CHEMISTRY
in the Community

Teacher's Edition

CHEMCOM
Fourth Edition

A Project of the American Chemical Society

W. H. FREEMAN AND COMPANY
NEW YORK

ChemCom Teacher's Edition

Fourth Edition Credits

American Chemical Society

Chief Editor: Henry Heikkinen

Revision Team: Laurie Langdon, Robert Milne, Angela Powers, Christine Gaudinski, Courtney Willis

Revision Assistants: Cassie McClure, Seth Willis

Teacher Edition: Lear Willis, Joseph Zisk

Ancillary Materials: Regis Goode, Mike Clemente, Ruth Leonard

Fourth Edition Editorial Advisory Board: Conrad L. Stanitski (Chair), Boris Berenfeld, Jack Collette, Robert Dayton, Ruth Leonard, Nina I. McClelland, George Miller, Adele Mouakad, Carlo Parravano, Kirk Soulé, Maria Walsh, Sylvia A. Ware (*ex officio*), Henry Heikkinen (*ex officio*)

ACS: Sylvia Ware, Janet Boese, Michael Tinnesand, Guy Belleman, Patti Galvan, Helen Herlocker, Beverly DeAngelo

ACS Safety Committee: Henry Clayton Ramsey (Chair), Wayne Wolsey, Kevin Joseph Edgar, Herbert Bryce

Technical Reviewers: Steve Cawthron, Kenneth Hughes, Susan C. Karr, Mary Kirchhoff, David Miller, Charles Poukish, Mary Ann Ryan, Tracy Williamson

Teacher Edition Reviewers: Drew Lanthrum, Karen Morris

Safety Consultants: Stanley Pine, Herbert Bryce

W. H. Freeman

Publisher: Michelle Russel Julet
Marketing Consultant: Arthur C. Germano
Marketing and Sales Director: Michael Saltzman
Project Editor: Rebecca Strehlow
Text Designer: Diana Blume
Cover Designers: Patricia McDermond, Diana Blume
Illustration Coordinator: Bill Page
Illustrations: Hudson River Studio, Network Graphics
Photo Editor: Vikii Wong
Photo Researchers: Elyse Reider, Tobi Zausner, Karen Barr
Production Coordinator: Susan Wein
Supplements and Multimedia Editor: Charles Van Wagner
Composition: Black Dot Graphics
Layout: Proof Positive/Farrowlyne Associates, Inc.
Manufacturing: Von Hoffman Press, Inc.

CD and Web Multimedia Package

Media Author: Roy Tasker, University of Western Sydney

Media Production, CADRE Design:

Media Producer: Ian Shakeshaft
Programmers: Ian Shakeshaft, Robert Bleeker
Animator: Nick McLeod
Research Assistant: Marianne Taylor
Graphic Designer: Janet Saunders

This material is based upon work supported by the National Science Foundation under Grant No. SED-88115424 and Grant No. MDR-8470104. Any opinions, findings, and conclusions or recommendations expressed in this publication are those of the authors and do not necessarily reflect the views of the National Science Foundation. Any mention of trade names does not imply endorsement by the National Science Foundation.

ISBN 0-7167-3918-6

Printed in the United States of America

Third Printing, 2002

CONTENTS

PREFACE

Welcome to the Teacher's Edition of *Chemistry in the Community,* Fourth Edition. This entirely new wraparound teacher's edition, along with the full *ChemCom* teaching package, offers complete, ongoing support for leading students through the *ChemCom* chemistry experience.

The Teacher's Edition

The wraparound Teacher's Edition, with reproductions of each Student Edition page, offers commentary when and where you can best take advantage of it.

Planning Guides lead in to each unit of the Teacher's Edition. Organized by section, these guide pages offer objectives, suggest class-time allotments, list special pedagogical features in the textbook, and identify relevant print and media program resources. Planning Guides also include materials lists to help you plan ahead for laboratory activities. When demonstrations are suggested in a *ChemCom* unit, the Planning Guide may provide detailed instructions and set-up photographs.

Commentary alongside Student Edition pages gives you point-of-use teaching strategies for students of various ability levels and learning styles. Preparation suggestions for laboratory activities appear where each lab begins. Teacher pages also include background material, supplemental activities, references to media, computer, and print supplements, and answers to all text questions and problems.

The Teacher's Edition concludes with an Appendix that contains duplicates of integral student handouts.

The Planning Guide lists student objectives for each section of the unit.

Section-by-section listings of specific print and media resources help you prepare class presentations.

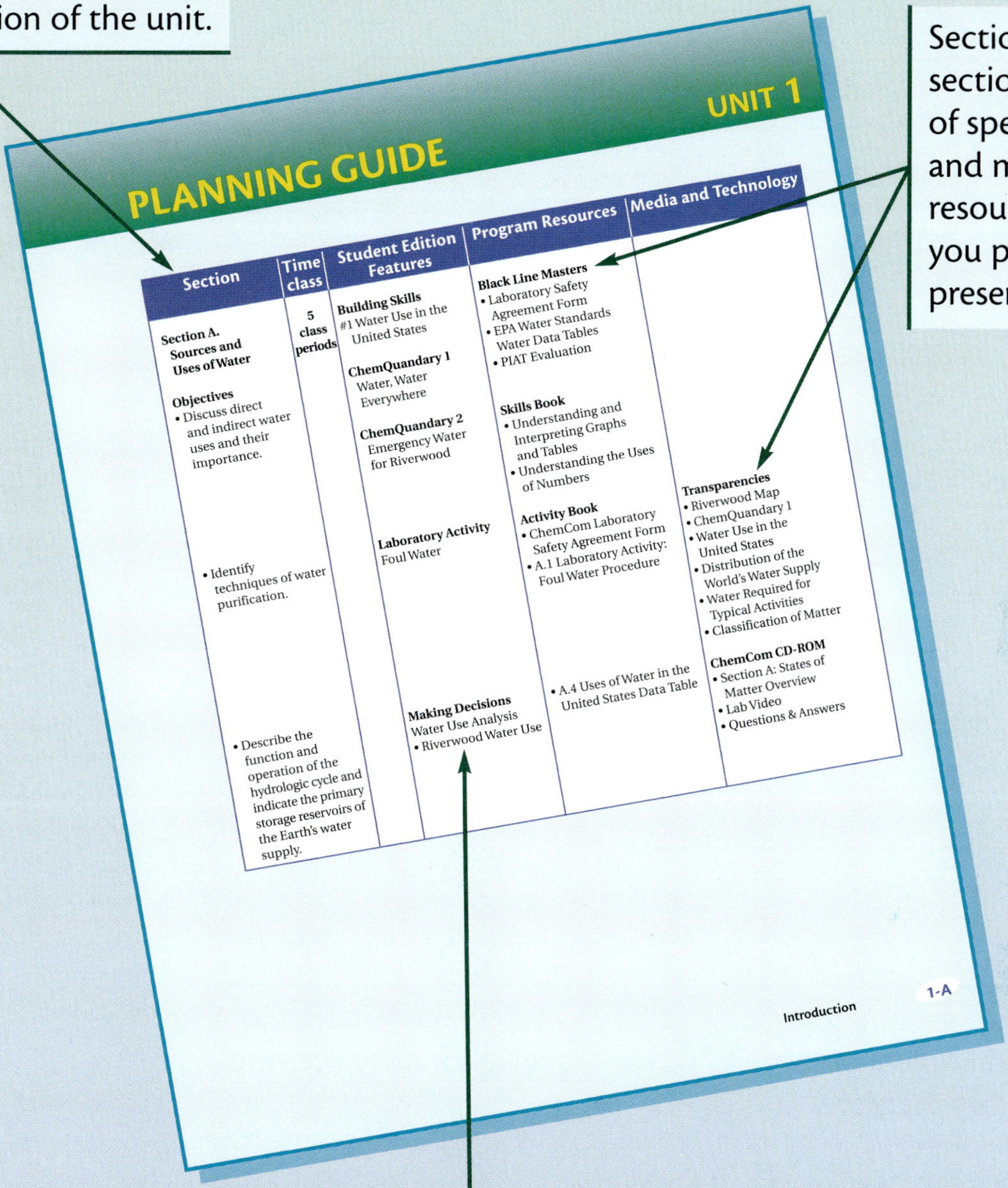

PLANNING GUIDE

UNIT 1

Section	Time class	Student Edition Features	Program Resources	Media and Technology
Section A. Sources and Uses of Water **Objectives** • Discuss direct and indirect water uses and their importance.	5 class periods	**Building Skills** #1 Water Use in the United States **ChemQuandary 1** Water, Water Everywhere **ChemQuandary 2** Emergency Water for Riverwood	**Black Line Masters** • Laboratory Safety Agreement Form • EPA Water Standards Water Data Tables • PIAT Evaluation **Skills Book** • Understanding and Interpreting Graphs and Tables • Understanding the Uses of Numbers	
• Identify techniques of water purification.		**Laboratory Activity** Foul Water	**Activity Book** • ChemCom Laboratory Safety Agreement Form • A.1 Laboratory Activity: Foul Water Procedure	**Transparencies** • Riverwood Map • ChemQuandary 1 • Water Use in the United States • Distribution of the World's Water Supply • Water Required for Typical Activities • Classification of Matter
• Describe the function and operation of the hydrologic cycle and indicate the primary storage reservoirs of the Earth's water supply.		**Making Decisions** Water Use Analysis • Riverwood Water Use	• A.4 Uses of Water in the United States Data Table	**ChemCom CD-ROM** • Section A: States of Matter Overview • Lab Video • Questions & Answers

Introduction 1-A

By section, the Planning Guide highlights Student-Edition features that promote problem solving, skill development, inquiry, and review.

Section-by-section orientation and teaching tips at point of use

Ideas for leading in to each activity

SECTION A

A.2 FOUL WATER
Laboratory Activity

This laboratory activity is designed to stimulate student interest in some of the practical problems associated with water quality. The goal of the activity is for students to purify a sample of "foul water" to the point at which it could be used to wash their hands. Do not extend the challenge to making the water suitable for drinking, because no laboratory-based materials should be swallowed. The activity introduces students to the techniques of filtration, absorption, and separation.

Time
50 minutes for student work
30 minutes for distillation demonstration

Materials (for a class of 24 working in pairs)
2–3 L of foul water (100 mL/pair) (See recipe below.)
12 100-mL graduated cylinders
24 150-mL beakers
12 funnels (73-mm inside diameter)
12 ring stands & clamps
12 clay triangles
12 pinch clamps
12 cups (paper or Styrofoam)
12 paper clips (for punching holes in cups)
12 filter papers (Whatman #1, 11 cm diameter)
12 125-mL Erlenmeyer flasks (for mixing charcoal and water)
150 g decolorizing charcoal
2 kg sand, cleaned
2 kg gravel, cleaned (small grind; roughly 5-mm diameter)
1 distillation apparatus
1 conductivity indicator
12 8–10-cm–long sections of rubber tubing

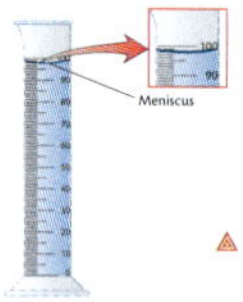

Figure 1 *To find the volume of liquid in a graduated cylinder, read the scale at the bottom of the curved part of the liquid (meniscus).*

The metric prefix milli- represents 1/1000th (0.001) of the unit indicated. Thus one milliliter (1 mL) is one-thousandth of a liter, or 0.001 L.

water ("foul water") sample as you can, producing water that is clean enough to use to rinse your hands. Even though most natural water samples are considerably less contaminated than your water sample, the challenge remains the same—to purify water supplies at low total cost to satisfy society's numerous needs.

A.2 FOUL WATER — Laboratory Activity

Introduction

Your objective is to purify a sample of "foul water," producing as much "clean water" as possible. **CAUTION:** *Do not test the purity of the water samples by drinking or tasting them.* You will use three water-purification procedures: (1) oil-water separation, (2) sand filtration, and (3) charcoal adsorption and filtration. In **filtration,** solid particles are sepa-rated from a liquid by passing the mixture through a material that retains the solid particles. The liquid collected after filtration is called the **filtrate.**

If you have not already done so, carefully read "Safety in the Laboratory," pages iv–v, before beginning the laboratory procedure.

Procedure

1. In your laboratory notebook, prepare a data table similar to the one shown here. Be sure to provide more space to write your entries.
2. Using a beaker, obtain approximately 100 mL (milliliters) of foul water from your teacher. Measure its volume precisely with a graduated cylinder. See Figure 1. Record the actual volume of the water sample in your data table.
3. Examine the properties of your sample: color, clarity, odor, and presence of oily or solid regions. Record your observations in the "Before treatment" row of your data table.

DATA TABLE

	Volume (mL)	Color	Clarity	Odor	Presence of Oil	Presence of Solids
Before treatment						
After oil-water separation						
After sand filtration						
After charcoal adsorption and filtration						

SAMPLE FOR REFERENCE ONLY

8 Unit 1 Water: Exploring Solutions

Foul water recipe

Estimate the amount of the following ingredients and add to an empty two-liter soft-drink bottle:
1/2 tablespoon table salt (NaCl)
1/2 teaspoon garlic powder
1/2 cup used coffee grounds
1/3 cup vegetable oil

Fill the bottle with tap water, cap tightly, and shake well. Save the excess for Section B.2 and the discussion of mixtures.

8

Oil-Water Separation

As you probably know, if oil and water are mixed and left undisturbed, the oil and water do not noticeably dissolve in each other. Instead, two layers form. Which layer floats on top of the other? Make careful observations in the following procedure to check your answer.

4. Place a funnel in a clay triangle supported by a ring and ring stand. See Figure 2. Attach a rubber tube to the funnel tip as shown.
5. Close the rubber tube by tightly pinching it with your fingers (or by using a metal pinch clamp). Gently swirl the foul-water sample for several seconds. Then immediately pour about half the sample into the funnel. Let it stand for a few seconds until the liquid layers separate. (Gentle tapping may encourage oil droplets to break free.)
6. Carefully open the tube, slowly releasing the lower liquid layer into an empty 150-mL beaker. Just as the lower layer has drained out, quickly close the rubber tube.
7. Drain the remaining layer into another 150-mL beaker.
8. Repeat Steps 5 through 7 for the other half of your sample, adding each liquid to the correct beaker. Which beaker contains the oily layer? How do you know?
9. Dispose of the oily layer as instructed by your teacher. Observe the properties of the remaining layer and measure its volume. Record your results. Save this water sample for the next procedure.
10. Wash the funnel with soap and water.

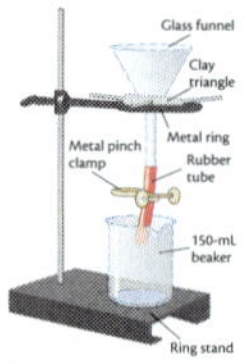

Figure 2 *Funnel in clay triangle.*

Sand Filtration

A **sand filter** traps and removes solid impurities—at least those particles too large to fit between sand grains—from a liquid.

11. Using a straightened paper clip, poke small holes in the bottom of a disposable cup. See Figure 3.
12. Add pre-moistened gravel and sand layers to the cup as shown in Figure 4. (The bottom layer of gravel prevents the sand from

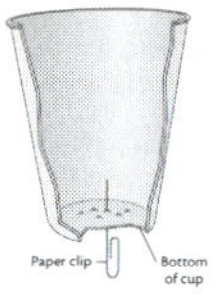

Figure 3 *Preparing a disposable cup.*

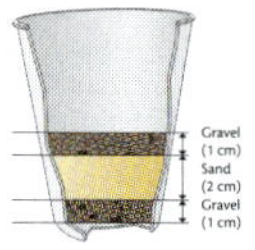

Figure 4 *Sand filtration.*

Section A Sources and Uses of Water 9

Safety

Students should wear safety goggles and aprons throughout the laboratory activity. During the post-lab distillation demonstration, the liquid should not become boiled to dryness, as the residue will be etched into the container, making cleaning difficult. More importantly, an overheated flask may break. About 1/4 of the original liquid should be left in the distillation flask. Water samples used in this activity should not be consumed.

SECTION A

Pre-Lab discussion

Because this is the first laboratory activity in the course, emphasize safety procedures, identification and use of safety equipment supplied in your classroom (such as the eyewash, shower, or fire extinguisher), and the general rules of safe laboratory conduct. It is recommended that Safety in the Laboratory, in the preliminary pages of the student text, be discussed. Review available safety videos and local safety policies, complete the safety compliance statement (in the Activity Workbook), and assess understanding with a safety quiz.

Explain that the protective goggle icon found at the beginning of each laboratory activity means students should wear goggles for the activity and heed any other safety precautions. Outline policies and procedures for students wearing contact lenses or glasses.

Display the foul-water mixture by pouring it into a large beaker. Review the purification methods used in the activity. Provide containers into which students can pour used sand, gravel, and charcoal.

Point out that data tables should be larger than the sample shown on page 8. Samples are only provided to give an idea of how to set up the table.

A review of how to construct a histogram may be appropriate before the Data Analysis to introduce the terms *range, mean,* and *median.*

Lab tips

- Dispense about 100 mL of foul water to each pair. Shake the mixture first to ensure homogeneity among samples. Remind students to record the actual volume they receive.

9

Thorough background for planning and implementing each laboratory activity

Safety information for all laboratory activities

Additional teacher laboratory material includes expected lab results and answers to all Student-Edition questions

Suggestions for additional activities

SECTION A

MEETING RAISES FISH-KILL CONCERNS

Chemists and other authorities report their progress in the investigation of the Snake River fish kill using such terms as *dissolved oxygen, suspended,* and *solubility.* In order to understand the fish-kill issue, students need to understand basic chemical concepts regarding the behavior of water.

ADDITIONAL ACTIVITIES

Students can maintain a "Chemistry in the News" bulletin board where local, regional, and national newspaper or magazine articles, cartoons, and so on, relating to the main idea of the unit being studied are displayed. Students should pay particular attention to actual reported fish kills. You may have students post these articles on the bulletin board or have them give a small report to the class. One example would be to have students type up a summary of the article in large print and post their summary along with the original article on the class bulletin board. Encourage multiple articles and post new articles on a regular basis to maintain interest.

Such activities help emphasize that Riverwood-type scenarios are not just stories but happen regularly. Chemistry-related issues appear consistently in both national and local news, and *ChemCom* topics focus on a wide variety of community settings and events.

MEETING RAISES FISH-KILL CONCERNS

BY CAROL SIMMONS
Riverwood News Staff Reporter

More than 300 concerned citizens, many "armed" with lists of questions, attended a Riverwood Town Hall public meeting last night to hear from the scientists investigating the Snake River fish kill.

Dr. Harold Schmidt, a scientist with the Environmental Protection Agency (EPA), expressed regret that the fishing tournament was canceled but strongly supported the town council's action, saying that it was the safest course in the long run. He reported that his laboratory is conducting further river-water tests.

Dr. Margaret Brooke, a State University water specialist, helped interpret information and answered questions. Local physician Dr. Jason Martingdale and Riverwood High School home economics teacher Alicia Green joined the speakers during the question-and-answer session that followed.

Dr. Brooke confirmed that preliminary water-sample analyses showed no likely cause for the fish kill. She reported that EPA chemists will collect water samples at hourly intervals today to look for any unusual fluctuations in dissolved-oxygen levels, as fish must take in adequate oxygen gas dissolved in water.

Concerning possible fish-kill causes, Dr. Brooke said that EPA scientists have been unable to identify any microorganisms in the fish that could have been responsible for their death. She concluded that "it must have been something dissolved or temporarily suspended in the water." If dissolved matter were involved, important considerations would include the relative amounts of various substances that can dissolve in water and the effect of water temperature on their solubility. She expressed confidence that further studies would shed more light.

Dr. Martingdale reassured citizens that "thus far, no illness reported by either physicians or the hospital can be linked to drinking water." Ms. Green offered water-conservation tips for housekeeping and cooking to make life easier for inconvenienced citizens. The information sheet that she distributed is available at Town Hall.

Mayor Edward Cisko confirmed that water supplies will again be trucked in from Mapleton today and expressed hope that the crisis will last no longer than three days.

Those attending the meeting appeared to accept the emergency situation with good spirits. "I'll never take my tap water for granted again," said Trudy Anderson, a Riverwood resident. "I thought scientists would have the answers," puzzled Robert Morgan, head of Morgan Enterprises. "They don't know either! There's certainly more involved in all this than I ever imagined."

22 Unit 1 Water: Exploring Solutions

SECTION B: A LOOK AT WATER AND ITS CONTAMINANTS

Section B returns to Riverwood's problem with another news report. To understand the issues, students need to know more about the chemistry of water and how substances interact with water. Some properties of water are given, and students learn to classify mixtures as suspensions, colloids, or solutions. A simple particulate model of matter is proposed. The universal symbolic language of chemistry is also introduced. Students then learn the difference between pure water and clean water, which leads them to laboratory testing for dissolved ions in water samples.

22

SECTION B

A LOOK AT WATER AND ITS CONTAMINANTS

As the preceding article indicates, scientists attribute the cause of the fish kill to something dissolved or suspended in the Snake River. What might those substances be? How can the search for the cause be narrowed further? Knowledge of the properties of water (and of substances that might be found in it) will aid in this task. To understand these properties, you will be introduced to matter at the particulate level. You will also begin to learn the language of chemistry and to use it to communicate with your classmates as you investigate the fish kill.

B.1 PHYSICAL PROPERTIES OF WATER

Water is a common substance—so common that it is usually taken for granted. You drink it, wash with it, swim in it, and sometimes grumble when it falls from the sky. But are you aware that water is one of the rarest and most unusual substances in the universe? As planetary space probes have gathered data, scientists have learned that the great abundance of water on Earth is unmatched by any planet or moon in our solar system. Earth is usually half-enveloped by water-laden clouds, as you can see in Figure 12. In addition, more than 70% of Earth's surface is covered by oceans having an average depth of more than three kilometers (two miles).

Kilo- (k) is the metric prefix meaning 1000. One kilometer (km) = 1000 meters (m).

Figure 12 *Earth as seen from space. What states of water can be observed in this winter scene?*

Section B A Look at Water and Its Contaminants 23

Many new chemical terms and concepts are introduced in Section B. It is important to remember that because these concepts will be revisited many times during the *ChemCom* course, it is not necessary to teach for mastery of each concept now. Many of the chemical concepts are in a spiral curriculum format. The Section Summary at the end of the unit provides important clues concerning the depth of knowledge required at this time.

With more information, students go back to the newspaper articles to separate fact from interpretation. Knowledge of chemistry provides a framework for asking questions about societal problems that may help lead to solutions.

SECTION B

B.1 PHYSICAL PROPERTIES OF WATER

This section focuses on the physical properties of water—density, physical state, surface tension, boiling and freezing points, and solvent characteristics. The important property of density is carefully investigated. To emphasize the unusual properties of water, begin the class session with some of the demonstrations described below. Ask students to interpret and explain the results of each demonstration.

Demonstration Ideas

1. Surface tension: Float a needle or a rubber band on the surface of the water. This can be done by gently lowering the needle onto the surface of the water using a bent paper clip. Show that more water can be added to a full glass of water. Fill a glass to the rim, then continue to add water drop by drop until it bulges well above the rim. The same effect can be demonstrated by filling a glass with water even with the rim, then asking students how many paper clips could be added to the same glass. Many more than they expect can be added due to the surface tension of the water clinging to the edge of the glass.
2. Density of water: Find the mass of 10.0 mL water in a 100-mL plastic graduated cylinder. Have the students calculate the density of liquid water. Then place the cylinder in the freezing compartment of a refrigerator. Find the new volume and have students calculate the density of ice.

23

Section overviews

Demonstration ideas

Answers follow questions for ease in assessing student progress

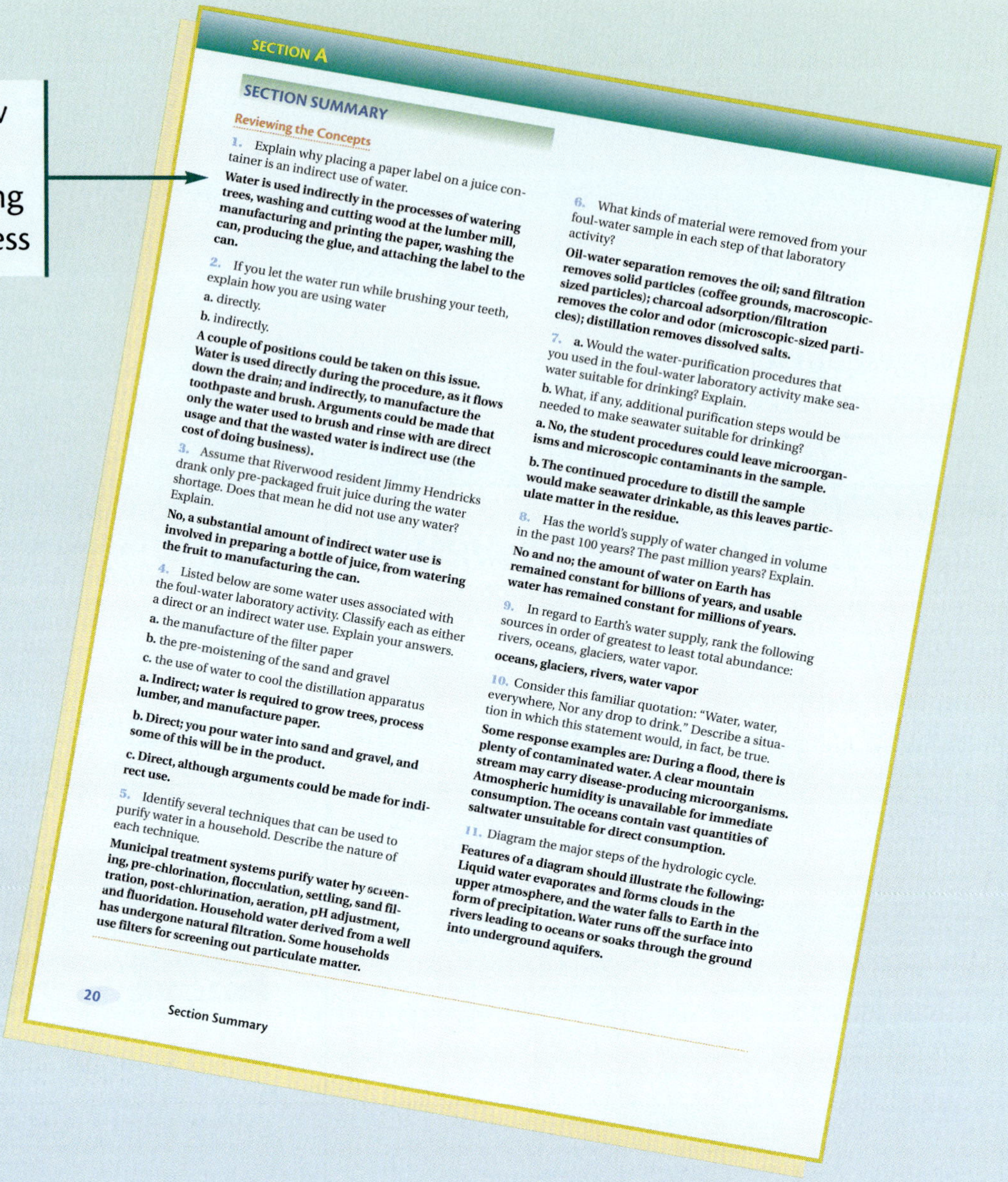

SECTION A

SECTION SUMMARY

Reviewing the Concepts

1. Explain why placing a paper label on a juice container is an indirect use of water.

Water is used indirectly in the processes of watering trees, washing and cutting wood at the lumber mill, manufacturing and printing the paper, washing the can, producing the glue, and attaching the label to the can.

2. If you let the water run while brushing your teeth, explain how you are using water

a. directly.

b. indirectly.

A couple of positions could be taken on this issue. Water is used directly during the procedure, as it flows down the drain; and indirectly, to manufacture the toothpaste and brush. Arguments could be made that only the water used to brush and rinse with are direct usage and that the wasted water is indirect use (the cost of doing business).

3. Assume that Riverwood resident Jimmy Hendricks drank only pre-packaged fruit juice during the water shortage. Does that mean he did not use any water? Explain.

No, a substantial amount of indirect water use is involved in preparing a bottle of juice, from watering the fruit to manufacturing the can.

4. Listed below are some water uses associated with the foul-water laboratory activity. Classify each as either a direct or an indirect water use. Explain your answers.

a. the manufacture of the filter paper

b. the pre-moistening of the sand and gravel

c. the use of water to cool the distillation apparatus

a. Indirect; water is required to grow trees, process lumber, and manufacture paper.

b. Direct; you pour water into sand and gravel, and some of this will be in the product.

c. Direct, although arguments could be made for indirect use.

5. Identify several techniques that can be used to purify water in a household. Describe the nature of each technique.

Municipal treatment systems purify water by screening, pre-chlorination, flocculation, settling, sand filtration, post-chlorination, aeration, pH adjustment, and fluoridation. Household water derived from a well has undergone natural filtration. Some households use filters for screening out particulate matter.

6. What kinds of material were removed from your foul-water sample in each step of that laboratory activity?

Oil-water separation removes the oil; sand filtration removes solid particles (coffee grounds, macroscopic-sized particles); charcoal adsorption/filtration removes the color and odor (microscopic-sized particles); distillation removes dissolved salts.

7. a. Would the water-purification procedures that you used in the foul-water laboratory activity make seawater suitable for drinking? Explain.

b. What, if any, additional purification steps would be needed to make seawater suitable for drinking?

a. No, the student procedures could leave microorganisms and microscopic contaminants in the sample.

b. The continued procedure to distill the sample would make seawater drinkable, as this leaves particulate matter in the residue.

8. Has the world's supply of water changed in volume in the past 100 years? The past million years? Explain.

No and no; the amount of water on Earth has remained constant for billions of years, and usable water has remained constant for millions of years.

9. In regard to Earth's water supply, rank the following sources in order of greatest to least total abundance: rivers, oceans, glaciers, water vapor.

oceans, glaciers, rivers, water vapor

10. Consider this familiar quotation: "Water, water, everywhere, Nor any drop to drink." Describe a situation in which this statement would, in fact, be true.

Some response examples are: During a flood, there is plenty of contaminated water. A clear mountain stream may carry disease-producing microorganisms. Atmospheric humidity is unavailable for immediate consumption. The oceans contain vast quantities of saltwater unsuitable for direct consumption.

11. Diagram the major steps of the hydrologic cycle.

Features of a diagram should illustrate the following: Liquid water evaporates and forms clouds in the upper atmosphere, and the water falls to Earth in the form of precipitation. Water runs off the surface into rivers leading to oceans or soaks through the ground into underground aquifers.

20 Section Summary

The Activities Workbook

Teachers of previous editions of *ChemCom* asked for ways to prevent potential damage to the textbook in the laboratory. They also wanted to minimize student laboratory preparation time and maximize laboratory table space. In response to their suggestions, a team of teachers has created the Activities Workbook to facilitate laboratory and classroom work while meeting science education standards. The result is a collection of perforated, removable activities pages for all the activities in the text, along with supplementary problem sets and assessment ideas. Not only does the Activities Workbook help teachers manage a busy classroom, it can help you meet standards, student needs, and school expectations.

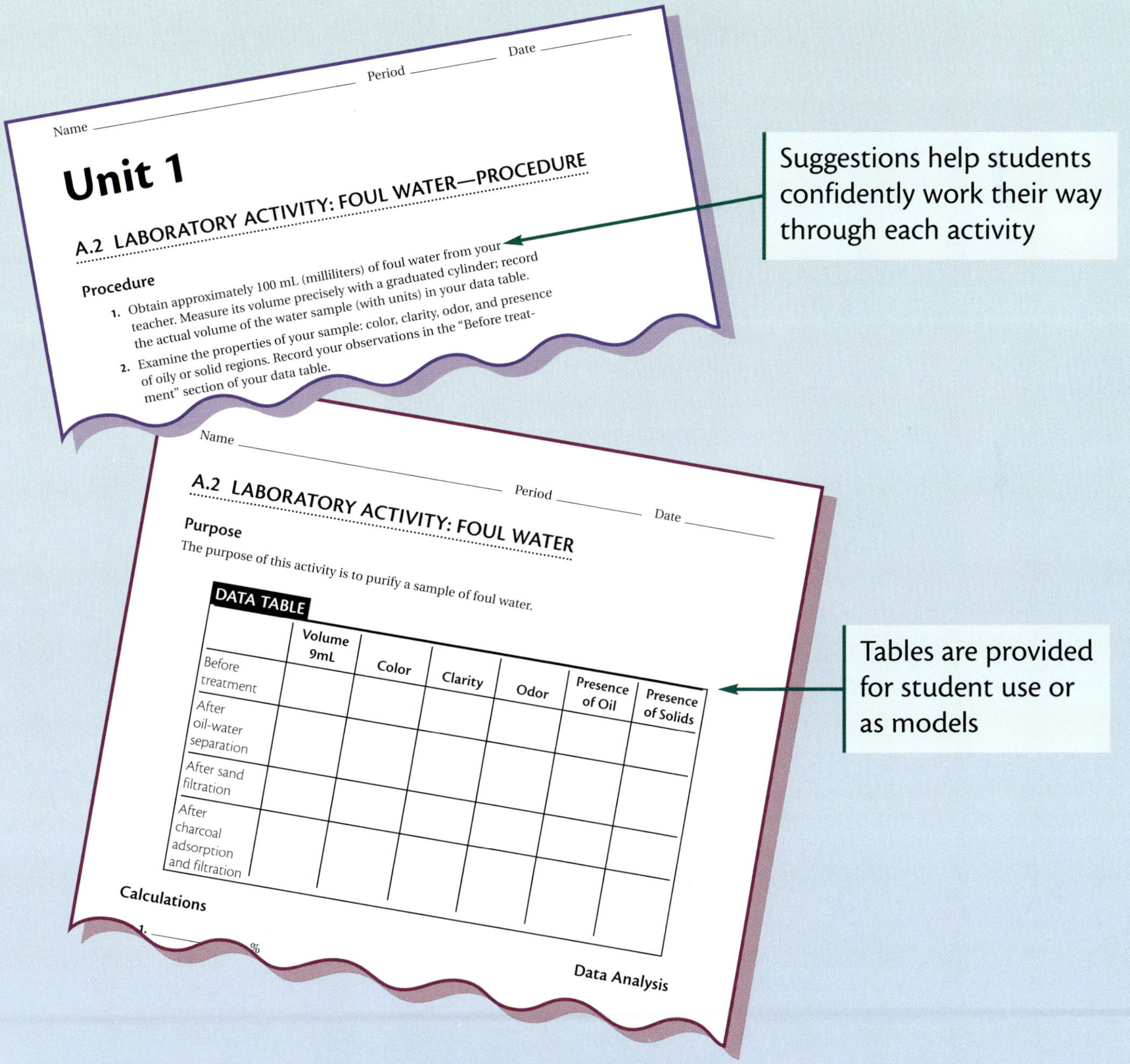

Name ________ Period ________ Date ________

Unit 1

A.2 LABORATORY ACTIVITY: FOUL WATER—PROCEDURE

Procedure

1. Obtain approximately 100 mL (milliliters) of foul water from your teacher. Measure its volume precisely with a graduated cylinder; record the actual volume of the water sample (with units) in your data table.
2. Examine the properties of your sample: color, clarity, odor, and presence of oily or solid regions. Record your observations in the "Before treatment" section of your data table.

Name ________ Period ________ Date ________

A.2 LABORATORY ACTIVITY: FOUL WATER

Purpose

The purpose of this activity is to purify a sample of foul water.

DATA TABLE

	Volume 9mL	Color	Clarity	Odor	Presence of Oil	Presence of Solids
Before treatment						
After oil-water separation						
After sand filtration						
After charcoal adsorption and filtration						

Calculations

1. ________ %

Data Analysis

The Skill Building Handbook

The *ChemCom* Skill Building Handbook provides teachers and students with lessons on the mathematics required for high school chemistry classes. Activities in the Skill Building Handbook guide students with diverse backgrounds toward understanding how scientific measurements can be applied in the search for understanding.

Students are introduced to significant figures and significant figures in calculations. They are given the opportunity to practice creating and interpreting graphs and tables. A separate section on equations and stoichiometry combines math skills with chemistry subject matter, offering in-depth work with the mole concept, mass-mole conversions, formulas, and limiting reactants.

CONTENTS

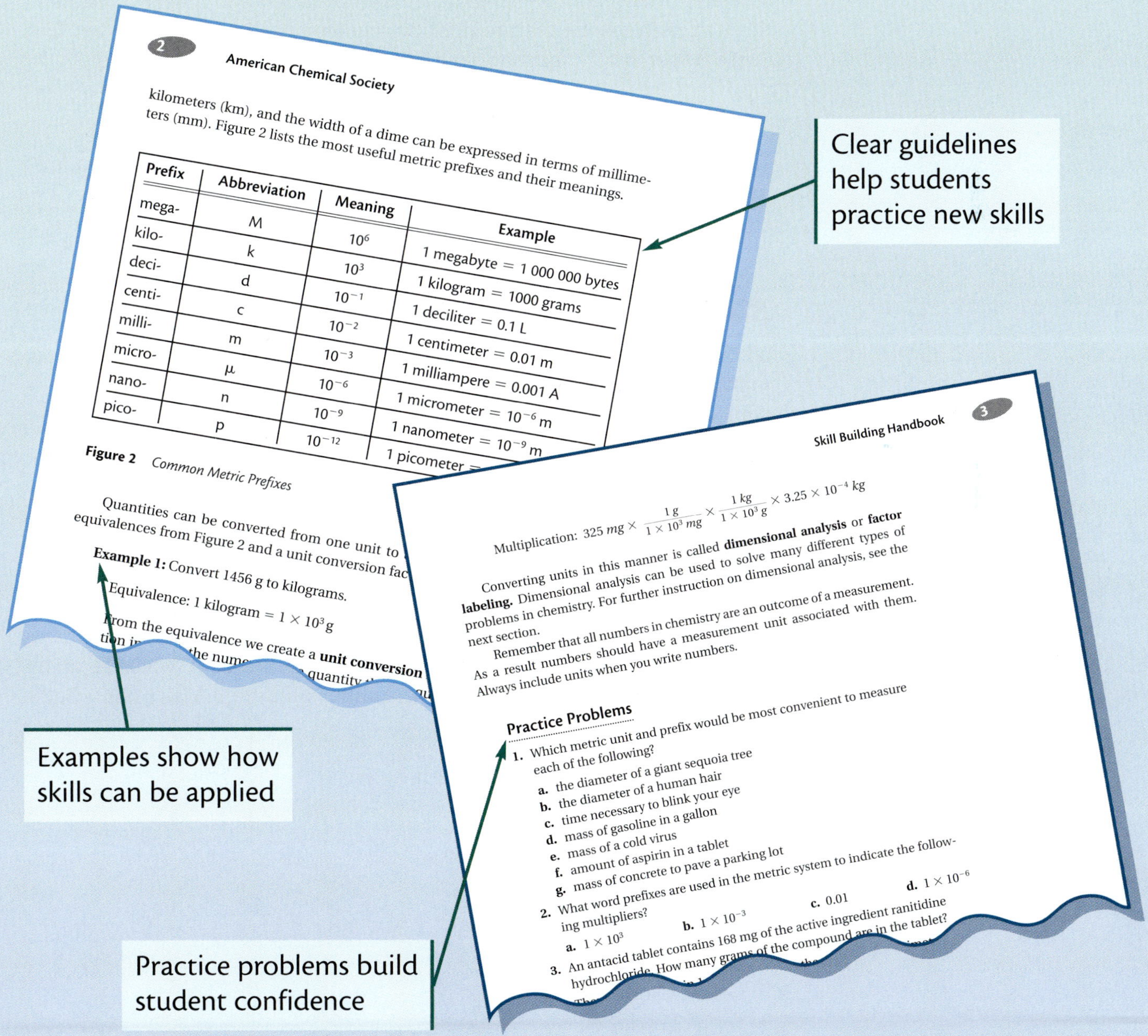

Clear guidelines help students practice new skills
Examples show how skills can be applied
Practice problems build student confidence

Assessment Tools

In print and computer versions, the *ChemCom* test bank contains hundreds of multiple-choice, short-answer, fill-in-the-blank, matching, and essay questions. A CD-ROM offers both Windows and Mac versions of the tests and is designed to give you the option of adding, editing, and resequencing questions.

The CD version of the test bank is also the access point for online testing with Diploma, a computerized test-bank package from Brownstone Research Group, in which teachers can create and administer exams on paper, over a network, and over the Internet. Questions include graphics, sound, multimedia, and interactivity. Security features allow you to restrict tests to specific computers or time blocks.

Online *ChemCom* quizzing makes it possible for you to use the Web to assess student comprehension. *ChemCom's* online quizzes, available through its Web site, feature multiple-choice questions on each text unit. Students get instant feedback and can take quizzes more than once.

Transparencies

To help you use cooperative visual learning techniques in the classroom, a package of 100 full-color transparencies introduces, clarifies, and reviews key topics. An accompanying booklet provides copies of each transparency for student or teacher use.

Black Line Copy Masters

A booklet of black line copy masters includes student handouts and assessment tools for "Putting It All Together" sections. All items may be photocopied for classroom use.

Media

Created exclusively for the Fourth Edition of *ChemCom*, media supplements give students a closer look at concepts that are traditionally difficult to visualize and comprehend. The *ChemCom* 4.0 CD-ROM offers highly interactive student features on selected topics, organized by text unit, plus a dynamic Periodic Table, a calculator, and useful electronic chemistry tools. The features and tools on the CD-ROM are also available at the *ChemCom* Web site, along with online quizzing. Visit the site at **www.whfreeman.com/chemcom**.

CHEMCOM AND THE NATIONAL SCIENCE EDUCATION STANDARDS[1]

The National Science Education Standards (NSES) outline what all students need to know, understand, and be able to do to be scientifically literate. No one course is expected to allow students to meet all of the standards. In fact, meeting the standards should be the result of the total school experience students have. As a textbook for a first-year chemistry course, we believe *ChemCom* provides excellent coverage of not only the physical science standards, but of parts of all the other content standards as well.

In the Standards document, for clarity of intent, each standard is accompanied by a section titled "A Guide to the Content Standard." It is this guide that describes the fundamental concepts and principles that underlie each standard. The relationship of *ChemCom* to these underlying concepts and principles is addressed in the following set of tables. For any component of each of the seven content standards, *ChemCom* may

- address *all* of the concepts and principles in the"Guide,"
- address *most* of the concepts and principles in the "Guide,"
- address *some* of the concepts and principles, or
- *not* address the concepts and principles.

The NSES Life Sciences Standard, while mainly focusing on the intellectual domain of the biological sciences, also includes important chemistry concepts that may not be taught in Biology I classes. Similarly, the Earth and Space Science Standard also includes chemistry concepts. In fact, modern chemistry is addressed throughout the seven content standards, a reflection of the central and useful character of chemical knowledge.

There is also a standard that addresses concepts and processes that provide a connection between and among different scientific disciplines. This Unifying Concepts and Processes Standard states:

As a result of activities in grades K–12, all students should develop understanding and abilities aligned with the following concepts and processes:

- *systems, order, and organization*
- *evidence, models, and explanation*
- *constancy, change, and measurement*
- *evolution and equilibrium*
- *form and function*

[1]National Research Council, 1996, National Science Education Standards. Washington, D.C., National Academy Press.

We encourage *ChemCom* teachers to work with other science teachers in their schools to ensure that these unifying concepts are addressed across the curricula. The *ChemCom* student text is rich with examples of these concepts as they apply to chemistry. One way to introduce these concepts might be to organize cross-disciplinary seminars that give students the opportunity to discuss each of these unifying concepts in the context of the particular sciences they are studying. We encourage you to consult the National Science Education Standards for additional information on these unifying concepts and processes.

<table>
<tr><th colspan="13">ChemCom and the Standards</th></tr>
<tr><th>Standards
(grades K-12)</th><th colspan="12">Components
(from "Guide to the Standards")</th></tr>
<tr><td>A. Inquiry</td><td colspan="6">1</td><td colspan="6">2</td></tr>
<tr><td>B. Physical Science</td><td colspan="2">1</td><td colspan="2">2</td><td colspan="2">3</td><td colspan="2">4</td><td colspan="2">5</td><td colspan="2">6</td></tr>
<tr><td>C. Life Science</td><td colspan="2">1</td><td colspan="2">2</td><td colspan="2">3</td><td colspan="2">4</td><td colspan="2">5</td><td colspan="2">6</td></tr>
<tr><td>D. Earth and Space Science</td><td colspan="3">1</td><td colspan="3">2</td><td colspan="3">3</td><td colspan="3">4</td></tr>
<tr><td>E. Science and Technology</td><td colspan="6">1</td><td colspan="6">2</td></tr>
<tr><td>F. Personal and Social Perspectives</td><td colspan="2">1</td><td colspan="2">2</td><td colspan="2">3</td><td colspan="2">4</td><td colspan="2">5</td><td colspan="2">6</td></tr>
<tr><td>G. History and Nature of Science</td><td colspan="4">1</td><td colspan="4">2</td><td colspan="4">3</td></tr>
</table>

Legend

addresses *all* fundamental concepts and principles in this section of the "Guide"

covers *most* fundamental concepts and principles

addresses *some* concepts and principles

section concepts *not* typically addressed in first-year chemistry courses

A. Inquiry Standard			
Components	***ChemCom* Level of Coverage**	**Activities in *ChemCom***	***ChemCom* Unit(s)**
1. Abilities Necessary to Do Scientific Inquiry	Addresses *all* concepts. Includes investigations, use of technology and math, using logic and evidence, modifying explanations, recognizing alternatives, communication and defending scientific arguments.	Labs, demonstrations, modeling matter, making decisions, building skills, math problems with graphs, word problems, library research, multimedia	Throughout
2. Understandings about Scientific Inquiry	Addresses *all* concepts. Includes use of math, logic, methods of scientists, purposes for research, development of science ideas.	Labs, demonstrations, modeling matter, making decisions, building skills, math problems with graphs, word problems, library research, multimedia	Throughout

B. Physical Science Standard			
Components	***ChemCom* Level of Coverage**	**Activities in *ChemCom***	***ChemCom* Unit(s)**
1. Structure of Atoms	Addresses *all* concepts. Includes atomic particles, fission and fusion, radioactive isotopes, radioactive series, nuclear pile. See also Standards A, F, G.	Labs, demonstrations, modeling matter, making decisions, building skills, math problems with graphs, word problems, library research, multimedia	1. Water 2. Materials 6. Atoms
2. Structure and Properties of Matter	Addresses *all* concepts. Includes Periodic Table and properties, ionic and covalent bonding, hydrogen bonding, physical and chemical properties, states of matter, organic structures, gas laws. See also Standards A, F.	Labs, demonstrations, modeling matter, making decisions, building skills, graphs, word problems, library research, multimedia	1. Water 2. Materials 3. Petroleum 4. Air 5. Industry 7. Food
3. Chemical Reactions	Addresses *all* concepts. Includes organic, redox, and acid/base reactions, reaction rates, catalysis, consumer, industrial, and environmental reactions, pH, moles, stoichiometry, enthalpy, activation energy. See also Standards A, F.	Labs, demonstrations, making decisions, building skills, activities sheet, math problems with graphs, word problems, library research, multimedia	1. Water 3. Petroleum 4. Air 5. Industry 7. Food
4. Motions and Forces	Concepts not typically addressed in first-year chemistry		
5. Conservation of Energy and Increase in Disorder	Addresses *all* concepts. Includes energy transfer, kinetic and potential energy, heat as random motion.	Labs, making decisions, building skills, math problems with graphs, word problems, library research, multimedia	3. Petroleum 4. Air 5. Industry 6. Atoms
6. Interaction of Energy and Matter	Addresses *all* concepts related to radiation.	Labs, making decisions, word problems, multimedia	4. Air 6. Atoms

C. Life Science Standard			
Components	***ChemCom* Concepts and Principles**	**Activities in *ChemCom***	***ChemCom* Unit(s)**
1. The Cell	Addresses *some* concepts. Includes chemical reactions in cells, proteins, enzymes, photosynthesis. See also Standards A, B, D, F.	Labs, making decisions, building skills, math problems, word problems, library research	3. Petroleum 7. Food
2. Molecular Basis of Heredity	Concepts not typically addressed in first-year chemistry		
3. Biological Evolution	Concepts not typically addressed in first-year chemistry		
4. Interdependence of Organisms	Addresses *most* concepts. Includes carbon and nitrogen cycles. See also Standards A, B, D, F.	Graphs, making decisions, word problems, library research	4. Air 5. Industry 7. Food
5. Matter, Energy, and Organization in Living Systems	Addresses *most* concepts. Includes food for energy and growth, biomolecules, ATP, cycle of energy and matter through living and nonliving systems. See also Standards A, B, D, F.	Labs, making decisions, building skills, word problems, library research	4. Air 5. Industry 7. Food
6. Behavior of Organisms	Concepts not typically addressed in first-year chemistry		

D. Earth and Space Science			
Components	***ChemCom* Concepts and Principles**	**Activities in *ChemCom***	***ChemCom* Unit(s)**
1. Energy in the Earth System	Addresses *some* concepts. Includes Earth's energy balance, atmosphere and climate, greenhouse effect. See also Standards A, B, C, F.	Making decisions, graphs, word problems, library research	4. Air
2. Geochemical Cycles	Addresses *most* concepts. Includes nonrenewable sources, conservation, carbon cycle. See also Standards A, B, F.	Labs, making decisions, graphs, word problems, library research	2. Materials 4. Air
3. Origin and Evolution of Earth System	Addresses *some* concepts. Includes radioactive isotopes for dating, half-lives. See also Standards A, B.	Labs, math problems, word problems, making decisions, building skills, library research	6. Atoms
4. Origin and Evolution of Universe	Addresses *some* concepts. Includes nuclear fusion reactions. See also Standard B.	Word problems, library research	6. Atoms

E. Science and Technology			
Components	***ChemCom* Concepts and Principles**	**Activities in *ChemCom***	***ChemCom* Unit(s)**
1. Abilities of Technological Design	Addresses *most* concepts. Includes evaluation and redesign of existing technologies.	Making decisions, word problems, library research	1. Water 2. Materials 3. Petroleum 4. Air 5. Industry 6. Atoms
2. Understandings about Science and Technology	Addresses *all* concepts. Includes chemical, petroleum, and nuclear technologies, nature of science and technology.	Labs, word problems, library research	2. Materials 3. Petroleum 5. Industry 6. Atoms

F. Science in Personal and Social Perspectives Standard			
Components	***ChemCom* Concepts and Principles**	**Activities in *ChemCom***	***ChemCom* Unit(s)**
1. Personal and Community Health	Addresses *some* concepts. Includes personal choice, food, chemistry and health.	Labs, making decisions, word problems, library research	1. Water, 6. Atoms 7. Food
2. Population Growth	Addresses *some* concepts. Includes chemical materials and petroleum.		2. Materials 3. Petroleum
3. Natural Resources	Addresses *all* concepts. Includes renewable and nonrenewable resources, uses and distribution of resources, recycling.	Labs, making decisions, graphs, math problems, word problems, library research	2. Materials 3. Petroleum 5. Industry
4. Environmental Quality	Addresses *all* concepts. Includes air and water quality issues and pollution prevention, nuclear wastes.	Labs, making decisions, math problems with graphs, word problems, library research	1. Water 2. Materials 4. Air 5. Industry
5. Natural and Human-Induced Hazards	Addresses *some* concepts. Includes risk analysis, water purity, and nuclear hazards.	Making decisions, graphs, word problems, library research	4. Air 6. Atoms
6. Science and Technology in Local, National, and Global Challenges	Addresses *all* concepts. Includes science, technology, political and social factors, complexity of decision making.	Labs, demonstrations, making decisions, math problems with graphs, word problems, library research	1. Water 2. Materials 3. Petroleum 4. Air 5. Industry 6. Atoms

G. History and Nature of Science Standard			
Components	***ChemCom* Concepts and Principles**	**Activities in *ChemCom***	***ChemCom* Unit(s)**
1. Science as a Human Endeavor	Addresses *most* concepts. Includes careers, interactions of scientists with society, work of scientists.	Making decisions, word problems, library research	Throughout
2. Nature of Scientific Knowledge	Addresses *most* concepts, but material tends to be less explicit than described in the Standards.	Labs, making decisions, math problems, word problems	Throughout
3. Historical Perspectives	Addresses *most* concepts. Includes atomic theory, nuclear chemistry, some molecular biology, some medical technology.	Labs, demonstrations, word problems, library research	2. Materials 3. Petroleum 4. Air 5. Industry 6. Atoms

CHEMICAL SAFETY, STORAGE, AND DISPOSAL REFERENCES

While no laboratory exercise is without some potential risk, most accidents can be prevented by basing laboratory practices and procedures on established chemical safety guidelines. The following references offer valuable background information. Since risk assessments are subject to change with new studies, it is important to keep up to date by reading professional journals (the *Journal of Chemical Education, The Science Teacher,* etc.) and learning about any guidelines and regulations issued in your local school district or state.

Aldrich Catalog Handbook of Fine Chemicals, Aldrich Chemical Co., P.O. Box 355, Milwaukee, WI 53201. 414-273-3850; 800-558-9160.

Chemical Catalog/Reference Manual, Flinn Scientific, Inc., P.O. Box 219, 770 Raddant Rd., Batavia, IL 60510. 630-879-6900; 800-452-1261.

Fire Protection Guide on Hazardous Materials, National Fire Protection Association, 11 Tracy Drive, Avon, MA 02322, 1994. 617-770-4543. National safety code as applied to laboratory fire prevention and protection.

The Merck Index, 12th Edition, Merck & Co., Inc., Whitehouse Station, NJ, 1996. Authoritative, brief descriptions of properties (including health hazards) of many substances.

NIOSH Pocket Guide to Chemical Hazards, U.S. Department of Health and Human Services, Public Health Service, Centers for Disease Control, National Institute for Occupational Safety and Health, Washington D.C., 1994. A pocket summary of hazardous properties of substances regulated by NIOSH.

School Science Laboratories: A Guide to Some Hazardous Substances, Council of State Science Supervisors and U.S. Consumer Products Safety Commission, Washington, D.C. 20207.

Committee on Chemical Safety (ACS), *Safety in Academic Chemistry Laboratories,* 6th edition. American Chemical Society, Washington, D.C. 20036, 1995.

School Science Safety Lecture Series/Secondary, Flinn Scientific, Inc., 770 Raddant Rd., Batavia, IL 60510, 1996. 630-879-6900; 800-452-1261. Overhead transparency masters, suggested student activities, student assessments.

Furr, A. Keith. *CRC Handbook of Laboratory Safety,* 4th edition , CRC Press, Boca Raton, FL, 1995.

Wood, Claire. *Safety in School Science Labs,* Flinn Scientific, Inc., P.O. Box 219, 770 Raddant Rd., Batavia, IL 60510. 630-879-6900, 800-452-1261.

Committee on Chemical Safety (ACS), *A Model Chemical Hygiene Plan for High Schools,* American Chemical Society, Washington, DC 20036, 1995. Product HS70A. 800-209-0423.

Prudent Practices in the Laboratory: Handling and Disposal of Chemicals, National Academy Press, Washington, D.C., 1995.

Starting With Safety: An Introduction for the Academic Chemistry Laboratory, American Chemical Society, Washington, D.C. 20036. 35-minute VHS videotape and teacher's guide. Product V5100. 800-209-0423.

Sourcebook, ChemSource, Inc. College of New Rochelle, New Rochelle, NY 10805, 1994. Sourcebook, Sourceview, and Guidebook to Preservice Use of ChemSource are available from the American Chemical Society, Washington, D.C. 20036. 800-209-0423 for orders.

INFORMATION ON ACIDS AND BASES

	Molar mass of reagent, g/mol	Approximate strength of concentrated reagent[a]	Molarity of concentrated reagent (M)	Milliliters of concentrated reagent needed to prepare one liter of 1 M solution[b]
Acetic acid, $HC_2H_3O_2$	60.5	99.8	17.4	57.5
Hydrochloric acid, HCl	36.46	37.2	12.1	82.5
Nitric acid, HNO_3	63.01	70.4	15.9	63.0
Sulfuric acid, H_2SO_4	98.07	96.0	18.0	55.5
Ammonium hydroxide NH_3 (aq)	17.03	28.0	14.5	69.0

[a]Representative value, mass percent

[b]Rounded to nearest 0.5 mL

CHEMICAL AND SCIENTIFIC EQUIPMENT SUPPLIERS

The following representative firms provide general supplies and equipment for all areas of science teaching as well as specific items for chemistry teaching.

Aldrich Chemical Company, P.O. Box 355, Milwaukee, WI 53201. 414-273-3850, 800-558-9160.

Boreal Laboratories, Ltd., 399 Vansickle Road, St. Catharines, Ontario, L2S 3T4, Canada. 905-984-3000, 800-387-9393.

Carolina Biological Supply Company, 2700 York Rd., Burlington, NC 27215. 910-584-0381, 800-334-5551. *www.carolina.com* Request Physical Science and Biological Science Catalogs.

Central Scientific Company, 911 Commerce Court, Buffalo Grove, IL 60089, 800-262-3626.

Connecticut Valley Biological Supply Company, Inc., 82 Valley Rd., P.O. Box 326, Southhampton, MA 01073. 413-527-4030, 800-628-7748.

Demco, Inc./Science Cap, 4801 Forest Run Road, Madison, WI 53704. 608-241-1201, 800-356-1200.

Edmund Scientific, 101 East Gloucester Pike, Barrington, NJ 08007. 609-573-6250.

Educational Innovations, Inc., 151 River Road, Cos Cob, CT 06807-2514. 203-629-6049, 888-912-7474. *www.teachersource.com*

Fisher Scientific Company, Educational Materials Division, 485 South Frontage Road, Burr Ridge, IL 60521. 630-655-4410. *www.fisheredu.com*

Flinn Scientific Inc., P.O. Box 219, Batavia, IL 60510. 630-879-6900, 800-452-1261. *www.flinnsci.com* The *Flinn Chemical Catalog* also provides information on chemical safety, storage, and disposal.

Frey Scientific, 100 Paragon Parkway, Mansfield, OH 44903. 419-589-1900, 800-225-3739.

Hach Company, P.O. Box 608, Loveland, CO 80539-0608. 800-227-4224. Test kits for environmental studies.

J.T. Baker, 222 Red School Lane, Phillipsburg, NJ 08865. 908-859-2151, 800-582-2537. *www. Jtbaker.com*

Lab Safety Supply, P.O. Box 1368, Janesville, WI 53547-1368. 608-754-2345. Specializes in safety equipment and supplies.

LaMotte Company, P.O. Box 329, Chestertown, MD 21620-0329. 410-778-3100, 800-344-3100. Test kits for environmental studies.

Markson, P.O. Box 1359, Hillsboro, OR 97123-9981. 800-528-5114. *www.marksson.com*

Micro Mole Scientific, 1312 North 15th Street, Pasco, WA 99301. 509-545-4904.

Nalgene Labware Division, P.O. Box 20365, Rochester, NY 14602-0365. 716-586-8800. Specializes in transparent and translucent plastic laboratory equipment.

NASCO, 901 Janesville Ave., Ft. Atkinson, WI 53538. 414-563-2446, 800-558-9595.

Ohaus Corporation, 29 Hanover Rd., Florham Park, NJ 07932. 201-377-9000, 800-672-7722.

S17 Science Supplies & Services Co. Ltd., PMB #231, 5 Columbia Drive, Niagara Falls, NY 19305-1275. 905-709-2033, 800-288-7151. *www.S17Science.com*

Sargent-Welch, 911 Commerce Court, P.O. Box 5229, Buffalo Grove, IL 60089-5229, 847-459-6625, 800-727-4368. *www.sargentwelch.com*

Science Kit & Boreal Laboratories, Inc., 777 E. Park Drive, Tonawanda, NY 14150-6782. 716-874-6020, 800-828-7777.

Vernier Software, 8565 SW Beaverton-Hillsdale Hwy., Portland, OR 97225-2429. 503-297-5317. *www.vernier.com*

Wards Natural Science Establishment, Inc. 5100 West Henrietta Rd., P.O. Box 92912, Rochester, NY 14692-9012. 716-359-2502, 800-962-2660.

COMBINED EXPENDABLE AND NON-EXPENDABLE ITEMS SUPPLY LIST

Quantities are for a class of 24 students working in pairs and completing all recommended *ChemCom* Laboratory Activities. For information concerning actual quantities needed for specific activities, see the supply list at the beginning of each unit of the Teachers Edition and the relevant laboratory activity materials list.

Quantity	Unit	Item	Unit-Activity
50	mL	Acetic acid	3-C.7
25	mL	Acetic acid, 0.1 M	4-C.6
0.3	mL	Acetic acid, conc.	1-B.8
25	mL	Acetic acid, glacial	4-C.6
100	mL	Alcohol (methyl, ethyl, or isopropyl)	6-B.8
300	mL	Alcohol, isopropyl (2-propanol)	3-A.2
4	Each	Alka Seltzer tablet	4-C.2
12	Set	Alpha, beta, and gamma radiation sets	6-B.4
25	mL	Aluminum nitrate, 0.1 M	4-C.2
25	mL	Ammonia, aqueous 0.1 M	4-C.6
25	mL	Ammonium acetate, 0.1 M	4-C.2
25	Gram	Ammonium chloride	1-C.7
25	mL	Ammonium chloride, 0.1 M	4-C.2
10	mL	Ammonium hydroxide, 14.8 M concentrated	4-D.6
120	mL	Ammonium molybdate-sulfuric reagent	5-A.5
85	Gram	Ammonium nitrate (NH_4NO_3) (alternate fertilizer)	5-A.3, 5-A.5
2.3	Gram	Amylase	7-C.5
4	Each	Apple	4-C.2, 7-C.3, 7-D.6
20	Gram	Ascorbic acid	5-A.5
12	Each	Aspirator adapters for faucets	4-B.6
12	Each	Balance	6-MM (Sect. C), 7-A.2
12	Each	Ball-&-stick or space-filling models	3-A.6, 3-A.7, 3-C.3
8	Each	Balloons, 4-in diameter	7-B.4
24	Each	Balloons, large	4-A.1
24	Each	Balloons, small	4-A.1
1.2	Gram	Barium chloride ($BaCl_2$)	5-A.3
1.2	Gram	Barium chloride dihydrate	1-B.7
25	mL	Barium hydroxide, 0.1 M	4-C.6
12	Each	Battery, 9 volt	2-A.3, 2-D.6
10	Each	Beaker, 50-mL	4-C.6

Quantity	Unit	Item	Unit-Activity
12	Each	Beaker, 100-mL	5-A.3
36	Each	Beaker, 400-mL	4-B.4, 5-A.5, 5-B.2
1	Each	Beaker, 250-mL	4-C.6
1	Each	Beaker, 600-mL	4-A.6
300	mL	Benedict's solution	7-C.5
12	Each	Boiling chips	3-A.2, 3-C.7
12.5	Gram	Boric acid, anhydrous	7-C.5
12	Each	Bottles, 8 oz. gas-connecting	7-D.5
12	Each	Bottles, squeeze (for alcohol)	6-B.8
6	mL	Bromothymol blue solution	4-B.6
60	mL	Buffer solution, pH 2	7-C.5
60	mL	Buffer solution, pH 4	7-C.5
60	mL	Buffer solution, pH 6	7-C.5
60	mL	Buffer solution, pH 8	7-C.5
60	mL	Buffer solution, pH 10	7-C.5
8	Gram	Calcium chloride dihydrate	1-B.7
40	Gram	Calcium chloride, anhydrous	5-B.2
1	Box	Calgon (sodium hexametaphosphate)	1-D.5
24	Each	Candles, large	3-B.3, 4-B.6
1	Lb	Candy, M&M's	7-D.5
1	Lb	Candy, Skittles	7-D.5
12	Each	Capillary tubes	4-A.6
5	Gram	Carbon, pencil lead	2-A.3
150	Gram	Charcoal, powdered	1-A.2
1	Pkg.	Chromatography paper, 1-in wide	7-D.5
25	mL	Citric acid, 0.1 M	4-C.6
10.5	Gram	Citric acid, hydrate	7-C.5
12	Each	Cloud chamber kit	6-B.8
12	Each	Cobalt-blue glass squares	5-A.3
2.37	Gram	Cobalt chloride ($CoCl_2 \cdot 6H_2O$)	5-B.2
1/2	Cup	Coffee grounds	1-A.2
12	Each	Coffee stirrers	4-C.8
20	Gram	Commercial fertilizer	5-A.5
1	Each	Conductivity apparatus	4-C.6
12	Each	Container or box (plastic or cardboard)	6-MM (Sect. C)
2	Each	Container, large glass or plastic	4-A.1
12	Each	Containers, sealed, opaque (possibly film canisters)	6-MM (Sect. A)
12	Gram	Copper powder	2-B.2
12	Each	Copper rods	2-B.3, 2-D.6
36	Cm	Copper strips, 0.25-cm wide	5-C.2
36	Cm	Copper strips, 0.5-cm wide	5-C.2
36	Cm	Copper strips, 1-cm wide	5-C.2
2	Meter	Copper wire, bare, 18 gauge	4-D.6
4	Gram	Copper(II) chloride dihydrate	2-A.3
11.2	Gram	Copper(II) nitrate ($Cu(NO_3)_2 \cdot 3H_2O$)	2-B.4, 5-C.2
58.3	Gram	Copper(II) sulfate pentahydrate	1-C.7, 2-D.6, 7-C.5

Quantity	Unit	Item	Unit-Activity
500	mL	Corn oil, yellow	4-A.6
12	Each	Crucible tongs	6-B.8
8	Each	Cultural tubes	7-B.4
36	Each	Cups, Styrofoam with lids	1-A.2, 4-B.4
12	Set	Dominos	6-MM (Sect. D)
1	Can	Drain cleaner	4-C.6
24	Each	Droppers	5-B.2
36	Each	Dropping bottle	3-C.7
10	Lb	Dry ice	6-B.8
24	Each	Electrical wires with alligator clips	5-C.2
20	Sets	Element data card	2-A.4
600	mL	Ethanol	1-C.7
100	mL	Ethanol, 95%	5-B.2
24	Pieces	Filter paper	2-C.5
108	Each	Filter paper, Whatman #1	1-A.2
12	Each	Flask, Erlenmeyer, 250-mL	4-B.6
12	Each	Flask, filter, 250-mL	4-B.6
2	Each	Flask, filter, 500-mL	4-D.6
1	Each	Food coloring, blue	7-D.5
1	Each	Food coloring, green	7-D.5
1	Each	Food coloring, red	7-D.5
1	Each	Food coloring, yellow	7-D.5
12	Each	Forceps	6-B.4, 6-B.8
12	Each	Funnels, glass, 7-cm diameter	4-B.6
1/2	Tbs	Garlic powder	1-A.2
12	Each	Gas burner	5-A.3, 5-A.5, 7-C.5
24	Each	Glass plates (about 5-cm square, 0.1 cm thick)	6-B.4
24	30 cm	Glass tube (90° bend)	3-A.2, 4-B.6
2	Each	Glasses, drinking	4-A.1
1	Box	Gloves, disposable	6-B.8
25	Gram	Glucose	1-C.7
12	Each	Graduated cylinder, 10-mL	5-A.5, 7-C.5, 7-D.2
1	Each	Graduated cylinder, 100-mL	4-D.6
12	Each	Graduated cylinders, 50- or 100-mL	5-A.5
2	Packages	Graph paper	2-A.6, 3-A.2, 6-B.4
2	Kg	Gravel	1-A.2
12	Each	Hammers, small	2-A.3
25	mL	Hexane	1-C.7
12	Each	Hot plate	3-C.7, 4-A.1, 4-A.7, 4-B.4, 5-A.3, 5-A.5, 5-B.2
1	Package	Hydrion paper (pH paper)	4-C.2

Quantity	Unit	Item	Unit-Activity
50	mL	Hydrochloric acid	2-A.3, 2-C.5
27	mL	Hydrochloric acid, 0.1 M HCl	4-C.8
5	mL	Hydrochloric acid, 1 M HCl	4-C.2
6	mL	Hydrochloric acid, 2 M HCl	5-A.3
50	mL	Hydrochloric acid, 3 M HCl	7-B.4
40	mL	Hydrochloric acid, 6.0 M HCl	4-C.2, 5-A.3
10	mL	Hydrochloric acid, 12 M concentrated	4-D.6
12	mL	Hydrogen peroxide, 3 %	7-C.3
18	Lb	Ice	1-C.3, 3-B.3, 4-A.1, 7-A.2
36	Each	Index cards, 3" x 5" (or cut as necessary)	3-B.3, 6-B.4, 7-A.2
25	Gram	Iodine	1-C.7
1	Gram	Iodine crystals	3-A.2
200	mL	Ion exchange resin	1-D.5
12	Cm	Iron (thin wire)	2-A.3
36	Cm	Iron metal strips or small iron nails	5-C.2
1.4	Gram	Iron(II) Sulfate ($FeSO_4$)	5-A.3
5	Gram	Iron(III) Nitrate ($Fe(NO_3)_3$)	5-A.3, 5-C.2
8	Gram	Iron(III) nitrate monohydrate	1-B.7
600	mL	Lamp oil	1-C.7
24	Each	Lead plates (about 5-cm square, 0.1cm thick)	6-B.4
1	Pkg	Legos	7-C.1
12	Each	Light, high intensity (bright flashlight)	6-B.8
1	Package	Litmus paper, blue	4-C.2
1	Package	Litmus paper, red	4-C.2
1	Vial	Litmus paper, red	5-A.3
0.5	Lb	Liver, raw	7-C.3
1.3	Gram	Magnesium nitrate ($Mg(NO_3)_2$)	5-C.2
10	Gram	Magnesium nitrate hexahydrate	2-B.4
121	Cm	Magnesium ribbon	2-A.3, 2-B.4, 4-C.2, 5-C.2, 7-B.4
24	Each	Marble chips ($CaCO_3$)	4-C.2
2	Box	Matches, long stem	2-D.6, 3-B.3, 4-D.6, 7-A.2
12	Each	Metal samples	4-A.6
12	Each	Meter sticks & rulers	4-A.1, 6-B.4
24	mL	Methyl alcohol (methanol)	3-C.7
72	Each	Micropipets	2-A.3, 2-B.4, 3-C.7, 7-D.5, 7-D.2
1	Each	Molecular model kit	7-C.1
12	Each	Nail, iron or zinc	2-D.6
25	Gram	Naphthalene flakes	1-C.7
12	Each	Nichrome or platinum wire loops (10-cm)	5-A.3

Quantity	Unit	Item	Unit-Activity
5	mL	Octyl alcohol	3-C.7
1/2	Cup	Oil, vegetable	1-A.2
24	Each	Paper clips	1-A.2, 7-A.2
1	Roll	Paper towels	4-A.7, 4-C.2
1	Each	Paramecium culture	4-C.2
12	Each	Paring knives	4-C.2
1236	Each	Pennies	2-D.3, 6-MM (Sect. C)
72	Each	Pennies, post-1982	6-MM (Sect. A)
72	Each	Pennies, pre-1982	6-MM (Sect. A)
5	mL	Pentyl alcohol (pentanol)	3-C.7
1	Each	Pin	4-A.1
180	Each	Pipets, Beral	1-B.7, 4-C.2, 4-C.8, 5-A.3, 5-C.2
1	Box	Plastic bags with sealing strips	4-C.2
1	Each	Plastic bottle, clear 1- or 2-liter	4-A.1
1	Each	Plastic bottle, clear 2-Liter	4-A.1
12	Each	Plastic, 2-cm square	4-A.1
35	Gram	Potassium chloride (KCl)	5-C.2
25	mL	Potassium hydroxide, 0.1M	4-C.6
1	Gram	Potassium iodate	7-D.2
1	Gram	Potassium iodide	7-D.2
0.5	Gram	Potassium nitrate (KNO_3)	5-A.3
25	mL	Potassium phosphate, 0.1 M	4-C.2
3.0	Gram	Potassium thiocyanate (KSCN)	1-B.7, 5-A.3
1	Each	Potato, Irish	7-C.3
12	Each	Radiation counter	6-B.4
12	Each	Radiation source (usually part of cloud chamber)	6-B.8
5	Each	Reagent bottles	3-C.7
1	Each	Refrigerator	7-C.5
12	Each	Ring stands with iron rings & wire gauze square	5-A.3, 5-A.5, 5-B.2, 7-A.3, 7-C.5
24	Each	Ring, clamps	7-A.3, 7-C.5
24	Each	Rubber aspirator hose, 30-cm	4-B.6
24	Each	Rubber bands, orthodontist type, small	4-A.6
36	Each	Rubber connectors	4-B.6
12	Each	Rulers, centimeter	4-A.6
5	Gram	Salicylic acid	3-C.7
2	Kg	Sand	1-A.2
5	Gram	Silicon (small lumps)	2-A.3
12.35	Gram	Silver nitrate	1-B.7, 2-B.4, 5-B.2
1	Bottle	Soap, liquid	1-D.5
25	mL	Sodium bicarbonate, 0.1 M	4-C.2
25	mL	Sodium bisulfate, 0.1 M	4-C.2

Quantity	Unit	Item	Unit-Activity
1.2	Gram	Sodium carbonate	1-D.5
25	mL	Sodium carbonate, 0.1 M	4-C.2
29	Gram	Sodium chloride (NaCl)	5-B.2, 1-C.7
25	mL	Sodium chloride, 0.1 M	4-C.2
173	Gram	Sodium citrate dihydrate	7-C.5
2.76	Gram	Sodium dihydrogen phosphate·H_2O	4-C.8
5.36	Gram	Sodium hydrogen phosphate·$7H_2O$	4-C.8
6	Gram	Sodium hydroxide (NaOH)	5-A.3
75	mL	Sodium hydroxide, 0.1 M	4-C.6
2	mL	Sodium hydroxide, 1 M NaOH	4-B.6
4.3	Gram	Sodium nitrate ($NaNO_3$)	5-A.3
25	mL	Sodium nitrate, 0.1 M	4-C.2
0.67	Gram	Sodium oxalate	1-B.7
382	Gram	Sodium phosphate (Na_3PO_4) (alternate fertilizer)	5-A.3, 5-A.5
1.6	Gram	Sodium sulfate (Na_2SO_4)	5-A.3
14.5	Gram	Sodium sulfite, Na_2SO_3	4-C.2, 7-D.6
36	Each	Soft drink cans, aluminum, with pull tabs	3-B.3, 4-A.1, 7-A.2
1	Bottle	Spray and Wash	2-B.4
150	mL	Starch sol, 0.1% (0.15 g in 150mL water)	7-C.5
100	Gram	Steel wool	2-D.3
12	Each	Stirring rods	5-A.5, 7-A.3
12	Each	Stoppers, cork (size 20)	7-A.2
12	Each	Stoppers, rubber, 1-hole to fit flask	4-B.6
1	Each	Stopper. rubber, 1-hole to fit graduated cylinder	4-D.6
60	Each	Stoppers, rubber to fit test tubes	7-C.5
12	Each	Stoppers, rubber, to fit bottles	7-D.5
12	Each	Straws, drinking, straight	4-B.6
120	Gram	Succinic acid	1-C.3
5	Gram	Sulfur (small lumps)	2-A.3
1	Liter	Sulfuric acid, 0.05 M	7-D.5
25	mL	Sulfuric acid, 0.1 M	4-C.6
5	mL	Sulfuric acid, conc.	3-C.7
24	mL	Sulfuric acid, conc. (6 M H_2SO_4)	5-A.3
1	Each	Syringe, large plastic with cap	4-A.1
1/2	Tbs	Table salt	1-A.2
1	Roll	Tape, transparent	4-D.6
1	Each	Telsa coil	4-D.6
60	Each	Test tubes (16 × 125 mm)	7-C.3, 7-C.5
60	Each	Test tubes (18 × 150 mm)	4-A.1, 5-A.3, 5-A.5, 5-B.2
24	Each	Test tubes (small)	5-A.3
12	Each	Test tube holders	7-C.5
24	Each	Test tube rack	5-A.5
12	Each	Thermometers	7-A.3
12	Each	Thermometers, alcohol (0 °C to 150 °C)	4-A.6
1	Each	Thistle tube (or long stem funnel)	4-D.6
5	Gram	Tin (mossy or foil)	2-A.3

Quantity	Unit	Item	Unit-Activity
3	Box	Toothpicks	4-C.8, 5-A.3, 7-D.5, 7-D.2
100	Gram	Trisodium carbonate, monohydrate	7-C.5
38	Gram	Trisodium phosphate, dodecahydrate	7-C.5
12	Each	U-tube or tygon tubing	D6
12	Each	Universal Indicator Color Chart	4-C.8
5	mL	Universal indicator	4-C.2, 4-C.8
25	Gram	Urea	1-C.7
84	Each	Vial, small (for metal samples)	2-A.3
1	Bottle	Vinegar, household	4-C.6
1	Each	Vitamin C tablets (100-mg)	7-D.2
12	Each	Voltmeter	5-C.2, 2-D.6
11.3	Liters	Water, distilled	1-C.7, 3-C.2, 5-A.5
1	Liter	Water, distilled or deionized	4-C.2, 4-C.8
3	Liters	Water, foul	1-A.2
12	Each	Wellplates, 24-well	4-C.8, 5-A.3, 5-C.2, 7-D.5
1	Each	Wire cutters	4-D.6
12	Each	Wire gauze	7-C.5
12	Each	Wire leads w/alligator clips (1 meter)	2-D.6, 4-D.6
1	Package	Wood splints	5-A.3
1	Pkg	Yeast cake	7-C.3
41	Gram	Zinc chloride	2-D.3
12	Gram	Zinc metal, mossy	2-A.3, 2-C.5
1.5	Gram	Zinc nitrate	5-C.2
10	Gram	Zinc nitrate hexahydrate ($Zn(NO_3)_2 \cdot 6H_2O$)	2-B.4
12	Cm	Zinc strip	2-B.4
36	Cm	Zinc strips, 0.25-cm wide	5-C.2
36	Cm	Zinc strips, 0.5-cm wide	5-C.2
36	Cm	Zinc strips, 1.0-cm wide	5-C.2
27	Gram	Zinc, granulated	2-D.3

PLANNING GUIDE UNIT 1

Section	Time class	Student Edition Features	Program Resources	Media and Technology
Section A. Sources and Uses of Water **Objectives** • Discuss direct and indirect water uses and their importance. • Identify techniques of water purification. • Describe the function and operation of the hydrologic cycle and indicate the primary storage reservoirs of the Earth's water supply.	**5 class periods**	**Building Skills** #1. Water Use in the United States **ChemQuandary** #1. Water, Water Everywhere #2. Emergency Water for Riverwood **Laboratory Activity** • Foul Water **Making Decisions** • Water Use Analysis • Riverwood Water Use	**Black Line Copy Masters** • Laboratory Safety Agreement Form • EPA Water Standards Water Data Tables • PIAT Evaluation **Skill Handbook** • Understanding and Interpreting Graphs and Tables • Understanding the Uses of Numbers **Activity Book** • ChemCom Laboratory Safety Agreement Form • A.2 Foul Water Procedure • A.3 Making Decisions: Uses of Water • A.6 Making Decisions Water Use Analysis	**Transparencies** • Riverwood Map • ChemQuandary 1 • Water Use in the United States • Distribution of the World's Water Supply • Water Required for Typical Activities • Classification of Matter **ChemCom CD-ROM** • Section A: States of Matter Overview • Laboratory Video • Questions & Answers

PLANNING GUIDE

Section	Time class	Student Edition Features	Program Resources	Media and Technology
Section B. A Look at Water and Its Contaminants **Objectives** • Classify matter in terms of elements, compounds, and mixtures: distinguish among the different types of mixtures (solutions, colloids, and suspensions). • Distinguish among a chemical symbol, formula, and a chemical equation. • Describe the three basic subatomic particles: protons, neutrons, and electrons. • Determine the formula and name of a simple ionic compound when provided with the anion's and cation's names and charges.	**7 class periods**	**Building Skills** #2. Density #3. Working With Symbols, Formulas, and Equations #4. Ionic Compounds **Laboratory Activity** • Water Testing **Making Decisions** • The Riverwood Water Mystery **Modeling Matter** • Pictures in the Mind **Chemistry at Work** • Environmental Cleanup: It's a Dirty Job . . . But That's the Point.	**Skills Book** • Understanding Equations and Stoichiometry: Understanding Chemical Equations **Activity Book** • B.6 Building Skills: Supplement: Ions and Ionic Compounds • B.7 Water Testing	**ChemCom CD-ROM** • Modeling Matter: Attraction Between Water Molecules • Molecular Views of Water • States of Matter • Laboratory Video **Videodisk** • B.7 Water Quality • B.7 Dealing With Hard Water **Transparency** • Common Ions

PLANNING GUIDE

Section	Time class	Student Edition Features	Program Resources	Media and Technology
Section C. Investigating the Cause of the Fish Kill **Objectives** • Use solubility curves to describe the effect of temperature on solubility, and calculate percent saturation. • Define the terms insoluble, unsaturated, saturated, and supersaturated, and calculate solution concentration. • Evaluate the risks of contaminants in our water supply, with particular attention to the heavy metal ions of lead and mercury. • Given the pH of a substance, classify it as acidic, basic, or neutral. • Describe the solubility of a molecular substance in water. • Use the polarity of water molecules to explain water's ability to dissolve ionic solid.	**5 class periods**	**Building Skills** #5. Solubility and Solubility Curve #6. Describing Solution Concentration **Laboratory Activity** • Constructing a Solubility Curve • Solvents **Modeling Matter** • Dissolving Ionic Compounds **Making Decisions** • Determining the Cause of the Fish Kill	**Activity Book** • C.1 Supplement: Solubility and Solubility Curves • C.2 Supplement: Solution Concentration • C.3 Solubility Curves: Alternative Procedure • C. 3 Constructing a Solubility Curve • C. 8 Solvents: Alternative Procedure **Teacher's Edition** • C.10 Additional ChemQuandary • C.11 Demonstration: Solubility of Gases vs. Temperature	**ChemCom CD-ROM** • C.1 Dissolving Ionic Compounds • C.7 Solubility of Molecular Substances • Laboratory Video • Questions & Answers **Transparencies** • Relationship Between Solute Solubility and Temperature • Polarity of Water and Formation of a Water Solution • Common Acids and Bases • Dissolved-Oxygen Requirement of Various Fish • Percent Composition of the Atmosphere

PLANNING GUIDE

Section	Time class	Student Edition Features	Program Resources	Media and Technology
Section D. Water Purification and Treatment **Objectives** • Compare and contrast natural and artificial water purification. • Assess the risks and benefits of water softening and chlorination.	**3 class periods**	**Building Skills** #7. Water Purification **ChemQuandary** • Bottled Water vs. Tap Water **Making Decisions** • D.4 Chlorination and THMs **Laboratory Activity** • D.5 Water Softening **Chemistry at Work** • Purifying Water Means More Than Going With the Flow	**Activity Book** • D.5 Water Softening	**ChemCom CD-ROM** • The Hydrologic Cycle Softening by Ion Exchange • Questions & Answers **Videodisk** • D.3 & D.4 Cleaning The Water Supply • Testing Water for Contamination **Transparencies** • The Hydrologic Cycle • Municipal Water Purification • U.S. Groundwater Hardness and Minerals Contributing to Water Hardness • Ion-Exchange Water-Softener Cycles
Putting It All Together	**2 class periods**		**Activity Book** • Riverwood Town • Laboratory Video Council Meeting Instructions and Responsibilities • Town Council Meeting Evaluation	**ChemCom CD-ROM** • Putting It All Together

SUPPLY LIST FOR UNIT 1

Expendable Items and Materials (per 24 students working in pairs)

Alphabetical by Item Name			
Quantity	**Unit**	**Item**	**Sections Used**
1.2	Grams	Barium chloride dihydrate	B.7
8.1	Grams	Calcium chloride dihydrate	B.7, D.5
1	Box	Calgon (sodium hexametaphosphate)	D.5
100	Grams	Cation exchange resin	D.5
150	Grams	Charcoal, decolorizing or aquarium charcoal	A.2
25	Grams	Copper(II) sulfate pentahydrate	C.8
1	Liter	Distilled water	C.8
600	mL	Ethanol	C.8
72	Pieces	Filter paper, Whatman #1	A.2, D.5
1/2	Teaspoon	Garlic powder	A.2
25	Grams	Glucose	C.8
2	Kilograms	Gravel, small grind	A.2
25	mL	Hexane	C.8
25	Grams	Iodine crystals	C.8
8	Grams	Iron(III) nitrate nonahydrate	B.7
1.4	Grams	Iron(II) sulfate heptahydrate	B.7
600	mL	Lamp oil	C.8
250	mL	Liquid soap solution	D.5
25	Grams	Naphthalene	C.8
12	Each	Paper clips	A.2
12	Each	Paper or Styrofoam cups	A.2
2.4	Grams	Potassium thiocyanate	B.7
2	Kilograms	Sand	A.2

SUPPLY LIST

Alphabetical by Item Name (continued)			
Quantity	**Unit**	**Item**	**Sections Used**
0.85	Grams	Silver nitrate	B.7
0.53	Grams	Sodium carbonate	B.7
1.24	Grams	Sodium carbonate monohydrate	D.5
25	Grams	Sodium chloride	C.8
0.67	Grams	Sodium oxalate	B.7
100	Grams	Succinic acid	C.3
1/2	Tablespoon	Table salt	A.2
25	Grams	Urea	C.8
1/2	Cup	Vegetable oil	A.2

By Section			
Quantity	**Unit**	**Item**	**Sections Used**
150	Grams	Charcoal, decolorizing	A.2
36	Pieces	Filter paper, Whatman #1	A.2
1/2	Teaspoon	Garlic powder	A.2
2	Kilogram	Gravel, small grind	A.2
12	Each	Paper clips	A.2
12	Each	Paper or Styrofoam cups	A.2
2	Kilogram	Sand	A.2
1/2	Tablespoon	Table salt	A.2
1/2	Cup	Vegetable oil	A.2
0.3	mL	Acetic acid, concentrated	B.7
1.2	Grams	Barium chloride dihydrate	B.7
5.9	Grams	Calcium chloride dihydrate	B.7
8	Grams	Iron(III) nitrate nonahydrate	B.7
1.4	Grams	Iron(II) sulfate heptahydrate	B.7

SUPPLY LIST

By Section (continued)			
Quantity	**Unit**	**Item**	**Sections Used**
2.4	Grams	Potassium thiocyanate	B.7
0.85	Grams	Silver nitrate	B.7
0.53	Grams	Sodium carbonate	B.7
0.67	Grams	Sodium oxalate	B.7
100	Grams	Succinic acid	C.3
600	mL	Ethanol	C.8
25	Grams	Copper(II) sulfate pentahydrate	C.8
1	Liter	Distilled water	C.8
25	Grams	Glucose	C.8
25	mL	Hexane	C.8
25	Grams	Iodine crystals	C.8
600	mL	Lamp oil	C.8
25	Grams	Naphthalene	C.8
25	Grams	Sodium chloride	C.8
25	Grams	Urea	C.8
2.2	Grams	Calcium chloride dihydrate	D.5
1	Box	Calgon (sodium hexametaphosphate)	D.5
100	Grams	Cation exchange resin	D.5
36	Pieces	Filter paper, Whatman #1	D.5
250	mL	Liquid soap solution	D.5
1.24	Grams	Sodium carbonate monohydrate	D.5

UNIT 1 WATER: EXPLORING SOLUTIONS

Introduction to Unit

The first *Chemistry in the Community* unit focuses on the fictional town of Riverwood and the problems with its water supply. To solve a fish-kill mystery, students explore the properties of water and aqueous solutions and examine procedures for obtaining adequate supplies of pure water. As students explore the Riverwood story, fundamental concepts related to formulas, nomenclature, and solubility are introduced and developed.

The unit demonstrates the *ChemCom* approach: create a need to know in the minds of the students via a science-technology-society (STS) problem or issue, and then develop the chemical concepts that help to resolve the issue. This is a departure from traditional chemistry courses, where students encounter the concepts first, and then use societal problems and issues to illustrate the use of the chemistry. It is important for you to remember that *ChemCom* is a spiral curriculum; when a concept is first introduced, it may contain only the information necessary to understand that issue or problem. For example, acids, bases, salts, and pH are introduced in this unit. However, it is unnecessary to teach them in greater depth at this time, because all of these topics will be revisited in subsequent units.

UNIT 1

WHAT separation techniques can be used to purify water?

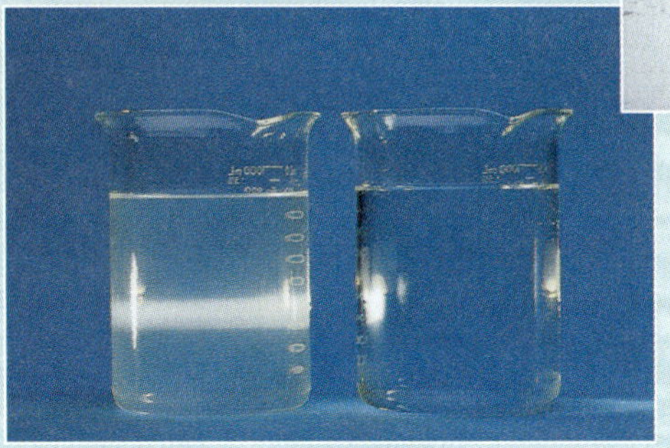

WHAT are the physical and chemical properties of water?

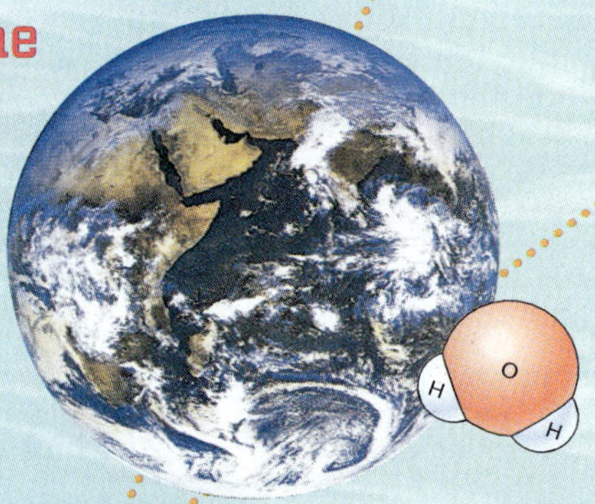

WATER: EXPLORING SOLUTIONS

WHY do some substances readily dissolve in water and others do not?

SECTION C **Investigating the Cause of the Fish Kill** (page 45)

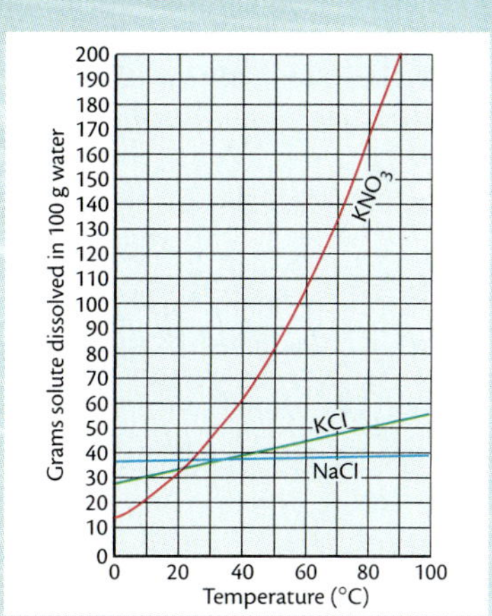

HOW do the properties of chlorine contribute to effective water treatment?

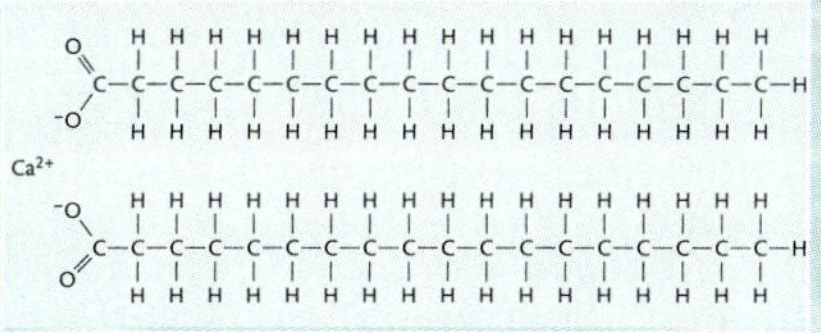

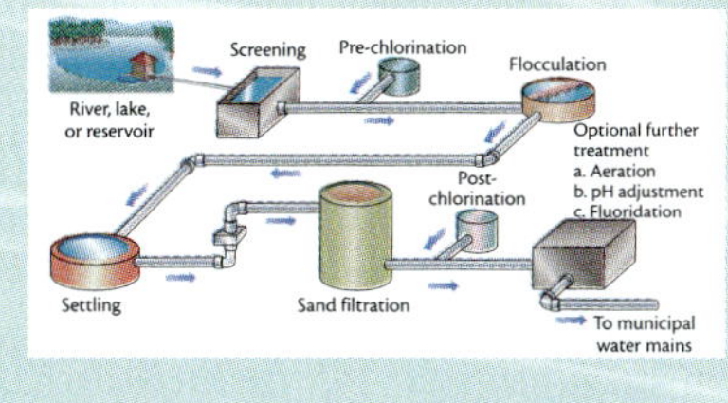

SECTION D **Water Purification and Treatment** (page 69)

Fish are dying in Riverwood's Snake River. Why? What are the consequences for the community? What is the solution to the mystery of the fish kill? Turn the page to find out more about this crisis and the importance of water in modern life.

CONCEPTUAL LEARNING GOALS OF UNIT 1

Unit 1 involves the students in exploring the chemistry of potential causes for a fish-kill scenario. The initial ChemQuandary poses questions to the students and gives the teacher an opportunity to assess prior knowledge and misconceptions, leading to the inquiry of water purification and a purification laboratory. The student is then guided through a "need to know" series of chemistry topics that adds to their knowledge in solving the fish-kill mystery. These topics include: matter, properties of matter, molecular structure, atomic structure, chemical formulas, nomenclature, ionic bonding, solubility of all states of matter, solution chemistry, heavy metals, pH, acids and bases, water purification, and distillation. These topics are introduced, and through the spiral curriculum methodology, most will be amplified and further developed later in the text. The unit culminates in a Town Council Meeting, which places in the hands of the students the evaluation process for deciding the cause of the fish kill, based on scientific data.

Section A: Setting the Stage Science, Technology, and Society (STS) Introduction

Two newspaper articles introduce students to a water-quality problem in Riverwood and set the stage for the major story line of the unit. Students follow the progress of Riverwood citizens in dealing with this issue. Encourage your students to note the difference between fact and opinion and to identify the kind of scientific information useful in solving such issues.

You may have students list the pertinent facts and opinions, and evaluate the comments made in the articles.

It is also worth noting that a large problem can often be solved more easily if it is first reduced to a series of smaller, more manageable subproblems. Students can identify some of the subproblems involved in the Snake River water-quality crisis.

FISH KILL CAUSES RIVERWOOD WATER EMERGENCY

SEVERE WATER RATIONING IN EFFECT

BY LORI KATZ
Riverwood News Staff Reporter

Citing possible health hazards, Mayor Edward Cisko announced today that Riverwood will stop withdrawing water from the Snake River and will temporarily shut down the city's water-treatment plant. Commencing at 6 P.M., river water will not be pumped to the plant for at least three days. And, if the cause of the fish kill has not been determined and corrected by that time, the shutdown will continue indefinitely.

During the pumping-station shutdown, water engineers and chemists from the County Sanitation Commission and Environmental Protection Agency (EPA) will investigate the cause of the major fish kill discovered yesterday. The fish kill extended from the base of Snake River Dam, located upstream from Riverwood, to the town's water-pumping station.

The initial alarm was sounded when Jane Abelson, 15, and Chad Wong, 16—both students at Riverwood High School—found many dead fish floating in a favorite fishing spot. "We thought maybe someone had poured poison into the reservoir," explained Wong.

Mary Steiner, Riverwood High School biology teacher, accompanied the students back to the river. "We hiked downstream for almost a mile. Dead fish of all kinds were washed up on the banks and caught in the rocks," Abelson reported.

Ms. Steiner contacted County Sanitation Commission officers, who immediately collected Snake River water samples for analysis. Chief Engineer Hal Cooper reported at last night's emergency meeting that the water samples appeared clear and odorless. However, he indicated some concern. "We can't say for sure that our water supply is safe until the reason for the fish kill is known. It's far better that we take no chances until then," Cooper advised.

Mayor Cisko canceled the community's "Fall Fish-In," scheduled to start Friday. No plans for rescheduling Riverwood's annual fishing tournament were announced. "Consensus at last night's emergency town council meeting was to start an investigation of the situation immediately," he said.

After five hours of often heated debate yesterday, the town council finally reached agreement to stop drawing water from the Snake River. Council member Henry McLatchen (also a Chamber of Commerce member) said the decision was highly emotional and unnecessary. He cited financial losses that motels and restaurants will suffer because of the

see Fish Kill, page 5

Alternative Approaches

ChemCom units provide the framework for student involvement in various local STS issues. Students can collect relevant newspaper articles and carry out individual projects related to local issues.

Fish Kill, from page 4

Fish-In cancellation, as well as potential loss of future tourism dollars due to adverse publicity. However, McLatchen and other council members sharing that view were outvoted by the majority, who expressed concern that the fish kill—the only one to occur in the Riverwood region in the past 30 years—may indicate a public health emergency.

Mayor Cisko assured residents that essential municipal services will not be affected by the crisis. Specifically, he promised to maintain fire department access to adequate supplies of water to meet fire-fighting needs.

Arrangements have been made to truck emergency drinking water from Mapleton. First water shipments are due to arrive in Riverwood by mid-morning tomorrow. Distribution points are listed in Part 2 of today's *Riverwood News*, along with guidelines on saving and using water during this emergency.

All Riverwood schools will be closed Monday through Wednesday. No other closings or cancellations have yet been announced. Local TV and radio will report new schedule changes as they are released.

A public meeting tonight at 8 P.M. at Town Hall features Dr. Margaret Brooke, a water expert at State University. She will answer questions concerning water safety and use. Brooke was invited by the County Sanitation Commission to help clarify the situation for concerned citizens.

Asked how long the water emergency would last, Dr. Brooke refused any speculation, saying that she first needed to talk to chemists conducting the investigation. EPA scientists, in addition to collecting and analyzing water samples, will examine dead fish in an effort to determine what was responsible for the killing. She reported that trends or irregularities in water-quality data from Snake River monitoring during the past two years also will play a part in the investigation.

5

SECTION A: SOURCES AND USES OF WATER

To help set the stage, have a brief open discussion on what the students believe may have caused the fish kill. Ask the students if they think the water is safe to drink. Remind them that in order to answer these questions, they will need to have a better understanding of chemistry.

A.1 THE INITIAL CHALLENGE

Indicate that the problems in Riverwood frequently happen elsewhere. You may have students do a search (newspaper and Internet) to see if this type of problem has occurred elsewhere recently. Emphasize the major themes of the unit.

Can we get enough water to supply our needs?

Can we get "pure" water?

TOWNSPEOPLE REACT TO FISH KILL AND ENSUING WATER CRISIS

BY JUAN HERNANDEZ
Riverwood News Staff Reporter

In a series of on-the-street interviews, Riverwood citizens expressed a variety of feelings earlier today about the crisis. "It doesn't bother me," said nine-year-old Jimmy Hendricks. "I'm just going to drink bottled water and pre-packaged fruit juice."

"I knew that eventually they'd pollute the river and kill the fish," complained Harmon Lewis, a lifelong resident of Fieldstone Acres, located east of Riverwood. Lewis, who traces his ancestry to original county settlers, still gets his water from a well and will be unaffected by the water crisis. He said that he plans to pump enough well water to supply the children's ward at Community Hospital if the emergency extends more than a few days.

Bob and Ruth Hardy, owners of Hardy's Ice Cream Parlor, expressed annoyance at the inconvenience but were reassured by council actions. They were eager to learn the reason for the fish kill and its possible effects on future water supplies.

The Hardys' daughter Toni, who loves to fish, was worried that late-season fishing would be ruined. Toni and her father won first prize in last year's angling competition.

Riverwood Motel owner Don Harris expressed concern for both the health of town residents and the loss of business due to the tournament cancellation. "I always earn reasonable income from this event and, without the revenue from the Fish-In, I may need a loan to pay bills in the spring."

The unexpected school vacation was "great," according to twelve-year-old David Price. Asked why he thought schools would be closed during the water shortage, Price said that all he could think of was that "the drinking fountains won't work."

Elmo Turner, whose residential landscaping has won Garden Club recognition for the past five years, felt reassured on one point. Because of the unusually wet summer, grass watering is unnecessary, and lawns are in no danger of drying as a result of current water rationing.

Dead fish were found washed up along the banks of the Snake River yesterday afternoon.

SECTION A SOURCES AND USES OF WATER

Riverwood will face at least a three-day water shortage. As the two newspaper articles indicate, the water emergency has aroused understandable concern among Riverwood citizens, town officials, and business owners. What caused the fish kill? Does the fish kill mean that Riverwood's water supply poses hazards to humans? In the pages that follow, you will monitor the town's progress in answering these questions as you learn more about water's important properties.

A.1 THE INITIAL CHALLENGE

Although Riverwood is imaginary, its problems are not. Residents of many actual communities have faced these problems and similar ones. In fact, two water-related challenges confront each of us every day. Can we get enough water to supply our needs? Can we get sufficiently pure water? These two questions are major themes of this unit, and their answers require an understanding of water's chemistry, uses, and importance.

The notion of water purity itself requires careful consideration. You will soon learn that the cost of producing a supply of water that is exactly "100%" pure is prohibitively high. Is that level of purity needed—or even desirable? Communities and regulatory agencies are entrusted with the responsibility of ensuring that water of sufficiently high quality for its intended uses is available at reasonable cost. How do they accomplish this task?

Even the apparently basic idea of "water use" presents some fascinating puzzles, as the following ChemQuandary illustrates.

CD-ROM WWW. Overview

This icon indicates an opportunity to explore further resources by using the ChemCom CD-Rom. You can also access the resources via the Web. See your teacher for further instructions. The appropriate location is printed near each icon.

CD-ROM WWW. Sources & Uses of Water

WATER, WATER EVERYWHERE — ChemQuandary 1

It takes 120 L of water to produce one 1.3-L can of fruit juice and about 450 L of water to place one boiled egg on your breakfast plate.

What explanation can you give for these two facts?

One liter (1 L) is a volume of approximately one quart.

As you just learned, one of the basic challenges that we face is related to the purity of the water we use. In the Laboratory Activity that follows, you will begin to address this challenge by trying to purify as much of a contaminated

CD-ROM WWW. Lab Video

WATER, WATER EVERYWHERE — ChemQuandary 1

This activity lends itself well to having students work in small cooperative groups (3 to 4 students). Have students make a list of the ways that water may be used in the production of an egg and of a can of juice.

Begin this brainstorming activity by asking students to list processes that are required before eggs or cans of juice can be available in a local grocery store. These include drinking water for the animals, using water for growing food for the chickens and for the trees that provide wood to build the chicken houses, for washing the eggs, and for producing the paper cartons in which eggs are sold.

To produce canned juice, the fruit trees must be watered and the fruit washed. What must also be taken into consideration is the amount of water used *indirectly* to produce the can that contains the juice. Digging the iron ore from the ground and moving it to the steel mill requires 4 L of water per can; the steel mill uses about 76 L of water to produce each can; and the rolling mill that fabricated the metal for the can consumes another 46 L of water. Thus, the total indirect volume of water used to make the juice can is about 126 L. Canned juice may help solve the short-term problem of water shortage in Riverwood, but it does not contribute to long-term water conservation. **Note: 1 quart = 0.946 liter.**

You may want to relate the analogy that the numbers given for one egg and a can of juice mean that eating two eggs and drinking 1.3 L of juice uses as much water as taking 10 showers, washing 10 loads of clothes or 25 loads of dishes, or flushing a toilet 100 times.

A.2 FOUL WATER

Laboratory Activity

This laboratory activity is designed to stimulate student interest in some of the practical problems associated with water quality. The goal of the activity is for students to purify a sample of "foul water" to the point at which it could be used to wash their hands. Do not extend the challenge to making the water suitable for drinking, because no laboratory-based materials should be swallowed. The activity introduces students to the techniques of filtration, absorption, and separation.

Time

50 minutes for student work

30 minutes for distillation demonstration

Materials (for a class of 24 working in pairs)

- 2–3 L of foul water (100 mL/pair) (See recipe below.)
- 12 100-mL graduated cylinders
- 24 150-mL beakers
- 12 funnels (73-mm inside diameter)
- 12 ring stands & clamps
- 12 clay triangles
- 12 pinch clamps
- 12 cups (paper or Styrofoam)
- 12 paper clips (for punching holes in cups)
- 12 filter papers (Whatman #1, 11 cm diameter)
- 12 125-mL Erlenmeyer flasks (for mixing charcoal and water)
- 150 g decolorizing charcoal
- 2 kg sand, cleaned
- 2 kg gravel, cleaned (small grind; roughly 5-mm diameter)
- 1 distillation apparatus
- 1 conductivity indicator
- 12 8–10-cm–long sections of rubber tubing

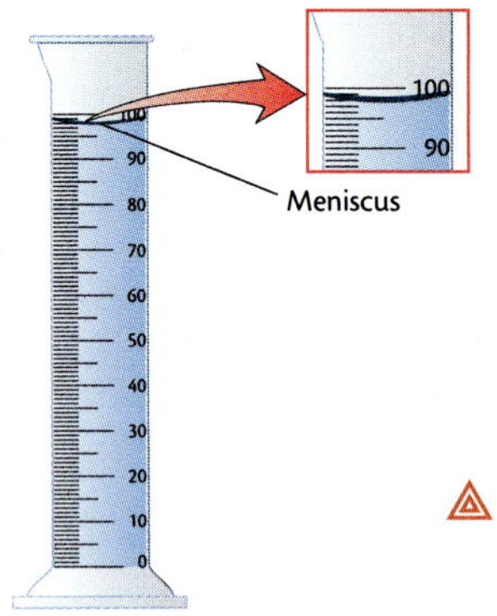

Figure 1 *To find the volume of liquid in a graduated cylinder, read the scale at the bottom of the curved part of the liquid (meniscus).*

water ("foul water") sample as you can, producing water that is clean enough to use to rinse your hands. Even though most natural water samples are considerably less contaminated than your water sample, the challenge remains the same—to purify water supplies at low total cost to satisfy society's numerous needs.

A.2 FOUL WATER

Laboratory Activity

Introduction

Your objective is to purify a sample of "foul water," producing as much "clean water" as possible. **CAUTION:** *Do not test the purity of the water samples by drinking or tasting them.* You will use three water-purification procedures: (1) oil-water separation, (2) sand filtration, and (3) charcoal adsorption and filtration. In **filtration,** solid particles are sepa-rated from a liquid by passing the mixture through a material that retains the solid particles. The liquid collected after filtration is called the **filtrate.**

If you have not already done so, carefully read "Safety in the Laboratory," pages iv–v, before beginning the laboratory procedure.

> The metric prefix *milli-* represents 1/1000th (0.001) of the unit indicated. Thus one milliliter (1 mL) is one-thousandth of a liter, or 0.001 L.

Procedure

1. In your laboratory notebook, prepare a data table similar to the one shown here. Be sure to provide more space to write your entries.
2. Using a beaker, obtain approximately 100 mL (milliliters) of foul water from your teacher. Measure its volume precisely with a graduated cylinder. See Figure 1. Record the actual volume of the water sample in your data table.
3. Examine the properties of your sample: color, clarity, odor, and presence of oily or solid regions. Record your observations in the "Before treatment" row of your data table.

DATA TABLE

	Volume (mL)	Color	Clarity	Odor	Presence of Oil	Presence of Solids
Before treatment						
After oil-water separation						
After sand filtration						
After charcoal adsorption and filtration						

Foul water recipe

Estimate the amount of the following ingredients and add to an empty two-liter soft-drink bottle:

1/2 tablespoon table salt (NaCl)

1/2 teaspoon garlic powder

1/2 cup used coffee grounds

1/3 cup vegetable oil

Fill the bottle with tap water, cap tightly, and shake well. Save the excess for Section B.2 and the discussion of mixtures.

Oil-Water Separation

As you probably know, if oil and water are mixed and left undisturbed, the oil and water do not noticeably dissolve in each other. Instead, two layers form. Which layer floats on top of the other? Make careful observations in the following procedure to check your answer.

4. Place a funnel in a clay triangle supported by a ring and ring stand. See Figure 2. Attach a rubber tube to the funnel tip as shown.
5. Close the rubber tube by tightly pinching it with your fingers (or by using a metal pinch clamp). Gently swirl the foul-water sample for several seconds. Then immediately pour about half the sample into the funnel. Let it stand for a few seconds until the liquid layers separate. (Gentle tapping may encourage oil droplets to break free.)
6. Carefully open the tube, slowly releasing the lower liquid layer into an empty 150-mL beaker. Just as the lower layer has drained out, quickly close the rubber tube.
7. Drain the remaining layer into another 150-mL beaker.
8. Repeat Steps 5 through 7 for the other half of your sample, adding each liquid to the correct beaker. Which beaker contains the oily layer? How do you know?
9. Dispose of the oily layer as instructed by your teacher. Observe the properties of the remaining layer and measure its volume. Record your results. Save this water sample for the next procedure.
10. Wash the funnel with soap and water.

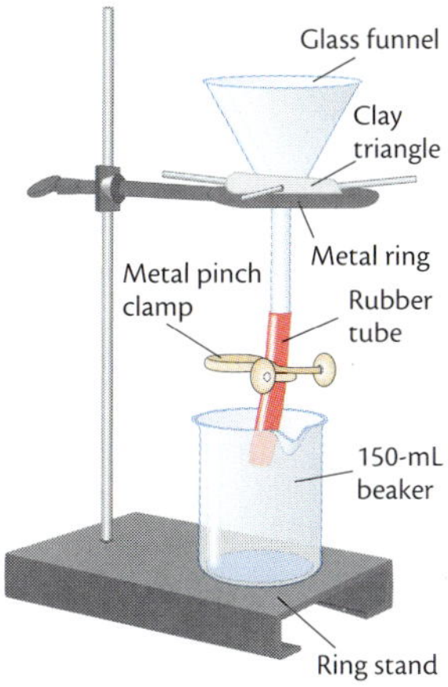

Figure 2 *Funnel in clay triangle.*

Sand Filtration

A **sand filter** traps and removes solid impurities—at least those particles too large to fit between sand grains—from a liquid.

11. Using a straightened paper clip, poke small holes in the bottom of a disposable cup. See Figure 3.
12. Add pre-moistened gravel and sand layers to the cup as shown in Figure 4. (The bottom layer of gravel prevents the sand from

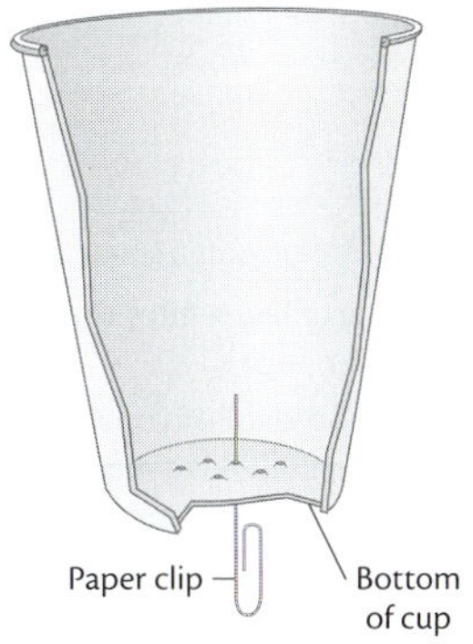

Figure 3 *Preparing a disposable cup.*

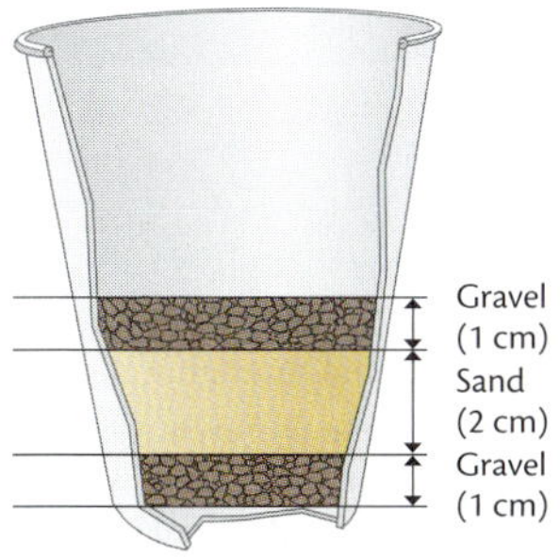

Figure 4 *Sand filtration.*

Pre-Lab discussion

Because this is the first laboratory activity in the course, emphasize safety procedures, identification and use of safety equipment supplied in your classroom (such as the eyewash, shower, or fire extinguisher), and the general rules of safe laboratory conduct. It is recommended that Safety in the Laboratory, in the preliminary pages of the student text, be discussed. Review available safety videos and local safety policies, complete the safety compliance statement (in the Activity Workbook), and assess understanding with a safety quiz.

Explain that the protective goggle icon found at the beginning of each laboratory activity means students should wear goggles for the activity and heed any other safety precautions. Outline policies and procedures for students wearing contact lenses or glasses.

Display the foul-water mixture by pouring it into a large beaker. Review the purification methods used in the activity. Provide containers into which students can pour used sand, gravel, and charcoal.

Point out that data tables should be larger than the sample shown on page 8. Samples are only provided to give an idea of how to set up the table.

A review of how to construct a histogram may be appropriate before the Data Analysis to introduce the terms *range, mean,* and *median.*

Safety

Students should wear safety goggles and aprons throughout the laboratory activity. During the post-lab distillation demonstration, the liquid should not become boiled to dryness, as the residue will be etched into the container, making cleaning difficult. More importantly, an overheated flask may break. About 1/4 of the original liquid should be left in the distillation flask. Water samples used in this activity should not be consumed.

Lab tips

- Dispense about 100 mL of foul water to each pair. Shake the mixture first to ensure homogeneity among samples. Remind students to record the actual volume they receive.

- Some teachers prefer to use plastic cups (instead of the paper or Styrofoam) so students can see the layers of sand and gravel. However, a sharper object (besides paper clips) is needed to puncture the small holes in plastic. You may need to use a hot paper clip if you use the hard clear plastic cups.
- Some forms of decolorizing charcoal (especially dust) are quite messy and can be irritating if inhaled. It is helpful to prepare a thick slurry of the charcoal with water beforehand, and have students use a spoonful of the mixture. Granular charcoal, used in aquariums, is a cheaper alternative than powdered charcoal. You may want to grind granular charcoal with a mortar and pestle prior to use. Activated charcoal works well if as much oil is removed as possible before adding the activated charcoal.
- An alternative to the funnel and rubber tubing method for removing oil is to have micropipettes siphoning the oil layer off the top of a beaker. Or, if available, use separatory funnels.
- Do not reveal the composition of the original foul-water sample until the Post-Lab Discussion.

Filtration does not remove the dissolved table salt; therefore, the final sample will conduct electricity. Compare the conductivity of the distilled water sample with the student-purified samples.

washing through the holes. The top layer of gravel keeps the sand from churning up when the water sample is poured into the cup.)

13. Gently pour the sample to be filtered into the cup. Catch the filtrate (filtered water) in a beaker as it drains through.
14. Dispose of the used sand and gravel according to your teacher's instructions. Do not pour any sand or gravel into the sink!
15. Observe the properties of the filtered water sample and measure its volume. Record your results. Save the filtered water sample for the next procedure.

Charcoal Adsorption and Filtration

Charcoal **adsorbs,** which means attracts and holds on its surface, many substances that could give water a bad taste, an odor, or a cloudy appearance. The pump system in a fish aquarium often includes a charcoal filter for this same purpose.

There are about 2.5 cm (centimeters) in an inch. You can "think metric" and make a good estimation of the length of a centimeter by realizing that a cassette audio cartridge or a piece of chalk has a thickness of about 1 cm.

16. Fold a piece of filter paper as shown in Figure 5.
17. Place the folded filter paper in a funnel. Hold the filter paper in position and wet it slightly so that it rests firmly against the base and sides of the funnel cone.
18. Place the funnel in a clay triangle supported by a ring. See Figure 2, page 9. Lower the ring so that the funnel stem extends 2 to 3 cm (centimeters) inside a 150-mL beaker.
19. Place one teaspoon of charcoal in a 125- or 250-mL Erlenmeyer flask.
20. Pour the water sample into the flask. Swirl the flask vigorously for several seconds. Then gently pour the liquid through the filter paper. Keep the liquid level below the top of the filter paper—liquid should not flow through the space between the filter paper and the funnel. (Can you explain why?)

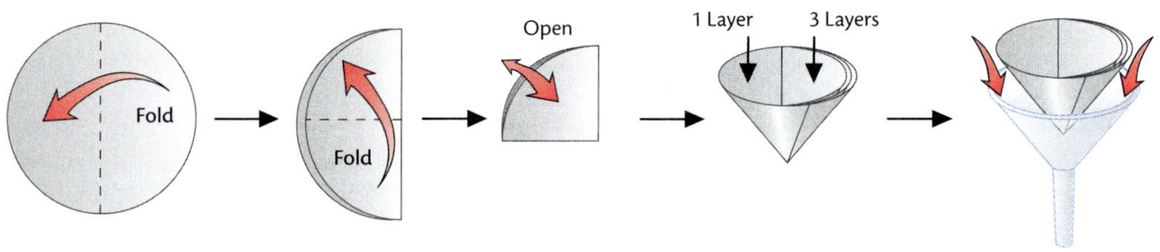

Figure 5 *Folding filter paper.*

21. If the filtrate is darkened by small charcoal particles, refilter the liquid through a clean piece of moistened filter paper.
22. When you are satisfied with the appearance and odor of your charcoal-filtered water sample, pour it into a graduated cylinder. Record the final volume and properties of the purified sample.
23. Follow your teacher's suggestions about saving your purified sample. Place used charcoal in the container provided by your teacher.
24. Wash your hands thoroughly before leaving the laboratory.

Expected Results

	Volume (mL)	Color	Clarity	Presence of Odor or Oil	Presence of Solids
Before treatment	80–100	dark	cloudy	yes	yes
Oil-water separation	75–95	dark	cloudy	less	yes
Sand filtration	70–90	dark	clearer	none	none
Charcoal filtration	65–85	lighter	clearer	none	none

Calculations

Record all calculations and answers in your laboratory notebook.

1. What percent of your original "foul water" sample did you recover as "purified water"?

$$\text{Percent of water purified} = \frac{\text{Volume of water purified}}{\text{Volume of foul-water sample}} \times 100\%$$

2. What volume of liquid did you lose during purification?
3. What percent of your original foul-water sample was lost during purification?

Data Analysis

Prepare to answer the following questions by compiling a list of the "percent of water purified" (percent recovery) values for water samples from each laboratory group in your class.

1. Construct a **histogram** showing the percent recovery obtained by all laboratory groups in your class. To do so, organize the data into equal subdivisions such as 90.0–99.9%, 80.0–89.9%, and so forth. Count the number of data points in each subdivision. Then use this number to represent the height of the appropriate bar on your histogram, as illustrated in Figure 6. In this histogram you can see that three data points fell between 70.0% and 79.9% in this example.
2. What was the largest percent recovery obtained by a laboratory group in your class? The smallest? The difference between the largest and smallest values in a data set is the **range** of those data points. What is the range of percent recovery data in your class?
3. What was the average percent recovery for your class? The **average** value is computed by adding all the values together and dividing by the total number of values. The result is also called the **mean** value.
4. The mean is one measure of what statisticians call central tendency—an expression of the most "typical" or "representative" value in a data set. Another important measure of central tendency is the **median,** or middle value. To find the median of your percent recovery data, list all values in either ascending or descending order.

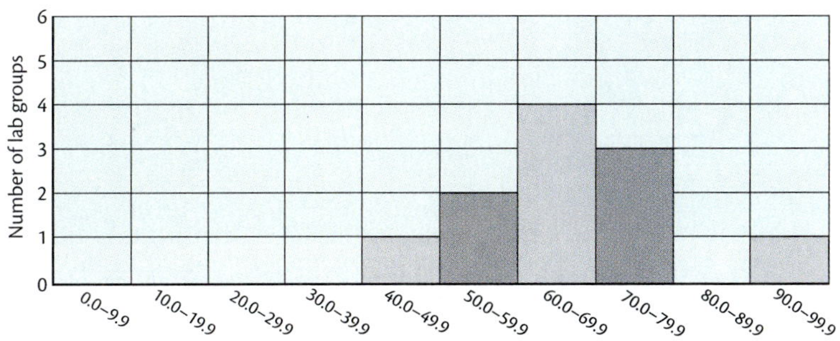

Figure 6 *Typical histogram of foul-water recovery data.*

Data analysis

Histograms and values will vary depending on the equipment and time spent completing the filtrations. Point out large variations in data and that groups following the same procedures and using the same equipment still had differences. This is one of the reasons for always repeating experiments and, generally, averaging the results.

1. See Figure 6 for typical histogram—most of the points will be in the 65 to 85% range.
2. Possible results: The largest percent recovery might be 90%, and the smallest percent recovery might be 50%. Thus the range would be 90% − 50% = 40%.
3. The average, or mean, is computed by adding all the values together and dividing by the total number of values. A typical mean is around 75%.
4. A median may be close to the mean but not necessarily. Median values help minimize the effect of extreme values, far off from the average.

This is a good opportunity to discuss the importance of data and how to view and display it. You may have students compare the value of the mean and the median and ask them which value best represents the data.

Calculations

1. Percent of water purified = $\frac{\text{Volume of water purified}}{\text{Volume of foul-water sample}} \times 100\%$
2. Volume lost = Volume of water before treatment − Volume of water after charcoal filtration
3. Percent lost = (volume lost)/(volume of foul-water sample) × 100%

Post-lab activities

Set up an apparatus to demonstrate distillation of water for the Post-Lab Activity (see Figure 3, Distillation of a Simple Mixture, in Unit 3: Petroleum, Section A, p. 179). The type of apparatus used will vary depending on the available equipment.

Check the electrical conductivity of the student samples and of distilled water. Compare the results. Distilled water acts as a control for this test. Using the distillation apparatus, distill a student group sample of purified water. Discard the initial 20% of the distillate, as it may contain dissolved gases, and allow about one fourth of the sample to remain in the distillation flask. Test the collected distillate for conductivity. Compare the results with previous tests, and relate the results to the usefulness of the purification procedures.

Many students think liquid samples are pure if they are clear and have no odor, but the presence of some contaminants cannot be detected with the unaided senses. The positive conductivity test on the sample before distillation should lead them to question their assumptions.

1. a. Heat the liquid sample (water in this case), channel the vapor (steam) through a cooling condenser, and collect the product.

   b. The first sample may contain dust and dirt that was in the equipment.

   c. Do not overheat the flask; there may be a high concentration of undesirable substances in this remaining portion of the liquid, which could also distill.

2. There is a substance in the purified sample that conducts electricity, which is not present after distillation.

Then find the value in the middle of the list—the point where there are as many data points above as below.

Consider this data set: 1 2 2 4 5 6 7

(1 2 2: three values lower; 4: median; 5 6 7: three values higher)

If there is an even number of data points, take the average of the two values nearest to the middle.

What is the median percent recovery for your class?

Post-Lab Activities

Seeing & Imagining Water (CD-ROM, WWW.)

1. Your teacher will demonstrate **distillation,** another technique for water purification.
   a. Write a description of the steps in distillation.
   b. Why did your teacher discard the first portion of distilled liquid?
   c. Why was some liquid purposely left in the distilling flask at the end of the demonstration?
2. Your teacher will organize the testing of the **electrical conductivity** of purified water samples obtained by your class. This test focuses on the presence of dissolved, electrically charged particles in the water. You will also compare the electrical conductivity of distilled water and tap water. (You will read more about electrically charged particles on page 32.) What do these test results suggest about the purity of your water sample?
3. Your teacher will test the clarity of the various water samples by passing a beam of light through each sample. Observe the results. The differences are due to the presence or absence of the **Tyndall effect.** See Figure 7. What does this test suggest about the purity of your water sample?

Figure 7 *Particles are suspended in the sample in the beaker on the left. The particles are too small to be seen but large enough to reflect light. This is the Tyndall effect. Particles in the solution in the beaker on the right are too small to reflect light. The Tyndall effect is also observable in nature.*

3. There may be some very small particles in water that are not seen until a beam of light is scattered by the particles.

The Tyndall effect may be tested by placing a piece of cardboard with a pencil-size hole on an overhead projector. Placing a beaker of water with a few drops of milk over the lit hole provides a good illustration. If the filtrate displays the Tyndall effect, the mixture is classified as a colloidal suspension. A laser may also be used—pocket laser pointers work especially well and are reasonably priced.

Questions

1. Is your "purified water" sample "pure" water? How do you know?
2. Suggest how you might compare the quality of your water sample with that of other laboratory groups. That is, how can the relative success of each laboratory group be judged? Why?
3. How would you improve the water-purification procedures you followed so that a higher percent of purified water could be recovered?
4. a. Estimate the total time you expended in purifying your water sample.
   b. In your view, did that time investment result in a large enough sample of sufficiently purified water?
   c. In the real world, it is often said that "time is money."
      i. If you spent twice as much time in purifying your sample, would that extra investment in time "pay off" in higher-quality water?
      ii. If you spent about ten times as much time?
   Explain your reasoning.
5. Municipal water-treatment plants do not use distillation to purify the water. Why?

A.3 USES OF WATER

Making Decisions

Keep a diary of water use in your home for three days. On a data table similar to the one shown here, record how often various water-use activities occur. Ask each family member to cooperate and help you.

DATA TABLE

Per Household	Day 1	Day 2	Day 3
Number of persons			
Number of baths			
Number of showers Average duration of showers (min)			
Number of toilet flushes			
Number of hand-washed loads of dishes			
Number of machine-washed loads of dishes			
Number of washing-machine loads of laundry			
Number of lawn/garden waterings Average duration of waterings (min)			
Number of car washes			
Number of cups used for cooking and drinking			
Number of times water runs in sink Average duration of running (min)			
Other uses and frequency			

Answers

1. There may still be some indications that the sample is not pure water. There most likely is some residual odor. Other indicators would be color, clarity, presence of immiscible liquids, and the presence of ions, as shown by the conductivity tests.
2. Line up samples and shine a bright light through them. Look for clarity, particles, and oil. Also check the odor of all samples. Tests may be extended to pH, conductivity, or spectrometric analysis.
3. Answers will vary: let the filtrations drip longer, premoisten the sand more, start with cleaner water, increase the number of filtrations, reduce the size of filter pores, use finer sand, or repeat the process.
4. a. Answers should vary between 5 to 20 minutes; however, some may have taken up to 40 minutes.

   b. Students should see that this procedure produces little water for the time spent.

   c. i. & ii. The process proceeds more slowly over time, so spending even ten times as long will not lead to corresponding yields. The demand for water determines the payoff. In municipal water treatment, where hundreds of thousands of gallons are produced daily, time is a significant factor. There is a trade-off between the quality of water and the cost to produce "pure" water. Many local water facilities publish an annual report, and it may add to the discussion to review one.
5. Distillation is too expensive for large volumes of water because of the time, energy, and equipment required. Some countries do use solar powered distillation plants to desalinate seawater.

A.3 USES OF WATER

Making Decisions

Throughout the *ChemCom* course, students will have many opportunities to explore the role of chemistry in their lives. This activity, which requires the participation of family members, generally elicits a positive response. The activity explores water use in the home and promotes awareness of the large quantities of water routinely used there. The activity also emphasizes information gathering as an important first step in making decisions. The data collected will be analyzed later in A.6 Water-Use Analysis: Making Decisions.

The time needed for data collection is three days. If possible, include one weekend day. Teachers may choose to extend the collection period to seven days, which can provide a better average computation. Provide students with a sample table, and illustrate how to record data in the table. Students often are unsure of how to estimate water usage. Teachers may want a student to measure the volume of a bathtub, but that may be impractical. A typical tub is 40 gallons up to the overflow drain. It is reasonable to expect a student to measure how much time it takes to fill a measured pitcher, then compute the shower's flow in gallons per minute. Typically, toilets use from 2 to 5 gallons per flush. All other water uses will vary in each household. It is the students' responsibility to make accurate estimates of how much water is used for each activity.

A.4 WATER SUPPLY AND DEMAND

The hydrologic (water) cycle is introduced. It is very likely that students are already familiar with the cycle. Consider assessing students' prior knowledge by setting up or displaying a water-cycle model and asking students to describe or discuss the steps involved. Key concepts to cover might include the source of heat that evaporates water, latent heat, vapor pressure, distillation, weather patterns, watersheds, surface flow, ground water, springs, geographical location, and any other variables that might change the path of the water.

This section discusses regional differences throughout the United States in sources and uses of water. It also refers back to indirect uses of water. Although students may have the impression that direct home use represents the major use of water, the fact is that growing, preparing, and manufacturing foods and consumer goods far exceed the more visible (direct) use of water by consumers. Most recent U.S. Geological Survey data indicate that the total water used by each person per day in the United States is 1340 gallons.

Make reference to ChemQuandary 1, as much water use is indirect.

Check the activities listed on the chart. If family members use water in other ways in the three-day-period, add those uses to your diary. Estimate the quantities of water used by each activity. You may be surprised by the large amount of water that you and your family use. Perhaps you are inclined to wonder what characteristics make water such a widely used and important substance.

A.4 WATER SUPPLY AND DEMAND

A trillion gallons is 1 000 000 000 000 gallons, or 10^{12} gallons.

Is so much water used in the United States that the nation is in danger of running out? The answer is both no and yes. The total water available is far more than enough. Each day, some 15 trillion liters (4 trillion gallons) of rain or snow falls in the United States. Only 10% is used by humans. The rest flows back into large bodies of water, evaporates into the air, and falls again as part of a perpetual **water cycle,** or **hydrologic cycle.** So that is the "no" part of the answer. However, the distribution of rain and snow in the United States does not necessarily correspond to regions of high water use. Figure 8 summarizes how available water is used in various geographic regions of the country, organized by six major water-use categories.

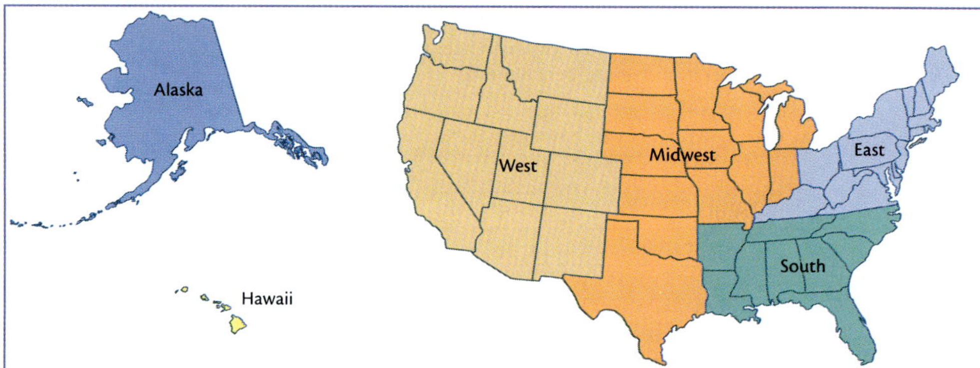

Percent of Water Used (%)

Purpose	Region					
	East	South	Midwest	West	Alaska	Hawaii
Irrigation	1	32	16	74	—	28
Other agricultural	—	2	1	1	—	—
Homes, offices	13	9	12	10	18	10
Manufacturing	6	6	8	1	17	2
Steam-electric	79	50	61	13	5	60
Mining	1	1	2	1	60	—

Figure 8 *Water use in the United States.*

WATER USE IN THE UNITED STATES

Building Skills 1

Refer to Figure 8 in answering these questions.

1. In the United States, what is the greatest single use of water in
   a. the East?
   b. the South?
   c. the Midwest?
   d. the West?
   e. Alaska?
   f. Hawaii?
2. Suggest reasons for differences in water-use priorities among these six U.S. regions.
3. Explain the differences in how water is used in the East and the West. Think about where most people live and where most of the nation's factories and farms are located. What other regional differences help explain patterns of water use?

An average family of four in the United States uses about 1200 liters (320 gallons) of water daily. That approximate volume represents direct, measurable water use. But in addition to that are many indirect, or hidden, uses of water that you probably have never considered. Each time you eat a slice of pizza, potato chips, or an egg, you are "using" water. Why? Because water was needed to grow and process the various components of each of these foods.

Consider again ChemQuandary 1 on page 7. At first glance, you probably thought that the volumes of water mentioned were absurdly large. How could so much water be needed to produce one boiled egg or one can of fruit juice? you might have asked. These two examples illustrate typical indirect (hidden) uses of water. The chicken that laid the egg needed drinking water. Water was used to grow the chicken's feed. Water was also used for various steps in the process that eventually brings the egg to your home. Even small quantities of water used for these and other purposes quickly add up when billions of eggs are involved!

In one of the Riverwood newspaper articles you read earlier, a youth was quoted as saying that he would drink bottled water and pre-packaged fruit juice until the water was turned back on. However, drinking fruit juice obtained from a grocery-store container involves the use of much more water than does drinking a glass of tap water. Why? Because the quantity of liquid in the container is insignificant when compared with the quantity of water used in making the container itself. The process of making a metal can, for example, is the source of the surprising 120 L of water mentioned on page 7! What examples of hidden water use associated with common materials do you encounter in daily life?

Although we depend on large quantities of water, we are relatively unaware of how much we use. This lack of awareness is understandable because water normally flows freely when taps are turned on—in Riverwood or in your home. Where does all this water come from? Check what you already know about sources of this plentiful supply.

WATER USE IN THE UNITED STATES

Building Skills 1

This activity provides practice in interpreting a figure to answer questions about regional water use.

Answers

1. a. steam-electric
   b. steam-electric
   c. steam-electric
   d. irrigation
   e. mining
   f. steam-electric
2. Possible reasons include a broad spectrum of uses for electricity, land forms (mountains, etc.), soil porosity, available precipitation, location of rivers and lakes, population density, climate for growing, and location of factories.
3. Because the East is more densely populated and requires a large amount of electrical power, 79% of the total water used in the East is for steam-electric, compared with 12.5% in the West. Because the East contains more manufacturing plants, 6% of the total water used in the East is for manufacturing, as compared with 1% in the West. In the East, only 0.5% of the water use is for irrigation, while the West and South use 74% and 32%, respectively.

ADDITIONAL BACKGROUND: INDIRECT WATER USE

Like energy, water is an essential behind-the-scenes ingredient in nearly every human endeavor. In the United States, where overall water use is over 5000 L/person/day, industrial processes—such as production of electrical energy, primary metals, and chemical products, petroleum refining, pulp and paper manufacturing, and food processing—account for about 51% of the use, agriculture for about 40%, and the municipal sector for about 9%. The point to make with students is that not only do they use large quantities of water directly (a U.S. average of about 316 L/person/day), but that even larger quantities are used on their behalf behind the scenes.

A.5 WHERE IS EARTH'S WATER?

Students are generally not surprised to learn that most of Earth's water supply (97%) is in the oceans. However, they may be puzzled to find that the next largest store of Earth's water is locked in glaciers and ice caps.

Model Earth's water supply with a 1-L beaker or graduated cylinder of water, pouring out various volumes into containers (with suitable explanatory talk) to illustrate relative distribution. For example: starting with an empty 1-L graduated cylinder, add 970 mL of water to indicate that 97% of all Earth's water comes from the oceans, add 21 mL for glaciers and ice caps, 6 mL for groundwater, 1 mL (or equivalent in drops) for lakes, a few drops for atmospheric moisture, and just a drop for rivers. Using blue water (or any other color) by adding food coloring makes for a more effective illustration.

Neither of these sources of water (glaciers or ice caps) is readily available. It would be very costly to change either to a useful form. Fortunately, many areas of the United States have an abundant water supply. Some cities, particularly in the arid parts of the West, already find it necessary to regulate water usage at times. Remind students that even though this country's total water-supply profile is favorable, there are local and regional problems. The fish-kill problem in Riverwood is a typical example. The Making Decisions sections in A.6 and A.7 challenge students to make thoughtful decisions about how water should be used during a shortage.

A.5 WHERE IS EARTH'S WATER?

States of Water CD-ROM WWW.

You are probably not surprised to learn that most of Earth's total water (97% of it, in fact) is stored in the oceans. However, the next largest global water-storage place is not as obvious. Do you know what it is? If you said rivers and lakes, you and many others agree—however, your answer is incorrect. The second largest quantity of water is stored in Earth's ice caps and glaciers. Figure 9 shows how the world's supply of water is distributed.

In some regions of the United States, water in its gaseous state is experienced as high humidity that contributes to summer discomfort.

As you know, water can be found in three different physical **states.** Water vapor in the air is in the **gaseous state.** Water is most easily identified in the **liquid state**—in lakes, rivers, oceans, clouds, and rain. Ice is a common example of water in the **solid state.** What other forms of "solid water" can you identify?

At present, most of the United States is fortunate to have abundant supplies of high-quality water. We turn on the tap, use what we need, and go about our daily routine—giving little thought to how that seemingly unlimited water supply manages to reach us.

If you live in a city or town, the water pipes in your home are linked to underground water pipes. These pipes bring water downhill from a reservoir or a water tower, usually located near the highest point in town, to all the faucets in the area. Water stored in the tower was cleaned and purified

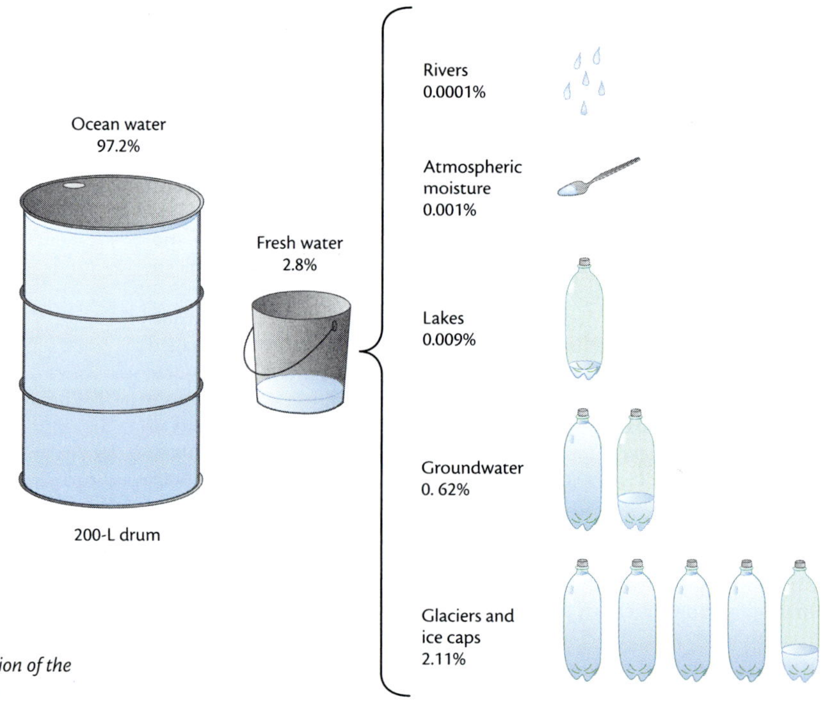

Figure 9 *Distribution of the world's water supply.*

at a water-treatment plant. It may have been pumped to the treatment plant from a reservoir, lake, or river. If your home's water supply originated in a river or other body of water, you are using **surface water.** If it originated in a well, you are using **groundwater.** Groundwater must be pumped to the surface.

If you live in a rural area, your home probably has its own water-supply system. A well with a pipe driven deep into an **aquifer** (a water-bearing layer of rock, sand, or gravel) pumps groundwater to the surface. A small pressurized tank holds the water before it enters your home's plumbing system.

About one-fifth of the nation's water supply is groundwater.

Not surprisingly, neither groundwater nor surface-water samples are completely pure. When water falls as precipitation (rain, snow, sleet, or hail) and joins a stream or when it seeps far into the soil to become groundwater, it picks up small amounts of dissolved gases, soil, and rock. However, these dissolved substances are rarely removed from water at the treatment plant or from well water. In the amounts normally found in water, they are harmless. In fact, some minerals found in water (such as iron, zinc, or calcium) are essential in small quantities to human health or may improve water's flavor.

When the water supply is shut off, as it was in Riverwood, it is usually shut off for a short time. But suppose a drought lasted several years. Or suppose the shortage was perpetual, as it is in some areas of the world. In such circumstances, what uses of water would you give up first? It is clear that the use of available water for survival purposes would have priority. Nonessential uses would probably be eliminated.

Now it is time to examine your water-use data. Refer to your completed water-use diary to find out how much water you use in your home and for what purposes. The table in Figure 10 on page 18 lists typical quantities of water used in the home for common purposes.

A.6 WATER-USE ANALYSIS — Making Decisions

Use the table in Figure 10 on page 18 to answer the following questions regarding your household's water use, as well as those of your classmates.

Questions

1. Estimate the total water volume (in liters) used by your household during the three days.
2. How much water (in liters) did one member of your household use, on average, in one day?

Compile the answers to Question 2 for all members of your class by creating a histogram. To review the construction of a histogram, see page 11. (*Hint:* The range in value for each histogram bar must be equal and should be chosen so that there are about ten total bars.)

3. What is the range of average daily personal water use within your class?
4. Calculate the mean and median values for the class data. Which do you think is more representative of the data set? That is, which is a better expression of central tendency for these data?

The table in Figure 10 indicates that a regular showerhead delivers 19 L each minute. In English units, that is 5 gallons each minute, or 25 gallons during a five-minute shower! To picture that volume of water, think of 25 gallons as 47 two-liter beverage bottles, or roughly twice the volume of gasoline in a normal automobile "fill up."

This section also traces tap water back through the pipes to its source, supporting the fact that Earth has a finite (though recyclable) supply of water. Students can investigate the source of water for their own community: reservoirs, water towers, pumping stations, wells, rivers, or other site-specific locations. A field trip to a municipal pumping station can be highly informative.

A.6 WATER-USE ANALYSIS
Making Decisions

In this activity, students analyze the data they collected earlier about daily use of water by their families. Students gain useful experience transforming raw data into more interpretable forms. The data processing includes finding the total volume of water used by the family during the survey, and then calculating the total water use per person, as well as the daily average water use. Students compare their results with individual water-use information provided in the unit.

The activity takes about 25 to 35 minutes. A data sheet for reporting the results may be helpful, and group work on calculations may help some students to complete them successfully. These calculations provide a good opportunity to stress the importance of keeping track of units associated with any quantities being computed. Answers will vary according to how accurately students recorded water use, location, and weather conditions.

An additional activity could be to compare the above calculated water use with that of an actual water bill. However, be prepared to convert the data in liters (from Making Decisions) to the units used in a typical water bill.

EMERGENCY WATER FOR RIVERWOOD

ChemQuandary 2

Students are required to compare collected data and make an estimate of the water requirement for the town of Riverwood.

Answers

1. Using the national average of 316 L/person/day × 19,500 of people × 3 days = amount of water needed (316 L/day is based on U.S. Geological Survey data)
2. Answers might include: weather conditions, industry or hospital needs, or other indirect water requirements that must be supported for the three days.
3. Water use in Riverwood is similar to national or student averages.

Figure 10 *Water required for typical activities.*

Water Required for Typical Activities

Activity	Volume of Water (L)
Bathing (per bath)	130
Showering (per min)	
Regular showerhead	19
Water-efficient showerhead	9
Cooking and drinking	
Per 10 cups of water	2
Flushing toilet (per flush)	
Conventional toilet	19
"Water saver" toilet	13
"Low flow" toilet	6
Watering lawn (per hour)	1130
Washing clothes (per load)	170
Washing dishes (per load)	
By hand (with water running)	114
By hand (washing and rinsing in dishpans)	19
By machine (full cycle)	61
By machine (short cycle)	26
Washing car (running hose)	680
Running water in sink (per min)	
Conventional faucet	19
Water-saving faucet	9

5. Compare your answer to Question 2 with the estimated total volume of water, 300 L, used daily by each person in the United States. What reasons can you give to explain any difference between your value and the national value?
6. Which is closer to the national average (mean) for daily water use by each person, your answer to Question 2 or the class average in Question 4? What reasons can you give to explain why that value is closer?

EMERGENCY WATER FOR RIVERWOOD

ChemQuandary 2

Recall that the Riverwood town council arranged to truck water from Mapleton to Riverwood for three days to meet the needs of Riverwood residents for drinking water and cooking water. The current population of Riverwood is about 19 500.

1. On the basis of the water-use data collected and analyzed by your class (pages 13–14 and 17–18), explain how you might estimate the total volume (in liters or gallons) of water actually trucked to Riverwood during the three days.

2. What additional information would help you to improve your water-volume estimate? Why?
3. What assumptions must you make to complete your estimate?

You are now quite aware of the amount of water that you use daily. Suppose you had to live with much less. How would you ration your water allowance for survival and comfort? This is exactly the question that Riverwood residents are confronting.

A.7 RIVERWOOD WATER USE

Making Decisions

Riverwood authorities have severely rationed your home water supply for three days while possible fish-kill causes are investigated. The County Sanitation Commission recommends cleaning and rinsing your bathtub, adding a tight stopper, and filling the tub with water. That water will be your family's total water supply for every use other than drinking and cooking for the three-day period. (Recall that water for drinking and cooking will be trucked in from Mapleton.)

Assuming that your household has just one tub of water, 150 L (40 gal), to use during these three days and considering the typical water uses listed here, answer the questions below.

- washing cars, floors, windows, pets
- bathing, showering, washing hair, washing hands
- washing clothes, dishes
- watering indoor plants, outdoor plants, lawn
- flushing toilets

CD-ROM WWW. Questions & Answers

1. Which water uses could you do without completely? What would be the consequences?
2. For which tasks could you reduce your water use? How?
3. Impurities added by using water for one particular use may not prevent its reuse for other purposes. For example, you might decide to save hand-washing water and use it later to bathe your pet dog.
   a. For which activities could you use such impure water?
   b. From which prior uses could this water be taken?

It should now be obvious that clean water is a valuable resource that must not be taken for granted. See Figure 11. Unfortunately, water is easily contaminated. In the next section, as Riverwood works on dealing with its water emergency, you will examine some of the causes of water contamination.

Figure 11 *Water-conservation arithmetic*

A.7 RIVERWOOD WATER USE

Making Decisions

This simulation involves water-use decisions that individuals might be required to make during a water shortage. Students may have experienced some actual short-term community emergency, such as a water-main break, that would increase the realism of this exercise. Students are challenged to plan how to use a limited supply (150 L) of water for maximum benefit. This activity is appropriate for group work and takes 15 to 20 minutes to complete.

Answers

1. Eliminate washing cars, windows, pets, and clothes, watering indoor and outdoor plants, and watering lawns. Some may answer bathing, showers, and washing hair! The consequences would be dirty cars, windows, and pets, dying plants, brown lawns, and increasing odor levels.
2. Reduce water for bathing, showering, washing clothes and dishes, and flushing the toilet. How would you reduce these? Take shorter showers, lower the bathwater level, wash dishes by hand, do not flush the toilet every time, and use low-flow toilets and showerheads.
3. a. for washing cars, flushing toilets, possibly watering plants and lawns

   b. from washing hands and from preparing food

It has been estimated that 30 to 50% of the water used in the United States is wasted. In the discussion, lead students to see how individual acts of conservation can lead to significant savings at the household level and even more at the community or national level, where the impact is multiplied by 260 million.

Explain the relevance of the Benjamin Franklin quotation (1746), "When the well is dry, we learn the worth of water." **Answer:** It is in the nature of economic systems to place a higher value on scarce commodities. Even though water is a renewable resource, it is possible to deplete and pollute it at a rate that exceeds natural chemical recycling processes.

SECTION SUMMARY

Reviewing the Concepts

1. Explain why placing a paper label on a juice container is an indirect use of water.

Water is used indirectly in the processes of watering trees, washing and cutting wood at the lumber mill, manufacturing and printing the paper, washing the can, producing the glue, and attaching the label to the can.

2. If you let the water run while brushing your teeth, explain how you are using water

a. directly.

b. indirectly.

A couple of positions could be taken on this issue. Water is used directly during the procedure, as it flows down the drain; and indirectly, to manufacture the toothpaste and brush. Arguments could be made that only the water used to brush and rinse with are direct usage and that the wasted water is indirect use (the cost of doing business).

3. Assume that Riverwood resident Jimmy Hendricks drank only pre-packaged fruit juice during the water shortage. Does that mean he did not use any water? Explain.

No, a substantial amount of indirect water use is involved in preparing a bottle of juice, from watering the fruit to manufacturing the can.

4. Listed below are some water uses associated with the foul-water laboratory activity. Classify each as either a direct or an indirect water use. Explain your answers.

a. the manufacture of the filter paper

b. the pre-moistening of the sand and gravel

c. the use of water to cool the distillation apparatus

a. Indirect; water is required to grow trees, process lumber, and manufacture paper.

b. Direct; you pour water into sand and gravel, and some of this will be in the product.

c. Direct, although arguments could be made for indirect use.

5. Identify several techniques that can be used to purify water in a household. Describe the nature of each technique.

Municipal treatment systems purify water by screening, pre-chlorination, flocculation, settling, sand filtration, post-chlorination, aeration, pH adjustment, and fluoridation. Household water derived from a well has undergone natural filtration. Some households use filters for screening out particulate matter.

6. What kinds of material were removed from your foul-water sample in each step of that laboratory activity?

Oil-water separation removes the oil; sand filtration removes solid particles (coffee grounds, macroscopic-sized particles); charcoal adsorption/filtration removes the color and odor (microscopic-sized particles); distillation removes dissolved salts.

7. **a.** Would the water-purification procedures that you used in the foul-water laboratory activity make seawater suitable for drinking? Explain.

b. What, if any, additional purification steps would be needed to make seawater suitable for drinking?

a. No, the student procedures could leave microorganisms and microscopic contaminants in the sample.

b. The continued procedure to distill the sample would make seawater drinkable, as this leaves particulate matter in the residue.

8. Has the world's supply of water changed in volume in the past 100 years? The past million years? Explain.

No and no; the amount of water on Earth has remained constant for billions of years, and usable water has remained constant for millions of years.

9. In regard to Earth's water supply, rank the following sources in order of greatest to least total abundance: rivers, oceans, glaciers, water vapor.

oceans, glaciers, rivers, water vapor

10. Consider this familiar quotation: "Water, water, everywhere, Nor any drop to drink." Describe a situation in which this statement would, in fact, be true.

Some response examples are: During a flood, there is plenty of contaminated water. A clear mountain stream may carry disease-producing microorganisms. Atmospheric humidity is unavailable for immediate consumption. The oceans contain vast quantities of saltwater unsuitable for direct consumption.

11. Diagram the major steps of the hydrologic cycle.

Features of a diagram should illustrate the following: Liquid water evaporates and forms clouds in the upper atmosphere, and the water falls to Earth in the form of precipitation. Water runs off the surface into rivers leading to oceans or soaks through the ground into underground aquifers.

Connecting the Concepts

12. Explain why it might be possible that a molecule of water that you drank today was once swallowed by a dinosaur.

The water used by dinosaurs remains on Earth to be used by us. The hydrological cycle is one method by which water is continuously redistributed around the planet. Atmospheric winds and ocean currents ensure that the water used today may be used again by ourselves, our neighbors, and our descendants.

13. You are marooned on a sandy island surrounded by ocean water. A stagnant, murky looking pond contains the only available water on the island. In your survival kit, you have the following items:

- one nylon jacket
- one plastic cup
- two plastic bags
- one length of rubber tubing
- one knife
- one 1-L bottle of liquid bleach
- one 5-L glass bottle
- one bag of salted peanuts

Describe a plan to produce drinkable water using only the items available.

Student answers may vary. Below are two suggested answers.

a. Steps: settle, decant, filter, collect, and disinfect. Construct a sand filtration device by using the knife to make a hole in the bottom of the cup. Fill about 2/3 of the cup with wet beach sand. Insert one end of the tube into the hole. Put the other end into the storage bottle. Collect water in one bag, allow time for mud to settle at bottom of the bag. Place the jacket (filter) over the cup, and pour off (decant) the water from the top of the bag into the cup. The water will go through the sand, then pass through the tube into the storage bottle. Add some bleach (about four drops per quart) to the storage bottle. Shake the bottle, place it in the sun (to further kill microorganisms) and increase the contact time of the sodium hypochlorite.
b. Make a distillation apparatus by digging a hole in a sunny area, fill it with water (sea or pond) to a level not above the cup, put the cup in the middle of the hole (stabilize by some method), drape the nylon jacket over the hole, and place something in the middle to cause the distilled vapor to condense and drip into the cup. This process will be slow and is not a great method for producing large amounts of drinkable water.

14. What would happen to Earth's hydrologic cycle if the evaporation of water suddenly stopped?

All the water would stay in oceans, rivers, and underground aquifers. No rain would occur to complete the natural purification cycle. The air would be very dry, and in many areas natural vegetation would die.

15. Consider these foul-water purification procedures: oil–water separation, sand filtration, charcoal adsorption and filtration, and distillation. Which procedure would be least practical to use for purifying a city's water supply? Why?

Distillation is too expensive, time-consuming, and energy consuming to be practical for the large-scale operations required by most municipalities.

16. A group of friends is planning a four-day backpacking hike. Some members of the group favor carrying their own bottled water; others wish to carry no water and instead buy bottled water in towns along the trail; still others want to purchase and carry portable filters for use with stream water.

a. What are the advantages and disadvantages of each option?

b. What additional information would the group need before making a decision?

a. You might suggest that students organize a response like the one in the following table:

Option	Advantages	Disadvantages
Carry Own Water	Convenient; no worry about supply; can be inexpensive	Heavy and bulky
Buying Bottled Water	No weight; no backpack space required	Must rely on availability, market hours and locations; can be expensive
Portable Filter	Little weight; removes micro-organisms	Streams must be near trails; takes time to produce desired quantities; some water will still have to be carried for immediate consumption

b. Additional Information Needed: How much water will be needed? What will it weigh? What size containers are needed to carry the water? Where are the stores that sell water located? What are their hours of operation? Do filters remove all disease-causing microorganisms? What is the rate of production of drinkable water by the filters? Where are streams located? What alternatives should be considered?

Extending the Concepts

17. A politician guarantees that if she is elected, every household will have 100% pure water in every tap. Evaluate this promise and predict the likelihood of its success.

Distillation to produce pure water is very expensive, and even then it would not be 100% pure. Some of the minerals and substances in the water are beneficial. Even if distilled water was produced, turbulent flow, temperature, and pressure would dissolve some air in the water so that it would no longer be 100% pure. Dissolved gases give water a pleasing taste; trace minerals are important for chemical reactions in our body. Other compounds are intentionally added to make the water safer and healthier. For example, fluorine is added to prevent tooth decay, and chlorine is added to keep harmful bacteria levels low.

18. Find out how much charcoal is used in a fish aquarium filter and how fast water flows through the filter. Estimate the volume of water that can be filtered by a kilogram of charcoal. How much charcoal would be needed to filter the daily water supply for Riverwood, population 19 500?

Students should check charcoal boxes and aquarium instructions for the recommended quantity of charcoal and for guidelines as to how often to change the charcoal. Research should include measuring the water flow rates through the filters and the amount of water purified for a given quantity of charcoal. Students need to determine what factor they can measure that will indicate when the charcoal in the filter has reached the end of its effectiveness. Once the data is gathered, the amount of charcoal required for the entire town can be extrapolated. The cost of charcoal probably differs by location, and teachers may have students do a cost comparison of filtration methods.

19. One unique characteristic of water is that it is present in all three physical states (solid, liquid, and gas) in the range of temperatures found on our planet. How would the hydrologic cycle be different if this were not true?

See the Extending Your Knowledge Overview below. Changing the temperature range of water would affect the climatology of the planet. Changing the hydrologic cycle would have a significant impact during each phase. Students may find it difficult to research all phases thoroughly, so the teacher may direct them to one particular phase of water. For example, suggest what might happen if water had a freezing point of −50 °C; or, perhaps a thematic approach, such as what might be observed if water did not make the crystalline lattice structures called snow.

20. Charcoal-filter materials are available in various sizes—from briquettes to fine powder. List the advantages and disadvantages of using either large charcoal pieces or small charcoal pieces for filtering.

The finer the powder, the greater the surface area available to adsorb odor and color. Charcoal that has been activated is designed especially to remove odor and color. Fine powder can be a lung irritant. Student study should include flow rates through various charcoal mediums and whether single or multiple filtering systems would be most advantageous for large quantities of water. Of course, cost is always a factor.

EXTENDING YOUR KNOWLEDGE OVERVIEW

The hydrologic cycle works because liquid water evaporates to form a lower-density gas, which rises into the atmosphere, cools, and forms clouds. Evaporation also leaves behind the dissolved particles present in impure water. Water in clouds may cool, producing higher-density liquid and solid states that fall back to Earth as precipitation.

The maximum and minimum temperatures on Earth are 45 °C and −40 °C, respectively. For the Moon, the values are 130 °C and −173 °C. On Venus the values are 480 °C and 90 °C. The large amount of water on Earth's surface helps to stabilize the surface temperature during Earth's rotation and revolution around the Sun. Compared to other common substances, water requires large quantities of heat energy for its temperature to increase or decrease. Also, a large quantity of energy is used or released when water changes from one physical state to another.

Heat of fusion is defined as the thermal energy needed to melt (convert a solid to a liquid) a specified quantity of a solid at its melting point.

Heat of vaporization is defined as the thermal energy needed to vaporize (convert to gas) a specified quantity of liquid at its boiling point.

Heat capacity refers to the quantity of thermal energy needed to change the temperature of a given sample of matter—without changing the phase—by one degree Celsius.

Compared with other common substances, water has an unusually high heat of fusion, heat of vaporization, and heat capacity. The next best candidate for a liquid that could support some type of life is ammonia.

Ammonia has much lower freezing and boiling points than water. Thus, even if ammonia were as good a solvent as water, an ammonia-based organism would have to function within a much narrower and lower temperature range (−78 °C to −33 °C).

Formula	Melting Point (°C)	Boiling Point (°C)	Heat of Fusion (J/g)	Heat of Vaporization (J/g)
H_2O	0	100	333	2260
NH_3	−78	−33	341	1370

MEETING RAISES FISH-KILL CONCERNS

Chemists and other authorities report their progress in the investigation of the Snake River fish kill using such terms as *dissolved oxygen, suspended,* and *solubility.* In order to understand the fish-kill issue, students need to understand basic chemical concepts regarding the behavior of water.

ADDITIONAL ACTIVITIES

Students can maintain a "Chemistry in the News" bulletin board where local, regional, and national newspaper or magazine articles, cartoons, and so on, relating to the main idea of the unit being studied are displayed. Students should pay particular attention to actual reported fish kills. You may have students post these articles on the bulletin board or have them give a small report to the class. One example would be to have students type up a summary of the article in large print and post their summary along with the original article on the class bulletin board. Encourage multiple articles and post new articles on a regular basis to maintain interest.

Such activities help emphasize that Riverwood-type scenarios are not just stories but happen regularly. Chemistry-related issues appear consistently in both national and local news, and *ChemCom* topics focus on a wide variety of community settings and events.

MEETING RAISES FISH-KILL CONCERNS

BY CAROL SIMMONS
Riverwood News Staff Reporter

More than 300 concerned citizens, many "armed" with lists of questions, attended a Riverwood Town Hall public meeting last night to hear from the scientists investigating the Snake River fish kill.

Dr. Harold Schmidt, a scientist with the Environmental Protection Agency (EPA), expressed regret that the fishing tournament was canceled but strongly supported the town council's action, saying that it was the safest course in the long run. He reported that his laboratory is conducting further river-water tests.

Dr. Margaret Brooke, a State University water specialist, helped interpret information and answered questions. Local physician Dr. Jason Martingdale and Riverwood High School home economics teacher Alicia Green joined the speakers during the question-and-answer session that followed.

Dr. Brooke confirmed that preliminary water-sample analyses showed no likely cause for the fish kill. She reported that EPA chemists will collect water samples at hourly intervals today to look for any unusual fluctuations in dissolved-oxygen levels, as fish must take in adequate oxygen gas dissolved in water.

Concerning possible fish-kill causes, Dr. Brooke said that EPA scientists have been unable to identify any microorganisms in the fish that could have been responsible for their death. She concluded that "it must have been something dissolved or temporarily suspended in the water." If dissolved matter were involved, important considerations would include the relative amounts of various substances that can dissolve in water and the effect of water temperature on their solubility. She expressed confidence that further studies would shed more light.

Dr. Martingdale reassured citizens that "thus far, no illness reported by either physicians or the hospital can be linked to drinking water." Ms. Green offered water-conservation tips for housekeeping and cooking to make life easier for inconvenienced citizens. The information sheet that she distributed is available at Town Hall.

Mayor Edward Cisko confirmed that water supplies will again be trucked in from Mapleton today and expressed hope that the crisis will last no longer than three days.

Those attending the meeting appeared to accept the emergency situation with good spirits. "I'll never take my tap water for granted again," said Trudy Anderson, a Riverwood resident. "I thought scientists would have the answers," puzzled Robert Morgan, head of Morgan Enterprises. "They don't know either! There's certainly more involved in all this than I ever imagined."

SECTION B: A LOOK AT WATER AND ITS CONTAMINANTS

Section B returns to Riverwood's problem with another news report. To understand the issues, students need to know more about the chemistry of water and how substances interact with water. Some properties of water are given, and students learn to classify mixtures as suspensions, colloids, or solutions. A simple particulate model of matter is proposed. The universal symbolic language of chemistry is also introduced. Students then learn the difference between pure water and clean water, which leads them to laboratory testing for dissolved ions in water samples.

SECTION **B**

A LOOK AT WATER AND ITS CONTAMINANTS

As the preceding article indicates, scientists attribute the cause of the fish kill to something dissolved or suspended in the Snake River. What might those substances be? How can the search for the cause be narrowed further? Knowledge of the properties of water (and of substances that might be found in it) will aid in this task. To understand these properties, you will be introduced to matter at the particulate level. You will also begin to learn the language of chemistry and to use it to communicate with your classmates as you investigate the fish kill.

B.1 PHYSICAL PROPERTIES OF WATER

Water is a common substance—so common that it is usually taken for granted. You drink it, wash with it, swim in it, and sometimes grumble when it falls from the sky. But are you aware that water is one of the rarest and most unusual substances in the universe? As planetary space probes have gathered data, scientists have learned that the great abundance of water on Earth is unmatched by any planet or moon in our solar system. Earth is usually half-enveloped by water-laden clouds, as you can see in Figure 12. In addition, more than 70% of Earth's surface is covered by oceans having an average depth of more than three kilometers (two miles).

Kilo- (k) is the metric prefix meaning 1000. One kilometer (km) = 1000 meters (m).

Figure 12 *Earth as seen from space. What states of water can be observed in this winter scene?*

B.1 PHYSICAL PROPERTIES OF WATER

This section focuses on the physical properties of water—density, physical state, surface tension, boiling and freezing points, and solvent characteristics. The important property of density is carefully investigated. To emphasize the unusual properties of water, begin the class session with some of the demonstrations described below. Ask students to interpret and explain the results of each demonstration.

Demonstration Ideas

1. Surface tension: Float a needle or a rubber band on the surface of the water. This can be done by gently lowering the needle onto the surface of the water using a bent paper clip. Show that more water can be added to a full glass of water. Fill a glass to the rim, then continue to add water drop by drop until it bulges well above the rim. The same effect can be demonstrated by filling a glass with water even with the rim, then asking students how many paper clips could be added to the same glass. Many more than they expect can be added due to the surface tension of the water clinging to the edge of the glass.
2. Density of water: Find the mass of 10.0 mL water in a 100-mL plastic graduated cylinder. Have the students calculate the density of liquid water. Then place the cylinder in the freezing compartment of a refrigerator. Find the new volume and have students calculate the density of ice.

Many new chemical terms and concepts are introduced in Section B. It is important to remember that because these concepts will be revisited many times during the *ChemCom* course, it is not necessary to teach for mastery of each concept now. Many of the chemical concepts are in a spiral curriculum format. The Section Summary at the end of the unit provides important clues concerning the depth of knowledge required at this time.

With more information, students go back to the newspaper articles to separate fact from interpretation. Knowledge of chemistry provides a framework for asking questions about societal problems that may help lead to solutions.

3. A density column: Carefully place the following materials in a large graduated cylinder (to show density in terms of g/mL) in this order: lead sinker (density 11.3), Karo syrup (density 1.36), glycerin (density 1.26), rubber stopper (density 1.2), ethylene glycol (density 1.11), water (density 1.0), plastic (density 0.9), oil (cooking or mineral, density 0.9), oak (density 0.9), alcohol (either ethanol or 2-propanol, density 0.79), and cork (density 0.2). To obtain discrete layers, pour the materials slowly down the side of the tilted graduated cylinder, or add each liquid layer using a long-stemmed funnel. Note that ethylene glycol is antifreeze. It can be useful to color the water with food coloring for an added effect.
4. Place an ice cube in a beaker of alcohol. The ice cube will sink rather than float because it is denser than the alcohol.

DENSITY Building Skills 2

Answers

1. The coffee grounds have the greater density, which is why they settle to the bottom; oil, having a density less than water, floats.
2. The density of solid ice is less than water, which is why an ice cube floats when placed in water. If this was reversed, ice would settle to the bottom, and in rivers and lakes of the colder climates the fish would no longer be able to bottom feed. Also, this might affect the overall water temperature of rivers and lakes in colder climates.
3. Mass and volume have to be determined. In the laboratory, a balance could be used to determine the mass. Its volume can be determined by measurement—length times width times height (if it is a regular-shaped object). If it is an irregular-shaped object, the water displacement method can be used to determine volume. Recommend summarizing how regular solids, irregular solids, liquids, and gases are measured for volume in computing a density.

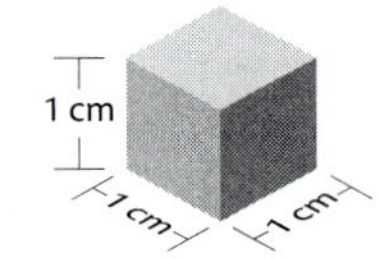

Figure 13 *One cubic centimeter (shown actual size).*

0 °C = 32 °F

Water is a form of matter. As you may recall from previous science courses, matter is anything that occupies space and has mass. All solids, liquids, and gases are classified as matter. Matter can be distinguished by its characteristic properties. Water has many important **physical properties,** properties that can be observed and measured without changing the chemical makeup of the substance. One physical property of matter is **density,** which is a measure of the mass of a material in a given volume. The density of water as a liquid is easy to remember. Because one milliliter (mL) of water has a mass of 1.00 g, the density of water is 1.00 g/mL. One milliliter of volume is exactly equal to one cubic centimeter (1 cm^3), which is pictured in Figure 13. Thus the density of water can also be reported as 1.00 g/cm^3. Another physical property of matter is freezing point. The freezing point of water is 0 °C at normal pressure. Can you think of other physical properties of water?

Water is the only ordinary liquid found naturally in our environment. Because so many substances dissolve readily in water, quite a few liquids are actually water solutions. Such water-based solutions are often called **aqueous solutions.** Even water that seems pure is never entirely so. Surface water contains dissolved minerals as well as other substances. Distilled water used in steam irons and car batteries contains dissolved gases from the atmosphere, as does rainwater.

Pure water is clear, colorless, odorless, and tasteless. The characteristic taste and slight odor of some tap-water samples are caused by substances dissolved in the water. You can confirm this by boiling and then refrigerating a sample of distilled water. When you compare its taste with the taste of chilled tap water, you may notice that "pure" distilled water tastes flat.

Water's physical properties, along with its chemical properties, distinguish it from other substances. In the following activity, you will compare the density of water with the density of some other common materials.

DENSITY — Building Skills 2

Most likely, you are already familiar with such physical properties of water as density, boiling point, and melting point. Use your experiences with water and other materials to answer the following questions concerning density.

1. In the foul-water laboratory activity (page 8), you observed that coffee grounds settled to the bottom of the water sample, whereas oil "floated" on top. Explain this observation in relation to the relative densities of coffee grounds, water, and oil.
2. How does the density of ice compare with that of liquid water? (*Hint:* Use your everyday experiences to answer this question.) What would happen to rivers and lakes (and fish) in cooler climates if the relative densities of ice and liquid water were reversed?
3. Suppose you were given a small cube of copper metal. What measurements would you need to make to determine its density? How would you make these measurements in the laboratory?

B.2 MIXTURES AND SOLUTIONS

How do you know when liquid water is not sufficiently pure? How can substances in water be separated and identified? Answers to these questions will be helpful in understanding and possibly solving the fish-kill mystery. But first you must learn how to recognize various types of mixtures.

When two or more substances combine yet retain their individual properties, the result is called a **mixture.** The foul water that you purified earlier is an example of a mixture because it contained coffee grounds, garlic powder, oil, and salt. As you discovered, the components of a mixture can be separated by physical means such as filtration and adsorption.

When you first examined your foul-water sample, did it look uniform throughout? Most likely, the coffee grounds had settled to the bottom and were not distributed evenly throughout the liquid. The foul water is an example of a **heterogeneous mixture** because its composition is not the same, or uniform, throughout. One type of heterogeneous mixture is called a **suspension** because the particles are large enough to settle out and can be separated by using a filter. Water plus coffee grounds and water plus small pepper particles are examples of suspensions.

A heterogeneous mixture's composition varies.

If the particles are smaller than those in a suspension, they may not settle out and thus may cause the water to appear cloudy. Recall what happened when your teacher shined a light through your sample of purified water. The scattering of the light, known as the Tyndall effect (see Figure 7, page 12), indicated that small, solid particles were still present in the water. This type of mixture is known as a **colloid.**

CD-ROM WWW. **Modeling Matter: Attraction Between Water Molecules**

A more familiar example of a colloid is milk, which contains small butterfat particles dispersed in water. These colloidal butterfat particles are not visible to the unaided eye; the mixture appears uniform throughout. Thus milk can be classified as homogeneous, which leads to the familiar term *homogenized milk.* Under high magnification, however, individual butterfat globules can be observed floating in the water. Milk no longer appears homogeneous. See Figure 14.

Particles smaller than colloidal particles also may be present in a mixture. When small amounts of table salt are mixed with water, as in your foul-water sample, the salt **dissolves** in the water. That is, the salt crystals separate into particles so small that they cannot be seen even at high

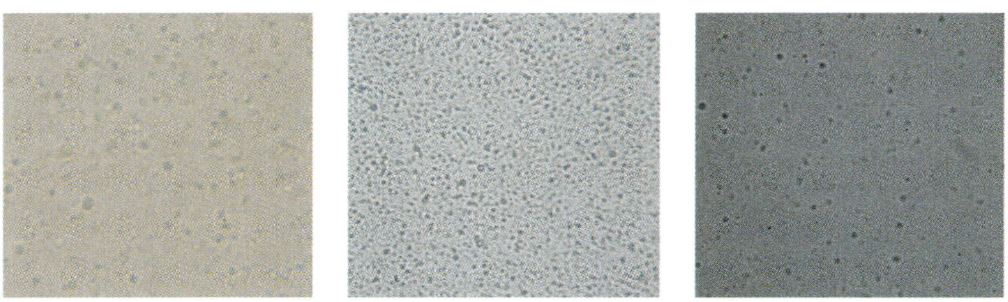

Figure 14 *Fat globules can be seen under magnification, so that milk no longer looks homogeneous.* Left: *Whole milk under 1000X magnification.* Center: *Whole milk under 400X magnification.* Right: *Non-fat milk under 400X magnification.*

B.2 MIXTURES AND SOLUTIONS

This section introduces water mixtures and their classification. A sample of the foul water used in Laboratory Activity A.2 is an effective visual aid. Dissolve table salt in water to demonstrate the formation of a solution.

In this section several important chemical concepts are introduced. It is likely that students have heard these terms before. It is suggested that you provide only enough information about these terms to help students understand the concepts of water solutions and their application to solving the fish-kill mystery. A graphic organizer may help in organizing these concepts. Using an organizational flowchart format, Matter is placed at the top, then Matter is divided into Pure Substances and Mixtures. Pure Substances is split into Elements and Compounds. Mixtures is split into Homogeneous and Heterogeneous. Homogeneous mixtures are Solutions. Generally, Heterogeneous mixtures are Colloids and Suspensions. It is recommended that the properties and characteristics of each category be reviewed.

Demonstrations

Several demonstrations are available to contrast the types of matter. A density column can be used to show a suspension and solution. A few drops of whole milk in a beaker of water can be used to display a colloidal suspension. The Chemical Sunset demonstration illustrates various aspects of solution and colloidal suspension mixtures. In *Chemical Demonstrations: A Sourcebook for Teachers, Volume 1,* Demonstration 96 on page 163, the procedure and reactions of the Chemical Sunset are explained. For large groups it is recommended that the solution be illuminated by an overhead projector. The white light projected through the solution illuminates a yellow-colored colloid; it later turns red, and finally it turns opaque. These chemical changes demonstrate the routine colors seen during a sunset, and the change from a solution to a colloidal suspension. The amount of colloidal sulfur increases, blocking more light, and producing various colors.

B.3 MOLECULAR VIEW OF WATER

Moving beyond directly observable properties of water, this section begins to build a model based on the idea of atoms and molecules. See the *ChemCom* CD-ROM for computer representations of water.

The purpose of this section is to introduce the idea of atoms and molecules; it is not necessary to go much beyond that goal. Even though chemical bonds are introduced, it is not appropriate at this time to go into the different types of chemical bonds. These topics will be revisited later in the text.

The term *matter* is presented with the assumption that students have prior knowledge of its definition and understanding. Remind students that everything around us is made of matter and that matter is anything that has mass and volume. Mass is the amount of material in an object, and volume is the amount of space an object occupies.

At the end of this section, have students evaluate the molecular representation of water, hydrogen and oxygen, in the formula H_2O. However, go beyond the formula and illustrate the V shape (not linear) of the water molecule. Suggest that this bent shape of the water molecule is significant. This shape is the key to understanding many of water's properties. Prompt students to think of how the shape and structure of a molecule play an important part in its properties. Ask questions such as: Is this what molecules look like? Why is water represented as having a V shape—why not linear? Does it make a difference? The term *polar molecule* will be introduced later in section C.4.

magnification. Nor do the particles exhibit the Tyndall effect when a light beam is passed through the mixture. These particles become uniformly mingled with the particles of water, producing a **homogeneous mixture,** or a mixture that is uniform throughout. All **solutions** are homogeneous mixtures. In a salt solution, the salt is the **solute** (the dissolved substance) and the water is the **solvent** (the dissolving agent). All solutions consist of one or more solutes and a solvent.

Molecular Views of Water CD-ROM WWW.

Evidence that something was still dissolved in your purified water sample came from the results of the conductivity test. The positive result (the bulb lit up) indicated that electrically charged particles were dissolved in the mixture.

B.3 MOLECULAR VIEW OF WATER

A compound can be broken down chemically into two or more simpler substances—either elements or new compounds. By definition, an element cannot be broken down into any simpler substances.

So far in this investigation of water, you have focused on properties observable with your unaided senses. In doing so, have you wondered why water's freezing point is 0 °C or why certain substances such as salt dissolve in water? To understand why water has its particular properties, you must investigate it at the level of its atoms and molecules.

All matter is composed of **atoms.** Atoms are often called the building blocks of matter. Matter that is made up of only one kind of atom is known as an **element.** For example, oxygen is considered an element because it is composed of only oxygen atoms. Because hydrogen gas contains only hydrogen atoms, it too is an element. Approximately 90 different elements are found in nature, each having its own unique type of atom and identifying properties.

What type of matter is water? Is it an element? A mixture? As you most likely know, water contains atoms of two elements—oxygen and hydrogen. Thus water cannot be classified as an element. And, because its properties are different from those of oxygen and hydrogen, water cannot be classified as a mixture either.

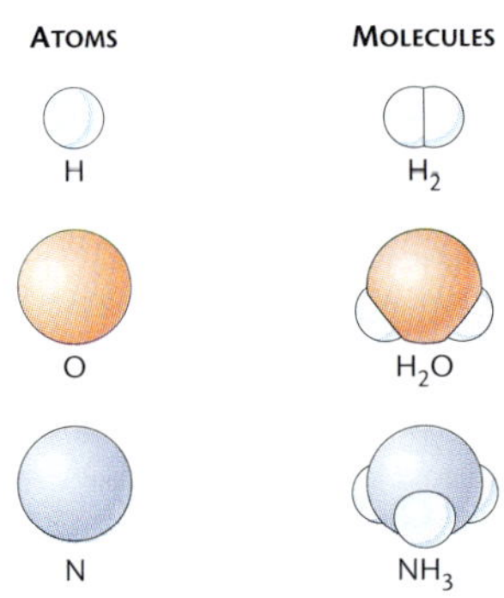

Figure 15 *On the left: hydrogen (H), oxygen (O), and nitrogen (N) atoms. On the right: hydrogen (H_2), water (H_2O), and ammonia (NH_3) molecules. Note the relative sizes of the atoms. Models similar to these are used throughout the textbook to depict atoms and molecules.*

Instead, water is an example of a **compound**—a substance composed of atoms of two or more elements linked together chemically in fixed proportions. To date, chemists have identified more than 18 million compounds. Compounds are represented by chemical formulas. In addition to water (H_2O), some other compounds and formulas with which you may be familiar include table salt (NaCl), ammonia (NH_3), baking soda ($NaHCO_3$), and chalk ($CaCO_3$).

Each element and compound is considered a **pure substance** because each has a uniform and definite composition as well as distinct properties. The smallest unit of a pure substance that retains the properties of that substance is a **molecule.** Atoms of a molecule are held together by **chemical bonds.** You can think of chemical bonds as the "glue" that holds atoms of a molecule together. One molecule of water is composed of two hydrogen atoms bonded to one oxygen atom, hence H_2O. An ammonia molecule (NH_3) contains three hydrogen atoms bonded to a nitrogen atom. Figure 15 shows representations of some atoms and molecules.

The following activity will give you a chance to apply an atomic and molecular view to a variety of common observations.

MODELING MATTER: PICTURES IN THE MIND

Students have an opportunity to illustrate substances on a molecular level, using visual learning. One method is to build some 3-D representations of molecules with marshmallows and toothpicks or other space-filling models. During exercise evaluation, it was discovered that many students had misconceptions about the nature of solids, liquids, and gases. In illustrations, gas particles would be correctly drawn as being very far apart, solids packed close together, but liquids were drawn as the intermediate between the two!

MODELING MATTER

PICTURES IN THE MIND

You live in a macroscopic world—a world filled with large-scale, readily observed things. As you experience the properties and behavior of bulk materials, you probably give little thought to the particulate world of atoms and molecules. If you wrap leftover cake in aluminum foil, it is unlikely that you think about how the individual aluminum atoms are arranged in the wrapping material. It is also unlikely that you consider what the mixture of molecules that make up air looks like as you breathe. And you seldom wonder about the behavior of atoms and molecules when you see water boiling or iron nails rusting.

Having a sense of how individual atoms and molecules might look and behave in elements, compounds, and mixtures can help you explain everyday phenomena. This activity will give you practice in observing, interpreting, evaluating, and creating visual models of matter at the particulate level.

To introduce you to these visualizations, consider this example: Suppose you want to draw a model of a homogeneous mixture of two gaseous compounds. You know that a homogeneous mixture is uniform throughout, so the two compounds should be intermingled and evenly distributed. You also know that compounds are composed of atoms of two or more different elements linked together by chemical bonds.

Suppose a molecule of one of the compounds contains two different atoms. To represent this molecule, you could draw two differently shaded or labeled circles to denote atoms of the two elements and a line connecting the atoms to denote a bond.

Suppose the other compound is composed of molecules that each contain three atoms, and that two of the atoms are of the same element. You now need to choose the order in which the atoms should be connected: the unique atom (Y) could be in the middle, X–Y–X, or on the end, X–X–Y. As long as you draw this imaginary compound in the same way every time, it does not matter which way you do it for this activity. However, the way in which atoms are connected in real compounds does, in fact, make a difference; X–Y–X is a different molecule from X–X–Y.

Examine the three models (a, b, and c) in the illustration. Which best represents a homogeneous mixture of the two compounds just described? You are correct if you said that b is the best visual model. The two types of molecules are uniformly mixed, and the atoms are shaded to indicate that they represent different elements. In a, the mixture is not homogeneous because the molecules are not uniformly mixed. Model c contains three different compounds instead of two. Notice that in a, bonded atoms in each molecule are connected by lines. In b and c, bonded atoms just touch each other. Both

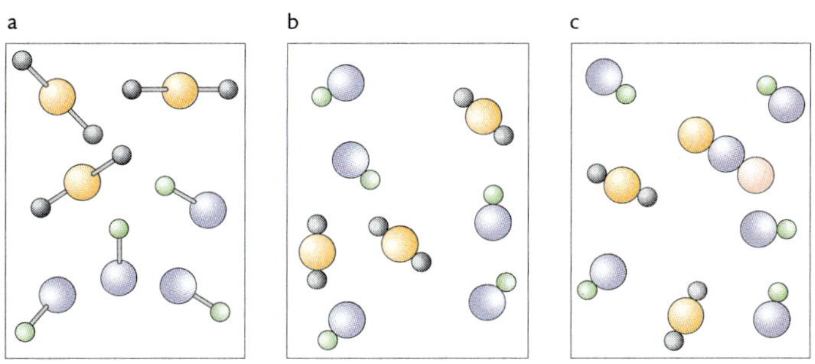

Answers

1. This drawing should have three distinct kinds of molecules. It is of little consequence how many atoms each molecule has. Real gases can be single atoms, such as in the noble gases; diatomic, as in oxygen and chlorine; or very complex, as in volatile organic compounds. The molecules must be spread far apart and uniformly distributed.

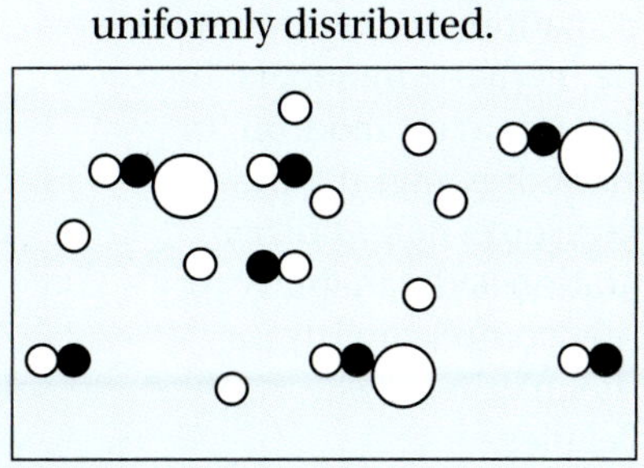

2. The model appears to represent a gaseous, heterogeneous mixture of two substances, each consisting of molecules with several atoms. In one angular molecule, there are two of one kind of atom bonded to a different atom in the middle. In the other, three atoms are bonded linearly to one different atom.

3. a. In a mixture of two gaseous elements, the particles should be spread far apart and mixed uniformly throughout the field, with several molecules of each of two gases. The gases could both be monatomic or diatomic; one of each is shown, and either is acceptable.

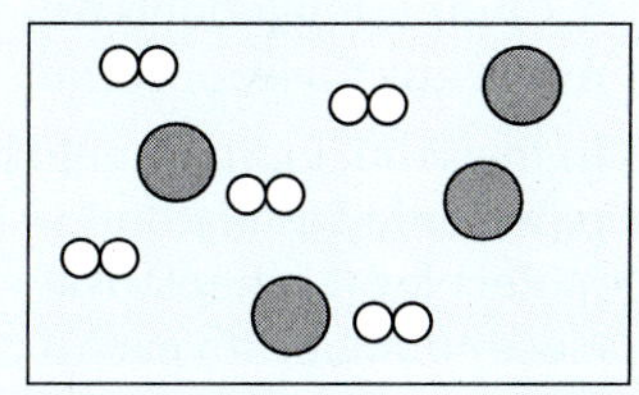

b. A two-atom compound of X and Z is linear. The only variation would be the size of the two atoms.

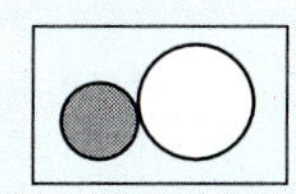

c. In a four-atom compound there are multiple choices as to how the student can connect the various parts. Any combination that connects a total of four atoms with at least one atom of X and Z included is satisfactory.

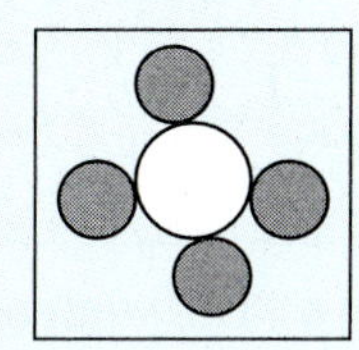

d. In this illustration, there should be significantly more of the two-atom molecules of L and R. The solute of D and T should be in the lesser amount. The solute particles should be homogeneously distributed throughout the solvent and separated from each other.

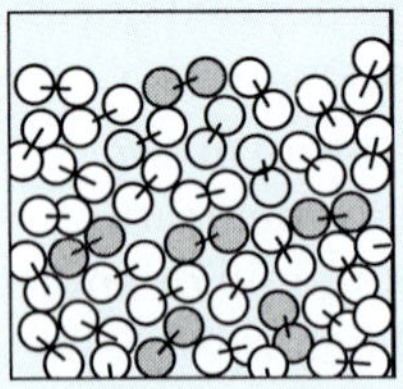

4. a. Students will find various results. The models should show the characteristics outlined above. The size, color, and total number matter very little.

b. Comparing models will reveal any misconceptions.

5. In the solid, iodine would be represented as diatomic molecules (I_2) in relatively close contact with each other. As a gas, they would still be diatomic, but spread much farther apart.

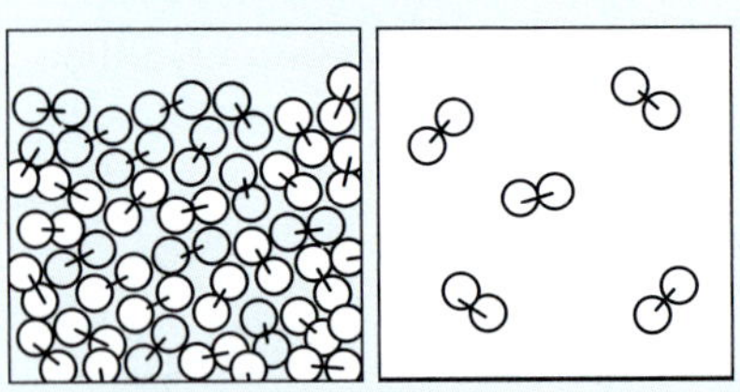

6. The student from Riverwood seems to have it about right, representing the element as a diatomic molecule, as in H_2 or O_2. It would also be acceptable to have the elemental gas as single free atoms, as in the noble gases, He and Ar. A gas-gas solution is expected to be homogeneously intermixed.

MODELING MATTER (cont'd)

representations are used by chemists; either one is acceptable in this activity.

Now it is your turn to create and evaluate various visual models of matter.

1. Draw a model of a homogeneous mixture composed of three different gaseous elements. Describe the key features of your drawing.
2. What kind of matter does the following model represent? Explain your answer.

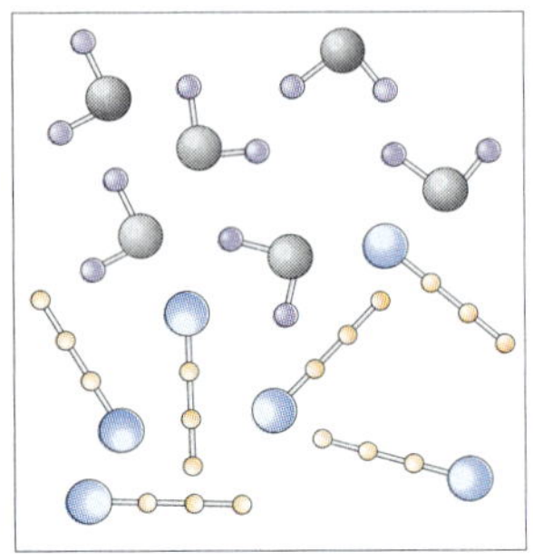

3. Draw a model of each of the following samples of matter. Write a description of key features of each model.
   a. a mixture of gaseous elements X and Z
   b. a two-atom compound of X and Z
   c. a four-atom compound of X and Z
   d. a solution composed of a solvent that is a two-atom compound of L and R, and a solute that is a compound composed of two atoms of D and one atom of T
4. One at a time, compare each visual model that you created in Question 3 with those of your classmates.
   a. Although the models may look a little different, does each set depict the same type of sample? Comment on any similarities and differences.
   b. Do the differences help or hinder your ability to visualize the type of matter being depicted?
5. The element iodine (I) has a greater density in the solid state than in the gaseous state. Draw models that depict and account for this difference at the atomic level. Iodine exists as a two-atom molecule.
6. A student in a chemistry class at Riverwood High School was asked to draw a model of a mixture composed of an element and a compound. Comment on the usefulness of the student's drawing.

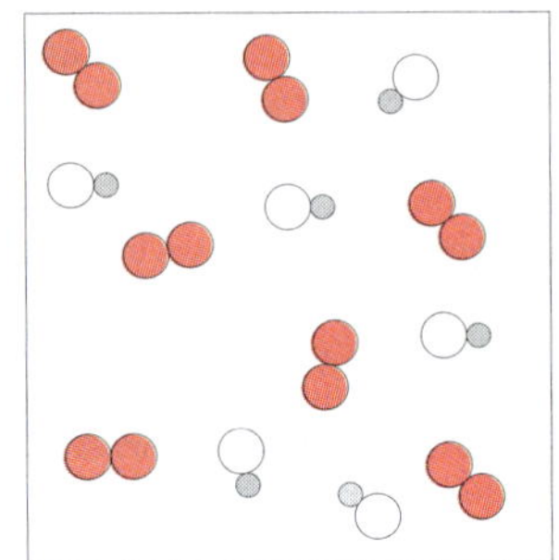

7. You have been interpreting and creating two-dimensional models of three-dimensional molecules.
   a. What are the limitations of two-dimensional representations?
   b. How can two-dimensional drawings be enhanced to show the features of three-dimensional atoms and molecules?
   c. Describe how you could make three-dimensional models from everyday materials.
8. a. How useful are models to you in visualizing matter at the particulate level?
   b. What characteristics do good models of matter have?

As you continue to study chemistry, you will encounter visual models of matter similar to those in this activity. When you see them, think about their usefulness as well as their possible limitations.

7. a. The flat representations sometimes give incorrect information. A good example is the methane molecule. In reality it is a tetrahedron. In student drawings it almost always looks flat. Another limitation is the inability to render complex twists or cross-links in large molecules.

b. A sense of three dimensions can be given with perspective, size, and shadowing. Changing the relative size of the molecules can add the perception of depth. By using perspective, parts of the molecule can be made to project out from the plane of the paper. Shadowing can illustrate the spherical nature of particles.

B.4 SYMBOLS, FORMULAS, AND EQUATIONS

An international "chemical language" for use in oral and written communication has been developed to represent atoms, elements, and compounds. The "letters" in this language's alphabet are **chemical symbols,** which are understood by scientists throughout the world. Each element is assigned a chemical symbol. Only the first letter of the symbol is capitalized; all other letters are lowercase. For example, C is the symbol for the element carbon and Ca is the symbol for the element calcium. Symbols for some common elements are listed in Figure 16.

All known elements are organized into the Periodic Table of the Elements, which is one of the most useful tools of chemists. As you continue your study of chemistry, you will learn more about this important table. For now, become familiar with this tool by locating each element listed in Figure 16 on the Periodic Table found on the inside back cover of this textbook. How many of these elements have you heard of before?

"Words" in the language of chemistry are composed of "letters" (elements) from the Periodic Table. Each "word" is a **chemical formula,** which represents a different chemical substance. In the chemical formula of a substance, a chemical symbol represents each element present. A **subscript** (a number written below the normal line of letters) indicates how many atoms of each element just to the left of the number are in one molecule or unit of the substance.

For example, as you already know, the chemical formula for water is H_2O. The subscript 2 indicates that each water molecule contains two hydrogen atoms. Each water molecule also contains one oxygen atom. However, the subscript 1 is understood and is therefore not included in a chemical formula. Here is another example. The chemical formula for propane, a compound commonly used as a fuel, is C_3H_8. What elements are present in propane, and how many atoms of each are there? You are correct if you said each propane molecule consists of three atoms of carbon and eight atoms of hydrogen.

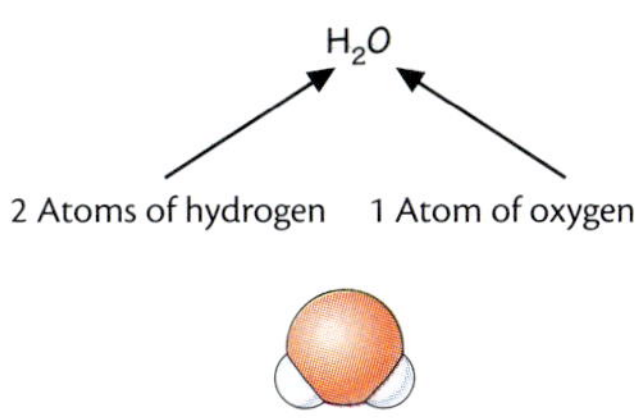

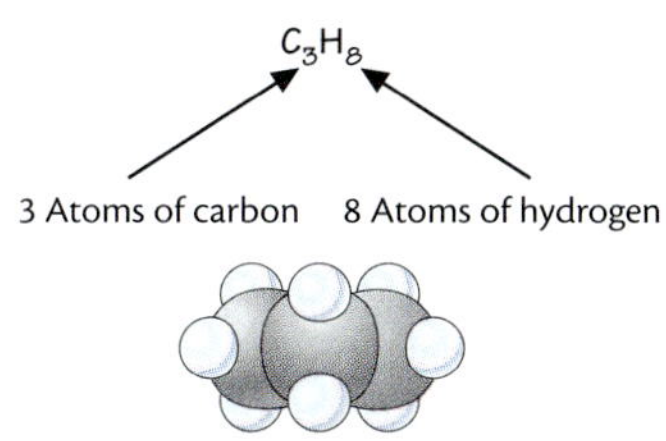

If formulas are the "words" in the language of chemistry, then **chemical equations** are the "sentences." Each chemical equation summarizes the details of a particular chemical reaction. **Chemical reactions** entail the breaking and forming of chemical bonds, causing atoms to become rearranged into new substances. These new substances have different properties from those of the original material(s).

Common Elements	
Name	Symbol
Aluminum	Al
Bromine	Br
Calcium	Ca
Carbon	C
Chlorine	Cl
Cobalt	Co
Copper	Cu
Gold	Au
Hydrogen	H
Iodine	I
Iron	Fe
Lead	Pb
Magnesium	Mg
Mercury	Hg
Nickel	Ni
Nitrogen	N
Oxygen	O
Phosphorus	P
Potassium	K
Silver	Ag
Sodium	Na
Sulfur	S
Tin	Sn

Figure 16 *Common elements.*

B.4 SYMBOLS, FORMULAS, AND EQUATIONS

Students are introduced to symbols, formulas, and equations and to the significance of subscripts and coefficients. Emphasize the importance of a common, worldwide chemical language. These topics are further developed in the Resources Unit. It is not necessary for students to memorize symbols or learn to balance equations now. Instead, show students where to find a listing of symbols and elements. Emphasize that an equation represents a chemical reaction that happens or that someone predicts might happen.

c. Students may suggest using gumdrops, marshmallows, clay, or any other common art modeling substance. Bonds could be represented by toothpicks, coffee stirrers, pipe cleaners, or other similar materials.

8. a. Visualizing the particulate level is hard to do. It is hoped students will find the models in this exercise helpful.

b. Good models reflect the true nature of the concept they represent. Models should be clear and easy to visualize and understand. Good models extend understanding by showing interactions and relationships.

See the *ChemCom* CD-ROM for computer model examples.

WORKING WITH SYMBOLS, FORMULAS, AND EQUATIONS

Building Skills 3

This activity reinforces the information just read and discussed by giving students practice in recognizing and interpreting symbols, formulas, and equations. Chemical equations are introduced in this section. The goal is to briefly illustrate reactions of chemicals and to introduce "atom inventory" balanced chemical equations. It is not necessary to explain the various types of chemical equations or how we balance these equations. Chemical equations will be covered in more depth in the Materials Unit.

Answers

1. a. i. phosphorus; ii. nickel; iii. copper; iv. cobalt; v. bromine; vi. potassium; vii. sodium; viii. iron

   b. phosphorus, nickel, cobalt, bromine

   c. Potassium, copper, sodium, and iron—all have Latin-name symbols, which might make them more international.

2. a. 2 hydrogen atoms, 2 oxygen atoms

   b. 1 calcium atom, 2 chlorine atoms

   c. 1 sodium atom, 1 hydrogen atom, 1 carbon atom, 3 oxygen atoms

   d. 2 hydrogen atoms, 1 sulfur atom, 4 oxygen atoms

3. a. Methane combusts in the presence of oxygen, yielding the reaction products of carbon dioxide and water.

   b. They are all compounds except oxygen.

   c. 1 carbon atom and 4 hydrogen atoms + 4 oxygen atoms = 1 carbon atom and 2 oxygen atoms + 4 hydrogen atoms and 2 oxygen atoms

   d. Illustrations will vary, but encourage a balanced number of atoms.

Extra items for practice

Ethylene glycol (car antifreeze): $C_2H_6O_2$—2 carbon atoms, 6 hydrogen atoms, 2 oxygen atoms; Ammonium phosphate (a fertilizer): $(NH_4)_3PO_4$—3 nitrogen atoms, 12 hydrogen atoms, 1 phosphorus, 4 oxygen atoms; Iron(III) oxide (common rust): Fe_2O_3—2 iron atoms, 3 oxygen atoms

Elements That Exist as Diatomic Molecules	
Element	**Formula**
Hydrogen	H_2
Nitrogen	N_2
Oxygen	O_2
Fluorine	F_2
Chlorine	Cl_2
Bromine	Br_2
Iodine	I_2

Figure 17 *These elements occur naturally as diatomic molecules.*

"GEN-U-INE DIATOMICS" can serve as a good memory device for all common diatomic elements. The names of the diatomic elements end in either GEN or INE, and U better remember them!

The chemical equation for the formation of water

$$2\,H_2 + O_2 \longrightarrow 2\,H_2O$$

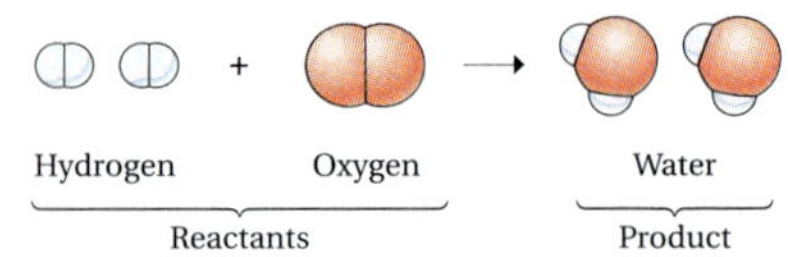

shows that two hydrogen molecules ($2\,H_2$) and one oxygen molecule (O_2) react to produce ($\rightarrow$) two molecules of water ($2\,H_2O$). The original (starting) substances in a chemical reaction are called the **reactants;** their formulas are always written on the left side of the arrow. The new substance or substances formed from the rearrangement of the reactant atoms are called **products;** their formulas are always written on the right side of the arrow. Note that this equation, like all chemical equations, is balanced—the total number of each type of atom (four H atoms and two O atoms) is the same for both reactants and products.

Perhaps you noticed that in the chemical equation for the formation of water the reactants hydrogen and oxygen are written with subscripts of 2 (H_2 and O_2). Do all elements have subscripts? Most uncombined elements in chemical equations are represented as single atoms (Cu, Fe, Na, and Mg, for example). A handful of elements are **diatomic molecules;** they exist as two bonded atoms of the same element. Oxygen and hydrogen are two examples of diatomic molecules. Figure 17 lists all the elements that exist as diatomic molecules at normal conditions. It will be helpful for you to remember these elements. Find the diatomic elements in the Periodic Table. Where are they located?

WORKING WITH SYMBOLS, FORMULAS, AND EQUATIONS

Building Skills 3

1. a. Name the element represented by each symbol below.

i. P	v. Br
ii. Ni	vi. K
iii. Cu	vii. Na
iv. Co	viii. Fe

   b. Which elements in Question 1a have symbols corresponding to their English names?

   c. Which is more likely to be the same throughout the world—the element's symbol or its name?

2. For each formula, name the elements present and give the number of atoms of each element.

a. H_2O_2	Hydrogen peroxide (antiseptic)
b. $CaCl_2$	Calcium chloride (de-icer for sidewalks)
c. $NaHCO_3$	Sodium hydrogen carbonate (baking soda)
d. H_2SO_4	Sulfuric acid (battery acid)

Look at the information available in a chemical equation:

$$N_2 + 3\,H_2 \longrightarrow 2\,NH_3$$

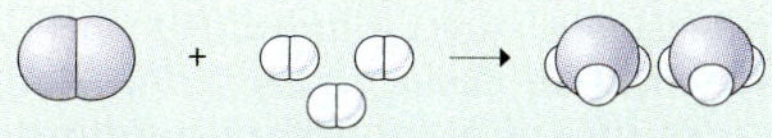

Nitrogen gas — Hydrogen gas — Ammonia gas

First, complete an "atom inventory" of this chemical equation:

N_2	+	$3\,H_2$	$\longrightarrow$	$2\,NH_3$
2 N atoms	+	6 H atoms	=	2 N atoms and 6 H atoms

Note that the total number of atoms of N (nitrogen) and H (hydrogen) remains unchanged during this chemical reaction.

Next, interpret the equation in terms of molecules:

N_2	+	$3\,H_2$	$\longrightarrow$	$2\,NH_3$
1 N_2 molecule		3 H_2 molecules	$\longrightarrow$	2 NH_3 molecules

Note that one molecule of N_2 reacts with three molecules of H_2 to produce two molecules of the compound NH_3, called ammonia. Also note that molecules of nitrogen (N_2) and hydrogen (H_2) are diatomic, whereas the ammonia molecule is composed of four atoms—one nitrogen atom and three hydrogen atoms.

3. The following chemical equation represents the burning of methane, CH_4, to form water and carbon dioxide:

$$CH_4 + 2\,O_2 \longrightarrow CO_2 + 2\,H_2O$$

a. Write a sentence describing the equation in terms of molecules.
b. Identify each molecule as either a compound or an element.
c. Complete an atom inventory for the equation.
d. Provide a visual model ("picture in your mind") of the chemical reaction. Let 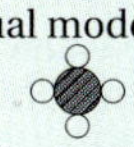 represent CH_4.

Let  represent CO_2.

Use the model of an H_2O molecule in Figure 15 (page 26) to draw a representation of H_2O similar to that of CH_4 and CO_2 shown here.

Household ammonia is made by dissolving gaseous ammonia in water.

B.5 THE ELECTRICAL NATURE OF MATTER

Previously, you were introduced to the concept of atoms and molecules. How do the atoms in molecules "stick" together to form bonds? Are atoms made up of even smaller particles? The answers to these questions require an understanding of the electrical properties of matter.

You have already experienced the electrical nature of matter, most probably without realizing it! Clothes often display "static cling" when they are taken from the dryer. The pieces of apparel stick firmly together and can

B.5 THE ELECTRICAL NATURE OF MATTER

The electrical nature of matter is discussed in terms of the interaction among charges, and in terms of neutrons, protons, and electrons. Water is described as a polar molecule whose structure helps explain certain observations, such as water's relatively high boiling point.

The demonstration of the two inflated balloons (that were rubbed against hair) helps to demonstrate the effects of charged particles. Another fascinating demonstration that can illustrate the electrical charges of water involves placing a charged comb or rod (a comb that was rubbed briskly through dry hair or on a wool sweater, or a static rod rubbed on various cloth materials) near (about 3 mm away) a narrow stream of falling water (e.g., water jet in a faucet or colored water from a burette against a contrasting background). The narrow stream will bead around the charged comb, indicating that water is an electrically charged substance.

This section provides more detail on the shape of a water molecule. It is important for students to understand that the molecular shape of a substance contributes to its properties. The term *partial charge* is briefly introduced. For now it is appropriate to indicate that a partial charge is different from a complete positive or negative charge.

B.6 IONS AND IONIC COMPOUNDS

The naming of ionic compounds and a simple method for writing formulas for ionic compounds are introduced and discussed very briefly. Be sure your students understand that ionic crystals are composed of regular arrangements of positive and negative ions (see Figure 19). The formula NaCl thus represents the one-to-one ratio of the number of cations and anions in the structure.

The concept of atoms combining together to form compounds is introduced, and connections are made to the electrical nature of matter. The main concept is reinforcing the idea that compounds are electrically neutral; the sum of all the ions is zero. In writing chemical formulas, use the sum method to determine the correct chemical formula. By using this method, students will develop a greater appreciation of ionic compounds. The Building Skills section that follows provides them with an opportunity to practice.

be separated only with effort. The shock that you sometimes receive after walking across a rug and touching a metal doorknob is another reminder of matter's electrical nature. And if two inflated balloons are rubbed against your hair, both balloons will attract your hair but repel each other, a phenomenon best seen when the humidity is low.

The electrical properties of matter can be summarized as follows:

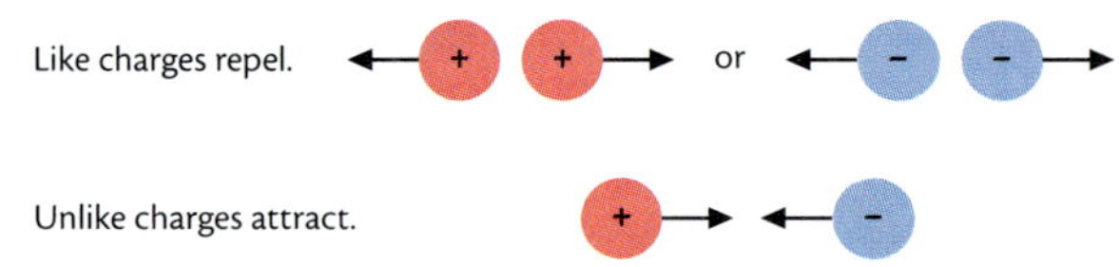

What are these positive and negative charges? How do they relate to the idea of atoms and molecules? The following points will be useful in answering these questions.

- Every electrically neutral (uncharged) atom contains equal numbers of positively charged particles called **protons** and negatively charged particles called **electrons.** In addition, most atoms contain one or more electrically neutral particles called **neutrons.**
- Positive–negative attractions between the protons in one atom and the electrons in another atom provide the "glue" that holds atoms together. This glue is the chemical bond that you read about on page 26.

States of Matter

These basic ideas will be used in later sections and in upcoming units to explain the properties of substances, the process of dissolving, and chemical bonding. Right now you will combine these ideas with your knowledge of atoms, chemical symbols, and chemical names to learn about a class of compounds that generally dissolve to some extent in water. It is possible that one or more of these compounds could be the cause of the fish kill.

B.6 IONS AND IONIC COMPOUNDS

Na	Electrically neutral sodium atom
Na^+	Sodium ion
Cl	Electrically neutral chlorine atom
Cl^-	Chloride ion
Na^+Cl^-	Sodium chloride (table salt)

Earlier in this unit (page 26), you learned about molecules. Molecules make up one type of compound. Another type of compound is composed of **ions,** which are charged atoms. Atoms can gain or lose electrons to form negative or positive ions, respectively. **Ionic compounds** are composed of positive and negative ions. An ionic compound has no net electrical charge; it is neutral because the positive and negative charges offset each other. The most familiar example of an ionic compound is table salt, sodium chloride (NaCl).

In solid ionic compounds, such as table salt, the ions are held together in crystals by attractions among the negative and positive charges. When an ionic compound dissolves in water, its individual ions separate from one another and disperse in the water. The designation (aq) following the symbol for an ion, as in Na^+(aq), means that the ions are in water (aqueous) solution.

When an atom gains one or more electrons (which have negative charge), the resulting negatively charged ion is called an **anion.** A positively charged ion, called a **cation,** results from an atom losing one or more electrons. An ion can be a single atom, such as a sodium ion (Na^+) or a chloride ion (Cl^-), or a group of bonded atoms, such as an ammonium ion (NH_4^+) or a nitrate ion (NO_3^-). An ion consisting of a group of bonded atoms is called a **polyatomic** (many-atom) **ion.** Figure 18 lists the formulas and names of common cations and anions.

Common Ions

Cations

1+ Charge		2+ Charge		3+ Charge	
Formula	**Name**	**Formula**	**Name**	**Formula**	**Name**
H^+	Hydrogen	Mg^{2+}	Magnesium	Al^{3+}	Aluminum
Na^+	Sodium	Ca^{2+}	Calcium	Fe^{3+}	Iron(III)*
K^+	Potassium	Ba^{2+}	Barium		
Cu^+	Copper(I)*	Zn^{2+}	Zinc		
Ag^+	Silver	Cd^{2+}	Cadmium		
NH_4^+	Ammonium	Hg^{2+}	Mercury(II)*		
		Cu^{2+}	Copper(II)*		
		Pb^{2+}	Lead(II)*		
		Fe^{2+}	Iron(II)*		

Anions

1− Charge		2− Charge		3− Charge	
Formula	**Name**	**Formula**	**Name**	**Formula**	**Name**
F^-	Fluoride	O^{2-}	Oxide	PO_4^{3-}	Phosphate
Cl^-	Chloride	S^{2-}	Sulfide		
Br^-	Bromide	SO_4^{2-}	Sulfate		
I^-	Iodide	SO_3^{2-}	Sulfite		
NO_3^-	Nitrate	CO_3^{2-}	Carbonate		
NO_2^-	Nitrite				
OH^-	Hydroxide				
HCO_3^-	Hydrogen carbonate (bicarbonate)				

*Some metals form ions that have one charge under certain conditions and a different charge under different conditions. To specify the charge for these metal ions, Roman numerals are used in parentheses after the metal's name.

Figure 18 *Common ions.*

Solid sodium chloride, NaCl(s), consists of equal numbers of positive sodium ions (Na^+) and negative chloride ions (Cl^-) arranged in three-dimensional networks called crystals. See Figure 19. The ionic compound calcium chloride, $CaCl_2$, presents a similar picture. However, unlike sodium ions, calcium ions (Ca^{2+}) each have a charge of 2+.

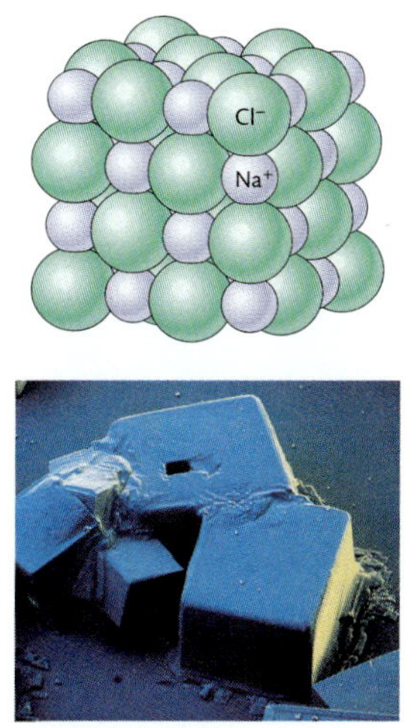

Figure 19 *Space-filling model of a sodium chloride (NaCl) crystal and a photo of magnified sodium chloride crystals.*

IONIC COMPOUNDS
Building Skills 4

This activity provides practice in recognizing, naming, and writing the formulas of ionic compounds. Remind students to acquire the ionic charges and partial names from the lists of ions available to them.

Note: Some students may use the common names for some compounds (baking soda or rust). Point out that such names give less information than the more systematic naming of both cation and anion.

The formula for sodium chloride is not ClNa (Rule 1); nor is it Na_2Cl_2 or Na_3Cl_3 (Rule 2).

The "Cl_2" in $CaCl_2$ represents two Cl^- anions—not a diatomic molecule of chlorine.

You can easily write formulas for ionic compounds by following two simple rules.

1. Write the cation first, then the anion.
2. The correct formula contains the fewest positive and negative ions needed to make the total electrical charge zero.

Why are the numbers of chloride ions different in sodium chloride (NaCl) and calcium chloride ($CaCl_2$)? In sodium chloride, the ion charges are 1+ and 1−. Because one ion of each type results in a total charge of zero, the formula for sodium chloride must be NaCl.

When cation and anion charges do not add up to zero, ions of either type must be added until the charges cancel. In calcium chloride, one calcium ion (Ca^{2+}) has a charge of 2+. Each chloride ion (Cl^-) has a charge of 1−; two Cl^- ions are needed to equal a charge of 2−. Thus two chloride ions (2 Cl^-) are needed for each calcium ion (Ca^{2+}). The subscript 2 written after chlorine's symbol in the formula indicates this. The formula for calcium chloride is $CaCl_2$. Using these rules, what is the formula for aluminum sulfide, an ionic compound made up of aluminum cations (Al^{3+}) and sulfide anions (S^{2-})?

Formulas for compounds containing polyatomic ions, such as Na_2CO_3 (sodium carbonate), follow these same basic rules. However, if more than one polyatomic ion is needed to bring the total charge to zero, the formula for the polyatomic ion is enclosed in parentheses before the needed subscript is added. Ammonium sulfate is composed of ammonium (NH_4^+) and sulfate (SO_4^{2-}) ions. Two ammonium cations with a total charge of 2+ are needed to match the 2− charge of the sulfate anion. Thus the formula for ammonium sulfate is $(NH_4)_2SO_4$.

The written name of an ionic compound is composed of two parts. The cation is named first, then the anion. As Figure 18 (page 33) suggests, many cations have the same name as their original elements. Anions composed of a single atom, however, have the last few letters of the element's name changed to the suffix *-ide*. For example, the anion formed from fluorine (F) is fluor*ide* (F^-). Thus KF is named potassium fluoride. The following activity will provide practice in naming and writing formulas for ionic compounds according to the universal language of chemistry.

IONIC COMPOUNDS
Building Skills 4

Prepare a data table similar to the one shown here that identifies the composition of each ionic compound described in Statements 2 through 7.

DATA TABLE

	Cation	Anion	Formula	Name
1.	K^+	Cl^-	KCl	Potassium chloride
	(Complete this chart for substances 2 through 7.)			
7.				

Completed Data Table

Cation	Anion	Formula	Name
1. K^+	Cl^-	KCl	potassium chloride
2. Ca^{2+}	SO_4^{2-}	$CaSO_4$	calcium sulfate
3. Ca^{2+}	PO_4^{3-}	$Ca_3(PO_4)_2$	calcium phosphate
4. NH_4^+	NO_3^-	NH_4NO_3	ammonium nitrate
5. Al^{3+}	SO_4^{2-}	$Al_2(SO_4)_3$	aluminum sulfate
6. Mg^{2+}	OH^-	$Mg(OH)_2$	magnesium hydroxide
7. Ca^{2+}	CO_3^{2-}	$CaCO_3$	calcium carbonate

Refer to Figure 18 on page 33 as needed to complete this activity. Potassium chloride, the primary ingredient in table-salt substitutes used by people on low-sodium diets, has been done as an example in the sample data table.

2. $CaSO_4$ is a component of plaster.
3. A substance composed of Ca^{2+} and PO_4^{3-} ions is found in some brands of phosphorus-containing fertilizer. This substance is also a major component of bones and teeth.
4. Ammonium nitrate, a rich source of nitrogen, is often used in fertilizer mixtures.
5. $Al_2(SO_4)_3$ is a compound that can be used to help purify water.
6. Magnesium hydroxide is called milk of magnesia when it is mixed with water.
7. Limestone and marble are two common forms of the compound calcium carbonate.

B.7 WATER TESTING

Laboratory Activity

Introduction

How can chemists detect and identify certain ions in water solutions? This activity will allow you to use a method that chemists, including those investigating the Riverwood fish kill, use to detect the presence of specific ions in water solutions.

The tests that you will perform in this activity are **confirming tests.** That is, a positive test confirms that the ion in question is present. In each confirming test, you will look for a change in solution color or for the appearance of an insoluble material called a **precipitate.** A negative test (no color or precipitate) does not necessarily mean that the ion in question is absent. The ion may simply be present in such a small amount that the test result may not be observed. Technologies are available to detect these very small amounts, however.

These tests are classified as qualitative tests, ones that identify the presence or absence of a particular substance in a sample. In contrast, quantitative tests determine the amount of a specific substance present in a sample. Both types of tests would most likely be used in determining the cause of the Snake River fish kill.

You will test for the presence of iron(III) (Fe^{3+}) and calcium (Ca^{2+}) cations, as well as chloride (Cl^-) and sulfate (SO_4^{2-}) anions. Although you are familiar with the names and symbols for Ca^{2+}, Cl^-, and SO_4^{2-}, the name for Fe^{3+} may look strange to you. Some elements can form cations with different charges. Iron atoms can lose either two electrons to form Fe^{2+} cations or three electrons to form Fe^{3+} cations. Thus the name "iron cation" is not descriptive enough; it does not distinguish between Fe^{2+} and Fe^{3+}. For this reason, Roman numerals are added to the name to indicate the charge on the ion. Examples of other cations that must include Roman numerals in their names are copper(I) and copper(II), and cobalt(II) and cobalt(III).

There are two types of iron cations: Fe^{2+} is designated Fe(II); Fe^{3+} is Fe(III).

B.7 WATER TESTING

Laboratory Activity

Students test water samples for the presence of dissolved ions—iron(III), calcium, chloride, and sulfate. The qualitative analysis procedures in this activity are based on a "confirming test" approach. Emphasize that a negative test does not necessarily mean that the ion was absent, only that it was not present in detectable quantities.

Natural water samples will be analyzed in this laboratory activity. Ocean water and bottled water are also good samples to test. It may be possible to simulate a sample of ocean water by dissolving sea salt in water.

A day or two before performing the lab, either you or your students need to collect water samples from local streams, rivers, or lakes. It will probably be more interesting to your students if you allow them to collect and analyze their own water samples and an unknown water sample.

Time

50 minutes

Materials (for a class of 24 working in pairs)

12 24-well microplates
120 Beral pipets

Test solutions

50 mL 0.5 M KSCN (potassium thiocyanate, 2.4 g/50 mL) [SCN^- ions]
50 mL 0.1 M $NaCO_3$ (sodium carbonate, 0.53 g/50 mL) [CO_3^{2-} ions]
50 mL 0.1 M $AgNO_3$ (silver nitrate, 0.85 g/50 mL) [Ag^+ ions]
50 mL 0.1 M $BaCl_2$ (barium chloride, 1.2 g $BaCl_2 \cdot 2H_2O$/50 mL) [Ba^{2+} ions]

Reference ion solutions

50 mL 0.1 M $Fe(NO_3)_3$ (8.0 g $Fe(NO_3)_3 \cdot 9\,H_2O$/50 mL) [Fe^{3+} ions]
50 mL 0.1 M $CaCl_2$ (5.9 g $CaCl_2 \cdot 2H_2O$/50 mL) [Ca^{2+} ions and Cl^- ions]
50 mL 0.1 M $FeSO_4$ (1.4 g $FeSO_4 \cdot 7H_2O$/50 mL) [SO_4^{2-} ions]

Safety

Students should wear safety goggles and aprons throughout the laboratory activity. Caution students that sodium oxalate is a poison and silver nitrate stains skin.

Advance preparation

Movement around the classroom during the activity is minimized if the pipets are labeled and filled with the solutions and distributed upside down in the wellplate. The same pipets and solutions can be used for several classes.

Pre-lab discussion

Repeat the philosophy behind this confirming test approach. Also emphasize the use of distilled water as a control. Remind students that although small quantities are used, it is necessary to exercise care in handling all materials in the laboratory. If your tap water is particularly pure, you may wish to "manufacture" your own tap water sample, which will contain at least one of the ions being tested.

Remind students to clean the microplates immediately after use.

Lab tips

If a test color is too dark, students may add small amounts of distilled water in order to see the color better. To help see faint precipitates, use scrap paper with black print.

You will perform each confirming test on several different water samples. The first solution will be a **reference solution**—one that contains the ion of interest. The second will be a **control**—a sample known not to contain the ion. The control in this activity is distilled water. The other solutions will be tap-water and natural-water samples that you or your teacher collected. These solutions may or may not contain the ion. To determine whether these solutions contain the ion, you will need to compare the results with your reference and control samples.

In your laboratory notebook, prepare four data tables (one for each ion) similar to the one shown. Add rows if you are testing more than one natural-water sample. Be certain to identify the source of each natural-water sample.

DATA TABLE: ____(Specify ion)____

Solution	Observations (color, precipitate, etc.)	Result (Is ion present?)
Reference		
Control		
Tap water		
Natural water from ____(source)____		

The following suggestions will help guide your ion analysis.

1. If the ion is in tap or natural water, it will probably be present in a smaller amount than in the same volume of reference solution. Thus the quantity of precipitate or color produced in the tap or natural water sample will be less than in the reference solution.
2. When completing an ion test, mix the contents of the well thoroughly, using a toothpick or small glass stirring rod. Do not use the same toothpick or stirring rod in other samples without first rinsing it and wiping it dry.
3. In a confirming test based on color change, so few color-producing ions may be present that it is difficult to determine if the reaction actually took place. Here are two ways to decide whether the expected color is actually present:

- Place a sheet of white paper behind or under the wellplate to make any color more visible.
- Compare the color of the control (distilled water) test with that of the sample. Distilled water does not contain any of the ions tested. So even a faint color in the tap or natural water confirms that the ion is present.

4. In a confirming test based on the formation of a precipitate, you may be uncertain whether a solid precipitate is present even after thoroughly mixing the solutions. Placing the wellplate on a black or dark surface often makes a precipitate more visible.

Procedure

The test procedures for each ion follow. If the ion of interest is present, a chemical reaction will take place, producing either a colored solution or a precipitate. The chemical equations are given for each ion.

Only ions that take part in the reaction are included in this type of equation.

Calcium Ion (Ca^{2+}) Test

$$\underset{\text{calcium ion}}{Ca^{2+}(aq)} + \underset{\text{carbonate ion}}{CO_3^{2-}(aq)} \longrightarrow \underset{\text{calciu}}{CaCO_3(s)}$$

carbonate

Follow these steps for each sample (Ca^{2+} reference, control, tap water, natural water):

1. Place 20 drops into a well of a 24-well wellplate.
2. Add three drops of sodium carbonate (Na_2CO_3) test solution to the well.
3. Record your observations, including the color and whether a precipitate formed.
4. Determine whether the ion is present and record your results.
5. Repeat for the remaining solutions.
6. Discard the contents of the wellplate as directed by your teacher.

Iron(III) Ion (Fe^{3+}) Test

$$\underset{\text{iron(III) ion}}{Fe^{3+}(aq)} + \underset{\text{thiocyanate ion}}{SCN^{-}(aq)} \longrightarrow \underset{\text{iron(III) thiocyanate ion}}{[FeSCN]^{2+}(aq)}$$

Follow these steps for each sample (Fe^{3+} reference, control, tap water, natural water):

1. Place 20 drops into a well of a 24-well wellplate.
2. Add one or two drops of potassium thiocyanate (KSCN) test solution to the well.
3. Record your observations, including the color and whether a precipitate formed.
4. Determine whether the ion is present and record your results.
5. Repeat for the remaining solutions.
6. Discard the contents of the wellplate as directed by your teacher.

Chloride Ion (Cl^-) Test

$$\underset{\text{chloride ion}}{Cl^{-}(aq)} + \underset{\text{silver ion}}{Ag^{+}(aq)} \longrightarrow \underset{\text{silver chloride}}{AgCl(s)}$$

Follow the same procedure as that for the Fe^{3+} ion, with the following changes:

- Use the Cl^- reference solution.
- In Step 2, add three drops of silver nitrate ($AgNO_3$) test solution instead of potassium thiocyanate (KSCN) test solution.

Expected Results

Reference Solution	Color	Precipitate
Fe^{3+}	red	no
Ca^{2+}	white	yes
Cl^-	white	yes
SO_4^{2-}	white	yes

Answers

1. a. The control represents the absence of the ion and therefore is useful in interpreting the experimental results.

   b. It is presumed to be essentially free of the ions detected by the tests.
2. The concentration of some ions may be too low to detect. The test for one ion may interfere with or mask the behavior of another ion. Interfering ions can give false-positive tests.
3. Ions may be present below levels detectable by this test.
4. Improper cleaning may leave stray ions behind that can contaminate and interfere with tests in the wellplate.

Post-lab discussion

Focus on student answers to questions. These questions develop analytical skills in interpreting laboratory data. Reinforce the idea that these kinds of chemical tests rely on the reference ion forming some sort of recognizable product.

B.8 PURE AND IMPURE WATER

This section distinguishes between pure water and clean water. It also discusses substances that enter our water supplies.

B.9 THE RIVERWOOD WATER MYSTERY

Making Decisions

This activity introduces students to an important aspect of a scientific approach to solving problems—distinguishing facts from opinions.

Sulfate Ion (SO_4^{2-}) Test

$$\underset{\text{sulfate ion}}{SO_4^{2-}(aq)} + \underset{\text{barium ion}}{Ba^{2+}(aq)} \longrightarrow \underset{\text{barium sulfate}}{BaSO_4(s)}$$

Follow the same procedure as that for the Fe^{3+} ion, with the following changes:

- Use the SO_4^{2-} reference solution.
- In Step 2, add three drops of barium chloride ($BaCl_2$) test solution instead of potassium thiocyanate (KSCN).

Questions

1. a. Why was a control used in each test?
   b. Why was distilled water chosen as the control?
2. Describe some difficulties associated with the use of qualitative tests.
3. These tests cannot absolutely confirm the absence of an ion. Why?
4. How might your observations have changed if you had not cleaned your wells or stirring rods thoroughly between each test?

B.8 PURE AND IMPURE WATER

Now that you have learned about water's properties and about some of the substances that can dissolve in water, you are ready to return to the problem of Riverwood's fish kill. Recall that various Riverwood residents had different ideas about the cause of the problem. For example, longtime resident Harmon Lewis was sure the cause was pollution of the river water. Which substances are regarded as pollutants, and which are harmless when dissolved in water?

What is the difference between clean and pure water?

Families in most U.S. cities and towns receive an abundant supply of clean, but not absolutely pure, water at an extremely low cost. You can check the water cost in your own area: If you use municipal water, your family's water bill will contain the current water cost per gallon. Divide that value by 3.8 (there are 3.8 liters in one gallon) to compute the current cost for one liter of water.

It is useless to insist on absolutely pure water. The cost of processing water to make it completely pure would be prohibitively high. And, even if costs were not a problem, it would still be impossible to have absolutely pure water. The atmospheric gases nitrogen (N_2), oxygen (O_2), and carbon dioxide (CO_2) will always dissolve in the water to some extent.

B.9 THE RIVERWOOD WATER MYSTERY

Making Decisions

Your teacher will divide the class into several different groups of students. Each group will complete this decision-making activity. Afterward, the entire class will compare and discuss the answers obtained by each group.

After listing facts and questions, students attempt to eliminate causes for the fish kill. Students should recognize that if the cause of the fish kill is water-related, it must be due to something suspended or dissolved in the water. This narrows the range of possible fish-kill causes.

Assign students to groups of four or five to work on this activity, after which the class can compare and discuss the groups' answers. Assign the rereading of all the newspaper articles in the unit as homework. Direct student groups to prepare answers to the three questions to share with the class. Suggest that groups might review their chemical knowledge from the previous sections to help them form questions.

At the beginning of this unit, you read newspaper articles describing the Riverwood fish kill and the reactions of several citizens to it. Among those interviewed were Harmon Lewis, a longtime resident of Riverwood, and Dr. Margaret Brooke, a water-systems scientist. These two people had very different reactions to the fish kill. An angry Harmon Lewis was certain that human activity—probably some sort of pollution—had caused the fish kill. Dr. Brooke refused to even speculate about the cause of the fish kill until she had conducted some tests.

Which of these two positions comes closer to your own reaction at this point? Complete the following activities to investigate the issue further.

1. Reread the fish-kill newspaper reports located at the beginning of Sections A and B. List all facts (not opinions) concerning the fish kill found in these articles. Scientists often refer to facts as data. **Data** are objective pieces of information. They do not include interpretation.
2. List at least five factual questions that you would want answered before you could decide on possible causes of the fish kill. Some typical questions might be: Do barges or commercial boats travel on the Snake River? Were any shipping accidents on the river reported recently?
3. Look over your two lists—one of facts and the other of questions.
   a. At this point, which possible fish-kill causes can you rule out as unlikely? Why?
   b. Can you suggest a probable cause? Be as specific as possible.

Later in this unit you will have an opportunity to test the reasoning that you used in answering these questions.

B.10 WHAT ARE THE POSSIBILITIES?

The activities that you just completed (gathering data, seeking patterns or regularities among the data, suggesting possible explanations or reasons to account for the data) are typical of the approaches scientists take in attempting to solve problems. Such scientific methods are a combination of systematic, step-by-step procedures and logic, as well as occasional hunches and guesses.

A fundamental yet difficult part of scientists' work is knowing what questions to ask. You have listed some questions that might be posed concerning the cause of the fish kill. Such questions help focus a scientist's thinking. Often a large problem can be reduced to several smaller problems or questions, each of which is more easily managed and solved.

The number of possible causes for the fish kill is large. Scientists investigating this problem must find ways to eliminate some causes and zero in on more promising ones. They try to either disprove all but one cause or produce conclusive proof in support of a specific cause.

As you recall, water analyst Brooke studied possible causes of the Snake River fish kill. She concluded that if the actual cause were water related, it would have to be due to something dissolved or suspended in the water.

CD-ROM WWW. Questions & Answers

In Section C, you will examine several categories of water-soluble substances and consider how they might be implicated in the fish kill. The mystery of the Riverwood fish kill will be confronted at last!

1. Possible facts or data:

   Dead trout and other fish were found floating from near the base of the Snake River Dam to the Riverwood water intake. Preliminary tests have been taken, looking for substances dissolved in the water that might be responsible for the fish kill. Early evidence seems to point to something dissolved in the water as a cause. The substance and source are still undetermined. The town council voted to shut down the water plant for three days.
2. Questions might be:

   What types of substances might be dissolved in water?

   Could contaminants enter below the dam?

   What possible emission sources are upstream from the dam?

   What happens to water as it goes over the dam? Do any additional substances dissolve in the water?

   What is the nature of the bubbles found in the fish?

   What substances might harm fish but not humans?

   Could the fish have died from a deficiency or overdose of a naturally occurring and necessary substance?
3. Answers will depend on the student questions. At this point, closure or a correct answer is not possible.

B.10 WHAT ARE THE POSSIBILITIES?

This section identifies the process just used by students as typical of the work of scientists. It is a bridge to Part C, where students investigate types of substances that can be dissolved or suspended in water and determine which ones are potentially harmful to aquatic life.

Knowledge of scientific facts, theories, laws, and principles helps the working scientist form questions or hypotheses, just as chemical knowledge helps students reject some possible causes as improbable.

Soon you will need to decide which evidence you will provide students for Section C and Putting It All Together. Four sets of data have been prepared, each with a different cause for the fish kill. Students will analyze the data and use the evidence in the data to determine the cause. You may want to use a different set of data with each class you teach. Or, you may want to vary the cause from one year to the next.

CHEMISTRY AT WORK

Environmental Cleanup: It's a Dirty Job . . . But That's the Point

Wayne Crayton spends his summers touring exotic islands in the Aleutian Islands chain off the coast of Alaska. But it's not just an adventure that he's embarked on. It's also his job.

As an Environmental Contaminants Specialist with the United States Army Corps of Engineers, Alaska District, Wayne investigates areas that were formerly used as military bases and fueling stations. Wayne and his teammates review and assess the damage (if there is any) that contaminants have done to key areas used by wildlife. Based on their findings, they then develop plans to fix the problems.

Wayne and his teammates recommend procedures for removing and treating contaminated soil and other material.

As part of its investigation planning, Wayne's team reviews information to determine what they're likely to find at a given site. For example, historical documents about the site will indicate whether the team members should be looking for petroleum residues or other contaminants; aerial photography and records from earlier investigations will help to identify specific areas that are potential sites of contamination.

At the site, the team collects soil, sediment, and water samples from the exact location where a contaminant was originally introduced to the environment, as well as samples from the area over which the contaminant might have spread. Wayne and his teammates may also collect small mammals or fish that have been exposed to the contaminants. After having collected the necessary samples, the team members return home quickly because some of the collected samples can degrade or change characteristics soon after collection.

Wayne works in an office in Anchorage during the rest of the year, analyzing and interpreting data and test results from the field investigations. He and his colleagues calculate concentrations of hazardous substances, including organochlorines, PCBs (polychlorinated biphenyls), pesticides, petroleum residues, and trace metals. Then they determine whether any of these substances present a risk to humans or the surrounding ecosystems.

Using these results, Wayne and his teammates recommend procedures for removing and treating contaminated soil and other material. In some situations, they decide that the best solution is to do nothing; the cleanup itself could destroy wetlands, disturb endangered wildlife, or have other negative effects on the environment. The Corps of Engineers uses the team's recommendations to direct the work of the contractor performing the actual cleanup.

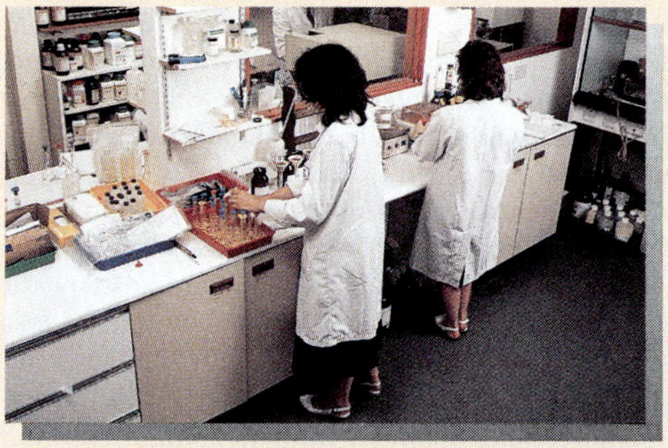

Solving Scientific Problems . . .

This icon indicates an opportunity to consult resources on the World Wide Web. See your teacher for further instructions.

Scientists often solve problems in unique ways—ways that are different from the methods used in other areas of academic research.

- Outline the problem-solving steps that the Environmental Contaminants Specialists use in planning their investigations as described in this article.
- Compare the steps used by these scientists with the steps that you have used in studying science.
- Conduct a World Wide Web search for any United States Army Corps of Engineers or United States Environmental Protection Agency investigations or projects that might be underway in your community.

41

SECTION SUMMARY

Reviewing the Concepts

1. When gasoline and water mix, they form two distinct layers. What do you need to know in order to determine which liquid will be found in the top layer?

Gasoline is less dense than water and insoluble, so it will form the top layer.

2. Identify each of the following materials as a solution, suspension, or colloid. Explain your choice in each case.

a. a medicine accompanied by the instructions "shake before using"

b. Italian salad dressing

c. mayonnaise

d. a cola soft drink

e. an oil-based paint

f. milk

D is a solution; a, b, and e are suspensions; c and f are colloids.

3. You notice beams of light passing into a darkened room through blinds on a window. Does this demonstrate that the room air is a solution, suspension, or colloid? Explain.

Suspension. The rays of sunlight are examples of the Tyndall effect, or scattering of the light. If the light beams hit suspended dust particles, they produce the Tyndall effect, which is most likely the case here.

4. Sketch a visual model on the molecular level that represents each of the following types of mixtures. Label and explain the features of each sketch.

a. a solution **c.** a colloid

b. a suspension

Student sketches might show: a. a solution as interspersed particles; b. a suspension as suspended particles with some solid particles on the bottom; and c. a colloid as particles suspended in between others.

5. **a.** Given a mixture, what steps would you follow to classify it as a solution, a suspension, or a colloid?

b. Describe how each step would help you to distinguish among the three types of mixtures.

a. and b. A solution is generally clear throughout, transmits light beams intact, does not exhibit the Tyndall effect, may have ionic-sized particles that cannot be seen visually or with a microscope and its particles do not settle to the bottom of the container. Suspensions have particles that will tend to settle, macroscopic particles that can be filtered from the mixture and seen with the eye or magnifying lens, and will tend to block or absorb light beams. A colloid exhibits the Tyndall effect and has micro-scopic particles that do not settle to the bottom of the container and can probably be seen with a microscope.

6. Using your knowledge of chemical symbols, classify each of the following substances as an element or a compound.

a. CO **c.** HCl **e.** $NaHCO_3$ **g.** I_2

b. Co **d.** Mg **f.** NO

B, d, and g are elements; the others are compounds.

7. Compare the physical properties of water (H_2O) with the physical properties of the elements of which it is composed.

At room temperature, water is a clear liquid; hydrogen and oxygen are colorless gases. Water boils at 100 °C and freezes at 0 °C; hydrogen boils at −252 C° and freezes at −259 °C; oxygen boils at −183 °C and freezes at −218 °C. The density of water is 1g/mL; hydrogen is 0.00008 g/mL and oxygen is 0.0013 g/mL. Both hydrogen and oxygen are flammable; water is not.

8. Look at the following drawings.

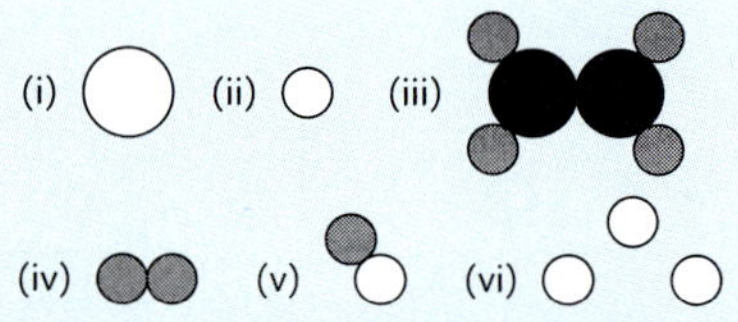

a. Which represent a pure element?

b. Which represent a compound?

a. i, ii, iv, and vi are elements

b. iii and v are compounds

9. Represent each chemical equation with drawings of the molecules and their component atoms. Use circles of different sizes or shading for each type of element.

a. $H_2(g) + Cl_2(g) \longrightarrow 2HCl(g)$

b. $2H_2O_2(aq) \longrightarrow 2H_2O(l) + O_2(g)$

Let  represent a hydrogen peroxide molecule, H_2O_2.

c. Using complete sentences, write a word equation for the chemical equations given in a and b. Include the numbers of molecules.

a. See text figure for similar drawings.

b. See text figure for similar drawings.

c. One molecule of diatomic hydrogen gas is added to one molecule of diatomic chlorine gas to yield two molecules of gaseous hydrochloric acid (or hydrogen chlorine).

Two molecules of aqueous hydrogen peroxide (or hydrogen dioxide) decompose to yield two molecules of liquid water and one molecule of oxygen gas.

10. Name the elements and list the number of atoms of each for the following substances.

a. phosphoric acid, H_3PO_4 (used in some soft drinks and to produce some fertilizers)

b. sodium hydroxide, NaOH (found in some drain cleaners)

c. sulfur dioxide, SO_2 (a by-product of burning most, if not all, types of coal)

a. 3 hydrogen, one phosphorus, 4 oxygen

b. one sodium, one oxygen, one hydrogen

c. one sulfur, two oxygen

11. Write chemical equations that represent the following word equations:

a. Baking soda ($NaHCO_3$) reacts with hydrochloric acid (HCl) to produce sodium chloride, water, and carbon dioxide.

b. During respiration, one molecule of glucose, $C_6H_{12}O_6$, combines with six molecules of oxygen to produce six molecules of carbon dioxide and six molecules of water.

a. $NaHCO_3 + HCl \rightarrow NaCl + H_2O + CO_2$

b. $C_6H_{12}O_6 + 6O_2 \rightarrow 6CO_2 + 6H_2O$

12. For each of the following elements, identify the number of protons or electrons needed for an electrically neutral atom.

a. Carbon:	6 protons	__ electrons
b. Aluminum:	__ protons	13 electrons
c. Lead:	82 protons	__ electrons
d. Chlorine:	__ protons	17 electrons

a. 6 b. 13 c. 82 d. 17

13. Decide whether each of the following atoms is electrically neutral.

a.	Oxygen:	16 protons	18 electrons
b.	Iron:	26 protons	24 electrons
c.	Silver:	47 protons	47 electrons
d.	Iodine:	53 protons	54 electrons

Only c is electrically neutral.

14. Write the symbol and show the charge (if any) on the following atoms or ions:

a. hydrogen with 1 proton and 1 electron

b. sodium with 11 protons and 10 electrons

c. chlorine with 17 protons and 18 electrons

d. aluminum with 13 protons and 10 electrons

a. H b. Na^+ c. Cl^- d. Al^{3+}

15. Indicate whether an Fe^{+3} ion would be attracted to or repelled from each particle in Question 14.

Fe^{+3} will be attracted to c, repelled by b and d, and will neither be attracted nor repelled by a.

16. **a.** Classify each of the following as an electrically neutral atom, an anion, or a cation.

i. O^{2-} ii. He iii. C iv. Ag^{+} v. Hg^{2+}

b. For each ion, indicate whether the electrical charge resulted from an atom gaining electrons, losing electrons, or neither.

a. ii and iii are neutral atoms; and i. is an anion; iv. and v. are cations.

b. i. gained electrons to become negatively charged; ii and iii are neutral; and iv. and v. lost electrons to become positively charged.

17. Write the name and formula for each compound that will be formed from the following combinations of cations and anions:

	OH^-	PO_4^{3-}	S^{2-}
Fe^{3+}	a.	b.	c.
K^+	d.	e.	f.
Ca^{2+}	g.	h.	i.

a. $Fe(OH)_3$ **b. $FePO_4$** **c. Fe_2S_3**

d. KOH **e. K_3PO_4** **f. K_2S**

g. $Ca(OH)_2$ **h. $Ca_3(PO_4)_2$** **i. CaS**

Connecting the Concepts

18. Explain the possible risks in failing to follow the direction "Shake before using" on the label of a medicine bottle.

The bottle contains a suspension; important ingredients may have settled to the bottom of the bottle. The bottle must be shaken to ensure that the patient receives a complete mixture of the compounds in the medication.

19. Why is it important that the symbols of the elements be internationally accepted?

Since chemicals for research, analysis, medications, and manufacturing are transported throughout the world, a common chemical language is necessary for accurate communication.

20. Draw a model of a solution in which water is the solvent and oxygen gas (O_2) is the solute.

The model would include water molecules fairly close together as a liquid, and diatomic oxygen molecules would be widely dispersed throughout.

21. An iron atom that has 26 protons and 23 electrons combines with an O^{2-} ion to form a compound.

a. What is the ionic charge on the iron atom?

b. Write the chemical formula for the compound.

a. $(+26) + (-23) = +3$ or gives Fe^{3+}

b. Fe_2O_3

Extending the Concepts

22. Is it possible to have a food product that is 100% "chemical free"? Explain.

No, all foods are composed of chemical compounds. Chemists ask the question, "What in the world is not made of chemicals?" Indeed all matter can be characterized as chemical substances. Most of the general public use the word *chemical* to refer to laboratory substances only.

23. Some elements in Figure 16 (page 29) have symbols that are not based on their modern names (such as K for potassium). Look up their historical names and explain the origin of their symbols.

Symbols for some elements are derived from the Latin name for the element.

Element	Symbol	Latin Origin
Gold	Au	Aurum; "shining dawn"
Iron	Fe	Ferrum
Lead	Pb	Plumbum
Mercury	Hg	Hydrargentum; "liquid silver"
Potassium	K	Kalium; first isolated by electrolysis in 1807
Silver	Ag	Argentum
Sodium	Na	Natrium; first isolated by electrolysis in 1807
Tin	Sn	Stannum

24. The symbols of elements (such as Na, Cu, and Cl) are accepted and used by chemists in all nations, regardless of the country's official language. However, the name of an element often depends on language. For example, the element N is nitrogen in English but azote in French. The element H is hydrogen in English but *Wasserstoff* in German. Investigate the names of some common elements in a foreign language of your choice. What are the meanings or origins of the foreign element names that you have found? How do those meanings or origins compare with those for the corresponding English element names?

Looking up the names of elements in foreign languages can be a good way to introduce interdisciplinary lessons. Recommend that some coordination be made with a foreign language teacher to review the history and structure of the foreign names of a few elements. Some examples: sodium (Na) in English is soda yet originated in Latin as *Natrium;* Copper (Cu) in Latin is *Cyprium* and in Late Latin is *Cuprum;* Chlorine (Cl) in Spanish is *Cloro,* in Greek is *Chlòros,* in French is *Chlore,* and in German is *Chlor.*

25. Investigate and report on why "100% pure water" would be unsuitable for long-term human consumption—even if taste were not a consideration.

Trace minerals are important for chemical reactions in our body. For example, fluorine prevents tooth decay, and iodine prevents goiter.

26. Using an encyclopedia or other reference, compare the maximum and minimum temperatures naturally found on the surfaces of Earth, the Moon, and Venus. The large amount of water on Earth serves to limit the natural temperature range on the planet. Suggest ways that water accomplishes this. To begin, find out what *heat of fusion, heat capacity,* and *heat of vaporization* mean.

See Extending Your Knowledge Overview at the end of Section A for information on these factors. The global ramifications of these factors lead to the weather and climatic conditions we live in.

27. Look up the normal freezing point, boiling point, heat of fusion, and heat of vaporization of ammonia (NH_3). If a planet's life forms were made up mostly of ammonia rather than water, what special survival problems might they face? What temperature range would an ammonia-based planet need to support "life"?

See Extending Your Knowledge Overview at the end of Section A.

EXTENDING YOUR KNOWLEDGE OVERVIEW

- Because 100 °C is equivalent to 180 °F, 1.8 °F = 1 °C or 1 °F = 0.56 °C, the following expression can be written:
  °F = (1.8 × °C) + 32. If this expression is solved for °C, we obtain °C = (°F − 32)/1.8.
- Because the density of oil is less than the density of liquid water, the oil will float on the water. An oil spill floating on water can ignite and burn, or it can block the sunlight from reaching the aquatic plants. It can slow the exchange of gases between the water and the atmosphere, and it can be blown by the winds and carried by the currents and tides to beaches. On the other hand, if oil were more dense than water, cleanup efforts would probably be more difficult.
- Symbols for some elements are derived from the Latin name for the element. Others originate from other cultures and languages.

Part B established that a possible cause of the Snake River fish kill was something dissolved in the water. Part C provides students with deeper understanding of the concepts of solubility and solutions, as well as the chemistry of ionic and molecular substances.

Part C begins by introducing basic terms and concepts related to solubility. After gaining a working knowledge of these ideas, students study several kinds of substances that dissolve in water. They first consider the importance of dissolved oxygen (DO) for the survival of aquatic life. To examine the possibility that a shortage of DO caused the fish kill, students find out how much oxygen is available to aquatic life and how much DO fish need to survive. This leads to questions about the influence of temperature on solubility. After examining and graphing the DO data from the Snake River, students rule out the possibility of DO shortages induced by thermal pollution and move on to consider other possibilities.

Acid contamination is then considered as another possible cause of the fish kill. Students learn to recognize common acids and bases from their formulas. Acid contamination is ruled out as a fish-kill possibility. Students next consider heavy-metal ions, and a decision-making activity leads them to rule this out as well. Part C ends by examining some molecular substances that dissolve in water.

In Part C you will present students with the Snake River data for the first time, and they will begin their analysis of the cause of the fish kill.

SECTION C

INVESTIGATING THE CAUSE OF THE FISH KILL

The challenge facing investigators of the Riverwood fish kill is to decide what in Snake River water was responsible for the crisis. In this section, you will learn about the process of dissolving, how solutions behave and are described, and what types of substances dissolve in water. What you learn will help to ensure that you have the knowledge and skills needed to evaluate the Riverwood data and to determine the cause of the fish kill.

C.1 SOLUBILITY OF SOLIDS

Could something dissolved in the Snake River have caused the fish kill? As you already know, a variety of substances can dissolve in natural waters. To determine whether any of these substances could be harmful to fish, you first need to know how solutions are formed and described. For example, how much of a certain solid substance will dissolve in a given amount of water?

Imagine preparing a water solution of potassium nitrate, KNO_3. What happens as you add a scoopful of solid, white potassium nitrate crystals to water in a beaker? As you stir the water, the solid crystals dissolve and disappear. The resulting solution remains colorless and clear. In this solution, water is the solvent, and potassium nitrate is the solute.

Remember: In a solution, the solvent is the dissolving agent and the solute is the dissolved substance.

What will happen if you add a second scoopful of potassium nitrate crystals to the beaker and stir? These crystals also may dissolve. However, if you continue adding potassium nitrate without adding more water, eventually some potassium nitrate crystals will remain undissolved on the bottom of the beaker, no matter how long you stir. The maximum quantity of a substance that will dissolve in a certain quantity of water (for example, 100 g) at a specified temperature is the **solubility** of that substance in water. In this example, the solubility of potassium nitrate might be expressed as "grams potassium nitrate per 100 g water" at a specified temperature.

Questions & Answers

From everyday experiences, you probably know that both the size of the solute crystals and the vigor and duration of stirring affect how long it takes for a sample of solute to dissolve at a given temperature. But do these factors affect how much substance will eventually dissolve? With enough time and stirring, will even more potassium nitrate dissolve in water? It turns out that the solubility of a substance in water is a characteristic of the substance and cannot be changed by any amount of stirring or time.

So what does affect the actual quantity of solute that dissolves in a given amount of solvent? As you can see from Figure 20 on page 46, the mass of

CD-ROM WWW. Dissolving Ionic Compounds

SECTION C: INVESTIGATING THE CAUSE OF THE FISH KILL

Introduction

Remind students that we are investigating the possible cause or causes of the Riverwood fish kill. The evidence points to something dissolved in the water that caused the problem. Water is a unique substance that can dissolve many solutes. To better investigate, we need to understand the solubility of substances in water.

C.1 SOLUBILITY OF SOLIDS

Section C.1 presents vocabulary and concepts related to solubility: solute, solvent, solution, saturation, and solubility curve. The temperature dependence of the solubility of gases and solids and the effect of pressure on gaseous solubility are explained. A solubility curve is used to help clarify the terms *saturated, unsaturated,* and *supersaturated.*

Demonstration ideas

Supersaturation can be demonstrated by dissolving 160 g sodium acetate trihydrate, $NaC_2H_3O_2{\cdot}3H_2O$, in 30 mL distilled water. Heat the mixture until all solid is dissolved. Allow the solution to cool to room temperature. Add a single crystal of sodium acetate to the solution, and observe the crystallization. The solution can be reused by reheating to dissolve the solid. Small reusable hand-warming packets contain the same compound and are reusable as well.

ADDITIONAL BACKGROUND INFORMATION

The solubility of a solute in a solvent (water in the present discussion) depends on the relative strengths of solute-solute and solute-solvent attractions. When a substance dissolves, bonds between solute molecules and between solvent molecules are broken. New bonds form between solute and solvent molecules. Breaking bonds always requires energy, and making bonds always produces energy.

For some solids, the overall dissolving reaction is endothermic because more energy is needed to overcome the solute-solute attractions than is released when the solute-solvent bonds are formed. In this case, adding thermal energy by raising the temperature will permit more solid to dissolve. This is reflected in the solubility curves in Figure 20. Pressure has no significant effect on the solubility of solids; grinding and stirring change the rate of dissolving but do not affect how much solid can dissolve.

In gases there are virtually no forces between solute molecules, and the process creates new bonds between solute and solvent molecules. The overall reaction produces energy. At higher temperatures less gas dissolves in a given quantity of water. The solubility of gases is also directly proportional to the pressure. This is demonstrated by observing canned or bottled soft drinks, in which the carbon dioxide is dissolved under as much as 10 atm of pressure. When the pressure is released by opening the container, bubbles of gas escape.

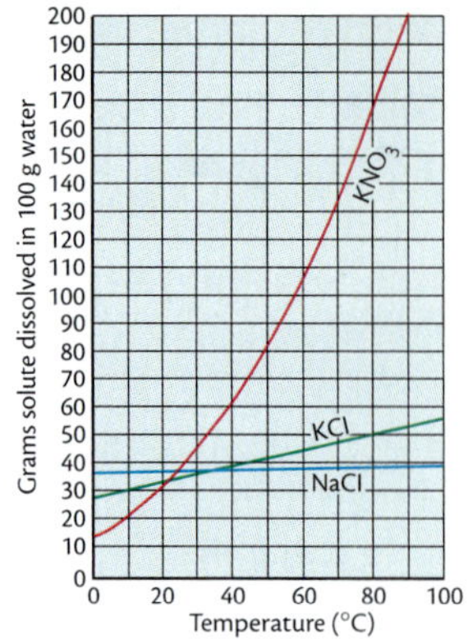

Figure 20 *Relationship between solute solubility in water and temperature.*

solid solute that will dissolve in 100 g water varies as the temperature of the water changes from 0 °C to 100 °C. The graphical representation of this relationship is called the solute's solubility curve.

Each point on the solubility curve indicates a solution in which the solvent contains as much dissolved solute as it normally can at that temperature. Such a solution is called a **saturated solution.** Thus each point on the solubility curve indicates a saturated solution. Look at the curve for potassium nitrate (KNO_3) in Figure 20. At 50 °C, how much potassium nitrate will dissolve in 100 g water to form a saturated solution? This value—80 g KNO_3 per 100 g water—is the solubility of potassium nitrate in 50 °C water. In contrast, the solubility of potassium nitrate in 20 °C water is only about 30 g KNO_3 per 100 g water. (Make sure that you are able to "read" this value on the graph.)

Note that the solubility curve for sodium chloride (NaCl) is nearly a horizontal line. What do you think this means about the solubility of sodium chloride as temperature changes? Compare the curve for sodium chloride with the curve for potassium nitrate (KNO_3), which rises steeply as temperature increases. You should be able to conclude that for some solutes, such as potassium nitrate (KNO_3), solubility in water is greatly affected by temperature, whereas for others, such as sodium chloride (NaCl), the change is only slight.

Now consider a solution containing 80 g potassium nitrate in 100 g water at 60 °C. Locate this point on the graph. Where does it fall with respect to the solubility curve? What does this tell you about the level of saturation of the solution? Because each point on the solubility curve represents a saturated solution, any point on a graph below a solubility curve must represent an unsaturated solution. An **unsaturated solution** is a solution that contains less dissolved solute than the solvent can normally hold at that temperature.

What would happen if you cooled this solution to 40 °C? (Follow the line representing 80 g to the left on the graph.) You might expect that some solid KNO_3 crystals would form and fall to the bottom of the beaker. In fact, this event is likely to occur. Sometimes, however, you can cool a saturated solution without forming any solid crystals, producing a solution that contains more solute than could usually be dissolved at that temperature. This type of solution is called a **supersaturated solution.** (Note that this new point lies above the solubility curve for potassium nitrate.) Agitating a supersaturated solution or adding a "seed" crystal to the solution often causes the "extra" solute to appear as solid crystals and settle to the bottom of the beaker, or precipitate. The remaining liquid then contains the amount of solute that represents a stable, saturated solution at that temperature.

One example of crystallization in a supersaturated solution may be familiar to you—the production of rock candy. A water solution is supersaturated with sugar. When seed crystals are added, they cause excess dissolved sugar to crystallize from the solution onto a string. Mineral deposits around a hot spring are another example of crystallization from a supersaturated solution. Water emerging from a hot spring is saturated with dissolved minerals. As the solution cools, it becomes supersaturated. The rocks over which the solution flows act as seed crystals, causing the formation of more mineral deposits.

SOLUBILITY AND SOLUBILITY CURVES

Building Skills 5

What is the solubility of potassium nitrate at 40 °C? The answer is found by using the solubility curve for potassium nitrate given in Figure 20. Locate the intersection of the potassium nitrate curve with the vertical line representing 40 °C. Follow the horizontal line to the left and read the value. The solubility of potassium nitrate in water at 40 °C is 60 g per 100 g water.

At what temperature will the solubility of potassium chloride be 25 g per 100 g water? Think of the space between 20 g and 30 g on the *y* axis in Figure 20 as divided into two equal parts, then follow an imaginary horizontal line at "25 g/100 g" to its intersection with the curve. Follow a vertical line down to the *x* axis. Because the line falls halfway between 10 °C and 20 °C, the desired temperature must be about 15 °C.

As you have seen, the solubility curve is quite useful when you are working with 100 g water. But what happens when you are working with other quantities of water? The solubility curve indicated that 60 g potassium nitrate will dissolve in 100 g water at 40 °C. How much potassium nitrate will dissolve in 150 g water at this temperature? You can "reason" the answer in the following way.

The amount of solvent (water) has increased from 100 g to 150 g—1.5 times as much solvent. That means that 1.5 times as much solute can be dissolved. Thus: $1.5 \times 60\text{ g} = 90\text{ g KNO}_3$.

The calculation can also be written as a simple proportion, which will give the same answer:

$$\frac{60\text{ g KNO}_3}{100\text{ g H}_2\text{O}} = \frac{x\text{ g KNO}_3}{150\text{ g H}_2\text{O}}$$

$$x\text{ g KNO}_3 = \frac{(60\text{ g KNO}_3)(150\ \cancel{\text{g H}_2\text{O}})}{(100\ \cancel{\text{g H}_2\text{O}})} = 90\text{ g KNO}_3$$

Refer to Figure 20 to answer the following questions.

1. a. What mass (in grams) of potassium nitrate (KNO_3) will dissolve in 100 g water at 60 °C?
   b. What mass (in grams) of potassium chloride (KCl) will dissolve in 100 g water at this temperature?
2. a. You dissolve 25 g potassium nitrate in 100 g water at 30 °C, producing an unsaturated solution. How much more potassium nitrate (in grams) must be added to form a saturated solution at 30 °C?
   b. What is the minimum mass (in grams) of 30 °C water needed to dissolve 25 g potassium nitrate?
3. a. A supersaturated solution of potassium nitrate is formed by adding 150 g KNO_3 to 100 g water, heating until the solute completely dissolves and then cooling the solution to 55 °C. If the solution is agitated, how much potassium nitrate will precipitate?
   b. How much 55 °C water would have to be added (to the original 100 g water) to just dissolve all of the KNO_3?

SOLUBILITY AND SOLUBILITY CURVES

Building Skills 5

The purpose of this activity is for students to learn to interpret solubility curves such as those in Figure 20. The answers can be found by using proportions or by thinking through the steps. This activity could be completed in small groups and used to supplement a class discussion.

Answers

1. a. 105 g
   b. 45 g
2. a. 45 g at saturation − 25 g in already = about 20 g to add
   b. $\frac{45\text{ g KNO}_3}{100\text{ g H}_2\text{O}} = \frac{25\text{ g KNO}_3}{x\text{ g H}_2\text{O}}$
   $x = 56$ g of water
3. a. 150 g KNO_3 − 95 g KNO_3 at 55 °C = 55 g of precipitate
   b. $\frac{95\text{ g KNO}_3}{100\text{ g H}_2\text{O}} = \frac{150\text{ g KNO}_3}{x\text{ g H}_2\text{O}}$
   $x = 158$ g of total water or add 58 g of water

C.2 SOLUTION CONCENTRATION

This section introduces percent as a way to express solution concentration and modifies that concept to include parts per hundred (pph), parts per thousand (ppt), and parts per million (ppm). Students may have read the term *ppm* in news articles about air and water pollution. Make sure students remember that the total mass of the solution also includes the mass of the solute. If possible, have students bring in beverages of various concentrations (i.e., fruit juices).

DESCRIBING SOLUTION CONCENTRATIONS

Building Skills 6

In this activity, students use the concept of percent to calculate concentrations of solutions.

Answers

1. a. Sugar

   b. Water
2. a. 17 g/(17 g + 183 g) × 100% = 8.5%

   b. 30 g/(30 g + 300 g) × 100% = 9.1%
3. a. 45 g/(45 g + 100 g) × 100% = 31%

   b. 45 g (45 g + 100 g + 155 g) = 15%

C.2 SOLUTION CONCENTRATION

The general terms saturated and unsaturated are not always adequate for describing the properties of solutions that contain different amounts of solute. A more precise description of the amount of solute in a solution is needed—an exact, numerical measure of concentration.

Solution concentration refers to the quantity of solute dissolved in a specific quantity of solvent or solution. You have already worked with one type of solution concentration expression: The water-solubility curves in Figure 20 (page 46) reported solution concentrations as the mass of a substance dissolved in a given mass of water.

Another way to express concentration is with percents. For example, dissolving 5 g table salt in 95 g water produces 100 g solution with a 5% salt concentration (by mass).

$$\frac{5 \text{ g salt}}{100 \text{ g solution}} \times 100\% = 5\% \text{ salt solution}$$

"Percent" means parts per hundred parts. So a 5% salt solution could also be reported as five parts per hundred of salt (5 pph salt). However, percent is much more commonly used.

$$\frac{5}{100} = \frac{50\,000}{1\,000\,000}$$

$$5\% \text{ (5 pph)} = 50\,000 \text{ ppm}$$

For solutions containing much smaller quantities of solute (as are found in many environmental water samples, including those from the Snake River), concentration units of **parts per million (ppm)** are sometimes useful. What is the concentration of the 5% salt solution expressed in ppm? Because 5% of 1 million is 50 000, a 5% salt solution is 50 000 parts per million.

Although you may not have realized it, the notion of concentration is part of daily life. For example, preparing beverages from concentrates, adding antifreeze to an automobile, and mixing pesticide or fertilizer solutions all require the use of solution concentrations. The following activity will help you review the concept of solution concentration, as well as gain experience with the chemist's use of this idea.

DESCRIBING SOLUTION CONCENTRATIONS

Building Skills 6

A common intravenous (abbreviated as IV) saline solution used in medicine contains 4.55 g NaCl dissolved in 495.45 g sterilized distilled water. Because a solution is a homogeneous mixture, the NaCl is distributed uniformly throughout the solution. What is the concentration of this solution, expressed as grams NaCl per 100 g solution?

The answer can be calculated in the following way:

$$\frac{4.55 \text{ g NaCl}}{4.55 \text{ g NaCl} + 495.45 \text{ g water}} = \frac{4.55 \text{ g NaCl}}{500 \text{ g solution}}$$

If 500 g solution contains 4.55 g NaCl, then you can determine the answer by calculating how much NaCl is contained in 100 g solution. So 100 g (or 1/5) of the solution will contain 1/5 of the total solute. One-fifth of the total solute is 0.91 g NaCl. Thus 100 g solution contains 0.91 g NaCl and 99.09 g water—1/5 as much as in the full 500-g solution:

$$4.55 \text{ g NaCl} \times 1/5 = 0.91 \text{ g NaCl}$$

4. This question may result in several types of answers. Simply keep adding solute until no more dissolves. It also provides a good way to extend the conceptual definition of *saturated* to the "real-world" notion of saturation. That is, you have a saturated solution when, after stirring, there is still some solute on the bottom of the container. Stirring can momentarily suspend additional solute and is necessary to ensure an even mixture. You could explain that you have a saturated solution of sugar in ice tea when no more sugar will dissolve in the tea and sugar falls to the bottom of the glass.

$$\frac{0.91\text{ g NaCl}}{0.91\text{ g NaCl} + 99.09\text{ g water}} = \frac{0.91\text{ g NaCl}}{100\text{ g solution}} = 0.91\%\text{ NaCl}$$

The concentration of this solution can be expressed as 0.91 g NaCl per 100 g solution. The solution is 0.91% NaCl by mass.

Now consider this example: One teaspoon of sucrose, which has a mass of 10 g, is dissolved in 240 g water. What is the concentration of the solution, expressed as grams sucrose per 100 g solution? As percent sucrose by mass?

Because the solution contains 10 g sucrose and 240 g water, it has a total mass of 250 g. A 100-g solution would contain 2/5 as much solute, or 4 g sucrose. Thus 100 g solution contains 4 g sucrose and 96 g water, a concentration of 4 g sucrose per 100 g solution. To determine the percent sucrose by mass,

$$\frac{10\text{ g sucrose}}{250\text{ g solution}} \times 100\% = \frac{4\text{ g sucrose}}{100\text{ g solution}} \times 100\% = 4\%\text{ sucrose by mass}$$

1. One teaspoon of sucrose is dissolved in a cup of water. Identify
   a. the solute.
   b. the solvent.
2. What is the concentration of each of the following solutions expressed as percent sucrose by mass?
   a. 17 g sucrose is dissolved in 183 g water.
   b. 30 g sucrose is dissolved in 300 g water.
3. A saturated solution of potassium chloride is prepared by adding 45 g KCl to 100 g water at 60 °C.
   a. What is the concentration of this solution?
   b. What would be the new concentration if 155 g water were added?
4. How would you prepare a "saturated solution" of potassium nitrate (KNO_3)?

Sucrose, $C_{12}H_{22}O_{11}$, is ordinary table sugar.

C.3 CONSTRUCTING A SOLUBILITY CURVE

Laboratory Activity

Introduction

You have seen and used solubility curves earlier in this unit (pages 46–47). In this activity, you will collect experimental data to construct a solubility curve for succinic acid ($C_4H_6O_4$), a molecular compound. Before you proceed, think about how your knowledge of solubility can help you gather data to construct a solubility curve.

- How can the properties of a saturated solution be used?
- What temperatures can you investigate?
- How many times should you repeat the procedure to be sure of your results?

Discuss these questions with your partner or laboratory group. Your teacher will then discuss with the class how data will be gathered and will demonstrate safe use of the equipment that will be used.

Safety Caution:

Remind students to remove the hot bath beaker from the hot plate or heating source before adding the succinic acid to the test tubes in the water bath. When succinic acid is heated to decomposition, its products can produce acrid smoke and irritating fumes. Students should handle the substance or its solutions with care. In particular, ensure that they do not allow any succinic acid crystals or solution to come into contact with the hot-plate surfaces or burner flames. For example, to avoid the risk of solution dripping onto the hot-place surface, before removing the stirring rod or thermometer from the test tube, students should "tap off" onto the tube's inner wall any drops of solution suspended on the stirring rod or thermometer.

Time

50 minutes

Materials (for a class of 24 working in pairs)

48 Pyrex® or Kimax® test tubes, 20 × 150 mm
24 glass stir rods
12 test-tube holders
12 alcohol thermometers
12 25-mL graduated cylinders
12 test-tube racks
12 400-mL beakers
6 hot plates
100 g succinic acid
ice
balances

C.3 CONSTRUCTING A SOLUBILITY CURVE

Laboratory Activity

Divide students into groups, with each group collecting solubility data at a different temperature. At each temperature, students should place a known mass of water into a suitable container (such as a test tube), then add solute to the container while stirring until solute remains on the bottom. The solution and undissolved solute can then be separated and the amount dissolved can be determined. Each point on the curve should have at least three trials so that the data can later be averaged reliably. Demonstrate how to set up a ring stand. Students may also need to be instructed in the use and care of balances.

Pre-lab discussion

Students should be reminded of the environmental context of this experiment, which is that the solubility of materials in water (e.g., agricultural runoff or industrial waste) is a function of temperature and that dissolved materials will have different environmental effects than undissolved materials. The terms *unsaturated, saturated,* and *supersaturated* should be defined within the context of this experiment.

Safety

Students should wear safety goggles and aprons throughout the laboratory activity. Students should be reminded that hot and cold glass is indistinguishable and that care should be taken in handling all glassware. If mercury thermometers are used, students should be told to leave the thermometer alone if it breaks and notify the instructor, who should have a mercury disposal kit at hand. Even with alcohol thermometers, students should be watched to ensure they do not use the thermometer as a stir rod. Succinic acid is toxic when ingested (LD_{50} 2.3 g/kg). Remind students to wash their hands at the end of the lab. The succinic acid should be reclaimed for future use, but it may also be washed down the drain with copious amounts of water.

Lab tips

Careful supervision is necessary during the laboratory activity due to the decanting of heated solutions and the use of hot plates. Before students start on the experiment, demonstrate how to decant a solution from a test tube using a glass stir rod. Have a test tube with approximately 4 g of succinic acid in it available so students can see how much material they need to dispense. It is suggested that each pair of students be assigned a particular temperature for which they will collect data. Each pair should report the average of two trials at one of the three temperatures—45 °C, 55 °C, or 65 °C—to the rest of the class. Use of temperatures lower than 45 °C may result in insufficient crystal formation. Use of temperatures higher than 65 °C poses a potential safety hazard. When students are finished, a beaker should be provided for reclaiming the succinic acid. The excess solution may be decanted or left to evaporate. The remaining crystals may be reused.

Expected Results

Temperature (°C)	Crystal Height (mm)
45	11
55	17
65	32

Safety

Keep the following precautions in mind while performing this laboratory procedure.

- The succinic acid that you will use is slightly toxic if ingested by mouth, so be sure to wash your hands thoroughly at the end of the laboratory.
- Never stir a liquid with a thermometer. Always use a stirring rod.
- Use insulated tongs or gloves to remove a hot beaker from a hot plate. Hot glass burns!
- Dispose of all wastes as directed by your teacher.

Procedure

1. To make a water bath, add approximately 300 mL water to a 400-mL beaker. Heat the beaker, with stirring, to either 45 °C, 55 °C, or 65 °C, as agreed to in your pre-lab class discussion. Ensure that the student team sharing your hot plate is investigating the same temperature. Carefully remove the beaker (using gloves or beaker tongs) when it reaches the desired temperature. NOTE: Do not allow the water-bath temperature to rise more than five degrees above the temperature that you have chosen. Return the beaker to the hot plate as needed to maintain the appropriate water-bath temperature.
2. Place between 4 g and 5 g succinic acid in each of two test tubes. **CAUTION:** *Be careful not to spill any of the succinic acid. If you do, clean up and dispose of the succinic acid as directed by your teacher.* Add 20.0 mL distilled water to each test tube.
3. Place each test tube in the water bath and take turns stirring the succinic acid solution with a glass stirring rod every 30 seconds for 7 minutes. Each minute, place the thermometer in the test tube and monitor the temperature of the succinic acid solution, ensuring that it is within 2 °C of the temperature that you have chosen.
4. At the end of 7 minutes, carefully decant the clear liquid from each test tube into a separate, empty test tube, as demonstrated by your teacher.
5. Carefully pour the hot water from the beaker into the sink and fill the beaker with water and ice.
6. Place the two test tubes containing the clear liquid in the ice bath for 2 minutes. Stir the liquid in each test tube gently once or twice. Remove the test tubes from the ice water. Allow the test tubes to sit at room temperature for 5 minutes. Observe each test tube carefully during that time. Record your observations.
7. Tap the side of each test tube and swirl the liquid once or twice to cause the crystals to settle evenly on the bottom of the test tubes.
8. Measure the height of crystals collected (in millimeters, mm). Have your partner(s) measure the crystal sample height and compare your results. Report the average crystal height for your two test tubes to your teacher.

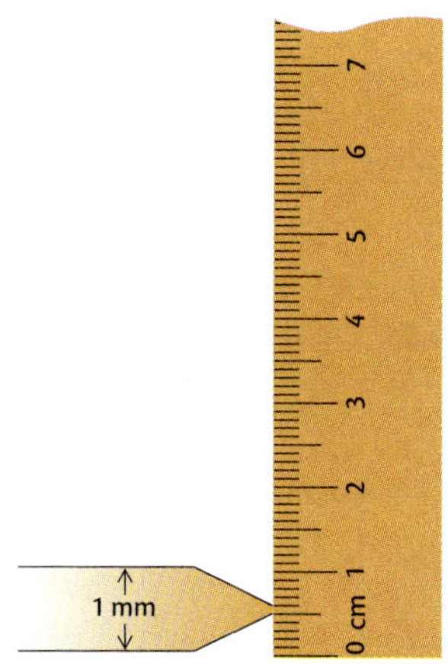

A millimeter is 1/10th of a centimeter.

9. Rinse the succinic acid crystals from the test tubes into a collection beaker designated by your teacher. Make sure that your laboratory area is clean.
10. Wash your hands thoroughly before leaving the laboratory.

Data Analysis

1. Find the mean crystal height obtained by your class for each temperature reported.
2. Plot the mean crystal height in millimeters (*y* axis) versus the water temperature in degrees Celsius (*x* axis).

Questions

1. Why is it important to collect data from more than one trial at a particular temperature?
2. How did you make use of the properties of a saturated solution at different temperatures?
3. Did all the succinic acid that originally dissolved in the water crystallize out of the solution? Explain your answer.
4. Given the pooled class data, did you have enough points to make a reliable solubility curve for succinic acid? Would the curve be good enough to make useful predictions about succinic acid solubility at temperatures not investigated in this activity? Explain your answer.
5. What procedures in this activity could lead to errors? How would each error affect your data?
6. Using your knowledge of solubility, propose a different procedure for gathering data to construct a solubility curve.

C.4 DISSOLVING IONIC COMPOUNDS

You have just investigated the process of a compound dissolving in water. What you observed is called a macroscopic phenomenon. However, what chemistry is primarily concerned with is what happens at the submicroscopic level—atomic and molecular phenomena that are not easily observed. As you have seen, temperature, agitation, and time all contribute to dissolving a solid material. But how do the atoms and molecules of the solute and solvent interact to make this happen?

Experiments suggest that water molecules are electrically **polar.** Although the entire water molecule is electrically neutral, the electrons are not evenly distributed in its structure. A polar molecule has an uneven distribution of electrical charge, which means that each molecule has a positive region on one end and a negative region on the other end. Evidence also suggests that a water molecule has a bent or V-shape, as illustrated in Figure 21 on page 52, rather than a linear, sticklike shape as in H–O–H. The "oxygen end" is an electrically negative region that has a greater

CD-ROM WWW. **Modeling Matter: Ionic Solutions**

4. Students should be able to interpolate using the data collected, but their answers should reflect that extrapolation is more problematic, as the solubility for temperatures outside the range investigated is completely unknown.
5. Factors students identify might include (but are not limited to): inaccurate temperature measurements, poor decanting technique, or failure of all the excess succinic acid to crystallize.
6. Accept reasonable answers that adequately describe solution preparation and isolation of dissolved solute.

C.4 DISSOLVING IONIC COMPOUNDS

Emphasize that dissolving can be viewed as requiring two steps—overcoming attractive forces among the particles of the solute (and solvent) and forming new attractive forces between solute and solvent particles. Careful examination of Figure 22 shows how the solution might look if we could see the ions and water molecules.

Explain that since water is a polar molecule, it can dissolve many ionic compounds. Remind students of the demonstration in which water bends around a charged object, indicating that water is an electrically charged substance (in B.5).

Answers

1. As the number of trials increases, the effects of random error decrease.
2. As the solutions cool, the solubility of succinic acid is lowered, and the excess succinic acid crystallizes.
3. There is still some succinic acid dissolved in the water at room temperature. The amount of crystal observed is the difference in the solubility of succinic acid at room temperature and the elevated temperatures used.

MODELING MATTER: DISSOLVING IONIC COMPOUNDS

You may want to stimulate the pictures in the mind by actually dissolving the KCl in 100 g of water at 70 °C.

Answers

1. a. At 50 °C, all 40 g of KCl will dissolve in 100 g of water. If the solution is cooled, you would expect to see the solute remain in solution until a temperature of 45 °C is reached. At that point, some of the solute is likely to begin coming out of solution. At 25 °C only 35 grams of KCl will dissolve, so we would expect that about 5 grams of solute would precipitate.

   b. The drawings (see below) should represent the solution being unsaturated at 50 °C and saturated at 40 °C. At any temperature below 40 °C, some precipitate would be expected (about 5 to 6 g at 25 °C). Precipitate would appear as some molecules of KCl clumped together at the bottom of the container, while most should be represented as still in solution.

2. a. The model would be unsaturated, just like the 50 °C model shown below.

   b. i. The model would still be unsaturated. With the total amount of water reduced, the concentration is higher, but still not saturated. The 20 g/75 g H_2O is equivalent to 27 g/100 g H_2O. This is well below the saturation point.

   ii. The amount of water would have to be reduced by half to 50 g in order to cause the KCl crystals to form. 20 g KCl/50 g H_2O is equivalent to 40 g KCl/100 g H_2O, the saturation point of KCl at 40 °C.

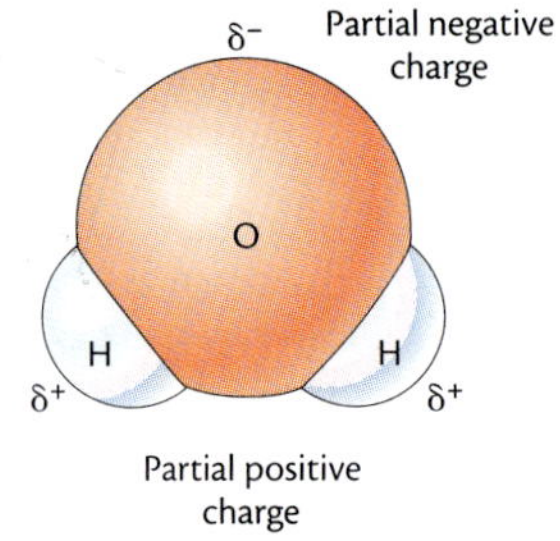

Figure 21 *Polarity of a water molecule. The δ+ and δ− indicate partial electrical charges.*

concentration of electrons (shown as δ−) compared with the two "hydrogen ends," which are electrically positive (shown as δ+). The Greek symbol δ (delta) means "partial"—thus partial plus and partial minus electrical charges are indicated. Because these charges balance, the molecule as a whole is electrically neutral.

Polar water molecules are attracted to other polar substances and to substances composed of electrically charged particles. These attractions make it possible for water to dissolve a great variety of substances.

One way to imagine the process of dissolving a substance in water is to liken it to a tug of war. Many solid substances, especially ionic compounds, are crystalline. In ionic crystals, positively charged cations are surrounded by negatively charged anions, with the anions likewise surrounded by cations. The crystal is held together by attractive forces between the cations and the anions. The substance will dissolve only if its ions are so strongly attracted to water molecules that the water "tugs" the ions from the crystal.

Water molecules are attracted to ions located on the surface of an ionic solid, as shown by the models in Figure 22a. The water molecule's negative (oxygen) end is attracted to the crystal's positive ions. The positive (hydrogen) ends of other water molecules are attracted to the negative ions of the crystal. When the attractive forces between the water molecules and the surface ions are strong enough, the bonds between the crystal and its surface ions become strained, and the ions may be pulled away from the crystal. Figure 22b uses models of water molecules and solute ions to illustrate the results of water "tugging" on solute ions. The detached ions become surrounded by water molecules, producing a water solution, as shown in Figure 22c.

> The water solubility of some ionic compounds can be extremely low indeed. For example, at room temperature, lead(II) sulfide, PbS, has a solubility of only about 10^{-14} g (0.00000000000001 g) per liter of water solution.

Using the description and illustrations of this process, can you determine what influences whether an ionic solid will dissolve? Because dissolving entails competition among three types of attractions—those between solvent and solute particles, between solvent particles themselves, and between particles within the solute crystals—the properties of both solute and solvent affect whether two substances will form a solution. Water is highly polar, so it will be effective at dissolving charged or ionic substances. However, if positive–negative attractions between cations and anions in the crystal are sufficiently strong, a particular ionic compound may be only slightly soluble in water.

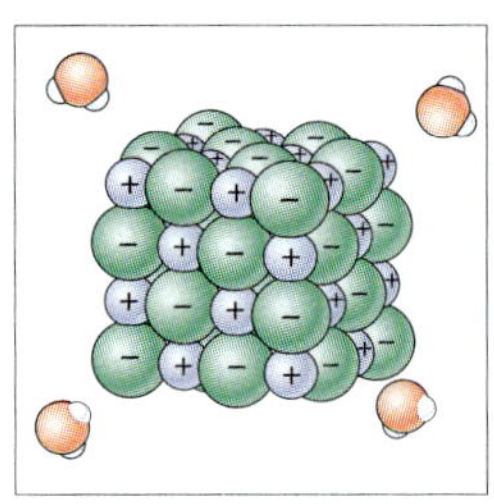

Figure 22a *Polar water molecules are attracted to the ions in an ionic crystal.*

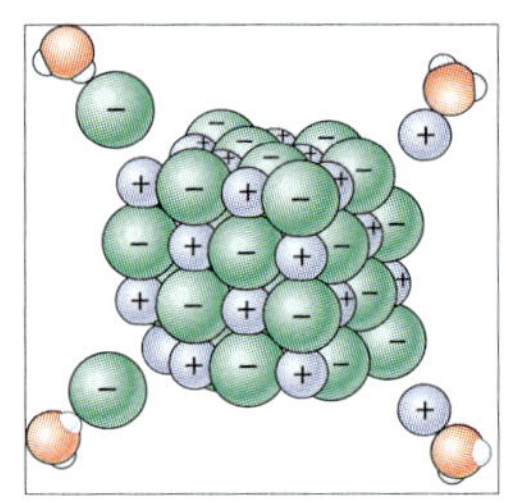

Figure 22b *Ions from the crystalline solid are pulled away by water molecules.*

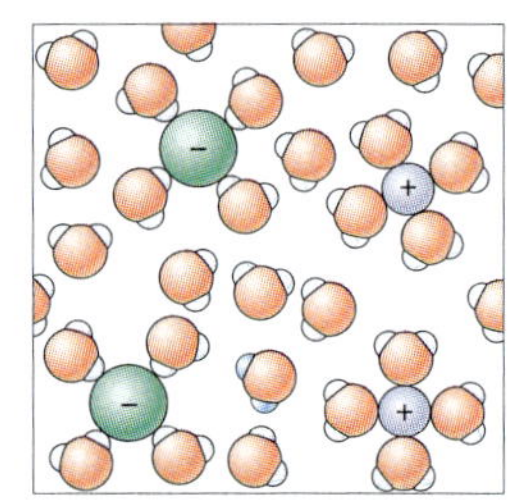

Figure 22c *A solution is formed when detached ions become surrounded by water molecules.*

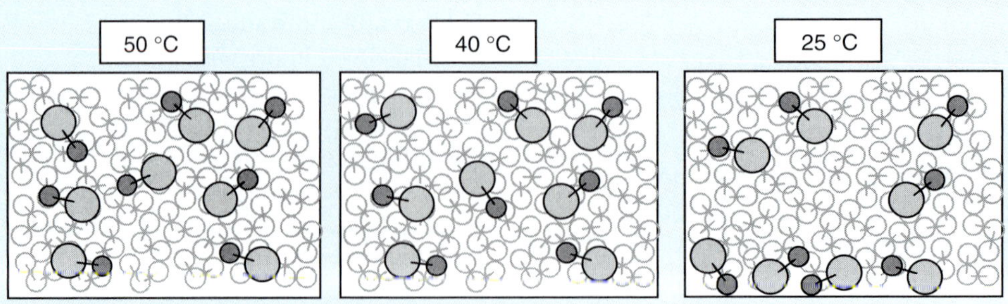

MODELING MATTER

DISSOLVING IONIC COMPOUNDS

You have now learned about solubility, solubility curves, and the process of dissolving ionic compounds in water. As part of these discussions, visual models, such as those presented in Figure 22 (page 52) have been used to describe the process of dissolving. In this activity, you will combine these models with your knowledge of solubility curves to create new models of ions dissolved in water.

1. Suppose you dissolved 40 g potassium chloride (KCl) in 100 g water at 50 °C. You then let the solution cool to room temperature, about 25 °C.
   a. What changes would you see in the beaker as the solution cooled? See Figure 23.
   b. Draw models of what the contents in the beaker would look like at the molecular level at 50 °C, 40 °C, and 25 °C. Keep these points in mind:

- You will need to consider whether the sample at each temperature is saturated, unsaturated, or supersaturated and draw the model accordingly. The solubility curve in Figure 23 will be helpful.
- It is impossible to draw all the ions and molecules in this sample. The ions and molecules that you draw will represent what is happening on a much larger scale.

2. An unsaturated solution will become more concentrated if you add more solute. Decreasing the total volume of water in the solution (such as by evaporation) also causes an increase in the solution's concentration. Consider a solution made by dissolving 20 g KCl in 100 g water at 40 °C.
   a. Draw a model of this solution.
   b. Suppose that while the solution was kept at 40 °C, 25% of the water evaporated.
      i. Draw a model of this solution and describe how it differs from the model of the original solution.
      ii. How much water must evaporate at this temperature to cause the first potassium chloride crystals to form?
3. A solution may be diluted (made less concentrated) by adding water.
   a. Draw a model of a solution containing 10 g KCl in 100 g water at 25 °C.
   b. Suppose you diluted this solution by adding another 100 g water with stirring. Draw a model of this new solution.
   c. Compare your drawings in 3a and 3b. What key feature is different in the two models? Why?

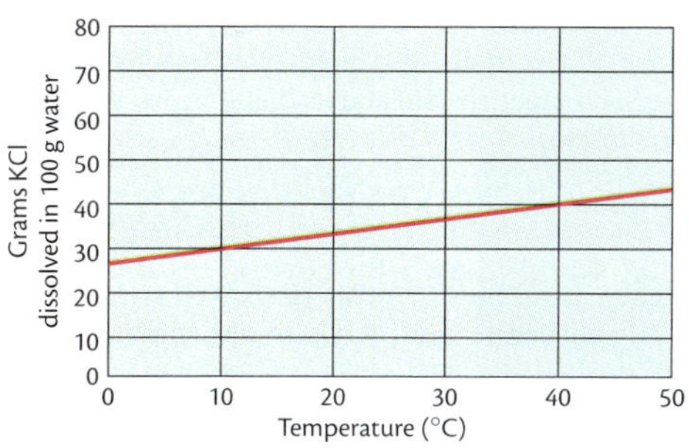

Figure 23 *Solubility curve for potassium chloride.*

Now that you know how solutions of ionic compounds are formed and described, it is time to consider some possible causes of the Riverwood fish kill. The information that you are about to read suggests some possible "culprits"—substances that can dissolve or be suspended in water and harm living things. These substances include heavy metals, acids and bases, molecular substances, and dissolved oxygen gas. Although all of these substances are normally found in natural water sources, the levels at which they are present can positively or negatively affect aquatic life.

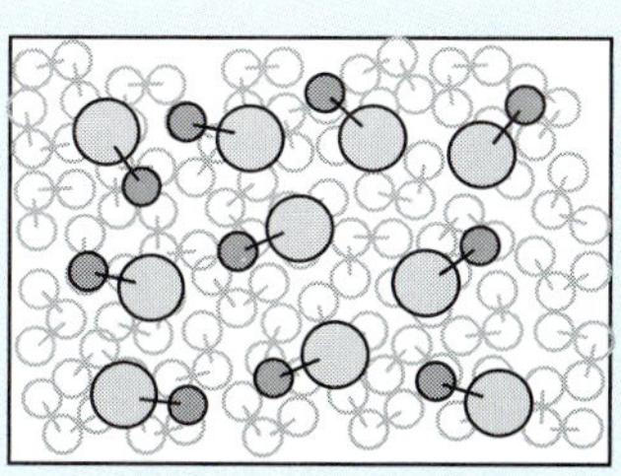

100 g water

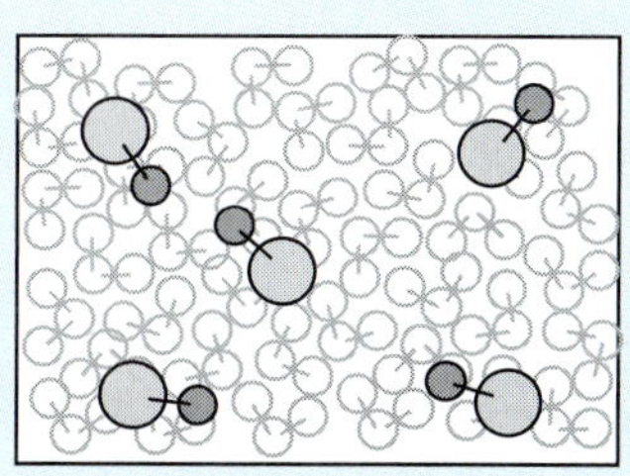

200 g water

3. a. See model below. The solution would be unsaturated and homogeneous. A solution with 10 g of KCl in 100 g of H_2O at 25 °C would allow all the solute to dissolve. A total of 35 g or so will dissolve under those conditions.

b. If the amount of solvent is doubled, the new solution should appear to be half as concentrated.

c. The appearance of the solute and solvent particles would be the same; only the relative numbers will change. The number of solute particles will be the same, but the number of solvent (water) molecules will double. Each drawing should show the solute particles homogeneously distributed, as neither solution is saturated.

INTRODUCTION TO POSSIBLE CAUSES

To help promote the "need to know" philosophy of *Chemistry in the Community,* the fish-kill investigation is divided into four possible causes. These causes include: heavy-metal ions, inappropriate pH, molecular substances, and dissolved oxygen levels. Students will gather information related to these causes and then make a determination on the cause of the fish problem. Depending on how the teacher organizes the final data, each class or group may come up with a different cause. However, keep in mind that students may come up with another cause or a combination of causes. If it can be chemically explained, this type of exploring should be encouraged.

C.5 INAPPROPRIATE HEAVY-METAL ION CONCENTRATIONS IN RIVER?

A candidate for responsibility for the fish kill is heavy-metal ion contamination. The cations of Pb and Hg are discussed in relation to their chemistry, commercial uses, and toxic effects on living organisms. The heavy metals are particularly dangerous because they accumulate in the body and have serious health consequences. Emphasize that metal ions, not metal atoms, are dissolved in water. Metal atoms *can* be found in the water as suspensions if they are scoured from the bottom sediments by strong water flow. Discuss the idea of allowable limits. (These are provided on the Snake River data at the end of this TE unit.)

C.5 INAPPROPRIATE HEAVY-METAL ION CONCENTRATIONS IN RIVER?

Many metal ions, such as iron(II) (Fe^{2+}), potassium (K^{+}), calcium (Ca^{2+}), and magnesium (Mg^{2+}), are essential to the health of humans and other organisms. For humans, these ions are obtained primarily from foods, but they may also be present in drinking water.

Not all metal ions that dissolve in water are beneficial, however. Some heavy-metal ions, called heavy metals because their atoms have greater mass than the masses of essential metallic elements, are harmful to humans and other organisms. Among the heavy-metal ions of greatest concern in water are cations of lead (Pb^{2+}) and mercury (Hg^{2+}). Lead and mercury are particularly likely to cause harm because they are widely used and dispersed in the environment. Heavy-metal ions are toxic because they bind to proteins in biological systems (such as your body), preventing the proteins from performing their intended tasks. As you might expect, because proteins play many important roles in body functioning, heavy-metal poisoning effects are severe. They include damage to the nervous system, brain, kidneys, and liver and even death.

The concentration of substances as they move through the food chain is known as bioaccumulation.

Unfortunately, heavy-metal ions are not removed as waste as they move up through the food chain. They become concentrated within the bodies of fish and shellfish, even when their abundance in the surrounding water is only a few parts per million. Such aquatic creatures then become hazardous for humans and other animals to consume.

In very low concentrations, heavy-metal ions are hard to detect in water and even more difficult and costly to remove. So how can heavy-metal poisoning be prevented? One of the easiest and most effective ways is to prevent the heavy-metal ions from entering water systems in the first place. This prevention can be accomplished by producing and using alternate materials that do not contain these ions and thus are not harmful to health or the environment. Such practices, which prevent pollution by eliminating the production and/or use of hazardous substances, are classified as examples of **green chemistry.** Such practices are applicable to heavy metals and to many other types of pollution.

Lead (Pb)

Locate lead on the Periodic Table at the back of your textbook. Compare its location with those of the essential metal ions that you just read about.

Lead is probably the heavy metal most familiar to you. Its symbol, Pb, is based on the element's original Latin name *plumbum*, also the source of the word "plumber."

Lead and lead compounds have been, and in some cases still are, used in pottery, automobile electrical storage batteries, solder, cooking vessels, pesticides, and paints. One compound of lead and oxygen, red lead (Pb_3O_4), is the primary ingredient in paint that protects bridges and other steel structures from corrosion.

Although lead water pipes were used in the United States in the early 1800s, they were replaced by iron pipes after it was discovered that water transported through lead pipes could cause lead poisoning. Romans constructed lead water pipes more than 2000 years ago; some of them are still in working condition. In modern homes, copper or plastic water pipes are used to prevent any contact between household water and lead.

Until the 1970s, the molecular compound tetraethyl lead, $Pb(C_2H_5)_4$, was added to gasoline to produce a better-burning automobile fuel. Unfortunately, the lead entered the atmosphere through automobile exhaust as lead oxide. Although the phaseout of leaded gasoline has reduced lead emissions, lead contamination remains in the soil surrounding heavily traveled roads. In some homes built before 1978 and not since repainted, the flaking of old leaded paint is another source of lead poisoning, particularly among children who may ingest the flaking paint.

Mercury (Hg)

Mercury is the only metallic element that is a liquid at room temperature. In fact, its symbol comes from the Latin *hydrargyrum,* meaning quick silver or liquid silver.

Locate mercury on the Periodic Table. Make a prediction about the locations of heavy metals on the Periodic Table.

Mercury has several important uses, some due specifically to its liquid state. It is an excellent electrical conductor, so it is used in "silent" light switches. It is also found in medical and weather thermometers, thermostats, mercury-vapor street lamps, fluorescent light bulbs, and some paints. Elemental mercury can be absorbed directly through the skin, and its vapor is quite hazardous to health. At room temperature, there will always be some mercury vapor present if liquid mercury is exposed to air, so any direct exposure to mercury is best avoided.

Because mercury compounds are toxic, they are useful in eliminating bacteria, fungi, and agricultural pests when used in antiseptics, fungicides, and pesticides. In the eighteenth and nineteenth centuries, mercury(II) nitrate, $Hg(NO_3)_2$, was used in making the felt hats popular at that time. After unintentionally absorbing this compound through their skin for several years, hat makers often suffered from mercury poisoning. Their symptoms included numbness, staggered walk, tunnel vision, and brain damage, thus giving rise to the expression "mad as a hatter."

The sudden release of a large amount of heavy-metal ions might cause a fish kill—depending on the particular metal ion, its concentration, the species of fish present, and other factors. Was such a release responsible for the Riverwood fish kill? As you read about other possible causes of the fish kill, keep in mind some questions that are relevant to all of the potential culprits. Is there a source of this substance along the Snake River near the site of the fish kill? What concentration of this solute would be toxic to various species of fish?

C.6 INAPPROPRIATE pH LEVELS IN RIVER?

You have likely heard the term pH used before, perhaps in connection with acid rain or hair shampoo. What is pH, and could it possibly help account for the fish kill in Riverwood? The **pH scale** is a convenient way to measure and report the acidic, basic, or chemically neutral character of a solution.

C.6 INAPPROPRIATE pH LEVELS IN RIVER?

Acidity and basicity play a major role in water quality and the survival of aquatic life. Students are introduced to general properties of acids and bases, and learn about some common acids and bases. The concepts of neutralization and the pH scale are also discussed briefly. The textbook's treatment of pH as a scale for distinguishing acids and bases is generally easily understood by students, who have encountered the term in advertisements (pH-balanced shampoo or deodorant) or in hot tub or swimming pool maintenance. The treatment of pH and acids and bases illustrates part of the *ChemCom* philosophy—the chemistry is introduced only as needed. Students will learn more about acids and bases later in the course.

DEMONSTRATION

Demonstrating the pH of various common household substances will help relate pH to students' lives. Common substances such as vinegar, citrus fruits, ammonia cleaner, baking soda, or antacids can be used to demonstrate pH. When testing these substances, use red and blue litmus paper, pH paper, and, if available, use a pH meter. Red cabbage juice contains a pigment that makes an excellent pH indicator. To extract the red pigment, boil the red cabbage in a small amount of water until the water turns a dark purple. Experiment with acids and bases using this liquid indicator. For a dramatic effect, place the red cabbage juice into a 1000-L graduated cylinder and add a few chunks of dry ice. As the dry ice (solid CO_2) dissolves, it will change the pH of the water indicator solution, thus making different colors.

Nearly all pH values are in the range from 0 to 14, although some extremely acidic or basic solutions may be outside this range. At room temperature, any pH values less than 7 indicate an acidic condition; the lower the pH, the more acidic the solution. Solutions with pH values greater than 7 are basic; the higher the pH, the more basic the solution. Basic solutions are also called alkaline solutions. Quantitatively, a change of one pH unit indicates a tenfold difference in acidity or alkalinity. For example, lemon juice, with a

Some Common Acids and Bases

Name	Formula	Use
Acids		
Acetic acid	$HC_2H_3O_2$	In vinegar (typically a 5% solution of acetic acid)
Carbonic acid	H_2CO_3	In carbonated soft drinks
Hydrochloric acid	HCl	Used in removing scale buildup from boilers and for cleaning materials
Nitric acid	HNO_3	Used in the manufacture of fertilizers, dyes, and explosives
Phosphoric acid	H_3PO_4	Added to some soft drinks to give a tart flavor; also used in the manufacture of fertilizers and detergents
Sulfuric acid	H_2SO_4	Largest-volume substance produced by chemical industry; present in automobile battery fluid
Bases		
Calcium hydroxide	$Ca(OH)_2$	Present in mortar, plaster, and cement; used in paper pulping and dehairing animal hides
Magnesium hydroxide	$Mg(OH)_2$	Active ingredient in milk of magnesia
Potassium hydroxide	KOH	Used in the manufacture of some liquid soaps
Sodium hydroxide	NaOH	A major industrial product; active ingredient in some drain and oven cleaners; used to convert animal fats into soap

Figure 24 *The name, formula, and common use of some familiar acids and bases.*

pH of about 2, is nearly ten times as acidic as soft drinks, which have a pH of about 3.

Acids and bases, some examples of which are listed in Figure 24, can also be identified by certain chemical properties. For example, the vegetable dye litmus turns blue in a basic solution and red in an acidic solution. Both acidic and basic solutions conduct electricity. Each type of solution has a distinctive taste and a distinctive feel on your skin. (**CAUTION:** *You should never test these sensory properties in the laboratory.*) In addition, concentrated acids and bases are able to react chemically with many other substances. You are probably familiar with the ability of acids and bases to corrode, or wear away, other materials. Corrosion is a type of chemical reaction.

Vinegar is an acid that you have tasted; that common kitchen ingredient is considered a dilute solution of acetic acid.

Most acid molecules have one or more hydrogen atoms that can be released rather easily in water solution. These "acidic" hydrogen atoms are usually written first in the formula for an acid. See Figure 24.

Many bases are ionic substances that include hydroxide ions (OH^-) in their structures. Sodium hydroxide, $NaOH$, and barium hydroxide, $Ba(OH)_2$, are two examples. Some bases, such as ammonia (NH_3) and baking soda (sodium bicarbonate, $NaHCO_3$), contain no OH^- ions but still produce basic solutions because they react with water to generate OH^- ions, as illustrated by the following equation.

$$NH_3 \quad + \quad H_2O \quad \rightarrow \quad NH_4^+ \quad + \quad OH^-$$

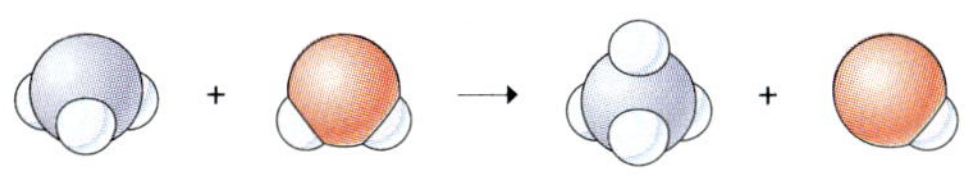

What about substances that display neither acidic nor basic characteristics? Chemists classify these substances as chemically neutral. Water, sodium chloride ($NaCl$), and table sugar (sucrose, $C_{12}H_{22}O_{11}$) are all examples of chemically neutral compounds.

At 25 °C, a pH of 7 indicates a chemically neutral solution. The pH values of some common materials are shown in Figure 25.

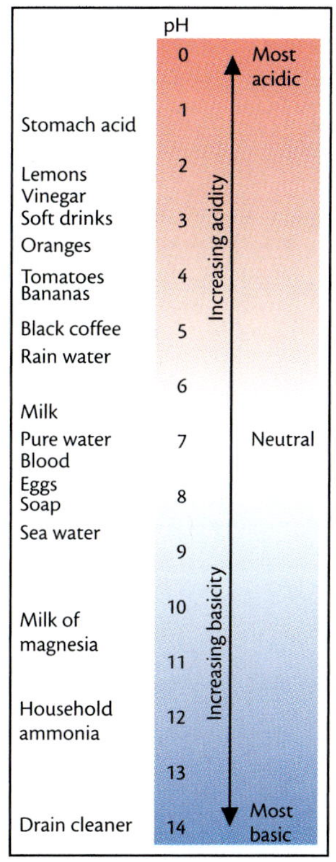

Figure 25 *The pH values of some common materials.*

As you can see in Figure 25, rainwater is naturally slightly acidic. This is because the atmosphere contains certain substances—carbon dioxide (CO_2) for one—that produce acidic solutions when dissolved in water. Both acidic and basic solutions have effects on living organisms—effects that depend on the pH of the water. When the pH of water is too low (meaning high acidity), fish-egg development is impaired, thus hampering the ability of fish to reproduce. Water solutions with low (acidic) pH values also tend to increase the concentrations of metal ions in natural waters by leaching the metals from surrounding soil. These metal ions can include aluminum ions (Al^{3+}), which are toxic to fish when present in sufficiently high concentration. High pH (basic contamination) is a problem for living organisms primarily because alkaline solutions are able to dissolve organic materials, including skin and scales.

The U.S. Environmental Protection Agency (EPA) requires drinking water to be within the pH range from 6.5 to 8.5. However, most fish can

57

tolerate a slightly wider pH range, from about 5 to 9, in lake or river water. Serious freshwater anglers try to catch fish in water between pH 6.5 and 8.2.

On a normal day, the pH of the water in the Snake River in Riverwood ranges between 7 and 8, nearly optimal for freshwater fishing. Could the pH have changed abruptly, killing the fish? Was acidic or basic contamination responsible for the Riverwood crisis?

C.7 INAPPROPRIATE MOLECULAR-SUBSTANCE CONCENTRATIONS IN RIVER?

Solubility of Molecular Substances CD-ROM WWW.

Until now, the types of substances considered suspects in the Riverwood mystery have been ionic substances, those that dissolve in water to release ions. Are there other types of substances that dissolve in water and possibly present a hazard to aquatic life? Some substances, such as sugar and ethanol, dissolve in water but not in the form of ions. These substances belong to a class of materials known as **molecular substances** because they are composed of molecules.

You investigated the solubility behavior of succinic acid earlier. See p. 49.

Unlike ionic substances, which are crystalline solids at normal conditions, molecular substances can be found as solids, liquids, or gases at room temperature. Some molecular substances such as oxygen (O_2) and carbon dioxide (CO_2) have little attraction between their molecules and are thus gases at normal conditions. Molecular substances such as ethanol (ethyl alcohol, C_2H_5OH) and water (H_2O) have larger between-molecule attractions, causing these "stickier" molecules to form liquids at normal conditions. Other molecular substances with even greater between-molecule attractions—succinic acid ($C_4H_6O_4$), for example—are solids at normal conditions. Their stronger attractive forces hold the molecules together more tightly, in effect determining in which state the substance will be found.

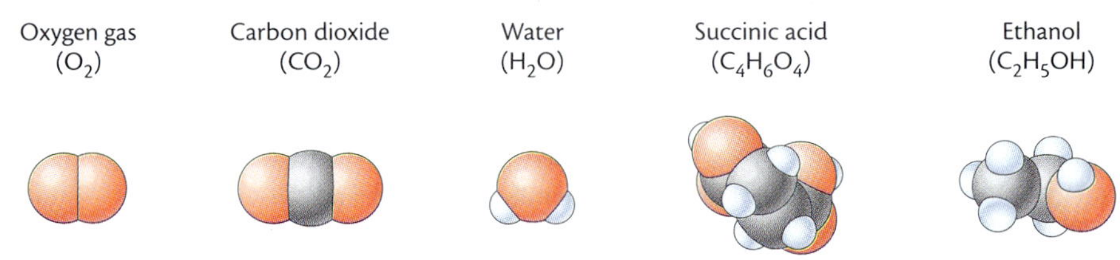

What determines how soluble a molecular substance will be in water? The attraction of a substance's molecules for each other compared with their attraction for water molecules plays a major part. But what causes these attractions? The distribution of electrical charge within molecules has a great deal to do with it.

Most molecular compounds contain atoms of nonmetallic elements. As you learned earlier (page 32), these atoms are linked together by the attraction of one atom's positively charged nucleus for another atom's negatively

C.7 INAPPROPRIATE MOLECULAR-SUBSTANCE CONCENTRATIONS IN RIVER?

After several possible explanations for the fish kill have been presented, the role of molecular substances is explored. This section gives a structural explanation as to why some substances dissolve in water while others do not. The explanation is based on polarity. The familiar saying, Oil and water don't mix, has a chemical basis in the generalization "like dissolves like." This means not only that polar solvents will dissolve polar solutes, but also that nonpolar solvents will dissolve nonpolar solutes.

Demonstration

An effective demonstration is the dissolving of Styrofoam (packing peanuts or cups) in a large beaker (1 L) that is partially filled with 100 mL acetone (fingernail polish remover). Enough of the Styrofoam quickly dissolves, reducing the volume of the peanuts or cup, to create a dramatic demonstration of solubility. Emphasis can be on rates of dissolving, solubility, and the types of molecules that would affect the dissolution rate.

C.8 SOLVENTS

Laboratory Activity

A discussion concerning toxic solutes that could possibly be found in the river may be a good introduction to this section. These could include pesticides, herbicides, fertilizers, preservatives, and paints. In addition, using the map of Riverwood, have a class discussion on which toxic substances might be found in the river based on the businesses located there.

In this lab activity, the students are given a set of solutes to test. The following set is one possibility: urea [$CO(NH_2)_2$], glucose ($C_6H_{12}O_6$), sodium chloride (NaCl), iodine (I_2), copper(II) sulfate ($CuSO_4$), calcium carbonate ($CaCO_3$), naphthalene ($C_{10}H_8$), ethanol (C_2H_5OH), and hexane (C_6H_{14}). Each of the above solutes was chosen for a particular reason. They exhibit a range of character from ionic to polar to nonpolar solutes. Additionally, the solvents can be tested as solutes.

Before having the students test the solubility of the solutes in water in Part II and in ethanol and lamp oil in Part III, have them draft an experimental plan. Students must think about a procedure that will control variables and give reasonable charged electrons. If differences in electron attraction between atoms are large enough, electrons move from one atom to another, forming ions. This is what often happens between a metallic atom and a nonmetallic atom when an ionic compound is formed. An atom's ability to attract shared electrons in its bonding within a substance is known as its **electronegativity.** In molecular substances, these differences in electron attraction, or electronegativities, are not large enough to cause ions to form, but they may cause the electrons to be unevenly distributed among the atoms.

You already know that the "oxygen end" of a water molecule is electrically negative compared with its positive "hydrogen end." That is, water molecules are polar and serve as the most common example of a polar solvent. Such charge separation (and resulting molecular polarity) is found in many molecules whose atoms have sufficiently different electronegativities.

Polar molecules tend to dissolve readily in polar solvents such as water. For example, water is a good solvent for sugar and ethanol, both composed of polar molecules. Similarly, nonpolar liquids are good solvents for other nonpolar molecules. Nonpolar cleaning fluids are used to "dry clean" clothes because they readily dissolve nonpolar body oils found in fabric. In contrast, nonpolar molecules (such as those of oil and gasoline) do not dissolve well in polar solvents (such as water or ethanol).

This pattern of solubility behavior—polar substances dissolving in polar solvents, nonpolar substances dissolving in nonpolar solvents—is often summarized in the generalization "Like dissolves like." This rule also explains why nonpolar liquids are usually ineffective in dissolving ionic and polar molecular substances.

Were dissolved molecular substances present in the Snake River water where the fish died? Most likely yes; at least in small amounts. Were they responsible for the fish kill? That depends on which molecular substances were present and at what concentrations. And that, in turn, depends on how each solute interacts with water's polar molecules.

In the following laboratory activity, you will investigate and compare the solubility behavior of some typical molecular and ionic substances.

Unfortunately, many nonpolar dry-cleaning solvents are damaging to both human health and the environment. However, the recent development of new technologies allows environmentally benign nonpolar solvents such as liquefied carbon dioxide to be used in the dry-cleaning industry.

Various molecular substances may normally be present at very low levels—so low that no harm to living things is observed.

C.8 SOLVENTS

Laboratory Activity

Introduction

The *Riverwood News* reported earlier that Dr. Brooke believes that a substance dissolved in the Snake River is one likely fish-kill cause. She based her judgment on her chemical knowledge and experiences with water and other substances. Dr. Brooke also has a general idea about which contaminating solutes she can initially rule out: those that cannot dissolve appreciably in water. Such background knowledge helps Dr. Brooke (and other chemists) reduce the number of water tests required in the laboratory.

What, then, do the terms "soluble" and "insoluble" mean? Is anything truly insoluble in water? It is likely that at least a few molecules or ions of any substance will dissolve in water. Thus the term "insoluble" actually refers to substances that are only very, very slightly soluble in water. Chalk,

Time

30 minutes

Materials (for a class of 24 working in pairs)

24 test tubes (assuming tubes are reused), 13 × 100 mm
12 test-tube racks
12 10-mL graduated cylinders
12 marking pencils
12 spatulas
12 medicine droppers
25 g urea, $CO(NH_2)_2$
25 g iodine, I_2
25 g ammonium chloride, NH_4Cl
25 g sodium chloride, NaCl
25 g glucose, $C_6H_{12}O_6$
25 g naphthalene flakes, $C_{10}H_8$
25 g copper(II) sulfate, $CuSO_4 \cdot 5H_2O$
25 mL hexane, C_6H_{14}
600 mL ethanol (denatured alcohol), C_2H_5OH
600 mL lamp oil
1 L distilled water, H_2O

Safety

Students should wear safety goggles and aprons throughout the laboratory activity. Direct skin contact with iodine is to be avoided. Consider demonstrating this part of the activity. Hexane, ethanol, and naphthalene are flammable. **No flames are allowed in the laboratory.**

Pre-Lab Discussion

It is important that students follow the procedures and use small solute samples to judge the solubility accurately. The gentle mixing is also important.

data on the solubility of each substance. An activity sheet in the Activity Workbook is provided to help organize these plans. Any plan that includes steps to have uniform sample size, mixing, and scale for determining relative solubility should be approved. Also scan for safe procedures.

CAUTION: *Hexane is highly flammable—ensure that there are no open flames in the room.*

Remind students not to touch any chemical, to use a wood splint to transfer solid samples, and to discard the splint in a teacher-designated container.

In this activity, students will explore the concept of "like dissolves like" by testing the solubility of several substances in water, ethanol, and lamp oil, examples of polar, slightly polar, and nonpolar solvents.

Lab Tips

Provide labeled containers for discarded hexane samples. After class, the hexane can be evaporated in the fume hood.

Mix powdered calcium carbonate in distilled water. Check conductivity of slurry vs. conductivity of distilled water for evidence of solubility.

Careful supervision is necessary during the laboratory activity because of the toxicity of iodine and naphthalene, and the flammability of hexane.

Part I answers

1. miscibility, amount of precipitate not dissolved
2. temperature, amounts of solute and solvent, stirring; to achieve a consistent test
3. Always mix chemicals together slowly, usually while stirring.

Part II answers

1. Naphthalene, hexane, and iodine, the water-insoluble solutes, would not be expected to dissolve in the river.
2. It is intended that the groups will report their results to the rest of the class. Differences should be discussed and debated, leading students to define the criteria for distinguishing solubility. Students may wonder if there is an objective method to determine solubility. This would be a good time to share with them how chemists determine and report solubility (or, the students themselves may suggest such a way).

   You will find a sample filled-in data table on page 1-68B of this Teacher Edition.

1.53 mg is equal to only 0.00153 g.

for example, is considered insoluble in water, even though 1.53 mg calcium carbonate ($CaCO_3$, the main ingredient in chalk) can dissolve in 100 g water at 25 °C.

In this laboratory activity, you will first investigate the solubilities of various molecular and ionic solutes in water. These solubility data, along with toxicity data, will help you rule out some solutes as likely causes of the fish kill. You will then test other solvents and examine the solubility data for any general patterns.

Your teacher will tell you which particular solutes you will investigate. List them in your laboratory notebook.

Part I: Designing a Procedure for Investigating Solubility in Water

Your teacher will direct you to discuss with either the whole class or your laboratory partner a procedure for testing the room-temperature solubilities of the substances listed in your laboratory notebook. (If you have performed solubility tests before, it may be useful to recall how you did them.) With your partner, design a step-by-step investigation that will allow you to determine whether each solute is soluble (S), slightly soluble (SS), or insoluble (I) in room-temperature water.

In this activity "insoluble" (I) means that the solubility is so low that no solute was observed to dissolve.

The following questions will help you design your procedure.

1. What particular observations will allow you to judge how well each solute dissolves in the polar solvent water? That is, how will you decide whether to classify a given solute as soluble, slightly soluble, or insoluble in water?
2. Which variables will need to be controlled? Why?
3. How should the solute and solvent be mixed—all at once or a little at a time? Why?

In designing your procedure, keep these concerns in mind.

- Avoid any direct contact of your skin with any solutes.
- Follow your teacher's directions for waste disposal.

When you and your partner have agreed on a written procedure, get it approved by your teacher. Construct a data table for your results and you are ready for Part II.

Part II: Investigating Solubility in Water

Use your approved procedure to investigate the solubility in water of the listed substances. Record the data in your data table.

Part III: Investigating Solubility in Ethanol and Lamp Oil

It is clear that the task of determining what may have caused the fish kill can be simplified somewhat by focusing efforts on substances that will dissolve appreciably in water. However, in dealing with other solubility-based problems, chemists sometimes find it helpful to use solvents other than water—ethanol and lamp oil serve that role in this activity.

You and your partner will investigate the solubility of some or all of the solutes from Part II in ethanol and lamp oil. You should also test the solubility of water in ethanol and in lamp oil. By gaining experiences with three

Part III answers

3. a. Ethanol acts very similar to water and hexane, dissolving the same substances.

   b. It also dissolves in lamp oil, and it appears to dissolve some of the same solutes as lamp oil.
4. a. and b. Answers will vary according to what students may have expected. Likely, they are unfamiliar with hexane and the results are new. You may want to confirm the molecular structure of water, alkanes, alcohols or organic solvents.

liquid solvents—lamp oil, ethanol, and water—you will be prepared to recognize some general patterns regarding solubility behavior.

Can you use the same procedure that you designed for Part II? If not, what parts of that procedure should be revised? In considering your Part III procedure, again keep the questions listed on page 60 in mind.

Have your proposed procedure approved by your teacher. Before you start the laboratory work, test your interpretation of the results of Part II by predicting what you think you will observe regarding solubility in each case. Include these predictions in your data table for Part III. Then collect and record your data for both solvents.

Wash your hands thoroughly before leaving the laboratory.

Questions

Part II

1. According to your data, which tested solutes are least likely to be dissolved in the Snake River? Why?
2. Compare your data with those of the rest of the class. Are there any differences? If so, how can those differences be explained?

Part III

3. a. How does the behavior of ethanol as a solvent compare with that of water?
   b. How does ethanol's behavior compare with that of lamp oil?
4. a. Were any of your solubility observations unexpected?
   b. If so, explain what you expected, why it was expected, and how your expectations compare with what you actually observed.
5. Based on your data, what general pattern of solubility behavior can you summarize and describe?
6. Predict the solubility behavior of each solid solute in:
   a. hexane, a liquid that is essentially insoluble in water.
   b. ethylene glycol, a liquid that is very soluble in ethanol.
7. a. Given that water is a polar solvent and lamp oil is a nonpolar solvent, classify each molecular solute tested as polar or nonpolar.
   b. How did you decide?
8. How useful is the rule "like dissolves like" for predicting solubility? Explain your answer on the basis of your results.
9. In Part II, water was the "solvent," but in Part III, water was a "solute."
   a. How can it be both?
   b. How do you know whether a substance is a solute or a solvent?

The familiar saying "oil and water do not mix" has a chemical basis!

You now know about the solubility of some solids and liquids in water. As you read on to learn about the behavior of gases in natural waters, consider the possibility that a dissolved gas was responsible for the Snake River catastrophe.

5. Generally, the substances soluble in water were insoluble in lamp oil, and vice versa.

6. a. The solutes that dissolved in lamp oil will also dissolve in hexane.

   b. Ethylene glycol (antifreeze) has similar solvent properties to ethanol and would have the same results.

7. a and b. Solvents being solute in water equates to polar, and insoluble equates to nonpolar. Likewise, in hexane, soluble equates to nonpolar, and insoluble equates to polar. Knowing that water is polar and insoluble in hexane leads to the conclusion that hexane is nonpolar. From there, polarity is deduced from the previous idea of "like dissolves like." You will probably need to tell students which solutes they tested were polar and nonpolar.

8. Good; from the results it should be clear that like dissolves like. If students did not get the expected results, the tests may need to be repeated under closer supervision.

9. a. In liquid mixtures, deciding which is the solute and which is the solvent can be confusing.

   b. According to convention, the liquid in greater quantity is the solvent and the liquid in lesser quantity is the solute.

Post-lab discussion

Center the post-lab discussion on the questions at the end of the activity. The newspaper clipping "Fish Kill Remains a Mystery" eliminates organic compounds from sources such as pesticides, fertilizers, and industrial waste as potential causes of the fish kill.

C.9 INAPPROPRIATE DISSOLVED OXYGEN LEVELS IN RIVER?

Continue to use the terms *solute, solvent,* and *saturated.* These terms from the solubility of solids also apply to gases. However, the relationship between the solubility of gases and temperature differs from that of the solubility of solids and temperature. In gases, the colder the liquid, the more gas that can dissolve in it. This is the opposite of what happens with the solubility of many solids. The relationship of gas solubility and pressure is explored. Most students are already familiar with the "fizz effect" of warm carbonated soft drinks and the effect of breaking the seal on a carbonated beverage.

Quick Demonstration

Open up two bottles of carbonated beverages, one very cold and one warm. Which bottle has the greatest release of gas?

ADDITIONAL CHEMQUANDARY

One 40-L aquarium is kept at 25 °C, while a second, identical aquarium is kept at 20 °C. Which aquarium can support a larger population of fish? Why?
Answer: Fish require dissolved oxygen to survive. The fish tank that is kept at the lower temperature (20 °C) will contain more dissolved oxygen to support a larger fish population (assuming all other factors are equal).

C.9 INAPPROPRIATE DISSOLVED OXYGEN LEVELS IN RIVER?

You have noted that the solubility of ionic and molecular solids in water usually increases when the water temperature is raised. Do gases behave similarly to solids in solution? Look at Figure 26, which shows the solubility curve for oxygen gas, plotted as milligrams oxygen dissolved per 1000 g water.

What is the solubility of oxygen gas in 20 °C water? In 40 °C water? As you can see, increasing the water temperature causes the gas to be less soluble! Note also the magnitude of the values for oxygen solubility. Compare these values with those for solid solutes as shown in Figure 20. At 20 °C, about 30 g potassium nitrate will dissolve in 100 g water. In contrast, only about 9 mg (0.009 g) oxygen gas will dissolve in ten times as much water—1000 g water—at this temperature. It should be clear that most gases are far less soluble in water than are many ionic solids.

When you considered the solubility of molecular and ionic solids in water, you found that solubility depended on two factors—temperature and the nature of the solvent. The solubility of a gas depends on these two factors as well. But it also depends on gas pressure. Referring to Figure 27, note what happens to oxygen's solubility as the pressure of oxygen gas above it is increased. Does more or less gas dissolve in the same amount of water? What happens if the gas pressure is doubled? For example, look at the solubility of O_2 at one atmosphere and at two atmospheres of pressure. Also consider the shape of the graph line. What type of relationship does a linear graph line indicate?

The metric unit for pressure is the pascal (Pa): 1 atmosphere (atm) = 101 325 Pa

As you have by now deduced, gas solubility in water is directly proportional to the pressure of that gaseous substance on the liquid. You see one effect of this proportionality every time you open a can or bottle of carbonated soft drink. As the gas pressure on the liquid is reduced by opening the container to the air, some dissolved carbon dioxide gas (CO_2) escapes from the liquid in a rush of bubbles.

Because there is not very much CO_2(g) present in air, carbon dioxide gas must be forced into the carbonated beverage at high pressure just before

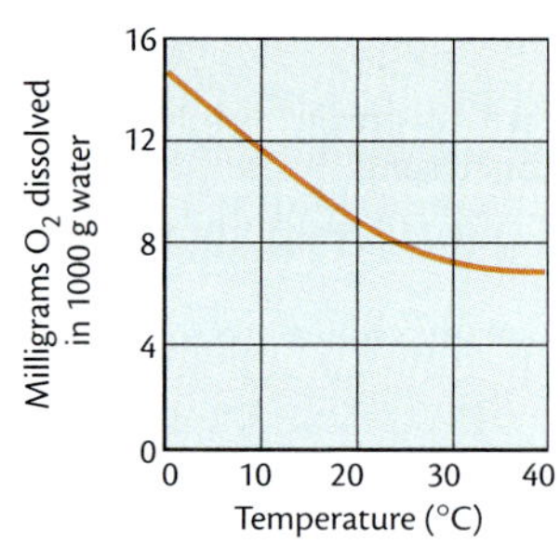

Figure 26 *Solubility curve for O_2 gas in water in contact with air.*

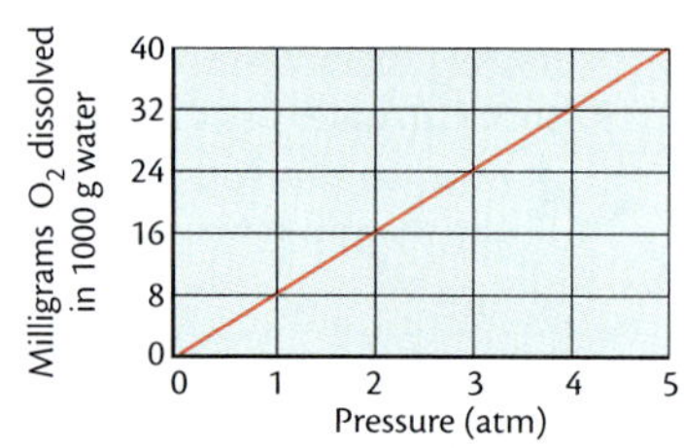

Figure 27 *Relationship between solubility and pressure of O_2 at 25 °C.*

the container is sealed. This increases the amount of carbon dioxide that can dissolve in the beverage. When the can or bottle is opened, CO_2 gas pressure on the liquid drops back to its usual low level. Dissolved carbon dioxide gas escapes from the liquid until it reaches its (lower) solubility at this lower pressure. When the fizzing stops, you might describe the beverage as having "gone flat." Actually, the excess carbon dioxide gas has simply escaped into the air; the resulting solution is still saturated with CO_2 at the new conditions.

C.10 TEMPERATURE, DISSOLVED OXYGEN, AND LIFE

CD-ROM WWW. **Temperature, Dissolved Oxygen & Life**

On the basis of what you know about the effect of temperature on solubility, you may be wondering if the temperature of the Snake River had something to do with the fish kill. As you just learned, water temperature affects how much oxygen gas can dissolve in the water. Various forms of aquatic life, including the many species of fish, have different requirements for the concentration of dissolved oxygen needed to survive. Figure 28 contains this information. How, then, does a change in the temperature of the natural waters affect the fish internally?

Fish are "cold-blooded" animals; their body temperatures rise and fall with the surrounding water temperature. If the water temperature rises, the body temperatures of fish also rise. This increase in body temperature in turn increases fish metabolism, a complex series of interrelated chemical reactions that keep fish alive. As these internal reaction rates speed up, the fish eat more, swim more, and require more dissolved oxygen. The rate of metabolism also increases for other aquatic organisms, such as aerobic bacteria, that compete with fish for dissolved oxygen.

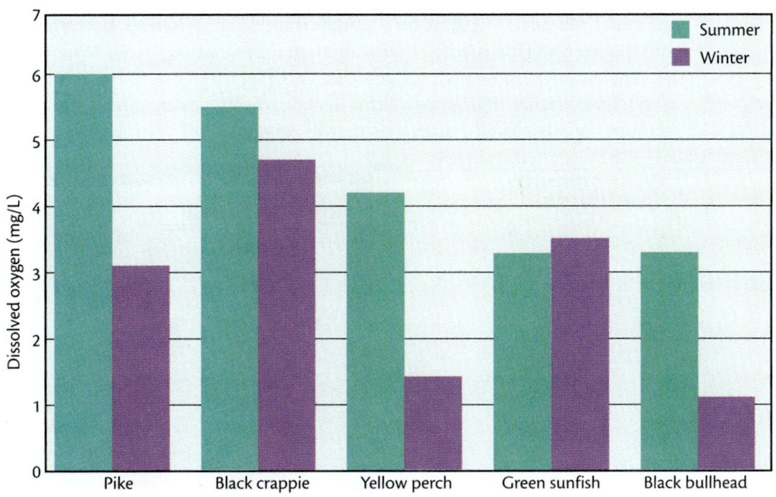

Figure 28 *Dissolved-oxygen requirements of various fish.*

C.10 TEMPERATURE, DISSOLVED OXYGEN, AND LIFE

The relationship between temperature and dissolved oxygen (DO) is important. The effect of temperature on chemical reactions is illustrated, and its relationship to biological functions in fish is also explored.

Remind students that there are several reasons for changes in lake or river water temperatures that can be traced to human activity. Sharp changes in water temperature may affect water organisms. Remind students that there is a minimum concentration of DO required for fish to survive. Often this value is stated as 4 ppm (parts per million).

Introduce the concept of percent saturation. This value incorporates oxygen levels at a given temperature, and it also illustrates that it is possible to have too much DO—a supersaturated solution. Many students have already heard of the "bends" that scuba divers might get if they surface too rapidly. This is an example of a supersaturated gas coming out of solution at a lower pressure. Students may be more familiar with the fish in their local area better than the fish listed in Figures 28 and 29.

A table of DO requirements for some common freshwater fish can be found on page 68B.

Demonstration

The effect of temperature on gas solubility can be demonstrated as follows: chill a bottle of carbonated drink (colored beverages show the result better) to just above 0 °C. Prepare three beakers of water. Have one at 0 °C (ice-water bath), one at room temperature, and the third at about 40 °C. Open the bottle and pour soda gently into three test tubes until full. Close each tube with a one-hole stopper and, with your finger over the stopper hole, invert the tube and immerse it in one of the water baths. Do the same with the other two. Having two students assist you, so that all three tubes are inverted and immersed simultaneously, is more effective. In the beakers at room temperature and 40 °C, carbon dioxide is released from the solution and can be seen as bubbles in the test tube. The relative numbers of bubbles show that the solubility of CO_2 is less at higher temperatures.

Figure 29 *Maximum water temperature tolerance in fish.*

Maximum Water Temperature Tolerance in Fish (24-hour Exposure)

Fish	Maximum Temperature °C	Maximum Temperature °F
Trout (brook, brown, rainbow)	24	75
Channel catfish	35	95
Lake herring (cisco)	25	77
Largemouth bass	34	93
Northern pike	30	86

As you can now see, an increase in water temperature affects fish both externally and internally. A long stretch of hot summer days sometimes results in large fish kills, in which hundreds of fish literally suffocate. Figure 29 summarizes the maximum water temperatures at which selected fish species can survive.

Sometimes hot summer days are not to blame. Often, high lake or river water temperatures can be traced to human activity. Many industries, such as electrical power generation, depend on natural bodies of water to cool heat-producing processes. Cool water is drawn from lakes or rivers into an industrial or power-generating plant, and devices called heat exchangers transfer thermal energy (heat energy) from the processing area to the cooling water. The heated water is then released back into the lakes or rivers, either immediately or after the water has partly cooled. If the water is too warm, it can upset the balance of life in lakes and rivers.

At this point, it should be clear that there is a lower limit on the amount of dissolved oxygen needed for fish to survive. Is there an upper limit? Can too much dissolved oxygen be a problem for fish? In nature, both oxygen and nitrogen gas are present in the air at all times. See Figure 30. When oxygen gas dissolves, so does nitrogen gas. This fact turns out to be significant when considering the upper limit for dissolved gases. When the total amount of dissolved gases—primarily oxygen and nitrogen—reaches

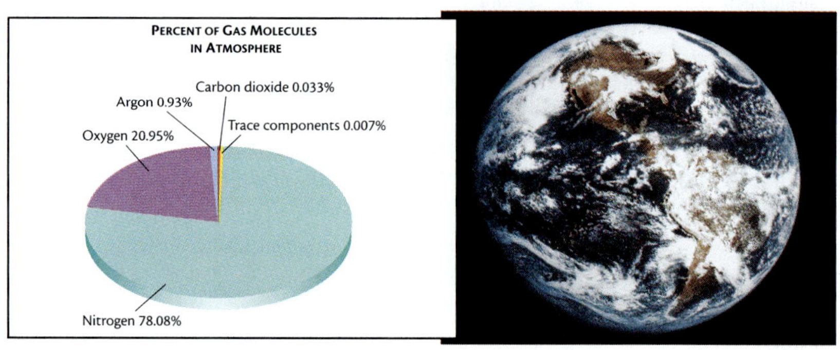

Figure 30 *The percent composition of the atmosphere—the envelope of gases that surrounds planet Earth.*

between 110% and 124% of saturation (a state of supersaturation), a condition called gas-bubble trauma may develop in fish.

This situation is dangerous because the supersaturated solution causes gas bubbles to form in the blood and tissues of fish. Oxygen-gas bubbles can be partially utilized by fish during metabolism, but nitrogen-gas bubbles block the flow of blood within the fish. This blockage results in the death of the fish within hours or days. Gas-bubble trauma can be diagnosed by noting gas bubbles in the gills of dead fish if they are dissected promptly after death. Supersaturation of water with nitrogen and oxygen gases can occur at the base of a dam or a hydroelectric project as released water forms "froth," trapping large quantities of air.

So, back to the original question: What caused the Riverwood fish kill? You have considered several possible causes. How will you determine which one was the actual cause? You will start by examining water-related data collected by scientists and engineers on the Snake River. From these data and what you have learned, you will make your own decision.

C.11 DETERMINING THE CAUSE OF THE FISH KILL

Making Decisions

Snake River watershed data have been collected and monitored since the early 1900s. Although some measurements and methods have changed during that time, excellent data have been gathered, particularly in recent years. Data are available regarding the following factors.

- water temperature and dissolved oxygen
- rainfall
- water flow
- dissolved molecular substances
- heavy metals
- pH
- nitrate and phosphate levels
- organic carbons

Your teacher will assign you to a group to study some of the data just listed. Each group will complete data-analysis procedures for its assigned data. After groups have finished their work, the class will share their analyses and draw conclusions about factors possibly related to the fish kill.

The following background information will help you complete the analysis of the data assigned to you by your teacher.

DATA ANALYSIS

Introduction

Interpreting graphs of environmental data requires a slightly different approach from the one you took in interpreting a solubility curve (page 47). Rather than seeking a predictable relationship, you will be looking for regularities or patterns among the values. Any major irregularity in the data may suggest a problem related to the water factor being evaluated. The following suggestions will help you prepare and interpret such graphs.

C.11 DETERMINING THE CAUSE OF THE FISH KILL

Making Decisions

This activity uses cooperative learning instructional strategies. Divide the class into small groups and give each group a full set of data. Students will analyze this data by graphing the information and looking for relationships and/or irregularities. After analyzing all the data, students are asked which cause appears to be the prime suspect. Have students ready to defend their reasoning. Plan to use cooperative learning strategies with this activity. For example, roles may be assigned to each student in the group. Such roles as material person, resource person, recorder, and presenter are effective in promoting positive interdependence.

You may want to consider having students use computer graphing programs to analyze the data. This is just one way to organize the analysis of the data; depending on your class size and organization, you may wish to take a different approach.

Four scenarios have been developed for evaluation by students. The scenarios are gas bubble trauma (too much dissolved oxygen in the river), pesticide contamination (molecular substances in the river), heavy-metal contamination (heavy metals from the sediment in the river), and not enough oxygen in the water.

The idea is that the teacher could vary which set of data is chosen for a class, period by period or year by year. This would prevent students from first period inappropriately sharing materials with the sixth period class. It is intended that only *one* of the data sets be used per period. (See Appendix A for background information.)

Answers

1 through 4. The graphs will vary according to the variables. Review the graphs provided on the *ChemCom* Web site for student accuracy. The site is located at www.whfreeman.com/chemcom, click on For Teachers on the left margin, then Supplemental Materials. The password is available from your sales representative. In general, it is very important for students to make valid interpretations of the data and to support their conclusions with evidence.

- Choose your scale so that the graph becomes large enough to fill most of the available space on the graph paper.
- Assign each regularly spaced division on the graph paper some convenient, constant value. The graph-paper line interval should have a value easily "divided by eye," such as 1, 2, 5, or 10, rather than a value such as 6, 7, 9, or 14.
- An axis scale does not have to start at "zero," particularly if the plotted values cluster in a narrow range not near zero.
- Label each axis with the quantity and unit being graphed.
- Plot each point. Then draw a small symbol around each point, like this: ⊙. If you plot more than one set of data on the same graph, distinguish each set of points by using a different color or geometric shape, such as: ⊡, ▽, or ◬.
- Give your graph a title that will readily convey its meaning and purpose to a reader.
- If you use technology—such as a graphing calculator or computer program—to prepare your graphs, ensure that you follow the guidelines just given. Different devices or programs have different ways of processing data. Choose the appropriate type of graph (scatter plot, bar graph, and so forth) for your data.

Procedure

1. Prepare a graph for each of your group's Snake River data sets. Label the *x*, or independent, axis with the consecutive months indicated in the data. Label the *y*, or dependent, axis with the water factor measured, accompanied by its units.
2. Plot each data point and connect the consecutive points with straight lines.

Questions

1. Is any pattern apparent in your group's plotted data?
2. Can you offer possible explanations for any pattern or irregularities that you detect?
3. Do you think the data analyzed by your group might help to account for the Snake River fish kill? Why? How?
4. Prepare to share your group's findings regarding Questions 1 through 3 in a class discussion. During the class discussion, take notes on key findings reported by each data-analysis group. Also note and record significant points raised in the data-analysis discussions.

Questions & Answers CD-ROM WWW.

Your class will reassemble several times during your study of Section D to discuss and consider implications of the water-analysis data that you have just processed. In particular, guided by the patterns and irregularities found in your analysis of Snake River data, your class will seek an explanation or scenario that accounts for the observed data and for the resulting Riverwood fish kill. Good luck!

SECTION SUMMARY

Reviewing the Concepts

1. If the solubility of sugar (sucrose) in water is 2.0 g/mL at room temperature, what is the maximum amount of sugar that will dissolve in 946 mL (1 quart) water?

946 mL × 2.0 g/mL = 1892 g

2. Explain why three teaspoons of sugar will completely dissolve in a serving of hot tea, but not in an equally sized serving of iced tea.

The solubility of sugar in water increases as the temperature of the water increases.

3. Rank the substances in Figure 20 (page 46) from most soluble to least soluble

a. at 20 °C. **a. NaCl, KCl, KNO_3**

b. at 80 °C. **b. KNO_3, KCl, NaCl**

4. A 35-g sample of ethanol is dissolved in 115 g water. What is the concentration of the ethanol, expressed as grams ethanol per 100 g solution?

100 g solution × 35 g ethanol / (115 g water + 35 g EtOH) = 23 g ethanol per 100 g of solution

5. Calculate the masses of water and sugar in a 55-g sugar solution that is labeled 20.0% sugar.

20.0% of 55 g = (.20) × (55 g solution) = 11 g sugar, and 55 g solution − 11 g sugar = 44 g water

6. Using the graph on page 46, answer the following questions about the solubility of potassium nitrate, KNO_3.

a. What is the maximum mass of KNO_3 that can dissolve in 100 g water if the water temperature is 20 °C?

b. At 30 °C, 55 g KNO_3 is dissolved in 100 g water. Is this solution saturated, unsaturated, or supersaturated?

c. A saturated solution of KNO_3 is formed in 100 g water at 75 °C. If the saturated solution is cooled to 40 °C, how many grams of solid KNO_3 should form?

a. about 32 g

b. At 30 °C, approximately 45 g of KNO_3 will dissolve in 100 g water; this solution is supersaturated.

c. At 75 °C, 150 g KNO_3 will dissolve; at 40 °C, about 60 g will dissolve; then 150 g – 60 g = 90 g of precipitate.

7. Draw a model that shows how molecules in liquid water generally arrange themselves relative to one another.

Drawing should show molecules close to each other, but not rigidly arranged as in an ice crystal. The polar OH bonds (hydrogen bonds) cause attractions between water molecules, which pull them into a spherical shape—much like beading on a clean surface. In the absence of the gravitational attraction of Earth, there is no force to pull them into a drop-like shape.

8. Why does table salt (NaCl) dissolve in water but not in cooking oil?

NaCl is composed of Na^+ and Cl^- ions. Polar water molecules will attract these charged particles. Cooking oil is nonpolar. There will be no attraction between nonpolar cooking oil molecules and the ions of NaCl.

9. a. What are heavy metals?

b. List some general effects of heavy-metal poisoning.

a. Heavy metals are metallic elements, such as lead, mercury, and cadmium, that are heavier than the essential elements, such as iron, calcium, and magnesium. Heavy metals dissolve in water as ions. Small amounts of heavy metals are toxic to humans and accumulate in the body. Heavy metals are also concentrated in the food chain.

b. Heavy metals damage the nervous system, kidneys, and liver, and cause mental retardation and even death. Heavy metals are especially toxic to infants and children.

10. What are some possible sources of human exposure to heavy metals?

Possible sources include gasoline fumes, aquifer water contaminated by waste disposal sites, old mine sites, tobacco smoke, streams polluted by industrial waste, and contaminated fish and foods. Vegetables, especially tomatoes and rice, concentrate dissolved cadmium from contaminated water sources and sewage sludge fertilizers. Mercury enters the aquatic food chain through bacteria, which form methyl mercury ions.

11. What items might be found in an urban landfill that could contribute heavy-metal ions to groundwater?

Accept all reasonable student answers, such as galvanized pipes, discarded paint, batteries, solder, land fill cover composed of incinerated sewage sludge, and pesticides.

12. Classify each sample as acidic, basic, or chemically neutral:

a. seawater (pH = 8.6) **a. basic**

b. drain cleaner (pH = 13.0) **b. basic**

c. vinegar (pH = 2.7) **c. acidic**

d. pure water (pH = 7.0) **d. neutral**

13. Which is more acidic, a tomato or a soft drink? (See Figure 25 on page 57.)

a soft drink

14. How many times more acidic is a solution of pH 2 than a solution of pH 4?

100 times (two powers of 10) or pH 2 = 0.01 pH 4 = 0.0001

15. Would you select ethanol, water, or lamp oil to dissolve a nonpolar molecular substance? Explain.

Lamp oil is a nonpolar organic molecule, which will dissolve a nonpolar molecular substance.

16. Explain the phrase "like dissolves like."

This is a "rule of thumb" to determine solubility. Polar substances and soluble ionic compounds are "like" because they are composed of charged particles or polar molecules that attract each other during the dissolving process. Nonpolar molecules dissolve "like" nonpolar molecules; due to their lack of charge, they mix during the dissolving process.

17. Explain why greasy dishes cannot be satisfactorily cleaned with pure water.

Water molecules are polar; grease is composed of nonpolar molecules.

18. As scuba divers descend, the pressure increases on the gases that they are breathing. How does the increasing pressure affect the amount of gas dissolved in their blood?

Increased pressure increases the amount of gas dissolved in body fluids.

19. Given your knowledge of gas solubility, explain why a bottle of warm cola produces more "fizz" when opened than a bottle of cold cola.

The solubility of carbon dioxide gas, the fizz in soda, increases as the temperature decreases and decreases as the temperature increases. This assumes that the pressure on both is the same.

Connecting the Concepts

20. At room temperature, C_2H_6 is a gas but C_2H_5OH is a liquid. Suggest an explanation for this difference.

The slightly polar structure of the alcohol encourages the molecules to be closer together or overlap, forming a product that is denser and liquid.

21. Predict the relative solubilities of C_2H_6 and C_2H_5OH in water.

The alcohol would be very soluble in water due to the polar nature of the molecule, and the gas would be less soluble.

22. From each of the following pairs, select the water source more likely to contain the higher concentration of dissolved oxygen. Give a reason for each choice.

a. a river with rapids or a calm lake

b. a lake in spring or the same lake in summer

c. a lake containing only sunfish or a lake containing only pike

a. Water in rapids has more opportunity to mix with the air (aeration).

b. In spring the lake is cooler, thus increasing the solubility of oxygen.

c. Overall, pike require more oxygen than sunfish to survive (although sunfish do require slightly more during the winter). Therefore, the "pike lake" probably has more dissolved oxygen.

Extending the Concepts

23. Read the label on a container of baking soda. Compare it with the label on a can of baking powder. Which one contains an acid ingredient? Suggest a reason for including the acid in the mixture.

Baking powder contains a dry powdered acid as well as baking soda; baking soda is 100% sodium bicarbonate. With an acid already present, baking powder will react with water to produce bubbles of CO_2. An acid such as vinegar (acetic acid) must be added in order for baking soda to produce bubbles of CO_2.

24. Describe how changes in solubility due to temperature could be used to separate two solid, water-soluble substances.

Dissolve the mixture of the two substances in a minimum amount of hot water. Reduce the temperature until one precipitates. Quickly filter the mixture and collect the less soluble substance in the filter paper. Further cool the filtrate until the second substance crystallizes (fractional crystallization).

25. The continued health of an aquarium depends on the balance of the solubilities of several substances. Investigate how this balance is maintained in a freshwater aquarium.

Solubility balance is maintained by adjusting the pH, aeration, and temperature of the water. Student answers will vary but should include regulations concerning the types of substances that can be discarded in their local landfill and the structure of landfills. Their report should include information about toxic waste pickup points.

26. The pH of rainwater is approximately 5.5. Rainwater flows into the ocean. The pH of ocean water, however, is approximately 8.7. Investigate reasons for the difference in pH.

The rainwater has interacted with gases in the atmosphere that make it slightly acidic. It gathers salts and minerals on the way to the ocean, where it is greatly diluted. The ocean contains basic salts, which also neutralize the acidic rainwater.

Possible data table for Laboratory Activity C.8, Part II (page 60)

Possible Data Table			
Solute	**Water**	**Ethanol**	**Lamp Oil**
Urea, $CO(NH_2)_2$	Soluble	Soluble	Insoluble
Ammonium chloride, NH_4Cl	Soluble	Soluble	Insoluble
Naphthalene, $C_{10}H_8$	Insoluble	Soluble	Soluble
Iodine, I_2	Insoluble	Soluble	Soluble
Copper(II) sulfate, $CuSO_4$	Soluble	Soluble	Insoluble
Ethanol, C_2H_5OH	Soluble	Soluble	Soluble
Sodium chloride, NaCl	Soluble	Soluble	Insoluble
Hexane	Insoluble	Soluble	Soluble
Water	Soluble	Soluble	Insoluble
Lamp oil	Insoluble	Soluble	Soluble

Additional data for Section C.10, Dissolved Oxygen (page 63)

Dissolved Oxygen Requirements of Some Additional Fish	
Fish Type	**DO Level in mg/L**
Salmon (chinook, coho, and sockeye)	3.9
Trout (rainbow, brown, and lake)	4.1
Northern pike	4.2
Largemouth bass	2.5
Channel catfish	3.2
Yellow perch	2.2

The average minimum dissolved oxygen level to maintain fish health is:
Cold-water species (trout, salmon)—7 mg/L
Warm-water species (bass, bluegill, catfish)—5 mg/L

SECTION D

WATER PURIFICATION AND TREATMENT

EDITORIAL
Attendance Urged at Special Council Meeting

A special town council meeting next Wednesday could result in important decisions affecting all citizens of Riverwood. The meeting will address two primary questions: Who is responsible for the fish kill? Who should pay the expenses of trucking water to Riverwood during the three-day water shutoff as well as any damages resulting from cancellation of the fishing tournament? These questions have financial consequences for all town taxpayers.

Those testifying at next week's public meeting include representatives of industry and agriculture, scientists taking part in the river-water analyses, and consulting engineers who have been studying the cause of the fish kill. Chamber of Commerce members representing Riverwood storeowners, representatives from the County Sanitation Commission, and officials of the Riverwood Taxpayer Association also will make presentations.

We urge you to attend and participate in this meeting. Many unanswered questions remain. Was the fish kill an "act of nature" or was it due to some human error? Was there negligence? Should the town's business community be compensated, at least in part, for financial losses resulting from the fish kill? If so, how should they be compensated and by whom? Who should pay for the drinking water brought to Riverwood? Can this situation be prevented in the future? If so, at what expense? Who will pay for it?

The *Riverwood News* will set aside part of its Letters to the Editor column in coming days for your comments on these questions and other matters related to the community's recent water crisis. For useful background information on water quality and treatment, we have prepared a special feature in today's paper that we think you will find useful.

Section D: Water Purification and Treatment Introduction

The editorial in the newspaper sets the stage for establishing responsibility for the Riverwood fish kill. In the previous section, students determined the possible cause or causes for the fish kill. The issue is whether the fish kill was an "act of nature" or whether there was some human error involved. If human activity is responsible, could this situation have been prevented?

D.1 NATURAL WATER PURIFICATION

The news article stresses the need for treated water and it reviews the three basic steps of nature's water-purification system. Help students to recognize that human-induced water purification and treatment methods resemble aspects of the hydrologic cycle. The problem of high concentration of dissolved minerals in some water supplies (hard water) is described briefly.

D.1 NATURAL WATER PURIFICATION

WATER PURIFICATION THROUGH HYDROLOGIC CYCLE

BY RITA HIDALGO
Riverwood News Staff Reporter

The residents of Riverwood share a sense of relief that the Snake River fish-kill mystery has been satisfactorily solved. However, ensuring the quality of Riverwood's water supply is a long-term commitment.

To act wisely about water use, whether in Riverwood or in other communities, residents must know how clean the water is and how it can be brought up to the quality required. It should not take an emergency or crisis to focus attention on these concerns.

How do water-treatment methods address threats similar to those investigated in the recent fish-kill crisis? This article provides details on how natural water-purification systems work to ensure the safety of community water supplies—particularly in light of potential threats in the form of water contamination. A companion article in today's *Riverwood News* looks at municipal water-treatment methods—procedures that mimic, in part, water-purification processes found in nature's water cycle.

Until the late 1800s, Americans obtained water from local ponds, wells, and rainwater holding tanks. Wastewater and even human wastes were discarded into holes, dry wells, or leaching cesspools (pits lined with broken stone). Some wastewater was simply dumped on the ground.

By 1880, about one-quarter of U.S. urban households had flush toilets; municipal sewer systems were soon constructed. However, as recently as 1909, sewer wastes were often released without treatment into lakes and streams, from which water supplies were drawn at other locations. Many community leaders believed that natural waters would purify themselves indefinitely.

Waterborne diseases increased as the concentration of intestinal bacteria in drinking water rose. As a result, water filtering and chlorinating of water supplies soon began. However, municipal sewage—the combined waterborne wastes of a community—remained generally untreated. Today, with larger quantities of sewage being generated and with extensive recreational use of natural waters, sewage treatment is part of every municipality's water-processing procedures.

Nature's water cycle, the hydrologic cycle, includes water-purification steps that address many potential threats to water quality. Thermal

see Water Purification, page 5

Water Purification, from page 3

energy from the Sun causes water to evaporate from oceans and other water sources, leaving behind any heavy metals, minerals, or molecular substances that were in the water.

This natural process accomplishes many of the same results as distillation. Water vapor rises, condenses into tiny droplets in clouds, and—depending on the temperature—eventually falls as rain or snow. Raindrops and snowflakes are nature's purest form of water, containing only dissolved atmospheric gases. However, human activities release a number of gases into the air, making present-day rain less pure than it used to be.

When raindrops strike soil, the rainwater collects additional impurities. Organic substances deposited by living creatures become suspended or dissolved in the rainwater. Located a few centimeters below the soil surface, bacteria feed on these substances, converting them into carbon dioxide, water, and other simple compounds. Such bacteria thus help repurify the water.

As water seeps farther into the ground, it usually passes through gravel, sand, and even rock. Waterborne bacteria and suspended matter are removed (filtered out). Thus three processes make up nature's water-purification system.

- **Evaporation,** then followed by **condensation,** removes nearly all dissolved substances.
- **Bacterial action** converts dissolved organic contaminants into a few simple compounds.
- **Filtration** through sand and gravel removes nearly all suspended matter.

Given appropriate conditions, people could depend solely on nature to purify their water. "Pure" rainwater is the best natural supply of clean water. If water seeping through the ground encountered enough bacteria for a long enough time, all natural organic contaminants could be removed. Flowing through sufficient sand and gravel would remove suspended matter from the water. However, nature's system cannot be overloaded if it is to work well.

If groundwater is slightly acidic (pH less than 7) and passes through rocks containing slightly soluble compounds such as magnesium and calcium minerals, a problem arises. Chemical reactions with these minerals may add substances to the water rather than removing them. In this case, the water may contain a relatively high concentration of dissolved minerals.

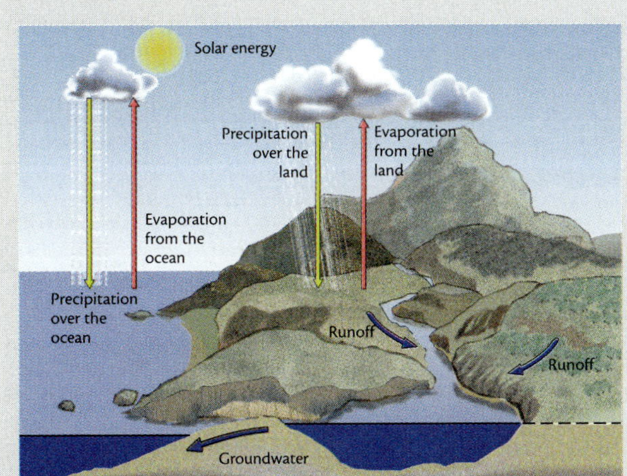

Your first laboratory activity in this unit (page 8) demonstrated that sand can act as a water filter.

CD-ROM WWW. Section A: The Hydrologic Cycle

Recall that the water cycle was first described on page 14.

D.2 MUNICIPAL WATER PURIFICATION

The news article reviews the major steps in the purification of a municipal water supply as shown in the diagram in the article. Direct students' attention to the diagram and the steps.

The process of adding a substance to our drinking water to benefit society is illustrated by the addition of fluoride ions to treated water. Students may have read articles about the controversy surrounding the addition of fluoride to treated water. Additional research is leading scientists to study the effects of increased soda consumption upon tooth decay, where the soda is high in sugar and low in fluoride. For the purpose of this section on water treatment, it is sufficient to mention that 1 ppm of fluoride ion is typically added to reduce tooth decay. Medical evidence for side effects is minimal, and the documented benefit for dental health is well established.

This is an excellent opportunity to organize a field trip to a local water treatment plant or a waste treatment plant. The steps mentioned in the text are found in most municipalities.

D.2 MUNICIPAL WATER PURIFICATION

WATER PURIFICATION THROUGH MUNICIPAL TREATMENT

BY RITA HIDALGO
Riverwood News Staff Reporter

Today, many rivers—such as the Snake River in Riverwood—are both a source of municipal water and a place to release wastewater (sewage). Therefore, water must be cleaned twice—once before its use, and again after its use. Pre-use cleaning, often called **water treatment,** takes place at a municipal filtration and treatment plant. It is the focus of this article.

The steps in a typical water-treatment process begin when intake water flows through metal screens that prevent fish, sticks, beverage containers, and other large objects from entering the water-treatment plant.

Chlorine, a powerful disinfecting agent, is added early in the water-treatment process to kill disease-causing organisms that may be present in the water. This step is called **pre-chlorination.** Then crystals of alum—aluminum sulfate,

see Municipal Treatment, page 6

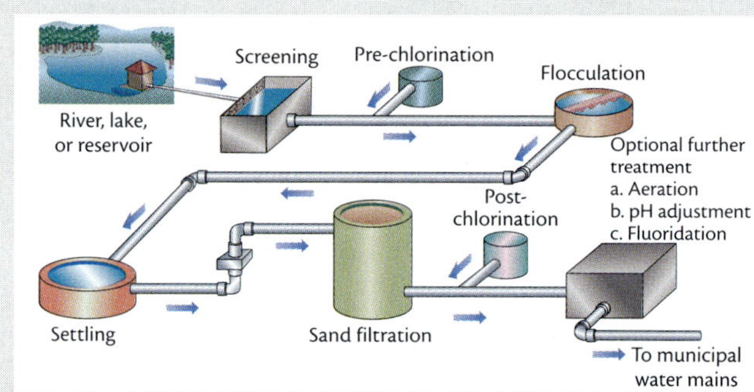

Municipal Treatment, from page 4

$Al_2(SO_4)_3$—and slaked lime—calcium hydroxide, $Ca(OH)_2$—are added to remove suspended particles such as colloidal clay from the water. These suspended particles can give water an unpleasant, murky appearance. The alum and slaked lime react to form aluminum hydroxide, $Al(OH)_3$, a sticky, jellylike material that traps and removes the suspended particles. This process is called **flocculation.** The aluminum hydroxide (holding trapped particles from the water) and other solids remaining in the water are allowed to settle to the tank bottom. Any remaining suspended materials that do not settle out are removed by filtering the water through sand.

In the **post-chlorination** step, the chlorine concentration in the water is adjusted to ensure that a low, but sufficient, concentration of residual chlorine remains in the water, thus protecting it from bacterial infestation.

Finally, depending on community regulations, one or more additional steps might take place before water leaves the municipal treatment plant. Sometimes water is sprayed into the air to remove odors and improve its taste. This process is known as **aeration.**

Water may sometimes be acidic enough to slowly dissolve metallic water pipes. This process not only shortens pipe life, but may also cause copper (Cu^{2+}) as well as cadmium (Cd^{2+}) and other undesirable ions to enter the water supply. Lime—calcium oxide, CaO, a basic substance—may be added to neutralize such acidic water, thus raising its pH to a proper level.

As much as about 1 ppm of fluoride ion (F^-) may be added to the treated water in a process called **fluoridation.** Even at this low concentration, fluoride ions can reduce tooth decay, as well as the number of cases of osteoporosis (bone-calcium loss) among older adults and hardening of the arteries.

WATER PURIFICATION

Building Skills 7

Refer to the two water-treatment articles by Ms. Hidalgo featured in the *Riverwood News* to answer these questions.

1. Compare natural water-treatment steps to the treatment steps found in municipal water systems.
   a. What are key similarities?
   b. What are key differences?
2. After reading the two water-treatment articles, a Riverwood resident wrote a letter to the *Riverwood News* proposing that the town's water-treatment plant be shut down. The reader pointed out that this would save taxpayers considerable money because "natural water treatment can take care of our needs just as well." Do you support the reader's proposal? Explain your answer.

Chlorine is probably the best known and most common substance used for water-treatment. It is found not only in community water supplies, but also in swimming pools. The following *Riverwood News* article provides background on chlorine's role in water treatment.

WATER PURIFICATION

Building Skills 7

Answers

1. a. An example of a similarity is well water being filtered through the ground (filtration).
   b. Evaporation is not a common step in municipal water treatment, but it is a step in the natural water treatment process.
2. Municipalities will not be able to keep up with the demand. With population increasing and the dumping of waste into our waterways, municipalities will not be able to get enough water from underground wells and rainwater collections. Without purification of water from waterways, underground wells will be drained dry in a relatively short time.

D.3 CHLORINATION OF WATER

The news article describes how chlorination of water offers both benefits and burdens. Students may have personal experience with chlorinated water in swimming pools or fish tanks. Encourage students to discuss these experiences, since they relate the chemistry to everyday activities.

The chlorination of our drinking water provides an opportunity to explore the history of diseases caused by polluted water, the history of social development around the location of freshwater, and the geography of watersheds. Students can research the history of social development due to contaminated drinking water.

A potential problem of chlorination is described in the production of trihalomethanes (THMs), which are known carcinogenic substances. You might want to challenge students to think of other ways we might be able to control waterborne disease without using chlorine. The unit ppb (parts per billion) is introduced; up to this point ppm (parts per million) was mostly used. Illustrate to the students that ppb is a very small amount and that 100 ppb is equal to 0.1 ppm.

ADDITIONAL BACKGROUND INFORMATION

On a worldwide basis, enough rain and snow fall each year to cover Earth's total land area to a depth of 83 centimeters. And yet, uneven distribution and inadequate technology make high-quality freshwater a major constraint on economic growth, food production, and general health in many nations. This is especially true of Asian and African countries, where periodic droughts and famines lead to the deaths of millions. Even when water is available in adequate quantities, it may be of poor quality. It has been estimated that 80% of all the world's sickness is caused by contaminated water and that persons with waterborne diseases occupy half of the world's hospital beds and die at a rate of 25 000 per day.

D.3 CHLORINATION OF WATER

CHLORINE IN PUBLIC WATER SUPPLIES

BY RITA HIDALGO
Riverwood News Staff Reporter

The single most common cause of human illness throughout the world is unhealthful water supplies. Without a doubt, chlorine added to public water supplies has helped to save countless lives by controlling waterborne diseases. When added to water, chlorine kills disease-producing microorganisms.

In most municipal water-treatment systems, **chlorination,** or chlorine addition, takes place in one of three ways.

- *Chlorine gas, Cl_2, is bubbled into the water.* Chlorine gas, which is a nonpolar substance, is not very soluble in water. It does react with water, however, to produce a water-soluble, chlorine-containing compound.
- *A water solution of sodium hypochlorite, NaOCl, is added to the water.*
- *Calcium hypochlorite, $Ca(OCl)_2$, is dissolved in the water.* Available as both a powder and small pellets, calcium hypochlorite is often used in swimming pools. It is also a component of some solid products sold as bleaching powder.

Laundry bleach is a sodium hypochlorite solution.

Regardless of how chlorination takes place, chemists believe that chlorine's most active form in water is hypochlorous acid, HOCl. This substance forms whenever chlorine, sodium hypochlorite, or calcium hypochlorite dissolves in water.

Unfortunately, there is a potential problem associated with adding chlorine to municipal water. Under some conditions, chlorine in water can react with organic compounds produced by decomposing animal and plant matter to form substances harmful to human health.

One group of such substances is known as the trihalomethanes (THMs). A common THM is chloroform, $CHCl_3$, which is carcinogenic, or cancer causing.

Because of concern about the possible health risks associated with THMs, the Environmental Protection Agency has placed a current limit of 100 ppb (parts per billion) on total THM concentration in municipal water-supply systems.

Possible risks associated with THMs must be balanced, of course, against the disease-prevention benefits of chlorinated water.

D.4 CHLORINATION AND THMs

Making Decisions

Several options are available to operators of municipal water-treatment plants that would help eliminate the possible THM health risks highlighted in the newspaper article that you just read. However, each method has its disadvantages.

- Treatment-plant water can be passed through an activated charcoal filter. Activated charcoal can remove most organic compounds from water, including THMs. *Disadvantage:* Charcoal filters are expensive to install and operate. Disposal also poses a problem because used filters cannot be easily cleaned of contamination. They must be replaced relatively often.
- Chlorine can be completely eliminated. Ozone (O_3) or ultraviolet light could be used to disinfect the water. *Disadvantage:* Neither ozone nor ultraviolet light protects the water once it leaves the treatment plant. Treated water can be infected by the subsequent addition of bacteria—through faulty water pipes, for example. In addition, ozone can pose toxic hazards if not handled and used properly.
- Pre-chlorination can be eliminated. Chlorine would be added only once, after the water has been filtered and much of the organic material removed. *Disadvantage:* The chlorine added in post-chlorination can still promote the formation of THMs, even if to a lesser extent than with pre-chlorination. Additionally, a decrease in chlorine concentration might allow bacterial growth in the water.

Your teacher will divide the class into working groups. Your group will be responsible for one of the three options just outlined. Answer the following questions.

1. Consider the alternative assigned to your group. Is this choice preferable to standard chlorination procedures? Explain your reasoning.
2. Can you suggest alternatives other than the three given here?

BOTTLED WATER VERSUS TAP WATER

ChemQuandary 3

When people do not like the taste of tap water, think it is unsafe to drink, or do not have access to other sources of fresh water, they may go to a vending machine or a market to buy bottled water. This bottled water often comes from a natural source, such as a mountain spring, or it may be processed at the bottling plant.

Is this water any better for you than tap water? Could it actually be more harmful? What determines water quality, and how can the risks and benefits of drinking water from various sources be assessed?

D.4 CHLORINATION AND THMs

Making Decisions

This Making Decisions points out that technological solutions to many problems create new problems. The formation of trihalomethanes (THMs), such as chloroform ($CHCl_3$) from the chlorination of water, is one example.

Alternative strategies such as ozone and ultraviolet light are introduced as a way to disinfect water. Mention that these methods are already in use in some public water supplies but that they do cost more. That brings up the question of how much we, as a society, are willing to pay for safe and clean water.

Assign three or four students to a group, and assign each group to evaluate one of the three alternative methods of water purification. Groups can present their reasoning to the class in oral reports.

Answers

1. Activated charcoal filters are expensive to install and operate on a large scale. Ozone or ultraviolet radiation disinfects water but does not provide added protection once the water leaves the plant. Eliminating the pre-chlorination step results in formation of smaller quantities of THMs. One disadvantage of this strategy is that the decreased chlorine concentration makes the water more susceptible to bacterial growth.
2. Other alternatives to chlorination include the use of chlorine dioxide, ozone, chloramines, UV irradiation, iodination, or some combination of these.

BOTTLED WATER VERSUS TAP WATER

ChemQuandary 3

A good way to introduce this activity is, with a bottle of spring water in hand, ask students if they believe this water to be pure, clean, and better than ordinary tap water. Then ask them to defend their answers. Encourage them to provide evidence or data to support their conclusion. Ask small groups of students to brainstorm for a few minutes and come up with three ways that we can perform tests to determine whether bottled spring water is better and safer to drink than regular tap water.

Answers

1. Possible answers might include bacteria, presence of ions, pH, and the presence of organic chemical substances.
2. Carefully measure the amount or concentration of the substances and know what levels are considered safe for humans (EPA safety limits). Any water (bottle or tap) that contains levels above the safety limit should be avoided.

D.5 WATER SOFTENING

Laboratory Activity

Students explore three ways to soften water.

Time

50 minutes

Materials (for 24 students working in pairs)

12 10-cm diameter funnels
12 ring stands
12 ring clamps
12 clay triangles
72 pieces of 11-cm filter paper (Whatman No. 1)
48 test tubes (18 × 150 mm)
24 test-tube racks
12 20-cm glass stirring rods
12 10-mL graduated cylinders
12 washing bottles
6 dropper bottles for sodium carbonate solution
Several burets for dispensing hard water
1.5 L hard water, containing 0.01 M Ca^{2+} (2.2 g $CaCl_2 \cdot 2H_2O$/1.5 L)
100 mL 0.1 M sodium carbonate in dropper bottles (1.2 g $Na_2CO_3 \cdot H_2O$ per 100 mL)
Liquid soap solution, Ivory or equivalent
12 wash bottles
1 L distilled water in wash bottles
1 lb sand
1 box Calgon (sodium hexametaphosphate)
100 g cation-exchange resin, ionic form Na^+ Dowex 50W-X8 or equivalent, 50–100 mesh

Safety

Students should wear safety goggles and aprons throughout the laboratory activity.

Advance preparation

Prepare the ion-exchange resin as a water slurry. Add 2 parts water to 1 part resin in a beaker. Decant excess water. Soak the resin overnight before it is used.

As usual, answering challenging questions requires gathering reliable data and information, weighing alternatives, and making informed decisions. Working with a partner, answer the following questions:

1. In your view, what are the most important two or three factors or considerations to analyze in deciding whether to drink tap water or bottled water?
2. For each factor listed in your answer to Question 1, what factual information would you need to gather to establish the advantages and disadvantages of drinking bottled water rather than tap water?

D.5 WATER SOFTENING

Laboratory Activity

Introduction

Water containing an excess of calcium (Ca^{2+}), magnesium (Mg^{2+}), or iron(III) (Fe^{3+}) ions is called **hard water.** Hard water does not form a soapy lather easily. River water usually contains low concentrations of hard-water ions. However, as groundwater flows over limestone, chalk, and other minerals containing calcium, magnesium, and iron, it often gains higher concentrations of these ions, thus producing hard water. See Figures 31 and 32.

In this laboratory activity, you will explore several ways of softening water by comparing the effectiveness of three water treatments for

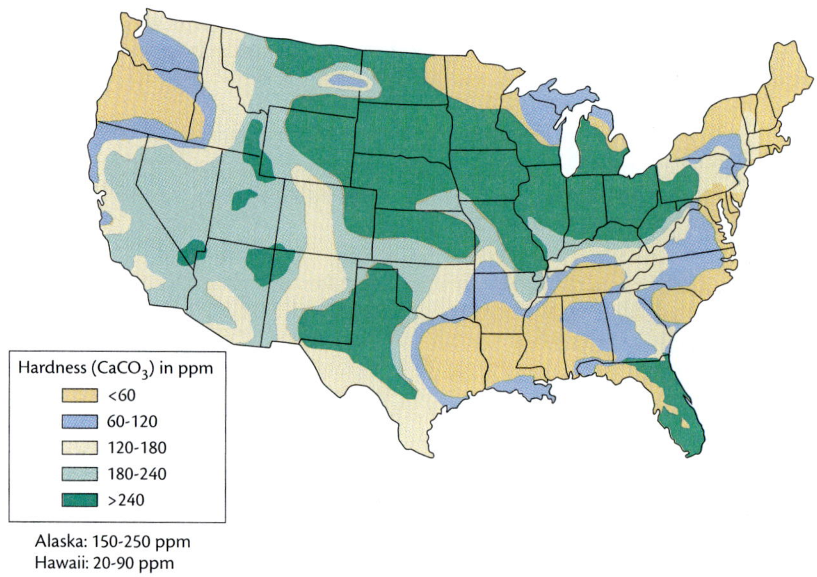

Figure 31 *U.S. groundwater hardness.*

Some Minerals Contributing to Water Hardness		
Mineral	**Chemical Composition**	**Chemical Formula**
Limestone or chalk	Calcium carbonate	$CaCO_3$
Magnesite	Magnesium carbonate	$MgCO_3$
Gypsum	Calcium sulfate	$CaSO_4 \cdot 2H_2O$
Dolomite	Calcium carbonate and magnesium carbonate combination	$CaCO_3 \cdot MgCO_3$

Figure 32 *The names, chemical composition, and chemical formulas for several minerals that contribute to water hardness.*

The "hardness" of water samples can be classified as follows:
Soft
< 120 ppm $CaCO_3$
Moderately hard
120–350 ppm $CaCO_3$
Very hard
> 350 ppm $CaCO_3$

removing calcium ions from a hard-water sample: sand filtration, treatment with Calgon, and treatment with an ion-exchange resin.

Calgon (which contains sodium hexametaphosphate, $Na_6P_6O_{18}$) and similar commercial products "remove" hard-water cations by causing them to become part of larger soluble anions. For example:

$$\underset{\text{Calcium ion from hard water}}{2\,Ca^{2+}(aq)} + \underset{\text{Hexametaphosphate ion}}{(P_6O_{18})^{6-}(aq)} \longrightarrow \underset{\text{Calcium hexametaphosphate ion}}{[Ca_2(P_6O_{18})]^{2-}(aq)}$$

Calgon also contains sodium carbonate, Na_2CO_3, which softens water by removing hard-water cations as precipitates such as calcium carbonate, $CaCO_3$. The equation for the reaction is shown below. Solid calcium carbonate particles are washed away with the rinse water.

$$\underset{\text{Calcium ion from hard water}}{Ca^{2+}(aq)} + \underset{\text{Carbonate ion from sodium carbonate}}{CO_3^{2-}(aq)} \longrightarrow \underset{\text{Calcium carbonate precipitate}}{CaCO_3(s)}$$

Another water-softening method relies on **ion exchange.** Hard water is passed through an ion-exchange resin such as those found in home water-softening units. The resin consists of millions of tiny, insoluble, porous beads capable of attracting cations. Cations causing water hardness are retained on the ion-exchange resin; cations that do not cause hardness (often Na^+) are released from the resin into the water to take the place of those that do. You will learn more about water-softening procedures after you have completed this laboratory activity.

Two laboratory tests will help you decide whether your hard-water sample has been softened. The first test is the reaction between hard-water calcium cations and carbonate anions (added as sodium carbonate, Na_2CO_3, solution) to form a calcium carbonate precipitate. The equation for this reaction is shown above. The second test is to observe the effect of adding soap to the water sample to form a lather.

Procedure

1. In your laboratory notebook, prepare a data table similar to the one shown on page 78.

Lab tips

The exchange resin contains 40 to 50% moisture and, thus, must be stored in sealed condition to avoid drying. The ion-exchange resin and sand can be reused from period to period. The resin can be regenerated for future use by washing it three times in a beaker with 1 M NaCl, and then three times with distilled water.

An alternative method to using liquid Ivory soap solution is to use a small amount of Ivory flakes. In using this method, place about 1/8 bar of Ivory soap in the microwave and heat it until it no longer expands. Allow it to cool and then crumble it into soap flakes. Use only a couple of small flakes in each test tube.

Purchase Calgon or equivalent water softeners. Make sure that it contains sodium hexametaphosphate for bath use, not sodium sesquicarbonate for dishes. This type of Calgon may be hard to find in some regions because of the phosphates.

Pre-lab discussion

Emphasize that hard water is caused by excess concentrations of calcium, magnesium, and/or iron(III) ions in water. Point out that water softening is done in many homes and will be discussed later.

DATA TABLE

Test	Filter Paper	Filter Paper and Sand	Filter Paper and Calgon	Filter Paper and Ion-Exchange Resin
Reaction with sodium carbonate (Na_2CO_3)				
Degree of cloudiness (turbidity) with Ivory soap				
Height of suds				

2. Prepare the equipment as shown in Figure 33. Lower the tip of each funnel stem into a test tube supported in a test-tube rack.
3. Fold four pieces of filter paper; insert one in each funnel. (In the following steps, funnels will be referenced, from left to right, as 1 to 4.)
4. Funnel 1 should contain only the filter paper; it serves as the control. (Hard-water ions in solution cannot be removed by filter paper.) Fill Funnel 2 one-third full of sand. Fill Funnel 3 one-third full of Calgon. Fill Funnel 4 one-third full of ion-exchange resin.
5. Pour about 5 mL of hard water into each funnel. Do not pour any water over the top of the filter paper or between the filter paper and the funnel wall.
6. Collect the filtrates in the test tubes. NOTE: The Calgon filtrate may appear blue because of other additives in the softener. They will not cause a problem. But if the filtrate appears cloudy, which means that some Calgon powder passed through the filter paper, use a new piece of filter paper and refilter the test-tube liquid.
7. Add 10 drops of sodium carbonate (Na_2CO_3) solution to each filtrate. Does a precipitate form? Record your observations. A cloudy precipitate indicates that the Ca^{2+} ion (a hard-water cation) was not removed.

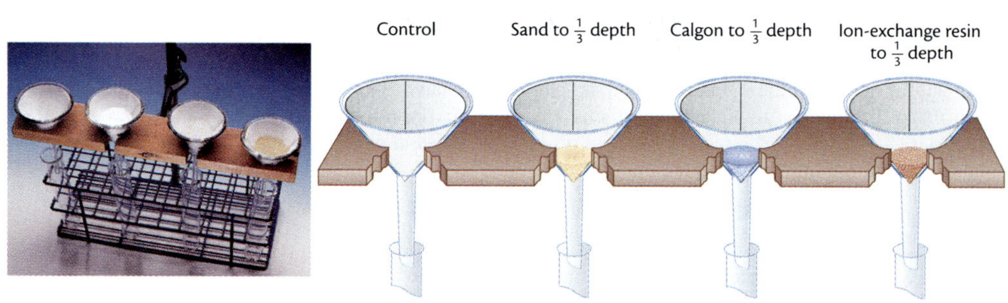

Figure 33 *Filtration setup.*

Expected Results

Reactant	Filter Paper	Filter Paper and Sand	Filter Paper and Calgon	Filter Paper and Ion-Exchange Resin
Sodium carbonate	Cloudy	Cloudy	No reaction	No reaction
Degree of cloudiness	Very clear	Very clear	Blue and turbid	Very turbid
Height of suds	None	None	1–2 mm	1–2 mm

8. Discard the test-tube solutions. Clean the test tubes thoroughly with tap water and rinse with distilled water. Place the test tubes back in the test-tube rack as in Step 2. Do not empty or clean the funnels; they will be used in the next step.
9. Pour another 5 mL of hard-water sample through each funnel, collecting the filtrates in the clean test tubes. Adjust test tube liquid heights, if necessary, to make all filtrate volumes equal.
10. Add one drop of Ivory liquid hand soap (not liquid detergent) to each test tube.
11. Stir each test tube gently. Wipe the stirring rod before inserting it into another test tube.
12. Compare the cloudiness, or **turbidity,** of the four soap solutions. Record your observations. The greater the turbidity, the greater the quantity of soap that dispersed. The quantity of dispersed soap determines the cleaning effectiveness of the solution.
13. Stopper each test tube and then shake vigorously, as demonstrated by your teacher. The more suds that form, the softer the water. Measure the height of suds in each test tube and record your observations.
14. Wash your hands thoroughly before leaving the laboratory.

Questions

1. Which was the most effective water-softening method? Suggest why this method worked best.
2. What relationship can you describe between the amount of hard-water ion (Ca^{2+}) remaining in the filtrate and the dispersion (cloudiness) of Ivory liquid hand soap?
3. What effect does this relationship have on the cleansing action of the soap?
4. Explain the advertising claim that Calgon prevents "bathtub ring." Base your answer on observations made in this laboratory activity.

D.6 WATER AND WATER SOFTENING

Hard water causes some common household problems. It interferes with the cleaning action of soap. As you observed, when soap mixes with soft water, it disperses to form a cloudy solution topped with a sudsy layer. In hard water, however, soap reacts with hard-water ions to form insoluble compounds (precipitates). These insoluble compounds appear as solid flakes or a sticky scum—the source of a bathtub ring. The precipitated soap is no longer available for cleaning. Worse yet, soap curd can deposit on clothes, skin, and hair. The structural formula for this objectionable substance, the product of the reaction of soap with calcium ions, is shown in Figure 34 on page 80.

If hydrogen carbonate (bicarbonate, HCO_3^-) ions are present in hard water, boiling the water causes solid calcium carbonate ($CaCO_3$) to form. The reaction removes undesirable calcium ions and thus softens the water.

4. Soaps react with the hard-water ions to form a precipitate in the water. When the water drains, this precipitate sticks to the sides of the tub, leaving a characteristic "ring." Softeners such as Calgon contain compounds that remove some of the ions responsible for hard water. The label on the box suggests that Calgon be used for laundry, bathing, hand washing, and general cleaning. Calgon will bind the positive ions found in hard water, but it does not form a precipitate with those ions. Thus, Calgon does not leave a "bathtub ring."

Post-lab discussion

Organize your post-lab discussion around the questions provided. Students may be asked to interpret their observations. Consider mentioning that the ion-exchange process is used in most home water-softener systems. Ask any students who happen to have these systems in their homes to explain the maintenance they require.

Note: Recently, some communities have banned salt-based ion-exchange systems, using mixed-bed H and OH resins instead. The mixed-bed resin appears to be equally effective, but not as easily recharged. Instead, the resin is "exchanged" by the company. Challenge your students to find out why salt-based resins are banned in some localities.

Answers

1. The ion-exchange resin should remove the hard-water ions best, but Calgon results may turn out better.
2. The higher the concentration of Ca^{2+} remaining in the filtrate (that is, the harder the water), the less soluble the soap.
3. The cleaning action of a soap depends on how dispersed or dissolved it is in the water. The more soap molecules that are bound to hard-water ions such as Ca^{2+}, the less there are available for cleaning.

D.6 WATER AND WATER SOFTENING

This section identifies sources of hard-water ions and explains common problems caused by hard water. It ends with a discussion of ion-exchange resins sometimes used to soften hard water.

ADDITIONAL BACKGROUND INFORMATION

The following equation summarizes the reaction that occurs in water softeners. R represents an ion-binding site in the resin:

$$2\ NaR + Ca^{2+} \rightarrow CaR_2 + 2\ Na^+$$

The resin is prepared by rinsing it thoroughly with solutions containing sodium ions, which bind to the anion sites. When hard water containing calcium or magnesium ions is filtered through the resin, the more highly charged ions have a greater attraction to the anion sites, replacing the sodium ions, which are then discharged with the flow of the liquid through the resin. The resin is regenerated by washing it with a concentrated sodium ion solution (sodium chloride solution). Hard water is prepared with these types of resins. In some parts of the United States, students may have water softeners in their homes. Encourage them to report on the operation of these devices.

Students will need to understand the concept of charge balance illustrated in this equation if they are asked to answer the Extending Your Knowledge questions. The notion is simply that the total electrical charge must remain constant on the resin and in the water. Here we see that two Na^+ ions are freed for each Ca^{2+} ion bound to the resin.

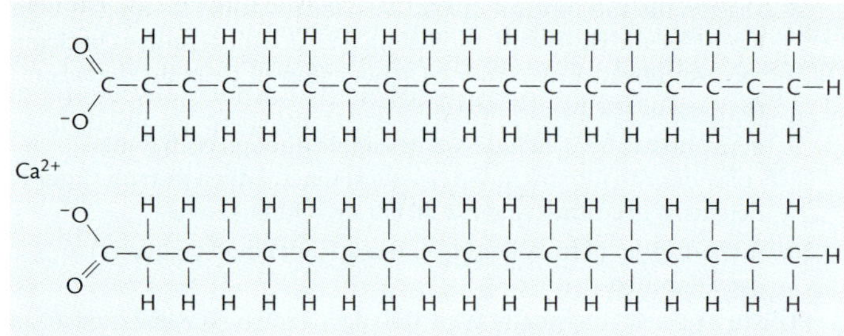

Figure 34 *Structural formula of a typical soap scum. This substance is calcium stearate.*

Hard Water & Soap Scum CD-ROM WWW.

However, the solid calcium carbonate produces rocklike scale inside tea kettles and household hot-water heaters. This scale (the same compound found in marble and limestone) acts as thermal insulation, partly blocking heat flow to the water. More time and energy are required to heat the water. Such deposits can also form in home water pipes. In older homes with this problem, water flow can be greatly reduced.

Fortunately, it is possible to soften hard water by removing some calcium, magnesium, or iron(III) ions. Adding sodium carbonate to hard water (as you did in the preceding laboratory activity) was an early method of softening water. Sodium carbonate (Na_2CO_3), known as washing soda, was commonly added to laundry water along with the clothes and soap. Hard-water ions, precipitated as calcium carbonate ($CaCO_3$) and magnesium carbonate ($MgCO_3$), were washed away in the rinse water. Water softeners in common use today include borax, trisodium phosphate, and sodium hexametaphosphate (Calgon). As you learned, Calgon does not tie up hard-water ions as a precipitate, but rather as a new, soluble ion that does not react with soap.

Figure 35 *In the 1960s and early 1970s, detergent molecules that were not decomposed by microorganisms caused some waterways to fill with sudsy foam, such as shown here. The development of biodegradable detergent molecules put an end to this unusual kind of water pollution.*

Most cleaning products sold today contain detergents rather than soap. Synthetic detergents act like soap but do not form insoluble compounds with hard-water ions. Unfortunately, many early detergents were not easily decomposed by bacteria in the environment—that is, they were not biodegradable. At times, "mountains" of foamy suds were observed in natural waterways. See Figure 35. These early detergents also contained phosphate ions (PO_4^{3-}) that encouraged extensive algae growth, choking lakes and streams. Because most of today's detergents are biodegradable and phosphate free, they do not cause these problems.

If you live in a hard-water region, your home plumbing may include a **water softener.** Hard water flows through a tank containing an ion-exchange resin similar to the one that you used in the water-softening laboratory activity. Initially, the resin is filled with sodium cations (Na^+). Calcium and magnesium cations in the hard water are attracted to the resin and become attached to it. At the same time, sodium cations leave the resin and dissolve in the water. Thus undesirable hard-water ions are exchanged for sodium ions, which do not react to form soap curd or water-pipe scale. Figure 36 illustrates this process.

It may not be economical to soften water of hardness less than about 200 ppm $CaCO_3$.

CD-ROM WWW. Water Softening by Ion Exchange

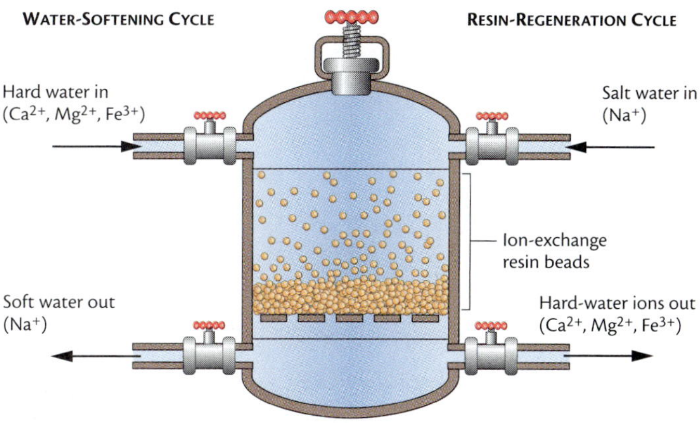

Figure 36 *Ion-exchange water-softener cycles.*

A water-softening unit typically uses from 5 to 6 pounds (2 to 3 kg) of sodium chloride (salt) for one regeneration.

As you might imagine, the resin eventually fills with hard-water ions and must be **regenerated.** Concentrated salt water (containing sodium ions and chloride ions) flows through the resin, replacing the hard-water ions held on the resin with sodium ions. Released hard-water ions wash down the drain with excess salt water. Because this process takes several hours, it is usually completed at night. After the resin has been regenerated, the softener is again ready to exchange ions with incoming hard water.

Water softeners are most often installed in individual homes. Other water treatment is done at a municipal level, both in Riverwood and in other cities.

CD-ROM WWW. Questions & Answers

ADDITIONAL CHEMQUANDARY

Ion-exchange water-softening units substitute sodium ions for hard-water ions such as Ca^{2+}. Does this create any new problems? **Answer:** Yes, because the water now contains higher concentrations of sodium ions. People with high blood pressure needing low-sodium intake must find alternate sources of drinking water. Many houses have small reverse osmosis units or other high-performance filtration systems, just for drinking water.

CHEMISTRY AT WORK

Purifying Water Means More Than Going with the Flow

When you drink from a water fountain, do you ever wonder where the water comes from? In some parts of the country, drinking water is provided by people such as **Phil Noe.**

"We have a limited supply of fresh water," says Phil, "so we've built a plant that lets us use water from our aquifers."

Phil is Production Manager at Island Water Association (IWA), which provides water for Florida's Sanibel and Captiva Islands. "We have a limited supply of fresh water," says Phil, "so we've built a plant that lets us use water from our aquifers." Aquifers are underground layers of permeable (porous) sand and limestone that contain large quantities of water.

After pumping the brackish, undrinkable water up from the Suwanee Aquifer to IWA's processing plant, Phil and his co-workers remove almost all of the salt and minerals, producing water that is more pure than many mountain streams. The process that accomplishes this task is known as reverse osmosis. The accompanying illustration is a comparison of osmosis with reverse osmosis. The amount of pressure applied in reverse osmosis must exceed the osmotic pressure, which tends to move the water from a region of higher vapor pressure to one of lower vapor pressure. Because osmotic pressure depends only on the concentration of the salt solution, it is known as a colligative property.

Aquifers are located at depths ranging between 700 and 900 feet. Water pumped up from these aquifers contains an average of 3000 ppm total dissolved solids (TDS). Feed water is pumped through hundreds of feet of pipes containing salt-filtering membranes. In the process, the purified water, called the permeate, separates from the salts and other dissolved solids. Eventually, the water is purified to levels as low as 100 ppm TDS.

Because Florida is subject to hurricanes and other tropical storms, Sanibel Island's water system maintains

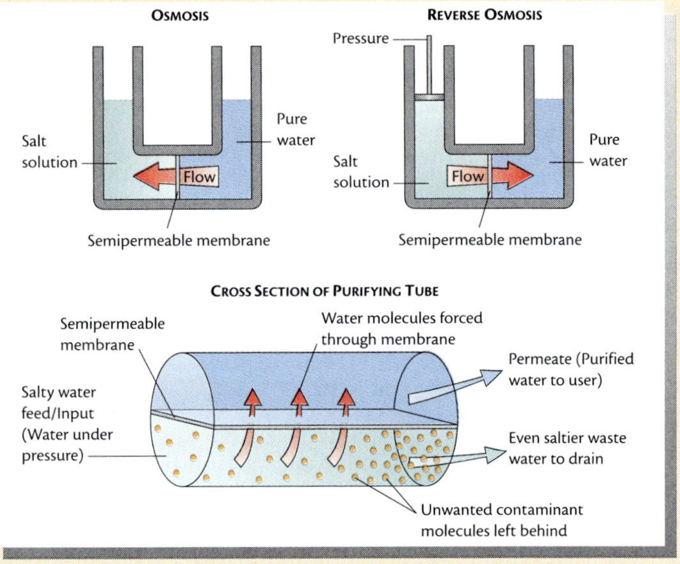

Osmosis occurs naturally when water in a dilute solution passes through a semipermeable membrane into a concentrated solution. In reverse osmosis, pressure must be applied to a concentrated solution to force water through a semipermeable membrane into a dilute solution.

82

After filtering to remove the larger solids, the water passes through equipment where reverse osmosis is used to produce purified water.

Saving up for a windy day.

15 million gallons of purified water in storage. As a result, island customers can manage without the plant in operation during brief periods of severe weather with no more inconvenience than several weeks without lawn sprinklers.

Eighty percent of IWA's feed water is converted into drinkable water. The remaining 20% is pumped out into the Gulf of Mexico. This water, a salt solution, is lower in salt and mineral concentrations than the gulf water itself and is harmless to marine life.

Web Search . . .

- Use the World Wide Web to search for other communities that rely on reverse osmosis for all or part of their water supply.
- Most communities use rivers, lakes, or wells to supply their water needs. Search the Web for information about municipal water-supply systems that use alternative water-purification processes in addition to reverse osmosis.
- What other kinds of water purification systems are in use in the United States?
- Use the Web to gather information about the cost of water for household use in various parts of the United States. Is there a connection between the cost of water and the location of the community? For example, is water more expensive in the desert than it is near the ocean?

Aquatic wildlife is unaffected by the salt water left over from the reverse osmosis process.

83

SECTION SUMMARY

Reviewing the Concepts

1. Explain how water can be purified through the actions of the hydrologic cycle.

Water is evaporated, separating it from its dissolved substances, suspended matter, and ions; water vapor condenses as precipitation, which is then filtered through sand and rock; some settling occurs in streams, rivers, and aquifers; rain's slightly acidic properties and ultraviolet radiation from the Sun minimize microorganism growth.

2. How are the properties of aluminum hydroxide related to the process of flocculation?

Aluminum hydroxide is a jelly-like material formed by the chemical reaction between aluminum sulfate (alum) and calcium hydroxide (slaked lime). It traps colloidal particles, such as clay, forming a heavy complex that settles to the bottom of the settling pond.

3. Why is calcium oxide (CaO) sometimes added in the final steps of municipal water treatment?

CaO, a basic substance, is used to neutralize water. Acidic water can dissolve metallic water pipes and release heavy metals such as cadmium to the water supply.

4. What advantages does chlorinated drinking water have over untreated water?

Chlorination kills disease-causing microorganisms that may be present in untreated water.

5. What are some disadvantages of using chlorinated water?

Some people dislike the taste and odor of chlorine in the water. Chlorinated water may kill aquarium fish, reacts with organic material in the water to form carcinogenic THMs, and cannot be used by people on kidney dialysis machines. Excessive chlorine can be hazardous to health.

6. Water from a clear mountain stream may require chlorination to make it safe for drinking. Explain why.

Even a clear mountain stream may contain disease-causing microorganisms or high concentrations of an unseen contaminant.

7. What is the origin of the expression "hard water"?

The term *hard* comes from the substantial concentrations of calcium and/or magnesium ions leading to calcareous water and/or hard deposits, the latter of which form on the inside of pipes or tanks.

8. When a sample of well water was mixed with a few drops of sodium carbonate solution, a precipitate formed. What does the formation of a precipitate indicate about the water sample?

It indicates the presence of Ca^{2+} ions, which form $CaCO_3$, a white precipitate.

9. Hard water often tastes better than distilled water. Explain.

Minerals and dissolved gases give a pleasant taste to water.

10. Which source in a given locality would probably have harder water, a well or a river? Explain.

A well. A river would tend to have a higher temperature. In a well, acidic water slowly dissolves metallic pipes, releasing iron ions. Warm water increases the solubility of hard-water ions, and scale forms in the standing water as the surface water evaporates and the concentration of hard-water ions increases.

Connecting the Concepts

11. A simple test for the hardness of water is to add soap to the sample and shake it. Explain how measuring the quantity of soap suds formed can be used to assess the hardness of water.

Hard-water ions bind with the soap molecules, forming precipitates that prevent the formation of suds.

12. Explain how hard water can interfere with the processes of a water-treatment plant.

Scale builds up in the pipes and heating units, reducing the flow of water.

Extending the Concepts

13. How much water would you have to drink to get your minimum daily requirement of calcium from water that contains 300 ppm calcium carbonate?

In the Food Unit, 1200 mg for 15- to 18-year-olds is listed as the Recommended Daily Allowance (RDA). Assuming that for every 100 parts of calcium carbonate the consumer gets out one calcium, then:

$$\textbf{1200 mg calcium} \times \frac{\textbf{1 kg}}{\textbf{10}^{6}\textbf{ mg}} \times \frac{\textbf{10}^{6}\textbf{ mg water}}{\textbf{300 mg calcium carbonate}} \times \frac{\textbf{100 parts calcium carbonate}}{\textbf{40 parts calcium}} = \textbf{4 kg water}$$

14. Explain why hard water stains in old sinks are found more often around hot-water faucets than around cold-water faucets.

The ions are more soluble in hot than in cold water, so the hot water faucet brings in water with a higher concentration of hard water ions, especially stain producing Fe^{3+}, then the ions are precipitated upon contacting the colder surface of the sink.

15. Investigate the reasons why sodium-based ion-exchange resins have been banned in some municipalities.

Sodium ion-exchange resins release concentrations of sodium ions, which may affect the health of some individuals.

EXTENDING YOUR KNOWLEDGE OVERVIEW

- Increased population growth has raised the demand for goods and services. The increased direct and indirect demand for "clean" water can no longer be met by nature's recycling system. New knowledge of the adverse health effects of poor-quality water has increased the trend toward regulation of water quality by government agencies. In addition, new chemical and instrumental tests can detect smaller concentrations of pollutants than was previously possible.
- Tertiary water treatment refers to a series of specialized chemical and physical processes that reduce the quantities of pollutants after primary and secondary treatment. The removal of organic wastes is typically the goal of tertiary water treatment. Sample procedures include activated carbon beds, chemical precipitation, ion exchange, electrodialysis, reverse osmosis, and distillation. The cost of tertiary water treatment can be high. Some communities are using natural wetlands for tertiary treatment.
- $Fe^{3+}(aq) + 3\ NaR \rightarrow FeR_3 + 3\ Na^+(aq)$. In this equation, R represents a cation-exchange resin site containing fixed negative charges, which interacts with mobile cations. Ion-exchange resins are macromolecules or insoluble polyelectrolytes having fixed charges distributed uniformly throughout the structure. During the ion-exchange reaction, the resins release sodium ions and bind the undesirable ion. Note that the charges are balanced as well as the numbers of atoms. See the *ChemCom* CD-ROM for an excellent animation of this process.
- About 50% of all U.S. residents (and more than 95% of all rural U.S. residents) rely on groundwater to supply their water needs. Twenty large U.S. cities rely exclusively on groundwater and twelve states use it for more than half of their public supplies. This was not addressed in the student text, but it makes an excellent topic for optional student reports. Refer students to the supplemental references.

PUTTING IT ALL TOGETHER

FISH KILL—FINDING THE SOLUTION

FISH KILL CAUSE FOUND

MEETING TONIGHT

BY ORLANI O'BRIEN
Riverwood News Staff Reporter

Mayor Edward Cisko announced at a news conference held early today that the cause of the fish kill in the Snake River has been determined. The details of the accidental cause will be released at a town council meeting tonight. As of today, levels of all dissolved materials in the river are normal and the water should be considered safe.

Accompanying Mayor Cisko was Dr. Harold Schmidt of the Environmental Protection Agency. Dr. Schmidt performed dissections on fish taken from the river within a few hours of their death. He also directed the team that analyzed accumulated river-water data in efforts that led to determining the cause of the fish kill. Dr. Schmidt gave assurances that the town's water supply is "fully safe to drink."

Mayor Cisko refused to elaborate on reasons for the accident. However, he invited the public to the special town council meeting scheduled for 8 P.M. tonight at Town Hall. The council will discuss events that caused the fish kill and how costs associated with the three-day water shutoff will be paid. Several area groups, as well as invited experts, plan to make presentations at tonight's meeting.

TOWN COUNCIL MEETING

Meeting Rules and Penalties for Rule Violations

CD-ROM WWW. Putting It All Together

1. The order of presentations is decided by council members and announced at the start of the meeting.
2. Each group will have a specified time for its presentation. Time cards will notify the speaker of time remaining.
3. If a member of another group interrupts a presentation, the offending group will be penalized 30 seconds for each interruption, to a maximum of one minute. If the group has already made its presentation, it will forfeit its rebuttal time.

PUTTING IT ALL TOGETHER: FISH KILL—FINDING THE SOLUTION

The newspaper article announces a town council meeting tonight to discuss who is responsible for the fish kill and who should pay for the costs associated with the problem.

At this simulated town council meeting, students will be given the opportunity to present data to support a particular point of view. If the class gets really involved, there may be some arguments and disagreements about how to solve this problem. This promotes presentation skills, using data to support an argument, expressing ideas in a scientific way, and realizing the civic opportunities available on a local level.

The simulated town council meeting addresses two questions: Who is responsible for the fish kill? Who should pay for the costs incurred because of the fish kill? The purpose of this culminating activity is to encourage students to use the chemistry they have learned to analyze real-world problems resembling those they will be called upon to consider as voting adults. The preparation for the town council meeting and the actual meeting should not require more than three 50-minute class periods. This time requirement could be reduced if some of the preparation work is assigned as homework.

Student groups prepare statements and develop presentations that express the viewpoint of the group. Individually, students are assigned to write an editorial or conduct an interview about the Riverwood problem. Both group and individual responsibilities are built into this activity.

This activity helps students recognize that while science does not provide absolute answers to social problems, scientific principles often can be used to help formulate possible solutions to these problems. Different people can reach different conclusions from the same facts, based upon their own perspectives or values. Science enables us to pose productive questions. Although scientific knowledge provides information about many of these problems, the decisions to take action must always involve additional considerations from social, economic, political, moral, and related perspectives. The activity is intentionally open-ended. There is no "right" answer for the class to reach, only one that is supported by the facts.

Pre-Activity Preparation and Discussion

Students can be assigned to one of eight "special interest" groups who will testify at the special meeting of the Riverwood Town Council. Adjust the number of groups according to your class size. Some suggestions on the approach various groups should take during the town meeting are included in the text, and additional information is included in Appendix A. Tell students to consider these suggestions only as a starting point.

You may wish to give students planning time to research and organize their presentations. Each group will have a limited time to present its position and to rebut the opposing position. Failure to stay within presentation time limits may result in loss of rebuttal time. Students assigned to the town council group will serve as timekeepers and manage the discussion.

Following the meeting, you may wish to have each group (or, alternatively, each student) write a letter to the editor of the Riverwood newspaper responding to questions posed regarding responsibility for the fish kill or the adequacy of the decisions made in the town council meeting. See the Activity Workbook for additional handouts for organizing this activity.

Figure 37 *View of a city council meeting in session in Austin, Texas. Open meetings such as this encourage citizen involvement in discussions of local issues and in related decision-making challenges.*

Town Council Members: Background Information

Your group is responsible for conducting the meeting in an orderly manner. Be prepared to:

1. Decide and announce the order of presentations at the meeting. Groups presenting factual information should be heard before groups voicing opinions.
2. Decide how the presentation area will be organized: where town council members controlling the meeting will be located, where the presenters will present, and where the groups and observers will be positioned.
3. Explain the meeting rules and the penalties for violating those rules.
4. Recognize each group at its assigned presentation time.
5. Enforce established presentation time limits by preparing time cards with "one minute," "30 seconds," and "time is up" written on them. These cards, placed in the speaker's line of sight, can serve as useful warnings.
6. Control the rebuttal discussions and open-forum speeches.
7. Summarize the options when testimony has been completed.
8. Conduct a vote of all town council members.
9. Report the results of the vote and future actions mandated by it.

Power Company Officials: Background Information

The power plant includes a dam and reservoir that ensure an adequate supply of cooling water. Your company's engineers control the rate of water release from the dam into the Snake River.

Normally, only relatively small volumes of water are released at any particular time. However, releasing large quantities of water from the dam is a standard way of preventing flooding. The last time such a large volume of water was released from the dam was 30 years ago. A fish kill was reported then, but the cause remained unknown. On that occasion, Riverwood and surrounding areas had experienced an unusually wet summer.

The dam, constructed in the 1930s, had the most current design of that time. Since then, its basic structure has not been modified.

Agricultural Cooperative Representatives: Background Information

Cooperative members in the Snake River valley include farmers and ranchers managing a variety of crops and livestock. Your cooperative assumes a proactive role in informing its members of current best practices and regulations regarding the use of agrochemicals and the management of wastes and runoff from fields and pastures.

Heavy rains present a problem for farmers. Although the rain is good for crops, it can wash away recently applied fertilizers and pesticides. This is not only expensive, but it can cause problems if these substances wash into the watershed.

Mining Company Representatives: Background Information

Riverwood began as a mining town on the Snake River, which provided early residents with a source of water. Your company intensely mined the hills surrounding Riverwood in the 1930s and 1940s. The important metals that came from this area included zinc and silver. The by-products of mining and processing the metal ores were collected in storage ponds built in accordance with the specifications and regulations of that time.

In seasons with average rainfall, the runoff from the waste ponds contains heavy-metal ions at levels within the values specified by your company's EPA permit. Your company monitors effluent values and keeps the ponds secured. Your company's structural engineers are responsible for upkeep of storage ponds at abandoned mine sites. However, during heavy rainfall, some underground settling in the mines and avalanches in hilly areas of the Snake River have been noted.

Scientists: Background Information

You are responsible for explaining how the analyzed data support the proposed cause of the fish kill. You should be prepared to explain what the data mean and why data fluctuations are noted from month to month or year to year. You may be called on to explain concepts such as pH, solubility of molecular and ionic substances, units of concentration, water-purification techniques, the hydrologic cycle, and other water-related concepts. It is important that you help council members and other attendees understand how the analyzed data document the cause of the fish kill.

Putting It All Together

Students should be assigned their roles in the town council meeting at least one class period before the activity begins. The town council members will organize and monitor the debate. Consider the personalities of the students you choose for these roles. Look for students with natural leadership ability.

If you wish to have smaller groups or more speakers, add other special interest groups to the meeting. For example, there could be two groups of engineers: one as a source of information, and the other as original designers of the dam. The Riverwood Corporation, with a fictitious product, could be represented. The only limits to this activity are time and your imagination!

Encourage students to bring appropriate props, such as hard hats or lab coats, to increase the authenticity of group identities. Each interest group could design a logo to be displayed on a badge or name tag symbolizing its identity. Encourage students to use their talents from other content areas—such as art, graphics, computer science, business, drama, and journalism—to create the town meeting atmosphere. If possible, rearrange the classroom tables and chairs to resemble a meeting room. A microphone adds realism to the setting. Videotaping the meeting lends an element of the theater, and students thoroughly enjoy seeing a recording of themselves.

Post-Activity Discussion

Following completion of this activity, conduct a class discussion regarding students' experiences in this Riverwood simulation. At this point students should abandon their assigned roles and critically analyze the entire decision-making process. This discussion is central to the success of the activity. Consider focusing on points such as these:

1. Informed decisions concerning any problem should take into account opposing views; don't jump to conclusions.
2. Ask students to identify points in the debate where chemistry knowledge was essential.
3. Point out the benefits and consequences of seeking consensus solutions to complex issues. Discourage students from viewing the debate in terms of winning or losing.
4. With students' help, highlight examples of factual, opinion-based, emotional, and logical arguments presented during the debate.
5. Point out that each group had worthwhile viewpoints to contribute to the debate. The final resolution is often a compromise among the different group positions.

The evaluation of student performance in this activity should include both individual and group grades. Group grades can be based upon the oral presentation or the simulated interview. It is helpful to have a written list of items to evaluate during the town council meeting. The list can be shared with the students before the debate begins. Items to consider include active participation in group work, effective use of facts, use of persuasive arguments, adherence to time requirements, and evidence of research beyond the text. Teachers will want to generate their own lists. A sample evaluation is available in the back of this book, the Blackline Master booklet, and in the Activity Book. Having students write letters to the editor permits you to evaluate individual performance.

Consulting Engineers: Background Information

Your consulting firm was hired to do a detailed examination of the cause of the fish kill. Your task was to determine whether accident, human error, negligence, or an unforeseen circumstance was responsible for the Riverwood crisis. In addition, you were asked to prepare scenarios or suggest improvements that would prevent recurring fish kills.

Your presentation should include the proposed solutions, the costs and benefits of each solution, and a detailed analysis of the cause. It is understood that you may not be familiar with cost analyses of major projects; however, you should try to make feasible estimates.

Chamber of Commerce Members: Background Information

Canceling the annual fishing tournament cost you and other Riverwood merchants a substantial sum of money. Close to one thousand out-of-town tournament participants were expected. Many would have rented rooms for at least two nights and eaten at local restaurants and fast-food establishments. In anticipation of this business, extra food supplies and support services were ordered. Fishing and sporting goods stores stockpiled extra fishing supplies. Some businesses have applied for short-term loans to help pay for their unsold inventories.

Local churches and the high school planned family social activities as revenue makers during the tournament weekend. For example, the school band scheduled a benefit concert that would have raised money to send band members to the spring band competition.

People are likely to remember the fish kill for many years. Tournament organizers predict that future fishing competition revenues in Riverwood will be substantially reduced due to this year's adverse publicity. Thus total financial losses resulting from the water emergency may be much higher than most current estimates predict.

You should be able to discuss how merchants and businesses were affected by this event and summarize the availability of support (as well as lack of support) to help resolve the issues.

County Sanitation Commission Members: Background Information

You are responsible for the protection and safety of the Snake River water supply. You are the group that completes most of the routine water testing for the supply of drinking water in Riverwood. It is important to know what the standards that specify the quality of drinking water mean and to explain how the water testing is done. You should be able to report maximum contamination levels (MCLs) for hazardous water contaminants. You should also know the allowable limits or expected ranges for other analyzed water data.

Riverwood Taxpayer Association Members: Background Information

Your organization is concerned about the financial effects of the fish kill on Riverwood citizens. Thus some of the important questions to be answered at the town council meeting should be addressed in your presentation. These questions include:

- Who will pay for the water brought into Riverwood during the water shutoff?
- Will taxes be increased to compensate local businesspeople for their financial losses? (Keep in mind that local merchants themselves are likely to be Riverwood taxpayers!)
- If the organization responsible for the fish kill takes measures to prevent its recurrence, will the costs be passed on to consumers? If so, how?

Because your presentation may be influenced by the testimony of other groups, you may find it useful to obtain their written briefs before the council meeting, if possible.

LOOKING BACK AND LOOKING AHEAD

The Riverwood water mystery is solved! In the end, scientific data provided the answer. And now human ingenuity will provide strategies for preventing the recurrence of such a crisis. In the course of solving the problem, the citizens of Riverwood learned about the water that they take for granted—abundant, clean water flowing steadily from their taps—and probably gained a greater appreciation for it.

Although Riverwood and its citizens exist only on the pages of this textbook, their water-quality crisis could be very real. The chemical facts, principles, and procedures that clarified their problem and its solution have applications in your own home and community.

Although the fish-kill mystery has been solved, your exploration of chemistry has only just begun. Many issues related to chemistry in the community remain, and water and its chemistry are only one part of a larger story.

LOOKING BACK AND LOOKING AHEAD

The Riverwood water mystery is solved! In the end, scientific measurements and data provided the answer. And now human ingenuity will provide strategies for preventing the recurrence of such a crisis. In the course of solving the problem, Riverwood citizens learned about the water that they take for granted—abundant, clean water flowing steadily from their taps—and probably gained a greater appreciation for it.

Although Riverwood and its citizens exist only on the pages of this book, their water-quality crisis could be very real. The chemical facts, principles, and procedures that clarified their problem and its solution have applications in your own home and community.

Many issues involving *Chemistry in the Community* remain, and water and its chemistry are only one part of a larger story. Several points are made in this section about the application of chemical facts, principles, and procedures in our own residences and communities. Remind students that the world around us is filled with chemistry and we have just begun our exploration of it.

PLANNING GUIDE UNIT 2

Section	Time class	Student Edition Features	Program Resources	Media and Technology
Section A. Why We Use What We Do **Objectives:** • Distinguish between chemical and physical properties and/or changes in matter when given specific examples. • Classify selected elements as metals, nonmetals, or metalloids based on observations of chemical and physical properties. • Use the Periodic Table to (a) predict physical and chemical properties of an element. (b) identify elements by their atomic masses and atomic numbers. (c) locate period and groups (families) of elements. • Determine properties of elements by the number and arrangement of the electrons in the atom.	**5 class periods**	**Building Skills** #1. Physical and Chemical Properties #2. Periodic Variation in Properties #3. Predicting Properties **Laboratory Activity** • Metal or Nonmetal? **Making Decisions** • Grouping the Elements • It's Only Money	**Activity Book** • A.3 Metals or Nonmetals • A.5 Building Skills: Periodic Variation in Properties • A.6 Periodic Table **Black Line Copy Masters** • Periodic Table • Periodic Properties • Metal Activity Series • Element Data Cards **Demonstrations** • Physical and Chemical Changes • Color Change: Chemical or Physical • Other Activities: Element of the Week	**CD-ROM** • Unit IIA: Why We Use What We Do **Transparencies** • Periodic Properties • Periodic Table

PLANNING GUIDE

Section	Time class	Student Edition Features	Program Resources	Media and Technology
Section B. Earth's Mineral Resources **Objectives:** • Describe the three major parts of Earth. • List and describe the factors that determine the feasibility of mining a particular ore at a particular site. • Explain why more active metals are more difficult to refine and process than less active metals. • Describe the process of oxidation-reduction. • Describe the three most common methods of separating a metal from its ore.	**5 class periods**	**Building Skills** #4. Properties and Uses of Copper #5. Trends In Metal Activity **Laboratory Activity** • Converting Copper • Relative Reactivities of Metals **ChemQuandary** #1. Discovery of Metals **Modeling Matter** • Electrons and Redox Processes	**Activity Book** • B.2 Converting Copper • B.4 Relative Reactivities of Metals	**CD-ROM** • Earth's Mineral Resources **Transparencies** • Earth's Composition • Production of Selected Materials Worldwide • Copper Cycle and Uses • Metal Activity Series

PLANNING GUIDE

Section	Time class	Student Edition Features	Program Resources	Media and Technology
Section C. Conservation **Objectives:** • State the Law of Conservation of Matter and apply the law by balancing chemical equations. • Write balanced chemical equations and relate them to the Law of Conservation of Matter. • Define the term mole and calculate the molar mass of a compound when provided with its formula and the atomic masses of its elements. • Calculate the percent composition by mass of a specified element in a given compound. • Distinguish between renewable and nonrenewable resources, and identify the methods of conserving our resources.	**8 class periods**	**Building Skills** #6. Accounting for Atoms #7. Writing Chemical Equations #8. Molar Mass #9. Percent Composition #10. Copper Life-Cycle Analysis **Laboratory Activity** • Retrieving Copper **Making Decisions** • Designing a Coin **Chemistry at Work** • Preserving the Past . . . for the Future	**Activity Book** • C.1 Balancing Equations • C.2 Balancing Equations • C.3 Molar Mass • C.4 Percent Composition and Conservation of Mass • C.5 Retreiving Copper **Skill Handbook** • Understanding Equation and Stoichiometry • Understanding Chemical Equations • Balancing Equations • What is a Mole? • Moles and Molar Mass • Mass-Mole Conversion • Calculate Percent Composition From a Formula	**CD-ROM** • Interpreting a Balanced Chemical Equation • Mole-Mass Conversion • Lab Video **Transparencies** • Aluminum Can Recycling • Waste Generated in the United States • Life Cycle of Materials

PLANNING GUIDE

Section	Time class	Student Edition Features	Program Resources	Media and Technology
Section D. Materials: Designing for Properties **Objectives:** • Describe the difference in the properties of an element due to the arrangement of the element's atoms. • Describe the difference between the properties of an element and the properties of its alloy. • Describe how the conductivity of a material can be changed. • Describe how the properties of a material may be changed by applying a coating.	**5 class periods**	**Building Skills** #11. Linking Properties to Structure #12. Alternatives to Metals **ChemQuandary** #2. A Case of Elusive Identity #3. Five Cents' Worth **Laboratory Activities** • Striking It Rich • Copper Plating **Chemistry at Work** • All That's Gold Doesn't Glitter . . . or Does It?	**Activity Book** • D.3 Striking It Rich • D.6 Copper Plating	**CD-ROM** • Unit II, Section D **Transparencies** • Common Alloy Compositions and Uses • Process of Electroplating
Putting It All Together Use the concepts of the properties of materials to design a new coin for a country.				**CD-ROM** • Putting It All Together

SUPPLY LIST FOR UNIT 2

Expendable Items and Unusual Materials (per 24 students working in pairs)

Alphabetical by Item Name			
Quantity	Unit	Item	Sections Used
12	Each	Battery, 9 volt	A.3, D.6
5	Grams	Carbon, pencil lead	A.3
10	Grams	Copper(II) nitrate trihydrate	B.4
16	Grams	Copper(II) sulfate pentahydrate	D.6
36	Each	Copper pennies	D.3
12	Grams	Copper powder	B.2
12	Each	Copper rods	B.2, D.6
4	Grams	Copper(II) chloride dihydrate	A.3
20	Sets	Element data card	A.4
24	Pieces	Filter paper	C.5
24	Sheets	Graph paper	A.6
50	mL	0.5 M Hydrochloric acid	A.3, C.5
12	Cm	Iron (thin wire)	A.3
10	Grams	Magnesium nitrate hexahydrate	B.4
18	Cm	Magnesium ribbon	A.3. B.4
12	Each	Matches/strikers	D.6
64	Each	Micropipets	A.3, B.4

SUPPLY LIST

Alphabetical by Item Name (continued)			
Quantity	Unit	Item	Sections Used
12	Each	Nail, iron or zinc	D.6
5	Grams	Silicon (small lumps)	A.3
7	Grams	Silver nitrate	B.4
12	Each	Small hammers	A.3
84	Each	Small vial (for metal samples)	A.3
1	Bottle	Spray & Wash	B.4
100	Grams	Steel wool	D.3
5	Grams	Sulfur (small lumps)	A.3
5	Grams	Tin(mossy or foil)	A.3
12	Each	U-Tube or Tygon Tubing	D.6
12	Each	Voltmeter	D.6
12	Each	Wire leads w/alligator clips	D.6
41	Grams	Zinc chloride	D.3
12	Grams	Zinc metal, mossy	A.3,C.5
10	Grams	Zinc nitrate hexahydrate	B.4
12	Cm	Zinc strip	B.4
27	Grams	Zinc, granulated	D.3

SUPPLY LIST

By Section			
Quantity	Unit	Item	Sections Used
5	Grams	Carbon, pencil lead	A.3
4	Grams	Copper(II) chloride dihydrate	A.3
12	Cm	Iron (thin wire)	A.3
5	Grams	Silicon (small lumps)	A.3
12	Each	Hammers, small	A.3
84	Each	Vials, small (for metal samples)	A.3
5	Grams	Sulfur (small lumps)	A.3
5	Grams	Tin (mossy or foil)	A.3
18	Cm	Magnesium ribbon	A.3, B.4
64	Each	Micropipets	A.3, B.4
50	mL	Hydrochloric acid	A.3, C.5
12	Grams	Zinc metal, mossy	A.3, C.5
12	Each	Battery, 9-volt	A.3, D.6
20	Sets	Element data card	A.4
24	Sheets	Graph paper	A.6
12	Grams	Copper powder	B.2
12	Each	Copper rods	B.3, D.6
10	Grams	Copper(II) nitrate trihydrate	B.4

SUPPLY LIST

By Section (continued)			
Quantity	Unit	Item	Sections Used
10	Grams	Magnesium nitrate hexahydrate	B.4
7	Grams	Silver nitrate	B.4
1	Bottle	Spray & Wash	B.4
10	Grams	Zinc nitrate hexahydrate	B.4
12	Cm	Zinc strip	B.4
24	Pieces	Filter paper	C.5
36	Each	Copper pennies	D.3
100	Grams	Steel wool	D.3
41	Grams	Zinc chloride	D.3
27	Grams	Zinc, granulated	D.3
16	Grams	Copper(II) sulfate pentahydrate	D.6
12	Each	Matches/strikers	D.6
12	Each	Nail, iron or zinc	D.6
12	Each	Voltmeter	D.6
12	Each	Wire leads w/alligator clips	D.6
12	Each	U-tube or tygon tubing	D.6

UNIT 2 MATERIALS: STRUCTURE AND USES

Introduction to Unit

Chemistry topics introduced include the language of chemical and physical properties, chemical symbols and formulas, the Periodic Table, reactivity of metals, oxidation-reduction, law of conservation of matter and energy, and writing and balancing chemical equations. Students will also use formulas to calculate molar mass and percent composition. Laboratory activities involve investigating properties of elements and comparing chemical reactivities of metals. In the PIAT of Unit 2, students are asked to design a coin.

UNIT 2

HOW can the chemical and physical properties of matter be explained?

SECTION A **Why We Use What We Do** (page 94)

WHERE are mineral resources found and how are they processed?

SECTION B **Earth's Mineral Resources** (page 112)

MATERIALS: STRUCTURE AND USES

SECTION C Conservation
(page 128)

WHAT information do chemical equations convey about matter and its changes?

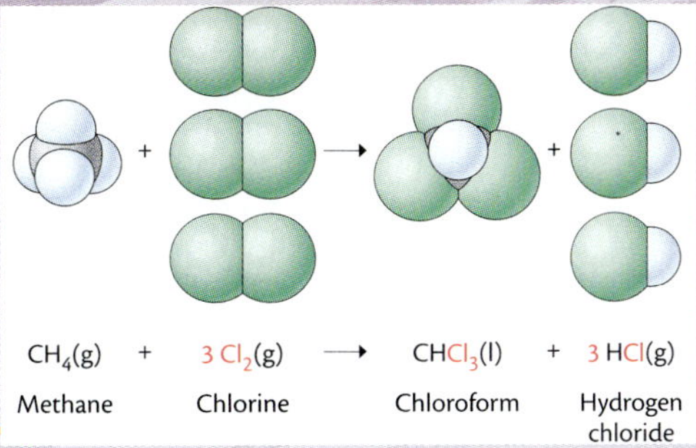

HOW can matter be modified to make it more useful?

SECTION D Materials: Designing for Properties
(page 153)

Your local Congresswoman has invited you and your classmates to submit a design for a new U.S. half-dollar coin. What will the coin's composition be? Why? What makes a particular material best suited for each intended use? Turn the page to learn the answers to these questions—and perhaps be the lucky contest winner.

IMPLEMENTING THE PROJECT OR THEME

Unit 2 continues the *ChemCom* approach—a need-to-know basis of a science-technology-society (STS) problem or issue. The story line for Unit 2 is about a competition to produce a sample of a proposed new half-dollar coin. The design competition is sponsored by Congresswoman Gonzales to stimulate interest in a bill authorizing the U.S. Mint to implement the design of such a coin. The students will use the information they learn in this unit to design the best possible coin. The procedures followed by the United States Mint are documented on their web site at *http://www.usmint.gov.*

THE HONORABLE MARIA GONZALES
United States House of Representatives

MEMORANDUM

TO: District 12 High School Principals and Chemistry Teachers
FROM: The Hon. Maria Gonzales
U.S. House of Representatives

SUBJECT: Coin-Design Competition

As you may know, I plan to introduce a Bill in the House of Representatives authorizing the production of a new half-dollar coin by the United States Mint. The purpose of this memorandum is to let you know about a contest my office is sponsoring for high-school chemistry students in your Congressional District. The competition involves proposing a possible design for the new coin. The winning design will be used as an example in House Committee deliberations and on the floor of the House as I seek support from colleagues for authorization of the new 50-cent coin.

Each high school within your Congressional District may submit one complete design. The team of students (or student) who creates the winning design will be honored, along with the rest of their chemistry class, at an open house at my local office. In addition, the winning student or team will travel with my staff to Washington, D.C., for the introduction of my Bill and for a public presentation of the suggested coin design.

A complete coin-design proposal must include the following information:

- Full name(s) and address(es) of the designer(s)
- Chemistry teacher's name and course title
- School name and phone number
- Coin diameter, thickness, and mass
- Detailed drawing or actual model of the coin, enlarged 5x for clarity
- Specifications for the composition of the coin's material
- Plans for obtaining or creating the materials used in the coin
- Two-page rationale for key decisions made in the coin's design

All completed proposals will be due in my office within six weeks of receipt of this memorandum.

MG/hs

92

Buoyed by the U.S. Mint's release of a new dollar coin and the success of the recent quarter-coin series featuring each state, Congressional Representative Maria Gonzales is planning to introduce a Bill authorizing the U.S. Mint to produce a new half-dollar coin.

Realizing that Congressional colleagues will request information about the proposed replacement, Representative Gonzales is sponsoring a contest for local high school chemistry students to propose a prototype design of the new coin.

As her memo suggests, every aspect of the half-dollar coin—from its appearance to its size and composition—is to be included in the designs. Because only one design can be submitted from each school, your class will have to decide which team's coin proposal will be selected for further screening in this competition.

As you go on to consider important features in your coin's design, you will learn about Earth's mineral resources and how they are used by society. You will learn why certain materials are used for particular new products—coins and other useful things—and how those materials are developed from available resources. Throughout this unit, keep in mind how such chemical knowledge can help guide your design of a new coin.

WHY WE USE WHAT WE DO

Introduction

Section A introduces students to the chemical and physical properties of the elements. Students discover patterns in the properties of the elements and that materials are selected for a particular use based upon their specific characteristics or properties. The Periodic Table is then used to predict properties of elements.

A.1 PROPERTIES MAKE THE DIFFERENCE

This section introduces chemical and physical properties and changes to determine those characteristics that make a substance suitable for minting coins. To further clarify the differences between chemical and physical change, perform the following minilab.

Minilab

Place a small sample of Cu in 1 M HCl; and a small sample of Zn in 1 M HCl. Observe differences in the two systems. Then notch the edge of a post-1982 penny with a file so that the inner layer is exposed. Place it in 3 M HCl for several days. The result is a hollow penny as the acid removes the more active zinc metal, leaving the copper shell.

DEMONSTRATIONS

To clarify the differences between physical and chemical properties and changes, the following demonstrations may be helpful. Ask students to observe the reactions carefully and record evidence for change before deciding whether the change is chemical or physical. Choose demonstrations from the following list or other sources you may have in your classroom.

SECTION A WHY WE USE WHAT WE DO

Every human-produced object, old or new, is made of materials selected for their specific characteristics, or properties. What makes a particular material best for a particular use? You can begin to answer this question by exploring some properties of materials.

A.1 PROPERTIES MAKE THE DIFFERENCE

Overview CD-ROM WWW.

In this unit, you will be considering the design of something you use every day—money in the form of coins. Throughout history, people have used many different items as money: beads, stones, printed paper, even precious metals, to name a few. What characteristics make an object suitable (useful) for manufacturing money or minting coins? How important is appearance? Size? What other important characteristics can you suggest?

As you already know, every substance has characteristic properties that distinguish it from other substances, thus allowing it to be identified. These characteristic properties include **physical properties,** or properties that can be determined without altering the chemical makeup of the material. Color, density, and odor are examples of physical properties. The ability of a material to undergo **physical changes,** such as melting, boiling, and bending, is often important in its use. Remember that in a physical change, the identity of the substance remains the same.

The **chemical properties** of a material often play important roles in its usefulness. Chemical properties relate to the tendency of a substance to undergo **chemical changes**—that is, to transform into new substances. Consider the common chemical change of rust forming on iron surfaces. The tendency of a material to rust, or react to form an oxide, is the chemical property that accounts for this chemical change. A chemical change can often be detected by observing the formation of a gas or solid, a color change, a change on the surface of a solid, or a temperature change (indicating that thermal energy has been absorbed or given off). Figure 1 illustrates some physical and chemical changes involving copper.

In the following activity, you will classify some characteristics of common materials as either physical or chemical properties.

PHYSICAL AND CHEMICAL PROPERTIES — Building Skills 1

Consider this statement: *Copper compounds are often blue or green.* Does the statement describe a physical or chemical property? To answer the question, first think about whether a change in the identity of a substance is involved. Has the substance been chemically changed? If the answer is

Physical changes

1. Gently heat a few crystals of iodine in a 250-mL flask with a loosely fitting cork. Heat very carefully until iodine vapors are just visible. Too much heating could cause the cork to become a projectile. **Perform this demonstration only if a hood is available.** Iodine vapors are toxic.
2. Place a peanut-sized piece of dry ice (CO_2) in an uninflated balloon, and tie the opening. As the dry ice warms, the balloon will expand.
3. Heat a bimetallic strip (20 cm or more in length) in a burner flame.
4. Shake and open a warm can of soft drink.
5. Dissolve a few crystals of any colored salt, such as copper(II) sulfate or potassium permanganate, in water.

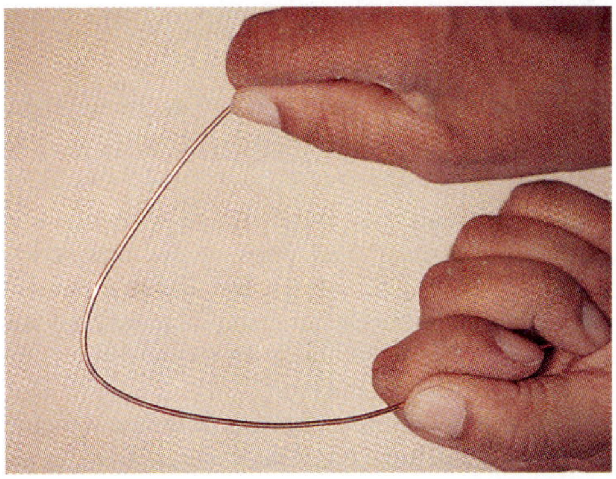

Figure 1 *Examples of physical and chemical changes involving copper.*

"no," then the statement describes a physical property; if the answer is "yes," then the description is of a chemical property.

Color is a characteristic physical property of many chemical compounds. A green copper compound in a jar on the shelf is not undergoing any change in its identity. Color, therefore, is a physical property. A change in color, however, often indicates a change in identity and thus a chemical change. For example, colored matter called litmus, derived from a plant-like organism called a lichen, turns from blue to pink when exposed to acid. This is a chemical change involving the chemical properties of litmus and acid.

Now consider this statement: *Oxygen gas supports the burning of wood.* Does the statement refer to a physical or chemical property of oxygen? If you apply the same key question—is there a change in the identity of the oxygen gas—you will arrive at the correct answer. The burning of wood—or combustion—involves chemical reactions between the wood and oxygen gas that change both reactants. The reaction products of ash, carbon dioxide, and water vapor are very different from wood and oxygen gas. Thus the statement refers to a chemical property of oxygen gas (as well as of wood).

Now it's your turn. Classify each statement as describing either a physical property or a chemical property. (*Hint:* Decide whether the chemical identity of the material does or does not change when the property is observed.)

1. Pure metals have a high luster (are shiny and reflect light).
2. The surfaces of some metals become dull when exposed to air.
3. Nitrogen gas, which is a relatively nonreactive element at room temperature, can form nitrogen oxides at the high temperatures of an operating automobile engine.
4. Milk turns sour if left too long at room temperature.
5. Diamonds are hard enough to be used on drill bits.
6. Metals are typically ductile (can be drawn into wires).
7. Bread dough increases in volume if it is allowed to "rise" before baking.

CD-ROM WWW. Why We Use What We Do

Chemical changes

1. There should be no open flames in the room during this demonstration. Place a few calcium turnings (calcium metal) in 200 mL of water in a 400-mL beaker. Hydrogen gas is evolved, which is made more noticeable if the beaker is placed on an overhead projector. Phenolphthalein added to the water causes the water to turn pink.
2. Burn a candle. This is especially effective if the candle is placed on a digital balance and the apparent loss of mass is noted.
3. Crack a raw egg into a beaker of boiling water. As an option, you can get the same effect by cracking a raw egg into 100 mL of concentrated HCl.
4. Demonstrate spontaneous endothermic reactions such as that between 15 g $Ba(OH)_2 \cdot 8H_2O$ and 15 g NH_4SCN. Place the reactants in a 250-mL flask; stopper lightly, shake, wet the outside of the bottom of the flask with water, and set on a small piece of wood.

   $Ba(SCN)_2 + NH_3 + H_2O$ are formed with a temperature drop sufficient to freeze the flask to the block. Soluble barium salts are very toxic; however, barium sulfate is very insoluble. The solid formed in this demonstration should be dissolved in a minimum of water and re-precipitated with 1.0 M sulfuric acid. Collect the precipitate of barium sulfate, place it in a plastic bottle, and cap the bottle. Then it can be safely discarded in a landfill.

PHYSICAL AND CHEMICAL PROPERTIES

Building Skills 1

In this activity, students practice recognizing the difference between physical and chemical properties. Chemical properties become evident when one substance is transformed into a new substance, whereas physical properties (such as melting, boiling, thermal and electrical conductivity, density, color, hardness) do not depend on the presence of another substance. In a physical change, the identity of the substance is not altered. Students should also realize that in considering a material for a particular use, its physical and chemical properties are important in determining the best substance for that use. They should also realize that cost and availability are important.

BUILDING SKILLS 1, cont.

Answers

1. Physical—no change in material, depends on nothing else.
2. Chemical—the metal reacts with the oxygen from the air.
3. Chemical—new substances are formed.
4. Chemical—material changes (taste), depends on bacteria in milk.
5. Physical—no change.
6. Physical—no change in identity, even though shape is changed.
7. Chemical—bread rising is due to the production of carbon dioxide by yeast; this involves a change of material.
8. Chemical—argon does not react with the filament.
9. Physical—no change in the metals.

A.2 THE CHEMICAL ELEMENTS

All the substances that make up our world are composed of around 100 chemical elements. These elements are grouped into three classes (metals, nonmetals, and metalloids) based upon the similarities and differences in their properties. The names and symbols of some of the elements are given in Figure 4. Remind students of the proper way to write a chemical symbol: the first letter is always capitalized, the second and third letter, if present, are lowercase. Discuss the derivation of certain symbols from the Latin name and give examples.

8. Argon gas, rather than air, is used in many lightbulbs to prevent oxidation of the metal filament.
9. Generally, metals are better conductors of heat and electricity than are nonmetals.

As you might imagine, many issues need to be considered when selecting materials for a specific use. A substance with properties well suited to a purpose may be either unavailable in sufficient quantity or too expensive. Or a substance may have undesirable physical or chemical properties that limit its use. In these and other situations, another material with most of the sought-after properties can often be found and used.

The cost of material is a particularly important issue when manufacturing coins and printed currency. Just imagine what would happen if the declared value of a coin were less than the cost of the metals contained in it. How would this affect the production and circulation of the coin? This situation actually occurred in the United States not too long ago. In the early 1980s, copper became too expensive to be used as the primary metal in pennies. In other words, the cost of the copper in a penny was greater than the value of the penny. A lower-cost replacement having similar properties was needed. Zinc, another metallic element, was chosen to replace most of the copper in all post-1982 pennies. Zinc is about as hard as copper and has a density (7.14 g/cm^3) close to that of copper (8.94 g/cm^3). Zinc is also readily available and less expensive than copper.

Unfortunately, zinc is more chemically reactive than copper. During World War II, copper metal was in short supply. To conserve that resource, zinc-plated steel pennies—known to coin collectors as "white cents" or "steel cents"—were created in 1943. The new pennies quickly corroded. As you can see in Figure 2, these pennies also looked considerably different from traditional copper pennies. Production of the zinc-plated pennies ended within a year.

Figure 2 *Zinc-copper and copper pennies (top); "new" and corroded zinc-steel pennies (bottom).*

The problems associated with the earlier zinc-plated pennies were solved in the early 1980s. In the new design, the properties of copper were used where they were most needed—on the coin's surface—and the properties of zinc where they were suitable—within the body of the coin. All post-1982 pennies are 97.5% zinc. They are composed of a zinc core surrounded by a thin layer of copper, added to increase the coin's durability and maintain its familiar appearance. Figure 3 shows a cross-section of a post-1982 penny.

Whether it is copper or zinc in a penny or tungsten in a lightbulb, the message is clear: Every substance has its own set of physical and chemical properties. But with millions of substances available, how can one identify the "best" substance(s) for a given need?

Fortunately, the list of possibilities can be greatly shortened. All substances are made of a relatively small number of building blocks—the atoms of the different chemical elements. Knowing the similarities and differences among atoms of elements and among combinations of those atoms can greatly simplify the challenge of matching a substance to its most appropriate uses.

ADDITIONAL ACTIVITIES

Color change: chemical or physical

Put some food coloring in water and have students state the type of change in the water. Use pH paper in an acid or a base and have students state the type of change on the paper.

Conclusion: A color change, by itself, is not an indication of a chemical change.

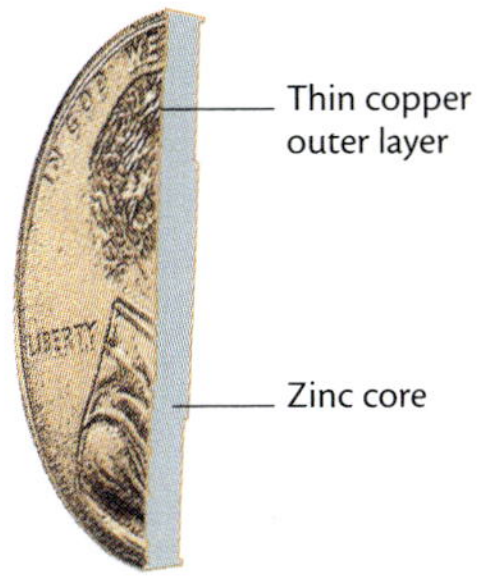

Figure 3 Cross-section showing composition of a post-1982 penny.

A.2 THE CHEMICAL ELEMENTS

CD-ROM WWW. …esenting …oms and Ions

You learned in Unit 1 that all matter is composed of atoms. One element differs from another because its atoms have properties that differ from those of all other elements. More than 100 chemical elements are known. The table in Figure 4 lists some common elements and their symbols. An alphabetical list of all the elements (names and symbols) can be found on page 108.

Elements can be grouped, or classified, in several ways according to similarities and differences in their properties. Two major classes are **metals** and **nonmetals**. Metals include such elements as iron, tin, zinc, and copper. Carbon and oxygen are examples of nonmetals. Everyday experience has given you some knowledge of metallic properties. The laboratory activity you will soon do will give you an opportunity to explore the properties of metals and nonmetals in more depth.

A relatively few elements called **metalloids** have properties that are intermediate to those of metals and nonmetals. That is, metalloids are like metals when it comes to some properties and like nonmetals when it comes to others. Examples of metalloids include silicon and germanium, both commonly used in the computer industry.

What properties of matter are used to distinguish metals, nonmetals, and metalloids? The next activity will help you find out.

Common Elements	
Name	Symbol
Aluminum	Al
Antimony	Sb
Argon	Ar
Barium	Ba
Beryllium	Be
Bismuth	Bi
Boron	B
Bromine	Br
Cadmium	Cd
Calcium	Ca
Carbon	C
Cesium	Cs
Chlorine	Cl
Chromium	Cr
Cobalt	Co
Copper	Cu
Fluorine	F
Gold	Au
Helium	He
Hydrogen	H
Iodine	I
Iron	Fe
Krypton	Kr
Lead	Pb
Lithium	Li
Magnesium	Mg
Manganese	Mn
Mercury	Hg
Neon	Ne
Nickel	Ni
Nitrogen	N
Oxygen	O
Phosphorus	P
Platinum	Pt
Potassium	K
Silicon	Si
Silver	Ag
Sodium	Na
Sulfur	S
Tin	Sn
Tungsten	W
Uranium	U
Zinc	Zn

Figure 4 *A table of common elements and their symbols.*

A.3 METAL OR NONMETAL? Laboratory Activity

Introduction

In this activity you will investigate several properties of seven elements and then decide whether each element is a metal, nonmetal, or metalloid. You will examine the color, luster, and form of each element and attempt to crush each sample with a hammer. You or your teacher (as a demonstration) will also test the substance's ability to conduct electricity. Finally, you will determine the reactivity of each element with two solutions: hydrochloric acid, HCl(aq), and copper(II) chloride, $CuCl_2$(aq).

ADDITIONAL ACTIVITIES

Dichotomous key

Assemble as many everyday products as possible that represent chemical elements. Household products are recommended, but the chemical supply room at school may be useful. (Examples: iron nail, copper pipe, silver flatware, gold-plated jewelry, coal, graphite, lead fishing weights, aluminum foil, magnesium ribbon, platinum wire, etc.). Have students develop a decision tree that ultimately separates every product into its own set of properties. For example, starting with products, the first division may be metals and nonmetals, then liquids or solids, and so on.

Element of the week

A popular activity for making elements more familiar to students is selecting an Element of the Week, by which each day has a small activity pertaining to that week's element. In a variation, each student "adopts" an element, prepares a report on it, and serves as the class expert on this element for the rest of the school year.

A.3 METAL OR NONMETAL? Laboratory Activity

In this laboratory activity, students will learn how to distinguish metallic from nonmetallic elements after examining their properties. Seven elements are tested for appearance, malleability, reaction with hydrochloric acid, reaction with copper (II) chloride solution, and electrical conductivity.

Time

50 minutes

Materials (for 24 students working in pairs)

several small hammers
conductivity apparatus
24 pipets
12 wellplates
7 small vials for samples of elements
12 forceps
25 mL 0.5 M hydrochloric acid solution (1 mL conc. HCl/25 mL)
25 mL 0.1 M copper (II) chloride solution (0.43 g $CuCl_2 \cdot 2H_2O$/25 mL)
The seven elements to be tested are:
Code A: 5 g Carbon (C) pencil lead
Code B: 12 cm magnesium (Mg) ribbon, cut to 1-cm lengths
Code C: 5 g silicon (Si), 12 small lumps
Code D: 5 g tin (Sn), 12 small pieces tin wire or tin foil
Code E: 5 g sulfur (S), 12 small lumps
Code F: 5 g iron (Fe), 1-cm pieces of thin wire (better for conductivity)
Code G: 15 g zinc (Zn), 12 small pieces mossy zinc or 12 1-cm pieces wire (better for conductivity)

Safety

Students should wear safety goggles and aprons throughout the laboratory activity. Remind students always to wear goggles.

Procedure

1. In your laboratory notebook, prepare a data table that has enough space to record the properties of the seven element samples, which have been coded with letters *a* to *g*.

DATA TABLE

Element	Appearance	Conductivity (optional)	Result of Crushing	Reaction with Acid	Reaction with $CuCl_2(aq)$
a.					
b.					
c.					
d.					
e.					
f.					
g.					

2. *Appearance:* Observe and record the appearance of each element, including physical properties such as color, luster, and form. You can record the form as crystalline (like table salt), noncrystalline (like baking soda), or metallic (like iron).
3. *Conductivity:* If an electrical conductivity apparatus is available, use it to test each sample. **CAUTION:** *Avoid touching the bare electrode tips with your hands; some may deliver an uncomfortable electric shock.* Touch both electrodes to the element sample, but do not allow the electrodes to touch each other. See Figure 5. If the bulb lights, even dimly, electricity is flowing through the sample. Such a material is called a **conductor.** If the bulb fails to light, the material is a **nonconductor.**

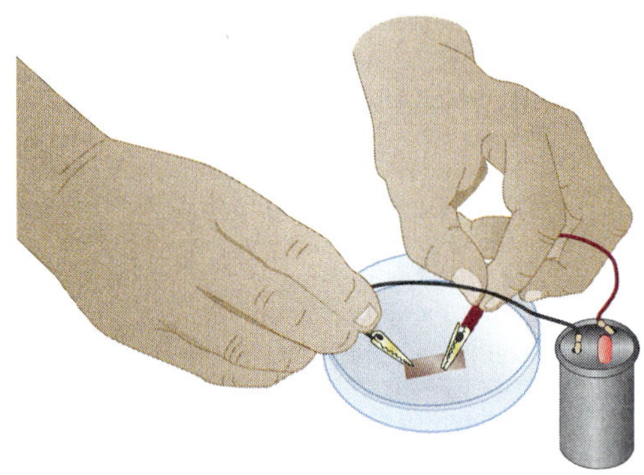

Figure 5 *Testing for conductivity.*

Advance preparation

Fill the pipets with 0.5 M HCl and 0.1 M $CuCl_2$. Wipe the stem of each pipet with paper towels. Cut the pencil lead, magnesium ribbon, tin foil, iron wire, and zinc wire to the appropriate lengths. Select small lumps of silicon and sulfur. Place the samples in small vials, labeled with codes as suggested above.

Pre-lab discussion

Review the differences between physical and chemical properties. Caution students not to handle any element samples with their hands. Review the procedures for using acids safely.

4. *Crushing:* Gently tap each element sample with a hammer. Based upon the results, decide whether the sample is **malleable,** which means it flattens without shattering when struck, or **brittle,** which means it shatters into pieces.
5. *Reactivity with acid.*
   a. Label seven wells of a clean wellplate *a* through *g*.
   b. Place a sample of each element in its appropriate well. The solid wire or ribbon samples provided by your teacher will be less than 1 cm in length. Other samples should be between 0.2 g and 0.4 g—you can estimate that as no larger than the size of a match head.
   c. Add 15 to 20 drops of 0.5 M HCl to each well that contains a sample. **CAUTION:** *0.5 M hydrochloric acid (HCl) can chemically attack skin if allowed to remain in contact for a long time. If any hydrochloric acid accidentally spills on you, ask a classmate to notify your teacher immediately. Wash the affected area immediately with tap water and continue for several minutes.*
   d. Observe and record each result. The formation of gas bubbles indicates that a chemical reaction has occurred. A change in the appearance of an element sample may also indicate a chemical reaction. Decide which elements reacted with the hydrochloric acid and which did not. Record these results.
   e. Discard the wellplate contents as instructed by your teacher.
6. *Reactivity with copper(II) chloride.*
   a. Repeat Steps 5a and 5b.
   b. Add 15 to 20 drops of 0.1 M copper(II) chloride ($CuCl_2$) to each sample.
   c. Observe each system for three to five minutes—changes may be slow. Decide which elements reacted with the copper(II) chloride and which did not. Recall the criteria you used in the acid test to determine if a reaction occurred. Record each result.
   d. Discard the wellplate contents as instructed by your teacher.
7. Wash your hands thoroughly before leaving the laboratory.

Questions

1. Classify each property tested in this activity as either a physical property or a chemical property.
2. Sort the seven coded elements into two groups based on similarities in their physical and chemical properties.
3. Which element(s) could fit into either group? Why?
4. Using the following information, classify each tested element as a metal, nonmetal, or metalloid.
   - Metals have a luster, are malleable, and conduct electricity.
   - Many metals react with acids; many metals also react with copper(II) chloride solution.
   - Nonmetals are usually dull in appearance, are brittle, and do not conduct electricity.
   - Metalloids have some properties of both metals and nonmetals.

Lab tips

- Do not allow students access to larger amounts than specified samples of active metals such as Mg. Some teachers have had experiences with students collecting and removing larger quantities of magnesium ribbon for unauthorized and highly dangerous "experimentation" outside of class.
- Consider having students "hammer" their samples on their metallic ring-stand base with a metal support ring; this spares the lab countertop and does not require hammers.
- A 9-V conductivity tester with light-emitting diodes is recommended. These testers are considered safe for student use. (See comment below.) If a 110-V apparatus is used, the teacher should demonstrate the activity.
- Copper turnings work well as a source of the metal for this activity.
- If there are fewer hammers than students or groups, demonstrate this part of the activity to the class.

Constructing a safe conductivity apparatus

Instructions for constructing a safe student conductivity apparatus from a 9-V battery, resistors, and LEDs available from electronic supply stores are found in the reference below. A commercial version is also available from Flinn Scientific.

Katz, D. and Willis, C., *Journal of Chemical Education*, 1994, p. 330

Expected Results

Code	Element	Appearance	Result of crushing	Conductivity (optional)	Reaction with acid	Reaction with $CuCl_2$(aq)
A	C	Gray, dull	Brittle	Yes	No	No
B	Mg	Shiny, silver	Malleable	Yes	Yes, bubbles	Yes, darkens
C	Si	Gray, shiny	Brittle	Yes	No	No
D	Sn	Shiny, silver	Malleable	Yes	Yes, bubbles	Yes, darkens
E	S	Yellow, dull	Brittle	No	No	No
F	Fe	Gray, shiny	Malleable	Yes	Yes, bubbles	Yes, darkens
G	Zn	Shiny, silver	Malleable	Yes	Yes, bubbles	Yes, darkens

Answers

1. Appearance — physical
   Crushing — physical
   Conductivity — physical
   Reaction with acid — chemical
   Reaction with $CuCl_{2(aq)}$ — chemical
2. Group 1: magnesium, zinc, iron, and tin; Group 2: carbon, silicon, and sulfur.
3. Silicon. It is shiny and conductive, but brittle.
4. Metals—magnesium, zinc, iron, and tin; nonmetals—carbon and sulfur; metalloids—silicon. Students may also classify carbon as a metalloid based on its conductivity. Ask students to consider the other properties of metals. Because carbon lacks the characteristic luster, it is not classified as a metal.

Post-lab discussion

Provide time for the student groups to compare their results and subsequent groupings of the elements. Discuss the criteria for assigning elements to groups. Allow students to resolve disagreements, but be ready to give suggestions if requested.

A.4 THE PERIODIC TABLE

Depending on your students' interest and ability, you may wish to point out that (unlike the 1800s atomic mass scale) the modern atomic mass scale of the elements is based on a particular kind of carbon atom (the isotope carbon-12), which is defined as having an atomic mass of exactly 12. This section describes the development of the Periodic Table as a search for patterns and regularities.

You have learned one classification scheme for elements: metals, nonmetals, and metalloids. The quantity of detailed information about the elements is enormous, however. And when choosing or designing materials for specific uses, the more information available about the elements (including similarities and differences among them), the better the decisions. How then is all the knowledge about the elements conveniently organized? You have already been introduced to the answer—the periodic table. Now you will explore its origins and gain greater understanding of the chemical information it conveys.

A.4 THE PERIODIC TABLE

By the mid-1800s, about 60 elements had been identified. Five of these were nonmetals that were gases at room temperature: hydrogen (H), oxygen (O), nitrogen (N), fluorine (F), and chlorine (Cl). Two liquid elements were also known, the metal mercury (Hg) and the nonmetal bromine (Br). The rest of the known elements were solids with widely differing properties.

In an effort to impose some organization on the information related to the elements, several scientists tried to place elements with similar properties near one another on a chart. Such an arrangement is called a **periodic table**. Dimitri Mendeleev, a Russian chemist, published a periodic table in 1869. A similar table is still used today. In some respects, the periodic table has a pattern that resembles a monthly calendar, in which weeks repeat on a regular (periodic) seven-day cycle.

The periodic tables of the 1800s were organized according to two characteristics of elements. It was known that atoms of different elements have different masses. For example, hydrogen atoms have the lowest mass, oxygen atoms are about 16 times more massive than hydrogen atoms, and sulfur atoms are about twice as massive as oxygen atoms (making them about 32 times more massive than hydrogen atoms). Based on such comparisons, an **atomic weight** was assigned to each element in Mendeleev's periodic table. This atomic weight then became one of the two criteria for arranging elements in the periodic table.

The other criterion for organizing elements in early periodic tables was their respective "combining capacity" with other elements, such as chlorine and oxygen. It was known that atoms of various elements differ in the way that they combine with another element. For example, one atom of potassium (K) or cesium (Cs) combines with only one atom of chlorine (Cl) to produce the compound KCl or CsCl. Such one-to-one compounds can be represented as ECl, where E stands for the Element combining with chlorine. However, one atom of magnesium (Mg) or strontium (Sr) combines with two atoms of chlorine to produce the compound $MgCl_2$ or $SrCl_2$, which can be represented in general terms as ECl_2. Atoms of other elements may combine with three or four chlorine atoms to produce compounds with the general formula ECl_3 or ECl_4.

In the first periodic tables, elements with similar chemical properties were placed in vertical groups (columns). Horizontal arrangements were based on increasing atomic weights of the elements. In the activity that

ADDITIONAL INFORMATION

Some everyday uses of the elements:
Carbon as graphite—steel making, lubricant, brake linings, pencils
Carbon as diamond—gemstone, abrasive
Carbon as activated carbon—decolorizing and purifying agent
Magnesium—high-strength/low-weight alloys, flares, fireworks
Silicon—transistors, semiconductors, solar cells
Tin—coating alloys (solder, bronze, and pewter)
Sulfur—producing sulfuric acid (no.1 industrial chemical by the amount produced), matches, gunpowder, vulcanized rubber
Iron—structural metal (steel)
Zinc—coating (galvanizing), alloys (brass), dry cell batteries

follows, you will develop a classification scheme for some elements in much the same way Mendeleev did.

A.5 GROUPING THE ELEMENTS Making Decisions

You will be given a set of 20 element data cards. Each card lists some properties of a particular element.

1. Arrange the cards in order of increasing atomic weight.
2. Place the cards in a number of different groups. Each group should include elements with similar properties. You might need to try several methods of grouping before you find one that makes sense to you.
3. Examine the cards within each group for any patterns. Arrange the cards within each group in some logical sequence. Again, trial and error may be a useful method for accomplishing this task.
4. Observe how particular element properties vary from group to group.
5. Arrange all the card groups into some logical sequence.
6. Decide on the most reasonable and useful patterns within and among card groups. Then tape the cards onto a sheet of paper to preserve your pattern for classroom discussion.

A.6 THE PATTERN OF ATOMIC NUMBERS

Creators of early periodic tables were unable to offer explanations for similarities in properties found among neighboring elements. For example, all of the elements in the leftmost column of the periodic table are very reactive metals. All the elements in the rightmost column are unreactive gases. The reason for these similarities, which was discovered about 50 years after Mendeleev's work, serves as the basis for the modern periodic table.

As you will recall from Unit 1, all atoms are composed of smaller particles, including equal numbers of positively charged protons and negatively charged electrons. The number of protons, called the **atomic number,** distinguishes atoms of different elements. For example, every sodium atom (and only a sodium atom) contains 11 protons. The atomic number of sodium is 11. Each carbon atom contains 6 protons. If the number of protons in an atom is 9, it is a fluorine atom; if 12, it is a magnesium atom. Thus the atomic number (number of protons) identifies every atom as a particular element.

Early periodic tables, much like the one you just constructed, used atomic weights to organize the elements. Although this method produces reasonable results for elements with relatively small atomic weights, it does not work well for more massive atoms. The reason for this is the existence of another small particle that makes up the atom, the electrically uncharged neutron. The total mass of an atom is largely determined by the combined number of protons and neutrons in its **nucleus.** The nucleus is a concentrated region of positive charge (from the protons) in the center of an atom. See Figure 6.

You can learn more about these subatomic particles in Unit 6.

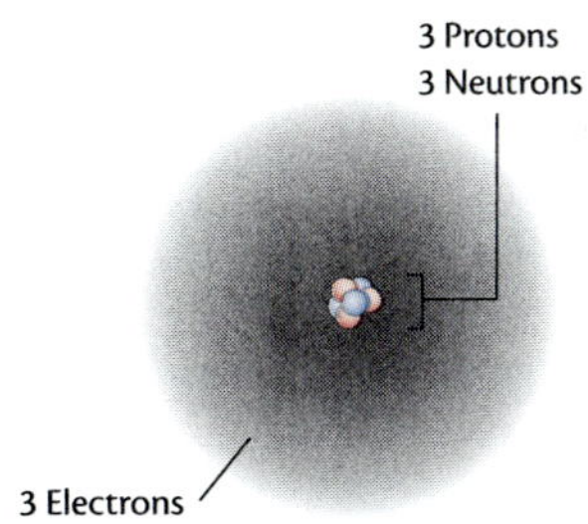

Figure 6 *Components of a particular atom of lithium (Li).*

A.5 GROUPING THE ELEMENTS

Making Decisions

Data cards containing selected chemical and physical properties of 20 elements are used to classify the elements according to similarities in their properties, an approach employed by Mendeleev (who worked with 63 elements and similar data). The cards can be duplicated to make a copy for each student. This works well as a group project in class. Groups can display their "periodic tables" and be ready to describe how patterns they used are useful in organizing and predicting.

A.6 THE PATTERN OF ATOMIC NUMBERS

After introducing the concept of atomic number, this section expands on the idea of grouping the elements and the relationship between atomic numbers and properties of elements.

PERIODIC VARIATIONS IN PROPERTIES

Building Skills 2

In this activity, students produce graphs illustrating the periodic relationships of the properties of elements. After discovering the cyclic (periodic) trends, students relate their interpretations to the modern Periodic Table. Such an approach was used in 1864 by Lothar Meyer, the second developer of the Periodic Table, when he plotted atomic volume versus atomic weight. Meyer was reluctant to use his scheme to predict undiscovered elements, not publishing his findings until 1870—one year after Mendeleev had published the pattern evident in compound formulas.

Students should begin by identifying the atomic number of each element by comparing the mass number and chemical properties of each to a periodic table. Mass number is the sum of the protons and neutrons in an atom. Mass number is also estimated by rounding atomic weight to the nearest whole number.

For example, the data card having a mass number of 1 would approximate the rounding of hydrogen's atomic weight, 1.008. The two cards with which students have difficulty both show a mass number of 40. They need to analyze the properties to decide which is argon (18) and which is calcium (20). Evaluating that one of the elements would be a gas at room temperature and chemically inert leads to identifying that data card as argon.

The total number of protons and neutrons in the nucleus of an atom is called the **mass number.** Electrons make up the rest of an atom, but because each electron is about 1/2000th the mass of a proton or neutron, the total mass of the electrons does not contribute significantly to the mass of an atom.

While all atoms of a particular element have the same number of protons, the number of neutrons can differ from atom to atom of an element. For example, carbon atoms always contain 6 protons, but they may contain 6, 7, or 8 neutrons. Thus individual carbon atoms can have mass numbers of 12, 13, or 14. Atoms with the same number of protons but different numbers of neutrons are called **isotopes.** In other words, isotopes are atoms of the same element with different mass numbers.

In the modern Periodic Table of the Elements, elements are placed in sequence according to their increasing atomic number. But because electrically neutral (uncharged) atoms contain equal numbers of protons and electrons, the Periodic Table is also sequenced by the number of electrons that the neutral atoms of each element contain.

Is there a connection between the atomic numbers used to organize the modern Periodic Table and the element properties used by nineteenth-century chemists to create their periodic tables? If there is, what is that connection? Continue reading to explore the relationship between atomic numbers and the properties of elements.

PERIODIC VARIATION IN PROPERTIES

Building Skills 2

Your teacher will assist you in identifying the atomic numbers of the 20 elements you considered earlier in the unit. Use these atomic numbers and information about each element's properties to prepare the two graphs described below. Look for patterns between atomic numbers and element properties as you construct the graphs.

Follow the graphing guidelines you learned in Unit 1 (page 66).

Graph 1. Trends in a chemical property.

1. On a sheet of graph paper, draw a set of axes and title the graph *Trends in a Chemical Property.*
2. Label the *x* axis *Atomic Number of E,* and number it from 1 to 20.
3. Label the *y* axis *Number of Oxygen Atoms per Atom of E.* Number it from 0 to 3; in increments of 0.5.
4. Construct a bar graph, as demonstrated in Figure 7, by plotting the oxide data from the element cards. For example, if no oxide forms, the height of the bar will be 0 because oxygen atoms do not form a compound with atoms of E. If E_2O (1 oxygen atom for 2 E atoms) is formed, the height of the bar is 0.5, which is the number of oxygen atoms for each E atom in the compound. Similarly, the heights of the bars for other oxides are 1 for EO, 1.5 for E_2O_3, 2 for EO_2, and 2.5 for E_2O_5. Do you understand why?

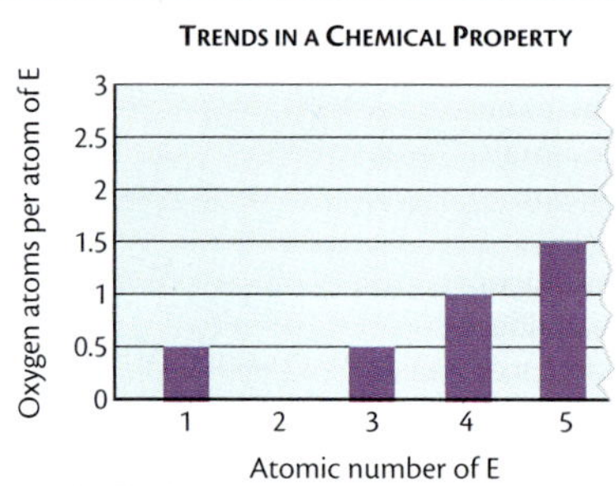

Figure 7 *Sample bar graph for oxide data.*

Mass number	Atomic number	Element	Symbol	Mass number	Atomic number	Element	Symbol
1	1	Hydrogen	H	12	6	Carbon	C
4	2	Helium	He	14	7	Nitrogen	N
7	3	Lithium	Li	16	8	Oxygen	O
9	4	Beryllium	Be	19	9	Fluorine	F
10	5	Boron	B	20	10	Neon	Ne

5. Label each bar with the actual symbol of the element E involved in that compound.

Graph 2. Trends in a physical property.

6. On a separate sheet of graph paper, draw a set of axes, and title the graph *Trends in a Physical Property.*
7. Label the *x* axis *Atomic Number* and number it from 1 to 20.
8. Label the *y* axis *Boiling Point (K)* and number it from 0 to 3000 K, as shown in Figure 8. Use as much of the space on the graph paper as possible to plot these kelvin temperatures.
9. Construct a graph as in Step 4, this time using the boiling point data from the element cards, as shown in Figure 8. Do not include data for the element with atomic number 6. The boiling point of this element (carbon) would be quite far off the graph.
10. Label each bar with the symbol of the element it represents.

Temperature in kelvins (K) is related to temperature in degrees Celsius (°C) by K = 273.15 + °C. You will learn more about the kelvin temperature scale in Unit 4.

QUESTIONS

1. Does either graph reveal a repeating, or cyclic, pattern? (*Hint:* Focus on elements represented by very large or very small bars.) Describe any patterns you observe.
2. Are these graphs consistent with patterns found in your earlier grouping of the elements? Explain.
3. Based on these two graphs, why is the chemist's organization of elements called a *periodic* table?
4. Where are elements with the highest oxide numbers located on the Periodic Table?
5. Where are elements with the highest boiling points located on the Periodic Table?
6. Explain any trends you noted in your answers to Questions 4 and 5.
7. Predict which element should have the lowest boiling point: selenium (Se), bromine (Br), or krypton (Kr). Explain how you decided.

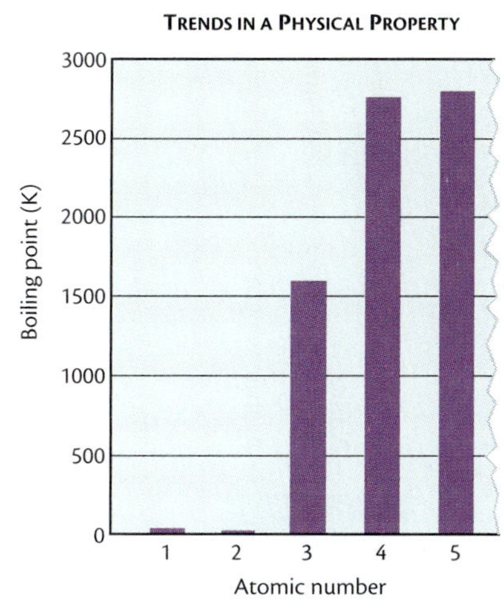

Figure 8 *Sample bar graph for boiling point data.*

When elements are listed in order of increasing atomic numbers and grouped according to similar properties, they form seven horizontal rows, called **periods.** This periodic relationship among elements is summarized in the modern Periodic Table, which you can see on page 104. To become more familiar with the Periodic Table, locate the 20 elements you grouped earlier. How do their relative positions compare with those shown on your chart?

Mass number	Atomic number	Element	Symbol	Mass number	Atomic number	Element	Symbol
23	11	Sodium	Na	32	16	Sulfur	S
24	12	Magnesium	Mg	36	17	Chlorine	Cl
27	13	Aluminum	Al	40	18	Argon	Ar
28	14	Silicon	Si	39	19	Potassium	K
31	15	Phosphorus	P	40	20	Calcium	Ca

Students should understand that a dash on a data card is interpreted as a "zero" (or indicates that the specified compound does not exist). If available, this activity could be completed using graphing calculators, or on computer graphing software.

As an alternate oxide analysis, students could plot the number of chlorine atoms of the chloride compound and melting point.

Additional or enrichment questions could ask students to predict the next elements to have high or low points in any of the graphs. It is also interesting to show the decreasing trend in boiling or melting peaks as atomic mass increases. This will prepare the students for predicting properties in Building Skills 3 (p. 105).

Answers

1. Both graphs show repeating patterns. Minimum values appear at the inert elements and maximum values for column-14 elements.
2. The relative positions of the elements should be the same on the Periodic Table and on the tables made by the students.
3. The elements repeat this pattern of physical characteristics after a certain interval or period. A table that orders the elements according to such repeating patterns is called a "periodic table."
4. The right side contains the highest oxide numbers.
5. The left side contains the highest boiling points.
6. As the oxide number increases, the boiling point decreases.
7. Krypton: Gases have lower boiling points than liquids or solids.

PERIODIC TABLE OF THE ELEMENTS

Key: 1 — Atomic number; Hydrogen — Name; H — Symbol; 1.008 — Atomic weight

Metals | Metalloids | Nonmetals

Period	Group 1 (1A)	2 (2A)	3 (3B)	4 (4B)	5 (5B)	6 (6B)	7 (7B)	8 (8B)	9 (8B)	10 (8B)	11 (1B)	12 (2B)	13 (3A)	14 (4A)	15 (5A)	16 (6A)	17 (7A)	18 (8A)
1	1 Hydrogen **H** 1.008																	2 Helium **He** 4.003
2	3 Lithium **Li** 6.94	4 Beryllium **Be** 9.01											5 Boron **B** 10.81	6 Carbon **C** 12.01	7 Nitrogen **N** 14.01	8 Oxygen **O** 16.00	9 Fluorine **F** 19.00	10 Neon **Ne** 20.18
3	11 Sodium **Na** 22.99	12 Magnesium **Mg** 24.31											13 Aluminum **Al** 26.98	14 Silicon **Si** 28.09	15 Phosphorus **P** 30.97	16 Sulfur **S** 32.07	17 Chlorine **Cl** 35.45	18 Argon **Ar** 39.95
4	19 Potassium **K** 39.10	20 Calcium **Ca** 40.08	21 Scandium **Sc** 44.96	22 Titanium **Ti** 47.87	23 Vanadium **V** 50.94	24 Chromium **Cr** 52.00	25 Manganese **Mn** 54.94	26 Iron **Fe** 55.85	27 Cobalt **Co** 58.93	28 Nickel **Ni** 58.69	29 Copper **Cu** 63.55	30 Zinc **Zn** 65.39	31 Gallium **Ga** 69.72	32 Germanium **Ge** 72.61	33 Arsenic **As** 74.92	34 Selenium **Se** 78.96	35 Bromine **Br** 79.90	36 Krypton **Kr** 83.80
5	37 Rubidium **Rb** 85.47	38 Strontium **Sr** 87.62	39 Yttrium **Y** 88.91	40 Zirconium **Zr** 91.22	41 Niobium **Nb** 92.91	42 Molybdenum **Mo** 95.94	43 Technetium **Tc** (98)	44 Ruthenium **Ru** 101.07	45 Rhodium **Rh** 102.91	46 Palladium **Pd** 106.42	47 Silver **Ag** 107.87	48 Cadmium **Cd** 112.41	49 Indium **In** 114.82	50 Tin **Sn** 118.71	51 Antimony **Sb** 121.76	52 Tellurium **Te** 127.60	53 Iodine **I** 126.90	54 Xenon **Xe** 131.29
6	55 Cesium **Cs** 132.91	56 Barium **Ba** 137.33	71 Lutetium **Lu** 174.97	72 Hafnium **Hf** 178.49	73 Tantalum **Ta** 180.95	74 Tungsten **W** 183.84	75 Rhenium **Re** 186.21	76 Osmium **Os** 190.23	77 Iridium **Ir** 192.22	78 Platinum **Pt** 195.08	79 Gold **Au** 196.97	80 Mercury **Hg** 200.59	81 Thallium **Tl** 204.38	82 Lead **Pb** 207.2	83 Bismuth **Bi** 208.98	84 Polonium **Po** (209)	85 Astatine **At** (210)	86 Radon **Rn** (222)
7	87 Francium **Fr** (223)	88 Radium **Ra** (226)	103 Lawrencium **Lr** (262)	104 Rutherfordium **Rf** (261)	105 Dubnium **Db** (262)	106 Seaborgium **Sg** (263)	107 Bohrium **Bh** (264)	108 Hassium **Hs** (265)	109 Meitnerium **Mt** (268)	110 Ununnilium **Uun** (269)	111 Unununium **Uuu** (272)	112 Ununbium **Uub** (277)		114 Ununquadium **Uuq** (285)		116 Ununhexium **Uuh** (289)		118 Ununoctium **Uuo** (293)

Lanthanides	57 Lanthanum **La** 138.91	58 Cerium **Ce** 140.12	59 Praseodymium **Pr** 140.91	60 Neodymium **Nd** 144.24	61 Promethium **Pm** (145)	62 Samarium **Sm** 150.36	63 Europium **Eu** 151.96	64 Gadolinium **Gd** 157.25	65 Terbium **Tb** 158.93	66 Dysprosium **Dy** 162.50	67 Holmium **Ho** 164.93	68 Erbium **Er** 167.26	69 Thulium **Tm** 168.93	70 Ytterbium **Yb** 173.04
Actinides	89 Actinium **Ac** (227)	90 Thorium **Th** 232.04	91 Protactinium **Pa** 231.04	92 Uranium **U** 238.03	93 Neptunium **Np** (237)	94 Plutonium **Pu** (244)	95 Americium **Am** (243)	96 Curium **Cm** (247)	97 Berkelium **Bk** (247)	98 Californium **Cf** (251)	99 Einsteinium **Es** (252)	100 Fermium **Fm** (257)	101 Mendelevium **Md** (258)	102 Nobelium **No** (259)

Each vertical column in the Periodic Table contains elements with similar properties. Each column is called a **group** or **family** of elements. For example, the lithium (Li) family, also called the **alkali metal** family, consists of the six elements (starting with lithium) in the first column at the left of the table. Each element (E) in this family is a highly reactive metal that forms an ECl chloride and E_2O oxide. By contrast, the helium family, at the right of the table, consists of very unreactive, and even inert, elements. Of the helium family, also called the **noble gas** family, only xenon (Xe) and krypton (Kr) are known to form compounds under normal conditions.

The arrangement of elements in the Periodic Table provides an orderly summary of the key characteristics of each element. By knowing the major properties of a certain chemical family, some of the behavior of any element in that family can be predicted. This can be very helpful in evaluating elements for certain uses.

CD-ROM WWW. **Trends & Variation in the Periodic Table**

Like sodium chloride (NaCl), all chlorides and oxides of lithium-family elements are ionic compounds.

PREDICTING PROPERTIES

Building Skills 3

Some of an element's properties can be estimated by averaging the respective properties of the elements located just above and just below it. This is how Mendeleev predicted the properties of several elements unknown in his time. He was so convinced of the existence of these elements that he left gaps in the periodic table for them, along with some of their predicted properties. When those elements were eventually discovered, they fit exactly as expected. Mendeleev's fame rests largely on the accuracy of these predictions.

For example, germanium (Ge) was not known when Mendeleev proposed his periodic table. However, in 1871 he predicted the existence of germanium, calling it ekasilicon. Given that the boiling points of silicon (Si) and tin (Sn) are 3267 °C and 2603 °C, respectively, the boiling point of germanium can be estimated.

Eka- means "standing next in order."

The three elements are in the same group in the periodic table. Germanium is below silicon and above tin. (You can verify this by locating these elements in the Periodic Table.) Calculating the average of the boiling points of silicon and tin gives

$$\frac{(3267\ °C)\ 1\ (2603\ °C)}{2}\ 5\ 2935\ °C$$

When germanium was discovered in 1886, its boiling point was found to be 2834 °C. The estimated boiling point of germanium, 2935 °C, is within 4% of its known boiling point. The periodic table helped guide Mendeleev (and now you) to a useful prediction.

Formulas for chemical compounds can also be predicted from relationships established in the Periodic Table. For example, carbon and oxygen form carbon dioxide (CO_2). What formula would be predicted for a compound of carbon and sulfur? The Periodic Table indicates that sulfur (S) and oxygen (O) are in the same family. Knowing that carbon and oxygen form CO_2, a logical—and correct—prediction would be CS_2 (carbon disulfide). Now it's your turn.

PREDICTING PROPERTIES

Building Skills 3

This activity demonstrates the usefulness of the Periodic Table in estimating properties of substances. Knowing the major properties of a chemical family allows one to predict (via interpolation and extrapolation) the likely behavior of elements in that family.

For further discussion, the properties of Sc and Ga may be compared to Mendeleev's prediction.

Display an element from various groups on the Periodic Table. Ask students to predict the properties of the other elements in that group.

Answers

1. Krypton has a predicted melting point (between −186 °C and −112 °C) of −149 °C (actual is 153 °C).
2. a. Rubidium has a predicted melting point (between 337 K and 302 K) of 320 K (actual melting point is 312 K).

   b. Sodium has a higher melting point than rubidium and potassium, at 97.8 °C.
3. $GeCl_4$
4. a. CF_4
   b. Al_2S_3
   c. KCl
   d. $CaBr_2$
   e. SrO

A.7 WHAT DETERMINES PROPERTIES?

The different numbers and arrangement of the electrons explain the reactivity among metals. The more reactive a metal, the greater its tendency to give up electrons. Students do not yet have the tools to understand much about electron structures, but they should begin to build the understanding that there is a relationship between structure and properties.

1. The element krypton (Kr) was not known in Mendeleev's time. Given that the boiling point of argon (Ar) is −186 °C and of xenon (Xe) −112 °C, estimate the boiling point of krypton.
2. a. Estimate the melting point of rubidium (Rb). The melting points of potassium (K) and cesium (Cs) are 337 K and 302 K, respectively.

   b. Would you expect the melting point of sodium (Na) to be higher or lower than that of rubidium (Rb)? Explain.
3. Mendeleev knew that silicon tetrachloride ($SiCl_4$) existed. Using his periodic table, he correctly predicted the existence of ekasilicon, an element just below silicon in the Periodic Table. Predict the formula for the compound formed by Mendeleev's ekasilicon and chlorine.
4. Here are formulas for several known compounds: NaI, $MgCl_2$, CaO, Al_2O_3, and CCl_4. Using that information, predict the formula for a compound formed from
   a. C and F.
   b. Al and S.
   c. K and Cl.
   d. Ca and Br.
   e. Sr and O.

A.7 WHAT DETERMINES PROPERTIES?

Properties of Metals (CD-ROM, WWW.)

What is responsible for differences in the number of chlorine atoms that react with a given atom—or in other properties that vary from element to element? Recall that atoms of different elements have different numbers of protons (atomic numbers). Therefore, atoms of different elements also have different numbers of electrons. Many properties of elements are determined largely by the number of electrons in their atoms and how these electrons are arranged.

You will learn more about electron arrangements in Unit 3.

A major difference between atoms of metals and nonmetals is that metal atoms lose some of their electrons much more easily than nonmetal atoms do. Under suitable conditions, one or more outer electrons of metal atoms may transfer to other atoms or ions. This is why metallic elements tend to form positive ions (cations).

Some physical properties of metals depend on attractions among their atoms. For example, stronger attractions among atoms of a metal result in higher melting points. The melting point of magnesium is 651 °C, whereas that of sodium is 98 °C. Thus attractions among the atoms in magnesium metal must be stronger than those in sodium metal.

Chemical and physical properties of nonmetals and compounds are also explained by the makeup of their atoms, ions, or molecules and by attractions among these particles. As you learned in Unit 1, the abnormally high melting and boiling points of water are due to the strong attractions among water molecules.

Several thousand new compounds are synthesized each year.

Understanding properties of atoms is the key to predicting and even manipulating the behavior of materials. Combined with a bit of imagination, this information allows chemists to find new uses for materials and to create new chemical compounds to meet specific needs.

A.8 IT'S ONLY MONEY

Making Decisions

Based on what you have learned so far, you can start to make some decisions about your design for the new coin. A good first step is to specify some characteristics that are necessary or desirable in the material you will use. Apply your knowledge of existing coins, as well as what you have learned about properties of elements, to answer these questions.

1. What physical properties must material chosen for the new coin possess?
2. What other physical properties are desirable?
3. What chemical properties are required of the coin's material?
4. What other chemical properties are desirable?
5. Which would make the best primary material for the new coin: a metal, nonmetal, or metalloid? Explain.
6. What assumptions did you make in order to answer the preceding questions?

Save your answers to these questions—they will help guide your coin-design work later in this unit.

CD-ROM WWW. Questions & Answers

A.8 IT'S ONLY MONEY

Making Decisions

Students should work in small groups to begin making some design decisions about the new coin that they will propose. Their answers to a series of questions given in the text can be the basis for your evaluation of each group. Some possible answers might be:

1. High melting point, cost, conductivity, malleable, identifiable color
2. Size to work in coin-op machines. Pleasing design
3. Resistance to corrosion, able to make alloys, bonds with composites
4. Compatible with potential coatings
5. Answers will vary. Solid metals will tend to be favored. Gaseous nonmetals are hardly feasible, and other nonmetals tend to lack a compatible structure. Metalloids tend to have at least one poor property for use as primary substance in a coin.

One idea for managing the data collected in answering these questions is to have students make a two-by-two matrix with "physical" and "chemical" properties on one side and "necessary" and "desirable" on the other (see sample chart below).

A recommended table to list characteristics for coin materials:

Properties	Necessary	Desirable
Physical		
Chemical		

The Elements (Values in parentheses are the mass numbers of the longest-lived isotopes.)

Element	Symbol	Atomic Number	Atomic Weight	Element	Symbol	Atomic Number	Atomic Weight
Actinium	Ac	89	(227)	Neodymium	Nd	60	144.24
Aluminum	Al	13	26.98	Neon	Ne	10	20.18
Americium	Am	95	(243)	Neptunium	Np	93	(237)
Antimony	Sb	51	121.76	Nickel	Ni	28	58.69
Argon	Ar	18	39.95	Niobium	Nb	41	92.91
Arsenic	As	33	74.92	Nitrogen	N	7	14.01
Astatine	At	85	(210)	Nobelium	No	102	(259)
Barium	Ba	56	137.33	Osmium	Os	76	190.23
Berkelium	Bk	97	(247)	Oxygen	O	8	16.00
Beryllium	Be	4	9.01	Palladium	Pd	46	106.42
Bismuth	Bi	83	208.98	Phosphorus	P	15	30.97
Bohrium	Bh	107	(264)	Platinum	Pt	78	195.08
Boron	B	5	10.81	Plutonium	Pu	94	(244)
Bromine	Br	35	79.90	Polonium	Po	84	(209)
Cadmium	Cd	48	112.41	Potassium	K	19	39.10
Calcium	Ca	20	40.08	Praseodymium	Pr	59	140.91
Californium	Cf	98	(251)	Promethium	Pm	61	(145)
Carbon	C	6	12.01	Protactinium	Pa	91	231.04
Cerium	Ce	58	140.12	Radium	Ra	88	(226)
Cesium	Cs	55	132.91	Radon	Rn	86	(222)
Chlorine	Cl	17	35.45	Rhenium	Re	75	186.21
Chromium	Cr	24	52.00	Rhodium	Rh	45	102.91
Cobalt	Co	27	58.93	Rubidium	Rb	37	85.47
Copper	Cu	29	63.55	Ruthenium	Ru	44	101.07
Curium	Cm	96	(247)	Rutherfordium	Rf	104	(261)
Dubnium	Db	105	(262)	Samarium	Sm	62	150.36
Dysprosium	Dy	66	162.50	Scandium	Sc	21	44.96
Einsteinium	Es	99	(252)	Seaborgium	Sg	106	(263)
Erbium	Er	68	167.26	Selenium	Se	34	78.96
Europium	Eu	63	151.96	Silicon	Si	14	28.09
Fermium	Fm	100	(257)	Silver	Ag	47	107.87
Fluorine	F	9	19.00	Sodium	Na	11	22.99
Francium	Fr	87	(223)	Strontium	Sr	38	87.62
Gadolinium	Gd	64	157.25	Sulfur	S	16	32.07
Gallium	Ga	31	69.72	Tantalum	Ta	73	180.95
Germanium	Ge	32	72.61	Technetium	Tc	43	(98)
Gold	Au	79	196.97	Tellurium	Te	52	127.60
Hafnium	Hf	72	178.49	Terbium	Tb	65	158.93
Hassium	Hs	108	(265)	Thallium	Tl	81	204.38
Helium	He	2	4.003	Thorium	Th	90	232.04
Holmium	Ho	67	164.93	Thulium	Tm	69	168.93
Hydrogen	H	1	1.008	Tin	Sn	50	118.71
Indium	In	49	114.82	Titanium	Ti	22	47.87
Iodine	I	53	126.90	Tungsten	W	74	183.84
Iridium	Ir	77	192.22	Ununnilium	Uun	110	(269)
Iron	Fe	26	55.85	Unununium	Uuu	111	(272)
Krypton	Kr	36	83.80	Ununbium	Uub	112	(277)
Lanthanum	La	57	138.91	Ununquadium	Uuq	114	(285)
Lawrencium	Lr	103	(262)	Ununhexium	Uuh	116	(289)
Lead	Pb	82	207.2	Ununoctium	Uuo	118	(293)
Lithium	Li	3	6.94	Uranium	U	92	238.03
Lutetium	Lu	71	174.97	Vanadium	V	23	50.94
Magnesium	Mg	12	24.31	Xenon	Xe	54	131.29
Manganese	Mn	25	54.94	Ytterbium	Yb	70	173.04
Meitnerium	Mt	109	(268)	Yttrium	Y	39	88.91
Mendelevium	Md	101	(258)	Zinc	Zn	30	65.39
Mercury	Hg	80	200.59	Zirconium	Zr	40	91.22
Molybdenum	Mo	42	95.94				

108

SECTION SUMMARY

Reviewing the Concepts

1. Classify each as a chemical or physical property:

a. Copper has a reddish brown color.

b. Iron may rust when left outdoors.

c. Carbon dioxide gas can extinguish a flame.

d. Molasses pours more slowly than water.

A and d are physical properties; b and c are chemical properties.

2. Classify each as a chemical or physical change:

a. A candle burns.

b. A carbonated beverage fizzes when the container is opened.

c. Hair curls as a result of a "perm."

d. As shoes wear out, holes appear in the soles.

A and c are chemical changes; b and d are physical changes.

3. **a.** List the steps you would complete in making chocolate-chip cookies.

b. Classify each step as involving either a chemical or a physical change.

Step 1: Measure and mix 3/4 c granulated sugar, 3/4 c packed brown sugar, 1 c margarine or butter, and 1 egg;
Step 2: Stir in 2 1/4 c flour, 1 tsp. baking soda, 1/2 tsp. salt;
Step 3: Stir in 2 c chopped nuts and one 12 oz. package chocolate chips;
Step 4: Drop by teaspoonfuls on ungreased cookie sheet and bake 8–10 minutes until light brown.

Steps 1–3 are physical changes; Step 4 is a chemical change.

4. Classify each property as characteristic of metals, nonmetals, or metalloids.

a. Shiny in appearance

b. Does not react with acids

c. Shatters easily

d. Dull in appearance but electrically conductive

A is metallic; b and c are nonmetallic, d is metalloid.

5. Classify each element as a metal, nonmetal, or metalloid.

a. tungsten

b. antimony

c. krypton

d. sodium

A and d are metals; b is a metalloid, and c is a nonmetal.

6. What would you expect to happen if you hit a sample of each of these elements with a hammer?

a. iodine

b. zirconium

c. phosphorus

d. nickel

A and c would break into pieces (nonmetals); b and d would flatten into sheets (metals).

7. What properties make nonmetals unsuitable for electric wiring?

Their brittleness and nonconductivity make them unsuitable.

8. Given the formulas AlN and $BeCl_2$, predict the formula for a compound containing

a. Mg and F **a. MgF_2**

b. Ga and P **b. GaP**

9. The melting points of oxygen and selenium are −218 °C and 221 °C, respectively. Estimate the melting point of sulfur.

1.5 °C (actual mp is 115.2 °C)

10. Would you expect the boiling point of chlorine to be higher or lower than that of iodine? Explain.

The boiling point of chlorine is lower because chlorine molecules have less mass and the forces of attraction between them (Van der Waals or dispersion) are less. Boiling is a function of changing the kinetic energy of a substance, Kinetic energy = $1/2mv^2$. Therefore, the substance with less mass would require less energy to boil. Also, iodine is closer to the metallic elements tending to be less non-metallic, or having a higher boiling point.

11. For medical reasons, people with high blood pressure are advised to limit the amount of sodium ions in their diet. Normal table salt (NaCl) is sometimes replaced by a substitute called Lite Salt (potassium chloride) for seasoning.

a. Write the formula for Lite Salt.

b. Why are the properties of Lite Salt similar to those of sodium chloride?

a. KCl

b. Sodium and potassium are in the alkali metals family and have similar chemical and physical properties. (Note: 0.396 g of Na per gram NaCl; 0.524 g of K per gram of KCl.)

12. Complete the table to the right for each electrically neutral atom.

Name	Protons	Neutrons	Electrons
a. ^{12}C	6	6	6
b. ^{13}C	6	7	6
^{41}Ca	c. 20	21	d. 20
e. Pt	f. 78	117	78
U	g. 92	146	h. 92

13. A student is asked to explain the formation of a lead(II) ion (Pb^{2+}) from an electrically neutral lead atom (Pb). The student says that a lead atom must have gained two protons to make the ion. How would you correct this student's explanation?

Two electrons have been lost. Electrons can be gained or lost by an elemental atom. If protons are gained or lost, the atom is that of a different element.

14. Supply the numbers of protons and neutrons for each of the isotopes in the table on the right.

Element	Mass Number	Number of Protons	Number of Neutrons
Mg	24	a. 12	b. 12
Mg	25	c. 12	d. 13
Mg	26	e. 12	f. 14

15. A scientist announces the discovery of a new element. The only characteristic given in the report is the element's mass number of 266. Is this information sufficient, by itself, to justify the claim of the discovery of a new element? Explain.

No, because the mass number does not identify the element. It would require that the atomic number be known.

16. Predict whether each element would be more likely to form an anion or a cation.

a.	sodium	a.	**cation**
b.	calcium	b.	**cation**
c.	fluorine	c.	**anion**
d.	oxygen	d.	**anion**
e.	lithium	e.	**cation**
f.	iodine	f.	**anion**

17. Noble gas elements rarely lose or gain electrons. What does this indicate about their chemical reactivity?

They are highly nonreactive.

Connecting the Concepts

18. Which pair is more similar chemically? Defend your choice.

a. copper metal and copper(II) ions

or

b. oxygen with mass number 16 and oxygen with mass number 18

Oxygen-16 and oxygen-18, because they are isotopes, having the same number of bonding electrons.

19. Three kinds of observations that may indicate a chemical change are listed below. However, a physical change may also result in each observation. Describe a possible chemical cause *and* a possible physical cause for each observation.

a. change in color

b. change in temperature

c. formation of a gas

a. physical: a wire becomes red hot; chemical: red litmus turns blue

b. physical: cooling due to evaporation of a liquid; chemical: burning candle

c. physical: carbon dioxide coming out of solution from a carbonated drink; chemical: Alka Selzer fizzing in water

20. The diameter of a magnesium ion (Mg^{2+}) is 156 pm (picometers, where 1 pm $=10^{-12}$ m); the diameter of a strontium ion (Sr^{2+}) is 254 pm. Estimate the diameter of a calcium ion (Ca^{2+}).

200 picometers (average of Mg and Sr diameters; actual is 187.6 pm)

21. Identify the element in the Periodic Table described by each statement:

a. This element is a member of a group of nonmetals. It forms anions with a -1 charge. It is in the same period as the metals used to form a penny.

b. This element is a metalloid. It is in the same period as the elements found in table salt.

a. bromine

b. silicon

22. Mendeleev arranged elements in his periodic table in order of their atomic weights. In the modern Periodic Table, however, elements are arranged in order of their atomic numbers. Cite two examples from the Periodic Table for which these two schemes would produce a different ordering of adjacent elements.

Ar (39.95) to K (39.10); Co (58.69) to Ni (58.69); Te (127.60) to I (126.90)

Extending the Concepts

23. How is mercury different from other metallic elements? Using outside resources, describe some applications that take advantage of the unique properties of metallic mercury.

Mercury is extremely dense and is the only metal that is a liquid at room temperature. It is used in mercury barometers, thermometers, manometers, electric relays, electrodes for certain batteries, and dental amalgams. Mercury is very toxic as a vapor and in many compounds, yet relatively safe in others.

24. Depending on how it is heated and cooled, iron can either be hard and brittle or malleable. Explain how the same metal can have both characteristics.

The rate of heating and cooling the iron determines the arrangement of atoms. This in turn determines its brittleness or malleability.

25. Construct a graph of the price per gram of an element versus its atomic number for each of the first twenty elements. Can the current cost of those elements be regarded as a periodic property? Explain. (*Hint*: Use a chemical supply catalog or the Web to locate the current price of each element.)

The cost per gram is not a periodic property. From the Flinn Chemical Catalog, the approximate cost per gram of a few substances is: Na ($ .42); Mg ($ .15); Al ($ .58); Si ($ .03); P ($1.90); S ($ .01).

26. Classify one or more pieces of jewelry you might possibly wear as being composed of metals, nonmetals, or metalloids.

Jewelry is made of metal because it is malleable; it can be drawn into wire and hammered into shape. Alloys are resistant to corrosion.

EARTH'S MINERAL RESOURCES

Introduction

Part B begins with an exploration of the chemical supplies of Earth, where they are located, and how society uses them. Copper is used as the representative metal for this section. The properties of copper that make it suitable for a particular use (coins) are discussed. The techniques used to refine a metal from its ore are explained with supporting lab activities. Oxidation-Reduction reactions (redox) are introduced to explain how metals are reduced from ores.

B.1 SOURCES AND USES OF METALS

Mineral resources are widely distributed among the nations of the world. Although the United States has a diversity of chemical resources, it must import much of its chemical resources to meet its needs. Since Earth's resources are limited, several factors are considered before it is decided to mine a particular metallic ore.

Use the chart *Worldwide Annual Production of Selected Metals* (Figure 10, p. 114) to discuss the need for trade among nations. (Example: Australia may trade aluminum with the United States for copper or lead.)

SECTION B

EARTH'S MINERAL RESOURCES

Among Earth's resources, metals—and the minerals from which they are extracted—have had long-standing importance for and use by humans. Those uses have ranged from toolmaking, energy transmission, and construction to works of art, decoration, and coin making. In this section, you will explore the properties and uses of minerals and metals. Using copper as a case study, you will learn about Earth's mineral resources and how they are converted to pure metals.

B.1 SOURCES AND USES OF METALS

Earth's Mineral Resources CD-ROM WWW.

Human needs for resources—whether to create a new coin, manufacture new clothing, construct a space-vehicle launch rocket, or supply fertilizer for food crops—must all be met by chemical supplies currently present on Earth. These supplies of resources are often cataloged by where they are found. The table in Figure 9 indicates the composition of Earth.

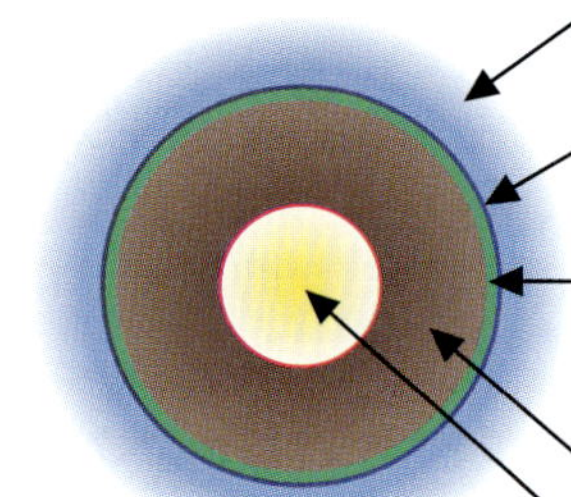

Earth's Composition

Layer of Planet	Thickness (Average)	Composition (Decreasing Order of Abundance)
Atmosphere	100 km	N_2 (78%), O_2 (21%), Ar (0.9%), He + Ne (<0.01%), variable amounts of H_2O, CO_2, etc
Hydrosphere	5 km	H_2O, and in ocean water approximately 3.5% NaCl and smaller amounts of Mg, S, Ca, and other elements as ions
Lithosphere: Crust	6400 km: Top 40 km	Silicates (compounds of metals, Si, and O atoms). Metals include Al, Na, Fe, Ca, Mg, K, and others Coal, oil, and natural gas Carbonates such as $CaCO_3$ Oxides such as Fe_2O_3 Sulfides such as PbS
Mantle	40–2900 km	Silicates of Mg and Fe
Core	2900 km to Earth's center	Fe and Ni

Figure 9 *Table of Earth's composition.*

Earth's atmosphere, hydrosphere, and outer layer of the lithosphere (the solid part of Earth) supply all resources for all human activities. The atmosphere provides nitrogen, oxygen, neon, argon, and a few other gases. From the hydrosphere come water and some dissolved minerals. The lithosphere provides the greatest variety of chemical resources. For example, petroleum and metal-bearing ores are found there. An **ore** is a naturally occurring rock or mineral that can be mined and from which it is profitable to extract a metal or other material. An ore contains a mixture of components. Of these, the most important are **minerals**—solid compounds containing the element or group of elements of interest.

The deepest mines on Earth barely scratch the surface of its crust. If Earth were the size of an apple, all accessible resources of the lithosphere would be located within the apple's skin. From this thin band of soil and rock, we obtain the major raw materials needed to build homes, automobiles, appliances, computers, videotapes, compact discs, and sports equipment—in fact, all manufactured objects.

As you can see from the table in Figure 10 on page 114, many of Earth's important resources are not uniformly distributed. There is no connection between a nation's supply of these resources and either its land area or its population. Quite often a particular region serves as the predominant supplier of certain metals vitally important to industry. For example, Africa holds much of the world's known reserves of chromium (80%), cobalt (54%), and manganese (61%).

The development of the United States as a major industrial nation has been facilitated, in part, because of the quantity and diversity of chemical resources found here. Yet in recent years, the United States has imported increasing amounts of certain vital chemical resources. For example, about 75% of the nation's tin (Sn) is imported.

The greatest challenge regarding mineral resources is deciding on the wisest uses of the available supplies. Immediate issues include some that have technical and economic implications. For example, is it worthwhile to mine a particular metallic ore at a certain site? The answer depends on several factors:

- amount of useful ore at the site
- percent of metal in the ore
- type of mining and processing needed to extract the metal from its ore
- distance of the mine from metal refining facility and markets
- metal's supply-versus-demand status

Copper, one of the materials you might be thinking of using in your coin design, provides a case study of a vital chemical resource. You will first consider worldwide sources of copper and how these copper-bearing materials are converted to pure copper. Later, you will explore some possible replacements for this resource.

Copper is one of the most familiar and widely used metals in modern society. Among all the elements, it is second only to silver in electrical conductivity. This property and its relatively low cost, corrosion resistance, and ductility (ease of being drawn into thin wires), make copper the world's most common metal for electrical wiring. Copper is also used to produce

Metal	Country	Percent production	Actual production (1000 metric tons)	World total production (1000 metric tons)	Year
Aluminum	United States	17%	3713	22 100	1998
	Russia	14%	3005		
	Canada	11%	2374		
	China	10%	2100		
	Australia	7%	1627		
Copper†	United States	15%	1720	11 100	1997
	Chile	13%	1389		
	Japan	12%	1350		
	China	9%	963		
	Russia	5%	600		
Iron ore○	United States	22%	65 900	305 300	1997
	Brazil	12%	37 300		
	Russia	11%	34 000		
	Ukraine	10%	32 000		
	Canada	9%	27 300		
Lead□	United States	22%	1450	5880	1998
	United Kingdom	6%	350		
	Germany	6%	335		
	France	5%	306		
	Japan	5%	302		
Nickel△	Russia	23%	260	1120	1997
	Canada	17%	190		
	New Caledonia	12%	137		
	Australia	11%	124		
	Indonesia	6%	72		
Silver△	Mexico	16%	2.7	16.4	1998
	United States	13%	2.1		
	Peru	12%	1.9		
	Australia	9%	1.5		
	China	9%	1.4		
Tin†	China	29%	61	213	1997
	Indonesia	19%	40		
	Malaysia	17%	36		
	Brazil	9%	19		
	Bolivia	8%	16		
Zinc†	China	18%	1500	3890	1997
	Canada	9%	743		
	Japan	8%	653		
	Republic of Korea	5%	390		
	Spain	5%	370		

† = world smelter production ○ = world pelletizing capacity □ = world refinery production △ = world mine production

Figure 10 *Production of selected metals worldwide.*

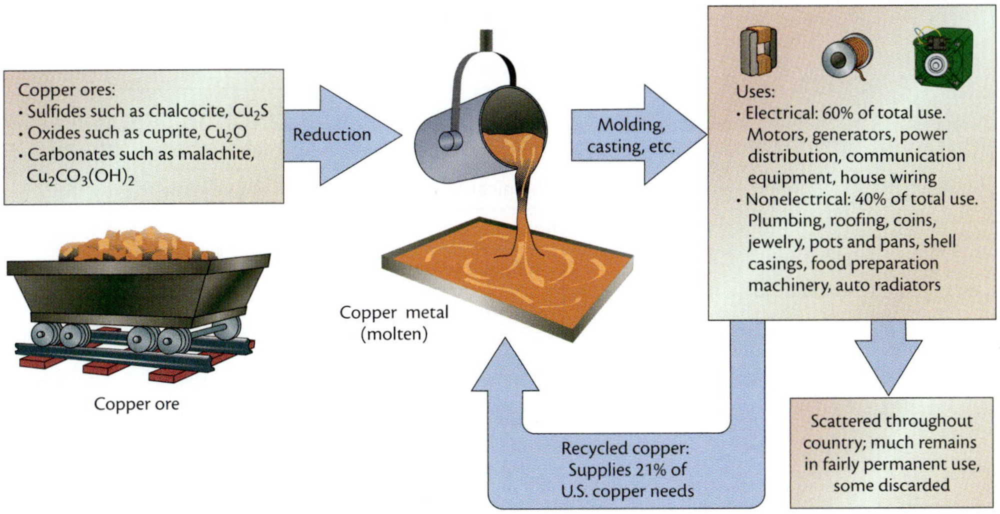

Figure 11 *The copper cycle and copper uses.*

brass, bronze, and other alloys, a variety of important copper-based compounds, jewelry, and works of art.

Alloys are discussed in more detail in Section D.

The first copper ores mined contained from 35% to 88% copper. Although such ores are no longer available, ores less rich in copper can be used. In fact, it is now economically possible to mine ores containing less than 1% copper. Copper ore is chemically processed to produce metallic copper, which is then transformed into a variety of useful materials. Figure 11 summarizes the copper cycle from sources to common uses to waste products.

Earth's accessible deposits of this valuable resource are destined to be depleted. Will future developments increase or decrease the need for copper? What copper substitutes are available? The following activity will help you address these questions.

PROPERTIES AND USES OF COPPER

Building Skills 4

Some of copper's properties are listed in the table in Figure 12 (page 116). Consider how these properties make copper suitable for uses depicted in Figure 11. For example, what properties make copper useful in electrical power generators? Copper's high electrical conductivity is certainly essential to this application. Copper's malleability and ductility are also important, making it possible to form copper wires and to wrap them in a generator. Corrosion resistance is also a benefit in such large, expensive equipment or in other applications.

PROPERTIES AND USES OF COPPER

Building Skills 4

Students connect copper's properties with its uses, and propose technological changes that could alter the use of copper.

Answers

1. Electrical uses: low reactivity, high electrical conductivity, malleability, and ductility; Non-electrical uses: attractive color (jewelry), malleability (anything requiring shaping), low reactivity and high thermal conductivity (pots and pans)
2. a. Increased recycling would decrease our need for imports (the United States imports 13% of its copper from Canada, Chile, Peru, and Zambia) and decrease the need for mining and refining lower-quality ores.
   b. Recycling has a limit. Not all copper can be recycled. House wiring usually must stay in place for the lifetime of the home, and the same is true for plumbing and roofing. The benefits of recycling usually reach a maximum in terms of economic profit before maximum recycling is reached, which reduces the impetus to improve recycling.
3. a. Removing pennies from circulation, decreasing the copper in other coins, and using credit cards would lower copper usage.
   b. Fiber optics, satellites, and cellular phones decrease the need for copper in communications wiring.
   c. More use of solar power could reduce power generation by turbines, reducing copper used in the wiring of the turbines.
   d. New conducting alloys and aluminum may decrease the demand for copper.

Properties of Copper	
Malleability and ductility	High
Electrical conductivity	High
Thermal conductivity	High
Chemical reactivity	Relatively low
Resistance to corrosion	High
Useful alloys formed	Bronze, brass, etc.
Color and luster	Reddish, shiny

Figure 12 *Properties of copper.*

1. Consider the remaining uses of copper listed in Figure 11. For each use, identify those particular properties that make copper an appropriate choice.
2. a. How would increased recycling of scrap copper affect future availability of this metal?
   b. Is there a limit to the role copper recycling can play? Why?
3. For each use listed below, describe a technological change that could decrease the demand for copper:
   a. coins
   b. communications
   c. power generation
   d. indoor electrical wiring

B.2 CONVERTING COPPER

Laboratory Activity

Introduction

You have seen many chemical reactions in your lifetime. Some, such as a fireworks display, are memorable. Others, such as the slow process of rusting, are far less dramatic. Have you ever stopped to think about what happens to the atoms involved in those reactions? Are the materials that made up the fireworks still there after they are launched into the sky and ignited? What about the iron that turns into rust?

In this laboratory activity, you will work with a powdered sample of elemental copper, a metal you may be considering for your coin design. As you observe its chemical behavior, think about whether its properties make it a good candidate for this use.

Procedure

1. Prepare a data table to record the masses you will determine in Steps 2 and 9.
2. Measure and record the mass of a clean, empty crucible. Add approximately 1 g copper powder to the crucible. Record the mass of the crucible with copper powder in it within the nearest 0.1 g. Find

B.2 CONVERTING COPPER

Laboratory Activity

Time

30 minutes

Materials (for a class of 24 students working in pairs)

12 crucibles
12 g copper powder
12 clay triangles
12 burners
12 spatulas

Safety

Students should wear safety goggles and aprons throughout the laboratory activity.

Students should avoid touching the hot crucible with their hands.

the actual mass of copper powder by subtracting the mass of the empty crucible from this value. Record the mass of copper powder.

3. Which properties of copper can be directly observed? Record your observations of the copper powder.
4. Set up the crucible, clay triangle, and burner as shown in Figure 13. The crucible lid should be slightly ajar.
5. Light the burner, and adjust it so that the flame tip just touches the bottom of the crucible.
6. Heat the crucible and its contents for two minutes. Remove the flame, and use a spatula to break up the solid in the crucible so that as much remaining copper metal is exposed as possible.

   CAUTION: *Avoid touching the hot crucible.*
7. Continue heating for about 10 minutes more, removing the flame and breaking up the solid every two to three minutes.
8. When you have finished the heating, extinguish the burner flame and allow the crucible and its contents to cool to room temperature. Answer Questions 1 and 2 while you are waiting.
9. After the crucible and its contents have cooled, find their mass. Use this value and the mass of the empty crucible to calculate the mass of the contents. Record these values in your data table.
10. Transfer your product to a clean 100-mL beaker. Label and store the beaker and product as indicated by your teacher.
11. Put away the other materials. Wash your hands thoroughly before leaving the laboratory.

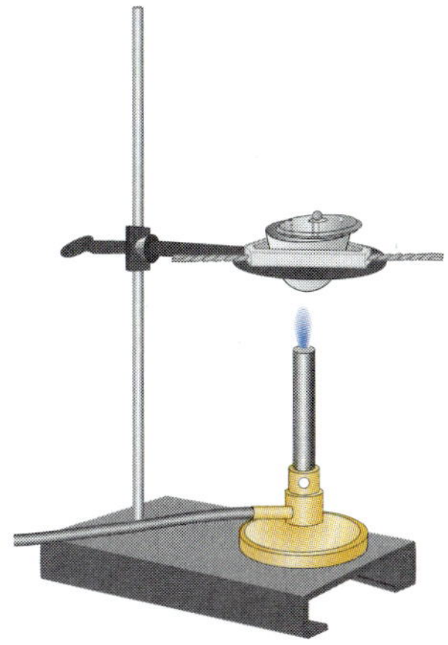

Figure 13 *Clay triangle holding a crucible over a burner.*

Questions

1. a. Were the changes you observed physical or chemical?
   b. How do you know?
2. a. Describe the changes you observed as you heated the copper.
   b. Did the copper atoms remain in the crucible? Explain.
3. a. What happened to the mass of the crucible contents after you heated the copper?
   b. Why do you think the mass changed?

B.3 METAL REACTIVITY

As you just observed, when copper metal is heated, it gradually reacts with oxygen in the air to produce a black substance. This is the equation for the reaction:

$$\underset{\text{Copper}}{2\,Cu(s)} + \underset{\text{Oxygen}}{O_2(g)} \longrightarrow \underset{\text{Copper(II) oxide}}{2\,CuO(s)}$$

Although copper reacts to form copper(II) oxide when heated, at room temperature the metal remains relatively unreactive in air. You are probably familiar with this fact from observing that copper wire and the copper surface on pennies do not turn black under normal conditions.

There are two common compounds of copper and oxygen—CuO and Cu_2O. Because the name "copper oxide" could be applied to both, a Roman numeral is added to indicate copper's ionic charge. Copper(I) oxide is Cu_2O because it contains Cu^+ ions; copper(II) oxide is CuO because it contains Cu^{2+} ions.

Answers

1. a. Chemical

   b. New product formed with different properties
2. a. Changed from reddish brown to a black powder

   b. Yes. Matter is neither created nor destroyed during a chemical reaction.
3. a. Mass increased

   b. Copper reacted with oxygen from the air

Post-lab discussion

To convince students that the powdered copper is actually copper metal despite its dull appearance, it is possible to "burnish" a sample of powdered copper to show its luster. With the flat surface of a metal spatula, press a small mound of powdered copper down on a piece of paper. Rub gently, then more firmly with the spatula blade. The characteristic shiny, metallic luster of copper will appear.

B.3 METAL REACTIVITY

We can compare the reactivities of elements by observing the relative ease with which different metals combine with oxygen. Gold, iron, and magnesium are discussed as examples of how a metal's reactivity help determine its utility for a specific purpose.

Lab tip

Check the quality of your supply of copper. Older supplies may have become oxidized over time. If the copper powder appears black, it is likely that it has oxidized already and should be replaced by a fresher supply for this activity. Copper turnings are not recommended as a substitute for the copper powder in this activity. Copper powder produces a more complete reaction in less time.

Save the CuO (black powder) for use in Laboratory Activity C.5.

B.4 RELATIVE REACTIVITIES OF METALS

Laboratory Activity

This activity emphasizes that different metals have varying degrees of reactivity. Students are to rank copper, magnesium, zinc, and silver according to observed chemical reactivities.

Time

50 minutes

Materials (for 24 students working in pairs)

12 24-well wellplates
48 pipets
60-mm magnesium ribbon cut into 5-mm lengths
60-mm zinc wire cut into 5-mm lengths
60-mm copper wire cut into 5-mm lengths
Silver metal (optional)
10 mL 0.2 M $Mg(NO_3)_2$ solution ($5.1\ g\ Mg(NO_3)_2 \cdot 6\ H_2O/100.0$ mL solution)
10 mL 0.2 M $Cu(NO_3)_2$ solution ($4.8\ g\ Cu(NO_3)_2 \cdot 3\ H_2O/100.0$ mL solution)
10 mL 0.2 M $Zn(NO_3)_2$ solution ($6.0\ g\ Zn(NO_3)_2 \cdot 6\ H_2O/100.0$ mL solution)
10 mL 0.2 M $AgNO_3$ solution ($3.4\ g\ AgNO_3/100.0$ mL solution)

Safety

Students should wear safety goggles and aprons throughout the laboratory activity. Caution students that silver nitrate solution stains skin (and clothing) and that direct contact with it should be avoided.

Keep a bottle of Spray & Wash in the lab to help remove any Ag^+ that may contact student hands.

Pre lab discussion

Observing the reactions of metals with other metallic solutions permits comparison of the reactivity of the metals. The more reactive the metal, the more likely it is to react with the solutions. Evidence for reaction includes formation of a solid, appearance or disappearance of the color of the solution, formation of a gas, or changes in temperature. Writing equations for these types of displacement reactions is emphasized in Section C of this unit.

Lab tips

Record keeping is easier if the wellplate is set up like the data table. Remind students to record the initial appearance of the metal samples (after Step 2) in their data table. Encourage students to observe immediately after the metal is added to the solution and a few minutes later. The magnesium/zinc nitrate reaction is difficult to detect. Very careful observations are necessary. Steel wool may be used in place of sandpaper or emery cloth to remove the oxide coatings.

Figure 14 *Magnesium metal and oxygen gas react so spectacularly that small samples of magnesium are used in some fireworks.*

Magnesium metal also reacts with oxygen gas. But unlike copper metal, magnesium heated in air ignites and produces a brief, blinding flash. See Figure 14. This is the equation for the reaction:

$$\underset{\text{Magnesium}}{2\ Mg(s)} + \underset{\text{Oxygen}}{O_2(g)} \longrightarrow \underset{\text{Magnesium oxide}}{2\ MgO(s)}$$

By contrast, gold (Au) does not react with any components in air; in particular, it does not react with oxygen gas, even at elevated temperatures. This is one reason why gold metal is highly prized in long-lasting, decorative objects, such as jewelry. Gold-plated electrical contacts, such as those used for automobile air bags and audio cable connectors, are very dependable because nonconducting oxides do not form on the contact surfaces.

Observing how readily a certain metal reacts with oxygen provides information about the metal's reactivity. If elements are ranked in relative order of their chemical reactivity, the ranking is called an **activity series**. Based on what you have just learned about gold and magnesium and what you already know about copper, how would you rank the three metals in terms of their relative chemical reactivity?

B.4 RELATIVE REACTIVITIES OF METALS

Laboratory Activity

Introduction

In this activity, you will investigate the reactions of the metals copper, magnesium, and zinc with solutions that each contain a metal cation. The four solutions you will use are copper(II) nitrate, $Cu(NO_3)_2$ (containing Cu^{2+});

magnesium nitrate, $Mg(NO_3)_2$ (containing Mg^{2+}); zinc nitrate, $Zn(NO_3)_2$ (containing Zn^{2+}); and silver nitrate, $AgNO_3$ (containing Ag^+).

Procedure

1. Devise an orderly procedure that will allow you to observe the reaction (if any) between each metal and each of the four ionic solutions. You will conduct each reaction in a separate well of your wellplate, using five drops of 0.2 M solution and a small strip of metal. How many different combinations of metals and solutions will you need to observe? How will you arrange things so you can complete your observations efficiently yet remain certain which metal and which solution are in each well?
2. Prepare a data table to help you organize the observations and results of the procedure you devise.
3. Obtain 5-mm strips of each of the three metals to be tested. Clean the surface of each metal strip by rubbing it with sandpaper or emery paper. Record observations of each metal's appearance.
4. Complete your planned procedure, writing your observations in your data table. **CAUTION:** *Avoid letting the $AgNO_3$ solution come in contact with skin or clothing as it causes dark, non-washable stains.* If no reaction is observed, write NR in the table. Record the observed changes if a reaction occurs.
5. Dispose of your solid samples and wellplate solutions as directed by your teacher.
6. Wash your hands thoroughly before leaving the laboratory.

Questions

1. Which metal reacted with the most solutions?
2. Which metal reacted with the fewest solutions?
3. With which of the solutions (if any) would you expect silver metal to react, if it were available to be tested?
4. List the metals (including silver) in order, placing the most reactive metal first (the one reacting with the most solutions) and the least reactive metal last (the one reacting with the fewest solutions).
5. Refer to your "metal activity series" list in Question 4. Write a brief explanation of why the outside surface of a penny is made of copper instead of zinc.
6. a. Which of the four metals mentioned in this laboratory activity might be an even better choice than copper for the outside surface of a penny? Why?
   b. Why do you think that metal is not used for that purpose?
7. Given your new knowledge about the relative chemical activities of these four metals,
   a. which metal is most likely to be found in an uncombined, or "free," (metallic) state in nature?
   b. which metal is least likely to be found chemically uncombined with other elements?

The sample data table shows the possibility of 16 (4 x 4) combinations of metals and solutions to be investigated. Two comments regarding implications of the data table:

a. Results for reaction of silver metal, (Ag), with the solutions can be given to students if you judge that silver is too expensive or it is unavailable for you to use in this investigation. However, small pieces of Ag can be obtained from most jewelers at low cost. Thus students can either investigate the silver reactions for themselves, or the silver-based combinations can be set up as a demonstration for students to view. One benefit, of course, is that the silver metal can be reused from year to year.

b. Interactions involving a metal with its own metal-ion solution—such as $Cu_{(s)}$ with $Cu(NO_3)_{2(aq)}$—have intentionally been left open for investigation by students on the sample data table. Based upon your teaching approach, you can either say nothing about that (thus allowing student teams to try those combinations or decide for themselves that those combinations do not need to be investigated), or ask students in your pre-lab discussion whether all combinations on the data table need to be investigated, thus guiding them to understand that some step-saving is possible.

Answers

1. Magnesium
2. Silver
3. None; silver would react with those metallic solutions with lower reactivity, such as platinum and gold.

Expected Results

Metal	Appearance	$Cu(NO_3)_2$	$Mg(NO_3)_2$	$Zn(NO_3)_2$	$AgNO_3$
Cu	Shiny, copper	—	NR	NR	Becomes silvery
Mg	Shiny, silver	Darkens	—	Flaky coating	Becomes silvery
Zn	Shiny, silver	Darkens	NR	—	Becomes silvery
Ag	Shiny, silver	NR	NR	NR	—

NR = no reaction

4. Magnesium, zinc, copper, silver.
5. Copper was chosen because it is lower than zinc in the activity series. In the presence of water, zinc reacts more readily with oxygen to form oxides.
6. a. Silver would have been a better choice than copper because of its lower chemical reactivity.

   b. Silver is more expensive than copper.
7. a. Silver is most likely to be found in the uncombined state because of its low reactivity.

   b. Magnesium is least likely to be found in the uncombined state because of its high reactivity.
8. a. Yes. The metal in a solution of its ions could be eliminated and still obtain all information needed to create a chemical reactivity chart.

   b. Cu in $Cu(NO_3)_2(aq)$, Mg in Mg $(NO_3)_2(aq)$, etc.

Post-lab discussion

In the post-lab discussion, resolve differences in results among student pairs. When students disagree, ask them for evidence supporting their claims. If necessary, repeat a reaction as a demonstration to show the differences.

B.5 METALS: PROPERTIES AND USES

The discovery that fire could be used to extract a metal from its ore led to the transformation of metals into useful materials. Students may ask how natural fibers served as components of ancient tools. You can point out applications such as wooden spears and tool handles, as well as fiber-based cords used to bind stone-cutting heads to handles.

Relative Reactivities of Metals CD-ROM WWW.

8. Reconsider your experimental design for this activity.
   a. Would it have been possible to eliminate one or more of the metal-solution combinations and still obtain all information needed to create chemical activity ratings for the metals?
   b. If so, which combination(s) could have been eliminated? Why?

Figure 15 *These stone, bronze, and iron tools represent three major ages of civilization.*

B.5 METALS: PROPERTIES AND USES

Humans have been described as toolmakers. Readily available stone, wood, and natural fibers were the earliest materials used to make useful tools—hammers, chisels, knives, spears, and grinding devices. It was the discovery that fire could transform materials in certain rocks into strong, malleable metals, however, that triggered a dramatic leap in the growth of civilization.

Gold and silver, found as free elements rather than in chemical combination with other elements, were probably the first metals used by humans. These metals were formed into decorative objects and, later, into coins. Their relative unreactivity made them excellent materials for those uses.

It is estimated that copper has been used to make tools, weapons, utensils, and decorations for about 10 000 years. Bronze, an alloy of copper and tin, was developed about 3800 B.C. Thus humans moved from the Stone Age into the Bronze Age. See Figure 15.

Eventually early people developed iron metallurgy, the extraction of iron from its ores. This led to the start of the Iron Age, more than 3000 years ago. In time, as humans learned more about chemistry and fire, a variety of metallic ores were transformed into increasingly useful metals.

DISCOVERY OF METALS — ChemQuandary 1

Copper, gold, and silver are far from being the most abundant metals on Earth. Aluminum, iron, and calcium, for example, are all much more plentiful. Why, then, were copper, gold, and silver among the first metallic elements discovered?

You have explored some of the chemistry of metals and know, for example, that copper metal is more reactive than silver but less reactive than magnesium. A more complete activity series is given in the table in Figure 16. The table also includes brief descriptions of common methods for retrieving each metal from its ore.

In this list the most reactive metallic elements are at the top; less reactive elements are closer to the bottom. An activity list can be used to predict whether certain reactions can be expected. For example, you observed in the laboratory that zinc metal, which is more reactive than copper, reacted with copper ions in solution. However, zinc metal did not react with magnesium ions in solution. Why? Because zinc is less reactive than magnesium.

Students should understand that although most elements are found in nature, they are not found in the uncombined metallic state. Metals are usually found in the form of a metallic ore.

It may be beneficial to students if a sample of an ore as it is found in nature is displayed along with a sample of the element in its metallic state.

DISCOVERY OF METALS
ChemQuandary 1

The concept of an activity series is introduced and related to the processes used to obtain a metal from its ore. Due to the low activity of copper, silver and gold are found uncombined in nature.

Metal Activity Series			
Element	Metal Ion(s) Found in Minerals	Process Used to Obtain the Metal	State of Metal Obtained
Lithium Potassium Calcium Sodium Magnesium Aluminum	Li^+ K^+ Ca^{2+} Na^+ Mg^{2+} Al^{3+}	Pass direct electric current through the molten mineral salt (electrometallurgy)	Li(s) K(s) Ca(s) Na(s) Mg(s) Al(s)
Manganese Zinc Chromium Iron	Mn^{2+} Zn^{2+} Cr^{3+}, Cr^{2+} Fe^{3+}, Fe^{2+}	Heat mineral with coke (C) or carbon monoxide (CO) (pyrometallurgy)	Mn(s) Zn(s) Cr(s) Fe(s)
Lead Copper Mercury Silver Platinum Gold	Pb^{2+} Cu^{2+}, Cu^+ Hg^{2+} Ag^+ Pt^{2+} Au^{3+}, Au^+	Heat (roast) mineral in air (pyrometallurgy) or find the element free (uncombined)	Pb(s) Cu(s) Hg(l) Ag(s) Pt(s) Au(s)

Figure 16 *Metal activity series (in decreasing order of reactivity).*

In general, a more reactive metallic element (higher in the activity series) will cause ions of a less reactive metallic element (lower in the activity series) to change to their corresponding metal.

TRENDS IN METAL ACTIVITY

Building Skills 5

Use the table in Figure 16 and the Periodic Table (page 104) to answer the following questions.

1. a. What trend in metallic reactivity is found from left to right across a horizontal row (period) of the Periodic Table? (*Hint:* Compare the reactivities of sodium, magnesium, and aluminum.)
   b. In which part of the Periodic Table are the most-reactive metals found?
   c. Which part of the Periodic Table contains the least-reactive metals?
2. a. Will iron (Fe) metal react with a solution of lead(II) nitrate, $Pb(NO_3)_2$?
   b. Will platinum (Pt) metal react with a lead(II) nitrate solution?
   c. Explain your answers to Questions 2a and 2b.
3. Use specific examples from the activity series in your answers to these two questions:
   a. Are least-reactive metals also the cheapest metals?
   b. If not, what other factor(s) might influence the market value of a metal?

TRENDS IN METAL ACTIVITY

Building Skills 5

This activity familiarizes students with the relative reactivity of metals, which determines whether or not certain chemical reactions may occur. It also explores the energy needed to free metals from their ores.

Answers

1. a. Metallic reactivity generally decreases from left to right across a period of the Periodic Table.

   b. The most-reactive metals are located on the left side.

   c. The least-reactive metals, such as copper, silver, and gold, are in the transition metal section.
2. a. Yes

   b. No

   c. Iron is more reactive than lead, according to the metal activity series, and will reduce lead ions from a solution. Platinum is less reactive than lead.
3. a. The least-reactive metals are not the cheapest, despite having the lowest production costs.

   b. Other factors include demand for the metal (low reactivity is one of the attractive features of gold, silver, and copper—all three are long-lasting) and overall availability (gold, silver, and copper are much less abundant than aluminum and iron on Earth).

B.6 MINING AND REFINING

Oxidation and reduction are defined, and the main techniques for separating metals from their compounds in ores are presented.

B.6 MINING AND REFINING

The process of converting a combined metal (usually a metal ion) in a mineral to a free metal involves a particular kind of chemical change. For example, the conversion of a copper(II) cation to an atom of copper metal requires two electrons.

Formation of Copper Metal

$$Cu^{2+} + 2e^- \longrightarrow Cu$$

Cu^{2+}	+	$2e^-$	⟶	Cu
Copper(II) ion				Copper metal

In general, to convert metal cations to neutral metal atoms, each cation must gain one or more electrons.

Chemists classify any chemical change in which a species gains one or more electrons as a **reduction.** Thus the conversion of copper(II) cations to copper metal is a reduction reaction. You can convince yourself of this fact by examining the equation above for that change. Chemists classify the reverse reaction, in which an ion or other species loses one or more electrons, as an **oxidation.** For example, under the right conditions copper atoms can be oxidized.

Formation of Copper(II) Ions

$$Cu \longrightarrow Cu^{2+} + 2e^-$$

Cu	⟶	Cu^{2+}	+	$2e^-$
Copper metal		Copper(II) ion		

Historically, "oxidation" referred to the chemical combination of a substance with oxygen, as the term itself suggests. Chemists now know that in nearly all cases in which oxygen combines with another element or compound, oxygen partially or fully removes one or more electrons from the other species. By today's definition, any reactant—be it oxygen or not—that causes a species to lose one or more electrons is said to cause that species to be oxidized.

One way to remember this is **OIL RIG: O**xidation **I**s **L**oss of electrons, **R**eduction **I**s **G**ain of electrons.

Whenever one species loses electrons, another species must simultaneously gain them. In other words, oxidation and reduction reactions never occur separately. Oxidation and reduction occur together in what chemists call **oxidation-reduction reactions** or, to use a common chemical nickname, **redox reactions.**

You have already observed redox reactions in the laboratory. In Laboratory Activity B.4 (pages 118–120), copper metal reacted with silver ions. Here is the oxidation-reduction reaction you observed:

$$Cu(s) + 2\,Ag^+(aq) \longrightarrow Cu^{2+}(aq) + 2\,Ag(s)$$

Cu(s)	+	2 Ag^+(aq)	⟶	Cu^{2+}(aq)	+	2 Ag(s)
Copper metal		Silver ion		Copper(II) ion		Silver metal

A Copper Redox Reaction

Each metallic copper atom (Cu) was oxidized (converted to a Cu^{2+} ion by losing two electrons) and each silver ion (Ag^+ from $AgNO_3$ solution) was reduced (converted to an Ag atom by gaining one electron).

In the same activity you found that copper ions could be recovered from solution as copper metal by allowing the copper ions to react with magnesium metal, an element more active than copper. Magnesium atoms were oxidized; copper ions were reduced. Do you see why?

Note that the total electrical charge on both sides of this equation is the same. Electrical charges—as well as atoms—must balance in a correctly written chemical equation.

$Cu^{2+}(aq)$	+	$Mg(s)$	$\longrightarrow$	$Cu(s)$	+	$Mg^{2+}(aq)$
Copper(II) ion		Magnesium metal		Copper metal		Magnesium ion

In some circumstances this reaction might be a useful way to obtain copper metal. However, as is often the case, the desired copper metal is gained at the expense of "using up" another highly desirable material—in this case, magnesium metal.

How do redox reactions occur? Many metallic elements are found in minerals in the form of cations because they combine readily with other elements to form ionic compounds. Obtaining a metal from its mineral requires energy and a source of electrons. A reacting chemical species that serves as the source of electrons is known as a **reducing agent**.

Look again at Figure 16 (page 121). The table highlights several techniques that are used to reduce metal cations—or, in other words, to supply one or more electrons to each cation. The specific technique chosen depends on the metal's reactivity and the availability of inexpensive reducing agents and energy sources.

Two major approaches summarized in the table are **electrometallurgy** and **pyrometallurgy.** As the table suggests, electrometallurgy involves using an electric current to supply electrons to metal ions, thus reducing them. This process is used when no adequate chemical reducing agents are available or when very high-purity metal is sought. Pyrometallurgy—the most important and oldest ore-processing method—involves the treatment of metals and their ores by heat, as in a blast furnace. Carbon (coke) and carbon monoxide are common reducing agents in pyrometallurgy. A more active metal can be used if neither of these will do the job.

A third approach to obtaining metals from their ions is the process called **hydrometallurgy**—the treatment of ores and other metal-containing materials by reactants in water solution. You used such a procedure when you investigated the reactivity of different metals in Laboratory Activity B.4. Hydrometallurgy is used to recover silver and gold from old mine tailings (the mined rock left after most of the sought mineral is removed) by a process known as leaching. As supplies of higher-grade ores become scarcer, it will become economically feasible to use hydrometallurgy and other "wet processes" on metal-bearing minerals that dissolve in water.

Several tips to consider in teaching about oxidation-reduction reactions

The metal with the most attraction (pull) for electrons is the least reactive—this can be related to the B.4 laboratory activity. That's why elemental gold is sometimes found in streams but metallic sodium is not. The meaning of the term *reduction* can be kept straight if it is associated with the notion that if an atom is reduced its electric charge is reduced.

Hints: OINK (Oxidation Increases the Number, the *K* doesn't stand for anything but it doesn't spell *oink* otherwise.) Red^2 (*Red*uction *red*uces the oxidation number)

Emphasize that oxidation and reduction always go together. Oxidation cannot occur without a simultaneous reduction process. The three reduction processes all require some sort of electron source—either electricity or a reducing agent that will lose electrons easily.

MODELING MATTER: ELECTRONS AND REDOX PROCESSES

Point out to students that in the chemical reactions depicted here it is possible to find copper (in the form of metallic Cu atoms) being oxidized, as well as copper (in the form of Cu^{2+} ions) being reduced. Given suitable conditions, electron-transfer processes can proceed in either direction.

The different number and arrangement of the electrons explain the difference in the reactivity of metals. The more reactive a metal, the greater its tendency to give up electrons. Electron dot structure is introduced. Students should realize that there is a relationship between electron structure and chemical and physical properties.

MODELING MATTER

ELECTRONS AND REDOX PROCESSES

The processes of oxidation (loss of one or more electrons) and reduction (gain of one or more electrons) can be clarified by visual representations of the events. To develop such representations, you will consider atoms of each of the metals you investigated in Laboratory Activity B.4.

First, however, a review of some key details about the composition of an atom is in order. Magnesium (Mg), an active metal, formed Mg^{2+} ions in several of the reactions. The atomic number of Mg is 12, indicating that an electrically neutral atom of magnesium contains 12 protons and 12 electrons. (Do you recall why those numbers must be equal for a neutral atom?)

If magnesium forms a Mg^{2+} ion, two negatively charged electrons must be removed from each magnesium atom. The bookkeeping involved in this change can be summarized this way:

Mg	$\longrightarrow$	Mg^{2+}	+	$2\,e^-$
12 protons (+) 12 electrons (−) Net charge: 0		12 protons (+) 10 electrons (−) Net charge: 2+		2 electrons (−) Net charge: 2−

To build a useful picture of this process in your mind, it is necessary to keep track of only the two electrons that each magnesium atom releases, rather than monitoring all 12 of the atom's available electrons. (In fact, in normal chemical reactions, a magnesium atom is not observed to release any of its other 10 electrons.)

Thus, for bookkeeping purposes, an atom of Mg will be depicted this way:

Mg: the symbol for the element with two dots attached.

Each dot represents one readily removable electron. The symbol Mg represents the remaining parts of a magnesium atom, including its other ten electrons. The resulting expression for Mg is called an **electron-dot structure,** or just a **dot structure.** The equation for the oxidation of Mg can be represented in electron-dot terms this way:

$$\text{Mg:} \longrightarrow Mg^{2+} + 2\,e^-$$

1. Construct a similar electron-dot expression for the change that occurred in Laboratory Activity B.4 when each of these events took place:
   a. An atom of zinc, Zn, was converted to a Zn^{2+} ion. (*Hint:* Zn has two readily removable electrons.)
   b. A silver ion, Ag^+, was converted to a metallic silver atom, Ag(s).
2. Apply the definitions of oxidation and reduction to your two equations in Question 1, and label each reaction appropriately.

Now consider one of the complete reactions that you observed in Laboratory Activity B.4. When you

MODELING MATTER

immersed a sample of copper metal, Cu, in silver nitrate solution, $AgNO_3$, a blue solution containing Cu^{2+} formed, as well as crystals of solid Ag. This is the reaction that occurred:

$$Cu(s) \quad + \quad 2\ Ag^{+}(aq) \longrightarrow Cu^{2+}(aq) \quad + \quad 2\ Ag(s)$$

Using dot structures, the reaction can be represented this way:

$$Cu\!: \ + \ Ag^{+} \ + \ Ag^{+} \longrightarrow Cu^{2+} \ + \ Ag\cdot \ + \ Ag\cdot$$

3. a. Which reactant (Cu or Ag^+) is oxidized?
   b. Which is reduced?
4. Why are two Ag^+ ions needed for each Cu(s) atom that reacts?

Each copper atom involved in this reaction loses two electrons. Thus copper atoms must be oxidized in the change. It is clear from the dot structures that those two electrons lost by copper are gained by two Ag^+ ions. So Ag^+ is the agent that caused the removal of electrons from Cu (resulting in the oxidation of Cu). The species involved in removing electrons from the reactant that is oxidized is called the **oxidizing agent**—in this case, Ag^+ ions.

5. a. Given that definition and explanation, what must be the reducing agent in the reaction between Cu(s) and Ag^+ ions?
   b. How would you define a reducing agent?

Now consider another reaction you observed in Laboratory Activity B.4:

$$Zn(s) \quad + \quad Cu^{2+}(aq) \longrightarrow Zn^{2+}(aq) \quad + \quad Cu(s)$$

6. Draw an electron-dot representation of this reaction.
7. a. Which reactant is oxidized?
   b. Which is reduced?
8. Identify the oxidizing agent and the reducing agent in this reaction.
9. Consider both of the oxidation-reduction reactions you analyzed in this exercise. What general features of an oxidation-reduction reaction would allow you to answer Questions 7 and 8 *without* drawing electron-dot representations?
10. Test your answer to Question 9 by considering a new oxidation-reduction reaction. Answer Questions 7 and 8 for this system:

$$Zn^{2+}(aq) \quad + \quad Mg(s) \longrightarrow Zn(s) \quad + \quad Mg^{2+}(aq)$$

Answers

1. a. $Zn \rightarrow Zn^{++} + 2e^-$ (oxidation)
   b. $Ag^+ + e^- \rightarrow Ag$ (reduction)
2. a. Zn metal was oxidized to the Zn^{++} ion (oxidation is a loss of electrons)
   b. Ag^+ ion was reduced to Ag metal (reduction is a gain of electrons)
3. Cu was oxidized.
   Ag was reduced.
4. Cu gives up two electrons—each Ag^+ can accept only one electron.
5. a. Copper is the reducing agent.
   b. The reducing agent is the substance that supplies the electron to the reactant that is reduced.
6. $Zn + Cu^{++} \rightarrow Zn^{++} + Cu$
7. a. Zinc is oxidized.
   b. Copper is reduced.
8. Zinc is the reducing agent.
   Copper is the oxidizing agent.
9. The substance oxidized is the reducing agent.
   The substance reduced is the oxidizing agent.
10. $Zn^{++} + Mg \rightarrow Zn + Mg^{++}$
    Zn is reduced (oxidizing agent).
    Mg is oxidized (reducing agent).

SECTION SUMMARY

Reviewing the Concepts

1. **a.** List and briefly describe the three major "parts" of Earth.

b. Which part serves as the main storehouse of chemical resources used in manufacturing consumer products?

a. Atmosphere: extends to 100 miles from Earth's surface and consists primarily of nitrogen and oxygen; Hydrosphere: consists of water and oceans that cover three-fourths of Earth; and Lithosphere: consists of Earth's crust, mantle and core

b. Earth's crust is the main source of Earth's resources.

2. List two resources typically found in each of the three major parts of Earth.

Some possible answers are:

Atmosphere: nitrogen and oxygen

Hydrosphere: sodium and magnesium

Lithosphere: petroleum; silicates, metals

3. According to information in Figure 10 on page 114, which of these four countries—the United States, Australia, China, or Brazil—produces the largest mass of these eight resources?

United States

4. Is there a connection between the distribution of mineral resources and the wealth of a nation? Explain.

The wealth of a nation depends on the availability of raw materials and resources that are desired at the time. For example, nations in the Middle East enjoy tremendous amounts of income from selling their vast amounts of petroleum. It could also be argued that human effort and industry are resources upon which a nation can capitalize, with Japan as an an example. Japan was largely dependent on importing materials to produce electronics and automobiles. Yet today, Japan's wealth is based largely on the ability to produce these goods in many places around the world.

5. What factors determine the feasibility of mining a particular metallic ore at a certain site?

The factors are: percent of metal in the ore; the type of mining and processing; the amount of useful ore; the distance of the mine to the refinery; and the metal's supply and demand.

6. A nineteenth-century gold mine, inactive for a hundred years, has recently reopened for further mining. What factors may have influenced the decision to reopen the mine?

Technology may have devised a more economical method of extracting the metal from its ore, or the price of the metal has increased substantially to make its mining profitable.

7. What is meant by referring to the amount of "useful ore" at a site?

A useful ore is one in which the percent of metal in a particular deposit makes it worth mining.

8. Why are active metals more difficult to process and refine?

The chemical bonds, which hold the metal to other atoms, require more energy to separate.

9. Based on your results from Laboratory Activity B.4, which metals involved in this activity would be the easiest to process? Why?

Copper and silver, due to their lower reactivity.

10. Why do most metals exist in nature as minerals rather than as pure metallic elements?

Most metals are chemically reactive and are found combined with other elements.

11. Consider these two equations. Which represents a reaction that is more likely to occur? Why?

a. $Zn^{2+}(aq) + 2\,Ag(s) \longrightarrow Zn(s) + 2\,Ag^{+}(aq)$

b. $2\,Ag^{+}(aq) + Zn(s) \longrightarrow 2\,Ag(s) + Zn^{2+}(aq)$

Zinc metal will reduce the silver ion to silver metal. Silver metal is not a strong enough reducing agent to reduce the zinc ion.

12. **a.** Why would it be a poor idea to stir a solution of lead(II) nitrate with an iron spoon?

b. Write a chemical equation to support your answer.

a. The iron would dissolve in the lead solution and the lead ions would deposit out of the solution as lead.

b. $\mathbf{Fe + Pb^{2+} \rightarrow Pb + Fe^{2+}}$

13. Write an equation for each of these processes.

a. the reduction of gold(III) ions

b. the oxidation of elemental vanadium to vanadium(II) ions

c. the oxidation of magnesium metal

a. $\mathbf{Au^{3+} + 3\ e^- \rightarrow Au}$

b. $\mathbf{V \rightarrow V^{2+} + 2\ e^-}$ **c.** $\mathbf{Mg \rightarrow Mg^{2+} + 2e^-}$

14. Identify each equation as representing either an oxidation or a reduction reaction.

a. $Fe^{2+} + 2\ e^- \rightarrow Fe$ **b.** $Cr \rightarrow Cr^{3+} + 3\ e^-$

c. $Al^{3+} + 3\ e^- \rightarrow Al$

a. reduction **b. oxidation** **c. reduction**

15. Consider the following equation:

$$Zn(s) + Ni^{2+}(aq) \longrightarrow Zn^{2+}(aq) + Ni(s)$$

a. Which reactant has been oxidized? Explain your choice.

b. Which reactant has been reduced? Explain your choice.

c. What is the reducing agent in this reaction?

a. Zinc metal was oxidized because it lost two electrons.

b. $\mathbf{Ni^{2+}}$ **was reduced because it gained two electrons.**

c. The reducing agent is zinc metal.

16. Explain how each process converts metal cations to metal atoms.

a. electrometallurgy

b. pyrometallurgy

c. hydrometallurgy

a. Electrometallurgy: cations gain electrons from the cathode or negatively charged electrode.

b. Pyrometallurgy: coke or carbon monoxide acts as a reducing agent to reduce the cation.

c. Hydrometallurgy: the cation is treated in solution with a more active metal that will reduce the cation.

17. What processes could you use to obtain these elements from their ores?

a. magnesium

b. lead

a. electrometallurgy

b. pyrometallurgy

Connecting the Concepts

18. How can a less active metal be used to prevent the corrosion of a more active metal?

Coating the more active element with a less active one inhibits the oxidation reaction by prohibiting contact with substances like air and water.

19. Large gold nuggets with masses of 45 kg (100 lb) or more have been discovered. What conditions might have allowed such large pieces of elemental gold to exist?

Gold may have been crystallized from molten solutions deep within Earth's crust in large pockets. The gold was protected from oxidizing agents.

20. There are thousands of tons of gold in sea water. Explain why it is unlikely that ocean water will ever be "mined" for gold.

The ions are too widely dispersed in solution to make it worthwhile.

21. In 1982, when the penny was converted from pure copper to copper and zinc, the outside surface was still coated with copper. List three reasons why this coating was used.

Copper has a relatively low reactivity, is inexpensive, develops a protective patina, and has a color that distinguishes the coin from others.

22. At one time, food cans were made with tin and soldered with lead. What kinds of health hazards were posed by this arrangement?

Lead or lead ions would seep into the food and make it toxic, causing some human disorders and birth defects.

23. Is there any connection between the process used to reduce a metal cation and the position of that element on the Periodic Table?

Yes. Alkali and alkaline earth metals as well as aluminum must be reduced electrochemically since they are so active. Transition elements are more easily reduced by pyrometallurgy or hydrometallurgy.

Extending the Concepts

24. Although aluminum is a more reactive metal than iron, it is often used for outdoor products. Investigate how this is possible.

Aluminum oxidizes to form a protective coating, which prevents it from further oxidation.

25. The uneven distribution of mineral resources sometimes affects relations between nations. Identify and describe one historical or current example of this fact.

The Gulf War was fought to protect Western interests in petroleum reserves in Saudi Arabia and Kuwait when Kuwait was invaded by Iraq.

26. What conclusions about materials can be drawn from a study of the substances used for currency in ancient civilizations? Explain your ideas by giving examples.

Valued materials available locally, such as beaver pelts, wampum, stones, and beads, were early forms of currency in North America. Silver and gold were used in ancient civilizations, since they were found uncombined in nature and their scarcity made them valuable. It would appear that early currency took the form of personally useful materials, and later took the form of easily traded, but less useful, metals and gems.

27. History documents that copper has been used by humans for 10 000 years, whereas aluminum has been used for only about 100 years. Suggest and explain some reasons for this difference.

The inexpensive method of extracting aluminum by electrolysis from its ore, bauxite, had not been invented until 1886 (by Hall and Herout).

28. The reactive metal aluminum is often used in containers for acidic beverages. Investigate and describe the technology that makes this possible.

Aluminum cans are coated on the inside with a thin layer of plastic.

29. What is a patina? Explain its value both aesthetically and chemically.

A patina is a product of corrosion forming a crust or film, usually on bronze or copper. The green-blue color is often aesthetically valued as being ornamental. The patina protects the metal underneath from further corrosion.

CONSERVATION

Introduction

This section introduces the law of conservation and carries that concept into methods for balancing chemical equations.

C.1. KEEPING TRACK OF ATOMS

- Each time a chemical equation is balanced, the Law of Conservation of Matter is illustrated.
- Emphasize that scientific laws are simple generalizations of observations.
- The equations for the reactions studied in *ChemCom* will be balanced using illustrations and formulas.
- Point out to students that coefficients should be written in front of a symbol or formulas.
- It is important to emphasize that a coefficient of one is never written but, rather, is understood.
- Have students write the symbols (s) for solid, (aq) for aqueous, (l) for liquids, or (g) for gases for any reaction that they may have performed in the lab and/or watched as a demonstration.

SECTION C

CONSERVATION

Conservation CD-ROM WWW.

In some chemical reactions, matter seems to be created. In others, matter seems to disappear. Neither actually occurs. In this section, you will learn what happens to atoms in chemical reactions and how the atoms can be tracked through chemical equations. This information will help you to consider the fate of Earth's resources as well as the materials and products developed from them.

C.1 KEEPING TRACK OF ATOMS

Many things people use daily seem to disappear. For example, fuel is depleted as an airplane flies from one destination to another. The ice cream you eat seemingly vanishes. Steel in automobile bodies rusts away. Although the original forms of such materials disappear when they are used, the atoms composing them remain.

Think for a moment about what happens to molecules of gasoline as they burn in an automobile engine. The carbon and hydrogen atoms that make up these molecules react with oxygen atoms in the air to form carbon monoxide (CO), carbon dioxide (CO_2), and water (H_2O). These products are released as exhaust and disperse in the atmosphere. Thus the atoms of carbon, hydrogen, and oxygen originally present in the gasoline and air have not been destroyed; rather, they have been rearranged into new molecules.

In short, "using things up" means changing materials chemically, not destroying them. The **law of conservation of matter**, like all scientific laws, summarizes what has been learned by careful observation of nature: In a chemical reaction, matter is neither created nor destroyed. Molecules can be converted and decomposed by chemical processes, but atoms are forever. In a chemical reaction, matter—at the level of individual atoms—is always fully accounted for.

Because chemical reactions cannot create or destroy atoms, chemical equations representing such reactions must always be balanced. What does a "balanced equation" mean? Recall your introduction to chemical equations in Unit 1 (pages 29–31). Formulas for the reactants are placed on the left of the arrow; formulas for the products are placed on the right. In a balanced chemical equation, the number of atoms of each element is the same on the reactant and product sides.

Consider the burning of coal as an example. Coal is mostly carbon (C). If carbon burns completely, it combines with oxygen gas (O_2) to produce carbon dioxide (CO_2). Here is a representation of the atoms and molecules involved in this reaction:

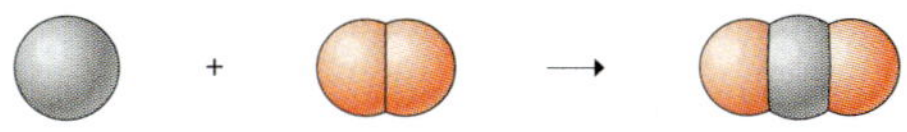

C	+	O_2	→	CO_2
1 Carbon atom (C)		1 Oxygen molecule (O_2)		1 Carbon dioxide molecule (CO_2)

Note that the numbers of carbon and oxygen atoms on the reactant side equal the respective numbers of carbon and oxygen atoms on the product side. This indicates that the equation is balanced.

The representation of the coal-burning reaction shows that one carbon atom reacts with one oxygen molecule to form one carbon dioxide molecule. Written with chemical formulas, the equation becomes

$C(s)$	$+$	$O_2(g)$	$\longrightarrow$	$CO_2(g)$
Carbon (in coal)	and	Oxygen gas	React to produce	Carbon dioxide gas

In writing chemical formulas for substances, the symbols for solid (s), liquid (l), and gas (g) are sometimes added. These symbols indicate the physical states of each substance under the conditions of the reaction.

The reaction in which you heated copper powder in air provides another example. Copper metal (Cu) reacts with oxygen gas (O_2) to form copper(II) oxide (CuO). Look at the representations below.

2 Copper atoms (Cu)	1 Oxygen molecule (O_2)	2 Copper(II) oxide formula units (CuO)

Interpreting a Balanced Chemical Equation (CD-ROM, WWW.)

Again, note that the numbers of copper and oxygen atoms on the reactant side equal the respective numbers of copper and oxygen atoms on the product side. Written as a chemical equation, the reaction is

$2\,Cu(s)$	$+$	$O_2(g)$	$\longrightarrow$	$2\,CuO(s)$
Copper metal	and	Oxygen gas	React to produce	Copper(II) oxide

You may have noticed that numbers have been placed in front of the copper and copper(II) oxide formulas. These numbers are called **coefficients**. Coefficients indicate the relative number of units of each substance involved in the chemical reaction. Reading this equation from left to right, you would say, "Two copper atoms react with one oxygen molecule to produce two formula units of copper(II) oxide."

It is standard not to write a coefficient of "1."

Why is the term "formula unit" used instead of "molecule"? Compounds of a metal and a nonmetal are ionic. (It might be helpful to review pages 32–34 of Unit 1.) These compounds are not found as individual molecules. Rather, they form large crystals made of ions. Chemists use the term **formula unit** when referring to the smallest unit of an ionic compound.

Look at the Periodic Table to recall which elements are metals and which are nonmetals.

In the following activity, you will practice recognizing and interpreting chemical equations. Then you will learn how to write them yourself.

ACCOUNTING FOR ATOMS

Building Skills 6

Students will learn to recognize a balanced equation. It should be pointed out that equations represent chemical reactions.

1. Answer provided.
2. a. Methane gas reacts with oxygen gas to produce carbon dioxide gas and water vapor.

   b.

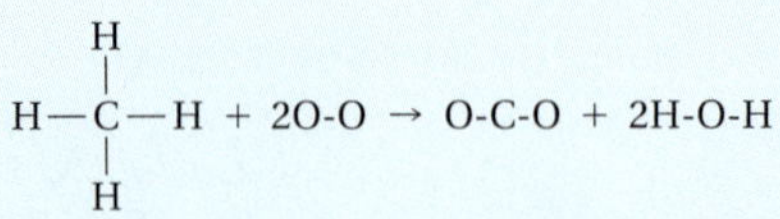

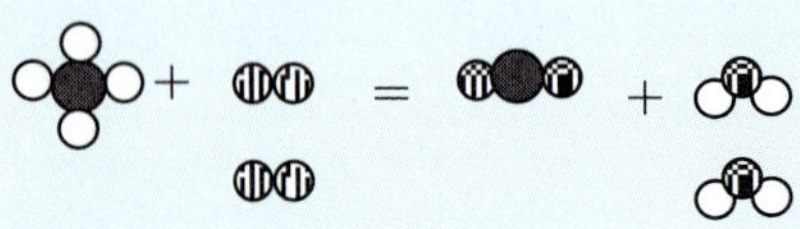

methane (CH_4) + oxygen (O_2) → carbon dioxide (CO_2) + water (H_2O)

c. Counting atoms

Reactants:	Products:
1 Carbon	1 Carbon
4 Hydrogen	4 Hydrogen
4 Oxygen	4 Oxygen

d. The respective numbers of carbon, hydrogen, and oxygen atoms are the same in the reactants and the products. The original statement is balanced.

This reaction produces heat, which can also be considered a product.

ACCOUNTING FOR ATOMS

Building Skills 6

For each chemical statement that follows,

a. interpret the statement in words,
b. draw a representation of the chemical statement (some structures are provided),
c. complete an atom inventory of the reactants and products, and
d. decide whether the expression—as written—is balanced.

To help guide your work, the first item is worked out.

1. The reaction between propane (C_3H_8) and oxygen gas (O_2) is a common source of heat for campers, recreational-vehicle users, and others using tanks of liquid propane fuel. A chemical statement showing the reactants and products is

$$C_3H_8(g) + O_2(g) \longrightarrow CO_2(g) + H_2O(g)$$

a. Interpreting this statement in words: "Propane gas reacts with oxygen gas to produce carbon dioxide gas and water vapor."

b. Using  to represent a propane molecule, the chemical statement can be represented as:

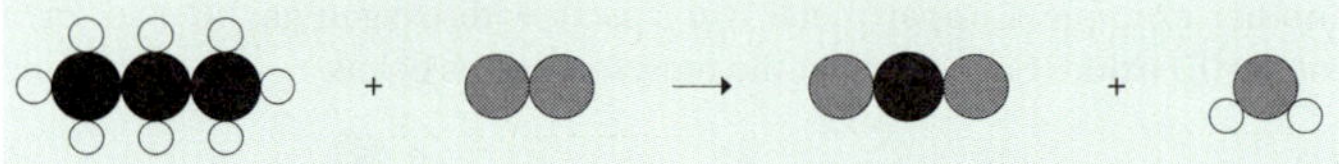

C_3H_8	+	O_2	⟶	CO_2	+	H_2O
1 Propane molecule (C_3H_8)		1 Oxygen molecule (O_2)		1 Carbon dioxide molecule (CO_2)		1 Water molecule (H_2O)

c. Counting the atoms on each side of the equation gives this atom inventory:

Reactant side	Product side
3 carbon atoms	1 carbon atom
8 hydrogen atoms	2 hydrogen atoms
2 oxygen atoms	3 oxygen atoms

d. The respective numbers of carbon, hydrogen, and oxygen atoms are different in reactants and products. Thus the original statement is not yet a properly written chemical equation. That is, it is not balanced.

2. Many people use natural gas to heat their homes. Natural gas contains methane (CH_4), which burns with oxygen (O_2) in air according to the equation

$$CH_4 + 2\,O_2 \longrightarrow CO_2 + 2\,H_2O$$

Use 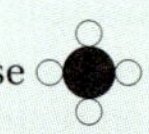 to represent a methane molecule.

3. When an acid reacts with a metal, hydrogen gas and an ionic compound are often formed. An expression for hydrobromic acid (HBr)

3. a. Hydrobromic acid reacts with magnesium to produce hydrogen gas and magnesium bromide.

   b. H-Br + Mg → H-H + Br-Mg-Br

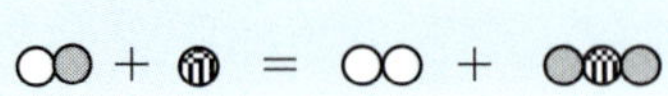

c.

Reactants:	Products:
1 Hydrogen	2 Hydrogen
1 Bromine	2 Bromine
1 Magnesium	1 Magnesium

d. The respective numbers of hydrogen and bromine atoms are different in reactants and products. The original statement is not yet a properly written chemical equation. It is not balanced.

reacting with magnesium metal to form hydrogen gas and magnesium bromide ($MgBr_2$) is:

$$HBr + Mg \longrightarrow H_2 + MgBr_2$$

Let 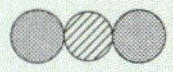 represent a formula unit of magnesium bromide ($MgBr_2$).

4. Hydrogen sulfide (H_2S) and metallic silver react in air to form silver sulfide (Ag_2S), commonly known as silver tarnish, and water:

$$4\,Ag + 4\,H_2S + O_2 \longrightarrow 2\,Ag_2S + 4\,H_2O$$

Let 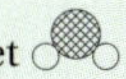 represent a hydrogen sulfide molecule and  represent a formula unit of silver sulfide.

Try Questions 5 and 6 without drawing representations. (Why might this be a good decision?)

5. Wood or paper can burn in air to form carbon dioxide and water vapor. One component that burns is cellulose, represented by $C_6H_{10}O_5$.

$$C_6H_{10}O_5 + 6\,O_2 \longrightarrow 6\,CO_2 + 5\,H_2O$$

6. Nitroglycerin, $C_3H_5(NO_3)_3$, the active component of dynamite, decomposes explosively to form N_2, O_2, CO_2, and water.

$$2\,C_3H_5(NO_3)_3 \longrightarrow 3\,N_2 + O_2 + 6\,CO_2 + 5\,H_2O$$

There are nine oxygen atoms in one $C_3H_5(NO_3)_3$ molecule. Can you see why?

C.2 NATURE'S CONSERVATION: BALANCED CHEMICAL EQUATIONS

The law of conservation of matter is based on the notion that Earth's basic "stuff"—its atoms—are indestructible. All changes observed in matter can be interpreted as rearrangements among atoms. Correctly written (balanced) chemical equations represent such changes. In the preceding activity, you practiced recognizing balanced chemical equations. Now you will learn how to write them.

As an example, consider the reaction of hydrogen gas with oxygen gas to produce gaseous water. First, write reactant formula(s) to the left of the arrow and product formula(s) to the right, keeping in mind that hydrogen and oxygen are diatomic (two-atom) molecules.

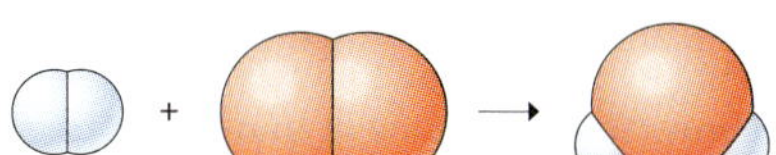

$H_2(g)$ + $O_2(g)$ ⟶ $H_2O(g)$

Check this expression by completing an atom inventory: Two hydrogen atoms appear on the left and two on the right. Hydrogen atoms are balanced. However, two oxygen atoms appear on the left and only one on the right. Because oxygen is not balanced, the expression requires additional work.

Under certain conditions, this reaction produces a violent explosion. When controlled, the reaction powers some types of rockets. Used in fuel cells, it generates electricity.

4. a. Solid silver reacts with hydrogen sulfide gas and oxygen gas to form solid silver sulfide and liquid water. (This is the reaction behind the tarnishing of silver and explains why silver vessels should not be used to heat sulfur-containing food such as eggs.)

b. 4 Ag + 4 H-S-H + O-O →
2 Ag-S-Ag + 4 H-O-H

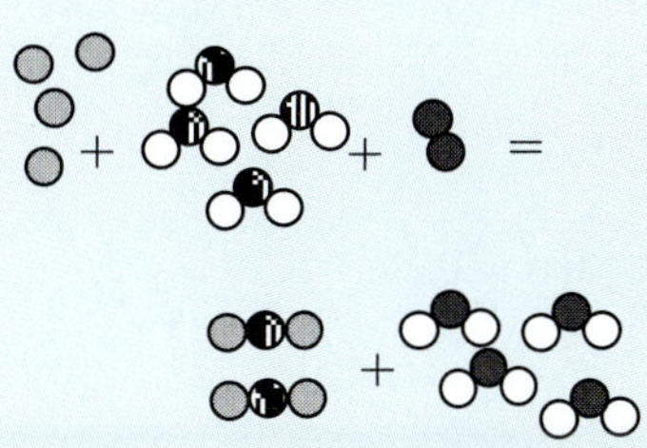

c. **Reactants:**	**Products:**
4 Silver	4 Silver
8 Hydrogen	8 Hydrogen
4 Sulfur	2 Sulfur
2 Oxygen	4 Oxygen

d. The respective numbers of sulfur and oxygen atoms are different in reactants and products. The equation is not balanced.

5. a. Cellulose reacts with oxygen gas to form carbon dioxide gas and liquid water.

b. Students are not expected to draw the structures. There is not always a need to draw the reaction structures if the reaction can be visualized by the student.

c. **Reactants:**	**Products:**
6 Carbon	6 Carbon
10 Hydrogen	10 Hydrogen
17 Oxygen	17 Oxygen

d. This is a balanced equation.

6. a. Nitroglycerin decomposes to form nitrogen gas, oxygen gas, carbon dioxide gas, and water vapor.

b. Drawing not required.

c. **Reactants:**	**Products:**
6 Carbon	6 Carbon
10 Hydrogen	10 Hydrogen
6 Nitrogen	6 Nitrogen
18 Oxygen	19 Oxygen

d. This is not a balanced equation.

C.2 NATURE'S CONSERVATION: BALANCED CHEMICAL EQUATIONS

The Law of Conservation is illustrated by giving step-by-step procedures used to balance chemical equations.

Stress to students that once an equation is written correctly, coefficients are used to balance atoms. Never change a subscript to balance an equation.

Demonstration

Place a 500 mL erlenmeyer flask containing a solution of NaI on a balance. Place a small test tube containing a solution of $Pb(NO_3)_2$ into the flask carefully, stopper the flask, and determine the mass. Invert the flask. Students should note that a chemical change ocurred, but the mass remained the same (Law of Conservation of Mass).

When balancing chemical equations, subscripts remain unchanged once the correct formulas have been written for the reactants and products. Coefficients must be adjusted instead.

Here is an *incorrect* way to complete the equation:

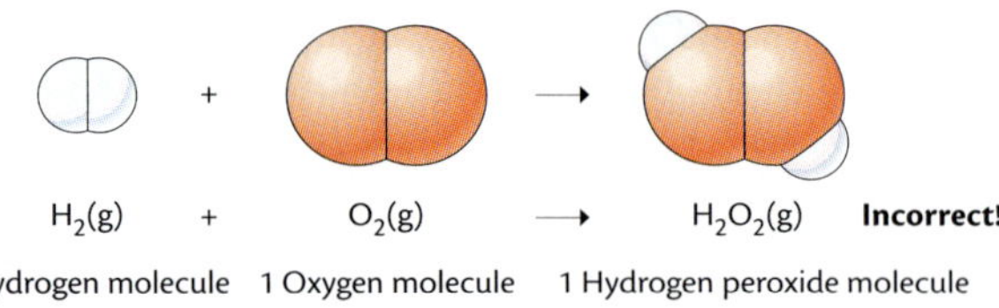

Although this chemical statement satisfies atom-inventory standards (two hydrogen and two oxygen atoms on both sides), the expression is wrong. By changing the subscript of O from 1 to 2 in the product, the identity of the product has been changed from water (H_2O) to hydrogen peroxide (H_2O_2). Because hydrogen peroxide is not produced in this reaction, the expression is incorrect.

Additional hydrogen, oxygen, and water molecules must be added to the appropriate side of the equation to balance the numbers of oxygen and hydrogen atoms. Another oxygen atom is needed on the product side to bring the number of oxygen atoms on both sides to two. Therefore, a water molecule is added:

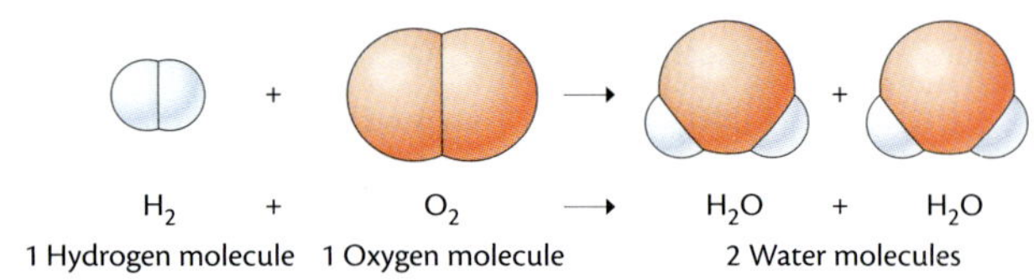

Good! Two oxygen atoms appear on each side of the equation. Unfortunately, there are now two hydrogen atoms on the left and four on the right—hydrogen atoms are no longer balanced. How can two hydrogen atoms be added to the reactant side? You are correct if you said by adding one hydrogen molecule:

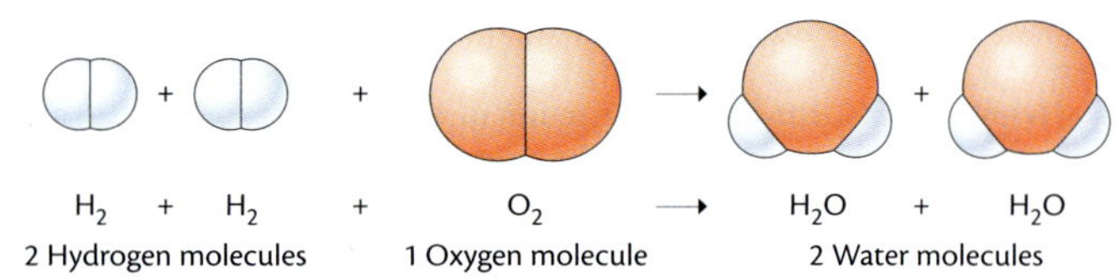

The atoms are balanced. Count them for yourself!

It is neither convenient nor efficient to draw representations for every chemical reaction. Thus this information is usually summarized in a chemical equation:

$$2\,H_2(g) + O_2(g) \longrightarrow 2\,H_2O(g)$$

The equation reads, "Two molecules of hydrogen gas react with one molecule of oxygen gas to produce two molecules of water vapor."

Here are some additional rules that may help you as you write chemical equations.

- If polyatomic ions, such as NO_3^- and CO_3^{2-}, appear as both reactants and products, treat them as units rather than balancing their atoms individually.
- If water is involved in the reaction, balance hydrogen and oxygen atoms last.
- Re-count all atoms after you think an equation is balanced—just to be sure!

WRITING CHEMICAL EQUATIONS

Building Skills 7

The reaction of methane gas (CH_4) with chlorine gas (Cl_2) occurs in sewage treatment plants and often in chlorinated water supplies. Common products are liquid chloroform ($CHCl_3$) and hydrogen chloride gas (HCl). A chemical statement describing this reaction follows below. As you can see from a quick glance, the equation is not balanced.

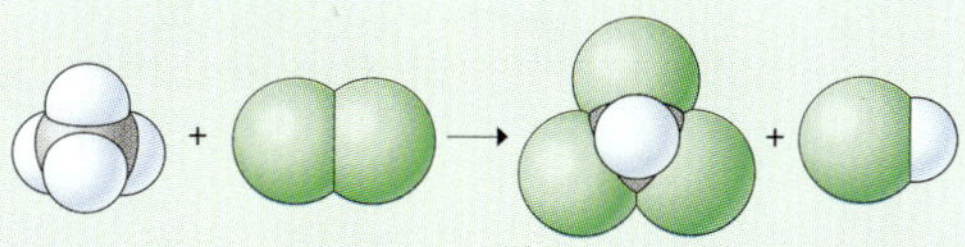

$CH_4(g)$ + $Cl_2(g)$ → $CHCl_3(l)$ + $HCl(g)$

Methane | Chlorine | Chloroform | Hydrogen chloride

To complete the chemical equation, you can follow this line of reasoning: One carbon atom appears on each side of the arrow, so carbon atoms balance.

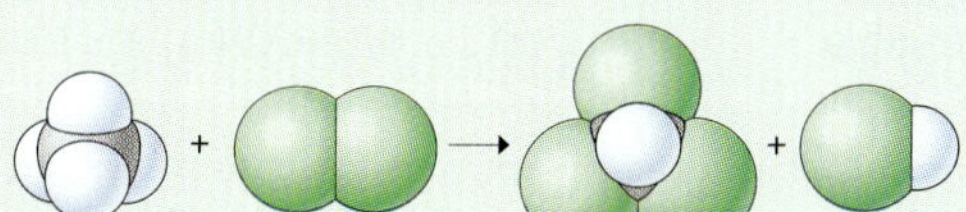

CH_4 + Cl_2 → $CHCl_3$ + HCl

Methane | Chlorine | Chloroform | Hydrogen chloride

Four hydrogen atoms are on the left, but only two on the right (one in $CHCl_3$, a second in HCl). To increase the number of hydrogen atoms on the product side, the coefficient of HCl must be adjusted. Because two

Chloroform is one of the THMs mentioned in Unit 1, page 74.

For convenience, the symbols (g) and (l) are removed. They will reappear in the final equation.

WRITING CHEMICAL EQUATIONS

Building Skills 7

Students practice balancing equations. Emphasize that equations represent chemical reactions. These equations represent particular aspects of actual resource processing or chemical manufacturing. Reinforce the idea that it is necessary to count atoms on both sides of the yield sign.

Answers

1. a. $2\ C(s) + O_2(g) \rightarrow 2\ CO(g)$

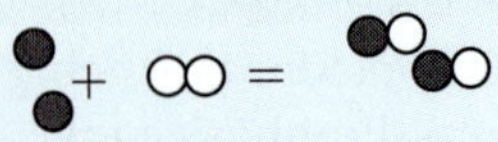

b. $Fe_2O_3(s) + 3\ CO(g) \rightarrow$
$2\ Fe(l) + 3\ CO_2(g)$

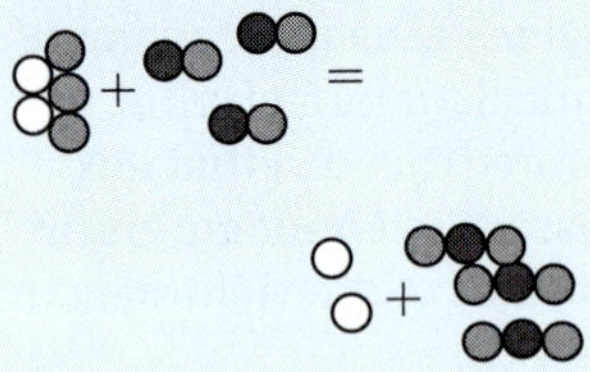

2. $2\ CuO(s) + C(s) \rightarrow 2\ Cu(s) + CO_2(g)$

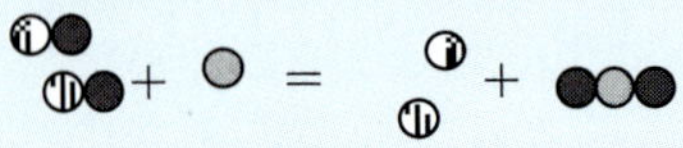

3. $4\ NH_3(g) + 7\ O_2(g) \rightarrow$
$4\ NO_2(g) + 6\ H_2O(l)$

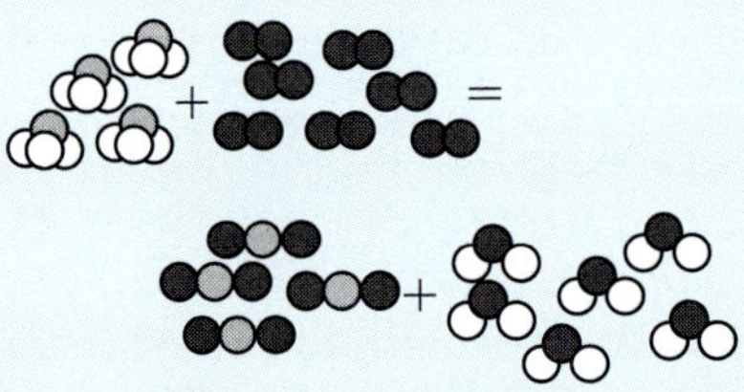

4. $2\ O_3(g) \rightarrow 3\ O_2(g)$

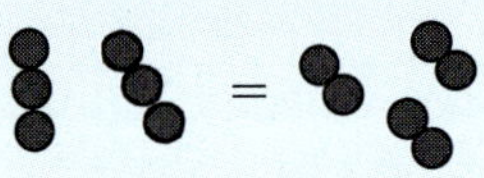

5. $Cu(s) + 2\ AgNO_3(aq) \rightarrow$
$Cu(NO_3)_2(aq) + 2\ Ag(s)$

6. $2\ C_8H_{18}(l) + 25\ O_2(g) \rightarrow$
$16\ CO_2(g) + 18\ H_2O(g)$

more hydrogens are needed on the right, the number of HCl molecules must be changed from 1 to 3. This gives four hydrogen atoms on the right:

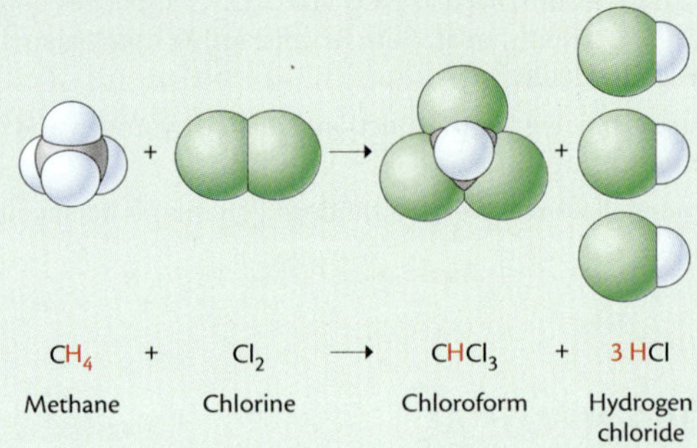

Now both carbon and hydrogen atoms are balanced. What about chlorine? Two chlorine atoms appear on the left and six on the right side. These six chlorine atoms (three in $CHCl_3$, three in 3 HCl) must have come from three chlorine (Cl_2) molecules. Thus 3 must be the coefficient of Cl_2.

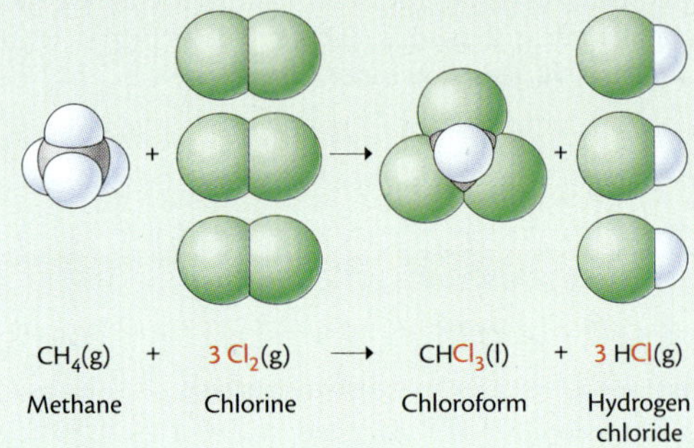

The chemical equation appears to be balanced. An atom inventory verifies that the equation is complete as written.

Reactant side	Product side
1 C atom	1 C atom
4 H atoms	4 H atoms (1 in $CHCl_3$, 3 in HCl)
6 Cl atoms (in 3 Cl_2)	6 Cl atoms (3 in $CHCl_3$, 3 in HCl)

Copy the following chemical expressions onto a separate sheet of paper, and balance each if needed. For Questions 1–4, draw a representation of your final equation to verify that it is balanced. Structures unfamiliar to you will be provided.

1. Two blast furnace reactions are used to obtain iron from its ore:
   a. $C(s) + O_2(g) \longrightarrow 2\ CO(g)$ b. $Fe_2O_3(s) + CO(g) \longrightarrow Fe(l) + 3\ CO_2(g)$

   Let 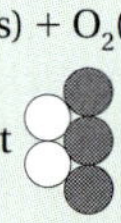 represent a formula unit of Fe_2O_3.

   Let  represent a molecule of CO_2.

2. The final step in the refining of a copper ore is:

$$CuO(s) + C(s) \longrightarrow Cu(s) + CO_2(g)$$

Let  represent a formula unit of CuO.

3. Ammonia (NH_3) in the soil reacts continuously with oxygen gas (O_2):

$$NH_3(g) + O_2(g) \longrightarrow NO_2(g) + H_2O(l)$$

4. Ozone (O_3) can decompose to form oxygen gas (O_2):

$$O_3(g) \longrightarrow O_2(g)$$

Let  represent an ozone molecule.

5. Copper metal reacts with silver nitrate solution to form copper(II) nitrate solution and silver metal:

$$Cu(s) + AgNO_3(aq) \longrightarrow Cu(NO_3)_2(aq) + Ag(s)$$

(*Hint:* Look at the formula for copper(II) nitrate, $Cu(NO_3)_2$. The subscript of two outside the parentheses indicates that this formula contains two nitrate (NO_3^-) anions. So one formula unit of $Cu(NO_3)_2$ contains one copper ion, two nitrogen atoms, and six oxygen atoms.)

6. Combustion of gasoline in an automobile engine can be represented by the burning of octane (C_8H_{18}):

$$C_8H_{18}(l) + O_2(g) \longrightarrow CO_2(g) + H_2O(g)$$

Recall that in writing chemical formulas for substances, the symbols (s), (l), and (g), for solid, liquid, and gas respectively, are sometimes added. The symbol (aq) means that the substance is dissolved in water. It is an aqueous solution.

C.3 ATOM, MOLECULE, AND ION INVENTORY

In Question 2 in the previous activity, you obtained the chemical equation

$$2\ CuO(s) + C(s) \longrightarrow 2\ Cu(s) + CO_2(g)$$

One interpretation of this equation is, as you know: *Two formula units of copper(II) oxide and one atom of carbon react to produce two atoms of copper and one molecule of carbon dioxide.* Although correct, this interpretation involves such small quantities of material that a reaction on that scale would be completely unnoticed. Such information would not be very useful, for example, to a metal refiner interested in how much carbon is needed to react with a certain large-scale amount of copper(II) oxide.

Chemists have devised a counting unit called the **mole** (symbolized mol) that helps solve the refiner's problem. You are familiar with other counting units such as "pair" and "dozen." The mole can be regarded as the chemist's "dozen." A pair of water molecules is two water molecules. One dozen water molecules is 12 water molecules. One mole of water molecules is 602 000 000 000 000 000 000 000 water molecules. This number—the number of particles (or "things") in one mole—is more conveniently written as 6.02×10^{23}. Either way, this is a very large number!

The number 6.02×10^{23} is called Avogadro's number.

C.3 ATOM, MOLECULE, AND ION INVENTORY

This section introduces the concepts of the mole and molar mass. Point out the relationship of moles to balanced equations. Emphasize that the mole is a counting unit. The concept of a mole is generally a difficult one for students; therefore, concrete activities are usually necessary to develop a thorough understanding. Emphasize the analogy of the "chemist's dozen." Point out that the reason for Avogadro's number being so large is that the units being counted (atoms and molecules) are so small. Analogies that help emphasize the enormity of the number 6.02 x 10^{23} include its approximate equality to the number of grains of sand on all the beaches of the world or the number of glasses of water contained in all the oceans.

The mole concept is based on the idea that if the unit mass of a particle is known, then the mass of a thousand, million, or billion particles can be calculated. Knowing the mass of a set number of particles allows us to "count by weighing."

After students have an understanding of the mole as the unit for counting atoms and molecules, build the connection between the mole and molar mass. Chemists count the number of particles in macroscopic samples of matter by weighing them, rather than actually counting the number of atoms, molecules, or ions. A display of one mole of a variety of common substances—sugar, salt, aluminum, copper, water—reinforces the idea that the mass of one mole varies with the substance.

Figure 17 *One mole each of copper, table salt, and water.*

To help you get a better idea of the size of a mole, consider this: Imagine stringing a mole of paper clips (6.02×10^{23} paper clips) together and wrapping the string around the world. It would circle the world about 400 trillion (4×10^{14}) times! And even if you connected a million paper clips each second, it would take you 190 million centuries to finish stringing one mole of paper clips.

As large as one mole of molecules seems, however, drinking that amount of water would leave you quite thirsty on a hot day. One mole of water is less than one-tenth of a cup of water—only 18 g (or 18 mL) of water. But that is why the mole is so useful in chemistry. It represents a number of atoms, molecules, or formula units large enough to be conveniently weighed or measured in the laboratory. Furthermore, the atomic weights of elements can be used to find the mass of one mole of any substance, a value known as the **molar mass** of a substance. Figure 17 shows a mole of several familiar substances.

Specific examples will help you to better understand this notion. Suppose you need to find the molar masses of carbon (C) and of copper (Cu). In other words, you want to know the mass of one mole of carbon atoms (6.02×10^{23} atoms) and one mole of copper atoms (6.02×10^{23} atoms). Rather than counting that collection of atoms onto a laboratory balance (good luck!), you can quickly get the answers from atomic-weight data. The atomic weight of each element is found on the Periodic Table. Carbon's atomic weight is 12.01; copper's is 63.55. If the unit "grams" is added to these values, the result is their molar mass:

Recall that mol is the symbol for the mole unit.

$$1 \text{ mol C} = 12.01 \text{ g C} \qquad 1 \text{ mol Cu} = 63.55 \text{ g Cu}$$

As you can now see, the mass (in grams) of one mole of an element's atoms equals the numerical value of the element's atomic weight. Any element's molar mass can be determined from the Periodic Table. The molar mass of a diatomic element is twice the mass given in the Periodic Table. Why?

The molar mass of a compound is the sum of the molar masses of its component atoms. For example, consider two compounds of interest to the copper-metal refiner—carbon dioxide (CO_2) and malachite ($Cu_2CO_3(OH)_2$).

One mole of CO_2 molecules contains one mole of C atoms and two moles of O atoms. Adding the molar mass of carbon and twice the molar mass of oxygen gives

$$1 \cancel{\text{mol C}} \times \frac{12.01 \text{ g C}}{1 \cancel{\text{mol C}}} = 12.01 \text{ g C}$$

$$2 \cancel{\text{mol O}} \times \frac{16.00 \text{ g O}}{1 \cancel{\text{mol O}}} = 32.00 \text{ g O}$$

$$\text{Molar mass of } CO_2 = (12.01 \text{ g} + 32.00 \text{ g}) = 44.01 \text{ g } CO_2$$

The molar mass of malachite is found in a similar way. However, the total atoms of each element in more complex compounds must be carefully counted.

One mole of malachite ($Cu_2CO_3(OH)_2$) contains 2 mol Cu, 1 mol C, 5 mol O (can you see where all five oxygen atoms are found in the formula?), and 2 mol H.

136

$$2\ \cancel{\text{mol Cu}} \times \frac{63.55\ \text{g Cu}}{1\ \cancel{\text{mol Cu}}} = 127.10\ \text{g Cu}$$

$$1\ \cancel{\text{mol C}} \times \frac{12.01\ \text{g C}}{1\ \cancel{\text{mol C}}} = 12.01\ \text{g C}$$

$$5\ \cancel{\text{mol O}} \times \frac{16.00\ \text{g O}}{1\ \cancel{\text{mol O}}} = 80.00\ \text{g O}$$

$$2\ \cancel{\text{mol H}} \times \frac{1.008\ \text{g H}}{1\ \cancel{\text{mol H}}} = 2.016\ \text{g H}$$

$$\text{Molar mass of } Cu_2CO_3(OH)_2 = 221.13\ \text{g } Cu_2CO_3(OH)_2$$

In summary, the molar mass of a compound is found by first multiplying the moles of each element in the formula by the molar mass of that element. Then all of the element masses are added together.

CD-ROM WWW. Modeling Matter: A Hypothetical Reaction

MOLAR MASSES — Building Skills 8

Find the molar mass of each substance:

1. The element nitrogen: N
2. Nitrogen gas: N_2
3. Sodium chloride (table salt): NaCl
4. Sucrose (table sugar): $C_{12}H_{22}O_{11}$
5. Chalcopyrite (a copper ore): $CuFeS_2$
6. Azurite (a copper ore): $Cu_3(CO_3)_2(OH)_2$ (*Hint:* Verify that this formula includes 8 moles of oxygen atoms.)

The term "one mole of nitrogen" is confusing. It could refer to one mole of N atoms or a mole of N_2 molecules. Thus it is important to specify which particular substance is involved.

How are chemical equations and molar masses useful in answering questions about large-scale reactions? The coefficients in a chemical equation show both the relative numbers of individual molecules (or formula units) of reactants and products and the relative numbers of moles of these same substances. Consider again the metal-refining example. The mole "counting unit" makes it easy to find the mass of carbon dioxide released during refining.

	2 CuO(s)	+	C(s)	⟶	2 Cu(s)	+	CO_2(g)
	2 Formula units CuO		1 Atom C		2 Atoms Cu		1 Molecule CO_2
Also:	2 mol CuO		1 mol C		2 mol Cu		1 mol CO_2

Thus for every two moles of CuO that react, one mole of CO_2 is produced. The molar masses of all four substances can be used to interpret the equation in terms of the masses involved:

2 CuO(s)	+	C(s)	⟶	2 Cu(s)	+	CO_2(g)
2 mol CuO		1 mol C		2 mol Cu		1 mol CO_2
159.10 g CuO		12.01 g C		127.10 g Cu		44.01 g CO_2

2 mol CuO, containing 2 mol Cu atoms and 2 mol O atoms, has a mass of 159.10 g.

$$2\ \cancel{\text{mol Cu}} \times \frac{63.55\ \text{g Cu}}{1\ \cancel{\text{mol Cu}}} = 127.10\ \text{g Cu}$$

Compare the total mass of the products to the total mass of the reactants. How does this illustrate the law of conservation of matter?

MOLAR MASSES
Building Skills 8

When students have an understanding of the concept of a mole, have them do these simple calculations. Caution students about the importance of parentheses in counting atoms and in calculating molar mass.

Answers

1. 14.01 g/mol
2. 14.01 g/mol × 2 = 28.02 g/mol
3. 22.99 g/mol + 35.45 g/mol = 58.44 g/mol
4. (12.01 g/mol × 12) + (1.008 g/mol x 22) + (16.00 g/mol × 11) = 342.30 g/mol
5. 63.55 g/mol + 55.85 g/mol + (32.01 g/mol × 2) = 183.42 g/mol
6. (63.55 g/mol × 3) + (12.01 g/mol × 2) + (16.00 g/mol × 8) + (1.008 g/mol × 2) = 344.69 g/mol

CHEMISTRY AT WORK

Preserving the Past . . . for the Future

Mary Striegel is the Materials Research Program Manager at the National Center for Preservation Technology and Training (NCPTT) in Natchitoches, Louisiana. NCPTT is an effort by the National Park Service to advance the use of modern technologies in the practice of historic preservation in the fields of archaeology, architecture, landscape architecture, materials conservation, and history.

Mary supervises scientists in the Center's Materials Research Program, where field tests are used to determine the ways that air pollution and acid rain affect historic buildings, outdoor works of art, and various other materials. In laboratory work, the Center's scientists isolate different pollutants in small chambers in order to investigate the way materials and preservation treatments interact. In case studies, the scientists observe the effects of pollutants on artifacts that have been exposed in the environment.

Scientist and art conservationists are combining their talents in an attempt to save these bronze treasures.

Preserving Art Treasures

Weathered to a powdery blue-green finish, bronze sculptures stand silently in city parks across the nation. Images of bygone heroes or memorials to noble historic struggles, the sculptures are gifts from previous generations. But bronze, an alloy made mostly of copper and tin, often corrodes in our modern industrial environment. Thus these links with the past are slowly being destroyed by the elements, pollution, acid rain, and changing temperatures. The problem is clearly evident on the pitted faces of the outdoor sculptures. Air pollution can ultimately lead to irregular patterns and decay of the metal—holes and streaks that are permanently etched into the piece.

But help may be on the way. Scientists and art conservationists are combining their talents in an attempt to save these bronze treasures.

The NCPTT has awarded a grant to researchers at North Dakota State University and the National Gallery of Art in Washington, D.C. The researchers'

138

task is to find more effective methods of testing and developing a coating system that will help outdoor bronze sculpture better resist corrosion. The coating, however, must not yellow or change the sculpture's appearance. Five new coating systems are currently being tested both with accelerated corrosion test methods and under natural exposure to corrosive environments.

The two coating systems that are showing the most promise in the initial stages of testing are each made of acrylic-urethane and wax. The researchers will try to develop a variety of successful coatings because what works well in a tropical environment may not be the best choice in a northern climate. Likewise, the coating that is most effective in a city with heavy industry may not be ideal for a rural location.

"Not only will the development and testing of these coating systems affect sculpture in our country, but there is the possibility it will impact how we take care of sculptures worldwide," Striegel said.

Photograph courtesy of NCPTT

How Does Science Happen?

1. Scientific research may result in advances that improve our ability to deal with problems. Science is a cooperative venture: discoveries are often made by various groups and individuals working toward a common goal. How does the research described here demonstrate the cooperative nature of science?
2. How can a coating work to protect an exposed surface? Describe some of the chemistry behind the use of coatings in this kind of preservation effort.
3. Use the Web to find out about other projects of the National Center for Preservation Technology and Training (NCPTT).

139

C.4 COMPOSITION OF MATERIALS

In considering an ore as a source of a metal, its percent composition is an important factor. The names and formulas of three ores of copper are given for consideration.

Hint: If your students are having problems with calculating percent composition, you might suggest it is parts per hundred. Help them remember how to set up percent problems: *Part over whole times 100* = percent.

C.4 COMPOSITION OF MATERIALS

One of the decisions you will make when designing your coin is whether to use only one material or a combination of materials. If your design uses more than one material, you will need to specify how much of each material will be present in the coin. The percent by mass of each material found in an item such as a coin is called its **percent composition**.

In Section A, you learned that the composition of the U.S. penny has changed several times. During 1943, it was made of zinc-coated steel. After this date and up until 1982, the penny was made mostly of copper. Since 1982, U.S. pennies have been made primarily of zinc. A post-1982 penny has a mass of 2.500 g and is composed of about 2.4375 g zinc and 0.0625 g copper. The percent composition of the penny can be found by dividing the mass of each constituent metal by the mass of the penny and multiplying by 100%:

$$\frac{2.4375\text{ g zinc}}{2.500\text{ g penny}} \times 100\% = 97.50\%\text{ zinc}$$

$$\frac{0.0625\text{ g copper}}{2.500\text{ g penny}} \times 100\% = 2.50\%\text{ copper}$$

The idea of percent composition also helps geologists describe how much metal or mineral is present in a particular ore. They can then evaluate whether the ore should be mined and also how it should be processed.

A compound's formula indicates the relative number of atoms of each element present in the substance. For example, one common commercial source of copper metal is the mineral chalcocite—copper(I) sulfide (Cu_2S). Its formula indicates that the mineral contains twice as many copper atoms as sulfur atoms. The formula also reveals how much copper can be extracted from a certain mass of the mineral—an important factor in copper mining and production.

How are the ideas of molar mass and percent composition useful in determining how much copper can be obtained from copper-containing minerals and ores? Some copper-containing minerals are listed in the table in Figure 18. The percent of copper in chalcocite can be found by applying what you know about molar masses.

The formula for chalcocite indicates that one mole of Cu_2S contains two moles of Cu, or 127.10 g Cu. The molar mass of Cu_2S is (2 × 63.55 g) + 32.07 g = 159.17 g. Therefore,

$$\%\text{ Cu} = \frac{\text{Mass of Cu}}{\text{Mass of Cu}_2\text{S}} \times 100 = \frac{127.10\text{ g Cu}}{159.17\text{ g Cu}_2\text{S}} \times 100 = 79.85\%\text{ Cu}$$

A similar calculation indicates that Cu_2S contains 20.15% sulfur. The sum of percent copper and percent sulfur equals 100.00%. Why?

Some Copper-Containing Minerals

Common Name	Formula
Chalcocite	Cu_2S
Chalcopyrite	$CuFeS_2$
Malachite	$Cu_2CO_3(OH)_2$

Figure 18 *Names and formulas for three copper-containing minerals.*

Knowing the percent composition of metal in a particular mineral is important in deciding whether a particular ore should be mined. What else needs to be considered? Suppose an ore is found to contain chalcocite, Cu_2S. Because nearly 80% of this mineral is composed of Cu (see calculation), it seems likely that this ore is worth mining for copper. However, another factor must be considered—the quantity of mineral contained in the ore. All other factors being equal, an ore that contains only 10% chalcocite would be less desirable as a copper source than one containing 50% chalcocite. Thus two factors must be taken into account to decide on the quality of a particular ore source: the percent mineral in the ore and the percent metal in the mineral.

Diagrams may be useful in understanding how these two percentage values relate to the total metal found in a particular ore. Consider an ore containing 10% chalcocite. Look at Figure 19. Suppose the rectangle in Figure 19a represents a piece of this ore. According to Figure 19b, 10% of the ore is composed of the mineral chalcocite. (One square is shaded with vertical stripes to show this.). You know that chalcocite itself is approximately 80% copper by mass. To represent this, 80% of the chalcocite square is shaded with horizontal stripes in Figure 19c. Now you can estimate visually how much copper is in this particular ore. How might similar diagrams represent an ore that is 50% chalcocite?

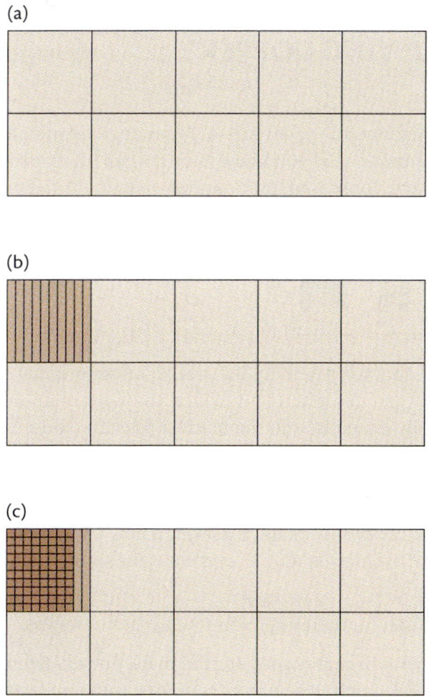

Figure 19 (a) *The rectangle represents the ore sample. Each square represents 10% of the sample.* (b) *One square is shaded with vertical stripes, representing an ore that is 10% chalcocite.* (c) *80% of the chalcocite is shaded with horizontal stripes. This represents the percent copper within chalcocite. Overall, only 8% of the ore is copper.*

PERCENT COMPOSITION

Building Skills 9

Students will demonstrate their ability to calculate the percent of a metal in its ore.

1. a. The molar mass of chalcopyrite, $CuFeS_2$, is $(63.55 \text{ g} + 55.85 \text{ g}) + (2 \times 32.00 \text{ g}) = 183.54 \text{ g}$

$$\% \text{ Cu} = \frac{\text{Mass of Cu}}{\text{Mass of CuFeS}_2} \times 100$$

$$= \frac{63.55 \text{ g Cu}}{183.54 \text{ g CuFeS}_2} \times 100$$

$$= 34.6\% \text{ Cu}$$

b. The molar mass of malachite, $Cu_2CO_3(OH)_2$ = 221.12 g

$$\% \text{ Cu} = \frac{127.10 \text{ g Cu}}{221.12 \text{ g CuCO}_3\text{(OH)}_2} \times 100 = 57.5\% \text{ Cu}$$

c. Chalcocite has the highest percent of copper and would be the most profitable.

2. $Cu_3(CO_3)_2(OH)_2$ contains 3 copper atoms, 2 carbon atoms, 2 hydrogen atoms, and 8 oxygen atoms

3. $$100 \text{ g} \times \frac{20 \text{ g}}{100 \text{ g}} \times 0.7984 = 15.97 \text{ g Cu}$$

$$100 \text{ g} \times \frac{30 \text{ g}}{100} \times 0.346 = 10.4 \text{ g Cu}$$

An ore of 20% chalcocite (Cu_2S) contains more copper.

4. Magnetite has the greater amount of iron. Assume a 1 Kg sample of each.

% Fe in hematite =

$$\frac{\text{Mass of Fe}}{\text{Mass of Fe}_2\text{O}_3} \times 100 =$$

$$\frac{111.7 \text{ g Fe}}{159.7 \text{ g Fe}_2\text{O}_3} \times 100 = 69.9\% \text{ Fe}$$

% Fe in magnetite =

$$\frac{\text{Mass of Fe}}{\text{Mass of Fe}_3\text{O}_4} \times 100 =$$

$$\frac{167.55 \text{ g Fe}}{231.55 \text{ g Fe}_3\text{O}_4} \times 100 = 72.4\% \text{ Fe}$$

The mass of Fe in magnetite, Fe_3O_4, is greater.

PERCENT COMPOSITION

Building Skills 9

Recall that the percent copper in chalcocite was calculated on page 140.

1. Use Figure 18 on page 140 to answer the following. In your calculations, assume that each mineral is present at the same concentration in an ore. Also assume that copper metal can be extracted from a given mass of any ore at the same cost.
   a. Calculate the percent copper in chalcopyrite.
   b. Calculate the percent copper in malachite.
   c. Which of the three minerals could be mined most profitably?
2. The chemical formula for the copper-containing mineral azurite is $Cu_3(CO_3)_2(OH)_2$. Complete an atom inventory for this compound. Use Building Skills 6 (page 130) as a model.
3. Suppose you have the choice either to mine ore that is 20% chalcocite (Cu_2S) or ore that is 30% chalcopyrite ($CuFeS_2$). Assuming all other factors are equal, which would you choose? Draw diagrams similar to those in Figure 19 on page 141 to support your answer.
4. Two common iron-containing minerals are hematite (Fe_2O_3) and magnetite (Fe_3O_4). If you had the same mass of each, which sample would contain the larger mass of iron? Support your answer with calculations.

C.5 RETRIEVING COPPER

Laboratory Activity

Introduction

In Laboratory Activity B.2 (pages 116–117), you heated metallic copper, producing a black powder that you know to be copper(II) oxide (CuO). Because atoms are always conserved in chemical reactions, the original copper atoms must still exist. In this laboratory activity, you will attempt to recover those atoms of metallic copper.

Procedure

Part I: Separating copper(II) oxide (CuO) from the sample

Most likely, in Laboratory Activity B.2, not all of the original copper powder reacted with oxygen gas when you heated the copper in air. Some copper metal is still likely mixed in with the black copper(II) oxide. The first steps in this activity involve separating this mixture into copper and copper(II) oxide. To do this, you will add dilute hydrochloric acid (HCl) to the black powder. Copper metal does not react with hydrochloric acid, so it will remain as a solid. The black copper(II) oxide, however, reacts with hydrochloric acid to produce copper(II) chloride ($CuCl_2$) and water, as shown in this equation:

$$\underset{\text{Copper(II) oxide}}{CuO(s)} + \underset{\text{Hydrochloric acid}}{2\,HCl(aq)} \longrightarrow \underset{\text{Copper(II) chloride}}{CuCl_2(aq)} + \underset{\text{Water}}{H_2O(l)}$$

Based on what you learned about mixtures on page 25, how would you classify the sample present in your beaker?

1. Obtain your beaker containing the black powder from Laboratory Activity B.2. Look closely at its contents. Is the material uniform throughout? If not, why not?

2. Add 50 mL of 1 M HCl to the beaker containing the copper oxide mixture. Record your observations of the solution. **CAUTION:** *Hydrochloric acid may damage your skin. If some HCl does spill on your skin, ask another student to notify your teacher immediately. Begin rinsing the affected area with tap water immediately.*
3. Gently heat the mixture to about 40 °C on a hot plate. Heat for 15 minutes, stirring every few minutes with a glass rod.
4. Remove the beaker from the heat source. Allow any unreacted copper metal remaining to settle to the bottom of the beaker. Then slowly decant the liquid into another 100-mL beaker.
5. Set aside the second beaker (containing the liquid) for Step 9.
6. Wash the solid copper remaining in the first beaker several times by swirling it gently with distilled water. Discard the liquid washings as instructed by your teacher.
7. Measure and record the mass of a piece of filter paper. Transfer the solid copper to the paper, and allow it to dry overnight.
8. When the sample and filter paper have dried, find the mass of the solid copper. This represents the portion of the original copper powder that failed to react to form CuO. Record this mass in your laboratory notebook.

Part II: Converting Copper(II) Chloride ($CuCl_2$) to Copper (Cu)

The final step is to convert the dissolved copper(II) chloride ($CuCl_2$) to copper metal. Perhaps you already have an idea how to do this. Recall the metal activity series that you devised in Laboratory Activity B.4 (pages 118–120). By adding solid metal samples to solutions containing other metal ions, you compared the chemical activity of each metal. Review your observations from that earlier laboratory activity. Which of the tested metals are more active than copper? Can you predict the result of placing one of those active metals in your copper(II) chloride solution? Check your prediction by performing these steps.

9. Obtain a watch glass that can completely cover the top of the beaker containing copper(II) chloride ($CuCl_2$) solution from Step 5. For each gram of copper powder that you started with in Laboratory Activity B.2, add about one gram of zinc metal to the $CuCl_2$ solution.
10. Immediately cover the beaker with the watch glass, and allow it to stand for several minutes. Record your observations.
11. After the reaction has subsided, remove the watch glass and gently dislodge solid copper that forms on the surfaces of the zinc pieces.
12. Continue to dislodge copper from the zinc until you are convinced that the zinc has stopped reacting with the solution. (How can you decide?) Then add 10 mL of 1 M HCl to the beaker and carefully remove any large pieces of solid zinc from the beaker with forceps. Replace the watch glass. Record your observations.
13. After a few minutes, carefully decant as much of the liquid as possible into another beaker.

⚠ Note warning about hydrochloric acid in Step 2.

C.5 RETRIEVING COPPER

Laboratory Activity

In this activity, students will retrieve the copper from copper (II) oxide that they produced in Laboratory Activity B.2 (p. 116).

Time

90 minutes

Materials (for a class of 24 students working in pairs)

CuO powder from activity B.2
1000 mL 1 M HCl
12 hot plates
12 stirring rods
12 thermometers
24 100 mL beakers
1 L distilled water
24 pieces filter paper
12 balances
12 watch glasses
12 forceps

Safety

Students should wear safety goggles and aprons during this lab activity.
Hydrochloric acid may damage your skin. Wash immediately with water.

Lab tip

Step 13: Students must wash the Cu thoroughly with distilled water to avoid reoxidation when product is left to dry overnight.

Post lab-discussion

You might have students calculate the percent of copper recovered from the copper(II) oxide powder.

Answers

1. a. All of the original copper did not react with the oxygen when heated.

   b. Expose more of the copper to air (increase surface area).

2. Answers will vary.

3. a. $Cu^{++} + 2e^- \rightarrow Cu$
   $Zn \rightarrow Zn^{2+} + 2e^-$

   b. i. Zinc

      ii. Copper

      iii. Zinc

      iv. Copper

4. a. Blue color became less intense.

   b. Zinc is more reactive than copper. Zinc replaced the copper ions in the solution.

   c. When the color of the solution disappears, all of the copper ions have been removed.

5. a. Resources used are H, Cl, O, and Zn.

   b. Zinc reacted with Cl^- to form $ZnCl_2$ solution. Hydrogen reacted with oxygen to form water.

14. Wash the solid copper several times with distilled water.
15. Transfer the copper to a preweighed piece of filter paper, and allow it to dry overnight.
16. When the sample and filter paper have dried, find the mass of copper metal. This represents the copper that you recovered from the copper(II) chloride solution.
17. Follow your teacher's instructions for disposing of waste materials.
18. Wash your hands thoroughly before leaving the laboratory.

Questions

1. In Laboratory Activity B.2 not all of the original copper powder reacted when you heated it in air.
   a. Why do you think the reaction was incomplete?
   b. How would you revise the procedure so that more copper(II) oxide could form?
2. a. In Laboratory Activity B.2 what mass of the original copper-powder sample reacted when you heated it? (*Hint:* Use the original mass of copper used in Laboratory Activity B.2 and the mass of copper residue found in Step 8 to calculate this.)
   b. What percent of the total copper-powder sample reacted?
3. In the reaction between copper(II) chloride ($CuCl_2$) solution and zinc metal, each Cu^{2+} ion gained two electrons to form an atom of copper metal. Each zinc metal atom lost two electrons to form a Zn^{2+} ion.
   a. Write a chemical equation that represents this process. (To review how, turn to pages 122–123.)
   b. Based on the equation you wrote in Question 3a, identify
      i. the species that was oxidized.
      ii. the species that was reduced.
      iii. the reducing agent.
      iv. the oxidizing agent.
4. Adding HCl to CuO resulted in the formation of a blue solution. This color is due to the presence of $Cu^{2+}(aq)$ ions. Consult your observations in answering these questions:
   a. Describe what happened to the solution color after you added zinc in Step 9.
   b. What caused the changes you observed in the solution?
   c. How can the color of the solution be used to indicate when the zinc metal has removed the Cu^{2+} ions in solution?
5. To recover Cu metal from the $CuCl_2$ solution, you had to use other resources.
   a. What resources were "used up" in this process?
   b. Where did each of them go?

C.6 CONSERVATION IN THE COMMUNITY

In some ways Earth is like a spaceship. The resources "on board" are all that are available to the inhabitants of the ship. Some resources—such as fresh water, air, fertile soil, plants, and animals—can eventually be replenished by natural processes. These resources are called **renewable resources.** As long as natural cycles are not disturbed too much, supplies of renewable resources can be maintained indefinitely. Other materials—such as metals, natural gas, coal, and petroleum—are considered **nonrenewable resources** because they cannot be readily replenished. If atoms are always conserved, why do some people say that a resource may be "running out"? Can a resource actually "run out"?

The answer can be found by first remembering that atoms are conserved in chemical processes, but molecules might not be. For example, the current production of new petroleum molecules in nature is very much slower than the current rate at which petroleum molecules are being burned to produce carbon dioxide, water, and other molecules. Thus the total inventory of petroleum on Earth is declining.

A resource—particularly a metal—can be depleted in another way. As you learned in Section B, profitable mining depends on finding an ore with at least some minimum metal content. This minimum level depends on the metal and its ore: from as low as 1% for copper or 0.001% for gold to as high as 30% for aluminum.

Once ores with high metal content are depleted, lower-grade ores with less metal content are processed. Meanwhile, atoms of the metal, once concentrated in rich deposits within limited parts of the world, gradually become spread out (dispersed) over wider areas of Earth. Eventually, the mining and extraction of certain metals may become prohibitively expensive for general use. At that time, for practical purposes the supply of that resource can be considered depleted.

Can such depletion scenarios be avoided? Can Earth's mineral resources be conserved? One strategy for conservation is to slow down the rate at which the resources are used. Part of this strategy includes rethinking personal and societal habits and practices involving resource use. Such rethinking can involve decisions such as whether it is better to use paper or plastic bags in grocery stores, as well as whether, as is done in some parts of the world, it is even better to encourage customers to bring their own reusable bags.

Rethinking can take the form of re-examining old assumptions, identifying resource-saving strategies, and, perhaps, uncovering new solutions to old problems. Possibly the most important part of rethinking resource conservation and management is to consider the most direct option—source reduction. That simply means decreasing the amount of resources used. The fewer resources used, the more that remain available for future generations.

Another approach is to replace a resource by finding substitute materials with similar properties, preferably materials from renewable resources. In addition, some manufactured items can be refurbished or repaired for reuse rather than sent to a landfill. Common examples include used car parts and printer cartridges, both of which can often be reused after rebuilding or refilling. Finally, certain items can be recycled, or gathered for

Resources such as petroleum are regarded as nonrenewable because they can be formed only over millions of years.

Minimum profitability levels for mining depend not only on the abundance of the metal in the ore but also on the metal's chemical activity. Less active metals (such as gold) can more easily be released from their compounds than can more active metals (such as aluminum). Thus richer ores are needed to make the mining of active metals profitable.

C.6 CONSERVATION IN THE COMMUNITY

The terms *renewable* and *nonrenewable* resources are explained in this section. This section explains that although atoms are always conserved, molecules are turned into a less useful form. "Running out" of a resource may mean that it is too spread out for economical extraction.

The term *conservation,* when used by the general public, often refers to management of resources. Conservation options that reduce resource demand and/or waste production include replacing, reusing, recycling, and rethinking.

reprocessing. Recycling allows the resources present in the items to be used again. Figure 20 illustrates the steps involved in recycling aluminum cans.

However, even the recycling process can create environmental problems. For example, aluminum scrap, excluding cans, may contain up to one percent magnesium. The magnesium must be removed before the aluminum is processed into other products. Conventional removal of magnesium requires the addition of chlorine gas, which requires special handling in the workplace and can contribute to air pollution. An alternative method uses ceramic oxides, a process that is safer to use and that eliminates chlorine emissions to the atmosphere.

In some nations most glass bottles are refilled, not discarded.

Ultimately, some items will be discarded if they are no longer wanted or needed. Each person in this country throws away an average of about 2 kg (4 lb) of unwanted materials (or waste) daily. Some products, such as yesterday's newspaper, become waste after they fulfill their initial purpose. Others, such as telephones and computers, become waste when they are discarded for newer models. Combined, materials directly discarded by U.S. citizens would fill the New Orleans Superdome from floor to ceiling twice each day.

As you can see in Figure 21, the largest fraction of municipal waste in the United States is paper. While recycling reduces this component some-

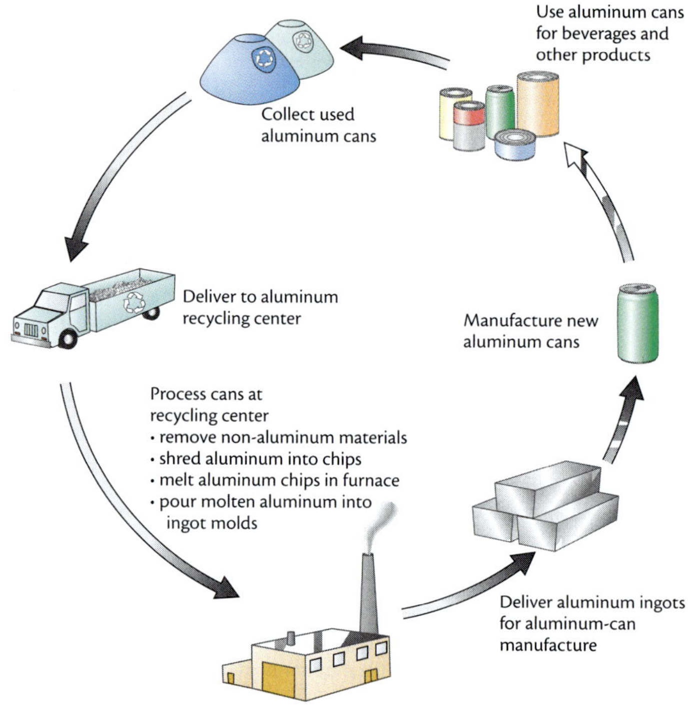

Figure 20 *Aluminum can recycling.*

146

what, the market for recycled paper is limited. Because this leaves a high proportion of combustibles in the waste stream, waste-to-energy plants have become an attractive option. More than 120 waste-to-energy plants currently operate in the United States, burning about 97 000 tons of solid waste per day. Each ton of garbage that serves as "fuel" in these plants produces about a third of the energy released by a similar quantity of coal.

Although waste-to-energy plants produce some fly ash and solid residues, such plants can allow the recycling of materials that otherwise would be disposed of as part of an unwanted product. In addition, waste-to-energy plants tend to increase recycling, both on-site and in the communities in which they are located.

Recycling, landfilling, and combustion for energy production are three options for the final step in the life cycle of a material—a topic you will now consider as you prepare to choose a material for your coin design.

WASTE GENERATED IN UNITED STATES IN 1997 BEFORE RECYCLING

Category	Weight (million tons)	Percent of Waste
Paper	83.8	38.6
Yard waste	27.7	12.8
Plastics	21.5	9.9
Metals	16.6	7.7
Wood	11.6	5.3
Food waste	21.9	10.1
Glass	12.0	5.5
Other	21.8	10.0

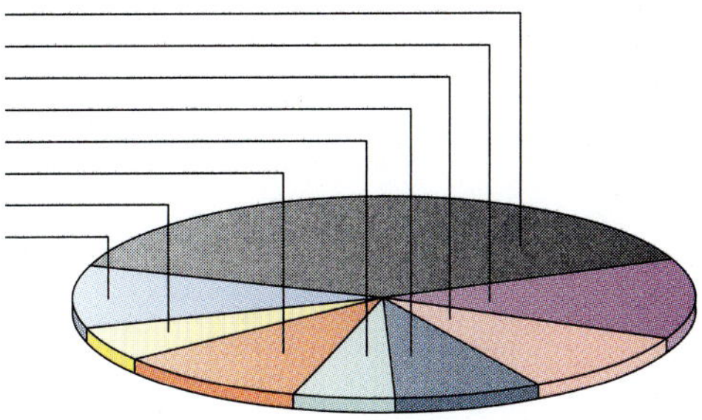

Total Generation: 217 million tons

WASTE GENERATED IN UNITED STATES IN 1997 AFTER RECYCLING

Category	Weight (million tons)	Percent of Waste
Paper	48.9	31.3
Yard waste	16.2	10.4
Plastics	20.4	13.0
Metals	10.1	6.4
Wood	11.0	7.0
Food waste	21.3	13.6
Glass	9.1	5.8
Other	19.3	12.3

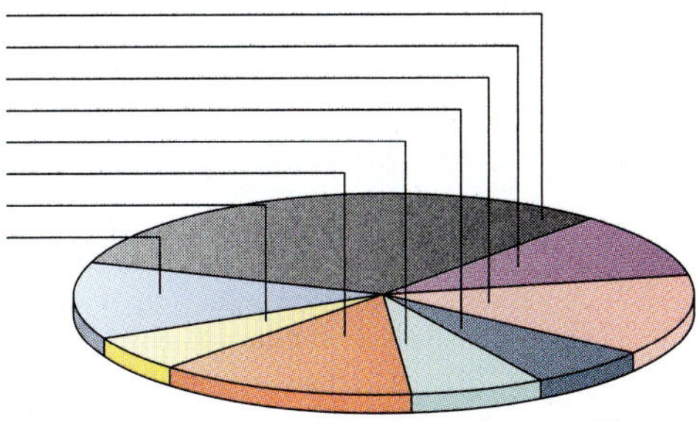

Source: Characterization of MSW in the U.S.: 1998 Update, U.S. EPA, Washington, DC

Total Discards: 156.3 million tons

Figure 21 *Composition of U.S. municipal waste stream before and after recycling.*

147

C.7 THE LIFE CYCLE OF A MATERIAL

In designing a new product, the full life cycle of the material must be considered, from mining to refining, designing a useful material, then recovering and reusing the material.

COPPER LIFE-CYCLE ANALYSIS

Building Skills 10

This activity lends itself well to small group work. Have student groups discuss the questions, then gather for a full-class discussion.

Answers

1. Refining.
2. Refining, designing.

C.7 THE LIFE CYCLE OF A MATERIAL

In designing a new product for human use, proper evaluation must include consideration of the full life cycle of the materials involved. Such a life cycle has several distinct stages. Raw materials are first obtained and then refined and synthesized into the desired material. That material is then used to make the product designed for a particular use. When the product is no longer useful, the materials may be recovered and reused, or they may end up scattered in landfills. Figure 22 illustrates this general life cycle.

In every step of the cycle, energy and resources are used. Laboratory Activities B.2 (Converting Copper) and C.5 (Retrieving Copper) are good examples of this fact. Recall that heat energy and chemical resources (hydrochloric acid and zinc metal) were used first to convert the copper metal to other substances and then to recover the copper metal. Because energy use and resource use impact economics as well as the environment, each step in the life cycle of a material becomes a factor to consider when a new product is designed.

The next activity will allow you to model this process for a familiar material—copper.

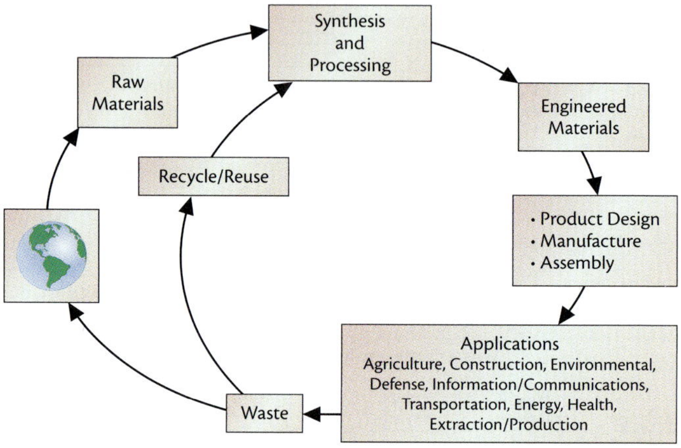

Figure 22 *Life cycle of materials.*

COPPER LIFE-CYCLE ANALYSIS

Building Skills 10

Imagine that you are involved in designing copper water pipes. Use Figure 11 (page 115) and Figure 22 to conduct a life-cycle analysis for the copper metal in your pipes. Consider how each step in the life cycle of copper will affect your final design by answering the following questions.

1. Which life-cycle steps consume significant quantities of energy?
2. Which steps (such as the reduction of a mineral in an ore to produce the metal) require the use of other materials?

3. How will you obtain the copper for your pipes?
4. Consider the transportation of materials in each step. How might this influence the design of your copper pipes?
5. Copper pipes may someday no longer be in use.
   a. What will happen to the copper in them?
   b. How can this issue be addressed when designing the pipes?
6. Consider your answers to each of the previous questions. How does each decision influence the cost of your copper pipes?

C.8 DESIGNING A COIN

Making Decisions

By now you have a good idea of what properties your half-dollar coin should possess. Perhaps you have even narrowed the range of materials still under consideration. To prepare for the final phase of your coin design, answer these questions.

1. Look at the list of required and desired physical and chemical properties you developed on page 107 of Section A. Revise and update your list using what you have learned since completing that activity.
2. Coins are traditionally made of metals. What alternative materials might be considered? What are the advantages or disadvantages of these alternatives?
3. Construct a table similar to the one shown below. Enter all necessary and desirable physical properties from your revised list in Question 1. Then list the materials you are considering for use in your coin. If you have not already done so, begin researching the properties of these materials. Indicate whether each material meets each criterion. For example, you might check the box if it does and place an "x" in the box if it does not. Construct a similar table for your list of necessary and desirable chemical properties.

	Materials				
	A	B	C	D	E
Necessary Properties					
Desirable Properties					

3. Reduce copper ore to copper metal.
4. For ease of transportation, copper pipes should be cut into shorter pieces that can be easily connected to each other rather than one long pipe.
5. a. The copper in the pipes may be recycled.
   b. The percent of copper in the pipe should be specified in the design. The properties of other metals combined with copper should have physical properties that will make it easy to separate from copper during the recycling process.
6. In each step, the cost of the copper pipe would likely increase.

C.8 DESIGNING A COIN

MAKING DECISIONS

Student groups will actually begin the design of their coin using all of the chemical principles discussed in this unit. Students do not need to complete their research at this point. They can work on these tables while studying Section D.

Answers

1. Student answers
2. Use a combination of different metals (alloys), plastics, or composites.
   Advantage: Use materials that are available. Disadvantage: May be difficult to duplicate.
3. Evaluate student chart.

4. a. Student answers

   b. Percent composition of the alloy

   c. Comparable to present coins

   d. 5 to 10 grams

5. a. They will disappear from circulation.

   b. Copper, silver, gold, and platinum

6. Costs of mining and refining the material

7. a. Mining new material or recycling

   b. Life cycle should be 50 years or longer

   c. Material should be reprocessed into a useful substance. Could the coin be refurbished or repaired for reuse?

   d. Consideration must be given to the composition of the coin. It should not contain any material that may cause damage to the environment during reprocessing.

4. Consider the design of your coin.
   a. What images are you considering for the obverse ("heads") and reverse ("tails") sides of the coin?
   b. What anticounterfeiting features are appropriate to include in your design?
   c. How large should this coin be?
   d. What should be its mass?
5. The cost of materials should be considered in your design.
   a. What would happen if the materials in the coin were worth more than the face value of the coin (fifty cents)?
   b. What particular materials have you already ruled out because of this concern?
6. What other costs need to be considered as you design your coin?

Questions & Answers CD-ROM WWW.

7. Consider the life cycle of your coin.
   a. How will the coin materials be obtained and processed?
   b. How long do you expect each coin to remain in circulation?
   c. What will happen to the materials when used coins are removed from circulation?
   d. How might impact on the environment be addressed at each stage of the coin's life cycle?

SECTION SUMMARY

Reviewing the Concepts

1. **a.** State the law of conservation of matter.

b. How is a scientific law, such as the law of conservation of matter, different from a law created by the government?

a. Matter can neither be created or destroyed, only changed in form. Earth's basic "stuff," its particles, is indestructible.

b. Governmental laws created by people are subject to change due to social and economic pressures. Governmental laws can be instituted and terminated by social whim. Scientific laws are fundamental explanations of nature that have been repeatedly observed and verified. Even explanations that are taken for granted are called theories until they are completely undisputed and pronounced as laws. Scientific laws are steadfast.

2. Complete atom inventories to determine if each of the following equations is balanced:

a. The preparation of tin(II) fluoride, an ingredient in some toothpastes (called "stannous fluoride" on some labels):

$$Sn(s) + HF(aq) \longrightarrow SnF_2(aq) + H_2(g)$$

Tin metal	Hydrofluoric acid	Tin(II) fluoride	Hydrogen gas

b. The synthesis of carborundum for sandpaper:

$$SiO_2(s) + C(s) \longrightarrow SiC(s) + CO(g)$$

Silicon dioxide (sand)	Carbon	Silicon carbide (carborundum)	Carbon monoxide

c. The reaction of an antacid with stomach acid:

$$Al(OH)_3(s) + 3\ HCl(aq) \longrightarrow AlCl_3(aq) + 3\ H_2O(l)$$

Aluminum hydroxide	Hydrochloric acid	Aluminum chloride	Water

a. Not balanced; hydrogen and fluorine atoms unbalanced

$$Sn(s) + HF(aq) \longrightarrow SnF_2(aq) + H_2(g)$$

Reactants	**Products**
1 tin	**1 tin**
1 hydrogen	**2 hydrogen**
1 fluorine	**2 fluorine**

b. Not balanced; oxygen atoms unbalanced

$$SiO_2(s) + C(s) \longrightarrow SiC(s) + CO(g)$$

Reactants	**Products**
1 silicon	**1 silicon**
1 carbon	**1 carbon**
2 oxygen	**1 oxygen**

c. Balanced

$$Al(OH)_3(s) + 3\ HCl(aq) \longrightarrow AlCl_3(aq) + 3\ H_2O(l)$$

Reactants	**Products**
1 aluminum	**1 aluminum**
6 hydrogen	**6 hydrogen**
3 oxygen	**3 oxygen**
3 chlorine	**3 chlorine**

3. Why are the phrases "using up" and "throwing away" inaccurate from a chemical viewpoint?

The atoms, elements, or basic materials are not destroyed or used up but changed in form during chemical reactions. The materials often transform into less useful products. It is impossible to throw anything away because Earth is self-contained and does not lose materials into space.

4. Find the molar mass of each substance.

a. oxygen gas, O_2 — **a. 32.00 g/mol**

b. ozone, O_3 — **b. 48.00 g/mol**

c. caffeine, $C_8H_{10}N_4O_2$ — **c. 194.20 g/mol**

d. a typical antacid, $Mg(OH)_2$ — **d. 58.32 g/mol**

e. aspirin, $C_9H_8O_4$ — **e. 180.46 g/mol**

5. Create an analogy that illustrates the mole concept. Refer to Figure 17 and the text in C.3, Atom, Molecule, and Ion Inventory (pages 135–137) for examples.

Possible example: If a mole of pennies was divided up evenly among the 300 million people in the United States, each person could spend $1000 per day for 50 years and still leave over 99.99% of the money to his or her heirs.

6. Balance each of the following chemical expressions:

a. Preparing phosphoric acid (used in making soft drinks, detergents, and other products) from calcium phosphate and sulfuric acid:

$\_\_Ca_3(PO_4)_2 + \_\_H_2SO_4 \longrightarrow \_\_H_3PO_4 + \_\_CaSO_4$

b. Preparing tungsten from one of its minerals:

$\_\_WO_3 + \_\_H_2 \longrightarrow \_\_W + \_\_H_2O$

c. Heating lead sulfide in air:

$\_\_PbS + \_\_O_2 \longrightarrow \_\_PbO + \_\_SO_2$

d. Burning gasoline:

$\_\_C_8H_{18} + \_\_O_2 \longrightarrow \_\_CO_2 + \_\_H_2O$

e. Rusting (oxidation) of iron metal:

$\_\_Fe + \_\_O_2 \longrightarrow \_\_Fe_2O_3$

a. $\underline{1}\ Ca_3(PO_4)_2(s) + \underline{3}\ H_2SO_4(aq) \longrightarrow \underline{2}\ H_3PO_4(aq) + \underline{3}\ CaSO_4(aq)$

b. $\underline{1}\ WO_3(s) + \underline{3}\ H_2(g) \longrightarrow W(s) + \underline{3}\ H_2O(l)$

c. $\underline{2}\ PbS(s) + \underline{3}\ O_2(g) \longrightarrow \underline{2}\ PbO(s) + \underline{2}\ SO_2(g)$

d. $\underline{2}\ C_8H_{18}(g) + \underline{25}\ O_2(g) \longrightarrow \underline{16}\ CO_2(g) + \underline{18}\ H_2O(g)$

e. $\underline{4}\ Fe + \underline{3}\ O_2 \longrightarrow \underline{2}\ Fe_2O_3$

7. A student is asked to balance this chemical expression:

$Na_2SO_4 + KCl \longrightarrow NaCl + K_2SO_4$

The student decides to balance the expression this way:

$Na_2SO_4 + K_2Cl \longrightarrow Na_2Cl + K_2SO_4$

a. Complete an atom inventory on the student's answer. Are all atoms conserved?

b. Did the student create a properly balanced chemical equation? Explain.

c. If your answer to Question 7b is "no," write a correctly balanced equation.

a. Yes

b. No

c. In balancing an equation, only the coefficients are manipulated:

$Na_2SO_4 + 2\ KCl \longrightarrow 2\ NaCl + K_2SO_4$

8. Find the percent metal (by mass) in each compound:

a. Ag_2S **a. silver sulfide—87.06% silver**

b. Al_2O_3 **b. aluminum oxide—52.93% aluminum**

c. $CaCO_3$ **c. calcium carbonate—40.04% calcium**

9. A 50.0-g sample of ore contains 5.00 g lead(II) sulfate ($PbSO_4$).

a. What is the percent lead metal (Pb) in $PbSO_4$?

b. What is the percent $PbSO_4$ in the ore sample?

c. What is the percent Pb in the total ore sample?

a. 68.32% **b. 10.0%** **c. 6.832%**

10. What is the difference between reusing and recycling? Give two examples of each, other than those presented in the textbook.

Reusing: material can be reused in its original form; for example, canning jars used year after year, garage sale treasures, antiques and used furniture items that are restored and reused.

Recycling: to collect certain items for reprocessing to allow the resources in them to be used again; for example, old fabric and clothing can be made into fiber for use in paper, and rubber tires can be recycled into material for flooring or construction material.

11. In addition to those found in this textbook, list four examples of

a. renewable resources.

b. nonrenewable resources.

a. renewable: biomass, plant and animal products like wood, paper, food, sand, ethanol fuel, or water, metals such as zinc and tungsten.

b. nonrenewable: nonmetals, natural diamonds and other precious stones, building stones such as marble and granite.

12. Classify each use as either recycling or reusing:

a. storing water in used juice bottles for an emergency

b. converting plastic milk containers into fibers used to weave clothing fabric

c. packing breakable items with shredded newspapers

A and c are reusing, b and d are recycling

Connecting the Concepts

13. Chromium minerals are found at three different mine sites in these forms:

a. Site 1: Chromite, $FeCr_2O_4$

b. Site 2: Crocoite, $PbCrO_4$

c. Site 3: Chrome ochre, Cr_2O_3

Based only on percent composition, at which site is chromium mining most feasible?

Site 1: 46.46% Cr, Site 2: 16.09% Cr, Site 3: 68.42% Cr

Site 3 has the highest percentage of chromium in chrome ochre and would be the most economically feasible.

14. Earth has been compared to a space station.

a. In what ways is this analogy useful?

b. In what ways is it misleading?

a. Earth may be viewed as a space station since Earth and its atmosphere essentially represent a closed system. The amount of resources is limited because matter can neither be lost nor gained. The analogy is useful in helping students realize that there is a finite amount of resources—water, fuel, raw materials—onboard, and no place to go to get more.

b. The analogy is a little misleading since Earth continually gains useful radiant energy from the Sun, while a space station more closely resembles an isolated (self-contained) system when it is in remote, unserviced space. A truly isolated system exchanges neither mass nor energy with its surroundings.

15. Describe at least two benefits of discarding less waste material and recycling more of it.

Discarding less waste decreases the waste storage problem, and recycling waste may lower the demand for resources resulting in a decrease in price. Other proven benefits include pollution reduction, energy savings, and the creation of new jobs.

16. One method of producing chromium metal includes, as the final step, the reaction of chromium(III) oxide with silicon at high temperature:

$$2\,Cr_2O_3(s) + 3\,Si(s) \longrightarrow 4\,Cr(s) + 3\,SiO_2(s)$$

a. How many moles of each reactant and product are shown in this chemical equation?

b. What mass (in grams) of each reactant and product is specified by this equation?

c. Show how this equation illustrates the law of conservation of matter.

a. reactants: 2 moles Cr_2O_3 and 3 moles of Si
products: 4 moles of Cr and 3 moles of SiO_2

b. reactants: 2 moles of Cr_2O_3 = 303.98 g; 3 moles Si = 84.26 g
products: 4 moles Cr = 207.98 g; 3 moles of SiO_2 = 180.25 g

c. The total mass of the reactants, 388.24 g equals the total mass of the products, 388.24 g.

Extending the Concepts

17. Why is aluminum metal more easily produced from recycled aluminum cans than from aluminum-containing materials such as clay, bauxite, or aluminum oxide ore?

17. Energy is used to separate the aluminum ore (bauxite) from unwanted minerals and dirt. A large quantity of electrical energy is needed to break the bonds in Al_2O_3 and reduce Al^{3+} ions to aluminum metal. The cost of electrical energy per mole of aluminum produced is very high.

Recycling aluminum from aluminum cans requires minimal energy (for melting and recasting) compared to the energy needed to process aluminum from its ores. It has been estimated that throwing away an aluminum beverage container wastes as much energy as pouring out such a can half-filled with gasoline; and the cost of recycling an aluminum can is approximately one-twentieth the cost of making a new can from ores.

18. In your laboratory experiences, you may encounter some compounds called hydrates. Examples of hydrates include $Na_2S_2O_3 \cdot 5\ H_2O$, $CaSO_4 \cdot 2\ H_2O$, and $Na_2CO_3 \cdot 10\ H_2O$.

a. Why are these compounds called "hydrates"?

b. Calculate the molar mass of each listed hydrate.

c. Calculate the percent composition of water in each hydrate.

d. Although hydrates contain significant amounts of water, they are found as solid substances at room temperature. How is this possible?

a. A hydrate is a compound that incorporates water molecules into its crystal structure.

b. $Na_2S_2O_3 \cdot 5\ H_2O = 248.18$ g/mol; $CaSO_4 \cdot 2\ H_2O = 172.17$ g/mol; $Na_2CO_3 \cdot 10\ H_2O = 286.139$ g/mol

c. $Na_2S_2O_3 \cdot 5\ H_2O = 36.29\%$; $CaSO_4 \cdot 2\ H_2O = 20.93\%$; $Na_2CO_3 \cdot 10\ H_2O = 62.96\%$

d. Hydrates retain a stable crystalline structure with the attached water molecules.

19. Atoms that presently make up your body may have once been part of Tyrannosaurus rex, Alexander the Great, or Cleopatra. Explain how this is possible.

Atoms are never destroyed under ordinary chemical conditions. The atoms making up the compounds of living organisms are recycled to be used again by new organisms when they die and decay.

SECTION D

MATERIALS: DESIGNING FOR PROPERTIES

As Earth's chemical resources continue to be extracted and used, issues of scarcity or cost, or other economic or political factors sometimes motivate a consideration of alternatives. One option is to find, modify, or create new materials to serve as substitutes. An ideal substitute satisfies three requirements: It is plentiful, it is inexpensive, and, of course, its useful properties match or exceed those of the original material. Alternative materials seldom meet these conditions completely. As a result, the benefits and burdens involved in each option must be weighed. This is true whether the design challenge involves a new coin, a better microwave oven, or a new computer storage device.

D.1 STRUCTURE AND PROPERTIES: ALLOTROPES

You have learned in your study of chemistry that elements are the fundamental building blocks of matter. You also know that each element—identified by a name, symbol, and particular populations of electrons and protons in its atoms—displays characteristic properties. For example, elements can be classified as metals, nonmetals, and metalloids. You are also aware that some elements, such as magnesium, are reactive and others, such as neon, are virtually inert. Given all of that, see if you can solve the following puzzle.

A CASE OF ELUSIVE IDENTITY — ChemQuandary 2

Imagine that you are given samples of three solids. You are told that each is a pure element.

You examine the first sample. It is a black solid that feels soft and a bit "greasy" to the touch. It leaves black marks when rubbed on the table. It is a useful lubricant and a fairly good conductor of electricity. A gram of the material sells for less than a penny.

You then inspect the second sample. It is a colorless, glasslike solid. It leaves deep scratch marks when rubbed on the table—in fact, it is among the hardest substances known. It is useful as an abrasive and as a coating for saw blades. It is a nonconductor of electricity. Depending on the quality of the solid piece, it can sell for $50 per gram or more than $20,000 per gram.

MATERIALS: DESIGNING FOR PROPERTIES

Introduction

Because the chemical resources from Earth are continuously being used up, it has become necessary to look for alternative materials to use as substitutes. The three requirements for a good substitute are that it must be plentiful, it must be inexpensive, and it must be equal to or better than the original material.

D.1 STRUCTURE AND PROPERTIES: ALLOTROPES

Review the properties and structure of elements before having students attempt to solve the ChemQuandary.

Activate and encourage your students' coin-design teams to complete their work during their study of Section D. They will make coin design presentations in PIAT. Set a tentative date for the presentations.

A CASE OF ELUSIVE IDENTITY

ChemQuandary 2

The three allotropes of carbon are introduced as clues to a puzzle, then explained. Emphasize that allotropes have the same atomic composition, but different structures and appearances. (Allotropes must be in the same physical state.)

Finally you look at the third sample. It is a fine, powdery solid made up of the roundest molecules found in nature. Although samples of the substance are present in prehistoric layers of Earth's crust, it was only discovered in 1985. Although the price of this substance is dropping, in pure form—gram-for-gram—it is currently more expensive than gold.

These are, indeed, three distinctly different substances. Yet you are told that all three samples are exactly the same element. In other words, each is composed only of atoms of one particular element—no impurities, no mixtures, no compounds.

How can atoms of one element make up such different materials? What additional information do you need to explain this? What element is being described?

Before you attempt to answer these questions, it will helpful for you to know that chemists recognize the three samples just described as allotropes of the same element. **Allotropes** are two or more forms of an element that have distinctly different physical or chemical properties.

To be considered allotropes, the forms of the element must be in the same state—either solid, liquid, or gas. Solid iron (Fe) and molten iron have distinctly different properties, but because they are in different states, they are not allotropes.

It is possible that some information in ChemQuandary 2 provided enough clues for you to identify the element involved: If you guessed carbon (C), you are correct. The identities of each of the three carbon allotropes can now be revealed.

The first sample, composed of a soft, black solid, is the carbon allotrope known as graphite. You encounter graphite either directly or indirectly nearly every day. It is a major component of pencil "lead." Due in part to its conductive properties, it is also found in many batteries.

The second sample is carbon in the form of diamond. Its melting point is among the highest of any element. Both natural and synthetic diamonds exist. The cost of a diamond depends in part on its quality and optical characteristics. When carefully selected and cut, natural diamonds have high decorative value, even though their chemical formula, C(s), is the same as the formula for the graphite in common pencil lead.

Because graphite and diamond are both pure carbon, they can burn in oxygen to produce carbon dioxide gas and thermal energy. Despite that, diamonds are not recommended as a heating fuel!

The third sample is one of many recently identified allotropes of carbon, a group of ball-like and even tubelike structures known as fullerenes. The particular fullerene described here is buckminsterfullerene, a 60-carbon hollow sphere (C_{60}), which resembles a soccer ball. Other fullerene molecules with formulas such as C_{70}, C_{240}, and C_{540} are known, as are structures composed of "rolled up" layers of carbon atoms in the form of hollow tubes termed nanotubes.

How do chemists account for the different properties of allotropes? The explanation lies in how the atoms of the element are linked and organized—that is, in the structure of the substances. Although the three allotropes of carbon are all composed only of carbon atoms, they have distinctly different atomic arrangements. The three allotrope models in Figure 23 illustrate this fact.

154

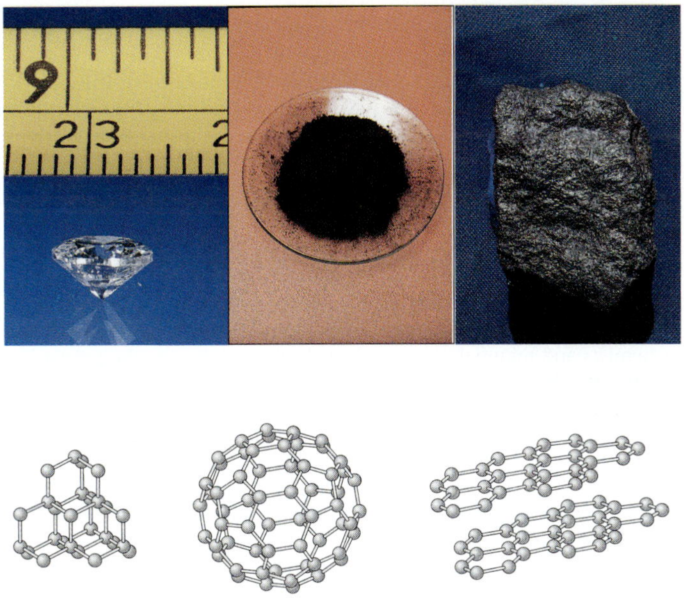

Figure 23 *Allotropic forms of carbon—their appearance and structures. From left to right, diamond, fullerene (in the form of buckminsterfullerene), and graphite. The structures shown of diamond and graphite are just small portions of three-dimensional structures that continue in every direction.*

LINKING PROPERTIES TO STRUCTURE

Building Skills 11

Based on the structural models shown in Figure 23, answer the following questions about some properties of carbon allotropes. The first question has been answered for you.

1. What feature of diamond's structure may account for its property as a hard, rigid substance, one that can scratch most other materials?

   Notice that every carbon atom in the interior of the diamond model is linked to four other carbon atoms by individual chemical bonds. These bonds have the effect of holding each carbon atom in place in a rigid three-dimensional structure. Each carbon-carbon linkage in diamond is a covalent bond involving the sharing of a pair of electrons. Thus "bending" or "denting" a diamond crystal would be extremely difficult because it would involve breaking a network of strong chemical bonds that locks each carbon atom in a fixed position. This accounts for diamond's rigidity and hardness.
2. How might the structure of diamond help explain why it is sometimes found in the form of large, single crystals?
3. What feature of graphite's structure might account for its usefulness as a lubricant?

CD-ROM WWW. **Structure and Properties: Allotropes**

LINKING PROPERTIES TO STRUCTURE

Building Skills 11

Answers

1. Rigid networks of tetrahedrons are linked into a strong, stable network.
2. The simple and regular structure provides a pattern that can be extended as large as can be, so long as the conditions and elemental carbon are available.
3. Planar sheets of molecules can slip by one another easily.

4. Balls of carbons would be very small and not linked to each other, appearing like powder.
5. Protective encasement of medicines.

D.2 MODIFYING PROPERTIES

Scientists and engineers are working together to find new and better materials. Two possible methods of changing the properties of materials are adding another substance or altering the methods through which the material is manufactured, or an entirely different substance may replace the material.

This section presents a variety of examples of how materials may be modified to change the properties necessary to meet specific needs and uses.

This section discusses the substitution of new materials for expensive or less available resources. Begin by asking students to describe changes in the materials used to make sports equipment (vaulting poles, golf clubs), medical equipment (heart valves, replacement joints, artificial skin), and automobiles (ceramic engines, plastic bumpers).

4. Why are fullerenes "powdery" as solids rather than composed of large-scale "chunks"?
5. A molecule of buckminsterfullerene (C_{60}) can be regarded as a hollow sphere. Chemists have demonstrated that it is possible to place an atom of another element inside the sphere. Can you think of any practical application for "carrying" atoms of another element inside fullerene molecules?

D.2 MODIFYING PROPERTIES

Throughout history—first by chance and more recently guided by science—humans have greatly extended the array of materials available for their use. Chemists and material scientists have learned to modify the properties of matter by physically blending or chemically combining two or more substances. Sometimes only slight changes in a material's properties are desired. At other times chemists create new materials with properties dramatically different from those of the constituent substances.

Black "lead" in a pencil is mainly graphite, a natural form of the element carbon. You have just learned that pure carbon can display distinctively different properties depending on how the carbon atoms are bonded in three-dimensional space. Thus carbon as graphite displays properties that are uniquely different from those of carbon as diamond or carbon as fullerene.

When first discovered, graphite was mistakenly identified as a form of lead, and the name stuck.

If pencil lead is mainly graphite, however, why are the properties of hard pencil lead (such as No. 4), which produces very light lines on paper, so different from the properties of soft writing lead (such as No. 1), which makes very broad, easily smudged lines? As you might suspect, the properties of pencil lead are controlled by the amount of another material that is mixed with the graphite. That second material is clay. Increasing the quantity of clay mixed with graphite produces harder pencil lead because less graphite can be rubbed off onto paper.

Examples of other materials designed or modified to meet specific needs abound. Clay, one of the most plentiful materials on this planet, is composed mainly of silicon and oxygen atoms and aluminum ions, along with magnesium, sodium, and potassium ions and water molecules. Early humans found that clay mixed with water, then molded and heated, formed useful ceramic products such as pottery and bricks.

In more recent times, researchers have used newly developed techniques and materials to produce engineering **ceramics**. Figure 24 compares the sources, processing, and products of conventional ceramics with newer, stronger engineering ceramics.

What properties of conventional ceramics made them useful in pottery and bricks? Characteristics such as hardness, rigidity, low chemical reactivity, and resistance to wear were certainly important. The main attractions of ceramics for future use, however, are their high melting points and their strength at high temperatures. Indeed, such ceramics have become attractive substitutes for steel in some applications. For example, diesel or turbine engines made of ceramics can operate at higher temperatures than metal engines. Such high-temperature engines run with increased efficiency, thus reducing fuel use.

Figure 24: *Conventional ceramic products such as bricks and pottery are made from clay. For example, kaolinite clay (top left) can be formed, by hand or machine, and fired in a kiln. Engineering ceramics, which may be higher melting, stronger, or less brittle than conventional ceramics, are produced from various minerals. They are designed for use in everything from the tiny resistors in computer circuit boards to huge jet turbines, both shown here.*

The major problem still facing researchers is that ceramics are brittle. They can fracture if exposed to rapid temperature changes, such as during hot engine cool-down. To avoid cracks in ceramic engine components (and resulting engine failure), scientists and engineers have developed precise manufacturing methods that carefully control the microstructure of the material. As these manufacturing practices become more economical, ceramics are likely to be more widely used in high-temperature applications.

Plastics have already replaced metals for many uses. These synthetic substances are composed of complex carbon-atom chains and rings with hydrogen and other atoms attached. Plastics generally are less dense and can be designed to be "springy" or resilient in situations where metals might become dented. Plastic automobile bodies are one example. The properties of certain plastics can be custom-made—in some cases without even changing the material's chemical composition. For example, polyethylene can be tailored to display relatively soft and pliable properties (as in a squeeze bottle for water) or crafted to be hard and brittle, almost glasslike in its behavior. Unfortunately, most plastics are made from petroleum, an important nonrenewable resource already in great demand as a fuel. You will learn more about petroleum in Unit 3.

157

ALTERNATIVES TO METALS
Building Skills 12

Answers

1. Wiring, motors, power distribution—aluminum

   Communication equipment, satellite transmissions—fiber optics

   Plumbing—plastic pipe

   Roofing—aluminum, fiberglass

   Coins—other alloys, more credit and debit cards, paper money

   Jewelry—silver, gold, chromium-plated nickel

   Pots and pans—aluminum, ceramics

   Shell casings—less copper, more zinc

   Food preparation machinery—stainless steel

   Auto radiators—aluminum, plastics, ceramics
2. Car bumpers, automobile engines, plumbing materials, sports gear, aircraft parts
3. Silver could replace copper in wiring, owing to its higher conductivity. This could also improve efficiency of computers, with lower resistance to electron flow increasing the speed of signal transmission.

D.3 STRIKING IT RICH
Laboratory Activity

This laboratory activity is intended to generate student interest in metallic resources and the chemistry involved in their transformation into useful materials. Students produce two types of brass, an alloy of zinc and copper. The final brass color may remind students of gold, and an intermediate type of brass may suggest the appearance of silver on the coin. Do not give away the anticipated results—let

Optical fibers developed by chemists, physicists, and engineers are well on their way to replacing conventional copper wires in phone- and data-transmission lines. Voice or electronic messages are sent through these thin, specially designed glass tubes of very pure silicon dioxide (SiO_2) as pulses of laser light. As many as 50 000 phone conversations or data transmissions can take place simultaneously in one glass fiber the thickness of a human hair. A typical 72-strand optical fiber ribbon can carry well over a million messages. The fiber's larger carrying capacity and noise-free characteristics outweigh its higher initial cost.

Chemists, chemical engineers, and materials scientists continue to find new and better alternatives to traditional materials for a variety of applications. Such custom-tailoring at the molecular level is possible because of chemical knowledge—knowledge of how the atomic composition of materials affects their observable properties and behavior.

Now it's your turn to consider possible uses of some alternative materials.

ALTERNATIVES TO METALS
Building Skills 12

1. Select four uses of copper from the list found in Figure 11 (page 115). For each use, suggest an alternative material that could serve the same purpose. Consider both conventional materials and possible new materials.
2. Suggest some common metallic items that might be replaced by ceramic or plastic versions.
3. Suppose silver became as common and inexpensive as copper. In what uses would silver most likely replace copper? Explain.

D.3 STRIKING IT RICH
Laboratory Activity

Introduction

Seeing is believing—or so it is said. In this activity, you will change the appearance of some pennies through chemical and heat treatment.

Procedure

1. In your laboratory notebook, prepare a data table similar to the one shown here, leaving plenty of room for your observations.

DATA TABLE

Condition	Appearance
Untreated penny	
Penny treated with Zn and $ZnCl_2$	
Penny treated with Zn and $ZnCl_2$ and heated in burner flame	

2. Obtain three pennies. Use steel wool to clean each penny until it is shiny. Record the appearance of the pennies.
3. Set aside one of the clean pennies to serve as a **control**—an untreated sample that can be compared later to the other two treated coins.
4. Weigh a 2.0-g to 2.2-g sample of granulated zinc (Zn) or zinc foil. Place it in a 250-mL beaker.
5. Use a graduated cylinder to measure 25 mL of 1 M zinc chloride ($ZnCl_2$) solution. Add the solution to the beaker containing the zinc metal. **CAUTION:** *Zinc chloride solution can damage skin. If any accidentally spills on you, ask a classmate to notify your teacher immediately; wash the affected area with tap water immediately.*
6. Cover the beaker with a watch glass and place it on a hot plate. Gently heat the solution until it just begins to bubble, then lower the heat to continue gentle bubbling. Do not allow the solution to boil vigorously or become heated to dryness. **CAUTION:** *Note the warning about zinc chloride solution in Step 5.*
7. Using forceps or tongs, carefully lower two clean pennies into the solution in the beaker. To avoid causing a splash, do not drop the coins into the solution. Put the watch glass on the beaker and keep the solution boiling gently for two to three minutes. You will notice a change in the appearance of the pennies during this time.
8. Using forceps or tongs, remove the two coins from the beaker. Rinse them under running tap water, then gently dry them with a paper towel. Set one treated coin aside for later comparisons.
9. Briefly heat the other treated, dried coin in the outer cone of a burner flame, holding it with forceps or tongs, as shown in Figure 25. Heat the coin only until you observe a color change, which will take 10 to 20 seconds. Do not overheat.
10. Immediately rinse the heated coin under running tap water, and gently dry it with a paper towel. Record your observations.
11. Observe and compare the appearance of the three pennies. Record your observations.
12. When finished, discard the used zinc chloride solution and the used zinc as directed by your teacher.
13. Wash your hands thoroughly before leaving the laboratory.

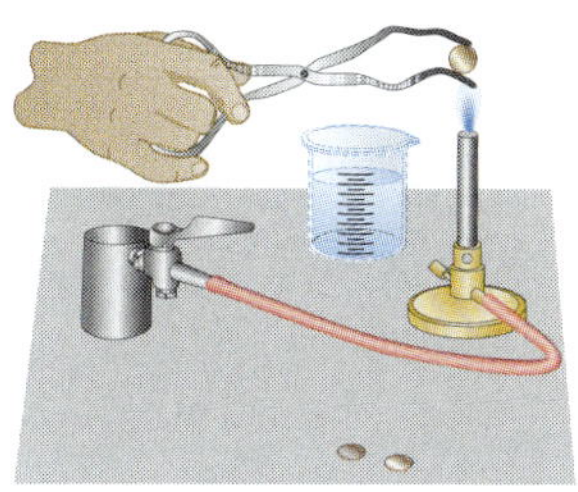

Figure 25 *Heating the treated penny.*

Questions

1. a. Compare the color of the three coins—untreated (the control), heated in zinc chloride solution only, and heated in zinc chloride solution and in a burner flame.
   b. Do the treated coins appear to be composed of metals other than copper? If so, explain.

students discover the changing appearance of the penny for themselves.

Success in this laboratory depends, in part, on the type of penny used in this activity. Best experience has been reported with new pennies—preferably uncirculated coins.

Time

30 minutes

Materials (for a class of 24 working in pairs)

24 protective goggles
36 250-mL beakers
12 25-mL graduated cylinders
12 crucible tongs
12 ring stands
24 ring clamps
12 watch glasses
12 wire gauze squares
12 gas burners or hot plates
12 matches or strikers
36 copper pennies
36 g granulated zinc (Zn) or zinc foil
350 mL of 1 M $ZnCl_2$ (136 g $ZnCl_2$/1 L solution)
2 L of distilled water
100 g steel wool

Safety

Students should wear safety goggles and aprons throughout the laboratory activity. Students must wear protective goggles for this and all other laboratory activities. Hot plates are safer than burners.

Advance preparation

Either pre- or post-1982 pennies can be used. However, since post-1982 pennies contain a much higher percent of zinc than pre-1982 pennies (97.6% vs. 5%), they are more likely to warp if overheated (melting point of copper: 1,083 °C, vs. zinc: 420 °C).

Pre-lab discussion

During a short pre-lab discussion, demonstrate how to operate a gas burner, as presented in the student text.

Lab tips

Remind students to refer to Figure 25 for the correct way to hold the coin in the flame.

Today's students may not know when pennies in regular circulation were 95% copper. Before 1982, a penny contained 2.95 g of copper and 0.15 g of zinc; currently, pennies contain only 0.06 g of copper and 2.46 g of zinc. Current prices of copper and zinc are found in the financial pages of most newspapers. If time permits, have students find the current prices of copper and zinc and calculate the value of a pre- and post-1982 penny. This information can be referenced for the final activity of this unit.

Answers

1. a. Untreated—copper color; heated in Zn and $ZnCl_2$ solution—shiny "silver" color; heated in Zn and $ZnCl_2$ solution and by flame—"gold" color

   b. Treated pennies resemble silver and gold.
2. By comparing physical properties like density (Zn: 7.1, Cu: 9.0, Ag: 10.5, Au: 19.3 g/mL), melting point (pure Zn: 420 °C, Cu: 1,083 °C, Ag: 962 °C, Au: 1,064 °C), and chemical reactivity (for example, zinc, copper, and silver all react with nitric acid, while gold does not).
3. Inexpensive substitute for silver or gold jewelry, cooking pan with different heat-conducting properties. (Student answers will vary.)
4. a. A solid solution of brass was produced; the copper atoms were dissolved in zinc.

   b. Yes, allowing the zinc to react with an acid such as HCl would remove the zinc and leave the copper layer. (Try this).

2. If someone claimed that a precious metal was produced in this activity, how would you decide whether the claim was correct?
3. Identify at least two practical uses for metallic changes similar to those you observed in this activity.
4. a. What happened to the copper atoms originally present in the treated pennies?

   b. Do you think the treated pennies could be converted back to ordinary coins? If so, what procedures would you use to accomplish this? **CAUTION:** *No laboratory work should be performed without your teacher's approval and direct supervision.*

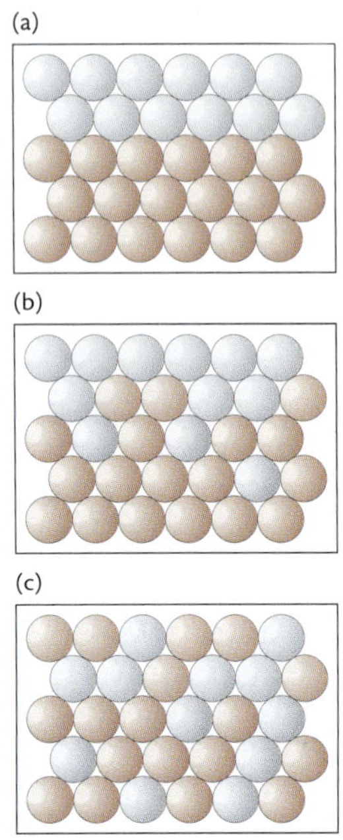

Figure 26 *Forming a brass outer layer on a penny. Brass samples contain mostly copper and between about 10% and 40% zinc (by mass).*

D.4 COMBINATIONS OF ELEMENTS: ALLOYS AND SEMICONDUCTORS

The laboratory activity you just completed demonstrated how metallic properties can be modified by creating an **alloy,** a solid combination of atoms of two or more metals. The immersion of a penny in hot zinc chloride solution produced a silvery alloy of zinc and copper called γ-brass. When you heated the penny in the burner flame, the zinc and copper atoms mixed more. The overall mixing of zinc and copper atoms to form brass is depicted in Figure 26. The resulting solid solution has a different concentration of zinc and copper and is known as α-brass. Brass has a golden color, unlike either copper or zinc. It is also harder than copper metal. Some other common alloys with familiar names are listed in Figure 27.

It is clear that one way to modify the properties of a particular metal is to form it into an alloy, just as you did when you produced a "gold-colored" penny. Often the results of alloying metals are unexpected, as you are about to discover.

FIVE CENTS' WORTH — ChemQuandary 3

A U.S. nickel coin is composed of an alloy of nickel and copper.

1. Based on your familiarity with that common five-cent coin, do you think the alloy in the coin contains more atoms of copper or more atoms of nickel? Why?

Now consider the facts: The nickel-coin alloy is composed of 25% nickel and 75% copper.

2. What does this suggest about a difference between an alloy (in this case a "solid solution" of copper and nickel atoms) and a simple mixture of powdered copper and powdered nickel?

In addition to alloys composed of "solid solutions" of two or more metals, other useful alloys include some that are well-defined compounds—that is, they have a constant, definite ratio of component metallic atoms. One example is Ni_3Al, a low-density, strong metallic alloy of nickel and

Post-lab discussion

Focus the discussion on the change in properties that occurs in the activity. Each substance has a unique set of properties, just as humans have unique sets of fingerprints. The change in properties in this activity allows us to predict that a change has occurred. Referring to the properties of a substance also permits identification of that substance. This idea of chemists as "matter detectives" occurs repeatedly throughout the *ChemCom* course.

Alloy and Composition	Examples (composition given in mass percent)	Comments
Brass Copper and zinc	Red brass 90% Cu, 10% Zn Yellow brass 67% Cu, 33% Zn Naval brass 60% Cu, 39% Zn, 1% Sn	Properties of brass vary with the proportion of copper and zinc and with the addition of small amounts of other elements. Brass is used for plumbing and lighting fixtures, rivets, screws, and ships.
Bronze Primarily copper with phosphorus, tin, zinc, and other elements	Coinage bronze 95% Cu, 4% Sn, 1% Zn Aluminum bronze 90% Cu, 10% Al Hardware bronze 89% Cu, 9% Zn, 2% Pb	Bronze is harder than brass. Its properties depend on the proportions of its components. Bronze is used for bearings, machine parts, telegraph wires, gunmetal, coins, medals, and bells.
Steel Primarily iron with carbon and small amounts of other elements	Steel 99% Fe, 1% C Nickel steel 96.5% Fe, 3.5% Ni Stainless steel 90–92% Fe, 0.4% Mn, <0.12% C, Cr (trace)	The properties of steel are often determined by the carbon content. High-carbon steel is hard and brittle; low- or medium-carbon steel can be welded and tooled. Steel is used for automobile and airplane parts, kitchen utensils, plumbing fixtures, and architectural decoration.
Other common alloys	Pewter 85% Sn, 6.8% Cu, 6% Bi, 1.7% Sb	Pewter is often used for figurines and other decorative objects.
	Mercury amalgams 50% Hg, 20% Ag, 16% Sn, 12% Cu, 2% Zn	Mercury amalgams have often been used for dental fillings.
	14 carat gold 58% Au, 14–28% Cu, 4–28% Ag	14-carat gold is popular in jewelry.
	White gold 90% Au, 10% Pd	White gold is also principally used for jewelry.

Figure 27 *Some common alloys and their uses.*

aluminum that is used as a component of jet aircraft engines. A very hard chromium-platinum alloy, Cr_3Pt, forms the basis of some commercial razor blade edges. And a special group of alloys, including the niobium-tin compound Nb_3Sn, displays superconductivity—the ability to conduct an electric current without any electrical resistance—if cooled to a sufficiently low temperature.

D.4 COMBINATION OF ELEMENTS: ALLOYS AND SEMICONDUCTORS

As a follow-up to the previous laboratory activity, students are introduced to the concept of an alloy (a solid solution of two or more metals). The properties of an alloy are different from that of the pure metals and difficult to predict.

FIVE CENTS' WORTH

ChemQuandary 3

Solid solutions such as alloys often have characteristics quite distinct from the component parts of the solution. It should not be a surprise that the five-cent coin does not look "75 percent copper." Have students consider a salt-water solution. The solution doesn't look or behave exactly like salt or water.

The internal composition of chemical resources may be changed to make them more useful for today's technology. Adding a small amount of other materials (five cents' worth) makes these changes. Alloys and semiconductor doping are two examples of these processes.

D.5 MODIFYING SURFACES

The performance of materials can be improved by coating (paint), electroplating, or adding a thin film. Each of these methods represents a positive interaction between science-technology and society.

This section provides a variety of concrete examples of how materials are modified to meet specific needs and uses.

A new coating being developed is a diamond coating, where a thin film of carbon atoms is applied to a surface. The coating resembles the structure of diamonds. This creates an exceptionally hard surface that may dramatically improve eyeglass lenses, cookware, and other products.

Combination of Elements: Alloys CD-ROM WWW.

Silicon Valley in California got its name from the element that is vital to the large number of computer-related industries located there.

Thus one important strategy for modifying the properties of metals is to produce alloys, which have properties that differ from those of the component elements. In fact, dramatic changes in the properties of a substance are possible when an extremely small amount of an element—as small as one atom per million atoms—is intermingled within a solid substance.

The metalloid known as silicon belongs in a class of materials called **semiconductors.** What does this term mean? You should recall that some elements (metals) are good conductors of electricity, whereas other elements (nonmetals) are not. Semiconductors, as the name implies, lie somewhere in between. In addition to silicon (Si), other elements and compounds that have semiconducting properties include germanium (Ge), tin (Sn), gallium arsenide (GaAs), and cadmium sulfide (CdS). Locate the elements identified as semiconductors on the Periodic Table (page 104). In what region of the table do they appear?

In the crystal structure of pure silicon, each atom is bonded to four other atoms. As a result, silicon's electrons are incapable of moving through the crystal. In other words, crystals of pure silicon display poor electrical conductivity at normal temperatures.

Although it sounds strange, adding certain impurities to pure silicon dramatically enhances its semiconductor properties. The impurities are atoms of other elements such as phosphorus, arsenic, aluminum, or gallium. This process, called **doping,** creates a situation in the solid silicon that allows charge carriers—either electrons or tiny regions of positive charge—to become mobilized within the crystal. This greatly improves silicon's semiconductor characteristics.

Most semiconductor devices (such as transistors and integrated circuits used in computers and other electronics) are based on solid crystals of silicon that contain intentionally added impurities. Those added atoms are responsible for the present silicon-based solid-state era, and, of course, for the existence of Silicon Valley itself.

Alloys and semiconductors involve changing the internal composition of a material to favorably affect its properties and uses. However, as you will now discover, there are other ways to modify or improve the properties of materials extracted from Earth's mineral resources.

D.5 MODIFYING SURFACES

In Section A you learned that skyrocketing copper prices in the early 1980s forced the United States to use zinc as the bulk metal in the penny. Copper was used to cover the coin's surface. This copper exterior not only preserves the appearance of the coin but also protects it from corrosion. The use of one material to protect the surface of another, less durable material is by no means new or unusual. A layer of protective material changes the properties of a manufactured product while allowing a less expensive or more available material to be used in the bulk of the item. As you continue your reading, you will learn about several types of surface treatments, both new and old, and how they can be used to enhance materials and products.

Coatings

The surface treatment that is probably most familiar to you is a **coating.** Coatings include paints, varnishes, and shellacs commonly applied to homes, cars, and many other products you buy and use. These materials are generally applied late in the manufacturing or construction process and are physically or chemically attached to the surface to be protected. Although coatings sink into pores in the base material, no bonding between the coating and the base takes place.

Paint, a typical example of a coating, consists of three components: pigment, solvent, and resin. The pigment provides color, the solvent allows application of the paint in liquid form, and the resin provides the desirable protective properties. Although paints have existed for centuries, growing concerns about the solvents released in their application and the heavy metals commonly found in their pigments have spurred the development of new formulations and methods of application. Of particular interest is the technique of powder coating, which eliminates the need for a solvent. See Figure 28. In this method, all components of the coating are blended and ground into an extremely fine powder. The product to be coated is then cleaned and sometimes pretreated. The powder is mixed with compressed air, pumped into spray guns, and given an electrical charge. The product is electrically grounded, allowing it to be coated by the charged powder. After coating, the product is heated to "cure" the powder into a tough shell. Many products, including bicycle and auto parts, are powder-coated to produce long-lasting, attractive surfaces.

A formulation, as used in industry, is much like a recipe for a product.

Plating

The process used to cover zinc with copper in making coinage involves using direct-current (DC) electricity, which causes redox reactions to occur between the metal and a metal-ion solution. This general process is known as **electroplating.** As you have already learned, cations of most metallic elements can be reduced. This fact is exploited in electroplating. For example,

Figure 28 *This bicycle's metal parts have been powder-coated to resist corrosion.*

CHEMISTRY AT WORK

All That's Gold Doesn't Glitter . . . Or Does It?

Did you ever wonder where the gold metal that makes up a high-school ring comes from? Or who is responsible for locating and recovering the material? **John Langhans** may have helped. He's a metallurgist with Barrick Goldstrike, a mining company in Elko, Nevada, that finds and recovers gold.

John's job is not an easy one. The normal concentration of gold in Earth's crust is only about 15 parts per billion (ppb). During the 1800s and early 1900s, miners looked for areas where gold was concentrated in the form of nuggets. More refined mining techniques enabled miners to locate deposits of finely divided gold. Such deposits account for most of today's gold mines.

John supervises laboratory personnel whose job it is to research and develop new methods for recovering gold. . .

Gold is often associated with certain types of rocks; quartz is an example. It is relatively easy to recover gold from these rock deposits. In the past decade, however, many of these deposits have been mined out, and the remaining gold is much harder to process.

When associated with sulfide minerals, gold is known as refractory ore. More sophisticated equipment and techniques must be employed to recover this gold. By using knowledge of chemistry and highly efficient recovery methods, John and his coworkers can extract gold from this material.

One method John's company uses is "autoclaving." In this process, the sulfide minerals are cooked under fairly high temperatures and pressures to dissolve the sulfides and expose the gold. The autoclaves work in a manner similar to that of a pressure cooker. You may be familiar with this piece of cookware if it is used to prepare meals in your home. Once the gold is exposed, a dilute basic solution containing cyanide ions (CN^-) is added to the ore to dissolve the gold. A gold cyanide complex $[Au(CN)_2]^-$ forms when gold atoms come in contact with the solution:

$$4\,Au + 8\,CN^- + 2\,H_2O + O_2 \longrightarrow 4\,[Au(CN)_2]^- + 4\,OH^-$$

Later, the gold is recovered through a process called electrowinning. Electrowinning uses an electrical current to transfer the gold onto steel wool. Finally, technicians heat the gold-plated steel wool to liquefy the gold, and pour the liquid gold into bar molds.

John supervises laboratory personnel whose job it is to research and develop new methods for recovering gold that are both safe and economical. He and his team also monitor the processing operations of mines to ensure that the environment is not degraded as a result of those operations.

164

Some Questions To Ponder . . .

Class rings are not 100% pure gold. Because pure gold is very soft, it is mixed with other metals (copper, silver, nickel, or zinc) to increase hardness and durability. The greater the amount of added metal, the lower the karat value. Pure gold is 24 karat. Most gold-containing rings are probably 12 karat, or about 50% gold.

Because of the specialized processes that Barrick Goldstrike uses, the company is able to profitably recover gold from refractory ore, even at grades as low as 5 grams gold per metric ton ore.

1. If a 12-karat-gold class ring has a mass of 11.2 g, how many metric tons of ore were processed in order to produce the gold to make the ring?
2. What is the value of the gold in the class ring if the market value of gold is about $10.50/g? (You may want to use the current actual price of gold, which is available in the business section of many newspapers or on the World Wide Web.)
3. During the 1800s, gold miners looked for gold in the form of nuggets. What property of gold allows it to be found in this form?

165

Figure 29 *Chrome plating can protect exposed steel surfaces.*

metal bumpers on trucks are often made of steel. In wet or snowy climates, the exposed steel would quickly corrode, or rust. Manufacturers protect the steel by coating it with chromium and nickel. See Figure 29.

Coins, too, have been plated with nickel for protection from corrosion. Writing half-reactions, which represent the reduction and oxidation parts of a redox process separately, can be helpful in understanding the plating process. The reduction half-reaction for plating nickel on steel can be written as follows:

$$Ni^{2+}(aq) + 2\,e^- \longrightarrow Ni(s)$$

But where do the electrons in this equation come from? Electroplating requires a power source, usually a battery or power supply when applied in the laboratory. The power source supplies electrons to the cathode (where reduction occurs) but it does not create those electrons. The ultimate source of the electrons in any electrochemical process is the anode. In electroplating, the anode is usually made of the metal to be plated. As metal cations are reduced, removed from the solution, and deposited on the object attached to the cathode, metal atoms at the anode are oxidized and dissolve into the plating solution. Figure 30 illustrates this process.

Thus the other half-reaction for the plating system is an oxidation, the reverse of the reaction shown above. Based on what you know about metal reactivities, would you expect this system of reactions to proceed spontaneously? Although systems that can plate metals without applied electric current have been developed, they must include a chemical reducing agent—which is sometimes more expensive than electric current.

Unlike coatings, platings are usually bonded to the surface of the bulk material through a metallic bond. This is desirable because it keeps the metal finish firmly attached to the object. After several layers of atoms have been deposited, the plating has the properties of the plating metal and can thus impart the properties for which it was selected. But what about the first few layers of atoms? Metallurgists have known for years that these layers tend to have properties distinctive from both the plating and the base materials. This knowledge helped lead to the development of another type of surface modification—thin films.

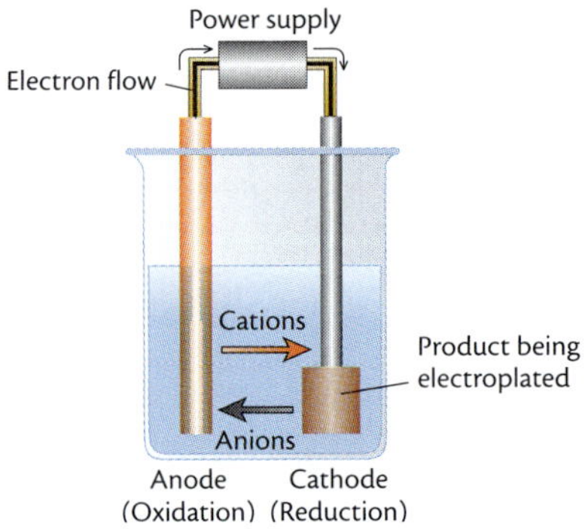

Figure 30 *The process of electroplating.*

166

Thin Films

Films are materials distinguished by the incredibly small thickness of their deposition on a surface. Modern thin films may be only one or two atoms or molecules thick! Films made of metal compounds, such as carbides and oxides, were originally developed to harden the edges of cutting tools. They are also used for decorative or protective purposes on many metal products. Alternating layers of these materials may enhance certain desired properties of the metal even more dramatically.

Thin metal films find many applications in the optics and electronics industries. For example, tungsten films interconnect circuits on microelectronic devices. Gold films can be deposited on glass or other surfaces to produce high-quality mirrors. Semiconductors also employ thin films in their construction.

Scientists are looking at the model provided by nature to develop new thin-films approaches. For example, seashells are created through the deposition of layers of inorganic and organic substances. Similar approaches are being used to create layers of engineering ceramics on surfaces.

The most common organic thin films are made of polymers, a subject you will explore in more detail in Unit 3. For now, however, it is important for you to know about the existence of polymer films because they have often been used to protect currency and important documents from counterfeiting and could be similarly used to protect coins. Many products you use daily are coated with polymers. Some common examples are automobile glass and the UV-protective and antiglare films on sunglasses. Why do you think it would be useful to have a thin film on automobile glass?

An interesting application of polymer thin films is in color-changing paint. Scientists have observed that the colors of butterflies result from the interference patterns of light caused by tiny fibers in their wings. Seeking to duplicate this effect, researchers have used both spun polyester and liquid-crystal polymers to develop car paint that appears to change color depending upon the angle from which it is viewed. Figure 31 illustrates this feature.

The development of thin films is a good example of the positive interaction of science and technology. As new discoveries are made, new applications result. Demands for better methods and more desirable properties lead to more basic research and subsequent applications. Because the properties of films are different from those of bulk materials, their study provides answers to important questions. The scope of their applications, meanwhile, continues to grow.

Figure 31 *The paint on this car changes color depending on the angle from which it is viewed.*

Time

30 minutes

Materials (for 24 students working in pairs)

12 U-tubes
12 buret clamps
12 ring stands
100 mL 0.75 M $CuSO_4$ (18.75 g $CuSO_4 \cdot 5\ H_2O$/100 mL solution) and 0.75 M H_2SO_4 (4.2 mL 18 M H_2SO_4/100 mL solution)
12 iron or zinc nails
12 wire leads with alligator clips attached
12 9-V batteries or power supply
12 copper metal strips
12 voltmeter or CBL (optional)

Safety

Students should wear safety goggles and aprons throughout the laboratory activity.

Always be careful when working with electricity, especially high-voltage power supplies.

CAUTION: *No lab activity should be performed without prior teacher approval.*

Advance preparation

U-tubes can be purchased, or 10-cm lengths of clear plastic tubing may be wrapped around the battery in the shape of a U. The batteries, connectors, and small alligator clips can all be purchased from local electronics supply stores.

Pre-lab discussion

Emphasize this as an oxidation-reduction reaction. Oxidation and reduction always go together. Stress the connection between this activity and previous experiences, the table of metal reactivity, and the concepts of oxidation and reduction.

D.6 COPPER PLATING

Laboratory Activity

This activity (page 168 in the Student Edition) introduces one technique used to modify the surface of materials. It can also be used to reduce metal from ores. Emphasize the fact that this process is endothermic, contributing to the higher cost of producing metals.

The copper-plating solution contains 0.75 M $CuSO_4$ and 0.75 M H_2SO_4. Students can use this procedure to plate keys, steel spoons, and so on, at your discretion.

Lab tips

Use fresh batteries. Rechargeable batteries are economical if you have time to recharge them between uses.

Answers

1. The positive electrode is the anode.
2. $Fe(s) \rightarrow Fe^{++} + 2\ e^-$

   $Zn(s) \rightarrow Zn^{++} + 2\ e^-$
3. The negative electrode is the cathode.
4. $Cu^{++} + 2\ e^- \rightarrow Cu(s)$
5. Anode—This should be made of the metal that will make up the plate. In this system it should be a copper strip. The copper will be oxidized and add to the copper ions available for reduction at the other electrode.
6. Cathode—The cathode should be the metal that is to be plated. Reduction occurs at the cathode. In this case the copper ions will be reduced on the surface of the nail or whatever metal object the student chooses to use.
7. Yes, everything in this laboratory can be scaled up to an industrial level. In fact, this very process is used as the final step in some copper refineries.

Post-lab discussion

Review the terminology of oxidation-reduction, and clarify the reactions at the cathode and anode in the questions. A helpful reminder is Red Cat (*Red*uction occurs at the *cat*hode.). Remind students of the energy cost of electrolysis. Point out that the more active elements of Groups I and II on the periodic chart, as well as aluminum, are commercially produced by electrolysis. Students should be able to explain the results of reversing the leads in the laboratory procedure.

D.6 COPPER PLATING

Laboratory Activity

Introduction

As you just learned, plating a metal requires the application of direct current, the presence of metal ions, and a suitable anode, usually made of the metal to be plated. In this activity, you will plate copper onto a nail. You will need the following materials:

- Beaker, U-tube, or transparent plastic (Tygon) tubing
- Copper plating solution

CAUTION: *Copper plating solutions are hazardous and corrosive.*

- Iron or zinc nail
- Copper metal strips
- 9-V battery or power supply

CAUTION: *Always be careful when working with electricity, especially high-voltage power supplies.*

- Wire leads with alligator clips
- Voltmeter or CBL kit (optional)

Procedure

Set up a system to deposit copper onto a nail. Keep in mind which metal should act as the anode and which as the cathode. Think about how electrons will flow. Test your setup. Record your results. If nothing happens, have your teacher check your setup. When you are finished with the activity, be sure to follow your teacher's instructions for disposal of wastes. Wash your hands thoroughly before leaving the laboratory.

Questions

Questions & Answers CD-ROM WWW.

1. What was the anode in this electrochemical cell?
2. Write an equation for the reaction that occurred at the anode.
3. What was the cathode in this cell?
4. Write an equation for the reaction that occurred at the cathode.
5. Does it matter what metal is used for the anode? Explain.
6. Does it matter what metal is used for the cathode? Explain.
7. Do you think this method would be useful for large-scale copper plating? Why or why not?

SECTION SUMMARY

Reviewing the Concepts

1. What is an allotrope?

An allotrope is a different form of an element that has different physical and chemical properties.

2. Name two substances that are allotropes of the same element and compare their properties.

Graphite and diamond are allotropes of the element carbon. Graphite is soft, slippery, black, and will conduct electricity. Diamond is the hardest substance known and will not conduct electricity.

3. A diamond, a chunk of coal, and your pencil lead contain the same substance. How do they differ in properties? What accounts for the differences in value of the items?

The abundance of coal and graphite make them cheap compared to rare diamonds. Diamonds are the hardest known substances and do not conduct electricity. Graphite is black, slippery, and conducts electricity. Coal is an impure form of carbon, and is a nonconductor.

4. What is an alloy?

An alloy is a solid solution of different metals.

5. Give examples of two alloys you use every day.

Examples include: stainless steel, brass, bronze, and 12-karat gold.

6. What is an advantage of using an alloy rather than a pure metal for a particular purpose?

A particular combination of metals in an alloy can alter the properties of the components to make the alloy more desirable than the metals themselves. For example, pure gold is soft; mixing it with other metals makes the gold alloy stronger for use in jewelry.

7. Where are the majority of elements that behave as semiconductors located on the Periodic Table?

The semiconductors occur in the metalloids and transition elements near the metalloids.

8. Would an unintended addition of a substance to a semiconductor be considered doping? Explain.

No, doping is the intentional addition to create a desired property. In this case, the substance would be an impurity.

9. Give three examples of how properties of materials can be modified by applying a coating.

Copper coating on newer pennies protects the inner zinc from corrosion.

Paint on a car body changes the color of the metal underneath it and protects it.

Metal coating on nonconductors, like plastic or glass, makes the products conductors.

Wax coating on milk cartons makes them waterproof.

10. Describe two ways in which electroplating is used in industry.

Steel bumpers on cars are electroplated with chromium and nickel to prevent rusting. Impure blister copper is purified by electroplating pure copper on the cathode, using the impure copper as the anode. Silver and gold are electroplated on cheaper metals in producing silver plate and "gold" jewelry.

11. How do coatings, electroplatings, and thin films differ?

Coatings are physically or chemically attached to the surface. Electroplating occurs when the metallic ions are reduced and electroplated on the object (cathode). Many layers of atoms are deposited and a metallic bond is formed between the plating material and the object. Thin films can be like electroplating, except they are only a few atoms thick.

12. Can a rusty car bumper be protected from further rusting by electroplating? Explain.

It would be better to electroplate the unrusted bumper since a metallic bond between the plating material and the bumper would last longer. Plating over rust is unlikely to form a durable bond.

Connecting the Concepts

13. Classify each major method of coating a surface as either a chemical or physical change. Explain your answers.

Coating—physical; no chemical bonds are formed between the base and coating. Electroplating—chemical; metal ions are reduced on the object and metallic bonds are formed between the plating material and the base. Thin films may be physical or chemical depending on how the thin film was applied.

14. Use the activity series in Figure 16 (page 121) to decide which metals would be easiest to use as electroplating material.

Metals low in reactivity would be most easily electroplated and used as electroplating material. The anode and the electroplating would be less reactive and likely to corrode.

15. New materials can successfully substitute for metals in some products. Explain why each of the following replacements can be made.

a. Ceramics replace steel in turbine engines.

b. Plastics replace steel in automobile bumpers.

c. Optical fibers replace copper in phone wires.

a. Ceramics have high melting points and can withstand high temperatures.

b. Plastics are resistant to denting, corrosion free, and cheaper.

c. Optical fibers can carry more voice and electronic messages with noise-free characteristics.

Extending the Concepts

16. Obtain information about various allotropic forms of the element sulfur.

a. Draw models depicting the major allotropes.

b. Compare the properties of these allotropes.

Amorphous—non-crystalline form; Rhombic—orthorhombic crystals form when sulfur cools below 95.5 °C; Monoclinic—needlelike crystals which occur when molten sulfur cools below 119 °C.

17. Define *synergy*. How can this term be used to describe properties of alloys? Give two examples.

Synergy: joint work; to work together; combined or cooperative action; or force greater than the sum of the parts. Alloys are made up of a mixture of two or more different metals, but have properties that are sometimes different than the properties of either of the metallic components alone. Some are stronger than their component metals alone, which demonstrates synergy. Thus, copper containing 2% beryllium by mass is much harder than pure copper. Sterling silver made up of 92.5% silver and 7.5% copper is harder than silver but soft enough to be made into jewelry and tableware. Some alloys are softer than their component metals. The presence of bismuth helps to soften a metal and lowers its melting point. An alloy of lead, tin, and bismuth is used to control water sprinklers employed in fire-extinguishing systems.

18. Use the Web and reference materials to find out why doping allows electrons to move more easily in semiconductors.

Semiconductors are very sensitive to impurities. The conductivity of silicon or germanium can be increased by a factor up to 10^6 by adding as little as 0.01% of an impurity. If small amounts of arsenic are added to a semiconductor, the arsenic can provide the extra electrons that allow the "doped" semiconductor to become a much better conductor of electricity.

MAKING MONEY

This section continues to develop the relationship between properties of metals and their uses, through consideration of the materials used for coins.

Choosing the best coin

Students are asked to work in small groups to design a metal coin. Using the concepts developed in this unit, each group designs a coin and shares the design with the class. The answers to a series of questions given in the text can be the basis for your evaluation of each group. (Section C.8, p. 149)

You might consider having students come up with the grading criteria as part of the discussion in Section C.8. If students know the expectations of an excellent design presentation in advance, they are much more likely to meet them. Students can be permitted to create their own grading rubric. Suggested grading categories might be: design adequacy of physical and chemical properties, appropriateness of design images, and design security.

PUTTING IT ALL TOGETHER

Making Money

Choosing the Best Coin

Putting It All Together CD-ROM WWW.

When a new product is created, the design process often includes proposals from several individuals or teams. A panel of experts or consumers then chooses the "best" of the proposals. Because each school can submit only one design for the new coin, you will follow a similar process in choosing the best proposal. Each team will present its coin design and findings to the class, which will serve as the selection committee. The following paragraphs summarize the essential elements of your design presentation.

Coin Model

Include a model or detailed drawing of the proposed coin. If you create a model, it does not have to be made of the specified material but should resemble it. Ensure that all models and designs are at least five times the actual coin size. In addition, specify the actual size, thickness, and mass of the coin.

Materials Design

Describe the material or materials chosen for the coin. Present your rationale for selecting those materials. Include an analysis of both necessary and desirable properties of the chosen material(s), as well as methods to discourage counterfeiting.

Material Source Data

Provide details on the source of the raw materials needed to produce the new coin. Include a discussion of how the materials will be mined and/or processed. Estimate the long-term availability of those resources.

Life Cycle Analysis

Present an analysis of the life cycle of the proposed coin. How long is the coin expected to last in general circulation? Will material(s) making up the coin be recycled or reused? Consider the disposal or reuse of the material(s) in the coin as part of its production cost. (See next item.)

Cost Analysis

Present and analyze production costs of the new coin. These involve factors such as location of resources, mining, production or processing (including energy costs), and distribution. Will the cost of materials be less than the value of the coin itself? Are overall production costs reasonable based upon the expected lifetime of the coin?

Looking Back and Looking Ahead

As you come to the end of this unit, it is appropriate to pause and reflect on what you have learned thus far. You have discovered some of the working language of chemistry (symbols, formulas, and equations), laboratory techniques, and major ideas in chemistry (such as periodicity, the law of conservation of matter, and atomic-molecular theory). This knowledge can help you better understand some important societal issues. Central among these issues is the use and management of Earth's chemical resources, including water, metals, petroleum, food, and air.

You have also explored other issues that enter into policy decisions concerning technological problems and challenges. Although chemistry is often a crucial ingredient in recognizing and resolving such issues, many problems are far too complex for a simple technological "fix." Issues of policy are not usually "either/or" situations, but often involve many dimensions and considerations. As a voting citizen, you will be concerned with a variety of issues that require some scientific understanding. Tough decisions may be needed. The remaining units of this textbook will continue to prepare you for this important responsibility.

Unit 3 deals with petroleum, a nonrenewable chemical resource so important that it deserves its own unit.

Looking back and looking ahead

Students have encountered chemical language (symbols, formulas, and equations), laboratory techniques, and basic chemical concepts that relate to societal problems and decisions associated with use of resources. In these decision-making tasks, chemistry remains an important ingredient. Chemical insight can assist in the definition and rational resolution of many complex problems facing this nation and the world today, but most social problems are too complex to have simple remedies.

This unit has gone beyond the Water unit in considering the issues associated with the general use of chemical resources. In contrast to water, which is definitely a renewable resource, metallic resources can be used up or scattered so widely that it is economically impractical to retrieve them. In the next unit, students will examine a nonrenewable resource (petroleum) in detail, exploring the problems involved in its wide use—now and in the future. Since it is impossible to recycle petroleum after it has been burned, students will explore alternatives to petroleum, consistent with the approach introduced in this Resources unit.

Section	Time class	Student Edition Features	Program Resources	Media and Technology
Section A. Petroleum—What Is It? **Objectives** • Describe the chemical makeup of petroleum, how it differs from other natural resources, and how it is refined. • Describe the process of fractional distillation and name the typical products manufactured from each fraction. • Represent the covalent bonding in hydrogen bonding by electron-dot, structural, and molecular formulas. • Write the general formulas for the first ten alkanes. • Define the term isomer and draw the structural formulas for the first three isomers of a given compound.	**5 class periods**	**Building Skills** #1. Hydrocarbon Boiling Points #2. Trends in Alkane Boiling Points #3. Boiling Points of Alkane Isomers **Laboratory Activity** • A.2 Separation by Distillation • A.6 Modeling Alkanes • A.7 Alkanes Revisited	**Activity Book** • A.2 Separation by Distillation Procedures • A.6 Modeling Alkanes • A.7 Alkanes Revisited • A.7 Naming Branched Alkanes • A.7 Electron Configuration **Black Line Copy Masters** • Distribution of World Petroleum, Population, and Consumption • Annual U.S. Consumption of Energy • Functional Groups I • Functional Groups II • PIAT Evaluation	**ChemCom CD-ROM** • Chemical Bonding • Modeling Alkanes **Transparencies** • Distribution of World Petroleum Reserves • Distillation of a Simple Mixture • Fractionating Tower • Hydrocarbon Boiling Points • First Ten Alkanes • Pentane and Octane Isomers

PLANNING GUIDE

Section	Time class	Student Edition Features	Program Resources	Media and Technology
Section B. Petroleum as an Energy Source **Objectives** • Explain endothermic and exothermic reactions in terms of bond breaking and bond forming. • Identify energy conversions and explain energy conversion efficiency. • Define the terms *heat of combustion* and *specific heat* and calculate energies of various combustion reactions. • Write balanced equations for the combustion of hydrocarbon fuels, including energy changes. • Explain the term octane number, state its relationship to grades of gasoline, and identify two ways of increasing octane numbers.	**5 class periods**	**Building Skills** #4. Wasted Energy #5. Energy Conversion Efficiency #6. Heats of Combustion **Laboratory Activity** • Combustion **Making Decisions** • Fuel Molecules in Transportation **ChemQuandary** #1. Using Fuel #2. A Burning Issue: Combustion **Chemistry at Work** • Oil's Well That Ends Well: The Work of a Petroleum Geologist	**Activity Book** • B.3 Combustion • B.3 Heats of Combustion Practice Problems **Skill Handbook** • Dimensional Analysis **Demonstrations** • B.1 Endothermic Reaction • B.1 Exothermic Reaction **Additional Activities** • B.3 Combustion • B.4 ChemQuandaries	**ChemCom CD-ROM** • Unit III: Section A **ChemCom Videodisk** • Energy From Fossil Fuels **Transparencies** • Formation of CO_2 and H_2O • Tracing Energy Conversions • Energy Use in an Automobile • Hydrocarbon Heats of Combustion

PLANNING GUIDE

Section	Time class	Student Edition Features	Program Resources	Media and Technology
Section C. Petroleum as a Building Source **Objectives** • Compare saturated and unsaturated hydrocarbons in terms of molecular models, formulas, structures, and physical and chemical properties. • Identify the functional groups for common alcohols, ethers, carboxylic acids, and esters. • Describe polymerization and give one example of addition and condensation reactions.	**5 class periods**	**Modeling Matter** • Polymer Structure and Properties **Making Decisions** • Builder Molecules in Transportation • Functional Groups **Laboratory Activity** • The Builders • Condensation **Chemistry at Work** • Optimizing Technology	**Activity Book** • C.3 The Builders • C.7 Condensation • C.7 IUPAC Rules • C.7 Functional Groups **Demonstrations** • C.1 Breakdown of Polystyrene • C.6 Synthesis of Nylon	**ChemCom CD-ROM** • Petroleum :Section III **ChemCom Videodisk** • Shaping Polymers • Creating Polymer Chains **Transparency** • Three Isomers of Butene

PLANNING GUIDE

Section	Time class	Student Edition Features	Program Resources	Media and Technology
Section D. Energy Alternatives to Petroleum **Objectives** • Describe major sources of energy and alternative sources of fuels and builder molecules for the future. • Describe alternative fuels that may be used to extend, supplement, or replace petroleum as Earth's primary fuel.	**3 class periods**	**Building Skills** #8. Fuel Sources Over the Years		**ChemCom CD-ROM** • Fuel Cells **ChemCom Videodisk** • Harnessing Alternative Energy Sources **Transparencies** • Annual U.S. Energy Consumption • Fuel Cell
Putting It All Together • Evaluate the ARL-600 TV ad, then write and produce an automobile commercial message.	**2 class periods**		**Activity Book** PIAT ad evaluation	

SUPPLY LIST FOR UNIT 3

Expendable Items and Unusual Materials (per 24 students working in pairs)

Alphabetical by Item Name			
Quantity	Unit	Item	Sections Used
300	mL	2-propanol (isopropyl alcohol)	A.2
50	mL	Acetic acid	C.7
12	Each	Ball & stick or space filling models	A.6, A.7, C.3
12	Each	Boiling chips	A.2, C.7
12	Each	Candles, paraffin	B.3
36	Each	Dropping bottle	C.7
12	Each	Glass tubing, 90 degree bend	A.2
12	Sheets	Graph paper	A.2
12	Each	Hot plate	C.7
6	Lbs	Ice	B.3
12	Each	Index cards	B.3
1	Gram	Iodine crystals	A.2
12	Each	Matches	B.3
24	mL	Methyl alcohol (methanol)	C.7
24	Each	Micropipets	C.7
5	mL	Octyl alcohol	C.7
5	mL	Pentyl alcohol (pentanol)	C.7
5	Each	Reagent bottles	C.7
5	Gram	Salicylic acid	C.7
12	Each	Soft drink cans w/pull tabs attached	B.3
5	mL	Sulfuric acid, conc	C.7
300	mL	Water, distilled	A.2

SUPPLY LIST

Expendable Items and Unusual Materials per 24 Students working in pairs.

By Section			
Quantity	**Unit**	**Item**	**Sections Used**
300	mL	2-propanol (isopropyl alcohol)	A.2
12	Each	Glass tubing, 90° bend	A.2
12	Sheets	Graph paper	A.2
1	Gram	Iodine crystals	A.2
12	Each	Boiling chips	A.2, C.7
12	Each	Ball-&-stick or space-filling models	A.6, A.7, C.3
12	Each	Candles, paraffin	B.3
6	Lbs	Ice	B.3
12	Each	Index cards	B.3
12	Each	Matches	B.3
12	Each	Soft drink cans with pull tabs attached	B.3
50	mL	Acetic acid	C.7
36	Each	Dropping bottle	C.7
12	Each	Hot plate	C.7
24	mL	Methyl alcohol (methanol)	C.7
24	Each	Micropipets	C.7
5	mL	Octyl alcohol	C.7
5	mL	Pentyl alcohol (pentanol)	C.7
5	Each	Reagent bottles	C.7
5	Grams	Salicylic acid	C.7
5	mL	Sulfuric acid, conc.	C.7
300	mL	Water distilled	CA.2

UNIT 3 PETROLEUM: BREAKING AND MAKING BONDS

Introduction to Unit

This unit looks at petroleum use throughout the world and at some of the issues that result from society's heavy reliance on petroleum products. Students investigate fractional distillation, energy, and organic chemistry during their development of an advertisement for an alternative vehicle.

Some additional activities for introducing sections are explained in the opening page of each section.

UNIT 3

WHAT are the chemical and physical properties of hydrocarbons?

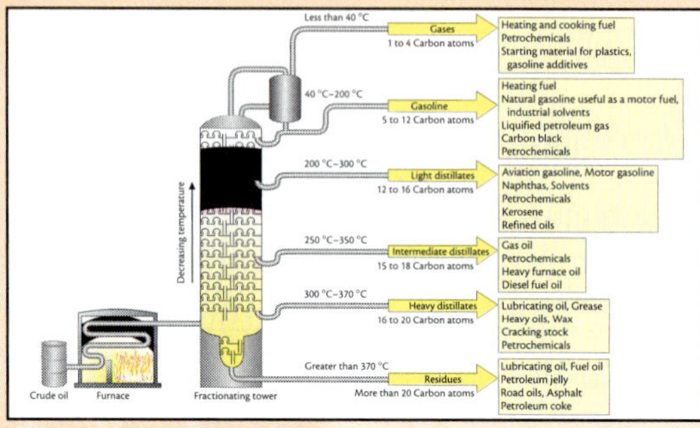

WHY do hydrocarbons make such good fuels?

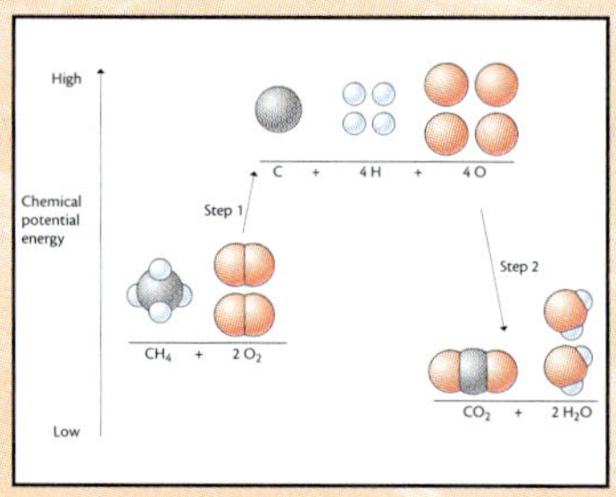

PETROLEUM: BREAKING AND MAKING BONDS

WHY are carbon-based molecules so versatile as chemical building blocks?

WHAT properties are important in considering substitutes for petroleum?

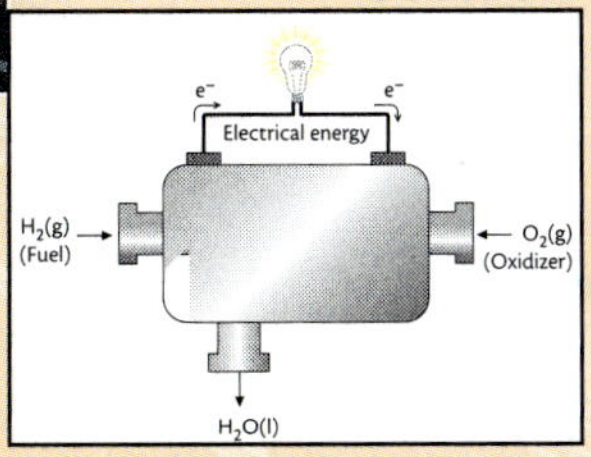

There is increased interest in alternative-energy transportation. Why? What are advantages and disadvantages to petroleum alternatives? How can the global supply of petroleum best be used? Turn the page to learn more about this energy-rich resource.

DEMONSTRATION

Display common products made from petroleum and samples of crude oil. Use these to discuss the nature of crude oil and the products derived from it. Easy examples are food containers, such as milk jugs, plastic wrap, or plastic grocery bags. Consider discussing sports equipment as well. Items such as roller blades and nylon sports jerseys might be of interest to students. Use the questions embedded in the student text as openers.

ADDITIONAL INTRODUCTION ACTIVITIES

Personal Petroleum Use Activity

1. Have students list ten personally important items made from petroleum.
2. Have students list items in the classroom that are not made from any petroleum.
3. Suppose there was a severe petroleum shortage.

   a. Which five petroleum-based products on your list would you be most willing to do without?

   b. Which five would you be least willing to give up?

Ask students: Were you surprised by all the items that are made from petroleum? Often people don't realize how important something is until they try to imagine life without it. Petroleum is such a large part of modern life we tend to take it for granted. We must realize, however, that the world is rapidly using up its limited supply of this nonrenewable resource. That leads to another important question: Will we have sufficient petroleum for our needs in the future?

A great video to show at this time is *FUEL-LESS*, 1996, American Petroleum Institute.

THE NEW ARL-600 TV COMMERCIAL—WORKING SCRIPT

Text/Script	Image/Sound
Clean . . .	(Video of driver and three passengers riding in the vehicle.)
Comfortable . . . ***Convenient . . .***	(The words "clean," "comfortable," "convenient" appear, in turn, as the announcer speaks them. "Clean" appears at the top of the screen; "Comfortable" appears in the middle; "Convenient" appears at the bottom. The "C" moves from the top to the bottom as the words appear.)
Those three words describe the new breakthrough personal vehicle, the ARL-600.	(Show the ARL-600 driving along a road as announcer reads line.)
Clean . . .	("Clean" appears on the screen and moves from left to right.)
Unlike petroleum-fueled vehicles, the ARL-600 is emission-free—no air pollutants, smoke, or smog. Why? Because the ARL-600 is directly powered by electricity, thanks to its new, high-capacity electric storage batteries. You drive guilt-free, knowing you're helping, not hurting, the environment.	(Show the ARL-600 accelerating [after sitting at a stoplight] next to gasoline-burning vehicles to show the difference in emissions.)
Comfortable . . .	("Comfortable" appears on the screen and moves from the top left corner to the bottom right corner.)
You and three others can easily ride in the ARL-600. And—thanks to electrical power—the ride is so quiet that driving the ARL-600 seems more like a walk in the park.	(Show four people riding in the ARL-600, then change the perspective to that of the driver with the window rolled down. The driver can hear birds chirping and children playing as she drives past a park.)
Convenient . . .	("Convenient" appears on the screen and moves from top center to bottom center.)

Imagine an end to stops at busy gas stations. How? You simply recharge your ARL-600's batteries when you return home, and you're ready for another day of gasoline-free driving.	(The ARL-600 passes a gas station, pulls up to a house, and the driver "plugs in" the vehicle.)
When fully charged, your electric-powered ARL-600 will take you wherever you want to go within a 90-mile range—school, work, or even soccer practice.	(Images of the ARL-600 in a variety of settings—shopping center, office parking lot, school, recreation area.)
One place you *won't* need to visit, however, is the auto-repair shop. Fewer moving parts mean less time and money spent on engine repairs.	(The ARL-600 drives past an auto-repair shop.)
And using your petroleum-free ARL-600 is easy on the pocketbook—as little as two cents per mile to operate. Compare that to more than five cents per mile for gasoline-burning vehicles.	(Image of a piggy bank being shaken with coins [mostly pennies] falling out onto the table.)
Help conserve petroleum resources! Visit your ARL-600 dealer for a test drive today.	(Side view of an ARL-600 at a dealer showroom with several people examining it. The "ARL-600" logo then appears below the vehicle.)

You live in a world of new products, devices, and materials. Whether presented on a billboard, displayed on television, featured in a magazine, or announced on the radio, every advertisement attempts to sell its product by informing the audience about specific features. The product may be "faster," "lighter," "easier to use," "newer," or "great-tasting"—to sample just a few of many common product claims. The advertisement for the ARL-600 is no exception; it highlights several energy- and fuel-related features of a new and (so it is claimed) "petroleum-free" vehicle.

In this unit, you will gain the knowledge and perspective you need to analyze the merits of the claims made about the ARL-600. For example, you will learn what petroleum is, what chemical and physical properties make it so useful, and how it is used. Then you can evaluate whether the ARL-600 is actually "petroleum-free" or not.

But you will learn more about petroleum than its use as a fuel. Many products you use every day are made from petroleum—an outstanding example is plastics. What is it that allows petroleum to be useful as both a fuel to burn and a building block from which to make many new substances? How long will known world reserves of petroleum last? What alternatives to petroleum are there?

As you learn about this valuable resource, consider the energy and fuel claims made in the ARL-600 advertisement. You will soon be invited to analyze those claims. Later in the unit you will have the opportunity to produce a design for a new-vehicle advertisement based on the knowledge you have gained.

PETROLEUM—WHAT IS IT?

Introduction

Section A examines the differences between petroleum and metallic ores. The separation and uses of the major components of petroleum are the main focus of this section. The characteristics of some key compounds found in petroleum are explored through a series of discovery-oriented activities, with emphasis on structure and properties. Models are used to explore the names and structures of common alkanes and the existence of structural isomers.

A.1 WHAT *IS* PETROLEUM?

Petroleum is a complex mixture of hydrocarbons when found in its natural state. Unlike metallic ores, which are typically compounds, petroleum is a mixture that must be refined before it can be used for building or burning.

Remind students that the unequal distribution of natural resources (petroleum and metallic ores) is responsible for much of the commercial trade among nations.

SECTION A

PETROLEUM—WHAT IS IT?

Introduction CD-ROM WWW.

The word "petroleum" is probably quite familiar to you. But do you know what petroleum is or what it is made of? Can you explain what properties make it useful for both burning and building? In this section you will explore the characteristics of some key compounds found in petroleum. Specifically, you will focus on their structure, bonding, and properties.

A.1 WHAT *IS* PETROLEUM?

The word "petroleum" comes from the Latin words *petr-* ("rock") and *oleum* ("oil").

Petroleum is a vitally important world resource. As pumped from underground, petroleum is known as crude oil, or "black gold." This liquid varies from colorless to greenish-brown to black, and may be as thin as water or as thick as soft tar. Crude oil cannot be used in its natural state. Instead, it is shipped by pipeline, ocean tanker, train, or barge to oil refineries, where it is separated into simpler mixtures. Some of these mixtures are ready for use, whereas others require further refinement. Refined petroleum is chiefly a mixture of various **hydrocarbons**—molecular compounds that contain atoms of the elements hydrogen and carbon only. Can you see how this class of compounds got its name?

Nearly 50% of the total energy needs of the United States are met by burning petroleum. Thus most petroleum is consumed as a fuel. Converted to gasoline, petroleum powers millions of automobiles in the United States, each traveling an average of 11 000 miles annually. Other petroleum-based fuels provide heat to homes and businesses, deliver energy to generate electricity, and propel diesel engines and jet aircraft.

But petroleum's importance goes beyond its use as just a fuel. Its other major use is as a raw material from which a stunning array of familiar and useful products are manufactured—from CDs, sports equipment, clothing, automobile parts, and carpeting to prescription drugs and artificial limbs. Based on your experiences with petroleum fuels and products, what percent of petroleum would you estimate is used for burning? For building? Can you identify other uses of petroleum? The answers in the next paragraph may surprise you.

What did you predict for the percent of petroleum used for burning? Fifty percent? Sixty percent? Astonishingly, 84% of petroleum is burned outright as fuel. Only about 7% is used for producing substances such as medications and plastics. The remaining 9% is used as lubricants, road-paving materials, and an assortment of miscellaneous products. For every gallon of petroleum that is used to produce useful products, more than five gallons are burned to release energy.

What happens to molecules in petroleum when they are burned or used in manufacturing? As in all chemical reactions, the atoms become rearranged to form new molecules. When hydrocarbons burn, they react with oxygen gas in the air to form carbon dioxide (CO_2) gas and water vapor.

WHO HAS THE OIL AND WHO USES IT? ACTIVITY

Figure 1 indicates where the world's known supplies of petroleum are located, along with the regional distribution of the world's population. The world's supplies, called *reserves,* can be tapped by available technology at costs consistent with current market prices.

Figure 1 also shows the regional distribution of the world's consumption of petroleum. Use the figure to answer these questions:

1. Which region has the most petroleum?
2. Which region has the most petroleum relative to its population?

These gases disperse in the air. The hydrocarbon fuel is used up; it will take millions of years for natural processes to replace it. Thus petroleum is a nonrenewable resource—much like the minerals you studied in Unit 2.

Like other resources, petroleum is not uniformly distributed around the world. Approximately 57% of the world's known crude oil reserves are located in just five Middle Eastern nations: Iran, Iraq, Kuwait, Saudi Arabia, and the United Arab Emirates. By contrast, the petroleum reserves of North America amount to only about 7% of the world's known supply. The distribution of crude oil reserves does not necessarily correspond to population or use of petroleum. For example, Asia, the Far East, and Oceania account for 60% of the world's population, but this region has only about 4% of the world's petroleum reserves. Figure 1 shows these global distributions.

You have just learned what petroleum is, what it is used for, and where it is found. Petroleum is actually a complex mixture of hydrocarbons that must be refined or separated into simpler mixtures in order to be useful. In the following activity, you will find out about this basic separation process as you investigate a simple mixture of two liquids.

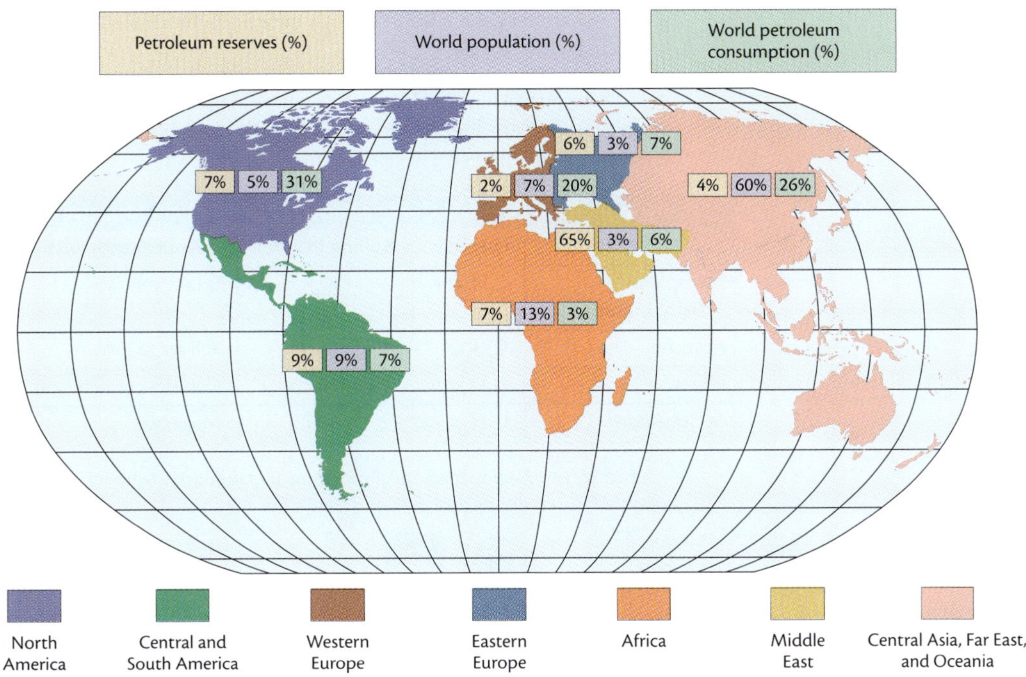

Figure 1 *Distribution of world's petroleum reserves, population, and consumption of petroleum.*

3. Which region has the least petroleum relative to its population?
4. Which regions consume a greater percentage of the world's supply of petroleum than they possess?
5. Which regions consume a smaller percentage of the world's supply of petroleum than they possess?
6. Which regions are likely to export petroleum?
7. Which regions are likely to import petroleum?
8. List several pairs of regions that might make petroleum trade agreements with each other. (A given region may be used more than once.)
9. Not all trade agreements that are possible actually exist. What factors might prevent trade relationships between regions?
10. What two regions are the largest petroleum consumers?
11. Why do these two regions consume such large quantities of petroleum?
12. a. Compare North America and the Middle East in terms of

    (i) population, (ii) petroleum reserves, and (iii) petroleum consumption.

    b. What possible consequences concerning petroleum use are suggested by these comparisons?

A.2 SEPARATION BY DISTILLATION

Laboratory Activity

Pre-lab discussion

This activity models the separation techniques used in petroleum refineries. Students should understand that petroleum is a mixture of many related substances that may be separated using fractional distillation.

The unknown liquid solution (a mixture of water and 2-propanol) is flammable.

Review graph construction and perhaps use computer graphing.

Time

50 minutes

Materials (for 24 students working in pairs)

24 100-mL beakers
12 100-mL graduated cylinders
12 125-mL Florence flasks
12 funnels
12 thermometers (–10 to 110 °C)
12 hot plates **(do not substitute a burner)**
24 ring stands
12 clamps
12 rubber stoppers with 2 holes to fit the tops of the flasks
12 condensers
12 corks
12 wooden splints
600 mL of distillation solution prepared by mixing 420 mL water and 180 mL 2-propanol (iso-propanol)
Boiling chips
Glass tube to connect the flask to the condenser (<90° bend)
Rubber tubing
Iodine crystals

A.2 SEPARATION BY DISTILLATION

Laboratory Activity

Introduction

You know that substances can often be separated by taking advantage of their different physical properties. One physical property commonly used to separate two liquids is density. But density will work only if the two substances are insoluble in each other, which is not the case with petroleum. Another physical property chemists often use is the boiling point of a substance. The separation of liquid substances according to their differing boiling points is called **distillation**.

As a liquid mixture is heated, the substance with the lower boiling point will vaporize first and leave the distillation flask; it will then be converted back to a liquid as it passes through a condenser—all before the second substance begins to boil. Each condensed liquid substance, called a **distillate**, can thus be collected separately.

As you might expect, heating the liquid mixture raises its temperature. However, once the first substance begins to boil and vaporize from the mixture, the temperature of the liquid remains steady until that component is completely distilled from the mixture. Continued heating will then cause the temperature to rise once again, this time until the second component begins to boil and distill.

In this activity, you will use distillation to separate a mixture of two liquids. Then you will identify the two substances in the mixture by comparing the observed distillation temperatures with the boiling points of several possible compounds listed in Figure 2.

The boiling points listed in Figure 2 are based on normal sea-level atmospheric pressure.

Figure 2 *Properties of substances that may be in the distillation mixture.*

Properties of Possible Components of Distillation Mixture

Substance	Formula	Boiling Pt. (°C)	Appearance with I_2
2-Propanol (rubbing alcohol)	C_3H_7OH	82.4	Bright yellow
Acetone	C_3H_6O	56.5	Yellow to brown
Water	H_2O	100	Colorless to light yellow
Cyclohexane	C_6H_{12}	80.7	Magenta

Before you begin, read the Procedure to familiarize yourself with the intended observations and measurements.

Procedure

1. Construct suitable data tables to record your observations and measurements.
2. Assemble an apparatus similar to that shown in Figure 3. Label two beakers Distillate 1 and Distillate 2.
3. Using a clean, dry graduated cylinder, measure a 50-mL sample of the distillation mixture. Pour it into the flask and add a boiling chip.

Safety

Students should wear safety goggles and aprons. The unknown liquid is flammable. No flames should be allowed in the laboratory. **Point out the hazards of heating unknown liquids that might be flammable. Never heat the flask to dryness.** Remind students that iodine must be treated with special care. Skin contact with iodine crystals can cause lesions, and iodine gas is intensely irritating to the skin. **Mercury thermometers should not be used.** Mercury spills that occur as a result of broken thermometers are not worth the risk.

Advance preparation

Construct and display a distillation setup (see Figure 2 in the student text). Demonstrate how to insert the thermometer and glass tube into the stopper

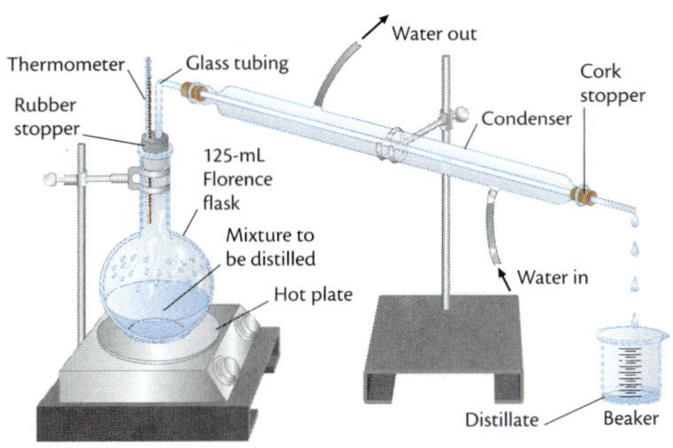

Figure 3 *Distillation of a simple mixture.*

4. Record your observations of the starting mixture.
5. Connect the flask to a condenser as indicated in Figure 3. Ensure that the hoses are attached to the condenser and water supply as shown. Position the Distillate 1 beaker at the outlet of the condenser so it will catch the distillate.
6. Ensure that all connections are tight and will not leak.
7. Turn on the water to the condenser and then turn on the hot plate to start gently heating the flask.
8. Record the temperature at which the first drop of distillate enters the beaker. Then continue to record the temperature every 30 seconds. Continue to heat the flask and collect distillate until the temperature begins to rise again. At this point, replace the Distillate 1 beaker with the Distillate 2 beaker.
9. Continue heating and recording the temperature every 30 seconds until the second substance just begins to distill. Record the temperature at which the first drop of second distillate enters the beaker. Collect 1 to 2 mL of the second distillate. **CAUTION:** *Do not allow all of the liquid to boil from the flask.*
10. Turn off the heat and allow the apparatus to cool. While the apparatus is cooling, test the relative solubility of solid iodine (I_2) in Distillate 1 and Distillate 2 by adding a small amount of iodine to each beaker and stirring. Record your observations.
11. Disassemble and clean the distillation apparatus and dispose of your distillates as directed by your teacher.
12. Wash your hands thoroughly before leaving the laboratory.

Questions

1. Plot your data on a graph of time (x axis) vs. temperature (y axis). Be sure to include all features of a correctly drawn graph. See page 66.
2. a. Using your graph, identify the temperature at which the first substance distilled and the temperature at which the second substance distilled.

   b. Because the liquid temperature does not change appreciably during distillation of a particular component, those portions of

using glycerin or another lubricant and hand protection. Be sure to grasp the thermometer close to the stopper. **You may wish to preassemble the pieces to reduce accidents.**

Lab tips

- Table 1 lists four possible compounds for students to consider as components of the original mixture.
  Use only a distillation solution that contains the 2-propanol–water mixture. This system poses the least hazard and provides good results. Do not reveal the composition of the mixture to the students. Allow them to use their observations to identify the mixture's components.

- **Check all connections and hose routing in each distillation apparatus before students begin work.** Remind students to keep the water pressure low to prevent the rubber tubing from separating from the condenser. A common mistake is connecting the inlet water to the upper connection of the condenser versus the lower one, in which case the condenser will not fill completely.

Expected results

The time will vary, but the boiling liquid should reach two temperature plateaus. The first, the boiling point of the 2-propanol, will be about 82° C; the second will be about 100° C, the boiling point of water. At higher altitudes the boiling points will be lower. The temperature of a pure substance does not change during the boiling process. However, sometimes mixtures act differently. This fact, along with errors in reading the thermometers, will cause the plateaus to vary. Iodine is more soluble in the first distillate.

Answers

1. Graphs should meet design criteria.
2. a. Temperatures should correspond to the plateaus at 82 and 100 °C.

   b. The plateaus should match the temperatures at which the first drops of distillate were collected.
3. b. The first distillate is 2-propanol, and the second is water.
4. a. Data will vary .

   b. Error in reading the thermometer, heating the mixture too rapidly.

5. Iodine is more soluble in 2-propanol than in water.
6. It could be distilled again. If two plateaus occur, it is not a pure substance. Another approach would be to see how close the boiling point comes to that of that of any of the pure substances in Figure 2. Another approach would be to measure the density and compare it with the known densities (which would have to be looked up) of the pure substances.
7. 2-propanol and cyclohexane, due to similar boiling points.
8. Similar, but with four plateaus, one for each liquid at its boiling point.

Post-lab discussion

Have students compare their conclusions regarding the components of the mixture. The discussion may focus on the contrasting physical and chemical properties of water and 2-propanol. Discuss the properties used to separate the mixture, namely volatility and boiling points.

Compare the results of the laboratory activity with fractional distillation of petroleum. In both, a mixture is separated by difference in boiling point. The laboratory sample involved two components, but crude oil contains hundreds of compounds, many with similar boiling points. Fractional distillation of crude oil generally results in a series of mixtures, each composed of similar molecular weight hydrocarbons.

the graph line should appear flat (horizontal). How well do these plateaus match the temperatures at which the first drops of each distillate were collected?

3. Using data in Figure 2 on page 178, identify each distillate sample.
4. Compile your data with the data of those students who distilled the same mixture.
   a. Find the mean and the mode for each distillate temperature.
   b. All laboratory teams did not obtain the same distillation temperatures. Why?
5. In which distillate was iodine more soluble? Explain.
6. What laboratory tests could you perform to decide whether the liquid left behind in the flask is a mixture or a pure substance?
7. Of the substances listed in Figure 2, which two would be most difficult to separate by distillation? Why?
8. a. What would a graph of time vs. temperature look like for the distillation of a mixture of all four substances listed in Figure 2?
   b. Sketch the graph and describe its features.

The statistical mode is the most frequently reported value in a set of data.

A.3 PETROLEUM REFINING

Unlike the simple laboratory mixture you investigated in the preceding activity, crude oil is a mixture of many compounds. Separating such a complex mixture requires the application of distillation techniques to large-scale oil refining. The refining process does not separate each compound contained in crude oil. Rather, it produces several distinctive mixtures called **fractions**. This process is known as **fractional distillation**. Compounds in each fraction have a particular range of boiling points and specific uses. Figure 4 illustrates the fractional distillation (fractionation) of crude oil.

First, the crude oil is heated to about 400 °C in a furnace. It is then pumped into the base of a distilling column (fractionating tower), which is usually more than 30 m (100 ft) tall. Many of the component substances of the heated crude oil vaporize. The temperature of the column is highest at the bottom and decreases toward the top. Trays are arranged at appropriate heights inside the column to collect the various fractions.

During distillation, the vaporized molecules move upward in the distilling column. The smaller, lighter molecules have low boiling points and either condense high in the column or are drawn off the top of the tower as gases. Fractions with higher boiling points contain larger molecules, which are more difficult to separate from one another and thus require more thermal energy to vaporize. These molecules condense in trays lower in the column. Substances with the highest boiling points never vaporize. These thick, or viscous, liquids—called bottoms—drain from the column's base. Each arrow in Figure 4 indicates the name of a particular fraction and its boiling-point range.

As you learn more about the characteristics of the fractions obtained from petroleum, think about how their products find uses in both traditional and electric vehicles.

Although the names given to various fractions and their boiling ranges may vary somewhat, crude oil refining always has the same general features.

A.3 PETROLEUM REFINING

Figure 4 clarifies several aspects of fractional distillation, which separates crude oil into components. The difference in temperature throughout the column promotes the evaporation, condensation, and subsequent separation of the various fractions. Each fraction boils at its normal boiling point.

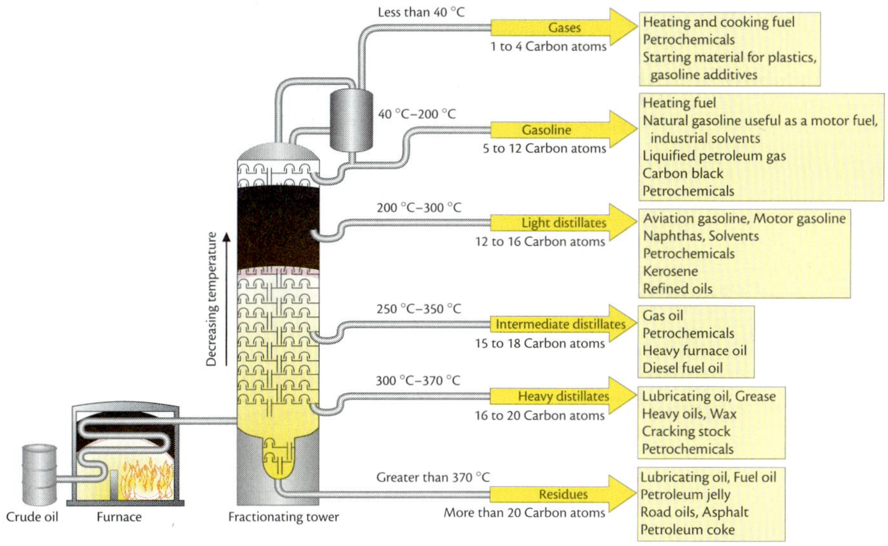

Figure 4 *A fractioning tower.*

A.4 A LOOK AT PETROLEUM'S MOLECULES

Petroleum's gaseous fraction contains compounds with low boiling points (less than 40 °C). These small hydrocarbon molecules, which contain from one to four carbon atoms, have low boiling points because they are only slightly attracted to each other or to other molecules in petroleum. Forces of attraction between molecules are called **intermolecular forces**. As a result of weak intermolecular forces, these small hydrocarbon molecules readily separate from each other and rise through the distillation column as gases.

Petroleum's liquid fractions—including gasoline, kerosene, and heavier oils—consist of molecules having from five to about twenty carbon atoms. Molecules with even more carbon atoms are found in the greasy solid fraction that does not vaporize. These thick, "sticky" compounds have the strongest intermolecular forces among all substances found in petroleum. It is not surprising that they are solids at room temperature.

Now complete the following activity to learn more about physical properties of hydrocarbons.

Just as "interstate highway" means a road that runs between states, intermolecular forces means forces between molecules.

HYDROCARBON BOILING POINTS — Building Skills 1

Chemists often gather and analyze data about the physical and chemical properties of substances. These data can be organized in many ways, but the most useful techniques uncover trends or patterns among the data.

Using the Visual

Show students the transparency of a fractionating tower. (See Figure 4, page 186).

A.4 A LOOK AT PETROLEUM'S MOLECULES

Fractional distillation separates petroleum into fractions. Each fraction consists of a mixture of hydrocarbons with boiling points that are close together. The difference in boiling point is directly related to the number of carbon atoms and the intermolecular forces between the carbon atoms.

HYDROCARBON BOILING POINTS
Building Skills 1

Chemists determine physical and chemical properties of substances and organize the data in ways that lead to the identification of useful patterns. These patterns often stimulate searches for an explanation of such regularities.

In this exercise, some students may organize the chemical information alphabetically. By having them do a few problems with tables arranged according to alphabetical or numerical rank, you may convince them that organizing the data according to boiling points provides a better method for arranging the properties of these compounds. Eventually, the naming system for organic compounds and its relationship to the number of carbons will reinforce this concept.

DEMONSTRATION

Dry ice (BP—78 °C) can be used to liquefy propane. A convenient setup is to attach a piece of copper tubing, with one end coiled or looped, to a piece of rubber tubing attached to a propane tank. If the coil is placed in a container of dry ice/acetone with the other end outside the container, liquid hydrocarbon can be collected in a beaker and observed to boil as it warms to room temperature. If liquid nitrogen (BP—196 °C) is used, methane or ethane from the laboratory gas jet can be liquefied.

Keep well away from all flames. Dispose of the distillate by allowing it to evaporate in a fume hood or in an open field.

Answers

1. a. The hydrocarbons in Figure 5 are alphabetized by name.

   b. Only in a very limited way. An alphabetical arrangement might be useful for locating a specific hydrocarbon in a handbook or index.

2. a. The boiling-point data would be better organized if the hydrocarbons were listed from high to low (or from low to high) boiling-point values.

   b. With lowest boiling point at top, the order would be:
   methane
   ethane
   propane
   butane
   pentane
   hexane
   heptane
   octane
   nonane
   decane

3. Methane, ethane, propane, butane. You may want to emphasize the point that substances that are gases at room temperature have already boiled at a lower temperature. Students often associate boiling with a high temperature.

4. Pentane

5. Decane has a higher boiling point than butane, indicating that the intermolecular forces between the decane molecules are greater than the intermolecular forces between butane molecules.

The development of the Periodic Table is an example of this approach. To refresh your memory about the Periodic Table, refer back to Section 2A.

Figure 5 *The boiling points of selected hydrocarbons.*

Hydrocarbon Boiling Points

Hydrocarbon	Boiling Point (°C)
Butane	−0.5
Decane	174.0
Ethane	−88.6
Heptane	98.4
Hexane	68.7
Methane	−161.7
Nonane	150.8
Octane	125.7
Pentane	36.1
Propane	−42.1

In a manner similar to the one you used earlier to predict a property of an unknown element, you can examine patterns among the boiling points of some hydrocarbons in order to make valuable predictions. Use the data found in Figure 5 to answer the following questions.

1. a. In what pattern or order are Figure 5 data organized?
   b. Is this a useful way to present the information? Explain.
2. You are searching for a trend or pattern among these boiling points.
   a. Propose a more useful way to arrange these data.
   b. Reorganize the data table based on your idea.

Use your reorganized data table to answer these questions:

3. Which substance(s) are gases (have already boiled) at room temperature (22 °C)?
4. Which substance(s) boil between 22 °C (room temperature) and 37 °C (body temperature)?
5. What can you infer about intermolecular forces among decane molecules compared to those in butane?

A.5 CHEMICAL BONDING

Chemical Bonding CD-ROM WWW.

Hydrocarbons and their derivatives are the focus of the branch of chemistry known as **organic chemistry**. These substances are called organic compounds because early chemists thought that living systems—plants or animals—were needed to produce them. However, chemists have known for more than 150 years how to make many organic compounds without any assistance from living systems. In fact, starting materials other than petroleum can be used to produce organic compounds. You will learn about some of these starting materials in Section C.

The carbon chain forms a framework to which a wide variety of other atoms can be attached.

In hydrocarbon molecules, carbon atoms are joined to form a backbone called a **carbon chain**. Hydrogen atoms are attached to the carbon backbone. Carbon's versatility in forming bonds helps to explain the abundance of different hydrocarbon compounds, as you will soon learn. Hydrocarbons

ADDITIONAL ChemQuandaries

1. Why would butane be a poor substitute for propane in the tanks of portable stoves used by campers?

Answer: Butane has a higher boiling point (0.5 °C vs. (−42.1 °C) than propane. At winter temperatures (and increased pressure), butane might liquefy. Since the stove works on the basis of pressurized gas, this would be a poor fuel choice. A better substitute might be either ethane or methane, because their boiling points are lower than propane.

can be regarded as "parents" of an even larger number of compounds that contain atoms of other elements attached to a carbon chain.

Electron Shells

How are atoms of carbon or other elements held to each other in compounds? The answer is closely related to the arrangement of electrons in atoms. You already know that atoms are made up of neutrons, protons, and electrons. Neutrons and protons are located in the small, dense, central region of the atom called the nucleus. Electrons occupy different **energy levels** in the space surrounding the nucleus. Similar energy levels are grouped into shells, each of which can hold only a certain maximum number of electrons. For example, the first shell surrounding the nucleus of an atom has a capacity of two electrons. The second shell can hold a maximum of eight electrons.

Consider an atom of helium (He), the first member of the noble-gas family. A helium atom has two protons (and two neutrons) in its nucleus and two electrons occupying the first, or innermost, shell. Because two is the maximum this shell can hold, the shell is completely filled.

The next noble gas, neon (Ne), has an atomic number of 10. This means that each neutral neon atom contains ten protons and ten electrons. Two electrons occupy (and fill) the first shell. The remaining eight electrons fill the second shell. In neon, each shell has reached its electron capacity.

Both helium and neon are chemically unreactive—their atoms do not combine with each other or with atoms of other elements to form compounds. By contrast, sodium (Na) atoms—with an atomic number of 11 and one more electron than neon atoms—are extremely reactive. Chemists explain sodium's reactivity as due to its tendency to lose that additional electron. Fluorine (F) atoms each have nine electrons—one less than neon atoms—and are also extremely reactive. Their reactivity is due to their tendency to gain an additional electron.

Noble-gas elements are essentially unreactive because their separate atoms already have filled electron shells. All but helium have eight electrons in their outer shells; helium needs only two to reach its first-shell maximum. A useful key to understanding the chemical behavior of many elements is to recognize that atoms with filled electron shells are particularly stable—that is, they are chemically unreactive. How does this guideline help to explain both the stability of noble-gas elements and the reactivity of elements such as sodium and fluorine?

When sodium metal reacts, sodium ions (Na^+) form. The +1 electrical charge indicates that each sodium atom has lost one electron. Each Na^+ ion contains eleven positively charged protons but only ten negatively charged electrons—thus the net +1 charge. With ten electrons, Na^+ possesses filled electron shells (two electrons in the first shell and eight in the second), just like a neon atom. Unlike sodium atoms, Na^+ ions are highly stable. In fact, the world's entire natural supply of sodium is found as Na^+ ions.

By losing an electron, each sodium atom has been oxidized. If you need to review this concept from Unit 2, see page 193.

Fluorine atoms react to form fluoride ions (F^-). The −1 electrical charge indicates that each electrically neutral fluorine atom has gained one electron. Each fluoride ion contains nine protons and ten electrons—a net −1 charge. The ten electrons in an F^- ion constitute the same electron population found in a neon atom. Once again, an element has reacted to attain the special stability associated with filled electron shells.

By gaining an electron, each fluorine atom has been reduced. See Unit 2, page 193, for a review of this concept.

A.5 CHEMICAL BONDING

The unique ability of carbon to form multiple bonds is responsible for the tremendous number of organic compounds. Ionic bonding, involving electron transfer, is contrasted with covalent bonds where electrons are shared. Electron-dot diagrams and structural formulas enable students to draw pictures of the compounds and lead them to the model building in the next activity.

Hint: A quick review of the electron-dot arrangement of carbon and hydrogen will be very helpful to students in constructing the structural formulas of hydrocarbons. Make sure that students understand the relationship between outer shell electrons and chemical reactivity. (Only the outermost electrons can react easily.) They should also realize that each (:) between the symbols in a structural formula represents a pair of electrons (one covalent bond).

Use a flannel board as a display to represent Lewis dot structures. Use small felt disks to represent electrons and large disks to represent atoms or ions. Colored transparency film on the overhead projector will give the same effect. (A one-hole puncher works well to make the smaller disks.)

2. If one looks at a see-through butane lighter, one finds that at least some of the butane is in the liquid state. Explain how this is possible if the lighter is at room temperature. (See also the second Extending the Concepts, question 35, page 194.)

Answer: The boiling point of a liquid also depends on pressure. Apparently, the butane is under pressure higher than 1 atm (where it boils at 0.5 °C), and its boiling point at this pressure is greater than room temperature. Therefore, some of the butane is a liquid.

Covalent Bonds

As you have just learned, electrons are either lost or gained in the formation of ionic substances. In molecular (non-ionic) substances, atoms achieve filled electron shells by sharing electrons rather than by losing or gaining electrons. Many molecular substances are composed of atoms of nonmetals that do not readily lose electrons. As you will see, the sharing of electrons between two nonmetallic atoms allows both atoms to complete their outer shells.

A hydrogen molecule (H_2) provides a simple example of electron sharing. Each hydrogen atom contains only one electron, so one more electron is needed to fill the first shell. Two hydrogen atoms can accomplish this if they each share their single electron. If an electron is represented by a dot (·), then the formation of a hydrogen molecule can be depicted this way:

$$H\cdot + \cdot H \longrightarrow H{:}H$$

The number of outer-shell electrons for any Group A element is equal to its group number in the Periodic Table. Thus Group 6 A elements each have six outer-shell electrons. The number of electrons in this section is also equal to the last digit in the 1–18 system.

The chemical bond formed between two atoms that share a pair of electrons is called a **single covalent bond**. Through such sharing, both atoms achieve the stability associated with complete electron shells. A carbon atom, atomic number 6, has six electrons—two in the first shell and four in the second shell. Only the electrons in the outer shell participate in chemical reactions. To fill the second shell to its capacity of eight, four more electrons are needed. These electrons can be obtained through covalent bonding.

Consider the simplest hydrocarbon molecule, methane (CH_4). In this molecule, each hydrogen atom shares its single electron with the carbon atom. Similarly, the carbon atom shares one of its four outer-shell electrons with each hydrogen atom. This arrangement is represented below.

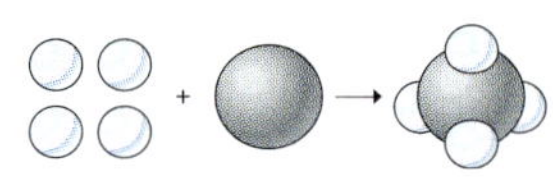

$$4\,H\cdot \;+\; \cdot\dot{\underset{\cdot}{C}}\cdot \longrightarrow \begin{matrix} & H & \\ & \because & \\ H{:} & C & {:}H \\ & \because & \\ & H & \end{matrix}$$

Electron-dot formulas are also referred to as Lewis structures. G.N. Lewis is given credit for laying the foundation of our current understanding of bond formation.

Here, as in the formula for a hydrogen molecule, dots surrounding each element's symbol represent the outer-shell electrons for that atom. Structures such as these are called **electron-dot formulas**. The two electrons in each covalent bond "belong" to both bonded atoms. Dots placed between the symbols of two atoms represent electrons that are shared by those atoms.

When determining the number of electrons associated with each atom, each shared electron in a covalent bond is "counted" twice, once for each element. For example, count the dots surrounding each atom in methane. You should notice that each hydrogen atom has a filled outer electron shell—two electrons in its first shell. The carbon atom also has a filled outer electron shell—eight electrons. Each hydrogen atom is associated with one

184

pair of electrons; the carbon atom has four pairs of electrons, or eight electrons.

For convenience, each pair of electrons in a covalent bond can be represented by a line drawn between the symbols of each atom. This yields another common representation of a covalently bonded substance called a **structural formula**.

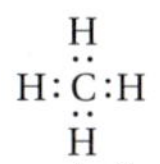

Electron-dot formula of methane, CH_4

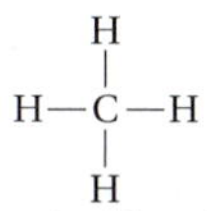

Structural formula of methane, CH_4

Although you can draw two-dimensional pictures of molecules on flat paper, assembling three-dimensional models gives a more accurate representation. Such an atomic model helps to predict a molecule's physical and chemical properties. The following activity provides an opportunity for you to assemble such models.

A.6 MODELING ALKANES

Laboratory Activity

Introduction

In this activity you will assemble models of several simple hydrocarbons. Your goal is to associate the three-dimensional shapes of these molecules with the names, formulas, and pictures used to represent them on paper.

Two types of molecular models are shown in Figure 6. Most likely, you will use ball-and-stick models. Each ball represents an atom, and each stick represents a single covalent bond (a shared electron pair) connecting two atoms. But of course molecules are not composed of ball-like atoms located at the ends of sticklike bonds. Experimental evidence shows that atoms are in contact with each other, much like what you see in space-filling models. However, ball-and-stick models are still useful because they can clearly represent the structure and geometry of molecules.

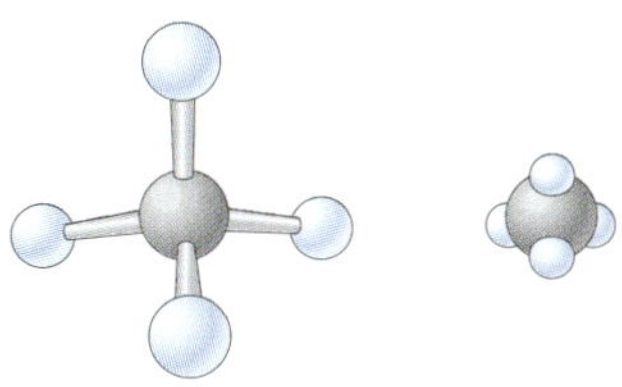

Figure 6 *Three-dimensional CH_4 models: ball-and-stick (left); space-filling (right).*

Look again at the electron-dot structure and structural formula for methane (CH_4) above. Methane, the simplest hydrocarbon, is the first member of a series of hydrocarbons known as **alkanes**. You will explore alkanes in this activity. Each carbon atom in an alkane forms single covalent bonds with four other atoms. Because each carbon atom is bonded to the maximum number of other atoms (four), alkanes are considered **saturated hydrocarbons**.

A.6 MODELING ALKANES

Laboratory Activity

Students will build three-dimensional models of methane, ethane, and propane. The general formula for alkanes is derived from the three-dimensional models. Students will practice writing electron-dot representations. Constructing models and diagrams enables students to see the differences among the three types of formulas.

Time

50 minutes

Materials (for 24 students working in pairs)

12 sets of space-filling or ball-and-stick models

Pre-lab discussion

One reason there are so many organic compounds is the unique bonding capability of carbon. Carbon can form four covalent bonds. This allows great flexibility in the way it can bond with itself and other elements.

Students should observe the bond angle of the C-H bond in a tetrahedral structure, such as in methane.

From the three-dimensional model, the "condensed" formula and the "structural" formulas of the hydrocarbons may be obtained.

Answers

1. No question—students assemble models.
2. a. The bond lines are positioned to indicate that an angle of the C-H bonds is 109.5° and shading is used to indicate depth.

   b. The pyramid structure. An equilateral triangle is formed with any three of the hydrogen atoms.

   c. Students' drawings should show the tetrahedral structure of methane.
3. a. 6 hydrogens

   b. 8 hydrogens

   c.

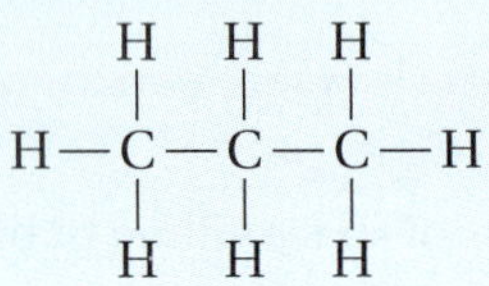

propane

4. a.

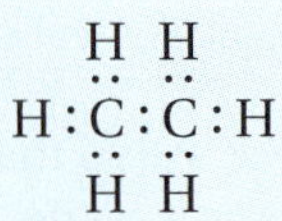

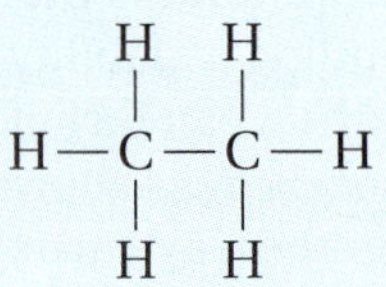

ethane

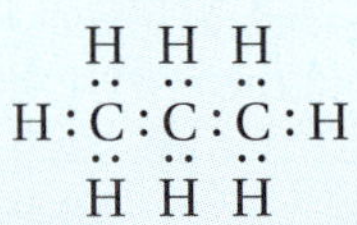

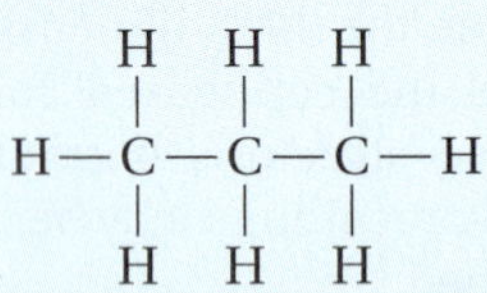

propane

   b. C_3H_8

Procedure

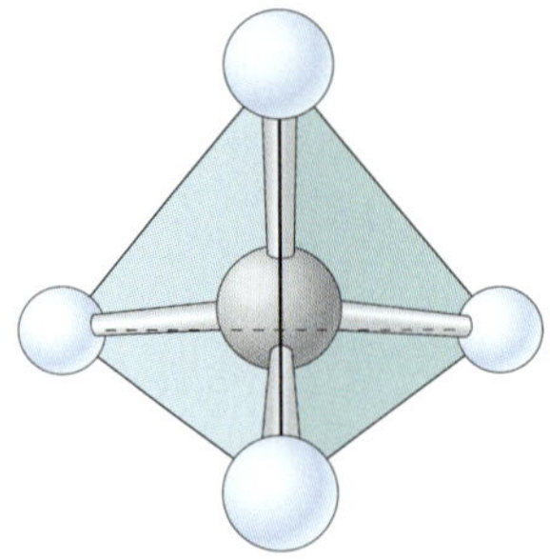

Figure 7 *The tetrahedral shape of a methane molecule.*

1. Assemble a model of methane (CH_4). Compare your model to the electron-dot and structural formulas on page 185. Note that the angles defined by bonds between atoms are not 90°, as you might think by looking at the structural formula. If you were to build a close-fitting box to surround a CH_4 molecule, the box would be shaped like a triangular pyramid, or a pyramid with a triangle as a base. A **tetrahedron** is the name given to this three-dimensional shape.

Why would the shape of a methane molecule be tetrahedral? Assume that the four pairs of electrons in the bonds surrounding the carbon atom—all with negative charges—repel one another. That is, the electron pairs stay as far away from one another as possible, arranging themselves so that they point to the corners of a tetrahedron. The angle formed by each C—H bond is 109.5°, a value that has been verified with several experimental methods. The angles are not 90°, as they would be if methane were flat. Verify this shape for yourself by arranging the atoms in your model.

2. Compare your three-dimensional model of methane to the representation of a tetrahedral molecule in Figure 7.
   a. How does the two-dimensional drawing (Figure 7) incorporate features that aid in visualizing the three-dimensional structure?
   b. Are there features of the two-dimensional figure that are difficult to translate into a three-dimensional structure? Explain.
   c. Translate your three-dimensional model into a two-dimensional drawing. Your drawing should convey the tetrahedral structure of methane.
3. Assemble models of a two-carbon and a three-carbon alkane molecule. Recall that each carbon atom in an alkane is bonded to four other atoms.
   a. How many hydrogen atoms are present in the two-carbon alkane?
   b. How many hydrogen atoms are present in the three-carbon alkane?
   c. Draw a ball-and-stick model, similar to the one in Figure 6 on page 185, of the three-carbon alkane.
4. a. Draw electron-dot and structural formulas for the two- and three-carbon alkanes.
   b. The molecular formulas of the first two alkanes are CH_4 and C_2H_6. What is the molecular formula of the third?

Examine your three-carbon alkane model and the structural formula you drew for it. Note that the middle carbon atom is attached to two hydrogen atoms, but the carbon atom at each end is attached to three hydrogen atoms. This molecule can be represented as CH_3—CH_2—CH_3, or $CH_3CH_2CH_3$. Formulas such as these provide convenient information about how atoms are arranged in molecules. For many purposes, such "condensed" formulas are more useful than molecular formulas such as C_3H_8.

Consider the formulas of the first few alkanes: CH_4, C_2H_6, and C_3H_8. Given the pattern represented by that series, try to predict the formula of the four-carbon alkane. If you answered C_4H_{10}, you are correct! The general molecular formula of all alkane molecules can be written as C_nH_{2n+2}, where

n is the number of carbon atoms in the molecule. So even without assembling a model, you can predict the formula of a five-carbon alkane: If $n = 5$, then $2n + 2 = 12$, and the formula is C_5H_{12}.

n can be any positive integer.

5. Using the general alkane formula, predict molecular formulas for the rest of the first ten alkanes. After doing this, compare your molecular formulas with the formulas given in Figure 8 to check your predictions.

The names of the first ten alkanes are also given in Figure 8. As you can see, each name is composed of a prefix, followed by *-ane* (designating an alk*ane*). The prefix indicates the number of carbon atoms in the backbone carbon chain. To a chemist, *meth-* means one carbon atom, *eth-* means two, *prop-* means three, and *but-* means four. For alkanes with five to ten carbon atoms, the prefix is derived from Greek—*pent-* for five, *hex-* for six, and so on.

6. Write structural formulas for butane and pentane.
7. a. Name the alkanes with these condensed formulas:
   (i) $CH_3CH_2CH_2CH_2CH_2CH_2CH_3$
   (ii) $CH_3CH_2CH_2CH_2CH_2CH_2CH_2CH_2CH_3$
   b. Write molecular formulas for the two alkanes in Question 7a.
8. a. Write the formula of an alkane containing 25 carbon atoms.
   b. Did you write the molecular formula or the condensed formula of this compound? Why?
9. Name the alkane having a molar mass of
   a. 30 g/mol.
   b. 58 g/mol.
   c. 114 g/mol.

CD-ROM WWW. Modeling Alkanes

Some Members of the Alkane Series

Name	Number of Carbons	Alkane Molecular Formulas: Short Version	Long Version
Methane	1	CH_4	CH_4
Ethane	2	C_2H_6	CH_3CH_3
Propane	3	C_3H_8	$CH_3CH_2CH_3$
Butane	4	C_4H_{10}	$CH_3CH_2CH_2CH_3$
Pentane	5	C_5H_{12}	$CH_3CH_2CH_2CH_2CH_3$
Hexane	6	C_6H_{14}	$CH_3CH_2CH_2CH_2CH_2CH_3$
Heptane	7	C_7H_{16}	$CH_3CH_2CH_2CH_2CH_2CH_2CH_3$
Octane	8	C_8H_{18}	$CH_3CH_2CH_2CH_2CH_2CH_2CH_2CH_3$
Nonane	9	C_9H_{20}	$CH_3CH_2CH_2CH_2CH_2CH_2CH_2CH_2CH_3$
Decane	10	$C_{10}H_{22}$	$CH_3CH_2CH_2CH_2CH_2CH_2CH_2CH_2CH_2CH_3$

Figure 8 *The first ten alkanes.*

7. a. (i) Heptane (ii) Nonane
   b. (i) C_7H_{16} (ii) C_9H_{20}
8. a. $C_{25}H_{52}$
   b. The condensed formula. The short version takes much less time and space; the long version would not fit on this page!
9. a. Ethane, C_2H_6: $2(12) + 6(1) = 30$ g/mol
   b. Butane, C_4H_{10}: $4(12) + 10(1) = 58$ g/mol
   c. Octane, C_8H_{18}: $8(12) + 18(1) = 114$ g/mol

Post-lab discussion

The enormous number and variety of hydrocarbons demand an orderly system of nomenclature. Learning the prefixes of the first ten alkanes allows students to name many compounds. To put a little humor in your discussion of organic nomenclature, consider giving students the following names for the alkanes C_{11} to C_{15}. (The actual IUPAC names are given in parentheses.)

C_{11} = Inane; (Undecane)
C_{12} = Nopane; (Dodecane)
C_{13} = Nogane; (Tridecane)
C_{14} = Rogane; (Tetradecane)
C_{15} = Windopane; (Pentadecane)

Stress the general formula for alkanes and the alternate ways of writing molecular formulas, as shown in Figure 8. Tell students that they do not have to choose a favorite way of representing compounds. Each version has its own advantages and purpose.

5. C_6H_{14}, C_7H_{16}, C_8H_{18}, C_9H_{20}, $C_{10}H_{22}$

6. 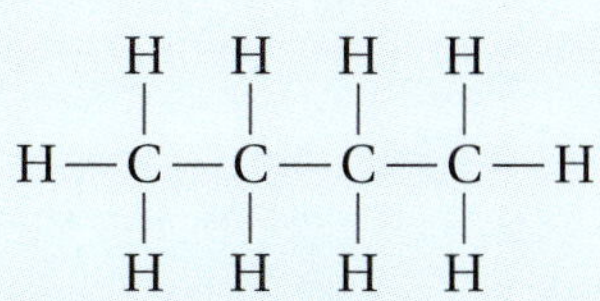

butane

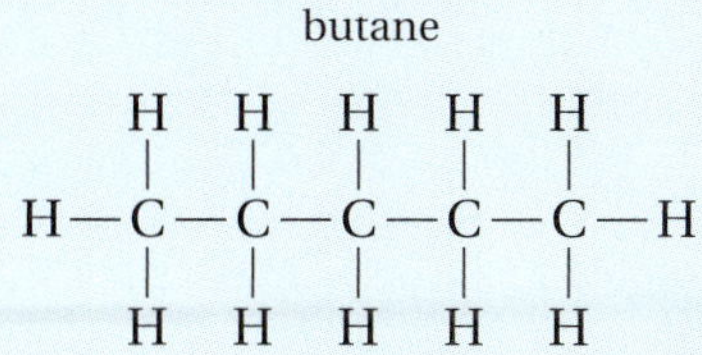

pentane

TRENDS IN ALKANE BOILING POINTS

Building Skills 2

Students are challenged to explain the observed boiling-point trend in the alkane homologous series and predict boiling points of larger hydrocarbons. Intermolecular forces in hydrocarbons are weak van der Waals forces that require relatively little energy to break. Discuss the chemical explanation for differences in boiling points, which is based on these intermolecular forces. Longer, and therefore less spherical, molecules have more surface area and more points of contact, and greater intermolecular attractions result.

Answers

1. Students' graphs should show a direct relationship.
2. Approximately 30 °C.
3. a, b. $C_{11}H_{24}$: 204 °C (actual: 196 °C), $C_{12}H_{26}$: 234 °C (actual: 216 °C), $C_{13}H_{28}$: 264 °C (actual: 235 °C)
4. The greater the number of carbons, the stronger the attractive forces.

TRENDS IN ALKANE BOILING POINTS

Building Skills 2

In the laboratory activity involving distillation, you used a technique that separates liquid mixtures according to boiling points of substances. You also know that the fractions of petroleum are separated based on their boiling points. Why do the fractions with the highest boiling points contain the largest molecules? Why are the smallest molecules found in the fractions with the lowest boiling points? In this activity, you will explore this trend in alkane boiling points.

Using data for the alkanes found in Figure 5 (page 182) and Figure 8 (page 187), prepare a graph of boiling points. The *x* axis scale should range from 1 to 13 carbon atoms (even though you will initially plot data for 1 to 10 carbon atoms). The *y* axis scale should extend from −200 °C to +250 °C.

1. Plot the data points. Draw a best-fit line through your data points according to these guidelines:
   - The line should follow the trend of your data points.
   - The data points should be equally distributed above and below the line.
   - The line should not extend past your data points.

   Figure 9 shows an example of a best-fit line.

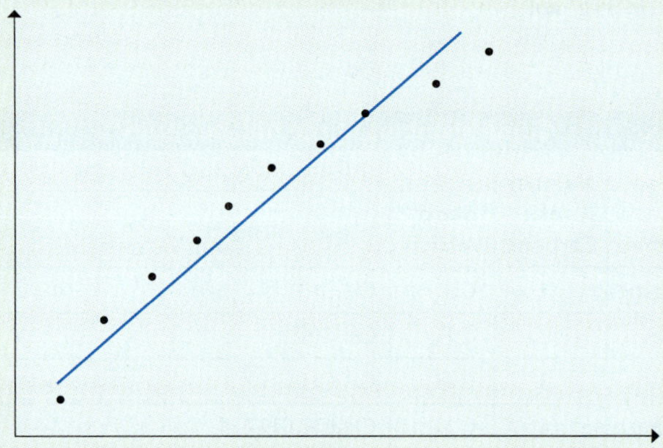

Figure 9 *An example of a best-fit line drawn through several data points.*

2. Estimate the average change in boiling point (in °C) when one carbon atom and two hydrogen atoms (—CH_2—) are added to a particular alkane chain.
3. The pattern of boiling points among the first ten alkanes allows you to predict boiling points for other alkanes.

a. Using your graph, estimate the boiling points of undecane ($C_{11}H_{24}$), dodecane ($C_{12}H_{26}$), and tridecane ($C_{13}H_{28}$). To do this, extend the trend of your graph line by drawing a dashed line from the graph line you drew for the first ten alkanes. This procedure is called **extrapolation**. Then read your predicted boiling points for C_{11}, C_{12}, and C_{13} alkanes on the *y* axis.
b. Compare your predicted boiling points to actual values provided by your teacher.

4. You learned that a substance's boiling point depends in part on its intermolecular forces, or attractions among its molecules. For the alkanes you have studied, what is the relationship between these attractions and the number of carbon atoms in each molecule?

CD-ROM WWW. Intermolecular Forces and Boiling Point

A.7 ALKANES REVISITED

Laboratory Activity

Introduction

The alkane molecules you have considered so far are **straight-chain alkanes**—each carbon atom is linked to only one or two other carbon atoms. In alkanes with four or more carbon atoms, many other arrangements of carbon atoms are possible. Alkanes in which one or more carbon atoms are linked to three or four other carbon atoms are called **branched-chain alkanes**. An alkane with four or more carbon atoms can have either a straight-chain or a branched-chain structure. In this activity you will use ball-and-stick molecular models to investigate such variations in alkane structures—variations that can lead to different properties.

A straight-chain structure:
C—C—C—C—C
A branched-chain structure:
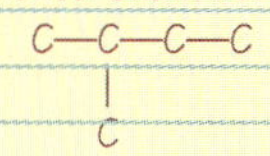

Procedure

1. Assemble a ball-and-stick model of a molecule with the formula C_4H_{10}. Compare your model with those built by others. How many different arrangements of atoms in the C_4H_{10} molecule can be constructed?

Molecules that have identical molecular formulas but different arrangements of atoms are called **isomers**. By comparing models, convince yourself that there are only two isomers of C_4H_{10}. The formation of isomers helps to explain the very large number of compounds that contain carbon chains or rings.

2. a. Draw an electron-dot formula for each C_4H_{10} isomer.
b. Write a structural formula for each C_4H_{10} isomer.

3. As you might expect, alkanes containing larger numbers of carbon atoms also have larger numbers of isomers. In fact, the number of different isomers increases rapidly as the number of carbon atoms

Time

50 minutes

Materials (for 24 students working in pairs)

12 space-filling or ball-and-stick model kits

Lab tips

Because building the isomers of hexane may require more atoms than found in a typical kit, students may need to form larger groups. The typical kit contains 12 to 14 carbon atoms, so if each team (with one kit) joins with two other teams (for a total of three kits/six students), each team can be asked to build as many unique isomers as it can in a given time period.

Answers

1. Two isomers of butane are possible.

2. a.
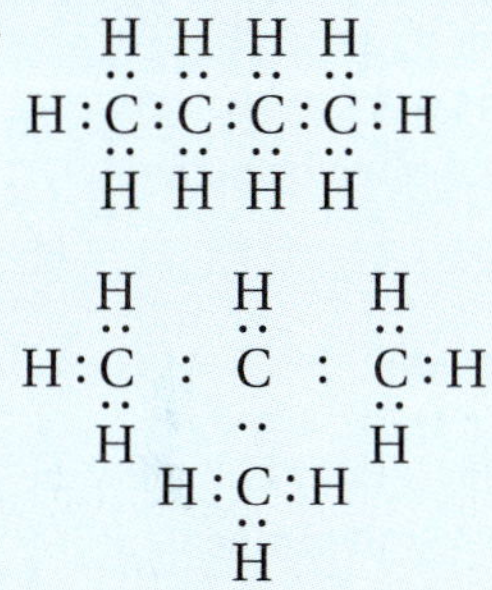

b.
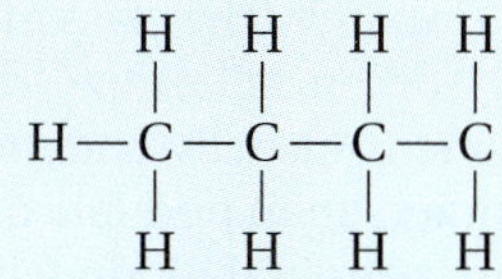
butane

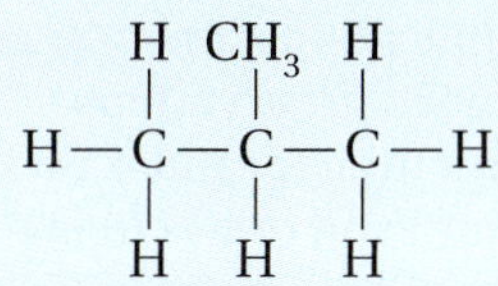
2-methylpropane

A.7 ALKANES REVISITED

Laboratory Activity

Students build models of C_4H_{10} and C_6H_{14} and find that in each case the atoms may be arranged in different ways (two and five, respectively), which introduces the concept of isomers. Students should realize that isomers exhibit different physical properties because they have different geometric arrangements of atoms.

Some students may confuse an isomer with an isotope, which was introduced in Unit 2. Stress the difference between them. Each isomer is a unique substance that has its own chemical and physical properties.

3. No. There are only three isomers of pentane possible: n-pentane, 2-methylbutane (isopentane), and 2,2-dimethyl propane (neopentane).

4. a.

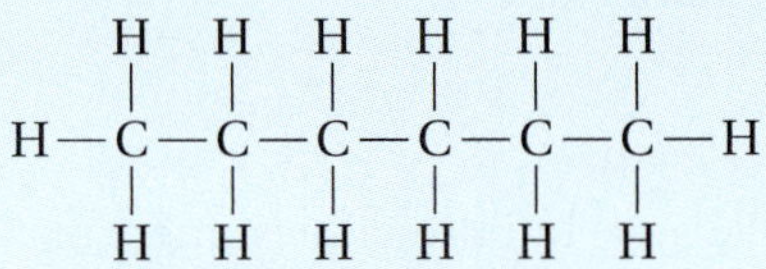

hexane

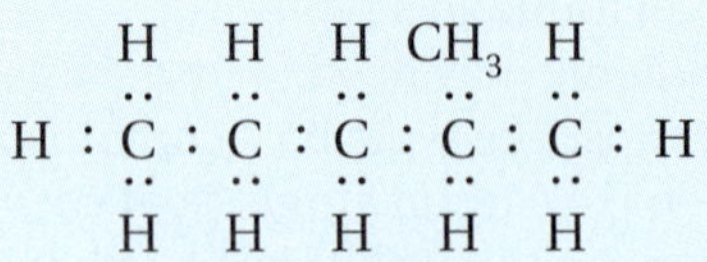

2-methylpentane

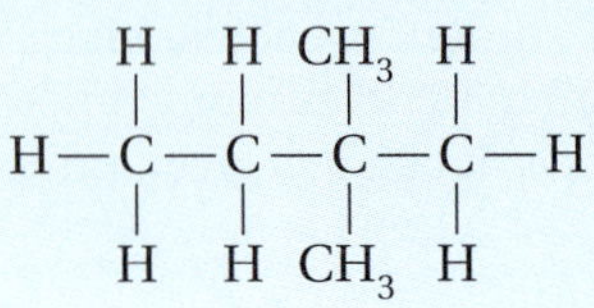

2,2-dimethylbutane

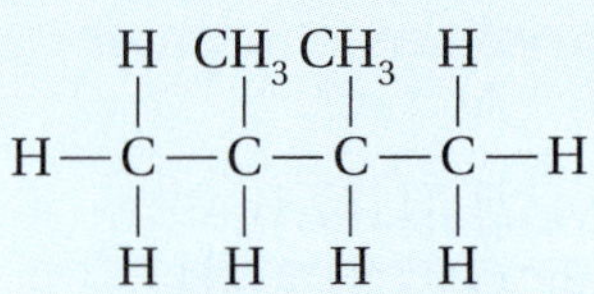

2,3-dimethylbutane

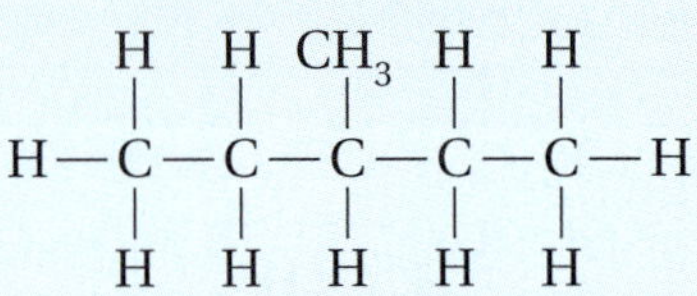

3-methylpentane

b. Hopefully the class found all isomers of hexane. Students usually find more isomers than possible due to variances in their drawings that are actually duplications. This a good time to draw various structural formulas on the board and draw circles around the longest continuous chain of carbons to visualize the attached groups.

increases. For example, chemists have identified three pentane (C_5H_{12}) isomers. Their structural formulas are shown in Figure 10. Try building these and other models. Are other pentane isomers possible?

Alkane Isomers

Alkane	Structural Formula	Boiling Point (°C)
C_5H_{12} isomers	$CH_3—CH_2—CH_2—CH_2—CH_3$	36.1
	$CH_3—CH(CH_3)—CH_2—CH_3$	27.8
	$CH_3—C(CH_3)_2—CH_3$	9.5
Some C_8H_{18} isomers	$CH_3—CH_2—CH_2—CH_2—CH_2—CH_2—CH_2—CH_3$	125.6
	$CH_3—CH_2—CH_2—CH_2—CH_2—CH(CH_3)—CH_3$	117.7
	$CH_3—CH(CH_3)—CH_2—C(CH_3)_2—CH_3$	99.2

Figure 10 *Some pentane and octane isomers.*

4. Now consider possible isomers of C_6H_{14}.
   a. Working with a partner, draw structural formulas for as many different C_6H_{14} isomers as possible. Compare your structures with those drawn by other groups.
   b. How many different C_6H_{14} isomers were found by your class?
5. Build models of one or more C_6H_{14} isomers, as assigned by your teacher.
   a. Compare the three-dimensional models built by your class with corresponding structures drawn on paper.
   b. Based on your examination of the three-dimensional models, how many different C_6H_{14} isomers are possible?

Because each isomer is a different substance, it has its own characteristic properties. In the next activity, you will examine boiling-point data for some alkane isomers.

5. a. Student models will vary. Some students may not recognize the same isomer on paper because of simple twists and bends possible in the model.

b. 5 isomers of hexane.

BOILING POINTS OF ALKANE ISOMERS

Building Skills 3

The purpose of this activity is to apply and test the idea that increased branching causes hydrocarbon boiling points to decrease. Students should see the relationship between the structure of an isomer and the boiling point of

BOILING POINTS OF ALKANE ISOMERS

Building Skills 3

You have already observed that boiling points of straight-chain alkanes are related to the number of carbon atoms in their molecules. Increased intermolecular forces are associated with the greater molecule-to-molecule contact possible for larger alkanes. Now consider the boiling points of some isomers.

1. Boiling points for two sets of isomers are listed in Figure 10 (page 190). Within a given set, how does the boiling point change as the extent of carbon-chain branching increases? Assign each of the following boiling points to the appropriate C_7H_{16} isomer: 98.4 °C, 92.0 °C, 79.2 °C.

a. $CH_3{-}CH_2{-}CH_2{-}CH_2{-}CH_2{-}CH_2{-}CH_3$

b. 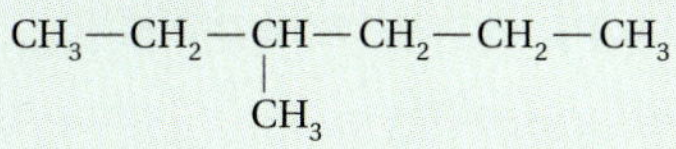

c. 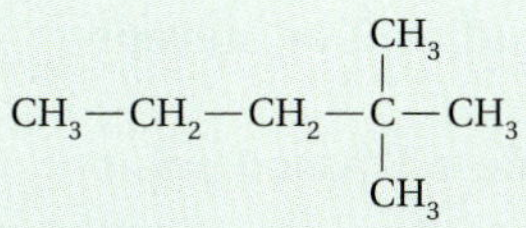

3. Here is the structural formula of a C_8H_{18} isomer:

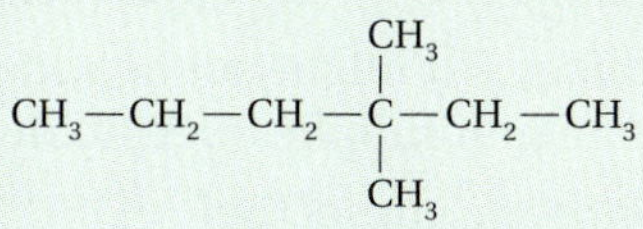

a. Compare it to each C_8H_{18} isomer listed in Figure 10. Predict whether it has a higher or lower boiling point than each of the other C_8H_{18} isomers.

b. Would the C_8H_{18} isomer shown here have a higher or lower boiling point than each of the three C_5H_{12} isomers shown in Figure 10?

4. How do you explain the boiling point trends that you observed in this activity?

Chemists and chemical engineers use information about molecular structures and boiling points to separate the complex mixture known as petroleum into a variety of useful substances, many of which you are quite familiar with. In Section B, you will learn how bonding helps to explain the use of petroleum as a fuel.

CD-ROM WWW. **Questions and Answers**

Answers

1. a. n-heptane, BP = 98.4 °C

b. 3-methylhexane, boiling point (BP) = 92.0 °C

c. 2,2-dimethylpentane, BP = 79.2 °C

3. a. Its actual boiling point is 112 °C, lower than octane and 2-methylheptane, and higher than 2,2,4-trimethylpentane.

b. One would expect it to have a higher boiling point than all the C_5H_{12} isomers, because of its larger size. (Some students may think that because it is branched, it might have a lower boiling point than unbranched pentane. In fact, its boiling point is 112 °C, compared with 36.1 °C for pentane.) Total molecular weight is also a factor.

4. The boiling points decrease with the increasing number of carbon side chains within an isomer series. The more compact shape affords fewer intermolecular contacts and that, in turn, makes these molecules separate more easily from one another.

the hydrocarbon. Intermolecular attraction decreases as the molecular structure approaches a more spherical shape (by increased branching), thus minimizing surface area contact between molecules. Knowledge of this provides an important introduction regarding altering fuels and octane numbers, as covered in Section C.

Students are not asked to name the isomers; emphasis should be placed on the arrangement of the carbon atoms. Each isomer must have the correct number of carbon and hydrogen atoms, as indicated in the molecular formula. Check student models to verify they are unique, not simply bent or twisted versions of the same isomer.

SECTION SUMMARY

Reviewing the Concepts

1. What is a hydrocarbon?

A hydrocarbon is a molecular compound made up of the elements hydrogen and carbon.

2. What does it mean to refine a natural resource?

To refine a natural resource means to remove the impurities from the raw material or separate the raw material into its useful components.

3. What is meant by saying that oil is "crude"?

Crude oil is the raw material as it is removed from the ground or the petroleum before refining.

4. What is the likelihood of discovering a pure form of petroleum that can be used directly as it is pumped from the ground? Explain your answer.

In chemistry, the word *pure* implies a substance that is of uniform composition. Petroleum is a complex mixture of many hydrocarbons with a wide range of boiling points. In order to be useful, petroleum must be separated into fractions consisting of hydrocarbons with boiling points close to one another. It is highly unlikely to find crude oil pure enough for a specific use.

5. About 16% of the petroleum used in the United States is used for "molecule building," producing nonfuel products that have a significant impact on everyday life. The remaining 84% of the petroleum is burned as fuel.

a. What is the average number of barrels of petroleum used daily in the United States for building (nonfuel) purposes?

b. How many barrels of petroleum are burned as fuel daily in the United States?

c. List four household items made from petroleum.

d. What materials could be substituted for each of these four items if petroleum were not available to make them?

a. Approximately 3 million barrels/day.

b. Approximately 15 million barrels/day.

c. The list will vary from plastic containers, fibers, medicines, detergents, adhesives, to cosmetics, etc.

d. Plastic food containers—glass or cardboard
Plastic film—wax paper
PVC pipe—copper pipe
Synthetic fibers—natural fibers such as wool, cotton, linen
Medicines—vegetable or herbal remedies
Detergents—soap
Adhesives—natural adhesives such as pitch
Laminated plastic countertops—granite, marble, wood
Plastic furniture—wood
Plastic housing for computers, radios, TVs, stereos—wood
Latex or acrylic paints—vegetable-oil-based paint

6. Name several fuels obtained from crude petroleum.

Natural gas, gasoline, kerosene, fuel oil, diesel fuel

7. List several products that might not be widely and easily available if petroleum supplies were to dwindle.

Gasoline, fuel oil, kerosene, diesel fuel

8. Rank the following hydrocarbons from their lowest boiling point to their highest: hexane (C_6H_{14}), methane (CH_4), pentane (C_5H_{12}), and octane (C_8H_{18}). Explain your rankings.

lowest BP	**methane**	**−161.7 °C**
	pentane	**36.1 °C**
	hexane	**68.7 °C**
highest BP	**octane**	**125.7 °C**

The boiling points increase due to increased forces of attraction between the molecules (called van der Waal's or dispersion forces). These forces increase with increasing mass and surface area.

9. Sketch the basic setup for a laboratory distillation. Label the key parts.

See Figure 3: A-2 Laboratory Activity: Separation by Distillation

10. Simple distillation is never sufficient to separate two liquids completely. Explain.

Since the molecular speeds of the molecules vary widely, some of the molecules of the higher boiling point fraction may vaporize before the other fraction has been completely condensed at its boiling point. Molecules from the end of the distillate of the lower boiling point fraction will have some of the molecules of the higher boiling point fraction mixed with them.

11. Explain why thermal energy is added at one point and removed at another point in the process of distillation.

Heat must be applied to the distilling flask to vaporize the lower boiling component. The vapor must be delivered to a condenser, where it is allowed to cool and condense back to a liquid in a separate collecting flask.

12. How does a fractional distillation differ from a simple distillation?

Simple distillation may consist of the separation of only two or three component substances. Fractional distillation of petroleum involves many components that are heated and vaporized. The vapor mixture rises and condenses on various trays in the fractionating column depending on the boiling points of the fractions. The lower boiling point fractions condense in the upper part of the tower, while the higher ones collect on trays below them. Each fraction is a mixture of hydrocarbons of similar boiling points, rather than one component, as in a simple distillation.

13. Petroleum fractions include light, intermediate, and heavy distillates and residues. List three useful products derived from each of these fractions.

Light: starting material for plastics, gasoline, gasoline additives, and solvents
Intermediate: kerosene, jet fuel, solvents, alcohols, dyes
Heavy distillates: fuel oil, diesel oil, lubricants, heavy oil
Residues: asphalt, paraffin, petroleum jelly, greases

14. Where in a distillation tower—top, middle, or bottom—would you expect the fraction with the highest boiling point range to be removed? Why?

The bottom fraction in the distillation tower would be the substance that is last to evaporate or having the highest boiling point.

15. After fractional distillation, each fraction is still a mixture. What must be done to further separate the components of each fraction?

Each fraction can be further separated by another fractional distillation.

16. What does each dot in an electron-dot diagram represent?

Each dot in an electron-dot diagram represents a bonding electron.

17. Each carbon atom has six electrons. Why does the electron-dot representation of carbon show only four dots?

The two electrons in the first energy level are not available for bonding.

18. Define the term "covalent bond."

A bond formed by sharing a pair of electrons between two atoms.

19. Draw an electron-dot diagram for a branched six-carbon hydrocarbon.

One example is 2-methylpentane.

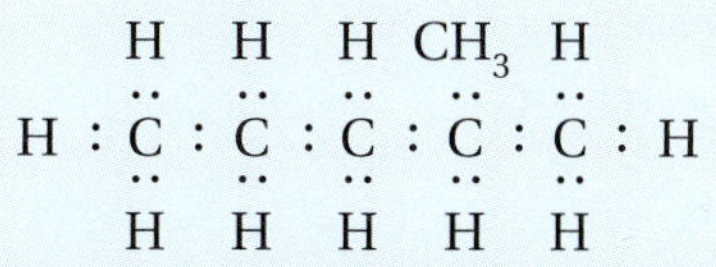

2-methylpentane

20. **a.** What additional information does a structural formula convey that a molecular formula does not?

b. In what ways is a structural formula an inadequate representation of a real molecule?

a. A structural formula shows the arrangement of atoms within the molecule, whereas the molecular formula only shows the number of atoms of each element in a molecule.

b. A structural formula can only show the molecule in two dimensions, whereas molecules are three-dimensional, and not as rigid as a structural formula might indicate.

21. Use the general molecular formula to write the molecular formula for an alkane containing:

a. 6 carbons b. 10 carbons

c. 16 carbons d. 25 carbons

a. C_6H_{14} b. $C_{10}H_{22}$ c. $C_{16}H_{34}$ d. $C_{25}H_{52}$

22. Alkanes are said to be saturated hydrocarbons. What does "saturated" mean?

It means that all of the bonds in the molecule are single covalent bonds.

23. What does *-ane* imply about the bonding arrangement in compounds such as hexane, butane, methane, and octane?

The suffix *-ane* implies that the carbon atoms are bonded in a chain of single covalent bonds.

24. Draw structural formulas for at least three isomers of C_9H_{20}.

There are many potential isomers of nonane. A few that students may draw are:

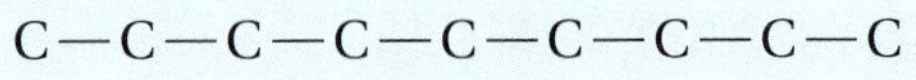

n-nonane

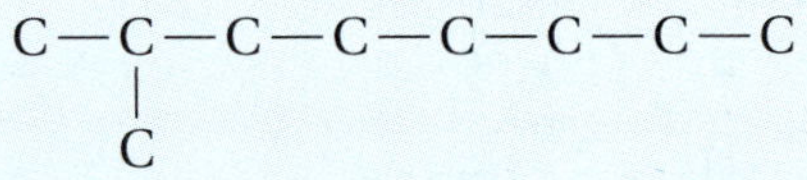

2-methyloctane

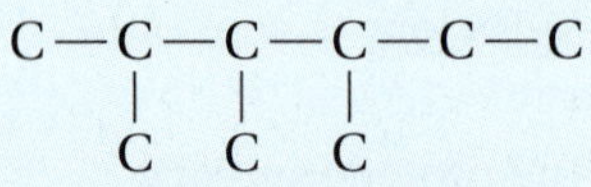

2,3,4-trimethylhexane

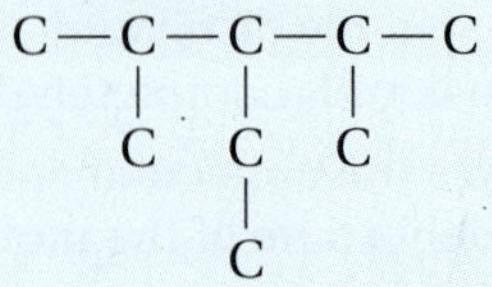

3-ethyl-2,4 dimethylpentane

25. What is the shortest-chain alkane that can demonstrate isomerism?

Butane can form the isomer isobutane, which is also named 2-methylpropane.

26. a. Draw two hexane isomers—one straight chain and one branched chain.

b. Which of the two would have the lower boiling point? Explain your choice.

a. There are four isomers of hexane: n-hexane, 2-methylpentane, 3-methylpentane, and 2,2-dimethylbutane. Two of these isomers are:

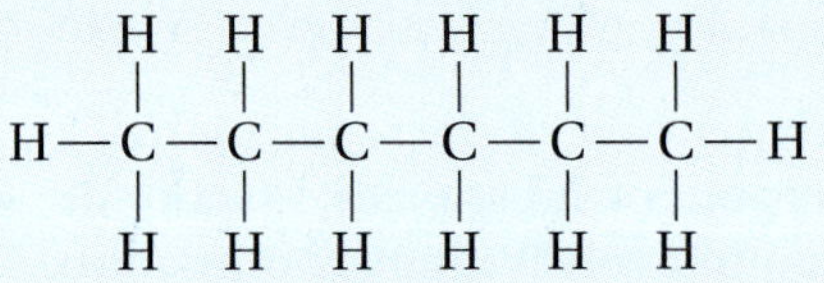

n-hexane

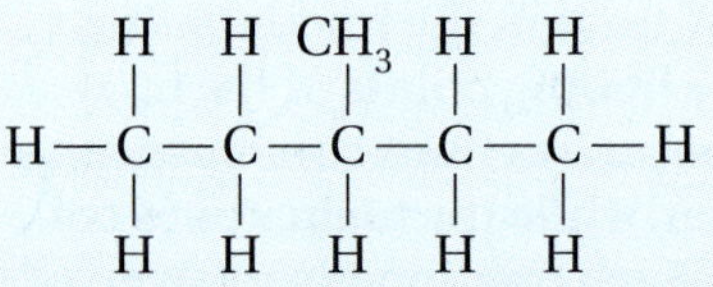

3-methylpentane

b. The branched chain hydrocarbon would have the lower boiling point. The branching shape of the molecule tends to be more compact or spherical. This decreases the surface area with the result that the intermolecular forces become weaker and are overcome at a lower temperature.

Connecting the Concepts

27. Why is petroleum considered a nonrenewable resource?

The total inventory of available petroleum on Earth is decreasing and is not replaceable or able to be replenished like water, plants, animals, or air.

28. In a fractionating tower, petroleum is generally heated to 400 °C. What would happen if it were heated to only 300 °C?

The fractions would condense in lower trays of the fractionating tower. The higher boiling fractions above 300 °C would not vaporize and condense. They would be left behind with the bottom residue.

29. The molar masses of methane (16 g/mol) and water (18 g/mol) are similar. At room temperature, methane is a gas and water is a liquid. Explain this difference.

Water has stronger forces of attraction between its molecules (hydrogen bonds) than methane (dispersion forces).

30. The traditional unit of volume for petroleum is one barrel, which contains 42 gallons. Assume that those 42 gallons provide 21 gallons of gasoline. How many barrels of petroleum does it take to operate an automobile for a year, assuming the auto travels 10 000 miles and goes 27 miles on a gallon of gas?

$$\frac{\textbf{10 000 mi}}{\textbf{1 yr}} \times \frac{\textbf{1 barrel}}{\textbf{21 gal gas}} \times \frac{\textbf{1 gal}}{\textbf{27 mi}} = \textbf{18 barrels/year}$$

31. Which mixture would be easier to separate by distillation—a mixture of pentane and octane or a mixture of pentane and a branched-chain octane isomer? Explain the reasoning behind your choice.

Pentane (36 °C) and octane (126 °C) would be easier to separate than pentane and a branched octane isomer (isooctane 116 °C) since they have a greater difference in boiling points.

Extending the Concepts

32. Is it likely that the composition of crude oil in Texas is the same as that of crude oil in Kuwait? Explain your answer.

The composition of petroleum from Texas and Kuwait would be very different, due to the conditions under which they were formed. All petroleum consists chiefly of saturated hydrocarbons, but may also contain unsaturated or aromatic derivatives and contaminants, depending upon the source.

33. Gasoline's composition, as blended by oil companies, varies in different parts of the nation.

Does the composition relate to the time of year? If so, what factors help to determine the composition of gasoline in various seasons?

The composition is blended to match the season and climate. The factors that determine the composition are viscosity and volatility. In cold conditions, more high-volatility hydrocarbon molecules are used in

gasoline, to promote proper fuel vaporization in the engine. Under warmer conditions, the blended gasoline contains more low-volatility hydrocarbon fractions, to avoid premature or too rapid vaporization.

34. What kind of petroleum trade relationship would be expected between North America and the Middle East? If other world regions become more industrialized and global petroleum supplies decrease, how might the North America–Middle East trade relationship change?

A good trade relationship is needed between North America and parts of the Middle East. The Middle Eastern countries, particularly the OPEC countries, have in the past been free to raise their prices, and oil-importing countries have been forced to pay the increases.

As other nations increase their use of Middle Eastern oil, the North America–Middle East trade relationship will depend on price increases. If it becomes too expensive to import oil from the Middle East, North America may turn to more locally available resources or search for petroleum substitutes. The United States now imports increasing amounts of petroleum from Mexico, South America, and Canada. Developing technology in other world regions might lead new nations to become oil exporters, providing another source of petroleum for North America, as well as increasing trade relations with those countries.

35. The hydrocarbon boiling points listed in Figure 5 (page 182) were measured under normal atmospheric conditions. How would those boiling points change if atmospheric pressure were increased? (*Hint:* Although butane is stored as a liquid in a butane lighter, it escapes through the lighter nozzle as a gas.)

The boiling points would increase with increased pressure. Butane, normally a gas at room temperature, is a liquid when held under pressure in a closed volume. The increased pressure raises butane's boiling point. When the valve is opened to the lower pressure of the atmosphere, the boiling point of butane becomes lower than room temperature, and the liquid vaporizes.

36. The two isomers of butane have different physical properties, as illustrated by their different boiling points. They also have different chemical properties. Explain how isomerism may contribute to differences in chemical behavior.

In isomers, the arrangements of the atoms in space are different, and branched alkanes burn more evenly than straight-chained alkanes.

37. What properties of petroleum make it an effective lubricant?

Petroleum is a very slippery, viscous liquid at room temperature, insoluble in water, and relatively inert.

38. When 1,2-ethanediol (ethylene glycol, also known as permanent antifreeze) is dissolved in water in an automobile's radiator, it helps keep the water from freezing. The permanent antifreeze-water solution has a lower freezing point than does pure water. Similarly, when an ionic substance such as table salt (NaCl) is dissolved in water, the solution freezes at a lower temperature than does pure water. Why is NaCl a highly undesirable additive for car radiators, whereas ethylene glycol is a suitable additive? (*Hint:* Compare the structure and chemical properties of these two substances.)

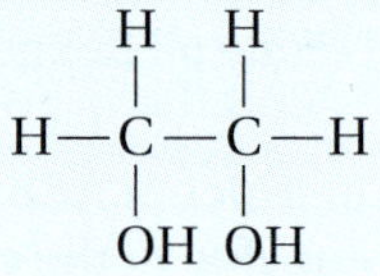

1,2-ethanediol
(ethylene glycol)

Sodium chloride ionizes in water. Salt water promotes corrosion, and although it will lower the freezing point of radiator fluid, it will contribute to the cooling system's early demise. Ethylene glycol is a molecular substance and is much less reactive with the radiator components. Ethylene glycol is also named 1,2-ethanediol.

SECTION B

PETROLEUM AS AN ENERGY SOURCE

People have used petroleum for almost 5000 years. The first oil well was drilled in the United States in 1859 in Pennsylvania. Since then, human life has been greatly altered by the increasing use of petroleum. In the following activity, you will begin to appreciate just how much everyday life is influenced by petroleum's role as an energy source—not just in powering automobiles and other vehicles, but in energizing modern society itself.

USING FUEL

ChemQuandary 1

Examine this textbook. It is a composite of a wide variety of materials—paper, colored inks, cardboard, coating material, and binding adhesive, to name a few. Indeed, a "materials story" could be written about this textbook—a story that would explore the origins of its component materials.

What is less obvious, perhaps, is that this textbook can also be analyzed in terms of energy: how energy (and the fuel from which the energy was produced) was involved in its manufacture, warehousing, and delivery to your school.

Complete an "energy trace" for your *ChemCom* textbook by following these steps. (a) Decide what general events must have occurred in the production of this textbook and its subsequent delivery to your school. (b) From where did the energy for each event come? In other words, what was the source of the needed energy? In your view, is the total cost of this textbook related more to the materials used (the paper, ink, adhesive, etc.) or to the energy required? Explain your reasoning.

B.1 ENERGY AND FOSSIL FUELS

Fossil fuels originate from biomolecules of prehistoric plants and animals. The energy released by burning these fuels represents energy originally captured from sunlight by prehistoric green plants during photosynthesis. Thus fossil fuels—petroleum, natural gas, and coal—can be thought of as forms of buried sunshine.

No one knows the exact origin of petroleum. Most evidence indicates that it originated from living matter in ancient seas some 500 million years ago. These species died and eventually became covered with sediments. Pressure, heat, and microbes converted what was once living matter into petroleum, which became trapped in porous rocks. It is likely that some petroleum is still being formed from sediments of dead matter. However, such a process is far too slow to consider petroleum a renewable resource.

Event	Type of Energy	Source
Release of electromagnetic radiation	Fusion	Sun
Tree absorbing radiation	Chemical	Plant synthesis
Tree making cellulose	Chemical	Plant synthesis
Processing tree to pulp to paper	Mechanical/chemical	Petroleum
Transporting paper to printer	Mechanical	Petroleum
Printing the textbook	Mechanical	Electrical/fossil fuel/petroleum
Transporting to school	Mechanical	Petroleum

PETROLEUM AS AN ENERGY SOURCE

Introduction

In Section B, students will tackle the "burning" issue of petroleum as a source of energy. Energy released from burning petroleum is described in terms of bond breaking and bond making. Students measure the energy released by burning a hydrocarbon in a simple calorimeter laboratory activity. The unit concludes with students writing an essay to confirm or dispute the claims made in the advertisement at the beginning of Unit 3.

USING FUEL

ChemQuandary 1

"Energy trace" for the *ChemCom* textbook.

Since the *ChemCom* textbook is made of paper, the textbook began with energy from the sun: fusion. See the table below.

B.1 ENERGY AND FOSSIL FUELS

This section identifies the sun as the primary source of energy on Earth and introduces the chemistry of energy storage and the energy release from petroleum molecules. Chemical reactions are discussed in terms of endothermic and exothermic processes.

Emphasize that the exact sequence of events involved in a reaction is often quite complex. Reactions go through a series of steps called the "reaction mechanism." Chemical equations simply show what we start with and what the end products are, but there is a lot more to the story. The story is simplified here; however, it does help in thinking about the energy of a reaction. Both of the reactions here illustrate that in an exothermic reaction, the energy necessary to break the bonds is less than that released when making bonds. The combustion of methane is exothermic overall. If a reaction is exothermic, then the reverse reaction is endothermic.

Have students practice writing the energy of reaction on the product or reactant side of the equation to represent an endothermic or exothermic reaction.

DEMONSTRATION IDEAS

1. The electrolysis of water, producing hydrogen and oxygen, demonstrates that the decomposition of water is endothermic. Collect a small quantity of hydrogen and oxygen in a test tube, and, igniting with a splint, demonstrate that the combination of hydrogen and oxygen is an exothermic reaction. (See *Chemical Demonstrations, A Sourcebook for Teachers,* vol. 1, Summerlin and Ealy, 1988, ACS Publications, for appropriate procedures and safety information on this demonstration.) The combustion of natural gas (mainly methane) in a gas burner also represents an exothermic process.
2. An endothermic reaction is used in cold packs. Place 100 mL of water into a large beaker and record the temperature. Quickly add 10–15 grams of ammonium nitrate to the water. Note the change in temperature as the solid dissolves. The dissolving of ammonium nitrate is endothermic. The temperature should decrease 6 to 9 °C. In this reaction, 600–900 calories are absorbed per 100 mL of water. A more accurate reading may be obtained by using an insulated or Styrofoam cup.

   Equation for the reaction:
   Heat + $NH_4NO_3(s) + H_2O(l) \rightarrow$
   $NH_4^+(aq) + NO_3^-(aq)$
3. Students are more familiar with exothermic reactions. Here is another example. Place 100 mL of water into a large beaker and record the temperature. Quickly add 10–15 grams of calcium chloride to the water. Note the change in temperature as the solid dissolves. The dissolving of calcium chloride is exothermic. The temperature should increase about 12 °C. The heat of solution for calcium chloride is about 117 calories per 100 mL of water. A more accurate reading may be obtained by using an insulated or Styrofoam cup. Discard the cup after the demonstration.

   Equation for the reaction:
   $CaCl_2(s) + H_2O(l) \rightarrow$
   $Ca^{2+}(aq) + 2\,Cl^-(aq)$ + Heat

A rock poised at the top of a hill has more potential energy—energy of position—than the same rock at the bottom of the hill.

Energy and Fossil Fuels (CD-ROM, WWW.)

Methane is a major component of natural gas.

Fossil-fuel energy is comparable in some ways to the energy stored in a wind-up toy race car. The "winding-up" energy that was originally supplied tightened a spring in the toy. Most of that energy is stored within the spring. That stored energy is a form of **potential energy**, or energy of position. As the car moves, the spring unwinds, providing energy to the moving parts. Energy related to motion is called **kinetic energy**. Thus the movement of the car is based on converting potential energy into kinetic energy. Eventually, the toy "winds down" to a lower-energy, more stable state and stops.

In a similar manner, chemical energy, which is another form of potential energy, is stored in the bonds within chemical compounds. When an energy-releasing chemical reaction occurs, as during the burning of a fuel, bonds break and reactant atoms reorganize to form new bonds. The process yields products with different and more stable arrangements of their atoms. That is, the products have less potential energy (chemical energy) than did the original reactants. Some of the energy stored in the reactants has been released in the form of heat and light.

The combustion, or burning, of methane (CH_4) gas illustrates such an energy-releasing reaction. It can be summarized this way:

CH_4	+	$2\,O_2$	$\longrightarrow$	CO_2	+	$2\,H_2O$	+	Energy
Methane gas		Oxygen gas		Carbon dioxide gas		Water		

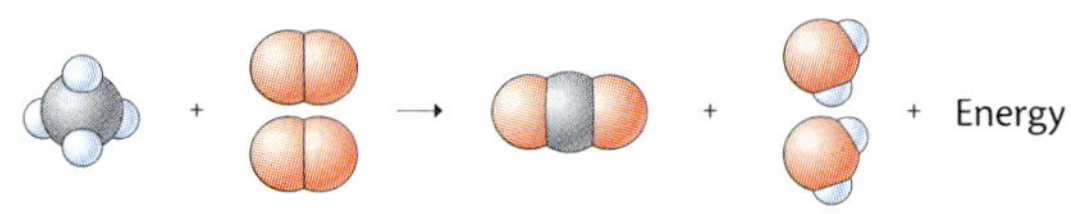

The reaction releases considerable thermal energy (heat). In fact, laboratory burner flames are based primarily on that reaction—performed, of course, under very controlled conditions. To gain a better understanding of the energy involved, imagine that the reaction takes place in two distinct steps—one involving bond-breaking and one involving bond-making.

In the first step, suppose that all the chemical bonds in one CH_4 molecule and two O_2 molecules are broken. (How many total chemical bonds need to be broken? Check the methane-burning reaction above to guide your thinking.) The result of this bond-breaking step is that separate atoms of carbon, hydrogen, and oxygen are produced. Such a bond-breaking step is an energy-requiring process, or an **endothermic** change. In an endothermic change, energy must be added to "pull apart" the atoms in each molecule. Thus energy appears as a reactant in Step 1.

An endothermic reaction can proceed only if energy is continuously supplied.

Step 1:
Energy + $CH_4 + 2\,O_2 \longrightarrow C + 4\,H + 4\,O$

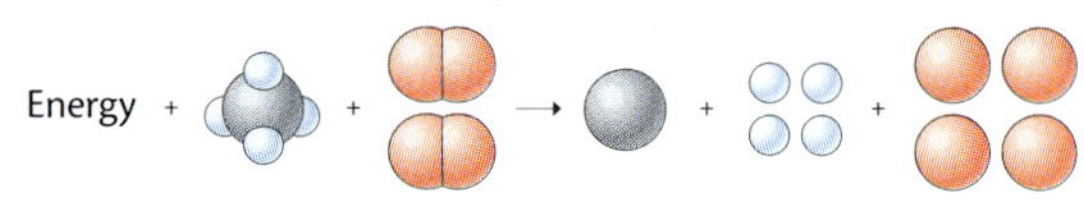

To complete the methane-burning reaction, suppose that the separated atoms now join to form the new bonds needed to make the product molecules: one CO_2 molecule and two H_2O molecules. The formation of chemical bonds is an energy-releasing process, or an **exothermic** change. Because energy is given off, it appears as a product in Step 2.

Step 2:

$$C + 4\,H + 4\,O \longrightarrow CO_2 + 2\,H_2O + Energy$$

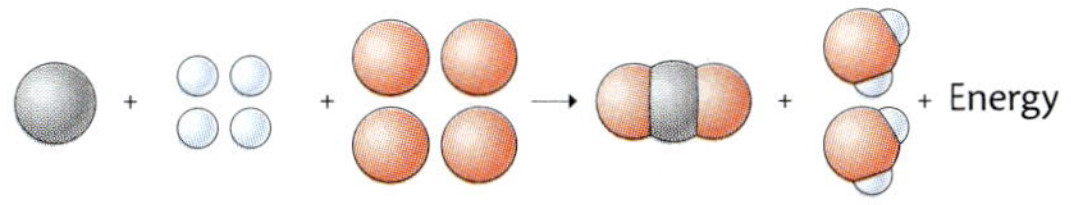

Although exothermic reactions—which release energy—can continue on their own, it often takes the addition of "starting energy" (activation energy) to initiate the chemical change. Thus a match must be "activated" by striking it on a matchbook cover before the exothermic burning reaction starts.

When methane burns, the energy released in forming carbon-oxygen bonds in CO_2 and hydrogen-oxygen bonds in H_2O is greater than the energy used to break the carbon-hydrogen bonds in CH_4 and the oxygen-oxygen bond in O_2. That is why the overall chemical change is exothermic. The complete energy summary for the burning of methane is shown in Figure 11.

Once an exothermic reaction begins, it releases energy until the reaction stops.

Whether an overall chemical reaction is exothermic or endothermic depends on how much energy is added (endothermic process) in the bond-breaking step and how much energy is given off (exothermic process) in the bond-making step. If more energy is given off than added, the overall process is exothermic. However, if more energy is added than is given off, the overall change is endothermic.

As a general principle, if a particular chemical reaction is exothermic, then the reverse reaction is endothermic. For example, the burning of hydrogen gas—involving the formation of water from hydrogen and oxygen—is exothermic.

$$2\,H_2 + O_2 \longrightarrow 2\,H_2O + Energy$$

Therefore, the separation of water into its elements—the reverse reaction—must be endothermic.

$$Energy + 2\,H_2O \longrightarrow 2\,H_2 + O_2$$

If a rock rolling downhill is an exothermic process, then pushing the same rock back uphill must be an endothermic process.

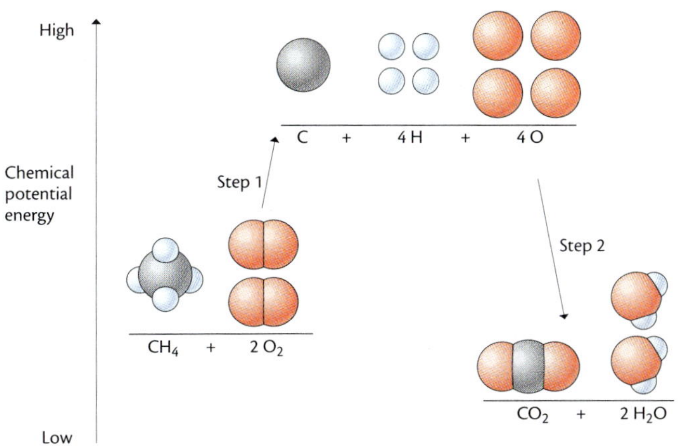

Figure 11 *The formation of carbon dioxide and water from methane and oxygen gases. In Step 1, bonds are broken, which is an endothermic process. Step 2, involving bond making, releases energy, so it is an exothermic process. Because more energy is released in Step 2 than is required in Step 1, the overall reaction is exothermic.*

CHEMISTRY AT WORK

Oil's Well That Ends Well: The Work of a Petroleum Geologist

Susan Landon is an independent petroleum geologist. Using her knowledge of chemistry, geology, geography, biology, and mathematics, she analyzes rocks and other geologic and geographic features for indications of likely oil and natural-gas reserves.

Susan searches for the folded or faulted terrain of Earth that often contains petroleum. Hydrocarbons (oil or natural gas) may be trapped in reservoirs in these areas. Using data about a given geographic area—such as satellite images, aerial photographs, topographic maps, and information about the rocks—Susan predicts the amount of oil or natural gas that may have been generated from organic matter in the rocks, and where it might occur.

If signs are good for a substantial petroleum deposit, Susan and her team work to find investors to fund further exploration. Such exploration often includes more detailed research, analysis, and drilling. During drilling, Susan and her colleagues examine rock samples and analyze data regarding electrical and other physical properties of the rocks encountered. Based on the findings, the group recommends either completing the well or plugging it.

In Susan's opinion, a good liberal arts background with a major in geology or a related science is excellent preparation for advanced studies in geology.

Wells? Now There's a Deep Subject!

Although there are many techniques for predicting the presence of underground oil or natural gas, the only way to know for sure is to drill a well.

The process of drilling a well begins with a drilling rig consisting of a derrick—a large, vertical structure made of metal that holds and supports a drilling pipe. Rotated by a motor, the drilling pipe ends in a bit that digs and scrapes down through soil and rock. Drilling fluid (usually a special mixture of water and other materials that is called "mud") is pumped down through the pipe to cool the bit and carry chips of rock back to the surface. A geologist can examine the chips and other material brought to the surface to determine the rock type and identify the presence of oil.

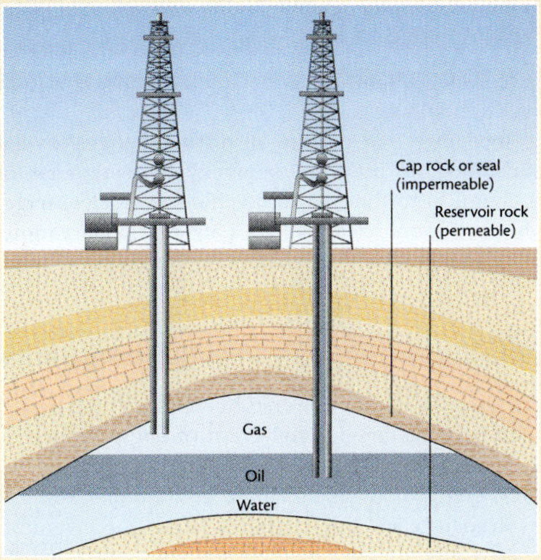

Although oil or natural gas sometimes flows to the surface spontaneously as a result of underground pressure, the more common procedure is to install a pump. Pumped from the well, the oil or natural gas is transported by truck or pipeline to a refinery. At the refinery, the oil or natural gas is processed and delivered to commercial and private consumers, who convert it into everything from polypropylene socks to carpeting to household heat.

A geologist can examine the chips . . . and identify the presence of oil.

B.2 ENERGY CONVERSION

The energy that is used on a daily basis may go through several conversions before it is actually used to do work. Energy is not "used up" in any of the conversions, but simply changed from one form to another. Like matter, the total amount of energy in the universe is conserved.

WASTED ENERGY

Building Skills 4

Emphasize that the automobile is a system that is made up of a series of interconnected and coordinated energy conversion devices. During the conversion process, about 75% of the energy is lost to the environment (wasted), generally as heat. The percent of the chemical energy (potential) actually converted into mechanical energy (kinetic) is a measure of the efficiency of the device. Point out that no device is 100% efficient. The automobile is only about 25% efficient.

The energy conversions that occur in Figure 13 are given below. The number and kind of conversions are not limited to the examples given.

- The defroster converts electrical energy into thermal energy.
- The generator converts mechanical energy into electrical energy.
- The engine converts chemical (potential) energy into thermal energy and mechanical energy.
- The gasoline in the tank represents chemical (potential) energy.

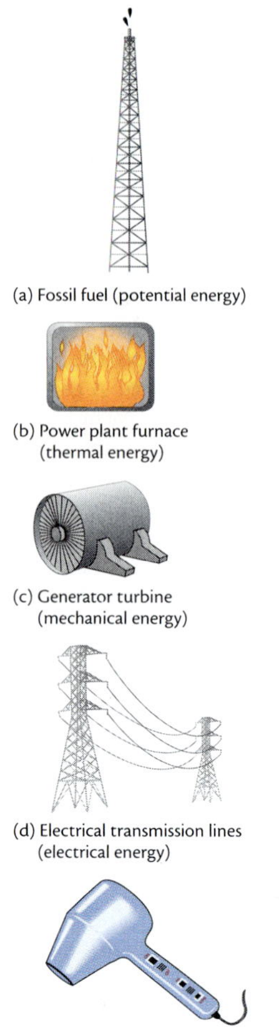

Figure 12 *Tracing energy conversions from source to final use.*

B.2 ENERGY CONVERSION

Scientists and engineers have increased the usefulness of energy released from burning fuels through devices that convert thermal energy into other forms of energy. In fact, much of the energy you use daily goes through several conversions before it reaches you.

Consider the energy-conversion steps involved in the operation of a hair dryer, which is illustrated in Figure 12. What detailed, step-by-step "energy story" can you infer from the illustration? You probably noted that in the first step, stored chemical energy (potential energy) in a fossil fuel (a) is released in a power-plant furnace, producing thermal energy (heat) (b). The thermal energy then converts water in a boiler to steam that spins turbines to generate electricity. Thus the power plant converts thermal energy to mechanical energy (a form of kinetic energy) (c) and then to electrical energy (d). When the electricity reaches the hair dryer, it is converted back to thermal energy to dry your hair, and also to mechanical energy (e) as a small fan blade spins to blow the hot air. Some sound energy is also produced, as any hair-dryer user knows!

It is important to recognize that despite all of the changes involved, energy is not really consumed, or "used up," in any of these steps. Its form simply changes from chemical to thermal to mechanical to electrical, for example. This concept is summarized by the **Law of Conservation of Energy**, which states that energy is neither created nor destroyed.

WASTED ENERGY

Building Skills 4

In one sense, an automobile—whether powered by electricity, gasoline, or even solar energy—can be considered a collection of energy-converting and energy-powered devices. Imagine that you are an "energy detective," and follow some typical energy conversions in an automobile. Name the type(s) of energy involved in each step required to get a conventional automobile window defogger to operate when the car is running. See Figure 13.

Although energy-converting devices have definitely increased the usefulness of petroleum and other fuels, some of these devices have problems associated with their use. For example, potentially harmful by-products may sometimes be generated by energy-converting devices. (You will read more about this issue later in the unit.) More fundamentally, some useful energy is always "lost" when energy is converted from one form to another. That is, no energy conversion is totally efficient; some energy always becomes unavailable to do useful work.

Consider an automobile with 100 units of chemical energy stored in its fuel tank as the mixture of molecules that make up gasoline. See Figure 14 on page 202. Even a well-tuned automobile converts only about 25% of that chemical energy (potential energy) to useful mechanical energy (kinetic energy). The remaining 75% of gasoline's chemical energy is lost to the surroundings as thermal energy (heat). The following activity will allow you to see what this means in terms of gasoline consumption and expense.

ENERGY CONVERSION EFFICIENCY

Building Skills 5

Assume that your family drives 225 miles each week and that the car can travel 27.0 miles on one gallon of gasoline. How much gasoline does the car use in one year?

Questions such as this can be answered by attaching proper units to all values, then multiplying and dividing them just as though they were arithmetic expressions.

For example, the information given in the problem can be expressed this way:

$$\frac{225 \text{ miles}}{1 \text{ week}} \quad \text{and} \quad \frac{27.0 \text{ miles}}{1 \text{ gal}}$$

or, if needed, as inverted expressions:

$$\frac{1 \text{ week}}{225 \text{ miles}} \quad \text{and} \quad \frac{1 \text{ gal}}{27.0 \text{ miles}}$$

Calculating the desired answer also involves using information you already know—there are 52 weeks in one year. You also know that the desired answer must have units of gallons per year (gal/y). When all of this information is combined, and care is taken to ensure that the units are multiplied and divided to produce "gallons per year," the following expression is formed:

$$\frac{225 \cancel{\text{miles}}}{1 \cancel{\text{week}}} \times \frac{1 \text{ gal}}{27.0 \cancel{\text{miles}}} \times \frac{52 \cancel{\text{weeks}}}{1 \text{ y}} = 433 \text{ gal/y}$$

Now answer these questions.

1. Assume that an automobile that averages 27.0 miles per gallon of gasoline travels 10 500 miles each year.
   a. How many gallons of gasoline will the car burn during the year's travels?
   b. If gasoline costs $1.35 per gallon, how much would be spent on gasoline in one year?
2. Assume your automobile engine uses only 25.0% of the energy released by burning gasoline.
   a. How many gallons of gasoline are wasted each year due to energy conversion inefficiency?
   b. How much does this wasted gasoline cost at $1.35 per gallon?
3. Suppose continued research leads to a car that travels 50.0 miles on one gallon of gasoline and whose engine is 50.0% efficient.
   a. In one year, how much gasoline would be saved compared to the first car?
   b. How much money would be saved assuming a gasoline price of $1.35 per gallon?

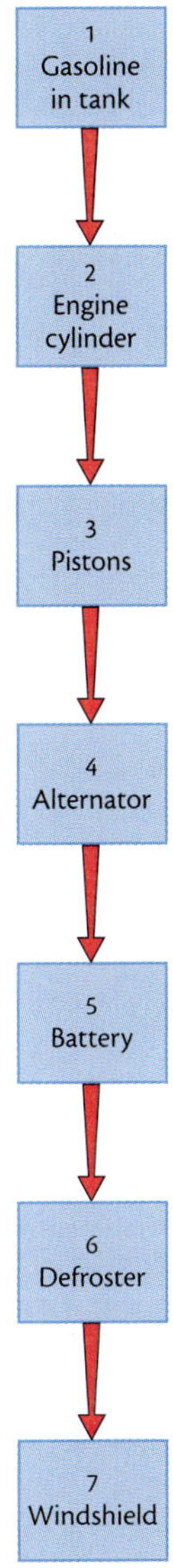

Figure 13 *Tracing energy conversions in an automobile.*

ENERGY CONVERSION EFFICIENCY

Building Skills 5

The energy loss associated with various energy conversions in the automobile is presented in the form of actual calculations. Students should be aware that the conversion of energy from chemical to heat to mechanical in any machine is usually less than 50% efficient. Generally, the greater the number of conversions, the lower the efficiency.

Representative calculations illustrate the monetary and resource losses resulting from inefficient energy conversion. The sample problems should be worked out with full explanations.

Answers

1. a. $\frac{10\,500 \text{ mi}}{\text{yr}} \times \frac{1 \text{ gal}}{27.0 \text{ mi}} = 389 \text{ gal}$

   b. $\frac{389 \text{ gal}}{\text{yr}} \times \frac{\$1.35}{\text{gal}} = \$525$

2. a. $\frac{389 \text{ gal}}{\text{yr}} \times 0.75 = 292 \text{ gal/yr}$

   b. $525 × 0.75 = $394 wasted

3. a. $\frac{10\,500 \text{ mi}}{\text{yr}} \times \frac{1 \text{ gal}}{50 \text{ mi}} = 210 \text{ gal}$

   389 gal − 210 gal = 179 gal saved

   b. 179 gal × $1.35 = $242 saved

B.3 COMBUSTION

Laboratory Activity

The concept of chemical combustion is introduced, including balanced equations that represent the release of energy. To make decisions about the use of fuels, one needs to know how to measure the amount of energy released by different fuels. Students will use a simple calorimeter to trap the energy from a burning candle and calculate the heat of combustion in kJ/g and kJ/mol.

Time

50 minutes

Materials (for 24 students working in pairs)

12 12-oz, empty soft drink cans (with pull-tabs)
12 ring stands
12 ring clamps
12 thermometers
12 stirring rods
12 index cards
balances
12 paraffin (represented as $C_{25}H_{52}$) candles
5 lb ice
matches

Safety

Students should wear safety goggles and aprons throughout the laboratory activity. Suggest to students that they should stir the water with a stirring rod, not a thermometer, to avoid breakage.

Caution students not to inhale carbon soot. If carbon soot gets on skin (it may be an irritant), wash it off immediately.

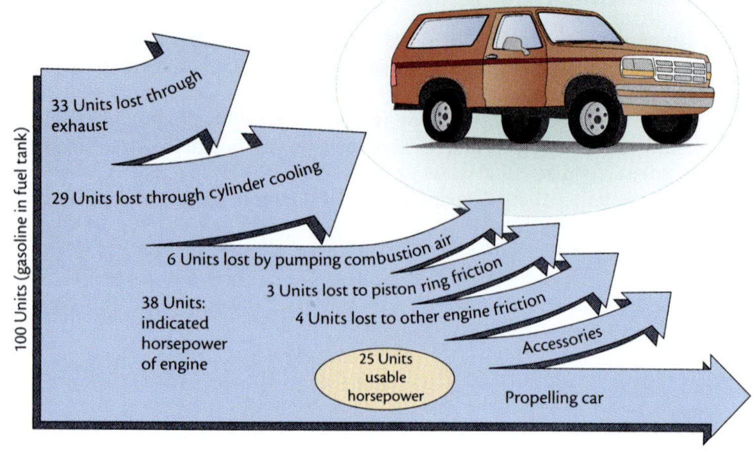

Figure 14 *Energy use in an automobile.*

Petroleum is neither limitless nor inexpensive. One way to maximize the benefits from available supplies of petroleum-based fuels is to reduce the total number of energy conversions the fuel undergoes on its way to its final use. Methods of increasing the efficiency of energy-conversion devices can also be sought.

Unfortunately, devices that convert chemical energy to heat and then to mechanical energy are typically less than 50% efficient. Solar cells, which convert solar energy to electrical energy, and fuel cells, which convert chemical energy to electrical energy, hold promise for either replacing petroleum or increasing the efficiency of its use.

Perhaps you have been wondering how the chemical energy in fuels is converted into heat and mechanical energy. What substances and chemical reactions are involved? How much energy is released? You will investigate these questions in the activity below.

B.3 COMBUSTION

Laboratory Activity

Introduction

You strike a match, and a hot, yellow flame appears. If you bring the flame close to a candlewick, the candle lights and burns. These events are so commonplace that you probably do not realize complex chemical reactions are at work.

Candle burning involves chemical reactions of the wax, which is composed of long-chain alkanes, with oxygen gas at elevated temperatures. Although many chemical reactions are involved in burning, or combustion,

Advance preparation

A day or two before the activity is scheduled, ask students to bring in from home empty aluminum soft drink cans (with pull-tabs attached). Flip up the tabs on the cans so that glass stirring rods can pass through (see Figure 15, page 204). This minimizes heat loss during heating, but makes stirring and adding the water a little more difficult.

Each class will need chilled water. If the tap water is not cool enough, add ice until the temperature is approximately 10 °C cooler than room temperature. Remove unmelted ice from the water.

Provide the students with a place on the chalkboard, poster board, or overhead transparency to record their value for the heat of combustion of candle wax in kJ/g and kJ/mol.

chemists simplify the process by usually focusing on the overall changes. For example, the burning of one component of candle wax, $C_{25}H_{52}$, can be summarized this way:

$$\underset{\text{Paraffin wax (Alkane)}}{C_{25}H_{52}(s)} + \underset{\text{Oxygen gas}}{38\ O_2(g)} \longrightarrow \underset{\text{Carbon dioxide gas}}{25\ CO_2(g)} + \underset{\text{Water vapor}}{26\ H_2O(g)} + \text{Energy}$$

As you already know, a burning candle gives off energy—the reaction is exothermic. Thus more energy must be liberated by forming bonds in the product molecules (carbon dioxide gas and water vapor) than was originally used to break the bonds in the wax and oxygen gas reactant molecules.

Fuels provide energy as they burn. But how much energy is released? How is the quantity of released energy measured? In this activity, you will measure the heat of combustion of a candle (paraffin wax) and compare this quantity with known values for other hydrocarbons. You will also investigate relationships between the quantity of thermal energy released when a hydrocarbon burns and the structure of the hydrocarbon.

Procedure

1. Read the entire procedure and prepare a suitable data table to record all of your specified measurements—masses, volumes, and temperatures.
2. Hold a lighted match near the base of a candle so that some melted wax falls onto a 3 × 5 index card. Immediately push the base of the candle into the melted wax. Hold the candle there for a moment to fasten it to the card.
3. Determine the combined mass of the candle and index card. Record the value.
4. Carefully measure (to the nearest milliliter) about 100 mL of chilled water. (The chilled water, provided by your teacher, should be 10 to 15 °C colder than room temperature.) Pour the 100-mL sample of chilled water into an empty soft-drink can.
5. Set up the apparatus as shown in Figure 15 on page 204, but do not light the candle yet! Adjust the can so the top of the candlewick is about 2 cm from the bottom of the can.
6. Measure both room temperature and the water temperature to the nearest 0.1 °C. Record these values.
7. Place the candle under the can of water. Light the candle. As the water heats, stir it gently.
8. As the candle burns and becomes shorter, you may need to lower the can so the flame remains just below the bottom of the can. **CAUTION:** *Lower the can with great care.*
9. Continue heating until the temperature rises as far above room temperature as it was below room temperature at the start. (For example, if the water is 15 °C before heating and room temperature is 25 °C, you would heat the water to 35 °C, which is 10 °C higher than room temperature.)

Pre-lab discussion

Introduce students to the SI energy unit, the joule (J). To raise the temperature of 1 g of water 1 °C, 4.2 J (or 1 calorie) is required. To provide additional perspective, explain that a 100-watt light bulb requires 6 000 J (6 kJ) of energy to operate for one minute.

Lab tips

Most student-quality thermometers can be read only to 0.5 °C.

Calculations

1. Volume of water × 1.0 g/mL = Mass of water in grams
2. Final temperature − Initial temperature = Total rise in temperature
3. Heat = Mass × Specific heat of substance × Change in temperature
4. Mass of paraffin wax burned = (Mass of candle + Index card before heating) − (Mass of candle + Index card after heating)
5. Here are some typical student results, but remember that values can vary widely:

Total thermal energy =

$$98.6 \text{ g water} \times \frac{4.2 \text{ J}}{\text{g} \times {}^\circ\text{C}} \times 21.2\ {}^\circ\text{C} = 8880 \text{ J}$$

Mass of candle consumed = 0.25 grams

Heat of combustion in J/g =

$$\frac{8880 \text{ J}}{0.25 \text{ grams}} = 36\,000 \text{ J/g}$$

Heat of combustion in kJ/g =

$$\frac{36\,000 \text{ J}}{\text{gram}} \times \frac{1 \text{ kJ}}{1000 \text{ J}} = 36 \text{ kJ/g}$$

Answers

1. The actual heat of combustion for propane (about 50.0 kJ/g) is higher than for paraffin (approximately 42.0 kJ/g). The relatively crude calorimeter used in this lab introduces many errors into the determination. Do not be surprised if the results vary greatly from team to team. Despite this, the results are often satisfactory, and this procedure effectively demonstrates the function of a calorimeter. The values determined in the laboratory activity will most likely be lower than the actual values.

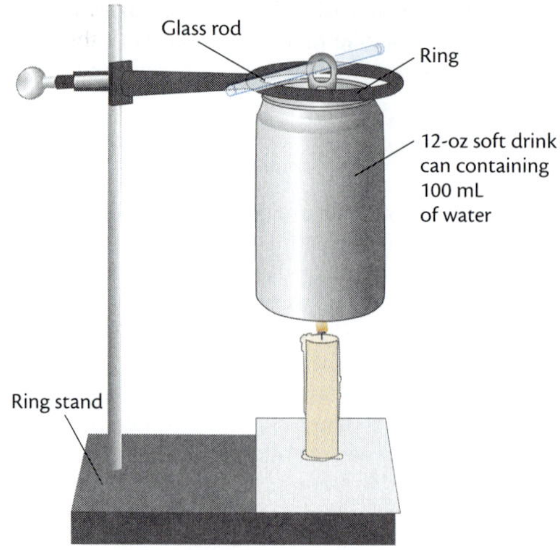

Figure 15 *Apparatus for determining heat of combustion.*

10. When the desired temperature is reached, extinguish the candle flame.
11. Continue stirring the water until its temperature stops rising. Record the highest temperature reached by the water.
12. Determine the mass of the cooled candle and index card, including all wax drippings.
13. Wash your hands thoroughly before leaving the laboratory.

Calculations

A characteristic property of a substance is the quantity of heat needed to raise the temperature of one gram of the substance by 1 °C. This value is called the **specific heat capacity** of the substance. The specific heat capacity of liquid water is about 4.2 J/(g·°C) (joules per gram per °C). This means that for each degree Celsius that liquid water is heated, each gram of the water absorbs 4.2 J of thermal energy.

> Specific heat capacity is sometimes shortened to the term "specific heat."

Suppose a 10.0-g water sample is heated from 25.0 °C to 30.0 °C, a temperature increase of 5.0 °C. How much thermal energy must have been added to the water?

> The exact specific heat capacity of water is 4.184 J/g. However, the rounded-off value of 4.2 J/g is adequate for and equally useful in this laboratory activity.

The answer can be reasoned this way. It takes 4.2 J to raise the temperature of 1 g water by 1 °C. In this example, however, there is 10 times more water and a temperature increase that is 5 times greater. Thus 10.0 × 5.0, or 50 times more thermal energy is needed. So, the specific heat must be multiplied by 50 to obtain the answer: 50 × 4.2 J = 210 J. It takes 210 J to increase the temperature of 10.0 g water by 5.0 °C.

The quantity of thermal energy given off when a certain amount of a substance burns is called the **heat of combustion**. The heat of combustion can be expressed as the thermal energy released when either one gram of

2. The actual molar heat of combustion is greater for paraffin (approximately 14 800 kJ/mol) than for propane (2200 kJ/mol). Student results should reflect a similar ordering of these two values, although their results will be lower than these accepted values because of heat losses.
3. Because each paraffin "molecule" on average contains 24 carbon–carbon bonds, while a propane molecule contains two carbon–carbon bonds, it is reasonable to expect that more energy would result from burning one mole of paraffin than from burning one mole of propane. Each additional —CH_2 unit produces one molecule of CO_2 and one of H_2O and additional heat.

substance burns or one mole of substance burns. If the amount of substance being burned is one mole, the quantity of thermal energy is called the **molar heat of combustion.** Using your laboratory data, you can calculate the heat of combustion of paraffin wax.

1. Calculate the mass of water heated. (*Hint:* The density of liquid water is 1.0 g/mL. Thus each milliliter of water has a mass of 1.0 g.)
2. Calculate the total rise in temperature of the water.
3. Calculate how much thermal energy was used to heat the water sample. Use values from the two preceding steps to reason out the answer, as illustrated in the sample problem.
4. Calculate the total mass of paraffin wax burned.
5. Calculate the heat of combustion of paraffin, expressed
   a. in units of joules per gram (J/g) of paraffin.
   b. in units of kJ/g.

   Hint: Assume that all the energy released by the burning paraffin wax is absorbed by the water. Divide the total thermal energy by the mass of paraffin burned to get the answer.

$$\text{Heat of combustion} = \frac{\text{thermal energy released (Step 3)}}{\text{mass of paraffin burned (Step 4)}}$$

(1 kJ = 1000 J)

Questions

Your teacher will collect your heat of combustion data, expressed in units of kilojoules per gram (kJ/g) of paraffin. Use the combined results of your class to determine a "best" estimate for this value. Will you decide to use the average value of the class results or the median value? Why? Use your selected value to answer the following questions.

1. How does your experimental heat of combustion (in kJ/g) for paraffin wax, $C_{25}H_{52}$, compare to the accepted heat of combustion for propane, C_3H_8? See Figure 16.

Heats of Combustion

Hydrocarbon	Formula	Heat of Combustion (kJ/g)	Molar Heat of Combustion (kJ/mol)
Methane	CH_4	55.6	890
Ethane	C_2H_6	52.0	1560
Propane	C_3H_8	50.0	2200
Butane	C_4H_{10}	49.3	2859
Pentane	C_5H_{12}	48.8	3510
Hexane	C_6H_{14}	48.2	4141
Heptane	C_7H_{16}	48.2	4817
Octane	C_8H_{18}	47.8	5450

Figure 16 *Hydrocarbon heats of combustion.*

improve accuracy of results. We have not gone that far in this simple lab, but it could be an extension for your advanced students.

b. Other experimental errors include inaccuracy in weighing, the burning wick's small contribution of energy to the total heat of combustion (its heat of combustion is probably not equal to that of paraffin), and incomplete combustion of paraffin. These factors also tend to lower the total quantity of heat measured. If time allows, refine the procedure based on the methods used by the students with the best results. Have all students repeat the experiment based on these refinements.

4. On a gram-for-gram basis, propane is a slightly better fuel. However, mole-for-mole (or molecule-for-molecule), paraffin is a much better fuel.
5. a. No. This is not a good assumption. Some thermal energy released during hydrocarbon burning will be lost to the surroundings directly from the flame, through the walls of the aluminum can, and from the water surface. This energy will not be accounted for in the calculations; thus, the experimental heat of combustion will be lower than the true value. It is possible to calculate an efficiency factor for an individual calorimeter. This takes into account all the heat losses and is used as a correction factor to

Post-lab discussion

The class discussion should focus on the posted student data. Some students' results may differ substantially from those of their classmates. This provides an excellent opportunity to suggest checking calculations carefully and looking for sources of experimental error. Observation of the setup (Figure 15, page 204) is helpful. It is particularly useful to look for heat loss, such as that absorbed by the can and that given off to the surrounding air.

This activity can and should be related to the more theoretical discussion of energy storage in chemical compounds and bond breaking and bond making in Section B.1.

ADDITIONAL ACTIVITIES

1. Have students make a list of advantages and disadvantages of using propane or paraffin to heat their homes. Topics that could be discussed are packaging, transport, system design, cost, and so on. This may help connect heat of combustion to a real-life scenario and also pertains to future activities.
2. Ask students for suggestions as to how a more accurate calorimeter could be designed. If you have access to a commercial calorimeter, demonstrate its design and function.
3. If alcohol burners are available, this activity might be extended to determine the heat of combustion of a liquid fuel such as hexane, methanol, or ethanol. Straight-chain hydrocarbons produce a lot of soot when used in an alcohol burner. Another extension would be to determine the heat content of a wood splint.

2. How do the molar heats of combustion (in kJ/mol) for paraffin and propane compare? (*Hint:* To make this comparison, first calculate the thermal energy released when one mole of paraffin burns. Because one mole of paraffin [$C_{25}H_{52}$] has a mass of 352 g, the molar heat of combustion will be 352 times greater than the heat of combustion expressed as kJ/g.)
3. Explain any differences noted between paraffin and propane in your answers to Questions 1 and 2. (*Hint:* Keep in mind the calculation you completed in Question 2.)
4. In your view, which hydrocarbon—paraffin or propane—is the better fuel? Explain your answer.
5. In calculating heats of combustion, you assumed that all thermal energy from the burning fuel went to heating the water.
   a. Is this a good assumption? Explain.
   b. What other laboratory conditions or assumptions might cause errors in your calculated values?

Energy from Combustion

B.4 USING HEATS OF COMBUSTION

With abundant oxygen gas and complete combustion, the burning of a hydrocarbon can be described by the equation

$$\text{Hydrocarbon} + \text{Oxygen gas} \longrightarrow \text{Carbon dioxide} + \text{Water} + \text{Thermal energy}$$

Energy is written as a product of the reaction because energy is released when a hydrocarbon burns. The combustion of a hydrocarbon is a highly exothermic reaction.

The equation for burning ethane (C_2H_6) is

$$2\ C_2H_6 + 7\ O_2 \longrightarrow 4\ CO_2 + 6\ H_2O + \textbf{?}\ \text{kJ thermal energy}$$

To complete this equation, the correct quantity of thermal energy involved must be included. Figure 16 indicates that ethane's molar heat of combustion is 1560 kJ/mol. That is, burning one mole of ethane releases 1560 kilojoules of energy. But according to the chemical equation above, two moles of ethane ($2\ C_2H_6$) are burned. Thermal energy must be "balanced" in terms of all other reactants and products. Thus the total thermal energy released will be twice that released when one mole of ethane burns:

$$2\ \cancel{\text{mol}} \times \frac{1560\ \text{kJ}}{1\ \cancel{\text{mol}}} = 3120\ \text{kJ}$$

The complete combustion equation for ethane is

$$2\ C_2H_6 + 7\ O_2 \longrightarrow 4\ CO_2 + 6\ H_2O + 3120\ \text{kJ}$$

As you found out in the preceding laboratory activity (and as is indicated in Figure 16), heats of combustion can also be expressed as energy produced when one gram of hydrocarbon burns (kJ/g). That information is very useful in determining how much energy is released when a certain mass of fuel is burned.

Any of the additional activities at left could also be used as a lab-practical exam for assessment.

For example, how much thermal energy would be produced by burning 12.0 g octane, C_8H_{18}? Figure 16 indicates that the burning of 1.00 g octane releases 47.8 kJ. Burning 12.0 times more octane will produce 12.0 times more thermal energy, or

$$12.0 \times 47.8 \text{ kJ} = 574 \text{ kJ}.$$

The calculation can also be written this way:

$$12.0 \cancel{\text{g octane}} \times \frac{47.8 \text{ kJ}}{1 \cancel{\text{g octane}}} = 574 \text{ kJ}$$

HEATS OF COMBUSTION

Building Skills 6

This activity will give you a better understanding of the energy involved in burning hydrocarbon fuels. Use Figure 16 (page 205) to answer the following questions. The first one is worked out as an example.

1. How much energy (in kilojoules) is released by completely burning 25.0 mol hexane, C_6H_{14}?

According to Figure 16, the molar heat of combustion of hexane is 4141 kJ. This means that 4141 kJ of energy is released when 1.00 mol hexane burns. So burning 25.0 times more fuel will produce 25.0 times more energy:

$$25.0 \cancel{\text{mol } C_6H_{14}} \times \frac{4141 \text{ kJ}}{1 \cancel{\text{mol } C_6H_{14}}} = 104\ 000 \text{ kJ}$$

Burning 25.0 mol hexane would thus release 104 000 kJ of thermal energy.

2. Write a chemical equation that includes thermal energy for the complete combustion of these alkanes:
   a. propane
   b. butane
3. Examine the data summarized in Figure 16.
   a. How does the trend in heats of combustion for hydrocarbons expressed as kJ/g compare with the trend expressed as kJ/mol?
   b. Assuming the trend applies to larger hydrocarbons, predict the heat of combustion for decane, $C_{10}H_{22}$, expressed as kJ/g decane and kJ/mol decane.
   c. Which prediction in Question 3b was easier? Why?
4. a. How much thermal energy is produced by burning two moles of octane?
   b. How much thermal energy is produced by burning one gallon of octane? (A gallon of octane has a volume of about 3.8 liters. The density of octane is 0.70 g/mL.)
   c. Suppose a car operates so inefficiently that only 16% of the thermal energy from burning fuel is converted to useful "wheel-turning" (mechanical) energy. How many kilojoules of useful energy would be stored in a 20.0-gallon tank of gasoline? (Assume that octane burning and gasoline burning produce the same results.)

HEATS OF COMBUSTION

Building Skills 6

The major use of petroleum is as a fuel for burning. The heat produced per gram is a primary factor in choosing appropriate types of hydrocarbons for burning.

Answers

1. Answer is worked in student text as an example.
2. $C_3H_8(g) + 5\,O_2(g) \rightarrow 3\,CO_2(g) + 4\,H_2O(g) + 2200 \text{ kJ}$

   b. $2\,C_4H_{10}(g) + 13\,O_2(g) \rightarrow 8\,CO_2(g) + 10\,H_2O(g) + 5718 \text{ kJ}$
3. a. The molar heats of combustion increase as the number of carbons increases (this is a linear relationship), while the heat of combustion per gram of hydrocarbon decreases slightly.

   b. The molar heat of combustion increases about 660 kJ per carbon atom in a hydrocarbon. The molar heat of combustion of decane should be 5450 kJ/mol + 2(660 kJ/mol) = 6770 kJ/mol. The heat of combustion per gram should be slightly less than that of octane. The exact number cannot be predicted, owing to the nonlinear relationship.

   c. The heat of combustion in kJ/mol was easier to predict, owing to the linear relationship.
4. a. $2 \text{ mol } C_8H_{18} \times \frac{5450 \text{ kJ}}{\text{mol}} = 10\ 900 \text{ kJ}$

   b. $1 \text{ gal } C_8H_{18} \times \frac{3.8 \text{ L}}{\text{gal}} \times \frac{1000 \text{ ml}}{\text{L}} \times \frac{.70 \text{ g}}{\text{ml}} \times \frac{47.8 \text{ kJ}}{\text{g}} = 127\ 000 \text{ kJ}$

   c. $20 \text{ gal} \times \frac{2660 \text{ g}}{\text{gal}} \times \frac{47.8 \text{ kJ}}{\text{g}} \times 0.16 = 410\ 000 \text{ kJ}$

B.4 USING HEATS OF COMBUSTION

Heats of combustion give students an opportunity to quantify the energy released in chemical reactions. Most charts give heat of reaction per mole. Therefore, if students are using two moles of the reactant, the heat of combustion must be multiplied by two.

5. a. $C(s) + O_2(g) \rightarrow CO_2(g) +$ 394 kJ

b. Octane (47.8 kJ/g) releases more energy per gram than carbon:

$$\frac{394 \text{ kJ}}{\text{mol}} \times \frac{1 \text{ mol C}}{12.01 \text{ g C}} = \frac{32.8 \text{ kJ}}{\text{g C}}$$

c. Carbon could be used in combustion reactions that do not require the vapor phase, such as charcoal briquettes for home barbecues.

d. Coal would not work in an automobile engine, which requires the fuel to be finely dispersed for rapid combustion.

ADDITIONAL ChemQuandaries

Imagine that you are planning a backpacking trip and wish to take along one of the first eight alkanes as a hydrocarbon fuel.

a. Which form of the data (kJ/mol or kJ/g) would be more useful in helping you pick the best fuel? Explain.

b. Ignoring the mass of the fuel container, which fuel would be most appropriate for your trip? Why?

c. In reality, what other factors would you need to consider if you selected one of the gaseous fuels?

Answers

a. Energy per mass (kJ/g) is more important because you have to carry it.

b. Methane because it releases the most energy per gram of substance.

c. Gaseous fuels such as methane must be kept in pressurized containers to be carried in practical quantities. Safety is an important factor to consider.

5. The molar heat of combustion of carbon contained in coal is 394 kJ/mol C.

a. Write a chemical equation for burning the carbon contained in coal. Include the thermal energy produced.

b. Gram for gram, which is the better fuel—carbon or octane? Explain your answer using calculations.

c. In what applications might coal replace petroleum-based fuel?

d. Describe one application in which coal would be a poor substitute for petroleum.

As automobile use grows around the world, demand for gasoline continues to increase rapidly. Because the gasoline fraction in a barrel of crude oil normally represents only about 18% of the total, researchers have been anxious to find a way to increase this yield. One promising method has been based on the discovery that it is possible to alter the structures of some petroleum hydrocarbons so that 47% of a barrel of crude oil can be converted to gasoline. This important chemical technique deserves further attention.

B.5 ALTERING FUELS

As you might expect, not all fractions of hydrocarbons obtained from petroleum are in equal demand or use at any particular time. The market for one petroleum fraction may be much less profitable than the market for another. For example, the invention of electric lightbulbs caused a rapid decline in the use of kerosene lanterns in the early 1900s. As a result, the kerosene fraction of petroleum, composed of hydrocarbon molecules with 12 to 16 carbon atoms, became a surplus commodity. On the other hand, automobiles dramatically increased the demand for the gasoline fraction (C_5 to C_{12}) that had earlier been discarded. The gasoline fraction could not be used safely in lanterns—it exploded!

Chemists and chemical engineers are adept at modifying or altering available chemical resources to meet new needs. Such alterations might involve converting less-useful materials to more-useful products, or—as in the case of kerosene in the early 1900s—converting a low-demand material into high-demand materials.

Cracking

Recall that the gasoline fraction obtained from crude oil refining includes hydrocarbons with 5 to 12 carbons per molecule.

By 1913, chemists had devised a process for converting larger molecules in kerosene into smaller, gasoline-sized molecules by heating the kerosene to 600 to 700 °C. The process of converting large hydrocarbon molecules into smaller ones through the application of heat and a catalyst is known as **cracking**. Through cracking, a 16-carbon molecule, for example, might produce two 8-carbon molecules:

$$C_{16}H_{34} \longrightarrow C_8H_{18} + C_8H_{16}$$

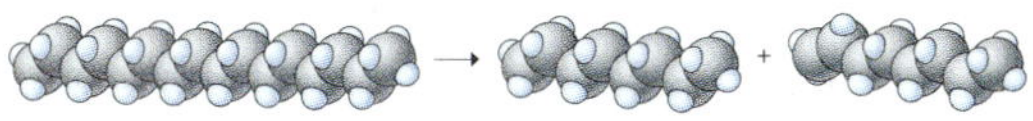

In practice, molecules with up to about 14 carbon atoms can be produced through cracking. Molecules with 5 to 12 carbon atoms are particularly useful in gasoline, which remains the most important commercial product of refining. Some C_1 to C_4 molecules (C_1, C_2, C_3, and C_4) produced in cracking are immediately burned, keeping the temperature high enough for more cracking to occur.

Today, more than a third of all crude oil undergoes cracking. The process has been improved by adding catalysts such as aluminosilicates. A **catalyst** increases the speed of a chemical reaction but is not itself used up. Catalytic cracking is more energy efficient because it occurs at a lower temperature—500 °C rather than 700 °C.

Gasoline is composed mainly of straight-chain alkanes, such as hexane (C_6H_{14}), heptane (C_7H_{16}), and octane (C_8H_{18}). As a result, it burns very rapidly. The rapid burning causes engine "pinging," or "knocking," and may contribute to engine problems. Branched-chain alkanes burn more satisfactorily in automobile engines; they do not ping as much. The structural isomer of octane shown below has excellent combustion properties in automobile engines. This octane isomer is known chemically as 2,2,4-trimethylpentane. Can you see how the name of the molecule is related to its structure? For convenience, this compound is frequently referred to by its common name, isooctane.

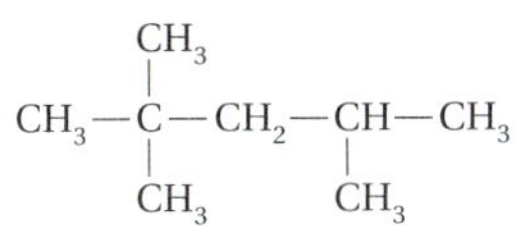

Isooctane, C_8H_{18}

Octane Rating

As you probably know, gasoline is sold in a variety of grades—and at corresponding prices. A common reference standard for gasoline quality is the octane scale. On this scale, isooctane, the branched-chain hydrocarbon you just learned about, is assigned an **octane number** of 100. Straight-chain heptane (C_7H_{16}), a fuel with very poor engine performance, is assigned an octane number of zero. Gasoline samples can be rated in comparison with isooctane and heptane. For example, a gasoline with knocking characteristics similar to a mixture of 87% isooctane and 13% heptane has an octane number of 87. The higher the octane number of a gasoline, the better its antiknock characteristics. Octane ratings in the high 80s and low 90s (85, 87, 92) are quite common, as a survey of nearby gas pumps, like those pictured in Figure 17 on page 210, will reveal.

Assigning an octane number of 100 to isooctane is arbitrary and does not mean that it has the highest possible octane number. In fact, several fuels burn more efficiently in engines than isooctane does and have octane numbers above 100.

B.5 ALTERING FUELS

The process of cracking or breaking large hydrocarbons into smaller molecules is presented. Fuel fractions can also be isomerized to produce branched-chained hydrocarbons that burn more satisfactorily in car engines. This section documents ways in which chemists have been able to engineer hydrocarbon molecules to fulfill particular needs. The older method of adding tetraethyl lead to boost octane ratings is contrasted with adding oxygen-containing fuels such as methanol and methyl-t-butyl ether (MTBE).

Students may be curious to know why gasoline is referred to as *unleaded,* when there is no *leaded* gasoline available. A brief discussion of the history of the addition of tetraethyl lead to boost the octane rating of gasoline prior to the mid-1970s would be beneficial at this point. A common misconception is that higher octane fuels are more powerful, since they are typically used in high-performance autos. In a way they are, but mostly the octane level prevents engine "knock." Knocking in engines is caused when the combustion of the air/fuel mixture inside the engine causes the piston to slap against the wall of the cylinder. Increasing the octane level causes a more controlled explosion (if that isn't a contradiction in terms!). This is especially important in high-performance engines, where the compression ratio is high, so much that it can even cause the air/fuel mixture to detonate on its own. This causes the "dieseling" effect in engines, where they continue to run even after you have turned off the ignition.

Sometimes there is a statement on gasoline pumps saying the octane ratings were obtained by the (R+M)/2 method. This means that the octane rating is an average of two different test values: (1) the research octane number (RON) for a test engine run under conditions simulating city driving; and (2) the motor octane number (MON) for the engine run under highway conditions. If the RON is 93 and the MON is 91, then the octane number is 92.

Gasoline that contains oxygen in addition to carbon and hydrogen delivers less energy per gallon than regular gasoline. These fuels improve the octane rating of the gasoline while reducing air pollutants. MTBE, an octane enhancer, has been used in attempts to reduce carbon monoxide in urban atmospheres. The Clean Air Act of 1992 requires that all oxygenated gasoline sold in at least the four winter months in the 41 most populated metropolitan areas (New York, Los Angeles, Washington, D.C., and so forth) must contain 2.7% oxygen by weight. Putting MTBE into gasoline brings the oxygen percentage of the mixture up to the required weight percent. Because the MTBE-enhanced fuel is partially oxygenated, it burns cleaner, producing carbon dioxide directly rather than carbon monoxide. However, MTBE is being reevaluated because of the unpleasant taste and odor it gives to ground water, even in minute concentrations. Therefore, scientists are looking at MTHE to replace MTBE. It has an octane rating of 87 and can be obtained from renewable resources.

Prior to the mid-1970s, the octane rating of gasoline was increased at low cost by adding a substance such as tetraethyl lead, $(C_2H_5)_4Pb$, to the fuel. This additive slowed down the burning of straight-chain gasoline molecules and added about three points to "leaded" fuel's octane rating. Unfortunately, lead from the treated gasoline was discharged into the atmosphere along with other vehicle exhaust products. Lead is harmful to the environment and, as a result, such lead-based gasoline additives are no longer used.

Oxygenated Fuels

The phaseout of lead-based gasoline additives meant that alternative octane-boosting supplements were required. A group of additives called **oxygenated fuels** have been blended with gasoline to enhance its octane rating. The molecules of these additives contain oxygen in addition to carbon and hydrogen.

Although oxygenated fuels actually deliver less energy per gallon than regular gasoline hydrocarbons do, their economic appeal stems from their ability to increase the octane number of gasoline while reducing exhaust-gas pollutants. In particular, oxygenated fuels encourage more complete combustion, producing lower emissions of air pollutants such as carbon monoxide (CO).

A common oxygenated fuel is methanol (methyl alcohol, CH_3OH), which is added to gasoline at distribution locations. In addition to its octane-boosting properties, methanol can be made from natural gas, coal, corn, or wood—a contribution toward conserving nonrenewable petroleum resources. A blend of 10% ethanol (ethyl alcohol, CH_3CH_2OH) and 90% gasoline, sometimes called **gasohol**, can be used as an oxygenated fuel in nearly all modern automobiles without associated engine adjustments or problems.

Figure 17 *Octane ratings are posted on gasoline pumps.*

Methyl tertiary-butyl ether, MTBE, which has an octane rating of 116, was initially introduced in the late 1970s as an octane-boosting fuel additive. In the 1990s, the role of MTBE as a pollution-reducing oxygenated fuel became increasingly important. In fact, MTBE became the most common oxygenated fuel additive in gasoline. At its peak, annual U.S. production of MTBE increased to more than 4 billion gallons (about 16 gallons per person), making it one of the top ten industrial chemicals.

By the late 1990s, however, evidence of contamination of groundwater and drinking water supplies due to seepage of MTBE from defective underground gasoline storage systems began to mount. MTBE dissolves readily in water, is resistant to microbial decomposition, and is difficult to remove in water-treatment processes. The unpleasant taste and odor that MTBE imparts to water, even at concentrations below those regarded as a public health concern, triggered consumer complaints. So although MTBE has been credited by the U.S. Environmental Protection Agency for substantial reductions in emissions of air pollutants from gasoline-powered vehicles in the 1990s, its reduced use or complete elimination as an oxygenated fuel is under active consideration, research, and policy debate.

A promising new fuel additive is MTHF, methyltetrahydrofuran. MTHF has an octane rating of 87, equal to that of regular unleaded gasoline, and the ability to increase the oxygenated level of the fuel. An added advantage of using MTHF is that it can be obtained from renewable resources, such as paper-mill waste products.

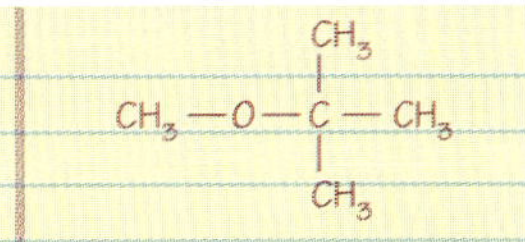

Methyl tertiary-butyl ether (MTBE).

A BURNING ISSUE — ChemQuandary 2

Methanol (CH_3OH) and ethanol (CH_3CH_2OH) are used as gasoline additives or substitutes. Their heats of combustion are 23 kJ/g and 30 kJ/g, respectively.

Consider the chemical formulas of methanol and ethanol. Gram for gram, why are heats of combustion of methanol and ethanol considerably less than those of any hydrocarbons discussed so far? See Figure 16, page 205.

Other octane-boosting strategies involve altering the structures of hydrocarbon molecules in petroleum. This works because branched-chain hydrocarbons burn more satisfactorily than straight-chain hydrocarbons. (Recall isooctane's octane number compared with that of heptane.) Straight-chain hydrocarbons are converted to branched-chain hydrocarbons by a process called isomerization. During isomerization, hydrocarbon vapor is heated with a catalyst:

$$CH_3-CH_2-CH_2-CH_2-CH_2-CH_3 \xrightarrow[\text{Catalyst}]{\text{Heat}} CH_3-\underset{\underset{}{}}{\overset{CH_3}{\overset{|}{CH}}}-\underset{\underset{CH_3}{|}}{CH}-CH_3$$

$C_6H_{14}(g)$ Straight-chain isomer → $C_6H_{14}(g)$ Branched-chain isomer

A BURNING ISSUE
ChemQuandary 2

The performance of gasoline as a fuel may be enhanced by additives such as methanol or ethanol because oxygen is already included in their structures. However, this "gasohol" mixture decreases fuel economy, because there is less energy per gallon of gasoline to propel the car.

Isomerization and cracking improve the burning of gasoline but also increase the cost, due to the extra fuel used during the manufacturing process.

The presence of oxygen in these two alcohols lowers their heats of combustion. It takes more energy to break the bonds to oxygen (relative to carbon), and less energy is released when new bonds are formed with oxygen.

B.6 FUEL MOLECULES IN TRANSPORTATION

Making Decisions

Answers

1. The manufacturer uses petroleum to manufacture the parts and in assembling the parts of the vehicle.
2. Petroleum is often used to generate the electricity on which the car runs.
3. Use other fossil fuels (such as coal and natural gas), solar energy, wind power, or nuclear reactors to generate the electricity.
4. Vehicle body parts (seat covers, floor covering, and so forth), tires, insulation on connectors, and dashboards are made from petroleum components.
5. Use natural fabrics (cotton, leather), natural rubber for tires and insulation, and wood or metal for dashboards.
6. Gasoline-free vehicles; no emission.
7. Omit claims of being "gasoline-free" and of having" no emissions." Instead, claim that the new car "reduces dependence on fossil fuels" and "reduces emissions from cars."

The branched-chain alkanes produced are blended with the C_5 to C_{12} molecules obtained from cracking and distillation, producing a high-quality gasoline.

Although cracked and isomerized molecules improve the burning of gasoline, they also increase its cost. One reason for this increase is the extra fuel used in manufacturing such gasoline.

B.6 FUEL MOLECULES IN TRANSPORTATION

Making Decisions

Now that you have examined the use of petroleum as a fuel, it is time to revisit the TV advertisement for the ARL-600 vehicle that opened this unit. First, use what you have learned about petroleum to answer the questions below. Then, write one or two paragraphs that critically evaluate the claims made in that advertisement.

1. In what ways does the manufacture of an ARL-600 use petroleum as a fuel?
2. In what ways does the operation of an ARL-600 use petroleum as a fuel?
3. What viable alternatives to petroleum can you suggest for fuel uses listed in Questions 1 and 2?
4. What direct, nonfuel uses of petroleum or its fractions are involved in the operation of an ARL-600?
5. What viable alternatives to petroleum can you suggest for the direct uses listed in Question 4?
6. What facts or concepts involving "energy from petroleum" in the ARL-600 advertisement are misleading? Why?
7. How would you correct the facts or concepts identified in Question 6?

Questions and Answers

CD-ROM WWW.

Section C will expand your knowledge of the "building" role of petroleum, focusing on it as a source of substances from which an impressive array of useful compounds and materials can be produced. This additional information will help you further decide about the accuracy of claims found in the new-car advertisement.

SECTION SUMMARY

Reviewing the Concepts

1. From a chemical viewpoint, why is petroleum sometimes considered "buried sunshine"?

Petroleum is derived from the decomposition of plants. The energy from the Sun was captured millions of years ago by green plants. This energy is stored in the bonds of the hydrocarbons that are left from the energy once stored in plants.

2. Why is the energy stored in petroleum more useful than solar energy for most applications? Give at least three reasons in your explanation.

Energy in petroleum is available and useable. For now, it is cheaper than solar energy. It can be transformed efficiently into heat and other forms of energy.

3. Write a word equation for photosynthesis. Include energy in your equation.

Carbon dioxide + Water + Energy → Glucose + Oxygen

4. Compare the equation for the burning of hydrocarbons to the equation for photosynthesis. How are they related to each other?

They are opposite reactions. Heat is a product of burning, whereas solar light energy is required for photosynthesis.

5. How were prehistoric green plants converted into today's petroleum reserves?

When plants die and decay, they are also largely transformed into carbon dioxide and water. However, under certain conditions, the glucose and other organic compounds that make up the plant only partially decompose, and the residue still contains substantial amounts of carbon and hydrogen. Such conditions arose at various times in the prehistoric past when vast quantities of plant life were buried beneath layers of sediment in swamps or the ocean bottom. There these remnants of vegetable matter were protected from atmospheric oxygen, and the decomposition process was halted. Under high pressure and temperature and over millions of years, these deposits were transmuted into petroleum.

6. Classify each of the following as an example of kinetic or potential energy.

a. a skateboard at the top of a hill

b. a charged battery

c. a rolling bowling ball

d. a gallon of gasoline

e. a waterfall

a. A skateboard at the top of a hill—potential

b. A charged battery—potential—chemical

c. A bowling ball rolling down the lane—kinetic

d. A gallon of gasoline—potential—chemical

e. Falling water—kinetic

7. Based on its structural formula, does a molecule of methane or a molecule of butane have more potential energy? Explain your answer.

Butane has more potential energy because it has more chemical bonds.

8. The burning of a candle is an exothermic reaction. Explain this fact in terms of the quantity of energy stored in the bonds of the reactants compared with the quantity of energy stored in the bonds of the products.

The total energy required to break the bonds in the reactants is less than the amount of energy given off by the formation of the bonds of the products. The excess energy is given off as heat and light.

9. Using Figure 11 on page 197 as a model, illustrate the energy change when hydrogen gas reacts with oxygen gas to produce thermal energy and water.

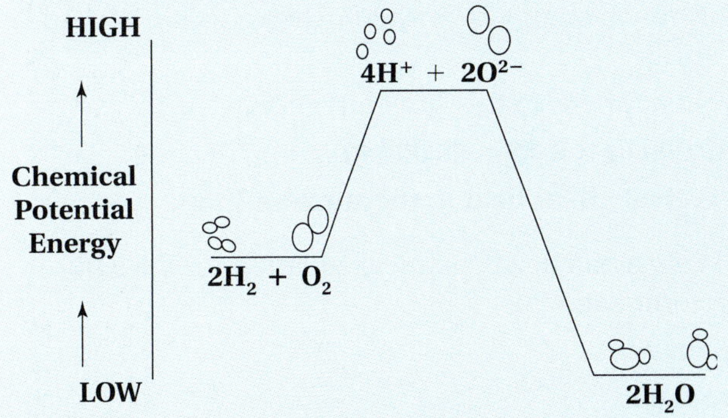

10. For each of the following reactions, determine whether bond-breaking or bond-making involves more energy.

a. burning wood in a campfire

b. activating a chemical "cold pack"

c. burning oil in a lamp

a. A campfire—making of bonds

b. Chemical "cold pack"—breaking of bonds

c. Oil lamp—making of bonds

11. Identify the energy conversions involved in making each of the following events possible.

a. riding a bike

b. illuminating a lightbulb

c. producing electricity in a wind-powered generator

d. running a gasoline-powered lawn mower

a. Riding a bike—chemical to mechanical to kinetic

b. Light bulb burning—electrical to heat and light

c. Wind-powered electrical generator—kinetic to mechanical to electrical

d. Gasoline-powered lawn mower—chemical to heat to mechanical to kinetic

12. In powering an automobile, 25% of the energy is said to be useful. What happens to the other 75% of the energy?

Seventy-five percent is lost to the surroundings as heat.

13. Explain what is meant by "energy conversion efficiency".

Energy efficiency represents the quantity of available energy that is converted into useful work expressed as a percent.

14. One gallon of gasoline produces about 132 000 kJ of energy when burned. Assume that an automobile is 25% efficient in converting this energy into useful work.

a. How much energy is "wasted" when a gallon of gasoline burns?

b. What happens to this "wasted" energy?

a. 132 000 kJ × 0.75 = 99 000 kJ

b. It is given off as heat to the surroundings.

15. Write a balanced chemical equation for the complete combustion of

a. methane

b. propane

c. octane

a. $CH_4 + 2\,O_2 \rightarrow CO_2 + 2\,H_2O$

b. $C_3H_8 + 5\,O_2 \rightarrow 3\,CO_2 + 4\,H_2O$

c. $2\,C_8H_{18} + 25\,O_2 \rightarrow 16\,CO_2 + 18\,H_2O$

16. When candle wax (a mixture of hydrocarbons) burns, it seems to disappear. What actually happens to the wax?

The hydrocarbon molecules during combustion change are dispelled into the atmosphere primarily as water vapor and carbon dioxide.

17. a. What are the products of the complete burning of octane?

b. What are the products of the complete burning of a mixture of C_8H_{18} and C_7H_{16}?

c. Compare and comment on your answers to Questions 17a and 17b.

a. Carbon dioxide, water vapor, and heat.

b. Carbon dioxide, water vapor, and heat.

c. The product of the complete burning of alkanes is always the same.

18. Water gas (a 50-50 mixture of CO and H_2) is made by the reaction of coal with steam. Because the United States has substantial coal reserves, water gas might serve as a substitute fuel for natural gas (composed mainly of methane, CH_4). Water gas burns according to this equation:

$$CO + H_2 + O_2 \longrightarrow CO_2 + H_2O + 525\text{ kJ}$$

a. How does water gas compare to methane in terms of thermal energy produced?

b. If a water gas mixture containing 10 mol CO and 10 mol H_2 were completely burned in O_2, how much thermal energy would be produced?

a. Methane is 891 kJ/mol; water gas is 525 kJ/mol. Water has about 60% of the energy per mole of methane.

b. 10 mol × 525 kJ/mol = 5250 kJ

19. How much energy is released when a 5600-g sample of water cools from 99 °C to 28 °C? The specific heat capacity of water is 4.18 J/(g·°C).

5600 g × 4.18 J/g·°C × (99 °C − 28 °C) × (1 kJ/1000 J) = 1660 kJ

20. The combustion of acetylene, C_2H_2 (used in a welder's torch), can be represented as:

$$2\,C_2H_2 + 5\,O_2 \longrightarrow 4\,CO_2 + 2\,H_2O + 2512\text{ kJ}$$

a. What is the molar heat of combustion of acetylene in kilojoules per mole?

b. If 12 mol acetylene burns fully, how much thermal energy will be produced?

a. 2512 kJ ÷ 2 mol = 1256 kJ/mol

b. 2512 kJ ÷ 2 mol × 12 mol = 15 070 kJ

21. a. List two factors that would help you decide which hydrocarbon fuel to use in a particular application.

b. Explain the importance of each factor.

a. The boiling point and heat of combustion.

b. Boiling point would identify the fuel's state during normal conditions and the type of container needed. The heat of combustion can be used to compare fuels.

22. In a laboratory activity, a student team measures the heat released by burning heptane (C_7H_{16}). Using the following data, calculate the molar heat of combustion of heptane in kJ/mol.

Mass of water	179.2 g
Initial water temperature	11.6 °C
Final water temperature	46.1 °C
Mass of heptane burned	0.585 g

179.2 g × (4.184 J/g °C) × (46.1 °C − 11.6 °C) = 25 867 J = 25.9 kJ
0.585 g × (1 mol/100.2 g) = 0.00584 mol
25.9 kJ ÷ 0.00584 mol = 4430 kJ/mol

23. List some molecules that might be formed when the following substances are cracked.

a. $C_{16}H_{34}$ b. $C_{18}H_{38}$

a. C_8H_{16} and C_8H_{18} or C_5H_{12} and $C_{11}H_{22}$
b. C_9H_{20} and C_9H_{18} or C_7H_{16} and $C_{11}H_{22}$

24. Why are catalysts used during the cracking process?

To lower the activation energy needed to break large hydrocarbon molecules down into smaller molecules.

25. Explain why only small amounts of catalysts are needed to crack large amounts of petroleum.

Catalysts are not used up in the cracking process; they just increase the rate of the reaction.

26. Explain the meaning of a fuel's octane rating.

Octane rating is a scale for measuring the antiknock characteristics of a gasoline.

27. A premium gasoline has an octane rating of 93. To what isooctane-heptane mixture does this rating compare?

Heptane is rated at 0 (poor antiknock behavior). Isooctane is rated at 100 (good antiknock behavior). The gas has the same antiknock behavior as a mixture of 93% isooctane and 7% heptane.

28. List two ways to increase a fuel's octane rating.

One can increase the percentage of branched-chain alkanes that burn more smoothly and increase the octane rating. Oxygenated fuels such as methanol and MTBE (methyl t-butylether) are blended with gasoline to enhance the octane rating.

29. Compare the molecular structures of octane and isooctane.

Isooctane is a branched-chain hydrocarbon. Octane is a straight-chain hydrocarbon.

30. How does the addition of oxygenating compounds affect a fuel's octane rating?

They increase the octane rating.

31. Why is tetraethyl lead no longer used as a gasoline additive?

The lead that was discharged into the atmosphere was a toxic pollutant.

Connecting the Concepts

32. In a laboratory activity, a student burns 4.2 g ethanol (C_2H_5OH). The molar heat of combustion of ethanol is 1366 kJ/mol.

a. How much energy is released?

b. The heat from this reaction is used to warm a 468-g sample of water in a calorimeter. The water temperature changes from 21 °C to 89 °C. How much energy is absorbed by the water? The specific heat capacity of water is 4.18 J/(g × °C).

c. Compare the quantity of heat released by the reaction to the quantity of heat absorbed by the calorimeter. What accounts for the difference?

a. 4.2 g × (1 mol/46.068 g) × (1366 kJ/mol) = 125 kJ

b. (468 g) × (4.18 J/g °C) × (89 °C – 21 °C) = 133 000 J = 133 kJ

c. (133 kJ ÷ 125 kJ) × 100 = 106%. This is an unusual lab result. Most often the amount of heat absorbed by the calorimeter is less than the amount given off by the reaction. Extra heat must have gone into the system from some other source; perhaps the room was quite warm on the day of the experiment. Another error could be that despite the measurements, the water was very cold to begin with and would absorb proportionately more heat from the surroundings. Finally, perhaps waste heat from another nearby experiment affected this setup.

33. Was the formation of petroleum the result of chemical or physical changes? Explain your answer.

The formation of petroleum is basically chemical, since plant material has to be broken down chemically by the action of microorganisms. Heat and pressure would promote this process.

34. You are given a choice of three fuels for use in heating your home—candles, butane, or gasoline. Based on their heats of combustion, which one would you choose? Why?

Per gram, butane would give off the most heat. It would be clean burning and convenient to use, like propane. Energy and money are wasted, and pollution in the form of CO or soot is given off.

Extending the Concepts

35. Powdered aluminum or magnesium metal releases considerable heat when burned. (The heats of combustion of aluminum and magnesium are 31 kJ/g and 25 kJ/g, respectively.) Would these highly energetic powdered metals be good fuel substitutes for petroleum? Explain.

Aluminum and magnesium are too expensive (per gram) as compared to petroleum; auto engines and home furnaces are not designed for solid fuels.

36. One theory concerning the extinction of the dinosaurs is that Earth was struck by a meteorite. The enormous cloud of dust and debris that resulted blotted out sunlight for a period of several years. How does this sequence of events explain the extinction of the dinosaurs?

Dinosaurs could not find enough plant life to eat and starved. One well-accepted theory is that a large meteor hit the Earth, sending up a huge amount of dust and debris into the upper atmosphere. This dust lingered for a long time, blocking sunlight and causing many plant species to die. This led to the prolonged winter-like weather, including cold and darkness, which in turn led to an inadequate food supply for the dinosaurs.

PETROLEUM AS A BUILDING SOURCE

Introduction

Section C focuses on the "building" theme for petroleum, with emphasis on the useful compounds and materials that can be produced from it. It describes alkenes, including *cis-trans* isomerism, and introduces the concepts of functional groups and aromatic compounds. Polymerization and condensation reactions are compared, and students prepare an ester by the condensation reaction of an alcohol with an organic acid.

C.1 CREATING NEW OPTIONS: PETROCHEMICALS

The variety of petrochemicals presented should give students some sense of the importance of the chemical industry. Alkanes are relatively unreactive, but the unsaturated (double) bond in alkenes is quite reactive, and therefore alkenes are used as starting materials for making many new molecules. Builder molecules can often be polymerized into macromolecules through addition reactions. Linking monomers to each other without loss of atoms forms addition polymers. These polymers may grow into chains containing 10 000 or more units. All monomers forming addition polymers have C=C bonds (carbon to carbon double bonds).

Students need not necessarily memorize names, formulas, or reactions, but they do need to understand that chemists can create many substances from petrochemicals.

Have students complete the Modeling Matter: Polymer Structure and Properties on page 219 as a cooperative group activity.

SECTION C

PETROLEUM AS A BUILDING SOURCE

Just as an architect uses available construction materials to design a building, a chemist—a "molecular architect"—uses available molecules to design new molecules. Architects must know about the structures and properties of common construction materials. Likewise, chemists must understand the structures and properties of their raw materials—the "builder molecules." In this section, you will explore the structures of common hydrocarbon builder molecules and some materials made from them.

C.1 CREATING NEW OPTIONS: PETROCHEMICALS

Until the early 1800s, all objects and materials used by humans were either created directly from what are called found materials, such as wood or stone, or crafted from metals, glass, and clays. Available fibers included cotton, wool, linen, and silk. All medicines and food additives came from natural sources. The only plastics were those made from wood (celluloid) and animal materials (shellac).

Today many common objects and materials created by the chemical industry are unlike anything seen or used by citizens of the 1800s or even the mid-1900s. Compounds produced from oil or natural gas are called **petrochemicals**. Some petrochemicals, such as detergents, pesticides, pharmaceuticals, and cosmetics, are used directly. Most petrochemicals, however, serve as raw materials in the production of other synthetic substances, particularly a wide range of plastics.

Plastics include paints, fabrics, rubber, insulation materials, foams, adhesives, molding, and structural materials. Worldwide production of petroleum-based plastics is more than four times that of aluminum products.

The astounding fact is that it takes relatively few builder molecules (small-molecule compounds) to make thousands of new substances. One particularly important builder molecule is ethene, C_2H_4, a hydrocarbon compound commonly called ethylene. The structural formula for ethene can be seen in the equation shown below. Because of the high reactivity of its double bond, ethene is readily transformed into many useful products. A simple example—the formation of ethanol (ethyl alcohol) from water and ethene—illustrates how ethene reacts.

Double bonds will be discussed in more detail on page 220.

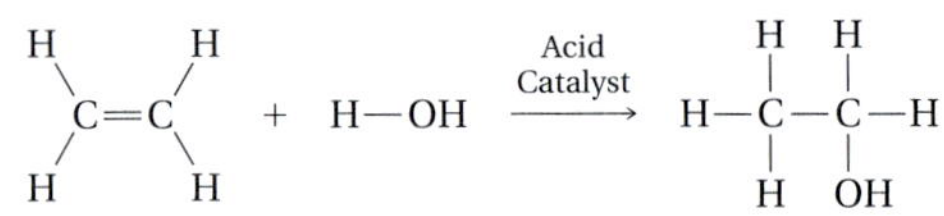

Ethene Water Ethanol

ADDITIONAL INFORMATION

See the table on page 233B for examples of commercial polymers.

In this reaction, the water molecule "adds" to the double-bonded carbon atoms by placing an H— on one carbon and an —OH group on the other. This type of chemical change is called an **addition reaction**.

Ethene can also undergo an addition reaction with itself. Because the added ethene molecule contains a double bond, another ethene molecule can be added, and so on. This creates a long-chain substance called polyethene, commonly known as polyethylene, a **polymer**. A polymer is a large molecule typically composed of 500 to 20 000 or more repeating units called **monomers**. In polyethene the repeating unit is an ethene (ethylene) monomer. The chemical reaction that produces polyethene can be written this way:

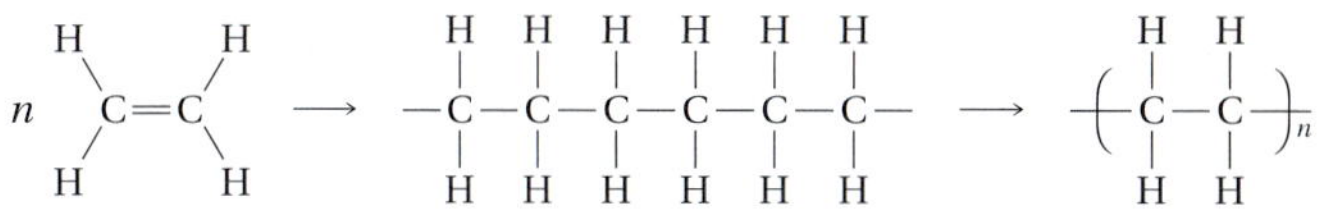

Ethene (Ethylene) — The growing polymer chain — Polyethene (Polyethylene)

Polymers formed in reactions such as this are called—sensibly enough—**addition polymers**. Polyethene, commonly used in grocery bags and packaging, is one of the most important addition polymers. The United States produces millions of kilograms of polyethene every year.

A great variety of addition polymers can be made from monomers that closely resemble ethene. The most common variation is to replace one or more of ethene's hydrogen atoms with an atom or atoms of another element. In the following examples, note the atom or atoms that replace hydrogen in each of these monomers and polymers.

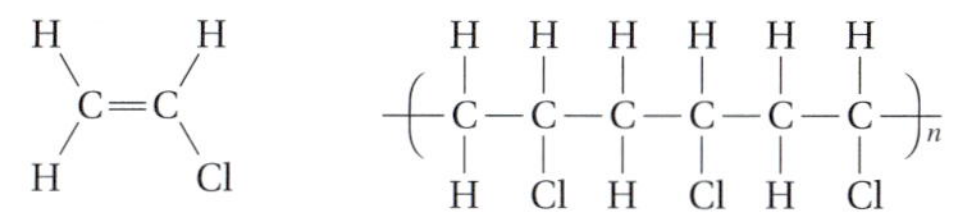

Vinyl chloride — Polyvinyl chloride

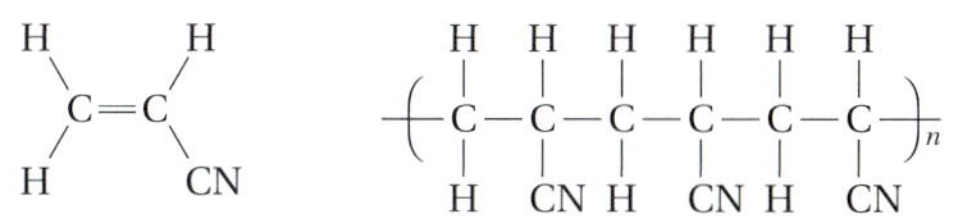

Acrylonitrile — Polyacrylonitrile

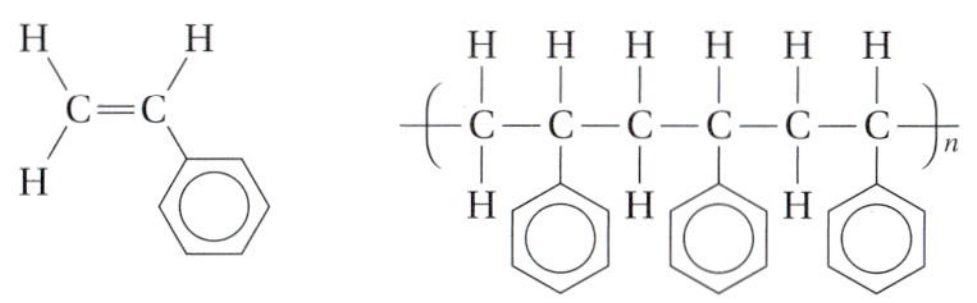

Styrene — Polystyrene

The atoms that compose the monomers dictate the properties of the polymer. Thus polyethene and polyvinyl chloride are fundamentally different materials. However, as with many modern materials, the properties of a polymer are often altered to meet a variety of needs and to produce a multitude of products.

Perhaps you are wondering why two names are shown for the same substances—ethene/ethylene and polyethene/polyethylene. Chemists worldwide have developed a system for naming chemical substances that makes it easy for them to communicate with one another. However, some substances were given names long before this system was universally accepted. Some "common names" of substances, such as ethylene and polyethylene, continue in wide use.

The hexagon-shaped ring depicted in styrene and polystyrene represents benzene, a substance that will be discussed on page 223.

DEMONSTRATION IDEAS

1. Students are always fascinated by the breakdown of polystyrene in acetone. This demonstration shows the characteristics of one polymer—polystyrene. Fill a transparent container, such as a large beaker or graduated cylinder, one-third full of acetone and begin adding Styrofoam pellets. You will certainly have no trouble getting volunteers to help. The vast quantity that can be broken down with bubbling occurs because the pellets are air-filled foam.
2. Fill a gallon-sized zip-closing bag with water. Seal it tightly. Then insert a sharpened pencil through both sides. The water does not leak out because of the elastic nature of the polymer in the bag (generally low density polyethylene). Many pencils can be poked into the bag at the same time, so long as they have smooth shafts. Do not try to use mechanical pencils or pens.

(a)

(b)

(c)
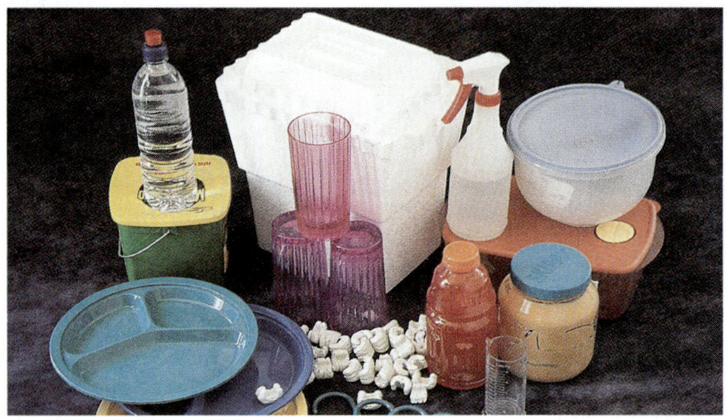

Figure 18: *Products commonly made from (a) polyvinyl chloride, (b) polyacrylonitrile, and (c) polystyrene.*

MODELING MATTER

POLYMER STRUCTURE AND PROPERTIES

Unmodified, the arrangement of covalent bonds in long, stringlike polymer molecules causes the molecules to coil loosely. A collection of polymer molecules (such as those in a piece of molten plastic) can intertwine, much like strands of cooked spaghetti. In this form the polymer is flexible and soft.

1. Using a pencil line on paper to represent a linear polymer, draw a collection of loosely coiled polymer molecules.

For most polymers like the ones you just drew, flexibility and ductility depend upon temperature. Ductility refers to the ability of a material to be drawn out, as into wires. When the material is warm, the polymer chains can slide past one another easily. However, the polymer becomes more rigid when it cools. Such polymers, called thermoplastics, are found in many everyday products.

The flexibility of a polymer can also be enhanced by adding molecules that act as internal lubricants between the polymer chains. For example, untreated polyvinyl chloride (PVC) is used in rigid pipes and house siding. With added lubricant, polyvinyl chloride becomes flexible enough to be used in raincoats and inflatable pool toys.

The reactions that form polymer chains can also take place perpendicular to the main chain, forming side chains. These polymers are called **branched polymers**. The branched form of polyethylene is shown below. The degree of branching can be controlled by adjusting reaction conditions.

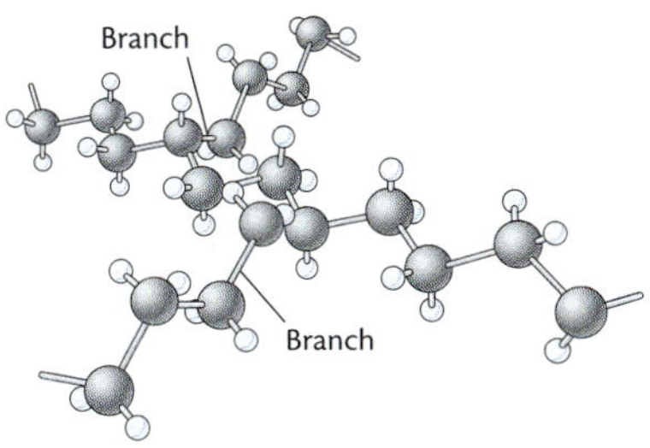

2. Draw at least two different models of branched chain polymers. Try to vary the representations—the forms of branched polymers can differ greatly.

Branching changes the properties of a polymer by affecting the ability of chains to slide past one another and by altering intermolecular forces.

Another way to alter the properties of polymers is through cross-linking. Polymer rigidity can be increased if the polymer chains are cross-linked so that they can no longer move or slide readily. You can see this for yourself if you compare the flexibility of a rubber band with that of a tire tread. Polymer cross-linking is much greater in the tire.

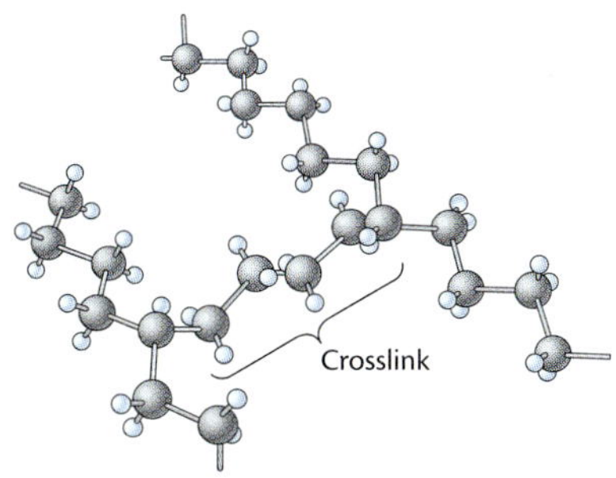

Above is cross-linked form of polyethylene.

3. Draw several linear polymer chains that have been cross-linked.
4. Draw several cross-linked, branched polymers. Then explain how cross-linking branched polymers compares with cross-linking linear polymers.

Polymer strength and toughness can also be controlled. To do this, polymer chains are arranged so that they lie in the same general direction. The aligned chains are then stretched until they uncoil. Polymers remaining uncoiled after this treatment make strong, tough films and fibers. Such materials include polyethene, used in everything from plastic grocery bags to artificial ice rinks, and polyacrylonitrile, found in fabrics.

5. Draw several aligned linear polymer chains.
6. Draw several aligned and cross-linked linear polymer chains.

MODELING MATTER: POLYMER STRUCTURE AND PROPERTIES

1.

2.

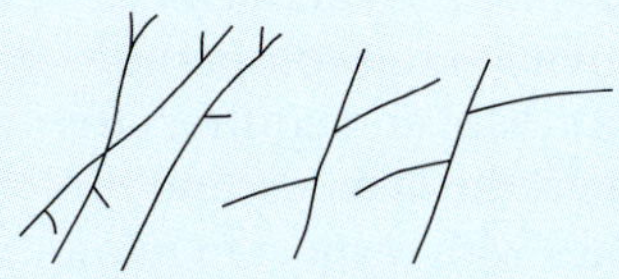

3.

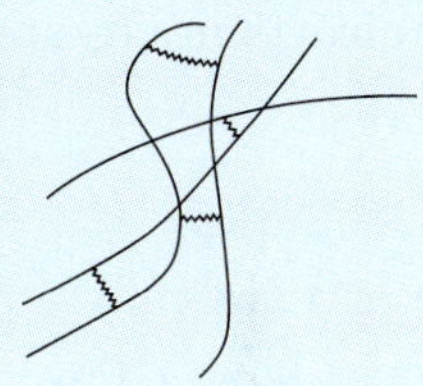

4.

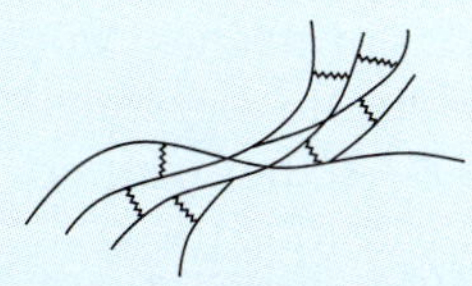

Branched polymers have many additional sites where cross-links can occur.

5.

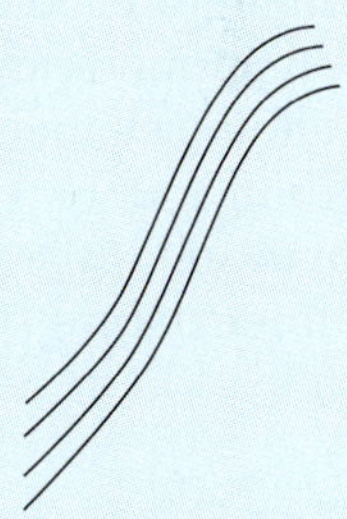

6.

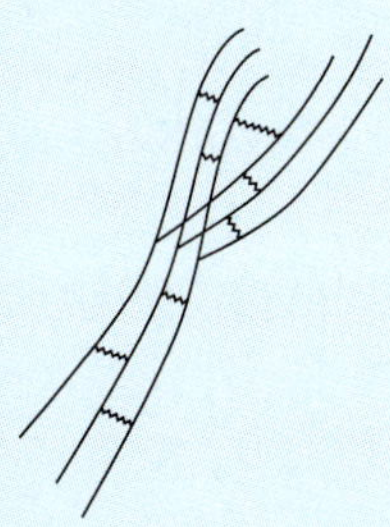

C.2 BEYOND ALKANES

The alkenes are unsaturated hydrocarbons that contain at least one carbon to carbon double covalent bond. Students are introduced to the possibility of including elements other than hydrogen and carbon in the organic compounds. The kind of atoms (nitrogen, oxygen, sulfur, or chlorine) and the arrangement of the atoms within the molecule determine the chemical reactivity of the compound.

Compare and contrast the alkanes, the alkenes, and the alkynes with respect to name (*-ane*, *-ene*, or *-yne*), general formula, chemical bonding, and saturation.

C.3 THE BUILDERS

Laboratory Activity

Students work with models to explore the bonding and characteristics of alkenes. Model kits must be able to represent double bonds—if they cannot, consider substituting the gumdrop/toothpick models. As the students construct alkene molecules, call attention to the lack of rotation around the double bond. Students are led to construct cyclic compounds and compounds containing elements in addition to carbon and hydrogen. Do not dwell excessively on nomenclature during this activity.

Time

50 minutes

Materials (for 24 students working in pairs)

12 sets of space-filling or ball-and-stick models

Polymers—3D Structures and Properties

CD-ROM WWW.

C.2 BEYOND ALKANES

Carbon is a versatile building-block atom. It can form bonds with other atoms in several different ways. As you learned in Section A, in alkane molecules each carbon atom is bonded to four other atoms. Compounds such as alkanes are called saturated hydrocarbons because each carbon atom forms as many single bonds as it can by bonding to four other atoms. In some hydrocarbon molecules, however, carbon atoms bond to three other atoms, not four. This series of hydrocarbons is called the **alkenes**. The simplest member of the alkenes was briefly introduced earlier in this section—ethene, C_2H_4.

The carbon-carbon bonding that characterizes the alkenes is a **double covalent bond**. In a double covalent bond, four electrons are shared between the bonding partners. Compounds containing carbon-carbon double bonds are described as **unsaturated**—not all carbon atoms are bonded to their full capacity with four other atoms. Because of their double bonds, alkenes are more chemically reactive—and therefore better builder molecules—than are alkanes.

The substituted alkenes make up another class of important builder molecules. In addition to carbon and hydrogen, these molecules contain one or more other elements, such as oxygen, nitrogen, chlorine, or sulfur. Adding atoms of other elements to hydrocarbon structures significantly changes their chemical reactivity.

Even molecules composed of the same elements can have quite different properties. For example, a molecule of ethanol (C_2H_6O)—also called grain alcohol—and a molecule of ethylene glycol ($C_2H_6O_2$)—often used as antifreeze—both contain carbon, hydrogen, and oxygen. The dramatic differences in their properties and uses result from the different arrangement of atoms in the two molecules.

Polymers—3D Formation and Chemical Structure

CD-ROM WWW.

C.3 THE BUILDERS

Laboratory Activity

Introduction

This activity, in which you will use models to simulate various arrangements of atoms, will help you to become more familiar with the alkenes and their polymers.

Procedure

Part 1: Alkenes

1. Examine the electron-dot and structural formulas for ethene, C_2H_4. Confirm that each atom has attained a filled outer shell of electrons.

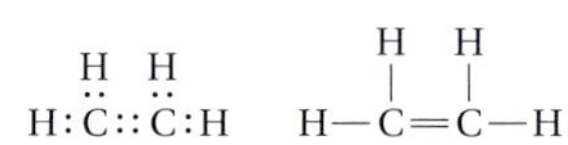

CH_2CH_2 or C_2H_4

Electron-dot formula   Structural formula   Molecular formula

The names of the alkenes follow a pattern much like that of the alkanes. The first three alkenes are ethene, propene, and butene. (Why is it impossible to have methene, a one-carbon alkene?) The same prefixes that you learned for alkanes are used to indicate the number of carbon atoms in the molecule's longest carbon chain. However, each alkene name ends in *-ene* instead of *-ane*.

2. Recall that the alkane general formula is C_nH_{2n+2}. Examine the molecular formulas of ethene (C_2H_4) and butene (C_4H_8). What general formula for alkenes do the molecular formulas suggest?
3. Assemble a model of an ethene molecule and a model of an ethane (C_2H_6) molecule. Compare the arrangements of atoms in the two models. Rotate the two carbon atoms in ethane about the single bond. Then try a similar rotation with ethene. What do you observe? Can you build a molecule in which you can perform a rotation about a double bond? Write a general rule to summarize your findings.
4. Build a model of butene (C_4H_8). Compare your model to those made by others. Remember that alkenes must contain a double bond.
   a. How many different arrangements of atoms in a C_4H_8 chain appear possible? Each arrangement represents a different substance—another example of isomers!
   b. Which structural formulas in Figure 19 correspond to models built by you or your classmates?

As with alkanes, alkenes are named according to the length of their longest carbon chain. The carbon atoms are numbered, beginning at the end of the chain closest to the double bond. The name of each isomer starts with the number assigned to the first double-bonded carbon atom. Look again at the butene isomer structures in Figure 19 and confirm this naming convention with 1-butene and 2-butene.

5. Does each of the following pairs represent isomers, or are they the same substance?
   a. $CH_2{=}CH{-}CH_2{-}CH_3$ or $CH_3{-}CH_2{-}CH{=}CH_2$
   b.

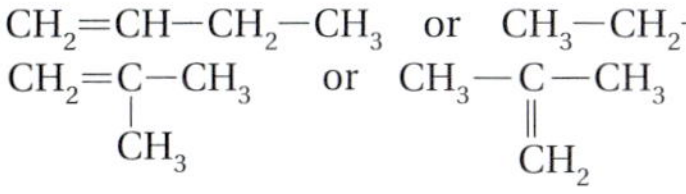

6. How many isomers of propene (C_3H_6) are there? Support your answer with the appropriate structure(s).
7. Are these two structures isomers or the same substance? Explain.

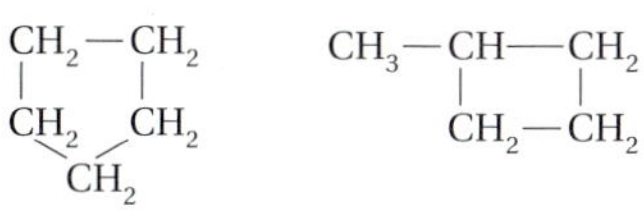

8. Based on your knowledge of molecules with single and double bonds between carbon atoms, assemble a model of a hydrocarbon molecule with a triple bond. Your completed model represents a member of the hydrocarbon series known as **alkynes**. Based on your understanding of how alkanes and alkenes are named, write structural formulas for
   a. ethyne, commonly called acetylene.
   b. 2-butyne.

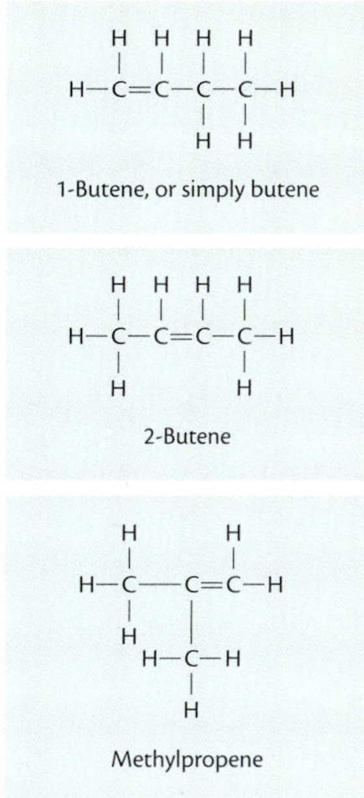

Figure 19: *Three isomers of butene, C_4H_8.*

Part 1. Alkenes

Electron-dot structures, structural formulas, condensed formulas, and systematic nomenclature are used to describe molecules with multiple bonds.

Answers

1. There must be at least two carbon atoms to form a C to C double bond.
2. C_nH_{2n}
3. Students should note that alkene molecules resist rotation about the double bond.
4. a. Four arrangements of C_4H_8 are possible, as well as cyclobutane.
   b. Students should recognize the 2-butene isomer; methylpropene is less easily recognized.
5. Each pair alone represents the same substance (not isomers). Some students may correctly point out, however, that the substance represented by the first pair (1-butene) is an isomer of the substance represented by the second pair (2-methylpropene).
6. Only one structure is possible for propene.
7. Isomers; same formula (C_5H_{10}) but different structures.
8. a. $H{-}C{\equiv}C{-}H$ (ethyne or acetylene)
   b. $CH_3{-}C{\equiv}C{-}CH_3$ (2-butyne)
9. Unsaturated; alkynes contain a triple carbon to carbon bond.

Part 2. Compounds of Carbon, Hydrogen, and Singly Bonded Oxygen

Students should find that the two isomers of C_2H_6O are easy to make and identify. They should recognize that the molecule with the -OH group would be more soluble in H_2O. If necessary, remind them that they have already tested this compound, ethanol, in the Water unit (Unit 1). The higher boiling point (80 °C) of ethanol compared to that of dimethyl ether (−24 °C) is explained by hydrogen bonding.

10. Students should make models using the nine atoms.

11. a. Two distinct structural isomers of C_2H_6O are possible.

    b. $CH_3{-}CH_2{-}O{-}H$ (ethanol) and $CH_3{-}O{-}CH_3$ (dimethyl ether).

    c. Yes, they have the same formula, but the atoms are arranged in a different order.

12. a. Ethanol is more soluble in water than dimethyl ether.

    b. Ethanol should have the higher boiling point.

Part 3. Alkene-based Polymers

13.

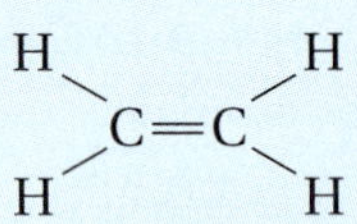

ethene

14.

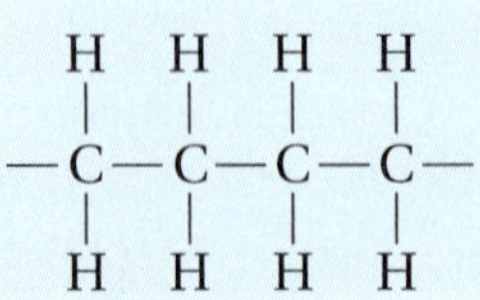

polyethylene

The double bond of the dimer was broken between carbon atoms.

15.

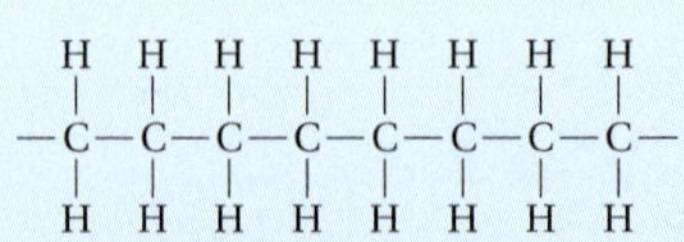

Polyethylene grows by chains of 2 carbons as each group connects.

16.

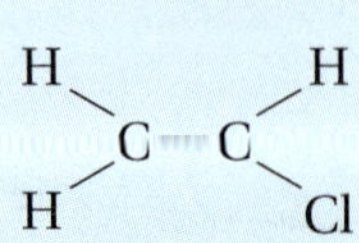

vinyl chloride

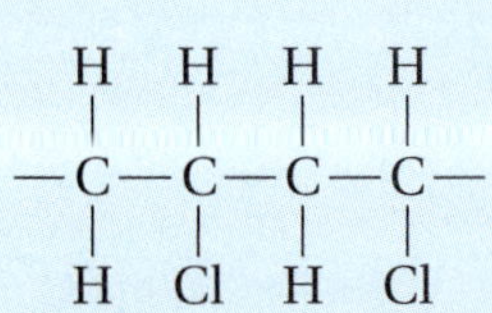

polyvinyl chloride

9. Are alkynes saturated or unsaturated hydrocarbons? Explain.

Part 2: Compounds of Carbon, Hydrogen, and Singly-Bonded Oxygen

10. Assemble as many different molecular models as possible using all nine of these atoms:

    2 carbon atoms (each forming four single bonds)
    6 hydrogen atoms (each forming a single bond)
    1 oxygen atom (forming two single bonds)

11. On paper draw a structural formula for each compound you have constructed, indicating how the nine atoms are connected. Compare your structures with those made by other classmates. After you are satisfied that all possible structures have been produced, answer these questions:
    a. How many distinct structures did you identify?
    b. What is the structural formula for each structure?
    c. Are all of these structures isomers? Explain.

12. Each compound you have identified possesses distinctly different physical and chemical properties.
    a. Recalling that "like dissolves like," which compound should be most soluble in water?
    b. Which compound should have the highest boiling point?

Part 3: Alkene-Based Polymers

In this part of the activity, you and your classmates will use models to simulate the formation of several addition polymers.

13. Build models of two ethene molecules.

14. Using information on page 217 as a guide, combine your two models into a dimer, or a two-monomer structure. What modifications in the monomer structure were necessary to accomplish this?

15. Combine your dimer with that of another lab team. Continue this process until your class has created a long-chain structure. Although your resulting molecular chain is not yet long enough to be regarded as a polymer, you have modeled the processes involved in creating a typical addition polymer.

16. Repeat Steps 13 and 14 for vinyl chloride and again for acrylonitrile. Then determine whether you can do the same for styrene. (See page 217 for the molecular structures for these substances.)

17. Assume the structures you built in Steps 15 and 16 became significantly larger. Give the name of each resulting polymer.

18. Are the chains you built linear or branched? How would this affect the properties of the polymer?

19. Make a polyethene chain that includes some cross-linking.
    a. Does this change the behavior of the model?
    b. How would the cross-linking change the properties of the polymer?

C.4 MORE BUILDER MOLECULES

So far, you have examined only a small part of the inventory of builder molecules that chemical architects have available. Now you will explore two important classes of compounds in which carbon atoms are joined in rings rather than in chain structures.

As a first step in doing this, picture a straight-chain hexane molecule, $CH_3-CH_2-CH_2-CH_2-CH_2-CH_3$. Next, remove one hydrogen atom from the carbon atom at each end. Then imagine those two carbon atoms bonding to each other. The result is the molecule known as cyclohexane.

$CH_3CH_2CH_2CH_2CH_2CH_3$

Hexane

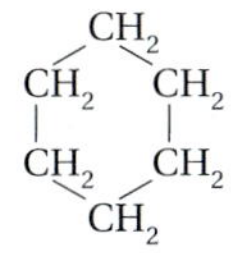

Cyclohexane

Cyclohexane is a starting material for making nylon, an important and familiar petrochemical polymer. Cyclohexane is representative of the **cycloalkanes**, which are saturated hydrocarbons made up of carbon atoms joined in rings.

Another important class of hydrocarbon builder molecules is the **aromatic compounds**. Unlike cycloalkanes, aromatic rings are unsaturated. Aromatic compounds have chemical properties distinctly different from those of the cycloalkanes and their derivatives. The structural formula of benzene (C_6H_6), the simplest aromatic compound, is shown below. In the representation on the right, each "corner" of the six-carbon (hexagonal) ring represents a carbon atom with its hydrogen atom.

> The pleasant odor of the first aromatic compounds discovered prompted their descriptive name.

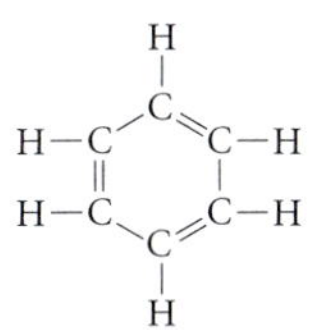

or

Although chemists who first investigated benzene proposed these structures, the chemical properties of the compound did not support their hypotheses. Recall that carbon-carbon double bonds (C=C) are usually very reactive. But benzene behaves as though it does not contain any such double bonds. A deeper understanding of chemical bonding was needed to explain benzene's puzzling structure.

Substantial experimental evidence indicates that all carbon-carbon bonds in benzene are identical. Thus its structure is not well represented by alternating single and double bonds. Instead, chemists often represent a benzene molecule in the following way:

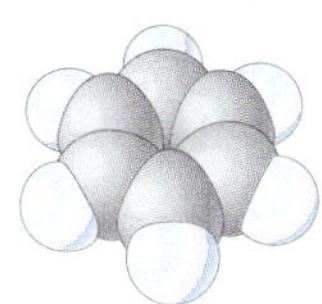

Benzene, C_6H_6

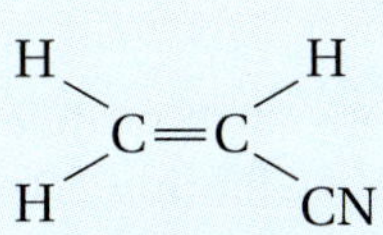

acrylonitrile

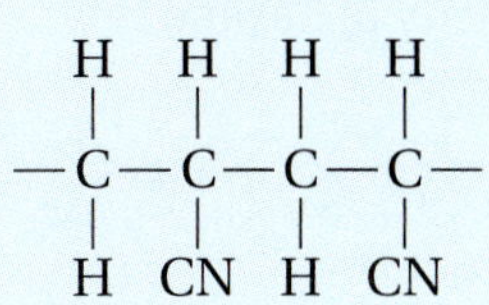

polyacrylonitrile

17. Polyvinyl chloride and polyacrylonitrile.

18. Linear; the branched chains would have different properties.

19. In cross-linked chains, the product will usually be more rigid and more resistant to heat. Cross-linked chains are not softened by heat, they simply char if exposed to flame. Cross-links cannot be remolded once they have been originally shaped.

ADDITIONAL ACTIVITY

Have students conduct research on petrochemicals made from the cycloalkanes and the aromatic hydrocarbons.

C.4 MORE BUILDER MOLECULES

The cycloalkanes are ring compounds like the aromatics, but the relative reactivity of the cycloalkanes have more in common with the linear alkanes. The special nature and reactivity of the benzene ring are contributed to the equal sharing of the electrons among all six-carbon atoms.

C.5 BUILDER MOLECULES CONTAINING OXYGEN

The functional groups responsible for alcohols, acids, and esters are diagrammed. It might be helpful for students to see that both alcohols and acids have the —OH functional group, but in esters the hydrogen is replaced by another carbon group. You may use the blackline masters on **Common Functional Groups** when discussing these structures. The specific functional group(s) found in a given organic compound determine its physical (solubility, boiling point, etc.) and chemical properties. Remind students that with the exception of combustion, unsubstituted alkanes are fairly unreactive. Molecules that contain the same functional group(s) share similar family properties.

If you have samples of some of these compounds, a safe way to test odors is to add a few drops of the liquid to a piece of cotton and then place the cotton in a stoppered test tube. The tubes can be circulated around the room with little risk of spilling or creating odors.

DEMONSTRATION IDEA

Condensation Polymers, p. 226

Synthesis of nylon: *Nylon* is a generic term used to represent synthetic polyamides. Nylons made from diamines and dibasic acids or diacid chlorides are designated by two numbers, the first representing the number of carbon atoms in the diamine, and the second the number of carbon atoms in the dibasic acid. The history of the development of nylon and its use in World War II make it a particularly interesting library research topic.

The inner circle represents the equal sharing of electrons among all six carbon atoms. The hexagonal ring represents the bonding of six carbon atoms to each other. Each corner in the hexagon is the location of one carbon atom and one hydrogen atom, thus accounting for benzene's formula, C_6H_6.

Although only small amounts of aromatic compounds are found in petroleum, large quantities are produced by petroleum fractionation and cracking. Benzene and other aromatic compounds are present in gasoline as octane enhancers; however, they are used primarily as builder molecules. Entire chemical industries (dye and drug manufacturing, in particular) are based on the unique chemistry of aromatic compounds.

C.5 BUILDER MOLECULES CONTAINING OXYGEN

In assembling molecular models with C, H, and O atoms, it is likely that you "discovered" one of the compounds depicted below:

$CH_3—OH$	$CH_3—CH_2—OH$
Methanol (methyl alcohol)	Ethanol (ethyl alcohol)

The —OH alcohol functional group should not be confused with the hydroxide anion, OH^-, found in ionic compounds such as sodium hydroxide, NaOH.

As you can see, each molecule has an —OH group attached to a carbon atom. This general structure is characteristic of a class of compounds known as **alcohols**. The —OH is recognized by chemists as a **functional group**—an atom or group of atoms that imparts characteristic properties to organic compounds. If the letter R is used to represent all of the molecule other than the functional group, then the general formula of an alcohol can be written as

$$R—OH$$

Any alcohol

In this formula, the line indicates a covalent bond between the oxygen of the OH group and an adjacent carbon atom in the molecule. In methanol (CH_3OH), the letter R represents CH_3—; in ethanol (CH_3CH_2OH), R represents CH_3CH_2—.

The formulas and structures of two common alcohols are shown below. What does R represent in each compound?

$CH_3CH_2CH_2OH$

1-Propanol

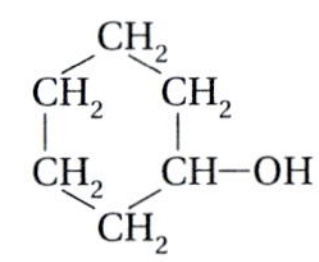

Cyclohexanol

(a)

Figure 20: *Familiar items that contain (a) alcohols and, on page 225, (b) carboxylic acids and (c) esters.*

Two other classes of oxygen-containing compounds, **carboxylic acids** and **esters**, are versatile and important builder molecules. The functional group in each class of compounds contains two oxygen atoms.

R—C(=O)—OH	R—C(=O)—OR
Carboxylic acid	Ester

Note that both classes of compounds have one oxygen atom double-bonded to a carbon atom and a second oxygen atom single-bonded to the

Materials

Solution 1: Gradually heat a reagent bottle of 1,6-diaminohexane (also called hexamethylenediamine, melting point (mp) = 42 °C) in a warm water bath until the solid begins to melt. Decant 2.9 g directly into a 100 mL beaker and dilute to 50 mL with 0.5 M NaOH solution (20 g NaOH per 1 L).

Solution 2: Prepare a 0.25 M diacid chloride solution by diluting 2.3 g of adipoyl dichloride to 50 mL with hexane (density of hexane is 0.66 g/cm³).

same carbon atom. Ethanoic acid (a carboxylic acid that is better known as acetic acid) and methyl ethanoate (an ester more commonly called methyl acetate) are examples of these two classes of compounds.

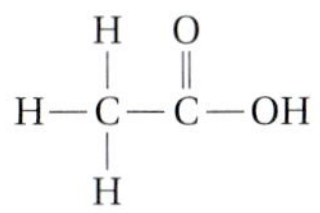

Ethanoic acid

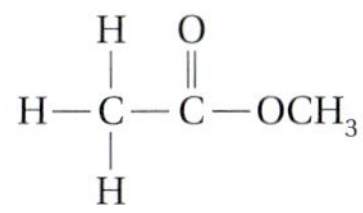

Methyl ethanoate

Some familiar things that include or involve the use of alcohols, carboxylic acids, and esters are shown in Figure 20. Builder molecules can be modified by adding other functional groups that include nitrogen, sulfur, or chlorine atoms. The rich variety of functional groups greatly expands the types of molecules that can ultimately be built by chemists.

(b)

(c)

Lab tips

- The reaction does not depend on the exact stoichiometry of the reactants, but the concentrations should be close to those indicated for optimal results.
- The aqueous diamine solution may be stored, but the acid chloride solution is somewhat unstable because of hydrolysis (only make up as much as you need on a given day). The pure acid chloride should be stored in a brown bottle and out of strong light under a hood.

Demonstration

Using equal volumes of the two solutions (as little as 5 mL of each can be sufficient), slowly pour Solution 2 down the side of the beaker, on top of Solution 1 (Solution 2 is less dense than Solution 1), so as to minimize agitation between the immiscible solutions.

Use forceps or a bent paperclip to hook the polymer film that forms at the interface of the two solutions, and slowly pull it from the center of the beaker. Continue pulling out the nylon "thread" until the solutions are exhausted. A thread 10–15 feet long is possible. If it breaks, you may be able to hook a new thread. Longer threads are possible if you avoid contact with the sides of the beaker.

Any excess solution should be placed in a solvent waste container, or stirred, and the resulting solid washed and disposed of in a solid waste container.

The reaction for this demonstration is shown at the bottom of page 226.

CAUTION: *The diamine, dichloride, and sodium hydroxide are all irritating to the skin, eyes, and respiratory system; using a hood is advised. It is also advisable to wear plastic or rubber gloves. The demonstration is quite safe if properly performed, but do not directly handle the product until it has been washed with ethanol or water.*

C.6 CONDENSATION POLYMERS

Condensation polymers are produced when two or more monomers react with each other to yield a larger hydrocarbon plus water. These reactions produce both natural and synthetic polymers. There must be at least two functional groups on each reacting monomer for the reaction to continue to form a chain. The monomers link together like cars of a railroad train or necklace beads.

The polymer that is used in soft drink containers (PET) may be recycled, or it may be broken down into monomers and repolymerized into high-quality recycled PET products. This property reduces the need for new petroleum-based builder molecules.

Emphasize that condensation reactions may be used to make both small and large polymers. Examples of condensation polymers are polyurethane foam, epoxy glue, nylon, dacron, peptides, and proteins.

C.6 CONDENSATION POLYMERS

Earlier you learned that many polymers can be formed from alkene-based monomers through addition reactions. Not all polymers are formed in this way, however. Natural polymers, such as proteins, starch, cellulose (in wood and paper), and synthetic polymers, including the familiar nylon and polyester, are also formed from monomers. But unlike addition polymers, these polymers are formed with the loss of simple molecules such as water when monomer units join. The term **condensation reaction** applies to this second type of polymer-making process, and the resulting product is called a **condensation polymer**.

One very common condensation polymer is polyethylene terephthalate (PET). This polymer is commonly used in large soft-drink containers. It also enjoys many other applications—as thin film for videotape and as the textile Dacron® for clothing, surgical tubing, and fiberfill. More than two million kilograms of PET are produced each year in the United States. Some examples of products made of PET are shown in Figure 21.

While some PET may be recycled, material that contains high quantities of additives, such as dyes or other polymers, is either landfilled or incinerated. The DuPont PetretecK process breaks down PET into its monomers, which can then be repolymerized into high-quality PET products. This technology decreases the need for new petroleum-based builder molecules as well as the quantity of PET being landfilled.

Condensation reactions can be used to make small molecules as well as polymers. In the next laboratory activity, you will use condensation reactions to produce esters. These reactions illustrate how organic compounds can be combined chemically to create new substances.

Figure 21 *Common uses of PET.*

Reaction for Demonstration

diamine + diacid chloride $\longrightarrow$

$$H_2N(CH_2)6NH_2 + Cl-\overset{\overset{\displaystyle O}{\|}}{C}-(CH_2)_8-\overset{\overset{\displaystyle O}{\|}}{C}-Cl \xrightarrow{NaOH}$$

polyamide + 2 HCl

$$-[\underset{}{\overset{\overset{\displaystyle H}{|}}{N}}-(CH_2)_6-\overset{\overset{\displaystyle H}{|}}{N}-\overset{\overset{\displaystyle O}{\|}}{C}-(CH_2)_8-\overset{\overset{\displaystyle O}{\|}}{C}]- + 2\,HCl$$

C.7 CONDENSATION

Laboratory Activity

Introduction

In this activity you will produce several petrochemicals through the reaction of an organic acid (an acid derived from a hydrocarbon) with an alcohol. The esters you will produce have familiar, pleasing fragrances. Many perfumes contain esters; the characteristic aromas of many herbs and fruits arise from esters contained in the plants.

Organic acids are also called carboxylic acids. See page 224.

One example of the formation of an ester is the production of methyl acetate from ethanoic acid (acetic acid) and methanol in the presence of sulfuric acid:

$$\underset{\text{Ethanoic acid}}{CH_3-\overset{\overset{\displaystyle O}{\|}}{C}-OH} + \underset{\text{Methanol}}{H-O-CH_3} \xrightarrow{H_2SO_4} \underset{\text{Methyl acetate}}{CH_3-\overset{\overset{\displaystyle O}{\|}}{C}-O-CH_3} + \underset{\text{Water}}{H-OH}$$

To emphasize the roles of functional groups in the formation of an ester, a general equation can be written using the R notation you learned earlier.

Recall that R stands for the "rest" of the molecule, or everything other than the functional group.

$$\underset{\text{Carboxylic acid}}{R-\overset{\overset{\displaystyle O}{\|}}{C}-OH} + \underset{\text{Alcohol}}{H-O-R} \xrightarrow{H_2SO_4} \underset{\text{Ester}}{R-\overset{\overset{\displaystyle O}{\|}}{C}-O-R} + \underset{\text{Water}}{H-OH}$$

Note how the functional groups of the acid and alcohol combine to form a water molecule, while the remaining atoms join to form an ester molecule. Sulfuric acid (H_2SO_4) acts as a catalyst—that is, it causes the chemical reaction to proceed faster without itself being used up.

Using a process similar to the reaction shown here, you will now produce the ester known as methyl salicylate.

Procedure

1. Prepare a water bath by adding about 50 mL tap water to a 100-mL beaker. Place the beaker on a hot plate, and heat the water until it is near boiling.
2. Obtain a small, clean test tube. Place 5 drops methanol into the tube. Next add 0.1 g salicylic acid. Then add 2 drops concentrated sulfuric acid to the tube. **CAUTION:** *Concentrated sulfuric acid will cause burns to skin or fabric. Add the acid slowly and very carefully.*
3. As you dispense these reagents, note their odors. **CAUTION:** *Do not directly sniff any reagents—some may irritate or burn nasal passages.* Record any odors you happen to note.
4. Place the test tube in the water bath you prepared in Step 1.
5. Using test-tube tongs, move the test tube slowly in the water bath in a small horizontal circle. Keep the tube in the water, and do not spill the contents. Note any color changes. Continue heating for three minutes.

C.7 CONDENSATION

Laboratory Activity

The synthesis of esters illustrates the observable changes produced when chemical reactions occur. The familiar odor of the products of this reaction makes this activity a pleasing experience.

Time

30 minutes

Materials (for 24 students working in pairs)

12 test tubes (15 × 125 mm)
12 10-mL graduated cylinders
12 150-mL beakers
12 test tube holders
12 test tube brushes
12 hot plates
12 dropping bottles for dispensing of organic reactants
50 mL concentrated sulfuric acid ($18\ M\ H_2SO_4$)
1 reagent bottle for dispensing $18\ M\ H_2SO_4$
24 mL methanol
6 g salicylic acid
boiling chips
50 mL pentyl alcohol
50 mL acetic acid
50 mL octyl alcohol

Safety

Students should wear safety goggles, aprons, and gloves throughout the laboratory activity. Demonstrate the appropriate way to smell chemicals. **Remind students to exercise extreme caution when using 18 M sulfuric acid. It is caustic and corrosive. Teachers may want to take charge of dispensing this themselves.**

Pre-lab discussion

Discuss the general condensation reaction for forming esters. Point out the role of chemists as molecular architects in designing molecules. The flavor and perfume industries are heavily dependent on chemically synthesized esters, which are either unavailable in nature or available only by expensive extraction from dilute sources. Many low-molecular-weight esters have pleasant fruity odors/tastes and are often responsible for the natural odors/flavors of ripe fruit, flowers, and other plant substances. The continued search for new odors/flavors and for a theory capable of correlating structure with odor/flavor is a potential area for future chemical careers. See the Post-lab discussion as well.

Lab tips

- The students will probably recognize the odor from its use in candy, perfumes, and so on, but caution them not to ingest any of the products.
- **Remind students to avoid the direct inhalation of any fumes—they should be gently fanned toward the nose.**

The boiling chip in the test tube minimizes erratic boiling, which can cause liquid to erupt out of the test tube. The hot water bath needs only to be above the boiling point of methanol (56 °C) and does not need to boil vigorously.

Many other combinations can be used if you have the available chemicals. (See Source Book, Volume 3, *Organic Chemistry*, pp. 5–10, ChemSource, Inc., 1994.)

6. If you have not noticed an odor after three minutes, remove the test tube from the water bath, hold the test tube away from you with the tongs, and wave your hand across the top of the test tube to waft any vapors toward your nose. Record observations regarding the odor of the product. Compare your observations with those of other class members.
7. Repeat the procedure using 20 drops pentyl alcohol, 20 drops acetic acid, and 2 drops sulfuric acid.
8. Repeat the procedure using 20 drops octyl alcohol, 20 drops acetic acid, and 2 drops sulfuric acid.
9. Dispose of your products as directed by your teacher.
10. Wash your hands thoroughly before leaving the laboratory.

Questions

1. Write the molecular formulas of the acid and alcohol from which you produced methyl salicylate. (*Hint:* You may need to look in a chemistry reference book.)
2. Write a chemical equation for the formation of methyl salicylate.
3. Repeat Questions 1 and 2 for the second ester you produced.
4. Describe the odors of the three esters produced in this laboratory activity.
5. Classify each of the following compounds as a carboxylic acid, an alcohol, or an ester:
   a. $CH_3-CH_2-CH_2-CH_2-OH$
   b. $CH_3-C(=O)-O-CH_2-CH_3$
   c. $CH_3-CH(-CH_2-C(=O)-OH)-CH_2-CH_3$
   d. $CH_3-C(=O)-OH$

C.8 BUILDER MOLECULES IN TRANSPORTATION

Making Decisions

In this section, you have had a chance to survey some of the many and varied roles of petroleum products as building blocks for countless everyday objects. You can easily find examples of products built from petroleum in your surroundings, products that were unknown even a few decades ago. Plastics have many advantages over traditional materials, including favorable strength-to-weight ratios and recyclability.

Expected Results

Acid	Alcohol	Odor
20 drops acetic acid	28 drops *n*-butyl alcohol	pear
10 drops acetic acid	20 drops octyl alcohol	orange
20 drops acetic acid	30 drops pentanol	banana
20 drops propanoic acid	15 drops ethanol	rum
0.4 g benzoic acid	20 drops ethanol	fruity

In the TV advertisement that opened this unit, a newly designed automobile was described as "petroleum-free." Is that possible? Is it plausible? In this activity, you will consider this claim and evaluate how builder molecules are used in automobile construction.

1. List several automobile components that are commonly made of polymers and plastics.
2. Which of the polymer parts you listed are made from petroleum builder molecules? (*Hint:* It may be easier—and faster—to determine which parts are not petroleum-based.)
3. Try to think of a replacement material (not another polymer or petrochemical) for each component on your list.
4. Decide whether there are any disadvantages associated with each replacement material you have suggested. List the most important disadvantages for each component, or write "none" if the replacement material provides similar properties at a comparable cost. A table such as the one shown here will help you to organize your ideas and answers.

CD-ROM WWW. **Questions and Answers**

Automobile Component	Petroleum Based? (Yes or No)	Potential Replacement Material	Disadvantages of Replacement Material

Post-lab discussion (C.7)

The post-lab discussion can continue the discussion of the reaction of the functional groups in the organic acids and alcohols that produce esters. You may use the transparency on **Esters** when discussing them. Emphasize that a seemingly small difference in molecular structure can make a very noticeable difference in properties. Have students note the contrasting odors of reactants and products. To reinforce this point, you may want to synthesize several other esters (see suggestion in Lab tips). Be sure to pass samples around the classroom and write out the specific reaction(s):

alcohol + organic acid $\xrightarrow{\text{acid catalyst}}$ ester + water

Be sure to distinguish organic acids (RCOOH) from common inorganic (or mineral) acids such as HCl, H_2SO_4, and HNO_3. Organic acids are typically weak acids and were originally obtained by isolation from plant and animal ("organic") sources. For example, formic acid (HCOOH/methanoic acid) was first obtained from the destructive distillation of ants, and acetic acid (CH_3COOH/ethanoic acid) from the aerobic fermentation of a mixture of cider and honey.

Students have probably smelled butyric acid ($CH_3CH_2CH_2COOH$/butanoic acid) in rancid butter or body odor. Today commercially used organic acids are typically synthesized from petroleum or coal tar, as are the alcohols with which they are reacted.

See page 230 for commentary on Section C.8.

Answers

1. Methanol, CH_3OH
   Salicylic acid (or 2-hydroxybenzoic acid), $C_7H_6O_3$
2. $CH_3OH + C_7H_6O_3 \rightarrow C_8H_8O_3 + H_2O$
3. $C_5H_{11}OH + CH_3COOH \rightarrow CH_3COOC_5H_{11} + H_2O$
   Pentyl alcohol + acetic acid → pentyl ethanoate + water
4. a. peppermint (methanol & salicylic acid)
   b. banana (pentyl alcohol & acetic acid)
   c. orange (octyl alcohol & acetic acid)
5. a. alcohol
   b. ester
   c. organic acid
   d. organic acid

C.8 BUILDER MOLECULES IN TRANSPORTATION

Making Decisions

In this culminating activity for Section C (pages 228–229), students may work in small groups to discuss the possibility and the plausibility of a petroleum-free vehicle.

Students' answers will vary. These are only suggested responses.

CHEMISTRY AT WORK

Optimizing Technology

A polymer is a large molecule consisting of a chain of small molecules joined together in a repeating fashion. Chemists develop polymers for use as "ingredients" in products with unique physical and chemical properties. Polymer products can be lightweight, hard, strong, and flexible. They can also have special thermal, electrical, and optical characteristics. Products made of polymers come from a variety of industries, including textile and industrial fiber, plastics, communication, packaging, and transportation.

The big boom in polymer chemistry occurred during the early 1900s with the advent of polymer materials such as polyethylene, nylon, and synthetic rubber. Today, the main focus of polymer chemistry is improving existing technologies and developing new synthetic and processing technologies that are more efficient.

Chemical testing to produce a new product with new properties begins in a chemical laboratory.

Polymer Chemistry: Fibers to Fit Your Needs!

Stacy Johnson is a Senior Research Chemist at the E.I. DuPont de Nemours and Company facility in Waynesboro, Virginia. Stacy's company makes LYCRA spandex, which is used in clothing such as underwear, hosiery, socks, swimwear, sportswear, and outerwear. LYCRA spandex is a strong elastomeric polymer that has the ability to stretch 400 to 800 times its original length without breaking. In the 1960s, LYCRA replaced much of the rubber commonly used in a variety of garments. Compared to rubber, LYCRA fibers have higher chemical stability and offer increased comfort and fit.

As a Senior Research Chemist, Stacy is responsible for developing new LYCRA products to meet the demands of the worldwide textile market. In the last decade, new chemical processes and additives have improved the properties of LYCRA, properties such as chlorine durability in swimwear and power enhancement for athletic performance in sportswear. In order to develop new products, Stacy must understand how market need and chemistry interact; that is, she must evaluate the various chemical routes to meeting that need and then implement the most effective one(s).

Automobile Component	Petroleum Based? (Yes or No)	Potential Replacement Material	Disadvantages of Replacement Material
Dashboard	Yes	Metal	Nonrenewable/expensive
Radio/CD player	Yes	Metal	Nonrenewable/expensive
Seat covers	Yes	Leather	Too expensive
Tires	Yes	Natural rubber	Scarce supply cannot meet demands

Chemical testing to produce a product with new properties begins in a chemical laboratory. An extensive "scale-up" process in which the same chemical reactions done in the laboratory are now reproduced in an industrial plant is needed to commercialize a new product. This process often requires several years of development. Thus an industrial chemist must not only know how to use chemistry to solve problems, but must also understand the larger picture that involves marketing and business.

Before earning a Ph.D. in chemistry at Pennsylvania State University, Stacy spent two years as a high school teacher in Athens, Texas. She feels that teaching was beneficial because through it she came to realize that chemistry, often an intimidating subject, can be made very exciting if it is explained in terms of its application to people's everyday lives. Stacy says, "Being a chemist is a very creative job. You are essentially building things with very tiny blocks that fit together in certain ways, and using the chemical rules and vocabulary you learned in school to describe what you built."

Preparing for a Career in Chemistry

1. Chemists are employed in a variety of fields. Conduct some research to find out what chemists working in the following areas do. (*Hint*: Go to the American Chemical Society Website at www.acs.org for help.)
   - Food and Flavor Chemistry
   - Forensic Chemistry
   - Science Writing
   - Chemical Education
   - Analytical Chemistry
   - Materials Science
   - Medicinal Chemistry
   - Biotechnology
   - Hazardous Waste Management
   - Environmental Chemistry
2. Science differs from technology. Review the definitions of the two words, and then identify several examples of each that are described in this Chemistry at Work feature.

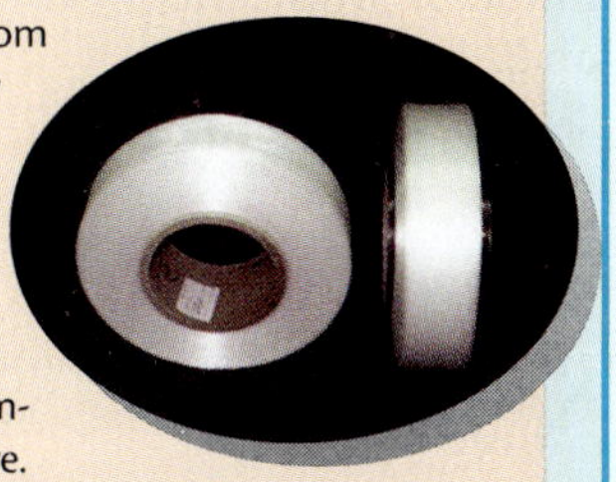

3. The rolls in the picture above are spools of LYCRA yarn. Imagine stretching a spandex fiber with your hands and then releasing it. What types of interactions must the polymer molecules in the fiber have with each other in order to stretch many times the original length and then return to the original size and shape?
4. In addition to synthetic polymers, natural polymers are also used in clothing. List several fabrics made from natural polymers.

SECTION SUMMARY

Reviewing the Concepts

Note: Provide students with the formula of propylene, H_2CCHCH_3 (1-propene) for question 8.

1. How many repeating units are found in each of the following structures?

a. a monomer	**a. Monomer: one**
b. a dimer	**b. Dimer: two**
c. a trimer	**c. Trimer: three**
d. a polymer	**d. Polymer: many**

2. List four examples of natural polymers and four examples of synthetic polymers.

Examples of natural polymers are starch, wool, silk, linen, cellulose, and cotton. Examples of synthetic polymers are synthetic fabrics, such as dacron, polyester, and nylon. Examples of plastics are polyethylene, Teflon, and PET.

3. What structural features make the properties of one polymer different from those of another?

The monomers that make up the polymer determine the properties.

4. List two techniques that can be used to modify the characteristics of a polymer. Explain each.

The addition of molecules that act as internal lubricants for the polymer chains makes the polymer more flexible. Untreated polyvinyl chloride (PVC) is used in rigid pipes and house siding. With added lubricant, polyvinyl chloride becomes flexible enough to be used in raincoats and shoes.

Polymer strength and toughness can be controlled by arranging the polymer chains so they lie generally in the same direction. Then the chains are stretched so that they uncoil. Strong films like the polyethene used in plastic bags or the polyacrylonitrile found in fabrics are examples.

5. Why is the term *unsaturated* used to describe the structures of alkynes and alkenes?

Alkenes and alkynes are characterized as unsaturated because they contain double and triple covalent bonds, which can undergo addition reactions.

6. Rank the following in order of decreasing reactivity: alkyne, alkane, and alkene. Explain your answer.

Increasing reactivity: alkane, alkene, and alkyne. The addition of atoms to double and triple covalent bonds occurs readily.

7. Draw an electron-dot diagram and structural formula for each of these molecules:

a. propane b. propene c. propyne

a.

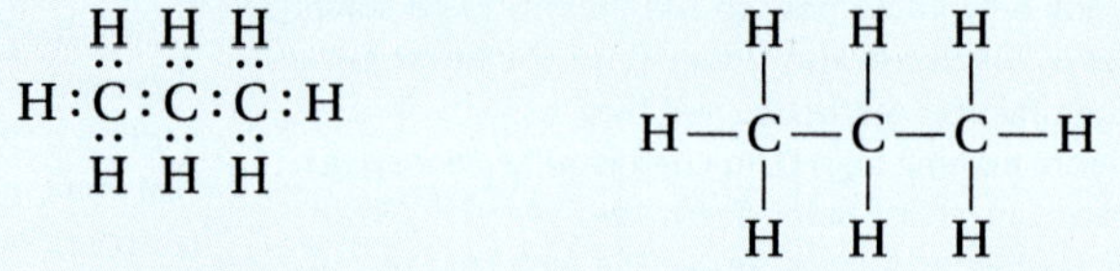

b.

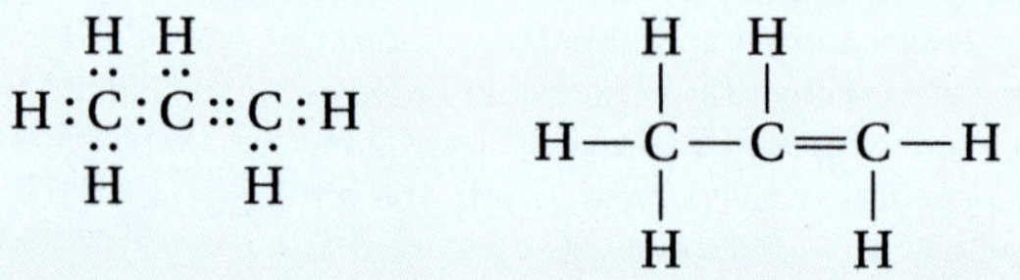

c.

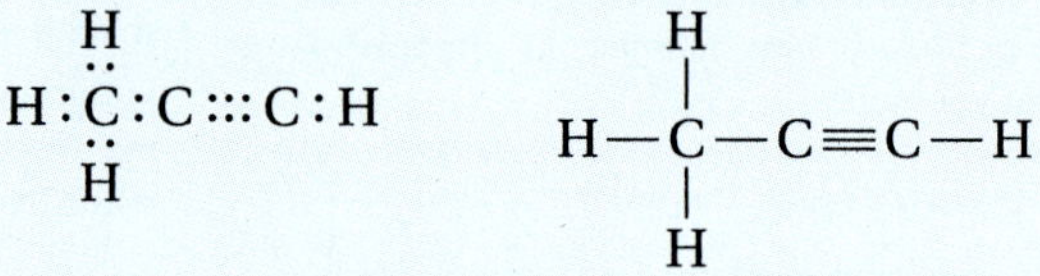

8. Use structural formulas to illustrate the polymerization that results in polypropylene.

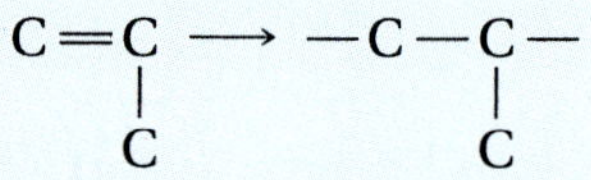

propylene (propene) polypropylene

9. Explain why alkanes cannot be used to make polymers.

Alkanes are saturated hydrocarbons. There is no double bond to be broken.

10. In what ways are cycloalkanes different from aromatic compounds?

Cycloalkanes are saturated hydrocarbons in which the carbon atoms are joined in a ring. Aromatic compounds are unsaturated hydrocarbons in which the carbon atoms are joined in a ring.

11. Draw a structural formula for both a saturated and an unsaturated form of C_4H_8.

Four arrangements of C_4H_8 are possible, cyclobutane is the only saturated isomer.

$H_3C-CH=CH-CH_3$

2-butene

$H_2C=CH-CH_2-CH_3$

1-butene

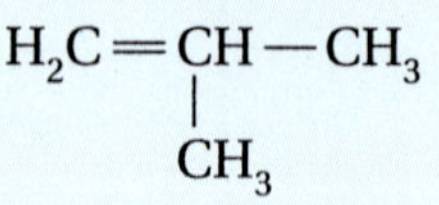

2-methylpropene

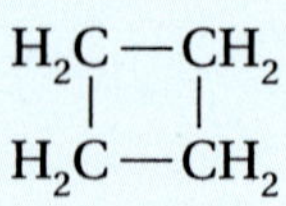

cyclobutane

12. How did aromatic compounds get their name?

Early discovery of aromatic compounds found that they had pleasant odors.

13. Why is the circle-within-a-hexagon representation of a benzene molecule a better model than the hexagon with alternating double bonds?

The electrons in covalent double bonds on the carbons are distributed equally around the ring.

14. a. Write the structural formula for a molecule containing at least three carbon atoms that represents (i) an alcohol, (ii) an organic acid, and (iii) an ester.

b. Circle the functional group in each structural formula.

c. Name each compound.

a. i. $CH_3—CH_2—CH_2—OH$

ii.

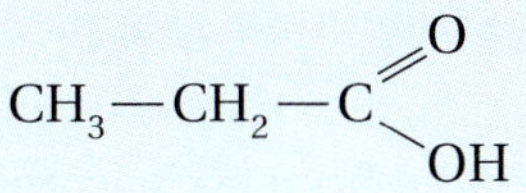

iii.

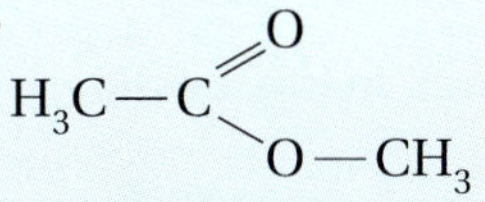

b. Students should circle the OH, COOH, and COO in the respective formulas.

c. Propanol, propanoic acid, methyl ethanoate (or ethanoic acid methyl ester)

15. What does the R stand for in ROH?

It is the rest of the molecule without the functional group.

16. Why is the word *condensation* used to describe the reaction that forms esters?

The H from the alcohol and the OH from the acid react to form water or condensation.

17. Draw the structural formula for the product of the following condensation reaction:

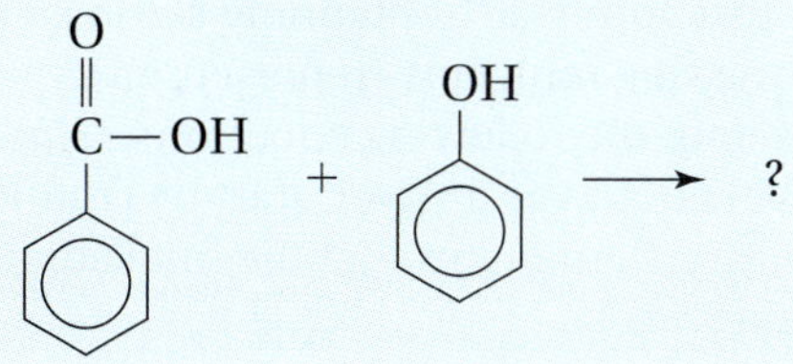

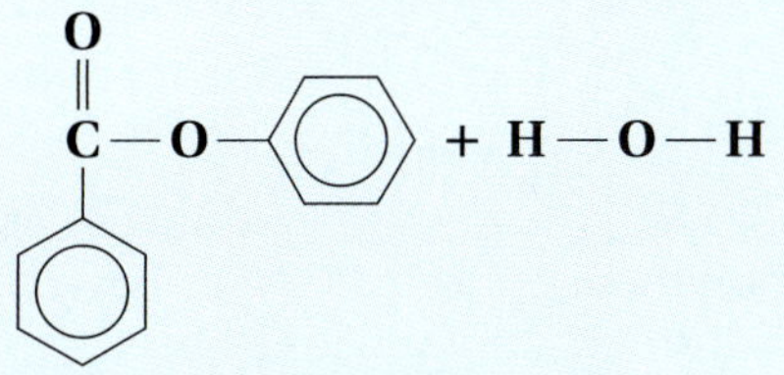

18. Acetic acid, CH_3COOH, and butyric acid, $CH_3(CH_2)_2COOH$, are common reactants in condensation reactions. Their structural formulas are shown below. Predict the products of a condensation reaction between each acid and each of the following alcohols by providing the name and structural formula for the product:

a. methanol (CH_3OH)

b. ethanol (C_2H_5OH)

c. propanol (C_3H_7OH).

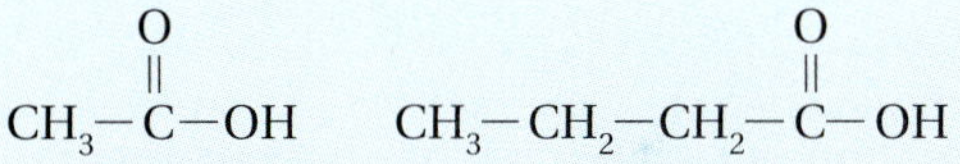

Acetic acid Butyric acid

For example, the product of a condensation reaction between acetic acid and ethanol is ethyl acetate.

Condensation products of Alcohols with Carboxylic Acids

Condensation Reactants	Acetic Acid CH_3COOH (Ethanoic Acid)	Butyric Acid (Butanoic Acid)
a. CH_3OH (methanol)	Methyl acetate, CH_3COOCH_3	Methyl butyrate, $C_3H_7COOCH_3$
b. C_2H_5OH (ethanol)	Ethyl acetate, $CH_3COOC_2H_5$	Ethyl butyrate, $C_3H_7COOC_2H_5$
c. C_3H_7OH (propanol)	Propyl acetate, $CH_3COOC_3H_7$	Propyl butyrate, $C_3H_7COOC_3H_7$

Connecting the Concept

19. Chemical synthesis is one of many branches of chemistry. Test your skill at planning syntheses by identifying the missing molecule (represented by a question mark) in each of the following equations. (*Hint:* If you are uncertain about the answer, start by completing an atom inventory. Remember that the final equation must be balanced.)

a. $CH_3{-}CH_3 + \textbf{?} \longrightarrow CH_3{-}CH_2Cl + HCl$

b.

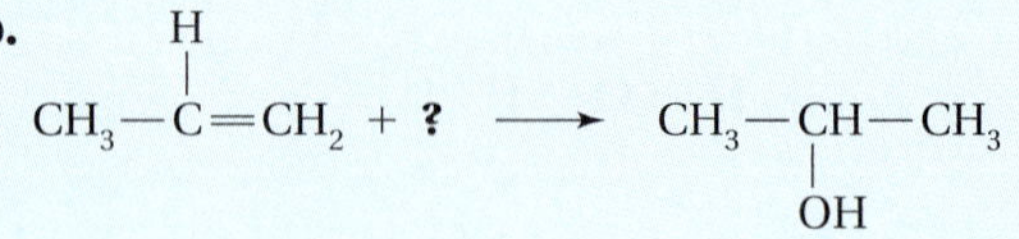

c.

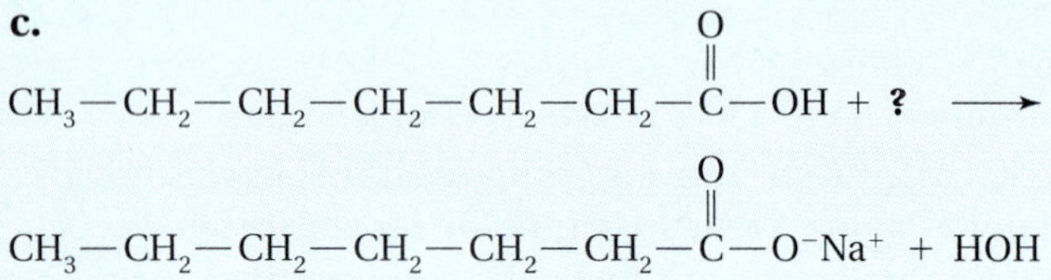

a. Cl_2 **b. HOH** **c. NaOH**

20. Using a molecular model to represent each molecule, explain the differences between benzene and cyclohexane.

Benzene is planar, whereas cyclohexane is bent.

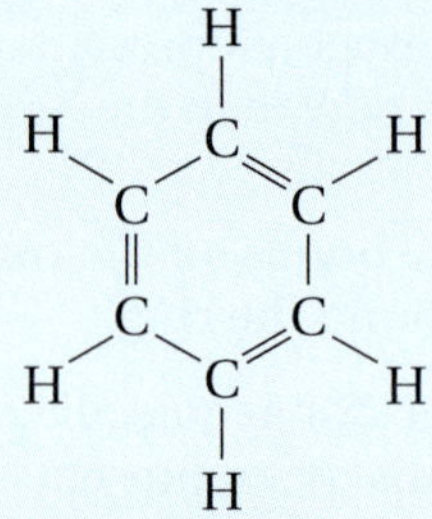

benzene

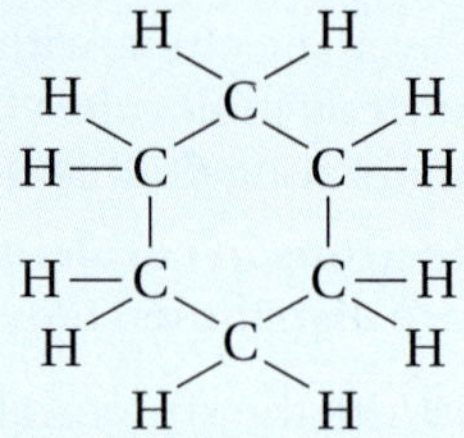

cyclohexane

21. There is a saying that "form follows function." How does this apply to the structure and properties of organic molecules?

The properties and use of an organic molecule depend on its structure and the functional groups in the molecule.

22. List five products manufactured from cyclic compounds.

Polystyrene, TNT (trinitrotoluene), nylon, kevlar, dacron, mylar, aspartame.

Extending the Concepts

23. Considering that petroleum is a natural resource, why aren't all petroleum-based polymers classified as natural?

The molecules found in these polymers have been altered chemically and are not found in nature.

24. Organic molecules that contain alcohol or carboxylic acid functional groups are often more soluble in water than similarly sized hydrocarbon molecules. Explain.

These functional groups make the molecule polar and will dissolve in polar molecules like water.

25. C_2H_6O has two isomers, one with a boiling point of 78 °C and the other with a boiling point of 224 °C.

a. Draw the structural formula for each isomer.

b. Which isomer has the higher boiling point? Explain your choice.

a.

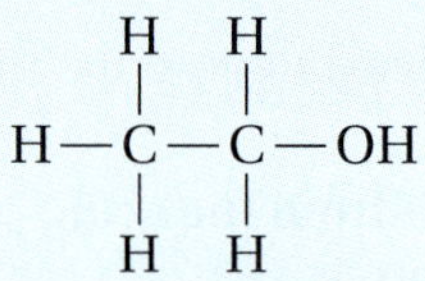

ethanol

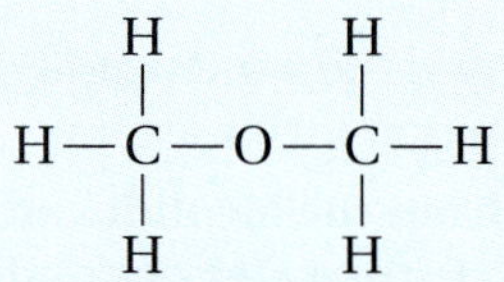

dimethyl ether

b. Ethanol would have a higher boiling point, due to the greater strength of its intermolecular bonds, whereas dimethyl ether is less polar than ethanol.

Some Important Commercial Polymers and Their Uses			
Polymer	**Repeating Unit(s) in Molecule**	**Monomer**	**Uses**
polyethylene	$-CH_2-CH_2-$	$CH_2{=}CH_2$	Films for packaging, molded toys, and housewares, plastic bottles, pipe and fittings
polyvinyl chloride	$-CH_2-CH(Cl)-$	$CH_2{=}CH(Cl)$	Pipe and fittings, film and sheeting (vinyl car tops), wire insulations, automotive parts, flooring, adhesives, coatings
polystyrene	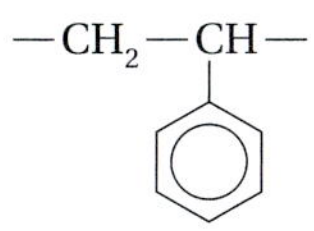	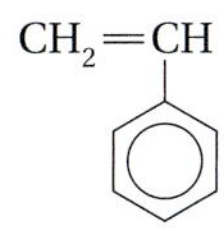	Packaging and containers, toys and recreational equipment, disposable food containers and utensils, brush handles, etc.
polypropylene	$-CH_2-CH(CH_3)-$	$CH_2{=}CH(CH_3)$	Fibers for carpeting, wool substitute, auto parts, rope, nets
polyacrylonitrile	$-CH_2-CH(CN)-$	$CH_2{=}CH(CN)$	Acrylic fiber, carpets, sweaters, single double knit clothing
polyester	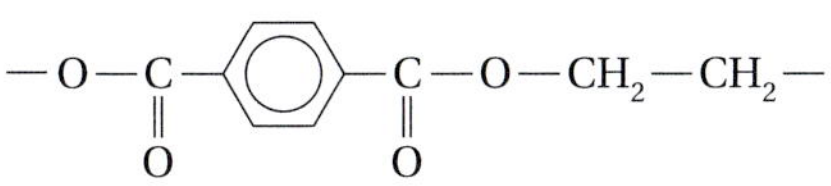	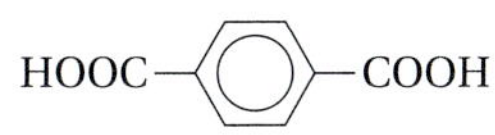 CH_2OH-CH_2OH	Textile fibers for clothing, sheets, curtains, towels, carpets
polyamide	$-C(=O)-(CH_2)_4-C(=O)-NH(CH_2)_6NH-$	$HOC(=O)(CH_2)_4C(=O)OH$; $H_2N(CH_2)_6NH_2$	Textile fibers, bristles, carpets, upholstery fabric, knitted goods, tire cord

ENERGY ALTERNATIVES TO PETROLEUM

Introduction

Section D explores various petroleum substitutes presently under consideration for use in place of the dwindling supply of nonrenewable fossil fuels. Alternative sources of fuel to power both commercial and personal vehicles are considered. Our reliance on petroleum-based products and our dependence upon petroleum as a major source of energy have demanded that we extend our research on how to supplement or replace petroleum as the primary fuel source. The promise of chemical technology is that economically and environmentally suitable alternatives to petroleum will be developed. Insights gained here lead to the culminating activity in Putting It All Together—the designing of a commercial for the EV-600 automobile.

Make the assignment for the commercial for the EV-600 vehicle, to allow time for library and/or Internet research.

D.1 ENERGY PAST, PRESENT, AND FUTURE

Earth's key sources of energy are explored, with the Sun as the primary source. Students should make the connection between the Sun and biomolecules. The growth of energy usage in our society has demanded changes in the kind of materials used as energy sources and the availability of the particular types of sources.

SECTION D ENERGY ALTERNATIVES TO PETROLEUM

Because petroleum is a nonrenewable resource, its total available inventory on Earth is finite. Thus other sources of energy must be found in order to meet the needs of modern society. In this section, you will explore several alternatives to petroleum and also learn about other ways to power personal and commercial vehicles.

The total energy used in the United States in one week is approximately 2×10^{18} J.

D.1 ENERGY: PAST, PRESENT, AND FUTURE

The Sun is our planet's primary energy source. Through photosynthesis, radiant energy from the Sun is stored as chemical energy. That is, green plants use the Sun's radiant energy to convert carbon dioxide and water into carbohydrates and oxygen gas. Thus green plants convert solar energy into chemical energy, which is stored in the bonds of carbohydrate molecules. When animals ingest and digest the plants, this chemical energy is released and used by the animals to form other organic molecules. Organic molecules found in plants and animals are called **biomolecules**.

Solar energy and energy stored in biomolecules are the key energy sources for life on Earth. Since the discovery of fire, human use of stored solar energy in wood, coal, and petroleum has been a major influence on civilization's development. In fact, the forms, availability, and cost of energy greatly influence how—and even where—people live.

In the past, abundant supplies of inexpensive energy were available. Until about 1850, wood, water, wind, and animal power satisfied the nation's slowly growing energy needs. Wood, then the predominant energy source, was readily available to most people, serving as an energy source for heating, cooking, and lighting. Water, wind, and animal power provided transportation and drove machinery and industrial processes. As demand for energy increased, primary fuel sources changed. Figure 22 illustrates how U.S. energy sources have changed since 1850. In the next activity, you will explore how energy supplies and fuel use have shifted in the United States during the past 150 years.

In 1850, the population of the United States was 23 million.

FUEL SOURCES OVER THE YEARS

Building Skills 8

As you can see from Figure 22, there has been a definite shift away from biomass (mainly wood, but also ethanol and waste) and toward fossil fuels—coal, petroleum, and natural gas. Use this figure to answer the questions that follow.

FUEL SOURCES OVER THE YEARS

Building Skills 8

In this activity, based on the graph in Figure 22 on page 235, students trace the growth of energy use and associate it with economic factors and the availability of particular types of energy sources. Note that basic graph-reading skills are practiced and enhanced in this activity.

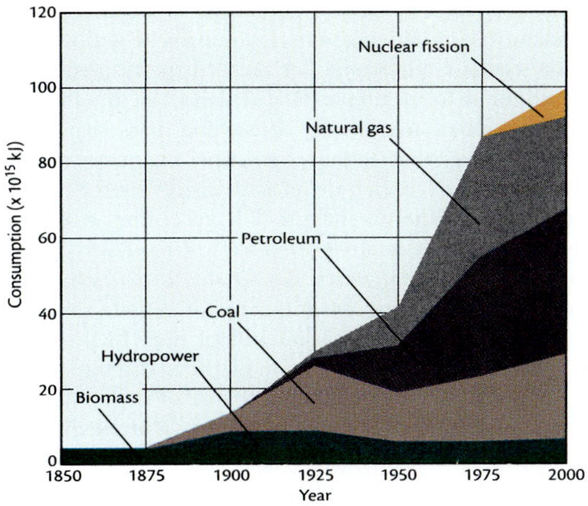

Figure 22: *Annual U.S. consumption of energy from various sources (1850–2000).*

1. a. Give the dates of the period during which biomass (mainly wood) supplied at least 50% of the nation's total energy.
   b. What were the chief modes of travel during that period?
   c. What factors might explain the decline in use of biomass after that period?
   d. What energy source increased in importance to replace biomass?
2. Compared with other energy sources, only a small quantity of petroleum was used as fuel before about the 1920s. What do you think petroleum's main uses might have been before that time?
3. Petroleum became increasingly important, and coal use began to decline, at about the same time.
   a. When did that occur?
   b. What could explain the growing use of petroleum after that date?
4. a. Which energy sources have become more important since 1975?
   b. What are the major uses of these energy sources?
5. a. Describe the trends in petroleum and coal use since 1975.
   b. What factors could explain these trends?

D.2 ALTERNATIVE ENERGY SOURCES

Everyday life in the United States requires considerable quantities of energy. As you just learned, the range of energy sources used in this country has indeed changed over time. As energy demands have accelerated, the nation has relied increasingly on nonrenewable fossil fuels—coal, petroleum, and natural gas. What is the future for fossil fuels, particularly petroleum?

Answers

1. a. Up to about 1910, wood supplied more than 50% of our energy needs.
   b. Trains were the chief form of long-distance transportation during this period, and horses were used for short distances.
   c. The decline in local supplies of wood and development of the internal combustion engine contributed to the demise of wood as a primary energy source.
   d. Coal was the next important energy source.
2. Oil burned in lamps was the most important use of petroleum prior to 1910.
3. a. Oil became an important energy source in the 1920s.
   b. Petroleum use increased as the number of automobiles increased, spurred by automobile assembly lines and mass production of cars.
4. a. The most recent energy source to enter the picture is nuclear energy.
   b. To generate electrical power.
5. a. Petroleum use continued to increase, although the rate of increase was somewhat less. This may be due to conservation efforts, including laws aimed at increasing the fuel efficiency of cars and trucks.
   b. More efficient automobiles, spurred by the oil embargo of the 1970s, and the increased use of coal and nuclear power to produce electricity.

D.2 ALTERNATIVE ENERGY SOURCES

This section investigates the possibility of using other options to supplement or replace petroleum as an energy source. The extraction of petroleum from shale, the liquification of coal, and "petroleum plantations" are some of the options presented to improve the use of petroleum.

Since all of the alternatives mentioned for building and burning petroleum have disadvantages, it is wise to consider conservation goals for extending the present supply.

Have students discuss the advantages and disadvantages of each alternative.

D.3 ALTERNATIVE-FUEL VEHICLES

Students examined alternatives to petroleum as a fuel in Section D.2. In this section, consideration is given to alternatives to the fuel used in vehicles. Factors that must be considered include overall cost of the vehicle, ease of refueling, emissions, and operational cost.

Have students evaluate each alternative fuel in a chart similar to the one below. This may require students to do some additional research on the Internet or using library resources.

The United States is a mobile society. More than 60% of U.S. petroleum is used for transportation. Although efforts to revitalize and improve public transportation systems merit attention, most experts predict our nation's citizens will continue to rely on personal vehicles well into the foreseeable future. And remember, even energy-conserving mass transit systems must have a fuel source. What options, then, does chemistry offer to extend, supplement, or even replace petroleum as an energy source?

Petroleum from tar sands and oil shale rock is an option with some promise. Major deposits of oil shale are located west of the Rocky Mountains. These rocks contain kerogen, which is partially formed oil. When the rocks are heated, kerogen decomposes into a material quite similar to crude oil. Unfortunately, vast quantities of sand or rock must be processed to recover this fuel. Moreover, current extraction methods use the equivalent of half a barrel of petroleum to produce one barrel of shale oil. Enormous amounts of water are also needed for processing—a problem where water is scarce.

A metric ton of oil shale typically contains the equivalent of 80 to 330 L of oil.

Because known coal reserves in the United States are much larger than known reserves of petroleum, another possible alternative to petroleum is a liquid fuel produced from coal. The technology for converting coal to liquid fuel (and also to builder molecules) has been available for decades, having been used in Germany more than 50 years ago. Current coal-to-liquid-fuel technology is well developed in the United States. However, the present cost of mining and converting coal to liquid fuel is considerably greater than that of producing the same quantity of fuel from petroleum. But if petroleum prices increase, obtaining liquid fuel from coal—itself a nonrenewable resource—may become a more attractive option.

Petroleum replacement candidates are not limited to other fossil fuels. Certain plants, including some 2000 varieties of the genus *Euphorbia*, capture and store solar energy as hydrocarbon compounds rather than as carbohydrates. These compounds may prove to be extractable and usable as a petroleum substitute. Other alternative energy sources currently in use or under investigation include hydropower (water power), nuclear fission and fusion, solar energy, wind energy, biomass, and geothermal energy. Alternate approaches include constructing more energy-efficient buildings, vehicles, and machines, as well as using alternative fuels. All of these initiatives are intended to further reduce the need to burn petroleum.

You have learned about some alternatives to using petroleum as a fuel source. In the next section, you will examine some specific fuel alternatives.

D.3 ALTERNATIVE-FUEL VEHICLES

As you now know, personal vehicles consume a significant portion of the petroleum burned for fuel. In recognition of the limited nature of petroleum as a resource and in consideration of the emissions produced by petroleum-burning engines, alternative-fuel vehicles are being developed, tested, and used. What are some of these fuels, and how are they used to propel vehicles? What are the advantages and disadvantages of the various

Alternative Vehicle	Fuel Type	Availability	Ease of Refueling	Efficiency	Environmental Impact (Emission)	Renewability
Compressed natural gas						
Electric cars						
Fuel cell						
Hybrid gasoline						

alternative fuels? The overview that follows will help you prepare your own automobile advertisement.

Compressed Natural Gas

Most passenger vehicles and buses can be converted to dual-fuel vehicles that run on either natural gas or gasoline. Natural gas, mainly methane (CH_4), is produced either from gas wells or during the processing of petroleum. Compressed and stored in high-pressure tanks, this product is commonly known as CNG (compressed natural gas). A refillable CNG tank, capable of powering an automobile up to 300 miles, can be comfortably installed in a car's trunk. Many CNG-powered vehicles are operating worldwide, particularly in government and mass transit fleets.

Among the advantages of CNG are wide availability and an 80% decrease (compared to gasoline) in carbon monoxide (CO) and nitrogen oxide (NO_x) emissions. However, refueling systems require a compressor, which increases the cost to $2000 to $4000 per vehicle.

Electric

Electric cars, including the fictitious ARL-600 vehicle featured in this unit, obtain their energy from a battery pack that is usually stored within the vehicle body and recharged with electricity at 120 or 240 volts. Powered by nickel-metal hydride (NiMH), lead-acid, or lithium-ion batteries, most electric vehicles can travel more than 100 miles on an eight-hour charge. The batteries provide energy to an electric motor, which turns the axle.

Figure 23 *An electric car being recharged.*

ADDITIONAL TEACHER RESOURCES

Synthetic fuels

To obtain fuel in a useful form requires a certain amount of energy for the equipment and for mining, pumping, and processing the fuel. Given present technology and oil prices, all the synthetic fuels have lower net energy values than the 50 to 1 ratio for domestic oil. This is the ratio of useful energy to the amount of energy required to produce the fuel.

Coal-derived fuels

Ordinary coal is composed primarily of carbon, with only about one hydrogen atom for every sixteen carbons. Because liquid and gaseous hydrocarbon fuels contain a much higher percent of hydrogen (methane, 4 H/1 C; octane, 2.25 H/1 C; heavy fuel oil, 1 H/6 C), temperature and pressure are used to break down the chemical structure of coal in order to increase the hydrogen to carbon ratio.

In World War II, when Germany was unable to import oil, coal-derived gasoline was used to fuel planes and tanks. More recently, South Africa's Sasol I plant has been producing oil from coal since 1960. The net energy value for coal-derived fuels is about 17:1.

Shale oil

Of the 1.8 trillion barrels of shale oil in Colorado, Wyoming, and Utah, about 600 billion are considered recoverable. Unfortunately, there are a number of problems with its extraction: (1) A great deal of energy is needed to extract kerogen, the oil-like component, while the high temperature required for processing the shale (approximately 500 °C) results in a net energy value of only about 6.5:1; (2) One to four barrels of water is needed to produce one barrel of oil; and (3) Much waste is generated (1.5 tons of shale generated per barrel of oil produced). Thus, at the present time, shale oil is expensive from both economic and environmental perspectives.

Alcohols

As partially oxidized hydrocarbons, alcohols have a lower energy content than an equal volume of gasoline. The chief advantages of using alcohols are that they are obtained from renewable resources and can be mixed with regular gasoline up to concentrations of 10 to 15%, without necessitating engine redesign. Methanol is made from wood, and, given methanol's high affinity for water, it is commonly used as a fuel-drying additive. Ethanol is made from grain (wheat or corn) and is less toxic and corrosive than methanol.

Although initially more expensive than gasoline-powered vehicles, electric vehicles do not require petroleum-based fuel, and their maintenance costs are lower. In addition, they do not produce any direct emissions. However, emissions may be produced some distance away at the electrical generating plant. And the limited lifetime of the batteries raises issues about recycling or disposal. The infrastructure for electric vehicles is largely in place in the form of the electrical power grid, but connections for recharging would need to be developed.

Fuel Cells

CD-ROM WWW.

Fuel Cell

An emerging option for providing electricity to power vehicles is the fuel cell. Although fuel cells were in use before 1840, they did not become a practical energy source until the 1960s, when they were used in the U.S. Space Program. Any fuel containing hydrogen (such as methanol or natural gas) can be used in a fuel cell, but only hydrogen (H_2) can be used directly.

As shown in Figure 24, one common form of fuel cell converts hydrogen fuel and oxygen gas from the air into electrical energy and water, which is its only emission. From a chemical viewpoint, such an operating fuel cell represents another way to release and harness the energy involved in "burning" hydrogen gas:

$$2\ H_2 + O_2 \rightarrow 2\ H_2O + \text{Electrical energy.}$$

The amount of hydroxide ions consumed at one electrode equals the amount produced at the other electrode. Overall, the concentration of hydroxide ions remains constant.

The fuel cell works by using porous electrodes that catalyze the reaction. In one common type of fuel cell, electrodes are in contact with a basic solution of potassium hydroxide, KOH(aq). At one electrode, hydrogen molecules (the fuel) react with hydroxide ions (OH^-), producing water and releasing electrons. The electrons flow from the fuel cell through an external circuit (where they do useful work), returning to the fuel cell at the other electrode. That second electrode catalyzes the reaction of oxygen gas (the oxidizer) with water and electrons to produce hydroxide ions, thus completing the electrical circuit. In the hydrogen-oxygen fuel cell, more water molecules are produced at one electrode than are consumed at the other.

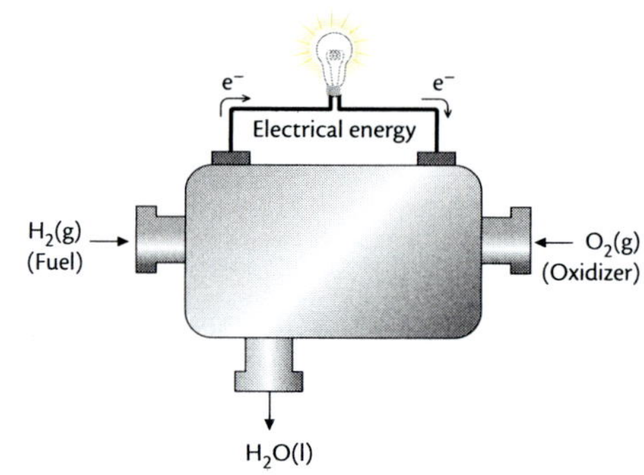

Figure 24 *The chemical reaction $2\ H_2 + O_2 \rightarrow 2\ H_2O$ + electrical energy takes place inside this fuel cell. See text.*

Fuel cells require no electrical recharging, eliminate or substantially reduce the release of air pollutants, and can obtain more useful power from a given quantity of fuel than can internal-combustion engines. However, challenges remain in developing fuel handling and processing options and reducing fuel-cell manufacturing and operating costs. The expense (in terms of high energy costs) required to obtain high-purity hydrogen (H_2) fuel for the type of fuel cell just described represents one of those key fuel-cell challenges.

Hybrid Gasoline-Electric

Some car designers believe that the hybrid gas-electric vehicle will best meet the needs of the consumer while reducing emissions and fuel costs. Hybrids come equipped with a small gasoline-burning engine as well as a battery-driven electric motor. The batteries are recharged while driving, partially through a conversion of braking friction into electricity. Hybrid vehicles typically achieve 70 miles per gallon and can travel more than 200 miles between fueling stops. Although hybrid vehicles produce the same kinds of emissions as fossil-fuel-burning vehicles, the quantity produced over a given distance is much smaller.

CD-ROM WWW. Questions and Answers

239

SECTION SUMMARY

Reviewing the Concepts

1. The supply of petroleum for both building and burning is limited. From the alternatives that have been discussed here, choose one for burning and one for building. Discuss the advantages and problems associated with the use of each substitute you have chosen.

For burning: Coal is readily available but burning it adds to pollution. Wood is readily available but requires special wood furnaces for heating homes. Natural gas is a good source of heat and clean burning, but it is expensive to build pipelines from the source.

For building: Liquid coal is a good source of carbon compounds but is expensive to process. Biomass requires land and other resources that may be expensive to obtain, but biomass can be a renewable source of useful hydrocarbons.

2. Some energy authorities recommend exploring ways to use more renewable energy sources, such as hydroelectric, solar, and wind power, as replacements for nonrenewable fossil fuels.

a. Why might this be a useful policy?

b. Which of these renewable sources is/are least likely to replace fossil fuels at this time? Why?

a. Substitutes will be in place as our sources of petroleum dwindle.

b. Hydroelectric, solar, or wind power would be least likely to be used, since they cannot be used directly for transportation fuels.

3. Consider coal, oil shale, and biomass as possible fuel substitutes for petroleum. Which do you think might play the most useful role in for the future? Explain.

Coal is in plentiful supply, but environmental problems are associated with its widespread use. The cost of converting coal to liquid fuel is currently greater than the cost of producing the same quantity of fuel from petroleum.

Petroleum can be extracted from oil shale and tar sand, the crude oil condensed from kerogen vapor. These shales contain a large supply of pre-oil. However, the technology for the process is not adequately developed, access to federally owned land containing the shale has not yet been secured, and the conversion processes require large amounts of water—already scarce in the western states where the oil shale and tar sands are found.

Petroleum can be extracted from some plants, implying the development of "petroleum plantations," yet larger-scale technology supporting this idea has not yet been demonstrated. Also, there would be the ethical question of whether it is justifiable to devote large tracts of land to "growing petroleum" when much of the world's population is starving.

4. **a.** Of the two major uses of petroleum—as a fuel and as a raw material—which one is likely to be curtailed first if petroleum supplies dwindle?

b. Give at least two reasons for your choice.

a. The burning of petroleum is most likely to be curtailed before building. The petrochemicals needed to produce detergents, drugs, plastics, and fibers will likely be given higher priority than certain uses of petroleum as fuel.

b. Burning uses much more petroleum. Since the total use for burning petroleum is such a huge number, even small improvements in efficiency or conservation yield large savings.

At least some of the building petroleum can be recycled, whereas the burning of petroleum results in a loss that cannot be recycled.

5. Although petroleum has been used for thousands of years, it is only in recent history that it has become a major energy source. List three technological factors that can explain this.

a. Increasing industrial technology that utilizes millions of machines and consumes large amounts of petroleum.

b. Increasing population and demand for electricity.

c. Mass production of the car by the invention of the automobile assembly line caused an increase in automobiles and fuel consumption.

6. List an advantage and a disadvantage of each of the following alternative power sources for vehicles.
a. compressed natural gas
b. electric
c. hydrogen fuel cell
d. hybrid gasoline-electric

	Advantage	Disadvantage
a. Compressed natural gas	**Reduces CO_2 emissions**	**Difficult to store fuel on vehicle**
b. Electric	**No direct emissions**	**Battery recycling**
c. Hydrogen fuel cell	**More efficient**	**Expensive**
d. Hybrid gasoline-electric	**Reduces emissions**	**Still not emission free**

7. Alternative-fuel vehicles are commercially available today. What factors might discourage people from purchasing them?

There are few refueling stations, and maintenance and repair may be difficult to obtain, since most repair shops are not equipped to deal with these vehicles.

The short range of these vehicles limits them to short trips before refueling.

Connecting the Concept

8. U.S. reserves of oil shale are approximately 87 quads, the equivalent of 150×10^9 barrels of oil. Suppose that this represented our total source of oil, from which we maintained production of 8 million barrels of oil daily.

a. How many years would this supply last?

b. The population of the United States is about 2.8×10^8 persons, and the United States uses about 24 barrels (about 1000 gallons) of oil per person per year. At that rate of consumption, how long would the oil shale reserves last?

c. Why is there a difference between the answers to Questions 8a and 8b?

a. 150×10^9 barrels $\times$ (1 day/8×10^6 barrels) $\times$ (1 yr/365 days) = 51.4 years.

b. 2.5×10^8 people $\times$ (24 barrels/person/year) = 6.0×10^9 barrels/yr.

oil shale = 150×10^9 barrels $\times$ (1 year/6.0×10^9 barrels) = 25 yr.

c. The potential production of petroleum from oil shale is not enough to keep up with current consumption.

9. We often consider alternative energy sources to be pollution free. Choose two alternative energy sources and evaluate them in terms of possible impact on the environment.

Alternate energy vehicles have much less of an impact on the environment than gasoline powered vehicles, since they produce less or no harmful emissions. However, the production of electricity to power the electric car produces harmful emissions at the electrical-generating plants, since many burn coal. Nuclear power plants produce radioactive waste that must be disposed of, as well as hot water, which can harm aquatic life.

10. What are some ways in which individuals could reduce their total consumption of petroleum products?

Increase fuel efficiency by keeping the car engine tuned up and by maintaining the proper tire pressure. Turn down the thermostat to 68° to lower fuel consumption, as well as to make sure the furnace is maintained for maximum efficiency. Switch off unnecessary electric lights. Promote research for "green car" technologies.

11. A friend claims that because she does not own a car, she does not consume any fossil fuels. Evaluate this claim.

Everyone who uses electricity originating from an oil- or coal-burning power plant consumes fossil fuels. Even without a car, one must use fossil fuels for heating or public transportation. The energy used to produce most consumer products, as well as food, requires fossil fuels for production, packaging, and transportation to markets. Fossil fuels serve as building materials for plastics, fabrics, dyes, drugs, building materials, and numerous other products.

Extending the Concepts

12. Describe several possible changes that would occur in your community if wind and solar power devices were installed on a large scale to provide power.

Large areas would be covered by numerous solar panels that rotate with the Sun and by large numbers of tall towers with windmills. The beauty of a rural landscape would be compromised. Placement of solar and wind generators may lead to political strife over who has access to the most desirable locations for generation of power.

13. How have the relatively low cost and easy availability of petroleum affected the search for and development of alternative energy sources?

These factors have slowed the research and development of alternative energy resources, because of a lack of support from the citizenry.

14. What characteristics of chemical substances do chemists seek in good petroleum substitutes? Why?

Desirable compounds should be readily available, with low expense in terms of energy to process the compounds, and with low environmental impact—no air or water pollution. The compounds should also not require the use of resources that meet other societal needs, such as food.

15. Using Internet or library resources, identify some state or federal programs that encourage development of alternative energy sources.

Through the Partnership for a New Generation of Vehicles, the federal government subsidizes participating companies that are developing "green car" technologies. The Department of Energy and its Office of Transportation have many programs and information concerning alternative fuel vehicles.

16. World experts disagree about how long fossil-fuel supplies will last. Research and evaluate some of the current opinions.

The world is not running out of oil, at least not yet. What our society does face in the near future is the end of abundant and cheap oil, on which all industrial nations depend. Before the next decade is over, the flood of conventional oil will crest, and production will enter a permanent decline.

More detailed information can be found in the March 1998 edition of *Scientific American*.

GETTING MOBILE

- This culminating activity has students design, present, and evaluate an advertisement using gasoline or one of the alternative fuels to power a vehicle.
- Encourage students to use technology in their presentations (computers, video, camcorders, and so forth).
- Commercials should reflect what students know about fuels.
- Allow time after each presentation for questions and answers.
- Commercial design should meet all of the criteria expressed in the students' textbook.

PUTTING IT ALL TOGETHER

Getting Mobile

If you are an average high school student in the United States, you have already viewed about a half-million television commercials, many of them for automobiles. As alternative-fuel vehicles become available, advertisements similar to the one that begins this unit may become more common.

To make informed, intelligent consumer decisions, it is important to analyze the information conveyed in such advertisements. You practiced this skill earlier in the unit as you evaluated the "petroleum-free" claim in the ARL-600 TV ad. Now you will further develop the skill by creating and defending your own product claims. You will write and produce your own automobile commercial message.

DESIGNING, PRESENTING, AND EVALUATING VEHICLE ADS

Each team of students will create and present an advertisement featuring an imaginary but plausible vehicle that uses a specific type of fuel. The fuels you can choose are electric, gasoline, hybrid gasoline-electric, or hydrogen in a fuel cell. Your vehicle must use only the selected fuel. Your commercial message will be presented to the class for analysis and comment based on concepts introduced and discussed thus far in this chemistry class.

Each commercial message must meet certain specifications. Use these specifications to guide development of your advertisement as well as evaluate the presentations of your classmates.

- **Time** The presentation of the message for your vehicle should take no more than 60 seconds of "air time."
- **Scientific Claims** All scientific claims must be accurate. Because the type of fuel (energy source) is the unique characteristic of your vehicle, you should highlight it and explain it briefly. If appropriate, compare your vehicle to those that depend solely on petroleum for fuel. Include the nature and implications of emissions released by your vehicle, as well as a reference to its fuel efficiency.
- **Comfort/Design/Safety Features** Special features can enhance customer appeal as well as challenge your design creativity. Invent a name and model for your vehicle, and give it one or more special features that you think are particularly significant.
- **Presentation** In addition to presenting a commercial message with accurate claims, your goal is to stimulate interest and vehicle sales. To accomplish this, your commercial presentation should be organized, visually stimulating, and motivating. Your script should be concise and to the point, while still presenting all necessary details.

LOOKING BACK AND LOOKING AHEAD

This unit has illustrated once again how chemical knowledge can inform personal and community decisions related to resource use and replacement. Thus far in your *ChemCom* studies, you have focused on three distinctive types of resources: water, minerals, and petroleum. The next unit explores the importance of another resouce—the thin, virtually invisible envelope of gases that surrounds Earth's surface. As you will discover, the properties and behavior of the atmosphere can be easily overlooked, but issues and concerns arising from human interaction with it merit attention.

LOOKING BACK AND LOOKING AHEAD

One of the primary objectives of *ChemCom* is to enable students to make wise personal and community decisions. This brief transitional section reflects on how the three types of natural resources—water, minerals, and petroleum—may influence their decisions. Each unit topic is vital to the survival of humans on Earth, as well as their comfort and well being. However, the next unit of study, "The Earth's Atmosphere," is of equal importance in helping them make wise decisions about environmental concerns and survival.

PLANNING GUIDE UNIT 4

Section	Time class	Student Edition Features	Program Resources	Media and Technology
Section A: Gases in the Atmosphere **Objectives:** • Earth's atmosphere is composed of a mixture of gases, primarily nitrogen, oxygen, and water vapor. • Pressure involves a force applied over a particular area. Air pressure is often measured in units of atmosphere (atm), pascals (Pa), or millimeters of mercury (mmHg). • The volume of a sample of gas in a flexible container decreases if external pressure increases. • Boyle's law describes this inverse relationship. • The volume of a sample of gas in a flexible container at constant pressure increases if its temperature is increased and decreases if its temperature is decreased. Charles' law describes this direct relationship. • The pressure exerted by a sample of gas in a rigid container increases if its temperature is increased and decreases if its temperature is decreased.	5 Class Periods	**Building Skills** #1. Graphing Atmospheric Data #2. Applications of Pressure #3. P-V Relationships #4. Charles' Law #5. T-V-P Relationships #6. Molar Volume and Reactions of Gases **Laboratory Activity** • A.1 Exploring Properties of Gases • A.6 Temperature-Volume Relationships **Modeling Matter** • Kinetic Molecular Theory	**Activity Book** • A.1 Exploring Properties of Gases • A.5 Boyle's Law: Supplement • A.6 Temperature-Volume Relationships • A.7 Charles' Law: Supplement • A.8 Temperature-Pressure Relationships: Supplement • A.9 Molar Volume: Supplement **Skill Handbook: Graphs** • XY-Plots	**ChemCom CD-ROM** • Unit IV: Section A **Transparencies** • Tropospheric Air • SI Units for Selected Physical Quantities and Pressure Exerted by a Brick • Boyle's Law • Temperature-Volume Measurements • Avogadro's Law

PLANNING GUIDE

Section	Time class	Student Edition Features	Program Resources	Media and Technology
• The kinetic molecular theory states that gases are composed of particles of negligible size that are in constant motion and engage in elastic collisions. The average kinetic of a gas sample is directly related to its temperature. • Equal numbers of gas molecules at the same temperature and pressure occupy the same volume. One mole of a gas at 0 °C and 1 atm pressure occupies a volume of 22.4 liters. • The coefficients in a balanced chemical equation involving gases indicate the relative volumes of gaseous reactants or products.				

PLANNING GUIDE

Section	Time class	Student Edition Features	Program Resources	Media and Technology
Section B: Radiation and Climate **Objectives** • Electromagnetic radiation includes x-rays, gamma rays, ultraviolet (UV), visible, and infrared (IR) radiation, radio waves, and microwaves. The energy transmitted by radiation varies according to its wavelength—the shorter the wavelength, the higher the energy. • Earth's atmosphere protects living organisms by absorbing and distributing solar energy. • Electromagnetic radiation can interact with matter to transfer energy.	5 Class Periods	**Building Skills** #7. Solar Radiation #8. Thermal Properties Of Materials #9. Trends in CO_2 Levels **ChemQuandary** #1. Always Harmful? #2. Grab Another Blanket! **Laboratory Activity** • B.4 Specific Heat Capacity • B.6 Carbon Dioxide Levels **Modeling Matter** • Incomplete Combustion **Making Decisions** • Planning for Greenhouse Gases	**Activity Book** • B.4 Specific Heat • B.6 Carbon Dioxide Levels **Skills Book: Graphs** • Additional Problems	**ChemCom CD-ROM** • Unit IV: Section B • Laboratory Videos **Transparencies** • The Parts of a Wave and the Solar Spectrum • Specific Heat Capacities of Common Materials • The Carbon Cycle
Section C: Acids in the Atmosphere **Objectives:** • Rain water is naturally acidic, but contaminants in the atmosphere can produce precipitation that is even more acidic than normal.	5 Class Periods	**Building Skills** #10. Acids and Bases #11. pH **Laboratory Activity** • C.2 Making Acid Rain • C.8 Buffers **Making Decisions** • C.3 Preventing Acid Rain • C.9 Lunar Habitation Module Prototype	**Activity Book** • C.2 Making Acid Rain • C.8 Buffers	**ChemCom CD-ROM** • Unit IV: Section C • Laboratory Videos **Transparencies** • Global Warming • SO_2 Emissions by Source and NO_2 Emissions by Source • Relationship among pH, Molar Concentration of H_30^+, and OH^-

PLANNING GUIDE

Section	Time class	Student Edition Features	Program Resources	Media and Technology
• Sulfur oxides and nitrogen oxides generated from natural and human sources contribute to acid rain. • Acids produce hydrogen (or hydronium) ions in water, while bases produce hydroxide ions. Strong acids and bases ionize completely; weak acids and bases ionize only partially. • Acidic solutions contain a higher concentration of hydrogen ions than hydroxide ions; basic solutions contain a higher concentration of hydroxide ions than hydrogen ions. Neutral solutions contain equal concentrations of hydrogen and hydroxide ions. • pH is a measure of the molar concentration of hydrogen ions in a solution. Solutions with pH 7.0 at room temperature are neutral, while those with a lower pH are acidic and those with a higher pH are basic. • Acidic precipitation can lower the pH of lakes and streams, which can adversely affect aquatic life.		**Modeling Matter** • Strong vs. Concentrated		

PLANNING GUIDE

Section	Time class	Student Edition Features	Program Resources	Media and Technology
• A buffered solution is capable of neutralizing limited amounts of either an acid or a base, thus resisting changes in pH.				
Section D: Air Pollution—Sources, Effects, and Solutions **Objectives** • Air pollution is a result of contributions from both primary and secondary sources. • Photochemical smog can intensify due to temperature inversions and adverse wind patterns. • Air pollution can be reduced by catalysts and other chemical technologies. • Chlorofluorocarbons (CFCs) can destroy stratosphere ozone.	5 Class Periods	**ChemQuandary** #3. Limiting Reactants and Scrubbers #4. Controlling Air Pollution **Laboratory Activity** • D.6 Cleansing Air **Making Decisions** • D.2 Identifying Major Air Contaminants • D.4 Vehicles and Smog	**Activity Book** • D.6 Cleansing Air **Skills Book: Equations and Stoichiometry** • Understanding Limiting Reactants	**ChemCom CD-ROM** • Unit IV: Section D • ChemCom Connections • Laboratory Video **Videodisk** • Controlling Air Pollution • The Air We Breathe • Monitoring Changes in the Earth's Atmosphere • Monitoring Air Quality **Transparencies** • Sources of U.S. Air Pollution • U.S. Pollutant Standards Index • A Temperature Inversion • Photochemical Smog Formation and Air Pollutants in the United States • Sulfur Dioxide Scrubber • Activation Energy Diagram • Stratospheric Ozone and Chlorine Monoxide Concentrations over Antarctica
Putting It All Together	2 Class Periods	**Putting It All Together** • Air Quality in the LHMP	**Black Line Copy Masters** • PIAT Evaluation	

D.6 CLEANSING AIR
Laboratory Activity

Part 1: The Electrostatic Precipitator

Materials for demonstration

100-mL glass graduated cylinder
18 gauge bare copper wire (exact gauge is not critical) (about 1 m)
1-hole rubber stopper to fit the graduated cylinder
wire cutters
transparent tape
Tesla coil
source of smoke (such as a 0.5 in^2 piece of heavy cardboard)

Safety

Teacher and students should wear goggles. Be aware of electric hazard due to high voltage. Keep the Tesla coil unplugged until ready for use and unplug immediately after use.

Procedure

1. Assemble the apparatus as shown in the diagram. If this demonstration is performed against a dark background, the reaction will be more visible.

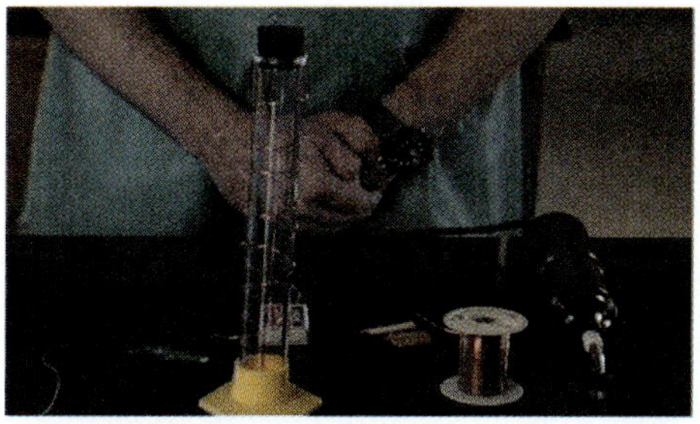

2. Insert the rubber stopper in a clean, dry graduated cylinder.
3. Wrap several turns of wire around the outside of the cylinder. Make a loop at the bottom end of the wire. Attach a wire from the loop to a water faucet to create a ground. **CAUTION:** *Be sure that the water supply line is metallic.*
4. Cut a straight piece of wire about 4 cm longer than the depth of the cylinder and make a small loop on one end. Insert the wire into the graduated cylinder through the stopper with the loop at the top. This wire should not touch the sides of the cylinder.
5. Remove the wire-stopper assembly, and ignite a small piece of heavy cardboard. Drop the burning cardboard into the cylinder and replace the wire-stopper assembly. (Again, be careful not to let the wire touch the sides of the cylinder.)
6. Plug in and turn up the Tesla coil.
7. Hold the Tesla coil as close as possible to the central wire in the cylinder, and allow a spark to jump to the top of the central wire. **CAUTION:** *Exercise extreme care with the Tesla coil.* Smoke should clear almost immediately.
8. Disassemble the apparatus and clean. The cylinder must be clean and dry before it can be used again.

Answers

1. The smoke disappears.
2. The copper rod produces a greenish color.

Part 2: Wet Scrubbing

The diagram in the text shows the flasks or bottles hooked up to a centralized vacuum system. The water-aspirator vacuum system also works if a centralized vacuum system is not available. The picture below shows a typical setup.

Materials for demonstration

1 pkg wide-range 1–12 (pH) paper
5 mL universal indicator solution
10 mL 12.0-M hydrochloric acid (concentrated HCl)
10 mL 14.8-M ammonium hydroxide (concentrated $NH_3(aq)$)
2 500-mL flasks
500-mL filter flask (500-mL flask can be used with 2-hole stopper). If using filter flask: 1-hole stopper to fit; 2 2-hole stoppers for other flasks
thistle tube (or long stem glass funnel)
2 straight pieces of glass tubing (to fit in first two flasks)
2 pieces rubber tubing; to connect first to second flask: 10 cm; to connect final flask to aspirator: 40 cm
aspirator adapters for sink faucet

Safety

Students should wear goggles throughout this activity.

Be sure there is adequate ventilation when using concentrated HCl and concentrated $NH_3(aq)$.

Wrap the flasks with clean tape, leaving space for viewing the contents.

Exercise caution when inserting the thistle tube into the stopper. Use glycerin to lubricate the stem, hold the stem close to the stopper, and protect your hand with a towel or multiple layers of paper toweling.

Procedure

1. Set up the wet-scrubbing apparatus as shown in the accompanying photo. Also refer to the diagram in the student text, Activity D.6.

2. Add 5–10 drops of universal indicator to each collecting bottle (or flask). Then fill each with water up to the neck, allowing just a small space for the inlet tube.
3. Place a moistened piece of pH paper between the stopper and glass rim so it hangs in the space between the stopper and the water level. (This will indicate the pH of the gases hovering above the water level.)
4. Add 10 mL of concentrated hydrochloric acid to the smoke-generating flask. Immediately stopper it.
5. Pour 10 mL of concentrated ammonium hydroxide through the thistle tube and immediately turn on the aspirator.
6. Direct students' attention to the color change in the water and pH paper for each flask.

SUPPLY LIST FOR UNIT 4

Expendable Items and Unusual Materials per 24 students working in pairs

Alphabetical by Item Name			
Quantity	Unit	Item	Sections Used
25	mL	Acetic acid, 0.1M (demo)	C.6
4	Each	Alka Seltzer tablet (demo)	C.2
25	mL	Aluminum nitrate, 0.1 M (demo)	C.2
12	Each	Aluminum soda can	A.1
25	mL	Ammonium acetate, 0.1 M (demo)	C.2
25	mL	Ammonium chloride, 0.1 M (demo)	C.2
10	mL	Ammonium hydroxide, 14.8 M concentrated	D.6
12	Each	Aspirator adapters for faucets	B.6
24	Each	Balloons, large	A.1
24	Each	Balloons, small	A.1
25	mL	Barium hydroxide, 0.1 M (demo)	C.6
1	250 mL	Beaker (demo)	C.6
10	50 mL	Beaker (demo)	C.6
12	Each	Beaker, 400 mL	B.4
1	Each	Beaker, 600 mL	A.6
6	mL	Bromothymol blue solution	B.6
12	Each	Candles, large	B.6
12	Each	Capillary tubes	A.6
25	mL	Citric acid, 0.1 M (demo)	C.6
12	Each	Coffee stirrers	C.8
1	Each	Conductivity apparatus (demo)	C.6
2	Each	Container, large glass or plastic	A.1
2	Meter	Copper wire, bare 18 gauge	D.6
500	mL	Corn oil, yellow	A.6
24	Each	Cups, Styrofoam with lids	B.4

SUPPLY LIST

Alphabetical by Item Name (continued)			
Quantity	Unit	Item	Sections Used
12	250 mL	Flask, Erlenmeyer	B.6
12	250 mL	Flask, filter	B.6
2	500 mL	Flask, filter	D.6
12	Each	Funnels, glass, 7-cm diameter	B.6
24	30 cm	Glass tube (90° bend)	B.6
2	Each	Glasses, drinking	A.1
1	Each	Graduated cylinder, 100-mL	D.6
12	Each	Hot plate	A.1, A.7, B.4
1	Package	Hydrion paper (pH paper)	C.2
12	mL	Hydrochloric acid, 0.1 M HCl	C.8
25	mL	Hydrochloric acid, 0.1 M (demo)	C.6
11	mL	Hydrochloric acid, 12 M concentrated	C.8, D.6
5	mL	Hydrochloric acid, 1 M HCl (demo)	C.2
25	mL	Hydrochloric acid, 6.0 M HCl	C.2
1	Bag	Ice, 5 pounds	A.1
1	Package	Litmus paper, blue	C.2
1	Package	Litmus paper, red	C.2
12	Cm	Magnesium ribbon	C.2
24	Each	Marble chips ($CaCO_3$)	C.2
1	Box	Matches	D.6
12	Each	Metal samples	A.6
1	Each	Meter stick	A.1
1	Roll	Paper towel	A.7, C.2
1	Each	Paramecium culture (demo)	C.2
12	Each	Paring knives	C.2

SUPPLY LIST

Alphabetical by Item Name (continued)			
Quantity	**Unit**	**Item**	**Sections Used**
1	Each	Pin	A.1
60	Each	Pipets, beral	C.2, C.8
1	Box	Plastic bags with sealing strips	C.2
1	Each	Plastic bottle, clear 1- or 2-liter	A.1
1	Each	Plastic bottle, clear 2- liter	A.1
12	Each	Plastic, 2-cm square	A.1
25	mL	Potassium hydroxide, 0.1 M (demo)	C.6
25	mL	Potassium phosphate, 0.1 M (demo)	C.2
2	Each	Red apples	C.2
24	Each	Rubber aspirator hose, 30 cm	B.6
1	Each	Rubber aspirator hose, 30 cm	D.6
24	Each	Rubber bands, orthodontist type, small	A.6
36	Each	Rubber connectors	B.6
1	Each	Rubber stopper, 1-hole to fit graduated cylinder	D.6
12	Each	Rubber stoppers, 1-hole to fit flask	B.6
12	Each	Rulers, centimeter	A.6
25	mL	Sodium bicarbonate, 0.1 M (demo)	C.2
25	mL	Sodium bisulfate, 0.1 M (demo)	C.2
25	mL	Sodium carbonate, 0.1 M (demo)	C.2
25	mL	Sodium chloride, 0.1 M (demo)	C.2
2.76	Grams	Sodium dihydrogen phosphate•H_2O	C.8
5.36	Grams	Sodium hydrogen phosphate•7 H_2O	C.8
25	mL	Sodium hydroxide, 0.1 M (demo)	C.6
1000	mL	Sodium hydroxide, 0.1 M NaOH	B.6, C.8
2	mL	Sodium hydroxide, 1 M NaOH	B.6

SUPPLY LIST

Alphabetical by Item Name (continued)			
Quantity	**Unit**	**Item**	**Sections Used**
25	mL	Sodium nitrate, 0.1 M (demo)	C.2
12	Grams	Sodium sulfite, Na_2SO_3	C.2
12	Each	Straws, drinking, straight	B.6
1	Each	Syringe, large plastic with cap	A.1
1	Roll	Tape, transparent	D.6
1	Each	Tesla coil	D.6
36	Each	Test tubes, 18×150 mm	A.1
12	Each	Thermometers, alcohol (0 °C to 150 °C)	A.6
1	Each	Thistle tube (or long stem funnel)	D.6
1	Box	Toothpicks	C.8
5	mL	Universal indicator	C.2, C.8
12	Each	Universal Indicator Color Chart	C.8
1	Liter	Water, distilled or deionized	C.2, C.8
12	Each	Wellplates, 24-well	C.8
1	Each	Wire cutters	D.6
1	Meter	Wire with alligator clips	D.6

SUPPLY LIST

By Section			
Quantity	**Unit**	**Item**	**Sections Used**
12	Each	Aluminum soda can	A.1
24	Each	Balloons, large	A.1
24	Each	Balloons, small	A.1
2	Each	Container, large glass or plastic	A.1
2	Each	Glasses, drinking	A.1
1	Bag (5-lb)	Ice	A.1
1	Each	Meter stick	A.1
1	Each	Pin	A.1
1	Each	Plastic bottle, clear, 1- or 2-liter	A.1
1	Each	Plastic bottle, clear, 2-liter	A.1
12	Each	Plastic, 2-cm square	A.1
1	Each	Syringe, large plastic with cap	A.1
36	Each	Test tubes (18 × 150 mm)	A.1
12	Each	Hot plate	A-1, A.7, B.4
1	Each	Beaker, 600 mL	A.6
12	Each	Capillary tubes	A.6
500	mL	Corn oil, yellow	A.6
12	Each	Metal samples	A.6
24	Each	Rubber bands, orthodontist type, small	A.6
12	Each	Rulers, centimeter	A.6
12	Each	Thermometers, alcohol (0 °C to 150 °C)	A.6
1	Roll	Paper towel	A.7, C.2

SUPPLY LIST

By Section (continued)			
Quantity	Unit	Item	Sections Used
12	Each	Beaker, 400-mL	B.4
24	Each	Cups, Styrofoam with lids	B.4
12	Each	Aspirator adapters for faucets	B.6
6	mL	Bromothymol blue solution	B.6
12	Each	Candles, large	B.6
12	250 mL	Flask, Erlenmeyer	B.6
12	250 mL	Flask, filter	B.6
12	Each	Funnels, glass, 7-cm diameter	B.6
24	30 cm	Glass tube (90° bend)	B.6
24	Each	Rubber aspirator hose, 30 cm	B.6
36	Each	Rubber connectors	B.6
12	Each	Rubber stoppers, 1-hole to fit flask	B.6
2	mL	Sodium hydroxide, 1 M NaOH	B.6
12	Each	Straws, drinking, straight	B.6
1000	mL	Sodium hydroxide, 0.1 M NaOH	B.6, C.8

SUPPLY LIST

By Section (continued)			
Quantity	Unit	Item	Sections Used
4	Each	Alka Seltzer tablet (demo)	C.2
25	mL	Aluminum nitrate, 0.1 M (demo)	C.2
25	mL	Aluminum acetate, 0.1 M (demo)	C.2
25	mL	Ammonium chloride, 0.1 M (demo)	C.2
1	Package	Hydrion paper (pH paper)	C.2
5	mL	Hydrochloric acid, 1 M HCl (demo)	C.2
25	mL	Hydrochloric acid, 6.0 M (demo)	C.2
1	Package	Litmus paper, blue	C.2
1	Package	Litmus paper, red	C.2
12	Cm	Magnesium ribbon	C.2
24	Each	Marble chips ($CaCO_3$)	C.2
1	Each	Paramecium culture (demo)	C.2
12	Each	Paring knives	C.2
1	Box	Plastic bags with sealing strips	C.2
25	mL	Potassium phosphate, 0.1 M (demo)	C.2
2	Each	Red apples	C.2
25	mL	Sodium bicarbonate, 0.1 M (demo)	C.2
25	mL	Sodium bisulfate, 0.1 M (demo)	C.2
25	mL	Sodium carbonate, 0.1 M (demo)	C.2
25	mL	Sodium chloride 0.1 M (demo)	C.2
25	mL	Sodium nitrate, 0.1 M (demo)	C.2
12	Grams	Sodium sulfite, Na_2SO_3	C.2
60	Each	Pipets, Beral	C.2, C.8
5	mL	Universal indicator	C.2, C.8
1	Liter	Water, distilled or deionized	C.2, C.8
25	mL	Acetic acid, 0.1 M (demo)	C.6

SUPPLY LIST

By Section (continued)			
Quantity	**Unit**	**Item**	**Sections Used**
25	mL	Barium hydroxide, 0.1 M (demo)	C.6
1	250 mL	Beaker (demo)	C.6
10	50 mL	Beaker (demo)	C.6
25	mL	Citric acid, 0.1 M (demo)	C.6
1	Each	Conductivity apparatus (demo)	C.6
1	Can	Drain cleaners (demo)	C.6
25	mL	Hydrochloric acid, 0.1 M (demo)	C.6
25	mL	Potassium hydroxide, 0.1 M (demo)	C.6
25	mL	Sodium hydroxide, 0.1 M (demo)	C.6
25	mL	Sulfuric acid, 0.1 M (demo)	C.6
12	Each	Coffee stirrers	C.8
12	mL	Hydrochloric acid, .1 M HCl	C.8
2.76	Grams	Sodium dihydrogen phosphate • H_2O	C.8
5.36	Grams	Sodium dihydrogen phosphate •7 H_2O	C.8
1	Box	Toothpicks	C.8
12	Each	Universal Indicator Color Chart	C.8
12	Each	Wellplates, 24-well	C.8

SUPPLY LIST

By Section (continued)			
Quantity	Unit	Item	Sections Used
10	mL	Ammonium hydroxide, 14.8 M concentrated	D.6
2	Meter	Copper wire, bare, 18 gauge	D.6
2	500 mL	Flask, filter	D.6
1	Each	Graduated cylinder, 100-mL	D.6
10	mL	Hydrochloric acid, 12 M concentrated	D.6
1	Box	Matches	D.6
1	Each	Rubber aspirator hose, 30 cm	D.6
1	Each	Rubber stopper, 1-hole to fit graduated cylinder	D.6
1	Roll	Tape, transparent	D.6
1	Each	Tesla coil	D.6
1	Each	Thistle tube (or long stem funnel)	D.6
1	Each	Wire cutters	D.6
1	Meter	Wire with alligator clips	D.6

UNIT 4
AIR QUALITY MANAGEMENT PLAN

Introduction to Unit

The Air Quality Management Plan for the Lunar Habitation Module Prototype (LHMP) Project serves as the organizing socio-technical issue for Unit 4. Student groups (4 to 5) should be assigned at the beginning of the unit with a tentative date of completion to allow time for research.

- Consideration should be given to the level of the students and to the type of schedule on which your school is operating (*e.g.*, block, semester, daily, and so forth). Student groups should be of varying academic abilities and motivational levels.
- Schedule class time at the end of the unit for oral presentations.
- To ensure that student groups are working throughout the unit, review interim reports at the end of each section. A check-off list (see end of unit) may help students organize their work and assure that they are covering all aspects of the project.
- Each team will prepare a proposal. Students should research how transportation, waste, energy, temperature, radiation, precipitation, humidity, pressure, and pollution might affect the inhabitants of the LHMP. This would include making some reasonable estimates of the types and amounts of gases involved. Once the teams have completed their preliminary proposals, each team should prepare an air quality proposal for presentation to the class.

UNIT 4

HOW does the composition of Earth's atmosphere affect its properties and behavior?

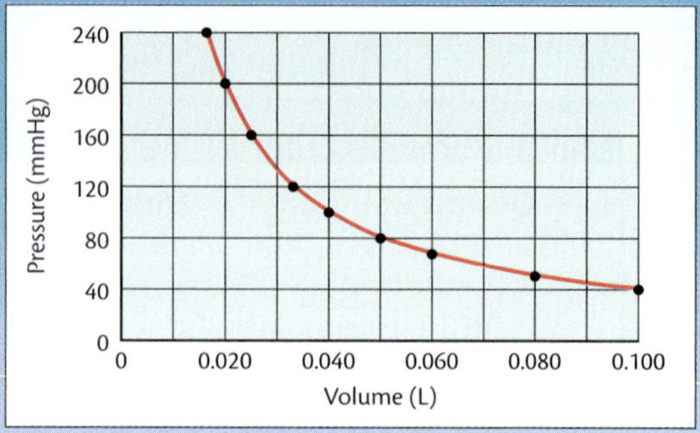

SECTION A Gases in the Atmosphere (page 248)

HOW does solar radiation influence conditions on Earth?

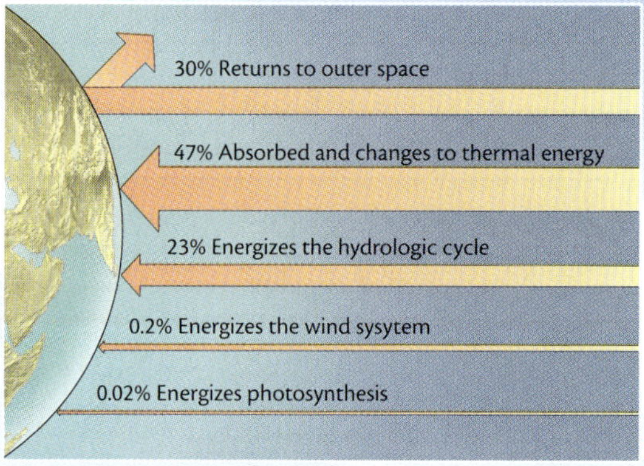

SECTION B Radiation and Climate (page 276)

AIR: CHEMISTRY AND THE ATMOSPHERE

WHAT are the major causes and consequences of acid rain?

SECTION C **Acids in the Atmosphere** (page 297)

HOW can air pollution be controlled?

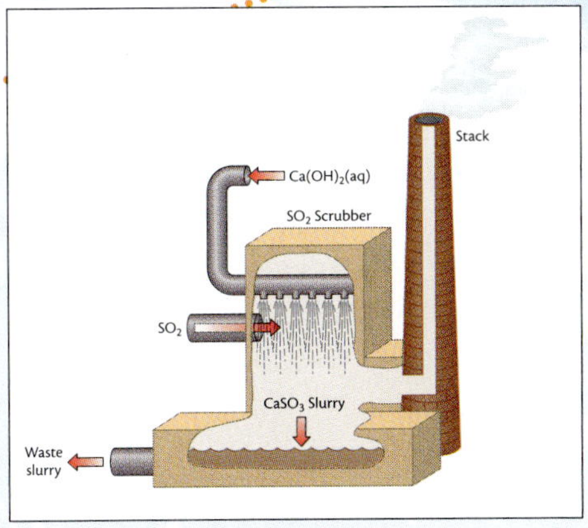

SECTION D **Air Pollution—Sources, Effects, and Solutions** (page 315)

You have just been invited to help design a self-contained module that may—at some future time—support human life on the Moon. What air-related features should the Earth-bound prototype incorporate? Turn the page to learn more about this challenging project.

GASES IN THE ATMOSPHERE

Introduction

The chemistry of the gases in the atmosphere is examined, with emphasis on the behavior of gases and the relationships among pressure, volume, and temperature. Some mathematical manipulation of the gas laws is included. Absolute zero and the kelvin temperature scale are introduced. Students will use the properties of atmospheric gases to determine how each will affect the living environment of the LHMP.

Opener CD-ROM WWW.

REQUEST FOR PROPOSALS

Air-Quality Management Plan for the Lunar Habitation Module Prototype (LHMP) Project

INTRODUCTION

The Lunar Habitation Association seeks proposals for air-quality management systems within its planned Lunar Habitation Module Prototype (LHMP) currently being prepared for construction.

The primary goal of the LHMP project is to develop on Earth's surface a self-contained human support system that later can be duplicated on the Moon's surface. As such, the system must be both self-contained and self-sustaining over extended time periods.

The LHMP will be constructed in a remote area in the southwestern United States. The domed structure will cover an area of 100 000 square meters with a maximum height of 50 meters. The LHMP will support a maximum of 50 people. The farthest point-to-point distance within the dome will be 350 meters. The dome will be constructed of a transparent polymer that allows transmission of visible and infrared radiation while reflecting almost all ultraviolet radiation. Within the dome will be an agricultural area, a residential zone, and a park area.

OBJECTIVES

Air-quality objectives for the LHMP focus on maintaining a suitable and livable atmosphere for human habitation by managing emissions of hydrocarbons, carbon dioxide, methane, volatile organics, particulates, ozone, and oxides of sulfur and nitrogen within the LHMP.

Proposals may address any of these issues related to an overall air-quality management plan:

- TRANSPORTATION: How does the movement of people and materials within the LHMP affect the quality of the air? What forms of transportation can be utilized to minimize these impacts?

- WASTE: How will the disposal, reuse, and/or recycling of human, food, and consumer wastes impact the air quality within the LHMP? What measures can be implemented to minimize these impacts?
- ENERGY: What impact will the production of energy sufficient to heat and power the LHMP and its life-support systems have on air quality within the LHMP? How can adverse effects be minimized?
- TEMPERATURE AND RADIATION: How will suitable temperatures be maintained within the LHMP? How will the effects of incoming radiation on the air and surfaces be controlled?
- PRESSURE: How will appropriate air pressure be sustained within the LHMP?
- PRECIPITATION AND HUMIDITY: How will an appropriate amount of precipitation (rain, mist, or fog) and an appropriate level of humidity be created within the LHMP?
- SMOG: What measures will be required to prevent or minimize the formation of photochemical smog within the LHMP?

PROPOSAL REQUIREMENTS

For each aspect addressed in the submitted plan, the proposal must include the following as appropriate:

- Identification of key air-quality issues and emissions
- Descriptions of the roles of the primary gases involved
- LHMP-monitoring schemes
- Recommended and alternative management options

FORM OF PROPOSAL

For each issue addressed in the submitted plan, a formal written proposal addressing all requirements must be submitted. Proposals should follow a scientific style of writing and should be double-spaced in a 12-point serif font.

Plans are being made to invite authors of selected proposals to deliver 15-minute presentations, including appropriate graphics and multimedia, to the Lunar Habitation Association.

A.1 EXPLORING PROPERTIES OF GASES

Laboratory Activity

Since atmospheric gases are colorless, odorless, and tasteless, students tend to think of them as "nothing." After completing this laboratory they should determine that gases, like solids and liquids, have definite physical properties.

You may want students to read through the procedures and write down their predictions prior to the laboratory session. This will reduce the time needed for completing the laboratory activity. Pre-lab procedures also assure that students will make their predictions before seeing other students working at the station.

The main focus of these activities is to raise questions and to motivate students to be more aware of the properties of gases.

Students may record their observations in a table similar to the table shown below.

Station Number	Prediction	Observation
↓	↓	↓

Time

50 minutes

Materials

18 medium to large balloons
Laboratory balance (accurate to .01 g)
Pin
Test tube
Meter stick
1-L beaker
Ring stand, clamp, wire gauze, and ring
Hot plate
Empty soft drink bottle
2 drinking glasses
Piece of plastic/Saran Wrap
Clean plastic bottle with lid (soda or detergent bottle)
Empty aluminum soft drink can
Large plastic or glass container

SECTION A GASES IN THE ATMOSPHERE

Gases in the Atmosphere (CD-ROM, WWW.)

You have been asked to generate an air-quality plan for the Lunar Habitation Module Prototype (LHMP). To do this, you will need to consider which gases (and how much of each) should be present and how each will affect the living environment in this closed system.

In addition to considering the specific gases that must be (as well as those that may be) inside the LHMP, it will be useful to know how these gases behave when there are changes in the conditions of temperature, volume, and pressure. As you know, solar radiation heats the Earth and everything on it, including atmospheric gases. How will radiation from the Sun affect a closed environment such as the LHMP? You will be able to answer this question and others by the end of Section A.

A.1 EXPLORING PROPERTIES OF GASES

Laboratory Activity

Introduction

Lab Video (CD-ROM, WWW.)

Because atmospheric gases are generally colorless, odorless, and tasteless, you might doubt that they are forms of matter; they appear to be "nothing." However, gases have definite physical and chemical properties, just as materials in the other two states of matter—solids and liquids—do. In this laboratory activity, you will perform a variety of experiments that illustrate some properties of air.

Before you start this activity, carefully read through the procedure. Decide what you think will happen at each of the nine laboratory stations, and write down your predictions.

Procedure

Nine stations have been set up around the laboratory. At each station you will perform the experiment indicated. The experiments can be done in any order. For each station:

- Reread the procedure.
- Review your prediction.
- Perform the experiment.
- Record your observations.
- Restore the station to its original condition.

When you have completed your work at all the stations, answer the questions at the end of the activity.

Station 1

1. Inflate a balloon and tie the end.
2. Place the inflated balloon on a balance, using a piece of tape to hold it in place.
3. Record the mass.
4. Remove the balloon from the balance. Use a pin to gently puncture the balloon near the neck and release most of the gas contained in it.
5. Place the deflated balloon on a balance (with the tape still attached) and record its mass.

Station 2

1. With its open end facing downward, lower an empty drinking glass into a larger container of water.
2. With the open end still under water, slowly tilt the glass.

Station 3

1. As shown in Figure 1, insert the rounded end of an unused balloon part way into an empty soft-drink bottle, stretching the balloon's neck over the mouth of the bottle.
2. Blow up the balloon so that it fills the bottle.
3. Remove and discard the used balloon.

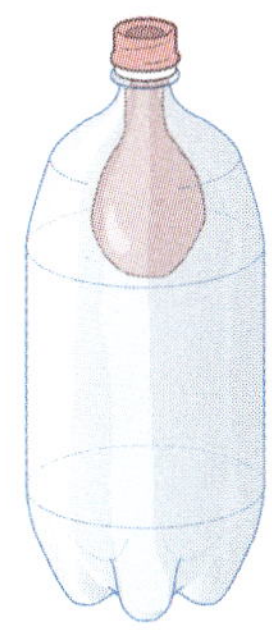

Figure 1 *Placement of balloon in bottle (Station 3).*

Station 4

1. Fill a test tube to the rim with water.
2. Cover the test tube rim with a piece of plastic.
3. Press down on the plastic to make a tight seal with the mouth of the test tube.
4. While continuing to press the plastic to the test tube, invert the test tube above a sink or a pan.
5. Without causing any jarring, gently remove your fingers from the piece of plastic.
6. Repeat the process with the test tube half-full of water.

Station 5

1. Locate the plastic bottle with a small hole in its side.
2. Cover the hole in the side of the bottle with your finger.
3. Fill the bottle with water.
4. Replace the cap tightly.
5. Holding the bottle over a sink, remove your finger from the hole.
6. Still holding the bottle over a sink, remove the cap.

Observations

Station 1: *Air has weight.* The balance should show a difference in mass of about 0.2 g between the filled balloon and the empty balloon. This demonstrates that seemingly "invisible" gases do possess weight, like other forms of matter. *Note:* The measurement does not allow for the buoyancy of the surrounding air.

Station 2: *Air occupies space.* The observation that the water does not enter the glass shows that air does take up space. When the glass is tilted, some air is allowed to escape and rises, indicating that air is less dense than the water.

Station 3: *Air has pressure and exerts pressure on both sides of a diaphragm.* Select a balloon that is easy to blow up. The class should notice that the balloon can be slightly blown up inside the bottle; however, it gets increasingly more difficult to blow up as it gets larger. Ask the class to explain the difficulty of this task.

Station 4: *Air exerts pressure in all directions.* Air pressure (acting upward against the covering) will support the weight of the water and prevent it from spilling out of the container. Repeat the process without filling the glass completely. Ask your students to account for the difference in the two cases.

Station 5: *Air exerts pressure in all directions.* The external pressure of the atmosphere prevents the water from coming out.

Pre-lab discussion

As students do the activity, ask them to

1. try to predict the outcome of the activity.
2. carefully observe the activity and, if necessary, account for differences between their predictions and the actual outcome.
3. try to identify the specific gas property that is being demonstrated, and why gases exhibit this property.

Station 6: *Pressure decreases as volume decreases. Pressure affects volume as well.* Remind students to be careful, as the can may be quite hot. The boiling process will force out much of the air that is inside the can and replace it with water vapor. Upon cooling, the water vapor will condense, leaving a partial vacuum inside the can. Atmospheric pressure outside the can will crush it.

Station 7: *Pressure is equal and opposite.* The water remains in the test tube as long as the mouth of the tube remains under water. With the tube half full of water students should be able to observe slight changes in the liquid level 0inside the tube when the tube is fully immersed and when it's lifted almost out of the water.

Station 8: *Pressure and volume are inversely related. As volume decreases, pressure increases. This is a demonstration of Boyle's Law. It also demonstrates that gases are compressible.*

Station 9: *Temperature and volume are directly related. This is a demonstration of Charles' law.*

Answers

1. # 1 and # 2 (air has both weight and volume).
2. #s 3, 4, 5, 6, and 7.
3. Student answers will be based on what they expected and what was observed.
4. a. Refer to the lab notes above.

   b. At all stations the role of air is to demonstrate the characteristic properties of gases.

   c. Particle models should all show air particles farther apart than particles of solid or liquid material in the experiment. The notion of the particles in each experiment should account for the pressure, volume, and temperature of the gas.
5. a. and b. Students' answers will vary. Some experiences that explain that air is matter and exerts pressure might be that of breath fogging, of leaning into the wind, or of headwind for a car or airplane.
6. a. When one draws on a single straw, pressure is decreased in that straw. The atmospheric pressure on the surface of the fluid remains the same. It is this difference in pressure that allows the fluid to flow upward. If two straws are used there is no pressure difference, and therefore the fluid does not rise. (If resourceful students use their tongues to close off the end of the second straw, they can accomplish the task.)

Figure 2 *Quickly invert the can containing boiling water into a container of ice water (Station 6).*

Station 6

1. Place about 10 mL of water in a clean, empty aluminum soft-drink can.
2. Place the can on a hot plate, and bring the water to a rapid boil.
3. Using tongs to handle the can, quickly remove it from the heat and immediately invert it into a container of ice-cold water. See Figure 2.

Station 7

1. Fill a test tube to the rim with water.
2. Cover the test tube opening with a piece of plastic.
3. While continuing to press the plastic to the test tube, invert the test tube and partially immerse it in a container of water.
4. Remove the piece of plastic.
5. Move the test tube up and down, keeping the opening of the test tube under water.
6. Repeat the process with the test tube half-full of water.

Station 8

1. Draw air into a syringe.
2. Seal the tip by placing the cap on the end.
3. Holding the cap in place, gently push the plunger down with your thumb.
4. Release the plunger.

Station 9

1. Inflate and tie off two unused balloons so that they are the same size—about the diameter of a small grapefruit.
2. Use tongs to submerge one inflated balloon in ice water.
3. Use tongs to submerge the other inflated balloon in hot tap water.

Questions

1. Which experiments are useful in demonstrating that air is matter? Explain.
2. Which experiments are useful in demonstrating that air exerts pressure? Explain.
3. Explain any differences between your predictions and the actual outcomes of the experiments.
4. For any two of the stations at which you performed experiments,
   a. describe your observations in detail.
   b. explain the role of air in the experiment.
   c. draw particle models showing the interactions between the gas particles in air and the other particles of matter in the experiment.

5. Describe an additional activity or experience you have had that demonstrates that
   a. air is matter.
   b. air exerts pressure.
6. Perform the following activity at home (or in the lunchroom at school): Put one end of a straw in a glass of water. Hold another straw outside the glass. Place the ends of both straws in your mouth and try to drink the water through the straw in the glass. See Figure 3.
   a. Describe what happens.
   b. Based on your observations, what makes it possible to drink liquid through a straw?

Figure 3 *What will happen when you try to drink water through the straw inside the glass?*

A.2 STRUCTURE OF THE ATMOSPHERE

Most of the atmosphere's mass and all of its weather are within 10 to 15 km of Earth's surface. This region is called the troposphere. Gases mix continuously in the troposphere, leading to a reasonably uniform composition around the world. The table in Figure 4 lists the components of tropospheric air. Chemical analysis shows that ancient air trapped in glacial ice has about the same chemical makeup as the current atmosphere, suggesting that there has been little change over very long periods of time.

As you know, air is a mixture of gases. Nitrogen is the most abundant gas, followed by oxygen, argon, and carbon dioxide. In addition to the gases listed in Figure 4, air samples can contain up to 5% water vapor, although in

The word *troposphere* comes from the Greek words *tropos*, which means turn, and *sphaira*, which means ball.

CD-ROM WWW. **Structure of the Atmosphere**

Tropospheric Air

Substance	Formula	Percent of Gas Molecules
Major components		
Nitrogen	N_2	78.08
Oxygen	O_2	20.95
Minor components		
Argon	Ar	0.93
Carbon dioxide	CO_2	0.033
Trace components		
Neon	Ne	0.0018
Ammonia	NH_3	0.0010
Helium	He	0.0005
Methane	CH_4	0.0002
Krypton	Kr	0.0001

Figure 4 *Gaseous components of dry tropospheric air.*

b. It is possible to drink liquid through a straw because of the difference in pressure inside and outside the straw.

Note: At this point, students are not expected to have correct molecular models for their observations. Question 4 serves as a starting point for students. Over the course of learning kinetic molecular theory and gas laws, students will develop a better understanding of these intermolecular interactions. At this point, they will be able to evaluate their ideas and think about the properties of gases.

A.2 STRUCTURE OF THE ATMOSPHERE

This section examines the composition of the first 10 to 12 km of the Earth's atmosphere. The chemical composition of the troposphere has remained constant over time. Students are asked to take an imaginary flight upward and to predict atmospheric conditions at various altitudes, which leads to the data used in the Building Skills activity that follows. (See *ChemCom* CD-ROM.)

A brief discussion of ppm and ppb may be useful at this point to help students visualize the minute concentrations of some components in air.

GRAPHING ATMOSPHERIC DATA
Building Skills 1

Encourage students to use their best techniques in constructing their graphs and to draw the best-fit lines. (Graphs may be computer generated.) Advise students that the data in Figure 5 do not include a warming trend that begins at 83 km, the start of the thermosphere, and goes out to 700 km.

1. Students' predictions will vary.
2. Students' graphs will vary.
3. a. Temperature decreases in a linear fashion to 12 km, increases linearly to 50 km, decreases again linearly to 80 km.

   b. Pressure decreases in a more consistent and regular pattern, following an exponential curve.
4. a. Fall. An increase in altitude means fewer molecules in a given volume of air, which will reduce the pressure.

   b. Rise. There are more molecules at the lower altitude, and a greater mass of air above; therefore there is more pressure.
5. a. i. Mass would decrease as altitude increased.

   ii. The number of molecules would decrease with increasing altitude.

   b. Direct relationship. Straight line.

   c. Mass depends only on the number of molecules. The relative composition of the atmosphere is fairly constant with altitude.
6. Students should identify on their graphs that the atmospheric layers have about the following thickness: troposphere 0–12 km, stratosphere 12–50 km, mesosphere 50–83 km, and thermosphere 83–700 km.

The amount of water vapor in air determines the humidity.

The only component of the atmosphere visible from space is condensed water vapor in the form of clouds.

most locations the range is 1–3%. Other gases naturally present in air at concentrations below 0.0001% (1 ppm) include hydrogen (H_2), xenon (Xe), ozone (O_3), nitrogen oxides (NO and NO_2), carbon monoxide (CO), and sulfur dioxide (SO_2). Human activity and natural phenomena such as volcanic eruptions can alter the concentrations of some trace gases and add other substances to air. This may lead to decreased air quality, as you will learn later in this unit.

Suppose it were possible for you to fly from Earth's surface to the farthest regions of the atmosphere. What do you think you would encounter as you traveled 5, 10, or 50 kilometers away from Earth? Would the air temperature change? Would you be surrounded by the same mixture of molecules as your altitude increased? Although you cannot take a trip such as this, you can find the answers to these and other questions about the atmosphere. Technology allows scientists to measure a range of air characteristics at different altitudes. In the following activity, you will analyze this type of data.

GRAPHING ATMOSPHERIC DATA
Building Skills 1

Figure 5 summarizes atmospheric data gathered at various altitudes. Use this information to answer the following questions.

1. Predict the shape of the graph line for a plot of
   a. temperature versus altitude.
   b. pressure versus altitude.

The unit mmHg—millimeters of mercury—is a common unit of gas pressure.

Atmospheric Data

Altitude (km)	Temp. (°C)	Pressure (mmHg)	Mass (g) of 1-L Sample	Total Molecules in 1-L Sample
0	20	760	1.20	250×10^{20}
5	−12	407	0.73	150×10^{20}
10	−45	218	0.41	90×10^{20}
12	−60	170	0.37	77×10^{20}
20	−53	62	0.13	27×10^{20}
30	−38	18	0.035	7×10^{20}
40	−18	5.1	0.009	2×10^{20}
50	2	1.5	0.003	0.5×10^{20}
60	−26	0.42	0.0007	0.2×10^{20}
80	−87	0.03	0.00007	0.02×10^{20}

Figure 5 *Atmospheric data at various altitudes.*

2. Prepare graphs according to the instructions below.
   a. For the first graph, plot temperature versus altitude data.
   b. For the second graph, plot pressure versus altitude data.
   c. Draw a best-fit line through the points of each graph. (Note that the line may be straight or curved.)
   d. Does the shape of either graph differ from what you predicted in Question 1? If so, how?
3. Compare the ways in which air temperature and air pressure change with increasing altitude.
   a. Which follows a more regular pattern?
   b. Try to explain this behavior.
4. Based on the graphed data, would you expect air pressure to rise or fall if you traveled from sea level (0 km) to
   a. Pike's Peak (4301 m above sea level)?
   b. Death Valley (86 m below sea level)?
5. a. Suppose you took one-liter samples of air at several altitudes. How would the following change?
      i. mass of the air sample
      ii. number of molecules in the air sample
   b. If you were to plot those two values (mass versus number of molecules) on a new graph, what would the plotted line look like?
   c. Why?
6. Scientists often characterize the atmosphere as having four layers: troposphere (nearest Earth), stratosphere, mesosphere, and thermosphere (outermost layer). Mark both graphs from Question 2 with lines at appropriate altitudes to indicate where you think the boundaries between the layers might be.

The term "pressure" has been used often in the previous pages. Although it is a common word used in daily conversation, what do scientists mean when they refer to pressure?

A.3 PRESSURE

CD-ROM WWW. Properties of Gases

In everyday language, the word "pressure" can have many meanings. Perhaps you use it to mean that you feel too busy or that you feel forced to behave in certain ways. The greater the pressure, the more "boxed in" you feel. To scientists, pressure also refers to force and space, but in quite different ways.

In science, pressure refers to the force applied to one unit of surface area:

$$\text{Pressure} = \frac{\text{Force}}{\text{Area}}$$

You can see from the equation that pressure is directly proportional to force and inversely proportional to area.

A.3 PRESSURE

The scientific meaning of "pressure" is introduced by describing several common examples. The relationship among pressure, force, and area is given by a mathematical equation. Make sure that students understand the difference between direct and inverse relationships.

A brief discussion of base and derived metric units may be beneficial to the student. Derived units compare two quantities, such as mass and volume (density) or distance and time (speed or velocity).

The SI units of both force and pressure are introduced. Atmospheric pressure data are given in both English and SI units, although students should use the pascal in their calculations.
1 atm = 760 torr = 760 mmHg = 29.92 in Hg = 101.3 kPa

To place this in a familiar context, think back to a common human experience: Someone steps on your foot. The pressure you feel depends on the two variables in the equation above—force and area. You will surely feel a difference in force between a 120-pound person stepping on your foot and a 180-pound person—assuming, of course, that both people have the same foot size and are wearing the same type of shoes. The larger the force (in this case, weight), the more pressure is exerted.

In the same way that a person exerts pressure on your foot, gas molecules exert pressure on their surroundings. You observed this in several experiments in Laboratory Activity A.1. See pages 248–250. For instance, at Station 4 you concluded that air pressure holds the piece of plastic against the water in an inverted test tube. How much pressure was the air exerting, and how could you measure it?

Many tires now also specify pressure in kilopascals (kPa).

You are probably familiar with several different units used to report pressure. For example, weather forecasters report barometric air pressure in inches of mercury. When you check air pressure in car or bicycle tires, most likely the gauge reads *p*ounds per *s*quare *i*nch, or *psi.* Other pressure units you may have already heard of are millimeters of mercury (mmHg), atmospheres (atm), and torr. With so many different units available for measuring and reporting pressure, how do scientists communicate with one another?

This issue of standardizing units of measurement is not a new one for scientists. Over the years, the scientific community has established standards upon which units are based. These include units of temperature, length, time, and mass. For example, the standard for length (the meter) is based on the distance traveled by light in a vacuum during 1/299 792 458 of a second.

The abbreviation SI comes from Système International de Unités, its French name.

Scientists have also agreed to use certain units when communicating results. This system, called the International System of Units, or SI, allows scientists from around the world to communicate with each other. Figure 6 lists some SI units. Note that some—such as mass and length—are **base units;** they express the fundamental physical quantities of the modernized metric system. Others, such as pressure, are **derived units**—they are formed by mathematically combining two or more base units.

Recall that pressure can be expressed as force per area. It is useful to understand the SI units of force and area before learning about the SI derived unit for pressure. The SI derived unit for force is the newton (N). To visualize a newton, imagine holding a personal-sized bar of soap (with a mass slightly greater than 100 g) in your open hand. The force that the bar of soap would exert downward on your hand is about one newton (1 N). The force (or weight) of any object depends on its mass and gravitational attraction. Thus an object will weigh much less on the Moon than on Earth.

The SI base unit of length is the meter; thus area is expressed as a derived unit—square meters (m^2). When a force of one newton is exerted over an area of one square meter, the pressure is equal to one pascal (Pa). The pascal is the SI derived unit of pressure. A pressure of one pascal is a relatively small value—it is roughly the pressure exerted downward on a

SI Units for Selected Physical Quantities		
Base Quantities	**Name**	**Symbol**
Length	meter	m
Mass	kilogram	kg
Time	second	s
Electric current	ampere	A
Thermodynamic temperature	kelvin	K
Amount of substance	mole	mol
Luminous intensity	candela	cd
Some Derived Quantities		
Area	square meter	m^2
	square centimeter	cm^2
Volume	cubic meter	m^3
	cubic centimeter	cm^3
Force	newton	N
Pressure	pascal	Pa

Figure 6 *SI units for selected physical quantities.*

slice of bread by a thin layer of butter spread on its top surface. Since a pascal is so small, the kilopascal (kPa), a unit a thousand times larger, is commonly used to express many pressures.

APPLICATIONS OF PRESSURE

Building Skills 2

Using what you have learned about pressure, answer these questions.

1. Figure 7 shows two bricks lying on the ground.
   a. Is each brick exerting the same force on the ground? Explain.
   b. Is each brick exerting the same pressure on the ground? Explain.

Figure 7 *Two similar bricks lie on the ground. Are they exerting the same pressure?*

APPLICATION OF PRESSURE

Building Skills 2

Repeating the demonstration of Section A.1 Station 7 at this time should reinforce the concept. Use test tubes of different height and/or an eudiometer tube to emphasize the importance of the height of the water column.

Answers

1. a. The force is the same in both cases, but the area over which the force is exerted is different.

   b. More pressure is exerted by the brick on its end, since it has less surface area in contact with the floor.

2. Pressure = force/area = 18 N/(.093 m × 0.055 m) = 18 N/0.005 m^2 = 3600 Pa

3. a. An axe.

   b. The cutting edge of an axe has a smaller area. Pressure is equal to force/area, therefore the smaller the area, the greater the pressure when the same force is used on each tool to chop the wood.

2. Calculate the pressure, in pascals, that the brick on the right side of the photo is exerting on the ground. It is exerting a force of 18 N, and the dimensions of the side in contact with the ground are 9.3 cm x 5.5 cm. (Remember, there are 100 cm in 1 m.)
3. You need to chop some wood to build a fire.
   a. Which tool—an axe or a hammer—are you likely to use?
   b. Explain your choice in terms of the concept of pressure.

You have just learned what pressure is and how it can be expressed in quantitative units. Perhaps you are wondering how pressure relates to the atmosphere. As you discovered in Laboratory Activity A.1, the atmosphere exerts a force on every object in it. On a typical day at sea level, air exerts a force of about 100 000 N on each square meter of your body—resulting in pressure of about 100 kilopascals (kPa). This pressure is also equal to one atmosphere (atm)—another unit commonly used by scientists and one with which you will become familiar. One atmosphere equals 101.3 kPa.

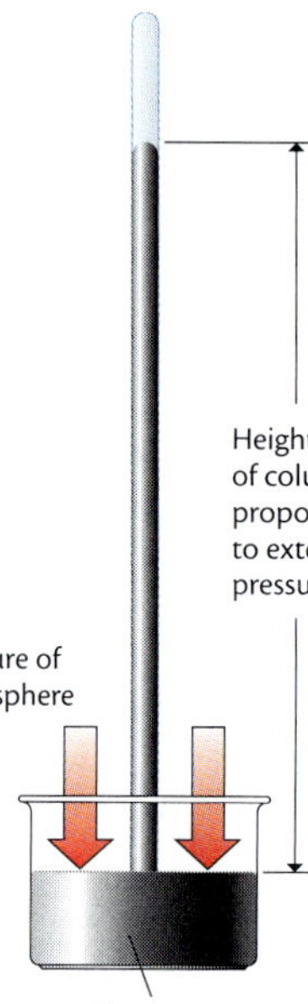

Figure 8 *A mercury barometer. On a typical day at sea level, the atmosphere will support a column of mercury 760 mm high. The pressure unit "atmosphere" is thus related to pressure in millimeters of mercury: 1 atm = 760 mmHg.*

Pressure can also be expressed in units of mmHg (millimeters of mercury) or inches of mercury. Such units suggest that air pressure can be measured as the height of a column of mercury. How is this possible? Recall that in one of the experiments you performed in Laboratory Activity A.1 (Station 7, page 250), you covered a test tube filled with water and inverted it into a container of water. You then uncovered the test tube. What did you observe? What force supported the weight of the column of water in the test tube?

Now imagine repeating the experiment with a taller test tube, and again with an even taller test tube. If the test tube were tall enough, at a certain height water would no longer fill the tube entirely when it was inverted in a container of water. This experiment was first performed in the mid-1600s. Scientists discovered that one atmosphere of air pressure could support a column of water only as tall as 10.3 m (33.9 ft). If the experiment is tried with even taller inverted tubes, the water still reaches only to the 10.3-m level.

Obviously, a **barometer** (a device that measures atmospheric pressure) based on a tube filled with water would be much too tall to be useful. So scientists replaced the water with mercury, a liquid 13.6 times denser than water. The resulting mercury column, illustrated in the barometer in Figure 8, is shorter than the water column by a factor of 13.6. Thus at one atmosphere of pressure, the mercury column has a height of 760 mm; 1 atm = 760 mmHg.

How does air pressure "support" the weight of a column of water or mercury? To answer this question, it will be helpful to develop a model for how gaseous atoms and molecules move and interact with each other and objects around them.

A.4 ATOMS AND MOLECULES IN MOTION

Your experiences with gases, liquids, and solids are at a sensory level—that is, you observe matter and its changes through your senses of sight, smell, touch, taste, and sound. In so doing, you arrive at some generalizations about the world around you. This is the first step in "doing science." Science does not stop with making and reporting observations, however. A more difficult task is crafting a theory that explains a majority of observations and that also predicts outcomes of experiments yet to be performed.

One such theory concerns the motion of atoms and molecules in solids, liquids, and gases. What do you already know about these states of matter from observing them? You know that solids have definite shapes and that their structures are rigid. Experience also tells you that liquids flow and—unlike solids—depend on a container to define their shape. Gases, likewise, do not have definite shapes. If you have ever blown up a balloon, then let go of it without tying it off, you also know that gases are very mobile.

Drawing from your observations, what would you expect about the movement of atoms and molecules in each state of matter? The atoms, molecules, and ions that make up solids are held tightly to one another. Although the particles vibrate, they are tightly packed and cannot move past one another. The molecules and ions within liquids are more mobile—they can move past each other, but attractive forces keep them from moving too far apart. The arrangement and motion of particles in solids and liquids are depicted in Figure 9.

Scientific laws describe the behavior (the "what") of nature, but do not provide explanations. Scientific theories and models offer "how"-type explanations of phenomena in the natural world.

Figure 9 *You can observe the properties of solids (left) and liquids (right) using your senses. Models help portray how atoms and molecules interact with each other and their surroundings.*

A.4 ATOMS AND MOLECULES IN MOTION

The differences among the relative motions of molecules in the solid, liquid, and gaseous states are discussed, with emphasis on the kinetic energy of gas molecules. The kinetic molecular theory is presented. The kinetic molecular theory is used to predict and explain the behavior of gases under different conditions of temperature and pressure.

Figure 10 *Particles of gaseous material (inside the balloons) are much farther apart than can be depicted in this visual model.*

You learned about attractive forces between molecules earlier. Refer to page 58 in Unit 1 if needed.

Most particles in the gaseous state are not as strongly attracted to each other. They move in straight-line paths at very high speeds and change direction only when they collide with each other or with another object. When the collisions occur with the walls of the container, the impact of the gaseous particles results in gas pressure. In contrast to particles in solids and liquids, particles in gases are very far apart. In fact, their size is negligible compared to the great distances separating them. Figure 10 illustrates the motion of particles in a gas. Compare this model of gases to the models of solids and liquids in Figure 9.

Gas particles move at speeds that are proportional to the quantity of kinetic energy they possess. **Kinetic energy,** the energy possessed by any moving object, is sometimes called "energy of motion." Its magnitude depends on the mass of the moving object and its velocity. Traveling at the same velocity, a more massive object has greater kinetic energy than a less massive object. Thus a baseball has much greater kinetic energy than a ping pong ball when both are thrown at the same velocity. If two objects have equal masses, the one traveling faster has greater kinetic energy. This helps explain the difference in damage to an automobile windshield when it is hit by a baseball thrown at 100 km/h compared to one tossed at 10 km/h.

The behavior of gas particles can be explained in the context of kinetic energy. This explanation is based on several postulates, some of which you have just read about. The postulates are as follows:

258

- Gases consist of tiny particles (atoms or molecules) whose size is negligible compared with the great distances separating them.
- Gas molecules are in constant, random motion. They often collide with each other and with the walls of their container or surrounding objects. Gas pressure is the result of molecular collisions with container walls and other objects.
- Molecular collisions are elastic. This means that although individual molecules in a gas sample may gain or lose kinetic energy, there is no gain or loss in total kinetic energy from these collisions.
- At a given temperature, molecules in a gas sample have a range of kinetic energies. However, the average kinetic energy of the molecules is constant and depends only on the temperature of the sample. Therefore, molecules of different gases at the same temperature have the same average kinetic energy. As temperature increases, so does the average velocity and kinetic energy of the gas molecules.

A postulate is an accepted statement used as the basis for further reasoning and study.

This set of postulates is the basis of the kinetic molecular theory (KMT), which can be used to explain observations of gas behavior. It can also be used to make predictions. For example, the theory can describe what will happen to the pressure of a gas sample if its volume is doubled, assuming no change in temperature. Read on to see how.

A.5 BOYLE'S LAW—P-V RELATIONSHIPS

Earlier, when you pushed down on a sealed syringe filled with gas, you observed that gases can easily be compressed—much more so than liquids or solids. In other words, the volume of a sample of gas is easily changed when an external pressure is applied to it.

You performed this experiment at Station 8 in the laboratory activity on page 250.

Consider your experience with the syringe. The more pressure you applied to the plunger, the smaller the volume of gas inside the syringe became. Consequently, the pressure of the gas sample below the plunger increased. Think of it this way: If the plunger rests in any given position, the external pressure pushing down on it must be equal to the pressure of the sample of gas below it.

Assume that the pressure of gas inside the syringe was 1 atm before you pushed on the plunger (at this point, only the atmosphere was exerting pressure on the plunger). If the volume of the gas in the syringe were changed to half of its original volume by pushing on the plunger, the pressure of the gas sample would be doubled. If the gas volume in the syringe were reduced to one-fourth of its original value, the gas pressure would become four times larger. In each case, the pressure of the gas inside the syringe increases by a proportional amount. Figure 11 (page 260) shows this relationship.

How does kinetic molecular theory explain this behavior? Gas molecules are in constant, random motion; gas pressure is the result of molecules colliding with the walls of their container. Assume that an external pressure is applied to a sample of gas at constant temperature. As the gas container becomes smaller, the gas-molecule collisions over a particular area of container wall increase. In fact, the number of collisions per unit area doubles—thus gas pressure doubles, as seen in Figure 12 on page 260.

A.5 BOYLE'S LAW—P-V RELATIONSHIPS

When the volume of a sample of a gas is reduced, while at the same time its temperature is held constant, the pressure will increase. This increase in pressure will result from the increased number of molecular collisions with the walls of the container. This inverse relationship between pressure and volume, at constant temperature, is a statement of Boyle's law.

Students may have difficulties with inverse relationships. It is helpful to have them use the "reason and ratio" method to predict whether the answer should be larger or smaller before they set up a ratio or begin their calculations.

Use class time to explain the example problems in the text and help students grasp the nature of this inverse relationship.

Consider demonstrating Boyle's law in action before asking students to work quantitative problems.

DEMONSTRATION IDEAS

A. Review Section A.1 Station 8: Use a plastic syringe to illustrate the relationship between pressure and volume.

B. Connect a eudiometer tube partly filled with water to a leveling bulb, via plastic tubing. When the level bulb is raised or lowered, the volume of the trapped gas in the tube changes. The volume of the gas in the tube decreases as the bulb is raised (increase in pressure) and increases when the bulb is lowered (decrease in pressure).

Figure 11 *As pressure increases, the volume of gas in the syringe decreases proportionally.*

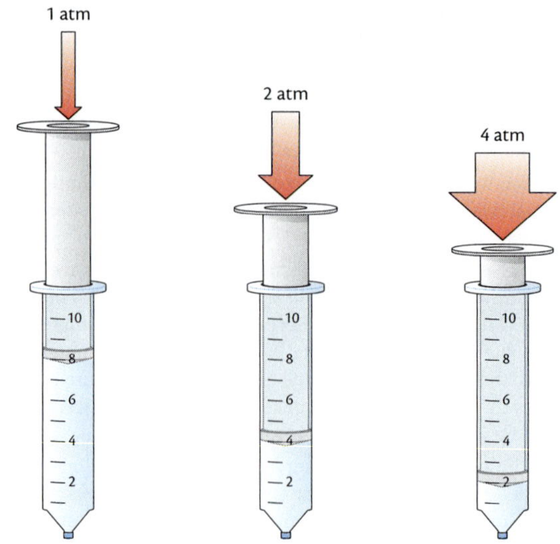

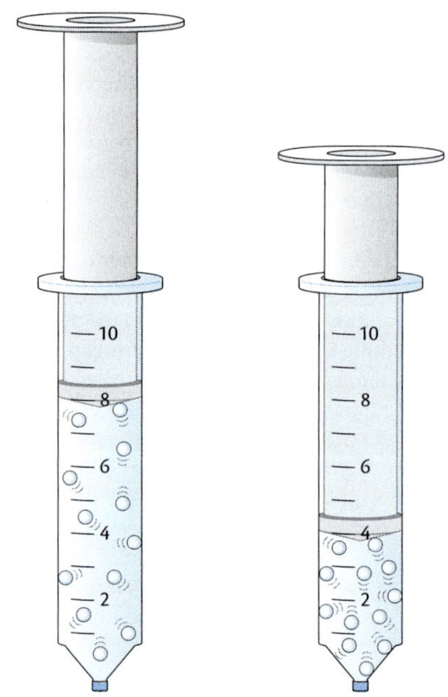

Figure 12 *As the volume of a sample of gas is reduced, the number of molecular collisions with the container wall—and thus the gas pressure—increases proportionally. The gas molecules in each syringe have been greatly magnified to depict the average distances among them, and thus to account for more frequent collisions with the container wall in the syringe on the right.*

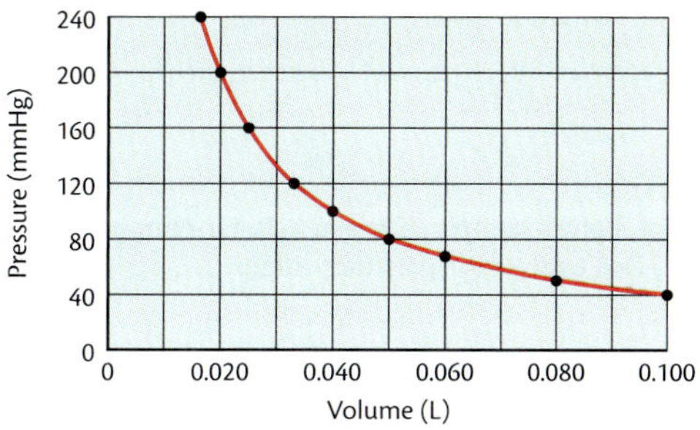

Figure 13 *The volume (V) of a gas sample, maintained at constant temperature, is inversely proportional to its pressure (P). Therefore the product P × V is constant. A plot of pressure vs. volume for any gas sample at constant temperature will be similar to this one.*

You have seen that the relationship for a gas sample between volume and pressure (at constant temperature) can be described in many ways. The changes in volume and pressure predicted by KMT for a particular gas sample can be described with words or pictures, such as those in Figures 11 and 12. It can also be described graphically, as shown in Figure 13. Note the shape of the curve; this indicates an inverse relationship between pressure and volume.

This relationship is useful for predicting a new gas pressure or volume resulting from a change in one of these variables. Again, consider the syringe in Figure 11. You know that initially, the gas occupies a volume of 8 mL and exerts a pressure of 1 atm. What would the pressure of the gas become if its volume were increased to 10 mL?

A "reason and ratio" method can be used to predict the resulting pressure. First, reason: If the volume increases from 8 mL to 10 mL, the pressure must decrease by a proportional quantity. Then, multiply the initial pressure (1 atm) by the appropriate volume ratio. The two possible volume ratios are:

$$\frac{8\ \text{mL}}{10\ \text{mL}} \quad \text{and} \quad \frac{10\ \text{mL}}{8\ \text{mL}}$$

Since your reasoning indicates that the pressure should decrease as the volume increases, use the ratio that is less than one. The calculation becomes:

$$1\ \text{atm} \times \frac{8\ \cancel{\text{mL}}}{10\ \cancel{\text{mL}}} = 0.8\ \text{atm}$$

If the volume of a particular sample of gas were to decrease (as in Figure 11), then its pressure would increase. In that case the volume ratio used must be greater than one. Similar reasoning applies to problems in which the initial and final pressures are known and the final volume is to be found. The "reason and ratio" method is useful, because you gain a sense of the answer before you do any calculations. When the answer doesn't fit with your reasoning, you can check to see whether you used the correct ratio.

The "reason and ratio" method can be applied to any gaseous system if you know three of the four values for pressure and volume. In fact, if P_1 and V_1 represent the initial pressure and volume of a sample of gas and P_2 and V_2 represent the final values for the same sample, then the calculation just completed becomes

$$P_1 \times \frac{V_1}{V_2} = P_2$$

SECTION A

P-V RELATIONSHIPS
Building Skills 3

Answers

1. a. Gases in the lungs, stomach, and sinuses and those dissolved in blood will expand under conditions of reduced pressure, causing conditions that can range from uncomfortable to life-threatening.

b. Carbon dioxide inside a sealed bottle or can of soda is under high pressure. When the container is opened, the gas expands and rushes out, due to lower external pressure. Some dissolved carbon dioxide also comes out of solution as pressure drops.

c. If the tennis ball cans are not pressurized, the pressurized gas in the balls themselves will leak out at a faster rate during storage. Tiny pores in the structural material of the balls will leak gas, just as helium leaks from an inflated balloon. The balls will also lose their "bounce."

d. Air is normally found in the middle ear, its volume dictated by the pressure outside the ear. A sudden decrease in this external pressure causes the air in the middle ear to expand rapidly and rush out the ear canal, into the throat cavity. (The "pop" is the releasing of the skin flap that covers the opening into the throat.)

2. a. i. Using the "reason and ratio" method the pressure on the inside of the can should be greater than 100 kPa. The 7 liters of gas would be compressed to fit into the 1/2-liter can.

ii. $\frac{7100 \text{ mL}}{492 \text{ mL}}$ or $\frac{492 \text{ mL}}{7100 \text{ mL}}$

iii. $V_1 = 7100$ mL, $P_1 = 100$ kPa, $V_2 = 492$ mL

$P_2 = \frac{7100 \text{ mL}}{492 \text{ mL}} \times 100 \text{ kPa} =$ 1440 kPa

b. $V_1P_1 = V_2P_2$

$7100 \text{ mL} \times 100 \text{ kPa} = 492 \text{ mL} \times P_2$

$P_2 = 1440$ kPa

Rearranging this equation gives

$$P_1 \times V_1 = P_2 \times V_2$$

or simply

$$P_1V_1 = P_2V_2$$

That is, for a given sample of gas at constant temperature, the product of its pressure and volume is a constant value:

$$P \times V = k$$

where k is a constant. This relationship is known as Boyle's law, named for the seventeenth-century English scientist who first proposed it, Robert Boyle. These general equations provide a way to obtain the same answers as in the "reason and ratio" method, as you can see below:

$$P_1 = 1 \text{ atm} \longrightarrow P_2 = ? \text{ atm}$$
$$V_1 = 8 \text{ mL} \longrightarrow V_2 = 10 \text{ mL}$$

$$P_1V_1 = P_2V_2$$
$$(1 \text{ atm})(8 \text{ mL}) = (P_2)(10 \text{ mL})$$

$$1 \text{ atm} \times \frac{8 \cancel{\text{mL}}}{10 \cancel{\text{mL}}} = P_2$$
$$0.8 \text{ atm} = P_2$$

In the following activity, you will apply Boyle's law to solve several pressure-volume problems.

P-V RELATIONSHIPS — Building Skills 3

1. Explain each statement.

a. Even if they have an ample supply of oxygen gas, airplane passengers experience discomfort if the cabin undergoes a drop in air pressure.

b. Carbonated soft drink cans "pop" when the container is opened.

c. Tennis balls are sold in pressurized containers.

d. When you climb a mountain or ride an elevator to the top of a tall building, your ears may "pop."

2. You can buy helium gas in small pressurized cans to inflate party balloons. Assume the container label indicates that the can delivers 7100 mL of helium at 100 kPa pressure. The volume of the can is 492 mL.

a. Find the pressure of the helium gas inside the can. Use the "reason and ratio" method. These steps will help guide your work:

i. Do you expect the pressure inside the can to be greater or less than 100 kPa? Explain.

ii. Write the two possible volume ratios (note that only one is valid).

iii. Calculate the gas pressure inside the can using the appropriate volume ratio.

b. Find the pressure of the helium gas inside the can, this time using the equation $P_1V_1 = P_2V_2$.

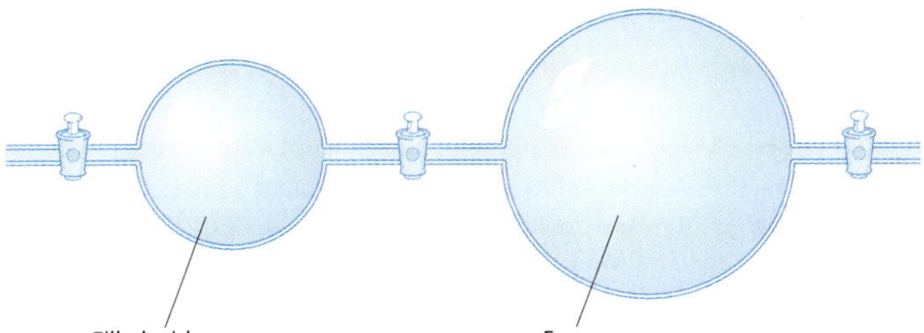

Figure 14 *Two glass bulbs are separated by a valve. The bulb on the left is filled with gas; the bulb on the right is empty. What will happen to the gas pressure after the middle valve has been opened?*

3. Two glass bulbs are separated by a valve as shown in Figure 14. The 0.50-L bulb on the left contains a gas sample at a pressure of 6.0 atm. The 1.7-L bulb on the right is evacuated—it does not contain any gas.
   a. What will happen to the total volume of the gas sample when the middle valve is opened?
   b. What will happen to the total pressure of the gas sample when the middle valve is opened?
   c. Calculate the pressure of the gas sample after the valve is opened.
   d. Draw a molecular model of the gas molecules before and after the valve is opened.
   e. Explain your model in terms of kinetic molecular theory.
4. Assume that the gaseous atmosphere held in the LHMP is inside of a sealed, leakproof, flexible dome. What would happen to the total volume of the LHMP atmosphere if a low-pressure weather system passed through the region where the lunar prototype was located?

In the problems you have encountered so far, gas temperature has remained constant. Does temperature affect gas volume? If so, how? The answers are important because many gas-related reactions in laboratory research, chemical processing, and food preparation occur at varying temperatures. You will begin to explore the relationship between gas temperature and volume by completing the following laboratory activity.

A.6 TEMPERATURE-VOLUME RELATIONSHIPS — Laboratory Activity

Introduction

Most matter expands when heated and contracts when cooled. As you know, gas samples expand and shrink to a much greater extent than either solids or liquids. In this activity you will investigate how temperature changes of a gas sample influence its volume—assuming pressure remains unchanged. To do this, you will heat a thin glass tube containing a trapped air sample and record changes in air volume as the sample cools.

CD-ROM WWW. Lab Video

4. The total volume would increase if a low pressure weather system passed through the region; thus, as the pressure lowers outside the LHMP the dome would expand, according to Boyle's law, if the temperature is constant.

A.6 TEMPERATURE-VOLUME RELATIONSHIPS
Laboratory Activity

This laboratory activity introduces students to the relationship between temperature and volume of a gas sample at constant pressure. Students observe the volume of a trapped air column at different temperatures. After making a graph of their temperature-volume data, the students are asked to predict the temperature that corresponds to zero volume. Students are introduced to the kelvin temperature scale. (273 K is read as 273 kelvin.)

Time

50 minutes

Materials (for a class of 24 working in pairs)

600-mL beaker and a hot plate or a small kitchen-type deep-fry appliance
300 mL yellow corn oil (to fill heating container deep enough to allow capillary tube immersion, the yellow color will enable students to see the fluid easily in the tubes)
24 small rubber bands, orthodontist's type, or rings cut from rubber tubing (2 bands/team)
12 capillary melting point tubes
12 non-mercury thermometers (capable of reading to 150 °C)
paper towels
12-cm rulers

3. a. The gas will expand to occupy the total volume of both bulbs.
   b. The total pressure will decrease.
   c. $V_1P_1 = V_2P_2$
   $0.5 \text{ L} \times 6.0 \text{ atm} = 2.2 \text{ L} \times P_2$
   $P_2 = 1.4 \text{ atm}$
   d. The model before the valve is opened would show gas molecules in the left glass bulb and none in the right bulb. The model after the valve is opened would show gas molecules distributed with equal but lower overall density in both bulbs.
   e. When the valve is opened, the high pressure gas molecules will flow to occupy the total volume of both containers.

Advance preparation

Obtain the rubber bands, or prepare from rubber tubing.

Note: Yellow corn oil is specified because it has given consistently satisfactory results. If you use some other type of oil, be sure you test it in this activity first. Results have been less satisfactory, for example, with mineral oil.

Safety

Students should wear safety goggles and aprons throughout the laboratory activity. Be especially careful with the hot oil, as it can cause serious burns. Do not heat the oil with a gas burner. Maintain the temperature below 130 °C.

Pre-lab discussion

This laboratory activity investigates the relationship between volume and temperature at constant pressure. The gas of interest is that trapped between the oil droplet and the closed end of the capillary tube. Point out that measuring the length of trapped air in the tube is valid because this length is proportional to volume.

Lab tips

Heat corn oil to 130 °C in a small deep-fry appliance or in a beaker on a hot plate. Since hot oil can be very hazardous, use one oil bath (under your personal control) for the entire class. Maintain the oil temperature as close to 130 °C as possible. Remind students to read and follow the directions carefully to avoid burns from hot oil.

Initially the capillary tubes should be fully immersed in oil for good results.

If the procedure is followed as written, students will record temperature-length data over a 90 °C to 100 °C range. This is necessary to provide a reasonable number of data points for extrapolation to zero volume. The first data point is at the hot-oil temperature and is the full length of the capillary tube. Measure the distance from the closed end of the tube to the oil plug, and record the temperature.

Expected results

The measured temperature-length data will yield reasonably linear plots. Experimental data have led to a median extrapolated temperature of −280 °C at zero volume, compared to the "expected" value of −273 °C.

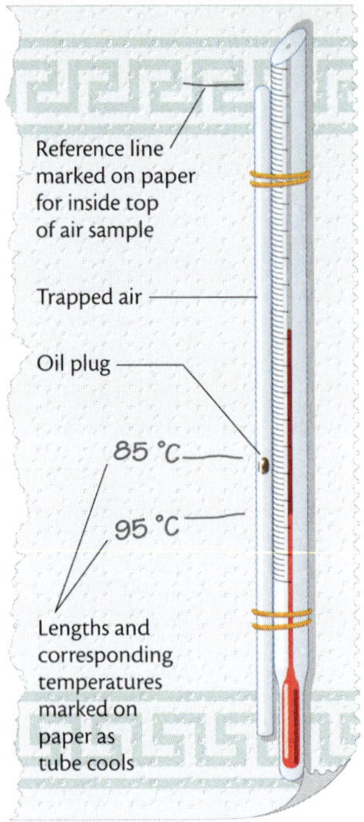

Figure 15 *Apparatus for studying how gas volume changes with temperature. The air trapped in this tube has cooled from 95 °C to 85 °C.*

Procedure

1. Using two small rubber bands, fasten a capillary tube to the lower end of a thermometer. See Figure 15. Place the open end of the tube closest to the thermometer bulb and 5–7 mm from its tip.
2. Immerse the tube and thermometer in a hot oil bath that has been prepared by your teacher. Be sure the entire capillary tube is immersed in oil. Wait for your tube and thermometer to reach the temperature of the oil (approximately 100 °C). Record the temperature of the bath.
3. After your tube and thermometer have reached constant temperature, lift them until only about one quarter of the capillary tube (open end down) is still in the oil bath. Pause for about 3 seconds to allow some oil to rise into the tube. Then quickly place the tube and thermometer on a paper towel (to avoid dripping) and carry them back to your desk. **CAUTION:** *Be careful not to touch the hot end of the thermometer or the drips of hot oil.*
4. Lay the tube and thermometer on a clean piece of paper towel on the desk. Make a reference line on the paper at the sealed end of the capillary tube. Also mark the upper end of the oil plug, as shown in Figure 15. Alongside this mark write the temperature corresponding to that air-column length.
5. As the temperature of the gas sample drops, make at least six marks representing the length of the air column trapped above the oil plug at various temperatures. Write the corresponding temperature next to each mark. Allow enough time so the temperature drops by 50–60 °C.
6. When the thermometer shows a steady temperature (near room temperature), make a final observation of length and temperature. Discard the tube and the rubber bands according to your teacher's instructions. Wipe the thermometer clean.
7. Measure the length (in centimeters) from each marked line to the mark for the sealed end of the tube. Record each length of the gas sample. Have your teacher check your data before you discard your paper towel.
8. Wash your hands thoroughly before leaving the laboratory.

Calculations

1. Prepare a sheet of graph paper for plotting your data, with length on the vertical axis and temperature on the horizontal axis. The *y* axis (length) should range from 0 to 10 cm; the *x* axis (temperature) should include values from −350 °C to 150 °C. Label your axes, and arrange the scales so the graph fills nearly the entire sheet.
2. Plot the length-temperature data for your air sample. Draw the best straight line through the plotted points. Using a dashed line, extend this straight graph line so that it intersects the *x* axis. Use your completed graph to answer the following questions.

Questions

1. At what temperature does your extended graph line intersect the x axis?
2. a. What would be the volume of your gas sample at that temperature?
   b. Why is that volume only theoretical?
3. Renumber the temperature scale on your graph, assigning the value zero to the temperature at which your plotted graph line intersects the x axis. The new scale expresses temperature in kelvins (K)—the kelvin temperature scale. One kelvin is the same size as one degree Celsius. However, unlike zero degrees Celsius, zero kelvins is the lowest temperature theoretically possible. It is called absolute zero.
4. Based on your graph, what temperature in kelvins (K) would correspond to 0 °C, the freezing point of water? What temperature would correspond to 100 °C, the normal boiling point of water?

Temperatures on the kelvin temperature scale are written as 10 K, for example, not 10 °K.

A.7 CHARLES' LAW

Your plotted data from the preceding laboratory activity indicate a relationship between volume and temperature for a sample of gas at constant pressure. What is that relationship? How might this relationship affect the LHMP? Why?

In the 1780s, French chemists (and hot-air balloonists) Jacques Charles and Joseph Gay-Lussac studied the changes in gas volume caused by temperature changes at constant pressure. Data for oxygen gas and nitrogen gas are shown in Figure 16. The plots for different gases and different sample sizes have different appearances. However, if all graph lines are extended to the x axis (an extrapolation), the lines meet at the same temperature. Lord Kelvin (an English scientist) used the work of Charles and Gay-Lussac

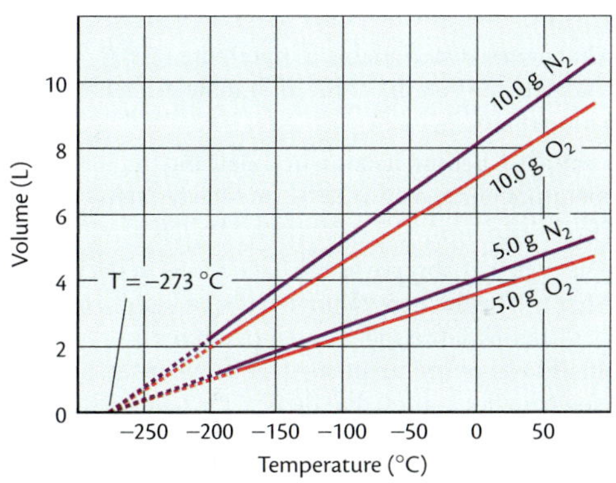

Figure 16 *Temperature-volume measurements of various gas samples at 1 atm pressure. Extrapolation has been made for temperatures below liquefaction.*

Calculations

1. Students should make an XY-scatterplot with the axes properly incremented and labeled.
2. Students should draw in a visual trend line using the plotted data. This line should extend beyond the data to intercept the x axis.

Answers

1. Student answers will vary but will be close to −273 °C.
2. a. Zero milliliters.

   b. The answer was obtained by extrapolation. The assumption here is that the gas will behave consistently as the temperature decreases. Real gases condense before absolute zero is reached. Once they condense, the volume will no longer change.
3. Students should re-label their graphs showing temperature in kelvins. (No answer required.)
4. 273 K, 373 K. (Student answers will vary but will be close to these numbers.)

Post-lab activity

Focus discussion of student data. This may be a good time to introduce how to compute percent error. Percent error = (|experimental value − actual value| ÷ actual value) × 100. Generally, students are within 10% of the value for absolute zero. An advanced activity may include computing the equation that mathematically represents the change in volume of an ideal gas as the temperature changes.

A.7 CHARLES' LAW

Students' data from the previous laboratory activity is used to establish the relationship between gas volume and kelvin temperature at constant pressure. The mathematical relationship results in a statement of Charles' law. Take time to reflect back and compare the graphs of Charles' law (direct relationship) to Boyle's law (inverse relationship).

T-V RELATIONSHIPS
Building Skills 4

Remind students that all of the following answers assume constant pressure.

1. The volume of the gas will decrease as its temperature decreases, indicating a direct relationship. It will decrease by a ratio of $\frac{253\text{ K}}{293\text{ K}}$.
2. The volume of the anesthetic gases would be greater inside the patient's body at the higher temperature. Volume and temperature are directly related. It will increase by a ratio of $\frac{310\text{ K}}{291\text{ K}}$.
3. a. The gas bubble would be larger. An increase in temperature would result in an increase in volume. Volume and temperature are directly related.

   b. $\frac{291\text{ K}}{415\text{ K}}$ or $\frac{415\text{ K}}{291\text{ K}}$

   c. $V_2 = 1.0\text{ mL} \times \frac{415\text{ K}}{291\text{ K}} =$ 1.4 mL
4. a. The volume of the balloon decreases, the air temperature in the car is lower than the air in the mall.

   b. $V_1/V_2 = T_1/T_2$
   $V_2 = V_1 \times T_2/T_1 =$
   $3.0\text{ L} \times 268\text{ K}/295\text{ K} = 2.7\text{ L}$

   c. No. When the balloon is placed into a room at 25 °C or 298 K, the volume of the gas in the balloon will return to its original volume of 3.0 L.

   d. The diagram should show the atoms closer together in the colder model.

   e. According to the kinetic molecular theory, as the temperature decreases, the speed of the molecules decreases, thus decreasing the volume of the gas.

K = °C + 273

to establish a simple mathematical temperature-volume relationship for gases. He based his relationship on his own new temperature scale.

Doubling the kelvin temperature of a gas sample doubles its volume. Reducing the kelvin temperature by one half causes the gas volume to decrease by one half, and so on. These relationships are summarized in Charles' law: At constant pressure and amount of gas, the volume (V) of a gas sample divided by its kelvin temperature (T) is a constant value:

$$\frac{V}{T} = k$$

where k is a constant. Relating the initial temperature and volume to the final temperature and volume gives:

$$\frac{V_1}{T_1} = \frac{V_2}{T_2}$$

This equation, as well as the "reason and ratio" method, is useful in predicting changes in gas volume and temperature at constant pressure.

T-V RELATIONSHIPS
Building Skills 4

Apply Charles' law in answering these questions.

1. What would happen to the volume of a balloon originally at 20 °C if you took it outdoors to a temperature of −20 °C?
2. In planning to administer an anesthetic gas to a patient, why must an anesthesiologist take into account the fact that during surgery the anesthetic gas is used both at room temperature (18 °C) and at the patient's body temperature (37 °C)?
3. An air bubble trapped in bread dough at room temperature (291 K) has a volume of 1.0 mL. The bread bakes at 415 K. Use the "reason and ratio" method to calculate the volume of the air bubble as the bread bakes. Follow these hints.
   a. Do you expect the gas volume to be larger or smaller than 1.0 mL? Explain.
   b. What are the two possible temperature ratios?
   c. Calculate the volume of the air bubble using the appropriate temperature ratio.
4. You buy a 3-L helium balloon in a mall and put it in a car. The temperature in the mall is 22 °C, and the temperature in the car is −5 °C.
   a. What will you observe as the balloon sits in the car?
   b. What will be the new volume of the balloon? (*Hint:* Remember that Charles' law applies to temperatures expressed in kelvins.)
   c. Should you return to the mall for a refill of the balloon? Explain.
   d. Sketch an illustration that depicts the helium atoms when the balloon was in the mall and when it was in the car.
   e. Explain your model using kinetic molecular theory.

A.8 TEMPERATURE-PRESSURE RELATIONSHIPS

Boyle's law states that volume and pressure are inversely related; as one increases, the other decreases. Charles' law states that volume and temperature are directly related; increasing one increases the other. Combine your knowledge of these relationships with what you learned about kinetic molecular theory to consider the possible relationship between the temperature (in kelvins) of a gas sample and its pressure.

If you are having difficulty doing this, picture a closed cylinder of gas, such as a deep-sea scuba tank (volume is constant). What would happen to the kinetic energies of the gas particles in the cylinder if you were to heat the tank? How would this affect the pressure?

If you concluded that raising the kelvin temperature of the gas should cause a corresponding increase in the gas pressure, you are correct. Increasing the temperature increases the kinetic energy (and thus the velocities) of the gas particles. Because the particles are traveling faster, the number of collisions with the container walls increases and the energy involved with each collision also increases. These effects cause an increase in the gas pressure. The mathematical expression for this relationship is:

$$\frac{P}{T} = k$$

where k is a constant. Particular changes in pressure and temperature at constant volume can be calculated using this equation:

$$\frac{P_1}{T_1} = \frac{P_2}{T_1}$$

You can also reason out the answer, using the correct pressure or temperature ratio. Regardless of method used, remember that the gas temperatures must be expressed in kelvins.

T-V-P RELATIONSHIPS — Building Skills 5

Solve the following problems using appropriate gas relationships.

1. a. If the kelvin temperature of a gas sample in a steel tank increases to three times its original value, what will happen to the pressure of the gas?
   b. Draw a molecular model that shows the movement of the molecules inside the tank at the two temperatures.
   c. Explain your model using kinetic molecular theory.
2. If a sample of gas is cooled at a constant pressure until it shrinks to one-fourth its initial volume, what change in its kelvin temperature must have occurred?
3. Explain why car owners in severe northern climates often add air to their tires in winter and release some air from the tires in summer.

CD-ROM WWW. Gas Laws

T-V-P RELATIONSHIPS

Building Skills 5

1. a. Pressure will increase by three times.

   b. Students' models should show equal numbers of particles in motion; the molecular motion at the higher-temperature condition should involve higher average velocities than at the cooler condition.

   c. As temperature increases, the molecular motion of the molecules will increase. This will increase both the number of collisions and the force of the collisions, resulting in an increase in pressure.
2. The absolute temperature (in kelvin) will decrease by a factor of 4; the absolute temperature will be 1/4 its original value.
3. Car owners add air to their tires in winter because the lower temperatures decrease the pressure of the enclosed volume of air. More air molecules are needed to ensure proper tire pressure. Air is released when summer arrives because the pressure from the enclosed volume of air increases as the temperature increases. Both under-inflation and over-inflation cause improper wear on the tires.

A.8 TEMPERATURE-PRESSURE RELATIONSHIPS

Students are asked to use their knowledge of Boyle's law and Charles' law to predict the relationship between temperature and pressure. This new relationship is also called Gay-Lussac's law.

Encourage students to think through the problems using the "reason and ratio" method before they use the mathematical relationship. Since the kinetic energies of the molecules increase with an increase in temperature, the pressure is also increased. Therefore, temperature and pressure are directly related if volume is held constant.

4. a. Changes in pressure and temperature change the volume.

   b. Since the particles are already closely packed, the volumes of liquids and solids remain nearly constant over a small range of temperatures and pressures.

5. When the balloon rises, the external pressure decreases, which should cause the volume to increase. Temperature also decreases, which should decrease the volume. The data in Figure 5 (page 252) show that the pressure changes are larger than the temperature changes. For example, from 0 km altitude to 10 km, the temperature changes 65 °C (65 K), a 22% change in absolute temperature, while the pressure changes from 760 mmHg to 218 mmHg, a change of 71%. Therefore, Boyle's law dominates, and the volume of the balloon will increase.

6. As the temperature of the gas in the can increases, the pressure of the gas will also increase, possibly causing the can to explode. This allows the volume to increase catastrophically.

7. a. The pressure of the gases inside the LHMP would be higher on hot days, unless the LHMP expands to equalize the pressure. The pressure will be lower on cool nights, or volume will decrease if the LHMP is flexible.

   b. The structure of the LHMP should be made of a flexible material to expand or contract with changes in temperature so that the pressure would remain fairly constant with changes in temperature.

Note: Section A.9 teacher commentary begins on page 270.

4. a. When the volume of a gas sample is measured, its pressure and temperature must also be specified. Why?
   b. That practice is normally not necessary for liquids or solids. Why?
5. Use kinetic molecular theory and gas laws to explain why a weather balloon steadily expands as it rises.
6. Why does the warning label on an aerosol can indicate not to dispose of the can in a fire?
7. Assume that the LHMP is enclosed in a transparent polymer structure.
   a. How will the gases inside the LHMP be affected by hot days and cool nights?
   b. What design features might be incorporated in the LHMP to reduce those heating and cooling effects?

A.9 IDEAL GASES AND MOLAR VOLUME

A gas that behaves under all conditions as the kinetic molecular theory predicts is called an **ideal gas**. At very high pressures or very low temperatures, real gases do not behave ideally. That is, the gas laws you have considered do not accurately describe gas behavior under such conditions.

At low temperatures, molecules move more slowly. As kinetic energy decreases, the weak attractions between molecules may become so significant that the gas condenses to a liquid, such as when air is liquefied. At high enough pressures—if the temperature is not too high—the gas molecules are pressed close enough together that these same forces of attraction may also cause a gas to condense. However, under usual conditions encountered in the atmosphere, most gas behavior approximates that of an ideal gas. Such gas behavior is well explained by the kinetic molecular theory.

Up until now, you have been considering the behavior of gases under differing conditions of volume, pressure, and temperature. One variable you have not yet fully considered is the size of the gas sample, expressed as the number of molecules contained in a particular gas volume. If you have the same volumes of oxygen gas, nitrogen gas, and carbon dioxide gas in three different balloons at the same temperature and pressure, how do the numbers of gas molecules compare? See Figure 18.

Figure 18 *These three balloons contain three different gases at the same temperature and pressure. Are the numbers of gas molecules in each the same or different?*

MODELING MATTER

KINETIC MOLECULAR THEORY

Previously, you modeled the structure of matter and changes in matter using pictures and symbols. Another way that matter can be modeled is through the use of analogies. An analogy can help you compare certain features of an abstract idea or theory to a situation that is familiar to you. Read the analogy provided here and answer the questions that follow it concerning the kinetic molecular theory of gases.

Imagine that a large group of dancers on an enclosed dance floor represents gas molecules bouncing around inside a container. The dancers move back and forth across the floor, but not off the floor. See Figure 17.

1. Decide which of the four variables—volume, temperature, pressure, or number of molecules—is most like each of the following. Explain your choices.
   a. the number of dancers
   b. the size of the room
   c. the beat of the music
   d. the number and force of collisions among the dancers
2. How does each of the following situations relate to what you have learned about gases and the kinetic molecular theory?
   a. The beat of the music and the number of dancers remain the same, but the size of the dance floor increases.
   b. The size of the dance floor and the number of dancers remain the same, but the beat of the music becomes faster.
   c. The size of the dance floor and the beat of the music are kept the same, but the number of dancers increases.
3. Suggest another dancing situation (similar to those in Question 2) that is analogous to the behavior of gases. Explain.
4. All analogies have limitations. For example, the dancing analogy fails to represent certain characteristics of gases. Gas molecules travel at very high velocities (on the order of 6000 km/h), whereas dancers move much more slowly.
   a. Suggest two other characteristics of gases that are not properly represented by the dancing analogy.
   b. Suggest two aspects of the dancing analogy that may be misleading concerning the behavior of gases.
5. Write your own analogy that is useful for modeling gas behavior.
   a. Identify the features of your analogy that relate to kinetic molecular theory and temperature-pressure-volume relationships of gases.
   b. Point out some key limitations in your analogy.

Figure 17 *Dancers on a dance floor.*

KINETIC MOLECULAR THEORY

MODELING MATTER

Students answer questions comparing a group of dancers to gas molecules.

Answers

1. a. number of molecules
   b. volume
   c. temperature
   d. pressure
2. a. This is similar to Boyle's law, in that at constant temperature (beat of music) and number of particles (number of dancers), if the volume (floor size) increases, the pressure (dancer collisions) should decrease.
   b. Gay-Lussac's law (pressure and temperature are directly related): If the volume (floor size) and the number of particles (number of dancers) remain constant, while the temperature (beat) increases, pressure (dancer collisions) should increase.
   c. If volume (floor size) and temperature (beat) are constant, but the number of molecules (dancers) increases, the pressure (collisions) should increase.
3. The size of the dance floor and the number of dancers remain the same, but the tempo of the music (temperature) decreases. Pressure (collisions) should decrease. (Many other possible student answers.)
4. a. The extreme small size of the gas molecules relative to the total volume of the gas sample. The relative sizes of the molecules and the space between them.
   b. Real dancers dance in pairs and tend to avoid colliding with other couples.
5. Students' analogies will vary.

A.9 IDEAL GASES AND MOLAR VOLUME

This section (beginning on page 268) leads students to make some general conclusions about the behavior of gases. Possible conclusions should include the following:

- The physical properties of all gases are nearly the same, provided the temperature is not too low and the pressure is not too high.
- Real gases at room temperature and standard atmospheric pressure behave nearly ideally since their molecules are so far apart.
- Avogadro's law and molar volume simplify the monitoring of gaseous reactions by measuring the volume of the gases involved. The law is based on the fact that the volume of a gas is directly proportional to the amount of gas (in moles).
- Volumes of gaseous reactions can be used experimentally in place of moles in problems, if temperature and pressure remain constant.
- An ideal gas is one in which the molecules occupy no volume and in which there is no attraction among molecules. Real gases at room temperature and standard atmospheric pressure can behave nearly ideally since their molecules are so far apart. When gases are under high pressure and low temperature, their molecules come closer together and no longer act ideally.

Note that students often have misconceptions about the concept of an ideal gas. They tend to regard ideal gases in the same category as "ideal home" or "ideal mate." Ideal gases are not ideal in the sense of being perfect or great, but in the sense that they have been idealized to simplify mathematical modeling.

Avogadro's complete name was Lorenzo Romano Amedeo Carlo Avogadro, conte di Quarequa e Cerreto.

That question was studied by the Italian lawyer and mathematical physics professor Amadeo Avogadro in the early 1800s. By making observations of gases similar to those you have completed, he proposed that equal volumes of gases at the same temperature and pressure contain the same number of molecules. This statement is commonly known as Avogadro's law.

At 0 °C and 1 atm, 22.4 L of gas, about the volume of three basketballs, contains one mole of molecules.

A consequence of Avogadro's law is that all gases have the same molar volume if they are at the same temperature and pressure. The **molar volume** is the volume occupied by one mole of a substance. At conditions of 0 °C and 1 atmosphere (1 atm), the molar volume of any gas is 22.4 L. There is no corresponding simple relationship between moles of various solids or liquids and their volumes.

The fact that all gases have the same molar volume under the same conditions simplifies thinking about reactions involving gases. For example, consider these equations:

$$N_2(g) + O_2(g) \longrightarrow 2\,NO(g)$$
$$2\,H_2(g) + O_2(g) \longrightarrow 2\,H_2O(g)$$
$$3\,H_2(g) + N_2(g) \longrightarrow 2\,NH_3(g)$$

You have learned that the coefficients in such equations indicate the relative numbers of molecules or moles of reactants and products. Using Avogadro's law (equal numbers of moles of gas occupy equal volumes), you can also interpret the coefficients in terms of gas volumes.

$$1\text{ volume } N_2(g) + 1\text{ volume } O_2(g) \longrightarrow 2\text{ volumes } NO(g)$$
$$2\text{ volumes } H_2(g) + 1\text{ volume } O_2(g) \longrightarrow 2\text{ volumes } H_2O(g)$$
$$3\text{ volumes } H_2(g) + 1\text{ volume } N_2(g) \longrightarrow 2\text{ volumes } NH_3(g)$$

The actual volumes could be expressed in any units that are convenient, such as liters (L) or cubic centimeters (cm^3). In the example above that represents the formation of water, you could combine 200 L $H_2(g)$ and 100 L $O_2(g)$ and expect to produce 200 L $H_2O(g)$, if all gases are measured at the same conditions. Note that, unlike mass, gas volumes are not necessarily conserved in a chemical reaction.

In this example, 200 L $H_2(g)$ and 100 L $O_2(g)$ do not "add up" to form 300 L of $H_2O(g)$.

These volume relationships allow chemists to monitor gaseous reactions by measuring the gas volumes involved. In the activity that follows, you will have an opportunity to check your understanding of this concept.

MOLAR VOLUME AND REACTIONS OF GASES

Building Skills 6

22.4 L = 1 mol gas at 0 °C and 1 atm.

1. What volume would be occupied at 0 °C and 1 atm by 3 mol $CO_2(g)$?
2. In a certain gaseous reaction, 2 mol NO react with 1 mol O_2:

$$2\,NO(g) + O_2(g) \longrightarrow 2\,NO_2(g)$$

   a. Given the same conditions of temperature and pressure, how many liters of $O_2(g)$ would react with 4 L of NO(g) gas?

Hint: 22.4 L of gas is about the volume of three basketballs.

Construct a box that is 28.1 cm × 28.1 cm × 28.1 cm, and label it "1 mol of any gas at 0 °C and 1 atm = 22.4 L," or display a molar volume box that may be purchased from one of the scientific supply companies.

b. How might a chemist use knowledge of molar volumes to monitor the progress of this reaction?

3. Toxic carbon monoxide (CO) gas is produced when fossil fuels such as petroleum burn without enough oxygen gas. The CO can eventually be converted to CO_2 in the atmosphere. Automobile catalytic converters are designed to speed up this conversion:

Carbon monoxide gas + Oxygen gas $\longrightarrow$ Carbon dioxide gas

a. Write the balanced equation for this conversion.
b. How many moles of oxygen gas would be needed to convert 50.0 mol of carbon monoxide to carbon dioxide?
c. What volume of oxygen gas would be needed to react with 968 L of carbon monoxide? (Assume both gases are at the same temperature and pressure.)

CD-ROM WWW. Questions & Answers

You now have a fundamental understanding of the behavior of all gases. In the next section, you will learn how radiation from the Sun interacts with gases in the atmosphere to produce a habitable climate on Earth.

Answers

1. $3.0 \text{ mol } CO_2 \times \frac{22.4 \text{ L}}{1 \text{ mol } CO_2} =$ $67 \text{ L } CO_2$

2. a. The ratio of moles NO to moles O_2 is 2:1; therefore, the ratio of liters NO to liters O_2 is also 2:1. Two liters O_2 would react with 4 L of NO.

   b. An example might be knowing how much volume to add for a reaction versus mass or amount in moles.

3. a. $2\ CO(g) + O_2(g) \rightarrow 2\ CO_2(g)$

   b. $50 \text{ mol CO} \times \frac{1 \text{ mol } O_2}{2 \text{ mol CO}} =$ $25 \text{ mol } O_2$

   c. $968 \text{ L CO} \times \frac{1 \text{ L } O_2}{2 \text{ L CO}} =$ $484 \text{ L } O_2$

MOLAR VOLUME AND REACTIONS OF GASES

Building Skills 6

This activity helps reinforce an important practical consequence of Avogadro's law. If all gases (at the same temperature and pressure) have the same molar volume, then in a chemical reaction involving gases, the coefficients indicate the relative volumes of reactants and products as well as the relative number of moles.

SECTION SUMMARY

Reviewing the Concepts

1. List the percentage of each of the most plentiful gases in the atmosphere.

The most abundant gases of the atmosphere are:

nitrogen	N_2	**78.08%**
oxygen	O_2	**20.95%**
argon	**Ar**	**0.93%**
carbon dioxide	CO_2	**0.033%**

2. Describe physical and chemical properties of the four most abundant gases that make up the atmosphere.

nitrogen—colorless, odorless, nonflammable

oxygen—colorless, odorless, supports combustion

argon—colorless, odorless, nonreactive

carbon dioxide—colorless, odorless, does not support combustion

3. Classify the mixture of gases in the atmosphere as a solution, a suspension, or a colloid. (*Hint:* See Unit 1, page 25.) Explain your answer.

The mixture of gases in the atmosphere is a solution. Each gas retains its chemical and physical properties, is homogeneous, and can be separated by physical means.

4. Compare the percent composition of gases in the atmosphere at sea level to the percent composition at high altitude.

The percent of gases in the atmosphere is essentially the same at the higher altitude as it is at the lower altitude. The percent of ozone increases at the higher altitudes.

5. **a.** Has the percent composition of the atmosphere changed significantly throughout human history?

b. What evidence can you cite for your answer?

a. No

b. The chemical composition of Earth's atmosphere is about the same today as in ancient air trapped in glacial ice. Many types of preserved samples indicate that air composition has not changed significantly.

6. It is much easier to slice a piece of cheese with the edge of a sharp knife than with the edge of a pencil. Explain this in terms of applied pressure.

The pressure on the cheese is greater with the sharp knife (smaller area). Pressure = *force/area.*

7. Vehicles that need to move easily over loose sand are often equipped with special tires. Would you expect these tires to be wide or thin? Explain your answer.

Vehicles that move over loose sand are equipped with wide tires. The greater the area of the tire the smaller the pressure.

8. U.S. weather reports on TV generally express air pressure in units of inches of mercury. Express a typical air pressure value in units of

a. atmospheres (atm).

b. kilopascals (kPa).

c. millimeters of mercury (mmHg).

a. 1 atm

b. 100 kPa (100 000 Pa)

c. 760 mmHg

9. Explain why a sealed bag of potato chips "puffs out" (increases in volume) if taken from sea level to a high mountain pass.

The sealed bag of potato chips inflates while going upward from sea level because the pressure on the inside of the bag remains constant while the atmospheric pressure on the outside decreases.

10. If the same bag of potato chips (see Question 9) were carried under water by a scuba diver, what would happen to the volume of the bag as it descended into deep water?

The bag would collapse because the pressure on the outside of the bag increases while the pressure on the inside remains constant (assuming temperature is constant).

11. A small quantity of the inert gas argon (Ar) is added to light bulbs to reduce the vaporization of tungsten (W) atoms from the solid filament. What volume of argon at a pressure of 760 mmHg is needed to fill a 0.21-L light bulb at a pressure of 1.30 mmHg? Assume the gas temperature remains constant.

$V_1 = 0.21\text{ L}, P_1 = 1.30\text{ mmHg}, P_2 = 760\text{ mmHg}$

$$V_1 = (P_2V_2)/P_1 = \frac{(1.30\,\text{mmHg} \times .21\text{ L})}{760\,\text{mmHg}} = 0.00036\text{ L} = .36\text{ L}$$

12. A 2.50-L balloon at 298 K contains butane gas, a fuel used in some camping stoves. At constant pressure, the temperature of the balloon is then lowered to 280 K. What will be the new volume of the balloon at this temperature?

$V_1 = 2.50\text{ L}, T_1 = 298\text{ K}, T_2 = 280\text{ K}$

$$V_2 = V_1T_2/T_1 = \frac{2.50\text{ L} \times 280\text{ K}}{298\text{ K}} = 2.35\text{ L}$$

13. What do you think a marshmallow, composed mainly of gas, would look like at temperatures close to absolute zero? Explain your answer.

According to Charles' law, temperature and volume are directly related. At a temperature close to absolute zero, the volume of the gas would approach zero. A marshmallow might appear shrunken and wrinkled, like a prune.

14. Experts recommend measuring automobile tire pressure when the tires are cold. Why is this better than measuring the tire pressure after driving for several hours?

Temperature and pressure are directly related. After driving for several hours, the temperature of the air in the tire would increase, resulting in an increase in pressure, since the volume of the tire remained constant.

15. Soccer players often notice that kicking the ball feels different on cold days than on warm days. Explain why this might be the case.

Temperature and pressure are directly related. On cold days, the pressure of the air on the inside of the ball would be less that it would be on warm days, if volume remains constant.

16. A steel tank at 300 K contains 0.285 L of gas at 1.92 atm pressure. The tank is capable of withstanding a maximum pressure of 5.76 atm. Assuming that doubling the kelvin temperature causes the internal pressure to double,

a. at what temperature will the tank burst?

b. will it burst if placed in a fire burning at 1275 K?

a. $T_1 = 300\text{ K}, P_1 = 1.92\text{ atm}, P_2 = 5.76\text{ atm}$

$$T_2 = 300\text{ K} \times \frac{5.76\text{ atm}}{1.92\text{ atm}} = 900\text{ K} = 627\text{ °C}$$

b. $T_1 = 300\text{ K}, P_1 = 1.92\text{ atm}, T_2 = 1275\text{ K}$

$$P_2 = 1.92\text{ atm} \times \frac{1275\text{ K}}{300\text{ K}} = 8.16\text{ atm}$$

Yes, the tank will explode at 1275 K (about 1000 °C).

17. A sample of gas in a rigid container is heated from 15 °C to 30 °C. By what proportion would you expect the gas pressure to increase?

Pressure and temperature are directly related. Therefore, if the temperatures are converted to kelvins, the change in volume would be a ratio of 303 K/288 K, or about 1.05 times the original volume. Students may make the mistake of saying that the Celsius temperature doubles and therefore, the pressure should also double.

18. Use the kinetic molecular theory to explain each of the following observations.

a. Decreasing the volume of a gas sample at constant temperature causes the gas pressure to increase.

b. At constant volume, the pressure of a gas sample changes when its temperature changes.

c. Filled balloons eventually become "flat" even when they are tightly tied shut.

d. Helium-filled balloons go "flat" faster than those inflated with air.

a. Decreasing the volume means that there is less space in which the molecules can move. This will cause molecules to hit the walls more often per unit time, causing an increase in pressure.

b. At higher temperatures, the kinetic energy of the molecules is greater because of their higher average velocities. This increased kinetic energy will cause more energetic and frequent collisions with the walls and thus higher pressure.

c. Given their high rate of motion and small size, gas molecules will eventually diffuse through the molecular structure of the balloon wall.

d. Helium molecules are smaller in size and exhibit a higher average velocity than molecules of air (N_2, O_2, CO_2). Helium molecules collide more frequently with the balloon walls and fit more easily between the molecular-scale openings.

19. Think of a pool table as a model of the behavior of gases.

a. How would you use pool balls to demonstrate the effect of an increase in the temperature of a gas sample?

b. How would you use this model to demonstrate the effect of an increase in the total number of gas molecules?

Each ball represents a molecule of the gas; the volume area of the pool table remains constant.

a. The balls would move faster to demonstrate the effect of the increase in temperature.

b. The total number of gas molecules may be demonstrated by increasing the number of balls on the table.

20. What volume would 8.0 g of helium (He) gas occupy at 0 °C and 1 atm?

$$8\text{ g He} \times \frac{1\text{ mol He}}{4.0\text{ g He}} \times \frac{22.4\text{ L He}}{1\text{ mol He}} = 44.8\text{ L He}$$

21. How many moles of gas molecules would you expect to find in a 2.0-L bottle of air at 0 °C and 1 atm?

$$\text{Total moles} = 2.0\text{ L} \times \frac{1\text{ mol}}{22.4\text{ L}} = 0.089\text{ mol gas}$$

22. **a.** How many moles of oxygen gas would be present in a 1.0-mL sample at 0 °C and 1 atm?

b. How many molecules of oxygen gas does your answer to Question 22a represent?

a. $\# \text{ moles} = 1.0 \text{ mL} \times \frac{1 \text{ L}}{1000 \text{ mL}} \times \frac{1 \text{ mol}}{22.4 \text{ L}} = 4.5 \times 10^{-5} \text{ mol}$

b. $\# \text{ molecules} = 4.5 \times 10^{-5} \text{ mol} \times \frac{6.02 \times 10^{23} \text{ molecules}}{1 \text{ mol}} = 2.7 \times 10^{19} \text{ molecules}$

23. This chemical equation represents the production of ammonia (NH_3) through the reaction of nitrogen gas with hydrogen gas:

$$N_2(g) + 3\ H_2(g) \longrightarrow 2\ NH_3(g)$$

a. If 1 mol $N_2(g)$ completely reacts with 3 mol $H_2(g)$ in a flexible container at constant temperature and pressure, would you expect the total gas volume to increase or decrease? Why?

b. How many moles of NH_3 would form if 2 mol N_2 reacted completely with hydrogen gas?

c. How many grams of hydrogen gas would be required to react completely with 2 mol N_2?

a. When one (1) mol N_2 reacts with 3 mol H_2 gas, the volume would decrease, starting at about 99.6 L and ending at 44.8 L. Unlike mass, gas volume is not conserved in a chemical reaction.

b. According to the balanced equation, 4 mol $NH_3(g)$ would be produced when 2 mol $N_2(g)$ reacts completely with $H_2(g)$.

c. 2 mol of $N_2(g)$ requires 6 mol $H_2(g)$ to react completely. $6 \text{ mol } H_2 \times \left(\frac{2\text{g } H_2}{1 \text{ mol } H_2}\right)$ 12g H_2.

24. In a particular chemical change, 1 L hydrogen gas (H_2) reacts with 1 L chlorine gas (Cl_2) to produce 2 L hydrogen chloride gas (HCl). All volumes are measured at the same conditions of temperature and pressure. Create a depiction of that chemical reaction, using

a. a sketch involving molecular models.

b. a written chemical equation.

a. Student models would be the same as those in question 9a. in Reviewing the Concepts of Unit 1, Section B.

b. $H_2(g) + Cl_2(g) \rightarrow 2\ HCl(g)$

Connecting the Concepts

25. If the pressure remains constant while the temperature of one mole of gas increases, would you expect the gas volume to increase or decrease? Explain your answer.

The gas volume would increase. Charles' law states that the volume and temperature are directly related if pressure remains constant.

26. Oxygen gas is essential to life as we know it. Earth's atmosphere contains approximately 21% oxygen gas. Would a higher concentration of atmospheric oxygen gas be desirable? Explain.

A higher atmospheric concentration of oxygen would cause more rapid corrosion (oxidation) of metals. Fires would be more common and difficult to control. Living organisms probably could gradually adapt to the metabolic effects of a higher oxygen concentration.

27. Based on your knowledge of the properties of gases, predict what will happen to the density of a 100-g helium gas sample when it is heated from room temperature to 60 °C at constant pressure.

Density = mass/volume. As temperature increases, volume will also increase at constant pressure. The mass will remain constant. Therefore the density will decrease.

28. If air at 25 °C has a density of 1.28 g/L, and your classroom has a volume of 2.0×10^5 L,

a. what is the mass of air in your classroom?

b. Assume that the room temperature is increased. How would that affect the total mass of air in the room

i. if the room were sealed?

ii. if the room were not sealed?

a. Mass of a roomful of air: 1.28 g/L × 2.00×10^5 L = 2.56×10^5 g

b. i. If the room is completely sealed, the mass of air in the room will remain constant.

ii. If the room is not sealed, the increased molecular motion would cause some of the air molecules to spread out to adjacent rooms.

Extending the Concepts

29. **a.** In what ways is an analogy involving an "ocean of gases" useful in thinking about the atmosphere?

b. In what ways does that analogy not apply to the behavior of the atmosphere?

a. Different environments are encountered at different heights. Pressure and temperature change with depth and height.

b. In some ways the atmosphere is more complex than the hydrosphere: (1) pressure changes exponentially, not linearly; (2) the temperature profile creates fairly permanent strata; and (3) its chemical composition varies with height and depth.

30. Some people suggest that they could dive to the bottom of a swimming pool and stay there, breathing through an empty garden hose that extends above the water surface. How would you argue against this plan?

The mass of the water pressing on the diver's lungs becomes so great, even in shallow water, that it quickly becomes impossible for the diver to draw air into the hose. The pressure exerted by the water is greater than the force human lungs can develop to draw in the air.

31. The gas behavior described by Boyle's law is a matter of life and death to scuba divers. On the surface of the water, the diver's lungs, tank, and body are at atmospheric pressure. However, under water a diver's body is under the combined pressure of the atmosphere and the water.

a. Why do scuba divers need pressurized tanks?

b. What would happen to the volume of the tank if it were not strong enough to withstand the pressure of the water outside?

c. Why is it necessary to exhale and rise slowly when ascending from the ocean depths?

d. How do the problems of a diver compare with those of a pilot climbing to a higher altitude in an unpressurized plane?

a. Pressurized tanks are needed for three purposes: to withstand the increased external pressure without being crushed, to enable sufficient air to be transported in a reasonable volume, and to be able to deliver the air at the higher pressures required by the depth of the water.

b. The volume would decrease (crushed).

c. If the diver does not exhale and rise slowly, the volume of the various gases trapped in the lungs and sinuses will rapidly expand, causing extreme pain and lung damage.

d. Divers experience increased pressures during the descent and, like pilots, experience reduced pressures during the ascent. In both cases, Boyle's law applies; pressurized tanks and regulators are used to deliver air at appropriate pressures, regardless of the pressure outside.

32. During extremely hot summer days, flights may not take off from an airport for safety reasons. In terms of Charles' law, suggest an explanation for this safety practice.

Aircraft require the laminar flow of dense air over the wings to create the pressure differentials needed to create lift. When the temperature increases, the volume of the air increases (according to Charles' law), lowering the density of air. This in turn reduces the lift created on the wing, requiring higher takeoff speeds (and lower abort speeds) to create the needed lift. These speeds can reach a point where the runway length will not permit safe takeoff.

33. The kelvin temperature scale is used extensively in theoretical and applied chemistry.

a. Find out how close scientists have approached a temperature of zero kelvins (0 K) in the laboratory setting.

b. Research results from the field of cryogenics (low-temperature chemistry and physics) have created many possible applications. Report on activities in this field.

Kelvin temperatures

a. Scientists have cooled cesium atoms to one-millionth of a degree of absolute zero.

b. Answers will vary. Students should use electronic and paper search techniques that are available in their library media center.

RADIATION AND CLIMATE

Introduction

Section B covers some technology issues associated with the atmosphere. The section begins with a discussion of the electromagnetic spectrum, with emphasis on ultraviolet radiation. The nature of energy input from the sun to the Earth is presented and the reasons for the greenhouse effect are explained. The energy exchanged at the Earth's surface clarifies some of the reasons for climatic conditions. Students gain some idea of the relative quantities of CO_2 produced in respiration and combustion and propose some changes that might lower CO_2 levels. The chemistry of the protection that the ozone layer provides us against ultraviolet radiation and the threat of certain chemicals of human origin to the ozone layer are presented.

B.1 SOLAR RADIATION

This section focuses on the nature of solar radiation and how it interacts with the components of the atmosphere. The energy liberated by fusion on the Sun is transferred to the Earth as electromagnetic radiation. All electromagnetic radiation

- travels at the speed of light (3.0×10^8 m/s).
- does not require a medium in which to travel.
- moves through space as bundles of energy.
- has an effect on living things that is determined by the energy of the radiation.

SECTION B

RADIATION AND CLIMATE

Solid carbon dioxide, $CO_2(s)$, is also known as Dry Ice.

Introduction CD-ROM WWW.

Without Earth's atmosphere, midday sunshine would heat a rock enough to fry an egg and the Sun's ultraviolet rays would burn exposed skin quickly; nights would be so cold that carbon dioxide gas would freeze to a solid. Because Earth's moon lacks an atmosphere, these extreme conditions are not science fiction; they exist there now.

The Sun's radiant energy and the gases that make up Earth's atmosphere combine to maintain a hospitable climate for life on the planet. Along with its vital role as a reservoir for both oxygen gas and carbon dioxide gas, the atmosphere provides protection from the Sun's ultraviolet rays. Additionally, the atmosphere plays an important part in moderating Earth's temperature by controlling how much solar radiation is trapped close to the surface of the planet.

In this section, you will learn about the nature of solar radiation, how it affects humans directly, and how it interacts with the components of the atmosphere.

Solar Radiation CD-ROM WWW.

B.1 SOLAR RADIATION

You can learn more about the process of fusion in Unit 6.

The enormous quantity of energy produced by the Sun, the main external source of energy for Earth, is a result of the fusion of hydrogen nuclei into helium. **Fusion** is the combining of two nuclei to form a new, heavier nucleus. Fusion occurs at high temperature and pressure and liberates an enormous quantity of energy, powering the Sun and other stars.

Some of the energy liberated in fusion is transferred from the Sun to Earth as **electromagnetic radiation,** which consists of a broad range of energetic emissions. It is useful to identify particular types of electromagnetic radiation, each representing a specific energy range, that together make up the **electromagnetic spectrum.** Figure 19 shows major types of electromagnetic radiation. Radio waves and microwaves are at the low-energy end of the spectrum; X-rays and gamma rays are at the high-energy end. Between these extremes are infrared and ultraviolet radiation and the familiar visible spectrum, the only range of electromagnetic radiation visible to the human eye.

The speed of light is about 3.0×10^8 m/s.

Electromagnetic radiation is composed of **photons,** or bundles of energy. Photons travel as waves and, as you might expect, move at the speed of light. Unlike sound waves or ocean waves, electromagnetic waves do not require a medium (or substance) for their movement. They can move through a vacuum.

Photons of electromagnetic radiation have energy ranges characteristic of the type of radiation involved (visible, ultraviolet, gamma, and so on). All waves, including the photon waves that make up electromagnetic radiation, involve oscillation. The rate of oscillation, or the number of waves that pass

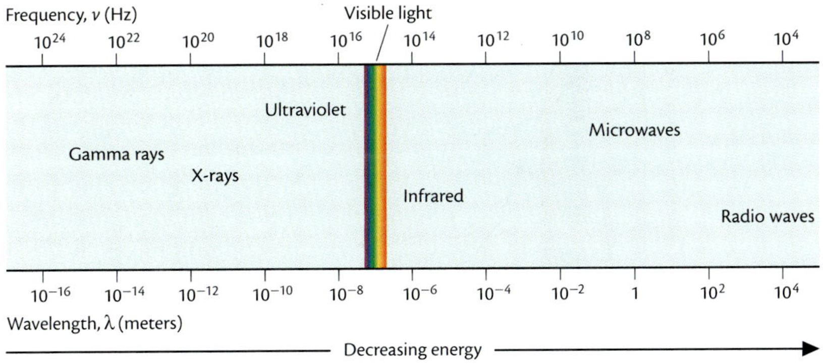

Figure 19 *The electromagnetic spectrum.*

a given reference point per second, is called **frequency.** The frequency of a wave is directly proportional to its energy—high-frequency radiation is also high-energy radiation.

Another characteristic of waves is wavelength, shown in Figure 20. The distance between the tops (or any corresponding part) of successive waves represents the **wavelength.** The wavelength and energy of a photon are inversely proportional; radiation with longer wavelengths is less energetic than radiation with shorter wavelengths.

Wavelength and frequency are inversely proportional.

Photons can transfer their energy when they interact with matter. The photon's energy—and thus its wavelength or frequency—largely determines the nature of its effect on living things and other types of matter.

The Sun emits radiant energy over a large portion of the electromagnetic spectrum. About 9% of this radiation is in the ultraviolet (UV) region,

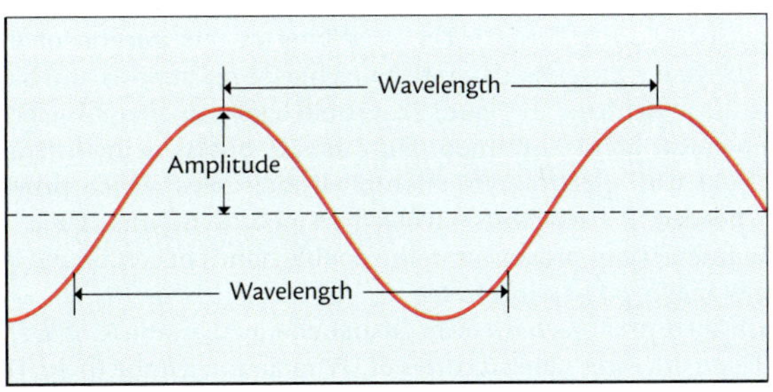

Figure 20 *Parts of a wave.*

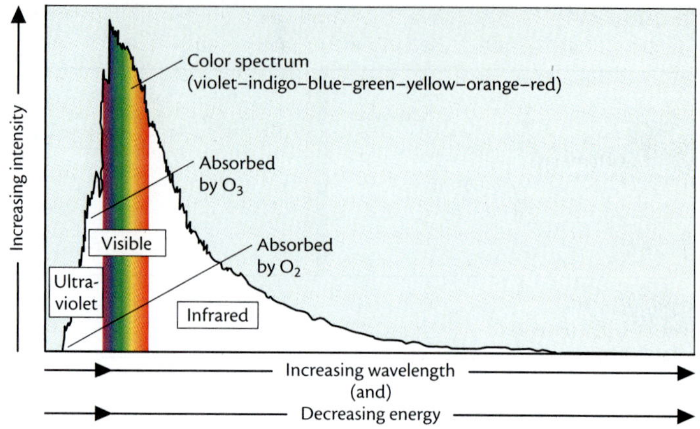

Figure 21 *The solar spectrum. Intensity, plotted on the y axis, is a measure of the quantity of radiation at a given wavelength.*

46% is in the visible region, and 45% is in the infrared (IR) region. The complete solar spectrum is shown in Figure 21. In the next few paragraphs you will be introduced in more detail to these three regions of the solar spectrum.

Infrared radiation consists of that portion of the electromagnetic spectrum just beyond the red end of the visible spectrum. Like microwaves, infrared radiation causes molecules to vibrate or rotate faster (depending on the wavelength of the radiation), which is observed as an increase in the temperature of the material. Most of the infrared portion of the Sun's radiation cannot reach Earth's surface because it is absorbed by CO_2 and gaseous H_2O in the atmosphere. However, some of the shorter-wavelength (higher-energy) solar radiation absorbed by Earth is transformed and reradiated as infrared energy. This reradiated energy is reflected back and retained by the atmosphere—a fact that will come into play later when the role of CO_2 (and certain other gases) in the atmosphere is considered.

In the 1860s, John Tyndall demonstrated the ability of CO_2 and gaseous H_2O to absorb heat radiation.

On a clear day, more than 90% of the visible portion of the solar spectrum reaches Earth's surface. The scattering of this portion of the Sun's energy (by water, air, and dust) is responsible for red sunsets and blue skies, such as those depicted in Figure 22. Visible light can energize electrons in some chemical bonds. An important example of this is the interaction of visible light with electrons in chlorophyll molecules, which provides the energy needed for reactions involved in photosynthesis. Such photon-electron interactions also occur in the double bonds of certain molecules in your eyes, making it possible to see.

Ultraviolet photons have even greater energy than those of visible radiation. There are three subcategories of UV radiation. Of the three, UV-A has the longest wavelengths and thus the lowest energy. UV-B has more energy, can cause sunburn, and with long-term exposure is linked to skin cancer.

278

Figure 22 *Some visible radiation from the Sun is scattered by Earth's atmosphere. This scattering creates the colors observed in the sky.*

UV-C radiation, the most energetic form of ultraviolet radiation, is useful for sterilization since it can kill bacteria and viruses. This is possible because UV-C photons have sufficient energy to break single covalent bonds. As a result, chemical changes can take place in materials exposed to some ultraviolet radiation—including damage to tissues of living organisms. UV-C consists of that portion of ultraviolet radiation with wavelengths shorter than 280 nm, UV-B wavelengths range from 280–320 nm, and UV-A has wavelengths longer than 320 nm.

Approximate Wavelengths:
IR: 10^3–10^4 nm
Visible: about 380–780 nm
UV: 180–380 nm.
(1 nm = 10^{-9} m)

UV-C is absorbed in the atmosphere before reaching Earth's surface. Most UV-B radiation (and much UV-A radiation) does not reach the surface of Earth either; it is absorbed by the stratospheric ozone layer (which you will learn more about later in this unit).

If all the UV radiation reaching the atmosphere actually continued to Earth's surface, it is likely that most life on Earth would be destroyed. Ultraviolet radiation, however, is not all bad. Humans and animals must have some exposure to sunlight, since vitamin D is produced when skin is exposed to moderate amounts of ultraviolet radiation.

Vitamin D is discussed further in Unit 7.

ALWAYS HARMFUL? — ChemQuandary 1

How true is the statement: "All radiation is harmful and should be avoided"?

B.2 EARTH'S ENERGY BALANCE

The mild average temperature (15 °C, or 59 °F) at Earth's surface is determined partly by the balance between the inward flow of energy from the Sun and the outward flow of transformed solar energy into space. However, certain properties of Earth also help determine how much thermal energy

ALWAYS HARMFUL?

ChemQuandary 1

"All radiation" includes the entire electromagnetic spectrum. It includes visible light, which is responsible for vision and is not considered harmful. Avoiding all radiation would mean isolation from radio waves, microwave equipment, all lighting sources, and infrared heaters, which is neither practical nor possible. This statement is obviously false.

B.2 EARTH'S ENERGY BALANCE

Earth's relatively stable temperature is explained by a balance between solar radiation entering the Earth's atmosphere and the outward flow of energy into space from the Earth. The re-radiation of visible light into lower energy infrared frequencies not only warms the atmosphere but also drives the hydrological system.

The interaction of infrared radiation between molecules in the atmosphere and outer space is known as the greenhouse effect. The role of carbon dioxide and water molecules in maintaining the average surface temperature capable of supporting life is discussed and compared to the average surface temperatures of Mars and Venus.

the planet can hold near its surface—where terrestrial life resides—and how much energy it radiates back into space. The combination of these factors helps create a balance of energy that leads to a hospitable climate on Earth.

Figure 23 shows the fate of solar radiation that enters Earth's atmosphere. About 30% of incoming solar radiation never reaches Earth's surface but is reflected directly back into space by clouds and particles in the atmosphere. Solar radiation is also reflected when it strikes such materials as snow, sand, or concrete at Earth's surface. In fact, visible light reflected in this way allows Earth's illuminated surface to be seen from space.

Of the remaining 70% of solar radiation that reaches Earth's surface, about two-thirds is absorbed, warming the atmosphere, oceans, and continents. Another one-third of this energy powers the hydrologic cycle—the continuous cycling of water into and out of the atmosphere by evaporation and condensation. This energy causes water to evaporate from the oceans, forming clouds, which then release the water as precipitation.

All objects with a temperature above absolute zero (0 K) radiate energy. The quantity of this radiated energy is directly related to an object's temperature in kelvins. Specifically, Earth's surface reradiates most absorbed solar radiation, but not necessarily at its original wavelengths. Some incoming UV and visible radiation is re-emitted at longer wavelengths—in the infrared region of the spectrum. This reradiated energy plays a major role in Earth's energy balance. Some molecules in the air absorb low-energy infrared photons rather than the original radiation, thus holding warmth in the atmosphere.

Chlorofluorocarbons are discussed in more detail later in this unit.

Carbon dioxide and water readily absorb infrared radiation, as do methane (CH_4), nitrous oxide (N_2O), and halogenated hydrocarbons such as CF_3Cl and other chlorofluorocarbons or CFCs. Because clouds consist of droplets of water or ice, they also absorb some infrared radiation. Energy absorbed by these molecules in the atmosphere is reradiated in all direc-

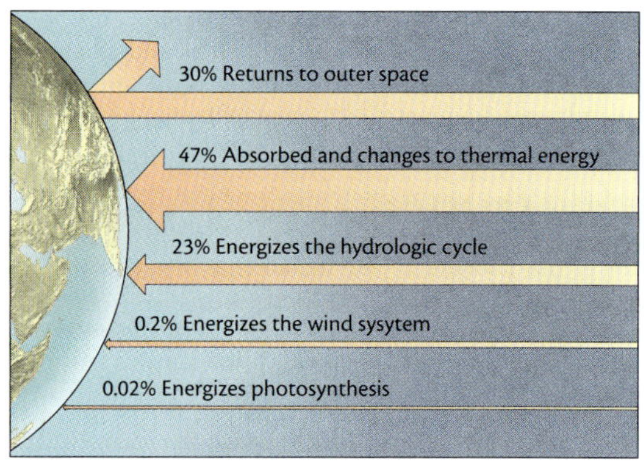

Figure 23 *The fate of incoming solar radiation.*

tions. Thus energy can pass back and forth between Earth's surface and molecules in the atmosphere many times before it finally escapes into outer space.

This trapping and returning of infrared radiation by carbon dioxide, water, and other atmospheric substances is known as the **greenhouse effect** because the process resembles the way thermal energy is held in a greenhouse on a sunny day. Atmospheric substances that effectively absorb infrared radiation are known as **greenhouse gases.**

Without water and carbon dioxide molecules in the atmosphere to absorb and reradiate thermal energy to Earth, the planet would have an average temperature of about −18 °C (0 °F). This temperature is quite close to the average surface temperature on Mars. At the other thermal extreme is the planet Venus, an example of a runaway greenhouse effect. Its atmosphere is composed of 96% carbon dioxide (and clouds made of sulfuric acid!), which prevents the escape of most infrared radiation. As a result, the average temperature on Venus (450 °C) is much higher than on Earth (15 °C). Although some of this difference is due to different planetary positions relative to the Sun, Venus (the second planet from the Sun) is actually hotter than Mercury (the planet nearest the Sun).

Check your understanding of the interaction of radiation with matter and the role of radiation in Earth's surface temperature by answering the following questions.

SOLAR RADIATION — Building Skills 7

1. Why is exposure to ultraviolet radiation potentially more harmful than exposure to infrared radiation?
2. Explain two essential roles played by visible radiation from the Sun.
3. Describe some possible consequences of increasing the concentration of carbon dioxide and other greenhouse gases in the troposphere.
4. Suppose Earth had a thinner (fewer gas molecules) atmosphere than it does now.
   a. How would average daytime temperatures be affected? Why?
   b. How would average nighttime temperatures be affected? Why?
5. Consider Figure 23. In what ways might incoming solar radiation within the LHMP differ from incoming solar radiation on the surface of Earth?

The interaction of solar energy with Earth's atmosphere is one factor in determining climates and weather. Radiant energy from the Sun warms Earth's land and water surfaces. Earth's warm surfaces, in turn, warm the air above them. As the warmer air expands, it is displaced by colder, denser air, causing the warmer air to rise. These movements of warm and cold air masses help create continuous air currents that drive the world's weather.

Answers

1. Ultraviolet radiation has shorter wavelengths than infrared radiation. The shorter the wavelength, the higher the energy of the radiation. Ultraviolet radiation has enough energy to disrupt chemical bonds and harm biological organisms.
2. Visible light activates photoreceptors in the eyes, making sight possible. Visible light also provides energy for photosynthesis reactions.
3. Increased concentrations of carbon dioxide may result in a warming trend, with subsequent side effects such as the melting of polar ice caps and the shifting of sites suitable for agriculture.
4. a. The average daytime temperature would be higher.

   b. The average nighttime temperature would be lower.

   If the atmosphere contained fewer molecules to absorb and re-radiate thermal energy to Earth, it would reach thermal balance at a lower temperature.
5. Incoming solar radiation within the LHMP would be different from that on the surface of Earth because within the LHMP, the radiation escapes back into space in a different fashion, as determined by the materials that make up the dome.

SOLAR RADIATION Building Skills 7

These questions are designed to focus attention on radiation from the sun and the factors that determine the temperature of the atmosphere.

Earth's climate and weather are influenced by (a) the interaction of solar radiation with the atmosphere, (b) the rotation of the Earth on its axis, (c) the revolution of the Earth around the sun, (d) the tilt of the Earth on its axis, and (e) the differing thermal properties of materials on the Earth's surface.

GRAB ANOTHER BLANKET!

ChemQuandary 2

On a clear night there are fewer clouds and thus fewer water molecules in the air to absorb and re-radiate thermal energy to Earth's surface.

B.3 AT EARTH'S SURFACE

The rate at which a substance absorbs energy is determined by its specific heat capacity and its reflectivity. A substance that becomes hot or cold quickly (gram for gram) has a lower specific heat capacity than one that does so more slowly. The high heat capacity of water enables it to moderate the temperature of land areas near oceans and to influence wind patterns.

The differing thermal properties of materials on Earth's surface represent another factor that influences climate.

Climate, which refers to the average or prevailing weather conditions in a region, is influenced by other factors. One factor is Earth's rotation on its axis, which causes day and night and influences wind patterns. Other significant factors are Earth's revolution around the Sun and tilt on its axis. The combination of these factors causes uneven distribution of solar radiation, which results in the four seasons, and thus influences climate.

GRAB ANOTHER BLANKET! ChemQuandary 2

On a clear night, the outside air temperature decreases much more quickly than on a cloudy night. Why is this?

B.3 AT EARTH'S SURFACE

If you live in the South or Southwest, or have visited there, you may have noticed that many cars are light-colored—both inside and out. Do you know why? A particular property of materials helps keep these vehicles cooler than those with darker surfaces. When radiation strikes a surface, some photons are absorbed—increasing the surface temperature—and some are reflected. The proportion of radiation that is reflected, expressed as the material's **reflectivity,** does not contribute to raising the object's temperature.

Light-colored surfaces reflect more radiation and therefore remain cooler than dark-colored surfaces. Clean snow, for example, reflects almost 95% of solar radiation, whereas forests reflect very little. Variations in the reflectivity of materials at Earth's surface help determine local surface temperatures. So, on a hot day, it is much more comfortable to walk barefoot across a lawn than across an asphalt parking lot. The lawn reflects 15–30% of the Sun's rays, whereas the black asphalt absorbs almost all the radiation.

Every material has a characteristic reflectivity and heat capacity, which together determine how much and how fast the material warms. As you have already learned (page 204), specific heat capacity is the quantity of thermal energy (heat) needed to raise the temperature of 1 g of a material by 1 °C.

In effect, specific heat capacity is a measure of a material's "storage capacity" for thermal energy. The lower a material's specific heat capacity, the more its temperature increases when a certain quantity of energy is added. The higher the specific heat capacity, the smaller the temperature increase will be for a given quantity of added energy. Thus materials with higher specific heat capacities can store more thermal energy. For instance, tin, with a specific heat capacity of 0.222 J/(g·°C), will show a larger temperature change in response to a certain quantity of energy input than paraffin, with a specific heat capacity of 2.9 J/(g·°C). Figure 24 summarizes the specific heat capacities of some other common materials.

Water's unique properties make it influential in the world's climates. One such property is its very high specific heat capacity, higher than that of most other common materials, as you see in Figure 24. Because of this, bodies of water can store large quantities of thermal energy and thus are slow to heat up or cool down. By contrast, land surfaces have much lower specific heat capacities; they heat up and cool down much more rapidly.

If you walked barefoot across an asphalt parking lot on a hot day, you would probably find walking on the painted stripes more comfortable than walking on the dark asphalt.

Specific Heat Capacities for Common Materials at 20 °C

Material	Specific Heat Capacity J/(g·°C)
air	1.00
aluminum	0.895
brass	0.380
carbon dioxide	0.832
copper	0.387
ethyl alcohol	2.45
gold	0.129
granite	0.803
iron	0.448
lead	0.128
silver	0.233
stainless steel	0.51
water	4.18
zinc	0.386

Figure 24 *Specific heat capacities.*

THERMAL PROPERTIES OF MATERIALS

Building Skills 8

1. Explain why each of the following occurs on a hot, sunny day.
   a. The chrome fender of a white convertible with its top down is cooler to the touch than its dark red seats.
   b. An asphalt sidewalk is warmer to the touch than a concrete sidewalk.
2. Use data in Figure 24 to explain why water is a better fluid for use in a hot pack than ethyl alcohol.
3. The surface of beach sand feels hotter than grass on a hot day. Which property is likely to be more responsible for this observation—specific heat capacity or reflectivity? Explain.
4. a. Which would you expect to be hotter, on average, in the summer, a city far away from any large body of water or a similar city located near a large body of water?
   b. Explain your choice.

THERMAL PROPERTIES OF MATERIALS

Building Skills 8

Students apply what they have learned about specific heat capacity and reflectivity to analyze common situations they may have experienced in their lives.

Answers

1. a. The chrome fender reflects more energy than the dark seat covers.
   b. The concrete is nearly white and reflects more light than the dark colored asphalt.
2. A given mass of water can store more heat energy and can release it over a longer period of time than the same mass of alcohol or other fluids having lower heat capacities.
3. Heat capacity is responsible for the higher temperature of sand. Although the sand is white and should reflect a lot of light energy, the sand's heat capacity allows the sand to get hotter than the grass.
4. a. A city far away from any large body of water will be hotter.
   b. Water has a moderating effect; therefore, the city located away from the water would be hotter. Water temperature rises and cools more slowly than that of air, soil, or asphalt. These temperature shifts between land and water lead to breezes. During the day there is a sea breeze, and at night there is a land breeze. Therefore, near the water the daytime temperatures do not get as high, and at night the temperatures do not drop as much as they do away from the water.
5. Water tends to have a more stable temperature. It can store more heat energy and release it over a longer period of time.

B.4 SPECIFIC HEAT CAPACITY

Laboratory Activity

Students use the specific heat capacity of a metal to identify it. The heat gained by the water must be equal to the heat lost by the metal (Law of Conservation of Energy). The heat capacity of the metal is calculated using the formula $E = mC\Delta T$.

Time

50 minutes

Materials

12 400-mL beakers
12 metal samples
12 balances
12 styrofoam cups
12 graduated cylinders
12 thermometers
12 hot plates
12 tongs

Safety

Students should wear safety goggles and aprons throughout the laboratory activity. Caution should be exercised in handling hot water.

Advance preparation

Metal samples may be selected from the following: lead (Pb); aluminum (Al); tin (Sn); copper (Cu); or zinc (Zn). Aluminum and zinc produce a larger temperature change to observe. Scientific supply companies sell the metals as density or specific heat kits. The physics department may have these kits available. Although not optimum, a 4-oz lead fishing weight may support immediate requirements. Often technology departments can also produce desired samples. The specific heat capacity of tin is 0.12.

5. A scientist designing an enclosed habitat specifies that a pond be placed within the habitat. Aside from aesthetic considerations, why might this be a good idea?

B.4 SPECIFIC HEAT CAPACITY

Laboratory Activity

Lab Video

Introduction

As you have just learned, one characteristic property of a material is its specific heat capacity. In this activity, you will use this property to determine the identity of a metal sample.

To do this, you will investigate the transfer of thermal energy from a hot sample of metal to cool water. The quantity of thermal energy gained by the water must be the same as the quantity of thermal energy lost by the metal. The specific heat capacity of the metal can be calculated using the following formula:

$$\text{Thermal energy } (E) = \text{Mass of material } (m) \times \text{Specific heat capacity of material } (C) \times \text{Change in temperature } (\Delta T)$$

or

$$E = m\,C\,\Delta T$$

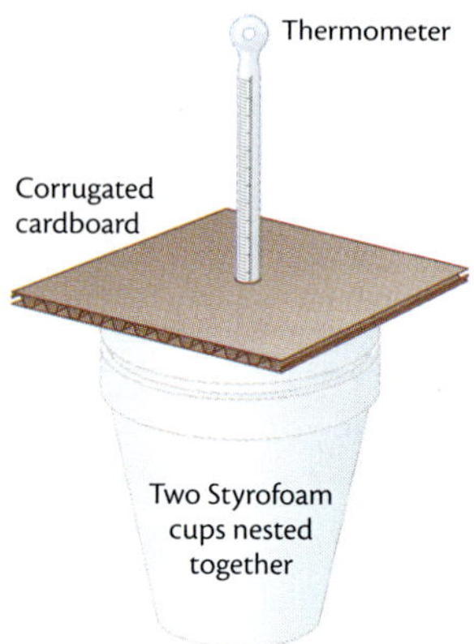

Figure 25 *Simple Styrofoam-cup calorimeter.*

Procedure

1. Half-fill a 250- or 400-mL beaker with water. Place the beaker on a hot plate and start heating.
2. Obtain a metal sample.
3. Determine and record the mass of the metal sample.
4. Place the metal sample into the water heating in the beaker. Allow the water to come to a boil, and keep it boiling for several minutes.
5. While the water is heating and boiling, obtain or set up a calorimeter. A simple version is shown in Figure 25.
6. Accurately measure a volume of water that represents about 60–80% of the volume of the calorimeter. Record this value. Add the water to the calorimeter.
7. Determine and record the temperature of the water in the calorimeter.
8. Determine and record the temperature of the water boiling on the hot plate.
9. Once the water has boiled for several minutes, use tongs to remove the metal sample and place it in the calorimeter.
10. Stir the water in the calorimeter, and record its temperature every 30 seconds until it reaches a maximum value and starts to drop.
11. Wash your hands thoroughly before leaving the laboratory.

Lab tip

Be sure that students remove the excess water from the metal before placing it into the water in the Styrofoam cup. Any water remaining on the metal could alter the specific heat capacity calculation.

Calculations

1. $E = \text{Mass}_{water} \times C_{water} \times \Delta T_{water}$
   Students will use this formula to compute the energy. A typical calculation might be:
   $$H = 200.0\text{ g} \times \frac{4.18\text{ J}}{\text{g}\cdot{}^\circ\text{C}} \times (22^\circ\text{C} - 20^\circ\text{C}) = 1670\text{ J}$$
2. $\text{Mass}_{water} \times C_{water} \times \Delta T_{water} = \text{Mass}_{metal} \times C_{metal} \times \Delta T_{metal}$

Calculations

1. Determine the quantity of thermal energy absorbed by the water in the calorimeter.
2. Assume that the thermal energy absorbed by the water is equivalent to the thermal energy released by the metal sample. Also assume that the metal sample was initially at the temperature of the boiling water. Calculate the specific heat capacity of the metal.

Questions

1. Using the calculated specific heat capacity value as well as data in Figure 24 (page 283), identify your metal sample.
2. The degree of certainty you can have about the identity of your metal sample depends in part on sources of error in the experiment. List and explain several potential sources of error in the experimental procedure that you followed.
3. Closely related to sources of error are assumptions made to simplify the experiment.
   a. List some assumptions you made in completing this experiment and its calculations.
   b. Evaluate each assumption in terms of how reasonable it is.
4. A recycling company decides to separate the various metals it receives by using the different specific heat capacities of the metals.
   a. Do you think this is a reasonable plan? Explain.
   b. Propose an alternative method of separating the metals.

Specific heat capacity and reflectivity of materials are two factors that affect energy balance at the surface of Earth. In the next section, you will learn more about the energy balance in the atmosphere as you follow the fate of global carbon.

B.5 THE CARBON CYCLE

More than 100 years ago it was suggested that a significant increase in the burning of fossil fuels might release enough carbon dioxide into the atmosphere to affect Earth's surface temperature. This suggestion was based on the idea that human activity can affect the processes in natural ecosystems, producing changes that might not always be beneficial. In particular, the use of fossil fuels by humans might perturb the natural movement of carbon within Earth's systems—the global **carbon cycle.**

Remember that a consequence of the Law of Conservation of Matter is that the numbers of atoms of all stable elements on Earth are essentially fixed. However, these atoms are found in many physical states and combined in many different compounds. In the carbon cycle, illustrated in Figures 26 and 27 (page 286), the different forms and compounds in which carbon atoms are found can be considered chemical reservoirs of carbon.

Answers

1. Students should be able to identify their metal sample. Typical experimental specific heat capacities are within 15% of the actual value.
2. Sources of error may include imprecise temperature and water measurements, heat absorbed by the thermometer, or heat lost from the calorimeter.
3. a. Assumptions:
   - The boiling temperature of the water is the same as the temperature of the metal.
   - All of the heat lost by the metal was gained by the water in the calorimeter.
   - No heat was exchanged between the calorimeter and the environment.

   b. Evaluations vary with the types of apparatus used.
4. a. To determine the specific heat capacities of the metals, each metal must be heated to a known temperature; then the quantity of heat the metal gives off to water must be measured. This is not a reasonable plan.

   b. Students will propose using such physical and chemical properties as magnetism, reactivity with acids, density, or appearance as the basis for separating the metals.

B.5 THE CARBON CYCLE

Although the number of carbon atoms on the Earth is fixed, the different forms and compounds of carbon are constantly being exchanged among various chemical reservoirs. The reservoirs of carbon include atmospheric CO_2 and CO_2 dissolved in lakes and oceans, limestone ($CaCO_3$) and organic molecules. Any change of carbon within its reservoir requires energy to be absorbed or released. The increases in the percentage of carbon dioxide in the Earth's atmosphere is due to a decrease in the amount of forest and the combustion of natural gas and gasoline.

Figure 26 *The carbon cycle—major pathways within the biosphere.*

These reservoirs include atmospheric CO_2, calcium carbonate ($CaCO_3$) in limestone, natural gas (CH_4), and organic molecules, to name a few. All movements within the carbon cycle—and thus among these reservoirs—either require energy or release energy. For instance, plants use CO_2 and energy from the Sun to form carbohydrates in photosynthesis. The carbohydrates are consumed by other organisms, and are eventually broken down or oxidized, releasing energy for use by those consuming organisms.

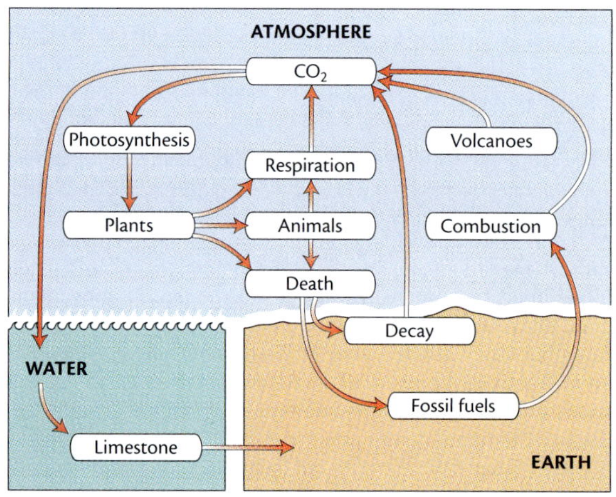

Figure 27 *The carbon cycle—relationships among major carbon reservoirs.*

286

The carbon used and circulated in photosynthesis represents only a tiny portion of the available global carbon. Gaseous CO_2 continually moves between the atmosphere and the oceans. In fact, 71% of Earth's carbon (in the form of CO_2) is contained in its oceans. Another 22% is trapped in fossil fuels and in carbonate rocks formed when dissolved CO_2 reacted with water to form carbonates, then sediments, then rocks. Dead organisms and terrestrial ecosystems (trees, crops, and other plants) each account for half of the remaining global carbon budget.

Without human activity, the distribution of carbon within its various reservoirs would remain relatively unchanged over time. Atmospheric CO_2 levels, however, have increased by about 30% since 1800. This increase can be explained as the result of several processes. For example, when limestone—a form of calcium carbonate ($CaCO_3$)—is decomposed to calcium oxide (CaO) to make concrete, CO_2 is released. Clearing forests removes vegetation that would ordinarily consume CO_2 through photosynthesis. The build-up of CO_2 is compounded when cuttings and scrap timber are then burned, releasing CO_2 into the atmosphere. Most significantly, burning fossil fuels releases CO_2 into the air, as these equations clearly illustrate:

Burning coal: $C(s) + O_2(g) \longrightarrow CO_2(g)$

Burning natural gas: $CH_4(g) + 2\,O_2(g) \longrightarrow CO_2(g) + 2\,H_2O(g)$

Burning gasoline: $2\,C_8H_{18}(g) + 25\,O_2(g) \longrightarrow 16\,CO_2(g) + 18\,H_2O(g)$

As you continue to learn about the role of CO_2 in the atmosphere, keep in mind processes that are likely to generate CO_2 within the LHMP and think about the impact of this added CO_2 on life in this closed system. Also consider how the carbon cycle within the LHMP may be different from Earth's global carbon-cycle system.

B.6 CARBON DIOXIDE LEVELS

Laboratory Activity

Introduction

Air usually has a very low concentration of CO_2. However, the CO_2 concentration in a small, enclosed space can be substantially increased by burning coal or petroleum, by allowing organic matter to decompose, or by accumulating a crowd of people or animals.

CD-ROM WWW. Lab Video

In this activity you will estimate and compare the amounts of CO_2 in several air samples. To do this, the air will be bubbled through water that contains an indicator, bromthymol blue. Carbon dioxide reacts with water to form carbonic acid:

$$CO_2(g) + H_2O(l) \longrightarrow H_2CO_3(aq)$$

As the concentration of carbonic acid in the bromthymol blue solution increases, the indicator color changes from blue to green and finally to yellow.

B.6 CARBON DIOXIDE LEVELS

Laboratory Activity

Students estimate the amount of carbon dioxide present in normal, combusted, and exhaled air by monitoring the change in acidity caused by CO_2 absorption in a solution. The indicator is bromthymol blue.

Time

50 minutes

Materials (for a class of 24 working in pairs)

12 small glass (not plastic) funnels (7-cm diameter)
12 250-mL filter flasks, wrapped with duct tape or towels
12 1-hole stoppers to fit filter flasks
12 aspirator adapters for faucets
12 small (30-cm) pieces of rubber aspirator hose
24 straight glass tubes
24 glass bends
24 ring stands and clamps
36 rubber connectors
36 test tubes (18 × 150 mm)
12 large candles
12 straws
bromthymol blue solution
12 250-mL Erlenmeyer flasks
0.1 M NaOH (4.0 g NaOH in 1000 mL solution, placed in each of six 15–20 mL dropper bottles)

Safety

Students should wear safety goggles and aprons throughout the laboratory activity. Insert the glass tubes into the rubber stoppers yourself to prevent student injury. If you decide to allow students to do this, emphasize the proper technique: using glycerin, protective towels, a close grip on the tube, and a twisting motion.

years have occurred since 1983. An international panel of climate scientists has predicted that under a business-as-usual scenario (in which no steps are taken to control the release of CO_2 and other greenhouse gases), global mean temperatures will rise about 0.3 °C each decade over the next century, leading to an increase of 1.0 °C to 3.5 °C by the year 2100. That prediction can be compared to the fact that Earth has warmed only 3 °C to 5 °C since the depths of the last Ice Age some 20 000 years ago.

The 1996 report of the United Nations Intergovernmental Panel on Climate Change states that Earth's average global surface air temperature has increased by 0.3 °C to 0.6 °C over the past 100 years. The 2000-page report was written by 500 climate scientists and reviewed by another 500 climate experts.

Greenhouse gases were first discussed on page 281.

Although most climate specialists agree that Earth's average temperature will increase about 1.0 to 3.5 °C over the next century, there is still disagreement about the causes of this warming. New data indicate that the lower atmosphere may not be warming at the same rate as Earth's surface, suggesting that factors other than the build-up of greenhouse gases may be responsible for the warming trend. Among these other factors is the circulation of the oceans. Ocean currents, such as El Niño and La Niña, transfer thermal energy within the oceans and can affect climates when this energy is transferred to land and atmosphere.

What are the projected effects of global warming? Based on the assumption of a temperature increase of 0.5 °C per decade, the oceans are predicted to rise about 5 cm each decade over the next century. This is due to the melting of polar ice caps and expansion of ocean water with increasing temperature. Such a change would produce an approximately 15-cm increase by 2030, a situation that would cause widespread flooding in major coastal cities throughout the world. The consequences of such flooding include serious health issues brought about by wastewater, agricultural runoff, and drinking water intermingling under flood conditions. Regional climate changes, including reduced summer precipitation and soil moisture in North America, would be expected. Climatic changes would not be uniform; the greatest temperature increases are predicted north of an imaginary line roughly connecting Sacramento, Denver, Kansas City, Cincinnati, and Philadelphia. Such a situation could have a major impact on food-growing areas. For example, northern regions might benefit from lengthened growing seasons, while southern growers would likely shift to crops that can benefit from warmer winters.

Unfortunately, predicted effects do not end there. Global warming would be expected to result in increases in extreme weather events such as floods, droughts, and heat waves. Why? A warmer atmosphere can hold more moisture, making big storms more likely but also drying out the land more quickly between storms. Another consequence of this trend would be a tendency toward shorter winters with more blizzards.

Scientists today understand the influence of CO_2 and other greenhouse gases on world climate better than they did even a decade ago. The use of sophisticated computer modeling has enhanced their research. The new understanding supports the idea of a surface global warming trend. In June

292

1992, representatives of more than 150 nations developed the Framework Convention on Climate Change, in which they agreed to develop policies and procedures to reduce emissions of greenhouse gases. The third meeting of parties to this agreement, held in Kyoto, Japan, in 1997, resulted in a protocol for addressing climate change. This protocol will enter into force 90 days after its ratification by at least 55 nations, including developed countries, which represent at least 55% of total 1990 CO_2 emissions. However, although 84 nations signed the protocol, many (including the United States) had not ratified it by early 2000.

The Kyoto Protocol sets ambitious goals for reducing greenhouse gas emissions. For industrialized nations, this task involves developing energy-efficient technologies, relying more heavily on renewable energy, and applying alternative processes that do not release greenhouse gases. For instance, an alternative process has been developed for producing the polystyrene foam used to make egg cartons and meat trays. This process uses CO_2 that would normally be released during the production of ammonia. The captured CO_2 replaces chlorofluorocarbons (CFCs) that had been used in this foam-producing process, thus reducing greenhouse gases in two ways. First, CO_2 that would normally be released by another industry is used. Second, CO_2 replaces CFCs, which have a global-warming potential 1700 times higher than that of CO_2.

TRENDS IN CO_2 LEVELS

Building Skills 9

The CO_2-level data given in Figure 30 show average measurements taken at the Mauna Loa Observatory in Hawaii. These data are also available on the World Wide Web. If possible, try to locate the most recent data before you start the following activity.

1. Graph the data in Figure 30. Prepare the *x* axis to include the years 1900 to 2050, and the *y* axis to represent CO_2 levels from 280 ppm to 600 ppm. Plot the data and draw a smooth curve to represent the trend in the plotted points.
2. Assuming the trend in your smooth curve will continue, extrapolate from your curve with a dashed line from the last year for which you have data to the year 2050.

You can now make and evaluate some predictions using the graph you have just completed.

3. What does your graph indicate about the general change in CO_2 levels since 1900?
4. Based on your extrapolation, predict CO_2 levels for
   a. next year.
   b. the year 2020.
   c. the year 2050.
5. a. Which prediction from Question 4 is likely to be the most accurate?
   b. Why?

CO_2 in the Atmosphere

Year	Approximate CO_2 Level (ppm by volume)
1900	287
1920	303
1930	310
1960	317
1965	320
1970	325
1972	328
1974	330
1976	332
1978	335
1980	338
1982	341
1984	344
1986	347
1988	351
1990	354
1992	356
1994	358
1996	363
1998	367

Figure 30 *Approximate carbon dioxide levels in the atmosphere, 1900–1998.*

TRENDS IN CO_2 LEVELS

Building Skills 9

This activity illustrates the use of trends to predict the future. Students plot known carbon dioxide concentrations by years and extrapolate to predict the carbon dioxide concentration in the year 2050.

Emphasize to students that a plot on a graph maybe either a curve or a straight line.

A transparency of the graph may be useful in helping students to extrapolate the results. During the discussion be sure to emphasize the limitations of extrapolation.

The web address for this data is **www:cdiac.esd.ornl.gov/ftp/maunaloa-co2/maunaloa.co2** and is easily located by searching for Mauna Loa CO_2.

Answers

1. No answer required.
2. Answer should be based on student graphs.
3. Carbon dioxide levels have increased since 1900.
4. a. approximately 370–380 ppm
   b. approximately 425 ppm
   c. 500 or more ppm
5. a. The predictions nearest the last measured data are more accurate than the one for 2050.
   b. The rate of carbon dioxide buildup in the atmosphere may change from previous rates. The further into the future we extrapolate, the greater the change in rate can become and the more uncertain our prediction.

6. Yes, the graph predicts the doubling of the carbon dioxide level in (b) the next 50 to 60 years.
7. a. The assumption is that there will be no drastic change in the processes that have produced the known data.

   b. The reliability of the prediction cannot be greater than the reliability of the data itself.
8. Carbon dioxide levels depend on the location, the type of industry, and the climate of the location. Mauna Loa is located on a tropical island with low industrial development but near active volcanoes.

B.8 PLANNING FOR GREENHOUSE GASES

Making Decisions

Students are asked to make some decisions on how to control the emissions of greenhouse gases on the LHMP.

Answers

1. a. methane: decomposition

   b. nitrous oxide: agricultural and industrial activities, internal combustion engines

   c. carbon dioxide: respiration, decomposition, combustion

   d. water vapor: respiration, decomposition, combustion
2. a. methane: could be used as a fuel

   b. nitrous oxide: reduce sources

   c. carbon dioxide: photosynthesis consumes CO_2

   d. water vapor: water for plants, maintain adequate humidity in LHMP.
3. The balance of energy flow depends on the relative amount of each gas in the system, as well as the physical structure of the LHMP.

6. a. Does your graph predict a doubling of the 1900 CO_2 level?
   b. If yes, in what year does this doubling occur?
7. a. What assumptions must you make when extrapolating from known data?
   b. How do these assumptions affect the accuracy of your predictions?
8. Why might the data for Mauna Loa be different from data for other locations around the planet?

B.8 PLANNING FOR GREENHOUSE GASES

Making Decisions

Later in this unit, you will be developing an air-quality management plan for a prototype Moon base—the LHMP described in the unit opener. One issue you will have to consider is the control of greenhouse-gas emissions. The following questions will help guide your approach to this aspect of the management plan.

1. Within the LHMP, what processes are likely to take place that would lead to the generation of each of these gaseous substances?
   a. methane
   b. nitrous oxide
   c. carbon dioxide
   d. water vapor
2. What processes within the LMHP could make use of or eliminate each gaseous compound in Question 1?
3. Based on your answers to Questions 1 and 2, would you expect the greenhouse effect to provide a stable, increasing, or decreasing temperature within the LHMP?

Questions & Answers CD-ROM WWW.

As you learned in Unit 1, water is naturally cleansed through processes in the hydrologic system. Similar cleansing of air takes place in the atmosphere—impurities are converted to more polar, water-soluble forms (hydrocarbons to CO_2 and H_2O, sulfur to SO_2, nitrogen to NO_2, for instance). These substances are then removed from the atmosphere through precipitation, or "raining out." The consequences and effects of these natural processes and their possible effects of human influences are discussed in Section C.

SECTION SUMMARY

Reviewing the Concepts

1. **a.** List the main types of electromagnetic radiation in order of increasing energy.

b. Describe how each type of radiation listed in your answer to Question 1a affects living things.

a. In order of increasing energy: infrared (least), visible (intermediate), and ultraviolet (highest).

b. Infrared radiation is important in heating the atmosphere, visible radiation provides energy for photosynthesis, and exposure to ultraviolet radiation is essential to vitamin D production.

2. Write an equation or a sentence describing the relationship between the wavelength of electromagnetic radiation and its energy.

The wavelength is inversely proportional to its energy. The shorter the wavelength, the higher the energy.

3. Why is the word *spectrum* a good descriptor of the types of energy found in electromagnetic radiation?

A "spectrum" implies a range. Each type of electromagnetic radiation covers a specific range of frequencies and wavelength. Exposure to X-rays is dangerous.

4. Describe some advantages and disadvantages of a hypothetical "flashlight" that would emit X-rays instead of visible light.

The hypothetical "flashlight" would not be visible to the human eye but could be used as an X-ray machine.

5. Why is visible light useful in photosynthesis, while other forms of electromagnetic radiation are not?

Visible light interacts with the chlorophyll molecules in green plants. This interaction provides the energy needed for the reactions involved in photosynthesis.

6. Ultraviolet light is used to sterilize chemistry-laboratory protective goggles. Why is it effective for this use?

The photons in ultraviolet light-C (UV-C) have sufficient energy to break single covalent bonds, which can kill bacteria and viruses. This same chemical change can cause damage to the tissues of other living organisms.

7. Rank the types of electromagnetic radiation in terms of how well each is absorbed by the atmosphere.

High-energy solar radiation penetrates the atmosphere, but low-energy infrared radiation is temporarily trapped.

8. Explain how the atmosphere absorbs and distributes solar energy, thus protecting life on Earth.

The stratosphere contains the ultraviolet-absorbing ozone that protects organisms. It also serves as a barrier that prevents the escape of molecules from the troposphere. The latter is accomplished by a temperature inversion between the troposphere and the stratosphere, which prevents molecules from passing between the layers.

9. **a.** Compare ocean water and beach sand in terms of how quickly each heats up when exposed to sunlight.

b. What properties of these two materials account for that difference in their behavior?

a. Beach sand heats up faster during the day and cools off faster at night than water.

b. The heat capacity of water is much greater than that of beach sand.

10. From a scientific viewpoint, why do many desert dwellers wear white or light-colored clothing?

The color white reflects all frequencies of visible light, whereas colored materials absorb some of the frequencies. Absorbed visible light is converted into molecular motion (thermal energy). Thus, wearing white will help keep one cooler.

11. Describe how atmospheric CO_2 and water help maintain moderate temperatures at Earth's surface.

Carbon dioxide and water trap lower-energy infrared radiation that is re-emitted by Earth. This is called the greenhouse effect. As a result, the Earth's average temperature is maintained at a higher level (15 °C) than would be the case without this insulating blanket.

12. List two ways in which human activities increase the amount of CO_2 in the atmosphere.

Burning fossil fuels and removing green plants from the surface of Earth increase the amount of carbon dioxide in the atmosphere. Burning appears to be the major factor.

13. How could the composition of the atmosphere be altered to produce

a. an increase in the average surface temperature of Earth?

b. a decrease in the average surface temperature of Earth?

a. An increase in the concentrations of carbon dioxide and water would result in a higher average surface temperature on the Earth.

b. A decrease in the concentrations of carbon dioxide and water would result in a lower average surface temperature on the Earth.

14. Explain why, on a sunny winter day, a greenhouse with transparent glass walls is much warmer than a structure with wooden walls.

The greenhouse allows light to enter but reflects infrared (heat) radiation from the materials inside the greenhouse. The wood walls block most incoming radiation and as a result no heating can occur.

15. What would be the effect of significant increases in carbon dioxide, methane, and water vapor in Earth's atmosphere?

A significant increase in carbon dioxide, methane, and water vapor in Earth's atmosphere would cause an increase in the surface temperature of Earth. This increase could cause CO_2 stored in ice, water, and the frozen floors of the northern forests to be released, thus increasing the amount of these gases in the atmosphere. The result could be an even greater increase in the surface temperature.

16. What natural and human processes might increase the amount of methane in the atmosphere?

The amount of methane in the atmosphere is increased by the decomposition of plants and animal wastes, the refining of fossil fuels, and the raising of livestock.

17. **a.** Sketch how a greenhouse works.

b. Sketch how the global greenhouse effect works.

The illustration should show the glass in a greenhouse allowing light to penetrate but infrared radiation being reflected back to the interior. The global system is much the same, except the upper atmosphere takes the role of the greenhouse glass.

Connecting the Concepts

18. How might each of the following conditions affect the level of your exposure to electromagnetic radiation?

a. living near a large industrial city

b. living at high altitude

c. living near the ocean or other large body of water

a. increase exposure to electromagnetic radiation

b. increase exposure to electromagnetic radiation

c. no effect

19. Explain hazards associated with not wearing suitable sunglasses at the beach on a sunny day.

Dark lenses that do not filter UV light can be especially dangerous. The dark filter causes the iris to open wide, allowing harmful UV light to enter.

20. How does the electromagnetic radiation that reaches the Moon's surface differ from the electromagnetic radiation that reaches Earth's surface?

The moon has no atmosphere, so there would be no absorption or reflection of any of the radiation by an atmosphere.

Extending the Concepts

21. Investigate how night-vision goggles work to allow objects to become visible in low-light conditions.

Night-vision goggles visualize infrared radiation as images. The goggles are sensitive to a broad spectrum of light from visible light to infrared. A special illuminator can increase the light available at the infrared end of the spectrum by converting radiation that is not visible to the human eye.

22. Some animals "see" a different range of the electromagnetic spectrum from the wavelengths to which the human eye responds. Investigate some examples, and suggest the survival value of these abilities.

One example: bees are sensitive to ultraviolet radiation, which may help them locate certain varieties of flowers to obtain food.

23. The presence of carbon dioxide in the atmosphere is only one part of the global carbon cycle. Investigate the current understanding of the carbon cycle, with particular attention to the role of oceans as a major "carbon sink."

Look for student explanations that reflect the fact that carbon is recycled constantly between living and nonliving systems (see Figure 26). Great amounts of carbon are stored in oceans as dissolved CO_2.

24. Scientists use deep-ice samples from the Antarctic to estimate the carbon dioxide concentrations in the atmosphere many thousands of years ago. Prepare a report on how the samples are analyzed to find this information.

Tiny bubbles of ancient atmospheric gas are locked in deep ice samples. Look for students to discuss the techniques of analyzing these samples.

SECTION C ACIDS IN THE ATMOSPHERE

During the 1960s and '70s, concern arose over dramatic declines in fish populations native to lakes in Scandinavia and eastern North America. In the 1980s, central European countries, notably Germany, reported large numbers of dead or dying trees. Although separated by many thousands of kilometers, these problems were and continue to be linked by the same cause: air contamination in the form of acid rain.

CD-ROM WWW. Acids in the Atmosphere

C.1 ACID RAIN

What is acid rain? In order to answer that question, it is first important to note that "acid rain" does not have to be rain—it can be any form of atmospheric precipitation. **Acid rain** is defined as fog, sleet, snow, or rain having a pH lower than about 5.6, the average pH of precipitation in the absence of air pollution. You probably recall that pure water is neutral, with a pH of 7. (See page 57.) Why, then, does natural precipitation have a pH less than 7?

The phrase "acid precipitation" is sometimes used instead of "acid rain," indicating that any form of precipitation—fog, sleet, snow, and rain—can be acidic.

In Unit 1, you learned that water samples—although they may be clean and healthful—are rarely pure. Even water that evaporates from Earth's surface recondenses and dissolves atmospheric gases such as nitrogen, oxygen, and carbon dioxide. The pH of rainwater is normally slightly acidic, at about 5.6, due mainly to the reaction of carbon dioxide with water to form carbonic acid:

$$CO_2(g) + H_2O(l) \longrightarrow H_2CO_3(aq)$$

This dilute solution of carbonic acid then falls to Earth in the form of rain, snow, or sleet.

Other natural events can contribute to the acidity of precipitation. Volcanic eruptions, forest fires, and lightning bolts produce sulfur dioxide (SO_2), sulfur trioxide (SO_3), and nitrogen dioxide (NO_2). These gases can then react with atmospheric water in much the same way that carbon dioxide does:

$$SO_2(g) + H_2O(l) \longrightarrow \underset{\text{Sulfurous acid}}{H_2SO_3(aq)}$$

$$SO_3(g) + H_2O(l) \longrightarrow \underset{\text{Sulfuric acid}}{H_2SO_4(aq)}$$

$$2\,NO_2(g) + H_2O(l) \longrightarrow \underset{\text{Nitric acid}}{HNO_3(aq)} + \underset{\text{Nitrous acid}}{HNO_2(aq)}$$

The actual processes that generate acidified precipitation are more complex than depicted here.

If all rain is naturally acidic, why is acid rain a problem? In the years since the beginning of the Industrial Revolution, fossil fuels such as coal, oil, and natural gas have been used as inexpensive energy sources. The combustion of fossil fuels not only heats homes, it also powers vehicles and

ACIDS IN THE ATMOSPHERE

Introduction

Section C explores the human impact on air pollution. Natural and human resources are compared as contributing factors of acid rain. The Arrhenius definitions of acid and bases are presented, along with equations showing the formation of the ions associated with each. Strong and weak acids and bases are compared in terms of the extent of ionization or dissociation and pH. The composition and function of buffers are discussed and tested in a laboratory activity.

C.1 ACID RAIN

Rain is naturally acidic (pH = 5.6) because the CO_2 gas in the atmosphere dissolves in water to form carbonic acid (H_2CO_3). However, NO_2 and SO_2 also contribute to acid rain. When these gases are dissolved by water droplets in the atmosphere, the gases cause the pH of atmospheric water to decrease thus becoming more acidic.

Acid rain may be any type of precipitation produced from the reaction between water and nonmetallic oxides. These oxides are emitted from power plants, industries, and the burning of fossil fuels. Acid rain may cause damage to natural ecosystems, as well as buildings and structures made of metals, marble and limestone due to the reaction between acid rain and the structures. Acid rain is difficult to control because it may be produced in one part of the world but, through the natural circulation of air, affect structures and ecosystems in another part.

The EPA website provides a great deal of information related to emissions and emissions trading.

Figure 31 *By 1997, this spruce-fir forest in North Carolina had been badly damaged by acid rain.*

generates electricity. Unfortunately, fossil fuels contain sulfur and nitrogen compounds that burn in air to produce sulfur oxides and nitrogen oxides. The result, as the equations on page 297 suggest, is the formation of acidic precipitation. The fact that the precipitation is much more acidic than normal leads to the problems associated with acid rain.

The sulfur oxides and nitrogen oxides emitted from power plants, various industries, and fossil-fuel-burning vehicles react with water vapor to form acids that lower the pH of rainwater—at times to as low as 4.5 to 4 in the Northeastern United States. (That pH range is about the acidity of orange juice.) Similar changes in pH have been observed for precipitation in Scandinavia, Central Europe, and other areas downwind from large industrial centers. Excessively acidic rain can lower the pH of lakes and streams, killing fish eggs and other aquatic life. Acid rain also damages plants by leaching minerals from the soil, preventing plants from using them. See Figure 31.

Problems caused by acid rain are not limited to effects on natural ecosystems. Most buildings and other structures contain metal, limestone, or concrete, all of which are materials susceptible to damage by acids. Statues and monuments (such as the Parthenon in Greece) that have stood for centuries now show signs of significant surface damage, due, in part, to acid rain. Acid attacks calcium carbonate in limestone, marble, and cement according to this equation:

$$\underset{\text{Sulfuric acid}}{H_2SO_4(aq)} + \underset{\text{Calcium carbonate}}{CaCO_3(s)} \longrightarrow \underset{\text{Calcium sulfate}}{CaSO_4(s)} + \underset{\text{Water}}{H_2O(l)} + \underset{\text{Carbon dioxide}}{CO_2(g)}$$

Calcium sulfate is much more soluble in water than is calcium carbonate. Thus as calcium sulfate forms, it washes away, uncovering fresh calcium carbonate that can react further with acid rain. See Figure 32.

The control of acid rain is made difficult because air pollution knows no political boundaries. Carried by air currents, acid rain often shows up hundreds of kilometers from the sources of pollution. For example, much of the acid rain that falls in Scandinavia has its origins in the industrial regions of Germany and the United Kingdom. The acid rain that falls on New England

298

Figure 32 *Acid rain has disfigured these limestone gargoyles on the Cathedral of Notre Dame in Paris, France.*

and southern Canada comes largely from the industrial Ohio Valley, in the midwestern United States.

The Clean Air Act Amendments of 1990 were enacted, in part, to address the issue of acid rain. The compliance plan created by the amendments imposed emissions restrictions on fossil-fueled power plants. As you can see in Figure 33, fuel combustion is the largest contributor to SO_2 emissions. To achieve these reductions, the U.S. Environmental Protection Agency (EPA) issued permits to electric utility plants, particularly the coal-burning power plants that have consistently been the largest sulfur dioxide emitters. The permits specify the maximum amount of SO_2 each plant can release annually. As a bonus for reducing emissions, plants with low emissions are allowed to auction or sell their excess permits to plants that produce more emissions. National SO_2 emissions decreased 12% between 1988 and 1997.

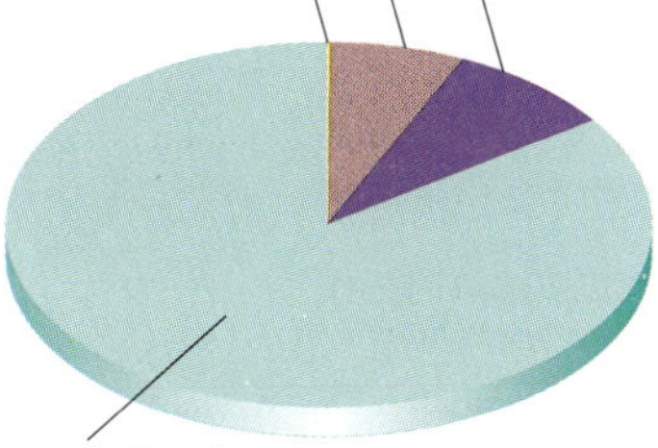

Figure 33 *SO_2 emissions by source, 1997. EPA data.*

299

There was a sharp decline in SO_2 emissions between 1994 and 1995, followed by a slight increase from 1995 to 1997. Is the increase a cause for concern? Probably not, since most electric utilities overcomplied in 1995. Instead of selling the emissions credits they earned, they used them to increase emissions slightly in the short term.

Reductions in SO_2 have been accomplished primarily by using more expensive, lower-sulfur coal. Between 1990 and 1995, annual U.S. sales of low to medium-sulfur coal rose by 103 million tons, while sales of higher-sulfur coal fell by 69 million tons. Some power plants have installed scrubbers that remove SO_2 from smokestack emissions, but the cost of this option is currently higher than that associated with using lower-sulfur coal.

Scrubbers are described in Section D.

Unlike SO_2 emissions, which are produced in the combustion of fuels containing sulfur, nitrogen oxides (NO_x) are formed by the reaction of atmospheric nitrogen gas with oxygen gas at the high temperatures produced by most forms of combustion—including internal combustion engines.

Nitrogen monoxide (NO)—also called nitric oxide—and nitrogen dioxide (NO_2) are sometimes referred to as NO_x (pronounced "nocks").

Figure 34 documents that combustion of fuel by power plants is joined by transportation as a major source of nitrogen oxide emissions. Thus, EPA permits for utility companies are a less effective strategy in reducing overall emissions of NO_x than for SO_2. Although the 1997 national average NO_2 concentration in air was 14% lower than in 1988, the total national NO_x emissions were only 1% lower. The EPA estimates that reducing NO_x emissions from fixed-location sources such as power plants is a cost-effective strategy. But the issue of NO_x emissions from mobile sources, such as automobiles and other modes of transportation, has yet to be addressed.

Have you seen any evidence of the effects of acid rain near where you live? In the following laboratory activity, you will have an opportunity to generate "acid rain" on a laboratory scale and observe some of its properties.

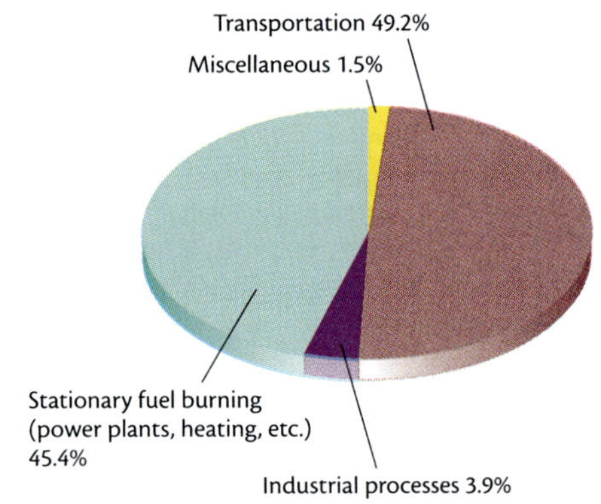

Figure 34 *NO_x emissions by source, 1997. EPA data*

C.2 MAKING ACID RAIN

Laboratory Activity

Introduction

In this activity you will generate a gas and dissolve it in water, creating a solution similar to acid rain. You will then observe the effects of this acidic solution on plant material (an apple skin), on a chemically active metal (magnesium, Mg), and on marble (calcium carbonate, $CaCO_3$). You will also use chemical indicators to evaluate the acidity of the solution.

CD-ROM WWW. Lab Video

Procedure

1. Place a 1-g sample of sodium sulfite (Na_2SO_3) in a test tube. Place the test tube in a 1-pint reclosable zip-seal bag, holding it upright from outside the bag.
2. Carefully fill a Beral pipet with 6 M hydrochloric acid (HCl). **CAUTION:** *6 M hydrochloric acid is corrosive. If you spill it on yourself or others, wash it off thoroughly and inform your teacher. Avoid breathing any HCl fumes.* Using a wash bottle, carefully rinse off any acid left on the outside of the pipet.
3. Without squeezing the bulb of the Beral pipet, place its stem inside the test tube containing the sodium sulfite. Carefully add 10 mL distilled water to the bag. Make sure that the distilled water does not come in contact with the hydrochloric acid or sodium sulfite. Carefully smooth the bag to remove most of the air from it. Seal the bag by closing the sealing strip. See Figure 35.
4. Slowly squeeze the Beral pipet through the outside of the bag so all the hydrochloric acid drips onto the solid sodium sulfite. Keep the

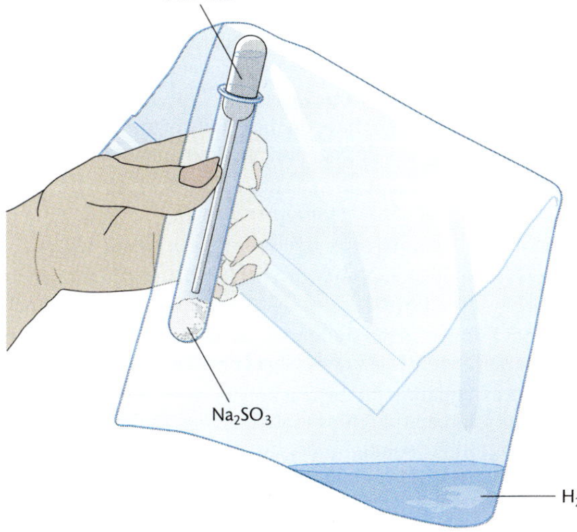

Figure 35 *Zip-seal bag arrangement for generating a laboratory sample of "acid rain." The Beral pipet and test tube are sealed inside the bag with the water sample, as shown.*

C.2 MAKING ACID RAIN

Laboratory Activity

The purpose of this activity is to make acid rain, then observe the effects of acid rain on plant material, an active metal, and marble.

Time

50 minutes

Materials (for a class of 24 working in pairs)

2 red apples
12 cm magnesium ribbon
12 zip-sealing pint-sized plastic bags
12 paring knives
36 Beral pipets
24 marble chips [calcium carbonate ($CaCO_3$)]
12 g sodium sulfite (Na_2SO_3)
25 mL 6.0 M HCl (hydrochloric acid)(50 mL 12 M HCl/100 mL solution)
1 pkg. Hydrion (pH) paper
1 pkg. red litmus paper
1 pkg. blue litmus paper
paper towels
distilled water

Safety

Students should wear safety goggles and aprons throughout the laboratory activity. Exercise caution when handling 6 M HCl. Acetone is highly flammable. Use only with proper ventilation.

Advance preparation

The red apple skins should be wiped with acetone to dissolve any protective wax, then rinsed with water and dried.

Lab tips

- Caution students to make sure the plastic bag is sealed quickly after adding required materials. The 6 M HCl is corrosive.
- You may want to peel the apples to avoid students handling paring knives.
- Orange peel may be substituted for apple peel.
- As marble chips may give poor results, small pieces of chalk may be substituted.
- If a fume hood is available, once the acid has been added, students should only open the bag in the hood.

Answers

1. $2HCl + Na_2SO_3 \rightarrow SO_2 + H_2O + 2NaCl$
2. The dissolved gas increased the acidity of the water. Sulfur dioxide reacts with water to form sulfurous acid and sulfuric acid. The solution turned blue litmus paper red, indicating an acid is present. Hydrion paper changed to the red-orange range, indicating an acidic solution.
3. $SO_2 + H_2O \rightarrow H_2SO_3$
4. The sulfur dioxide produced in this chemical reaction is identical to the SO_2 released from combustion. It combines with water to form an acid.
5. Students' reports of pH will vary but will probably be lower.
6. Because the marble in statues reacts with acid, and because the iron in the steel girders reacts with acid, the statue and bridge would deteriorate.
7. a. $H_2SO_3 + CaCO_3 \rightarrow CaSO_3 + H_2O + CO_2$

   b. The $CaSO_3$ is a soluble compound that will wash away. This is how marble structures are diminished.

test tube upright so the contents do not spill out. Keep the bag sealed.

5. Allow the reaction in the test tube to proceed for 1 to 2 minutes, gently tapping the test tube every few seconds. After the reaction in the test tube has stopped, gently swirl the water in the bottom of the bag for another 1 to 2 minutes. Do not swirl the water so vigorously that it mixes with the contents of the test tube.
6. Carefully open a top corner of the bag and, using a clean, dry Beral pipet, transfer three pipets of liquid from the bottom of the bag to a clean, dry test tube. Reseal the bag. Label the test tube "A."
7. Peel two pieces of fresh apple skin, approximately 2 cm long, and place them on a paper towel. Add 4 to 5 drops of solution from test tube A to the outside of one piece of apple skin. Add 4 to 5 drops of distilled water to the second piece of apple skin. (The second apple skin sample serves as a control. Why?) After 3 minutes, observe the two apple-skin samples; record your observations.
8. Place one drop of distilled water each on a fresh piece of red litmus paper, blue litmus paper, and pH paper. Record your observations.
9. Repeat Step 8, except use solution from test tube A in place of the distilled water. Record your observations.
10. Place a 1-cm length of magnesium ribbon in a separate clean, dry test tube. Add one pipet of the solution from test tube A. Observe the reaction for at least 3 minutes. Record your observations.
11. Add two small marble chips (calcium carbonate, $CaCO_3$) to the solution remaining in test tube A. Observe the marble chips for at least 3 minutes; record your observations.
12. Dispose of all remaining solutions as directed by your teacher.
13. Wash your hands thoroughly before leaving the laboratory.

Questions

1. The gas formed in the test tube by the reaction of hydrochloric acid and sodium sulfite was sulfur dioxide, SO_2. The other products formed were water and NaCl(aq). Write a balanced equation for the reaction that produced SO_2.
2. What effect did this gas have on the acidity of the distilled water placed inside the plastic bag? Support your answer with data.
3. Write a chemical equation that shows how "acid rain" (H_2SO_3) was produced from SO_2 inside the zip-seal bag.
4. Describe how this activity models the production of acid rain.
5. As you know, precipitation with a pH less than 5.6 is defined as acid rain. How does the pH of your solution compare to this value?
6. If a liquid similar to the solution in the plastic bag moistened a marble statue or steel girders, what effect might it have?
7. a. Write an equation for the reaction between your "acid rain" (H_2SO_3) and marble chips ($CaCO_3$). (*Hint:* Carbon dioxide and calcium sulfite solution were two of three products formed.)

   b. Explain how that equation relates to the effect of acid rain on marble statues and building materials.

Post-lab discussion

Focus on the experimental results and the answers to the questions.

DEMONSTRATION

Place a drop of paramecium (or mixed) culture on a microscope slide. Pond water, if available, is an interesting alternative. Allow student groups to examine the drop under the microscope and record their observations. Add three drops of the acid rain solution prepared in the zip-sealing plastic bag.

CHEMISTRY AT WORK

Using Chemistry to Bring Criminals to Justice

For a moment, no one in the courtroom moved. Aware that everyone was staring at her, Susan could hear the sound of her own heart beating.

Then the prosecutor continued. "Ms. Ragudo, please tell this court," he said, holding up the small plastic bag of white powder, "the results of your analysis of the substance labeled Exhibit A."

For many people, the thought of testifying in a criminal court case is enough to make them break out in a sweat. But for **Susan Ragudo,** it's just another part of her job as a forensic scientist employed by the Commonwealth of Virginia. Forensic science—the use of science in criminal investigations—can describe many types of inquiry. Susan's specialized area involves analyzing crime scene material for evidence of illegal substances.

To identify an unknown white substance, for instance, Susan begins with a color test, introducing the substance to a reagent. Depending on the resulting color change, Susan knows what the substance most probably is. For example, the drug heroin will turn purple, cocaine will turn blue, and so on. Next, Susan uses thin-layer chromatography (TLC) to compare the substance to a known sample of the suspected drug. This narrows Susan's choices even more and also tells her whether one or more substances are present. Finally Susan tests her conclusion by using gas chromatography/mass spectroscopy. Then she returns the substance—and her analysis data—to the law enforcement officers.

Susan explains that in her field it is essential to know one's way around a laboratory. It's also important to have an analytical mind—to be able to "zero in on" a substance logically, by conducting and interpreting tests. Susan adds that she must also be comfortable speaking in public, because she is sometimes required to testify in court about her findings.

Calling All Sleuths

Susan Ragudo uses chemistry to help determine whether substances are illegal drugs. In what other types of criminal investigation could chemistry be helpful? What types of tests could be used? Write a story about a crime in which everyone overlooks a small chemical clue—everyone but you, that is. Don't be bashful—make yourself a hero, and explain how you used a basic understanding of chemistry to unravel the crime and convict the criminal!

Truth vs. Evidence

1. Some chemical tests can conclusively prove what a substance is *not;* it is often much harder to establish what it *is*. Explain why one type of proof is more difficult than the other.
2. Most tests are quite reliable but still may only be 95% or 99% accurate. Given that no test is likely to be 100% accurate, how sure must evidence be for you to accept it as "the truth?" Explain your answer.
3. Juries sit in judgment regarding the facts in a criminal case. Who serves as the jury in deciding what scientific research results are regarded as true?

303

C.3 PREVENTING ACID RAIN

Making Decisions

1. Sulfur dioxide and sulfur trioxide are formed from burning coal and other fossil fuels, often in power plants. Nitrogen oxides are generally the result of automobile engines.
   - **a.** The sulfur oxides are easier to control because there are relatively few power-generating plants and it is easy to target environmental regulations and emission levels at them.
   - **b.** It is hard to implement anti-pollution measures that may result in higher energy costs and lower supplies for consumers.
2. a. Subsidizing the cost of pollution prevention through tax dollars can help, as can subsidizing research into more efficient processes.
   - **b.** Where the fines can be made greater than the profit from creating pollution and in regard to any emissions that present an acute danger
3. You would need to add up the total cost of production, including both direct and indirect costs, such as the production of the batteries and the electricity.

C.3 PREVENTING ACID RAIN — Making Decisions

You have learned about the roles of sulfur oxides and nitrogen oxides in forming acid rain, as well as current efforts to control these emissions. Use this information to answer the following questions.

1. a. In what ways is the generation of sulfur oxides easier to control than the generation of nitrogen oxides?
   b. In what ways is it more difficult?
2. a. Describe incentives you think could be offered by the government to help encourage reduction of these emissions.
   b. Which emissions might be reduced more effectively through the use of penalties?
3. In terms of decreasing total release of sulfur oxides and nitrogen oxides into the atmosphere, how could you decide whether an electric vehicle that uses electricity from a fossil-fuel-burning power plant is preferable to a gasoline-powered vehicle?

What Is an Acidic Solution? CD-ROM WWW.

Arrhenius also proposed that an increase in atmospheric carbon dioxide may increase global temperatures.

C.4 STRUCTURE DETERMINES FUNCTION

In the late nineteenth century the Swedish chemist Svante Arrhenius defined an acid as any substance that generates hydrogen ions (H^+) when dissolved in water. He defined a base as a substance that generates hydroxide ions (OH^-) in water. In aqueous solutions, the hydrogen ion released by the acid is bonded to water. Such a combination can be represented as $H^+(aq)$, $H(H_2O)^+$, or, more commonly, $H_3O^+(aq)$, the **hydronium ion:**

$$H^+(aq) + H_2O(l) \longrightarrow H_3O^+(aq)$$

Consider the example of carbon dioxide, $CO_2(g)$, dissolving in water to produce carbonic acid, $H_2CO_3(aq)$:

$$H_2O + CO_2(g) \longrightarrow H_2CO_3(aq)$$

The transfer of a hydrogen ion, H^+, from the carbonic acid to water produces a bicarbonate ion and a hydronium ion, which accounts for the acidity of the solution:

$$H_2O(l) + H_2CO_3(aq) \longrightarrow H_3O^+(aq) + HCO_3^-(aq)$$

Most common acids are substances that contain one or more hydrogen atoms that can be released as $H^+(aq)$ in water. The remainder of the original acid molecule—with hydrogen's single electron still attached—becomes an anion. As shown in the following equations for two common acids, the dissolved acid produces hydronium ions, $H_3O^+(aq)$, and an aqueous anion.

Nitric acid: $H_2O + HNO_3(aq) \longrightarrow H_3O^+(aq) + NO_3^-(aq)$

Sulfuric acid: $H_2O + H_2SO_4(aq) \longrightarrow H_3O^+(aq) + HSO_4^-(aq)$

Bases generally contain a cation and the hydroxide anion, OH^-. When such bases dissolve in water, the cations and hydroxide ions separate and

C.4 STRUCTURE DETERMINES FUNCTION

Acids and bases are defined in terms of the ions generated when they are dissolved in water: the hydronium ion (H_3O^+) for acids and the (OH^-) hydroxide ion for bases.

Equations are given for the formation of acids and bases in water. The basic and acidic properties of a solution disappear and the substance is neutralized. A neutralization reaction results in the formation of an ionic compound (a salt) and water when a suitable amount of a base is added to an acid. The reactions are presented as net ionic equations.

disperse uniformly through the solution. The hydroxide ions (OH^-) give basic solutions (also referred to as **alkaline** solutions) their characteristic properties:

Potassium hydroxide: $KOH(s) \longrightarrow K^+(aq) + OH^-(aq)$
Barium hydroxide: $Ba(OH)_2(s) \longrightarrow Ba^{2+}(aq) + 2\ OH^-(aq)$

Acidic solutions are characterized by an excess of $H_3O^+(aq)$, while alkaline solutions characterized by an excess of $OH^-(aq)$. Pure water is a neutral substance (pH = 7.0 at 25 °C) containing equal amounts of $H_3O^+(aq)$ and $OH^-(aq)$.

Mixing equal amounts of $H_3O^+(aq)$ and $OH^-(aq)$ results in a **neutralization** reaction. When these ions react with each other, water is produced:

$$H_3O^+(aq) + OH^-(aq) \longrightarrow 2\ H_2O(l)$$

Both acidic and basic characteristics are destroyed by this reaction; pure (neutral) water is produced, which contains small, equal amounts of $H_3O^+(aq)$ and $OH^-(aq)$.

Consider the neutralization of hydrochloric acid, HCl(aq), with sodium hydroxide, NaOH(aq), according to this equation:

$$HCl(aq) + NaOH(aq) \longrightarrow H_2O(l) + NaCl(aq)$$

As is true for all neutralization reactions, this reaction produces water and a salt (an ionic compound).

ACIDS AND BASES — Building Skills 10

1. Hydrochloric acid (HCl) is an acid found in gastric juice (stomach fluid).
   a. Write an equation for the formation of ions when this acid dissolves in water.
   b. Name each ion.
2. Below are three acids that may be found in acid rain. For each, give the name of the acid and write an equation indicating its reaction with water, forming ions.
   a. HNO_3—formed from the reaction of water with nitrogen oxides produced in car engines and lightning bolts
   b. H_2CO_3—bears the major responsibility for the natural acidity of precipitation
   c. H_2SO_3—formed from the reaction of water with sulfur dioxide produced by the combustion of sulfur-containing fuels
3. Each base listed below has commercial and industrial applications. Give the name of each and write an equation showing the ions liberated when it dissolves in water.
   a. $Mg(OH)_2$—the active ingredient in milk of magnesia, an antacid
   b. $Al(OH)_3$—a compound used to bind dyes to fabrics
   c. NaOH—a compound often found in drain and oven cleaners

ACIDS AND BASES
Building Skills 10

Students are given an opportunity to practice writing ionization and dissociation equations.

Answers

1. a. $HCl(aq) + H_2O(l) \rightarrow H_3O^+(aq) + Cl^-(aq)$
   b. hydrochloric acid + water → hydronium ion + chloride ion
2. a. $HNO_3(aq) + H_2O(l) \rightarrow H_3O^+(aq) + NO_3^-(aq)$
   Nitric acid + water → hydronium ion + nitrate ion
   b. $H_2CO_3(aq) + H_2O(l) \rightarrow H_3O^+(aq) + HCO_3^-$
   carbonic acid + water → hydronium ion + hydrogen carbonate ion
   c. $H_2SO_3(aq) + H_2O(l) \rightarrow H_3O^+(aq) + HSO_3^-(aq)$
   sulfurous acid + water → hydronium ion + hydrogen sulfite ion
3. a. $Mg(OH)_2(s) \rightarrow Mg^{2+}(aq) + 2\ OH^-(aq)$
   magnesium hydroxide → magnesium ion + hydroxide ion
   b. $Al\ (OH)_3(s) \rightarrow Al^{3+}(aq) + 3\ OH^-(aq)$
   aluminum hydroxide → aluminum ion + hydroxide ion
   c. $NaOH(s) \rightarrow Na^+(aq) + OH^-(aq)$
   Sodium hydroxide → sodium ion + hydroxide ion.
4. a. $2\ H^+(aq) + 2Cl^-(aq) + Mg^{2+}(aq) + 2\ OH^-(aq) \rightarrow 2\ H_2O(l) + Mg^{2+}(aq) + 2Cl^-(aq)$
   Hydrogen ion + chloride ion + magnesium ion + hydroxide ion → water + magnesium ion + chloride ion

C.5 pH AND ACIDITY

The pH concept, introduced in the Water Unit, is explored more thoroughly in this section. Students should recognize that the smaller the pH number, the greater the acidity. As pH decreases, the concentration of the $H^+(H_3O^+)$ increases. As pH increases, the concentration of the OH^- also increases. At room temperature, pH + pOH = 14.

Emphasis should also be placed on the importance of a change of one power of ten in the concentration of the hydronium ion (H_3O^+) for each unit of pH. One unit change in pH represents a tenfold change in acidity and alkalinity.

A solution with pH = 0 is very acidic; in fact, the solution would have a H^+ concentration of 1 M:

$$pH = -\log[H^+] = \log(1) = 0.$$

It would be helpful to prepare one or more solutions as a demonstration of molarity as a unit of concentration. Some examples: 29 g of NaCl in 500 mL of solution to make a 1.0 M solution of NaCl or 10.1 g KNO_3 in 100 mL solution to make a 1.0 M solution of KNO_3.

4. A person takes a dose of the antacid magnesium hydroxide, $Mg(OH)_2$, to relieve excess stomach acid, HCl(aq).

a. Write the total equation for the neutralization of HCl(aq) with $Mg(OH)_2$(aq).

b. Name each product shown in the chemical equation you wrote to answer Question 4a.

C.5 pH AND ACIDITY

What Is pH? CD-ROM WWW.

Water and all of its solutions contain both hydronium ions (H_3O^+) and hydroxide ions (OH^-).

- In pure water and neutral solutions, the concentrations of these ions are equal, but very small.
- In acidic solutions, the hydronium ion concentration is larger than the hydroxide ion concentration. In very acidic solutions the hydronium ion concentration is much, much larger than the hydroxide concentration.
- In basic solutions, the hydroxide ion concentration is larger than the hydronium ion concentration. In very basic solutions, the hydroxide ion concentration is much, much larger than the hydronium concentration.

Chemists often express solution concentration in terms of the amount of solute dissolved in a particular volume of solution. When the amount of solute (in moles) is divided by the volume of the solution (in liters), the resulting value is called the **molar concentration,** or **molarity** (symbolized by M). Figure 36 illustrates how a solution of a particular molar concentration (molarity) can be prepared in the laboratory.

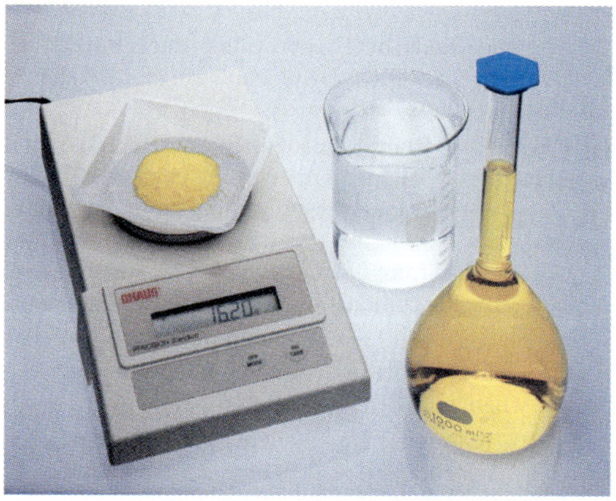

Figure 36 *Preparation of a 0.1000-M solution of sodium chromate, Na_2CrO_4. The molar mass of Na_2CrO_4 is 162.0 g/mol. A 0.1000-M sample of that solute (16.20 g) is dissolved in enough water to produce 1.000 L of final solution. A line etched into the neck of the volumetric flask in the photo precisely indicates that volume measurement.*

Expressed in symbolic terms,

$$M = \frac{\text{mol solute}}{\text{L solution}}$$

Thus if 2.0 mol sodium chloride (NaCl) were dissolved in 1.0 L of total solution, the molar concentration of the solution would be:

$$\frac{2.0 \text{ mol NaCl}}{1.0 \text{ L solution}} = 2.0 \text{ M NaCl}$$

A solution containing 1.0 mol NaCl in 0.50 L of solution is also a 2.0 M NaCl solution. If 0.16 mol NaCl were dissolved in 2.0 L of solution, the molar concentration would be:

$$\frac{0.16 \text{ mol NaCl}}{2.0 \text{ L solution}} = 0.080 \text{ M NaCl}$$

The concept of pH is related to the concentration of hydronium ions (H_3O^+) found in water or aqueous solutions. The term pH stands for "power of hydronium ion," where "power" refers to the mathematical power (exponent) of 10 in the hydronium ion's molar concentration.

For example, a solution containing 0.001 mol H_3O^+ (1×10^{-3} mol H_3O^+) per liter has an H_3O^+ concentration of 0.001 M and a pH of 3. The pH value can be interpreted as the exponent on the power of 10 with its sign changed. Thus a solution with 0.000 000 01 mol H_3O^+ (1×10^{-8} mol H_3O^+) per liter (0.000 000 01 M) would have a pH of 8. Figure 37 shows how pH is related to the molar concentration of $H^+(H_3O^+)$ and OH^-.

As you have just learned, the pH scale expresses acidity and alkalinity using powers of 10. Thus lemon juice, pH 2, is 10 times more acidic than a soft drink at pH 3, which in turn is 10 000 times more acidic than pure water at pH 7 (four steps farther up the pH scale). Similarly, a solution with pH 11 is 100 times more basic than a solution with pH 9.

Using similar reasoning, a solution with a pOH of 2 would contain 1×10^{-2} mol OH^- (0.01 mol OH^-) per liter.

pH

Building Skills 11

1. Using the values in Figure 37, describe the mathematical relationship between pH and the $H_3O^+(H^+)$ concentration, as well as between the H_3O^+ and OH^- concentrations at 25 °C.

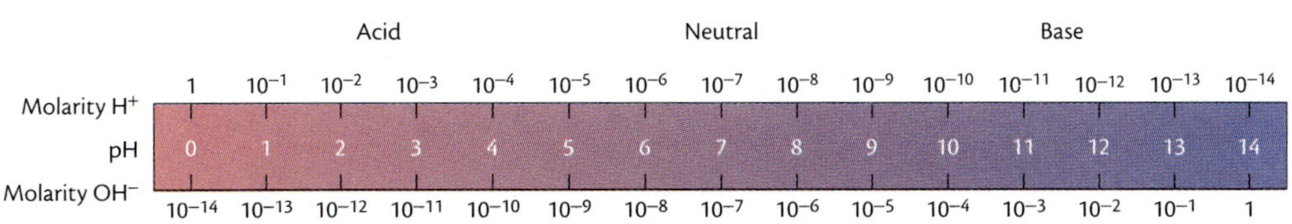

Figure 37 *The relationships among pH, molar concentration of H3O+, and molar concentration of OH− at 25 °C.*

		Hydronium ion	Hydroxide ion (OH^-)
2.	a. acidic	1×10^{-6} M	1×10^{-8} M
	b. acidic	1×10^{-1} M	1×10^{-13} M
	c. basic	1×10^{-13} M	1×10^{-1} M
	d. acidic	1×10^{-3} M	1×10^{-11} M
	e. neutral	1×10^{-7} M	1×10^{-7} M
	f. basic	1×10^{-11} M	1×10^{-3} M

3. A cola drink is 1000 (10^3) times more acidic than milk.
4. The Wheeling rainfall was 10 000 (10^4) times more acidic than normal. Students should see that a pH of 1.5 represents a H_3O^+ concentration between .01 M and 0.1 M (actually .032 M) and that a pH of 5.5 represents a H_3O^+ concentration between .0000001 M and .000001 M (actually .00000032 M). The ratio is then 10 000. Mathematically, $\frac{10^{-1.5}}{10^{-5.5}} = \frac{1}{10^{-4}} = 10^{8} = 10\,000$.
5. a. Aqueous solutions that might be present in the LHMP are solutions of NO_x, SO_x and CO_2 produced by combustion, respiration, and decomposition reactions.

   b. The pH of each of these solutions must be monitored because each solution may pose potential problems due to its acidic characteristics.

pH

Building Skills 12

1. pH	Hydronium Ion Concentration	Hydroxide Ion Concentration
1	.1 M	.0000000000001 M
4	.0001 M	.0000000001 M
6	.000001 M	.00000001 M
9	.000000001 M	.00001 M
12	.000000000001 M	.01 M

As the concentration of H_3O^+ decreases, the pH increases. As the acidity increases, the pH decreases. As concentration of H_3O^+ decreases, the OH^- concentration increases. As the concentration of base increases, the pH also increases.

6. a. 1.0 mol NaOH was spilled.

b. Concentration of NaOH =

$$\frac{1 \text{ mol NaOH}}{1 \times 10^6 \text{ L}} = 0.000001 \text{ M}$$

NaOH

c. Since the concentration of OH^- is 0.000001 (10^{-6}), Figure 37 indicates that pOH = 6 and pH = 8 for that solution.

C.6 STRENGTHS OF ACIDS AND BASES

The strength of an acid or a base is determined by the degree of ionization in an aqueous solution. Emphasize that *strong* and *weak* are not synonymous with concentrated and dilute. Strong and weak refer to the extent of ionization. Strong acids and bases ionize completely in solution to form hydronium and hydroxide ions, whereas weak acids and bases are only slightly ionized in solution.

Demonstration

Acid and bases conduct an electric current. Use a conductivity apparatus to compare conductivity of strong-acid and weak-acid solutions. This demonstration should help establish the definition of strong and weak when used to describe acid-base properties.

Materials

conductivity apparatus
250-mL beaker
6 beakers, 50 mL
0.1 M acetic acid
0.1 M barium hydroxide
0.1 M hydrochloric acid
0.1 M citric acid
0.1 M sodium hydroxide
0.1 M potassium hydroxide

Safety

Students should wear aprons and goggles throughout the activity.

2. Some common aqueous solutions and their typical pH values are listed below. Classify each as acidic, basic, or neutral, and determine the hydronium ion and hydroxide ion concentrations in each.

a. Milk, pH = 6
b. Stomach fluid, pH = 1
c. A solution of household lye, pH = 13
d. A cola drink, pH = 3
e. Sugar dissolved in pure water, pH = 7
f. Household ammonia, pH = 11

3. How many times more acidic is a cola drink than milk? (See information in Question 2.)

4. In 1979 a rainstorm over Wheeling, West Virginia, produced rainfall with a pH of 1.5. Compared to rainfall at pH 5.5, how many times more acidic was the Wheeling rainfall?

5. a. Identify some aqueous solutions that might be present in the Lunar Habitation Module Prototype (LHMP).
b. For which of these solutions would pH monitoring be necessary?

6. A worker accidentally spills 40 g of NaOH into the water storage container for the LHMP. The container holds 1×10^6 L of water.
a. How many moles of NaOH were spilled?
b. What is the molar concentration of the resulting solution? (*Hint:* This is also the concentration of OH^- in the solution.)
c. Use Figure 37 to find the pH of this solution.

C.6 STRENGTHS OF ACIDS AND BASES

Acid Concentration and Strength CD-ROM WWW.

In everyday discussions of solutions, "strong" tends to be associated with the idea of "concentrated," while "weak" often means "diluted." However, to a chemist, the words "strong" and "weak" in relation to acid and base solutions have very different meanings. Acids and bases are classified as strong or weak according to the extent to which they form ions in solution. An acid is considered strong if it ionizes completely—meaning that every dissolved molecule reacts to produce a hydronium ion and an anion. A base is considered strong if every dissolved molecule or unit produces a hydroxide ion and a cation. The total number of dissolved molecules or units does not matter. Strong means complete ionization, while weak means only partial ionization.

When a strong acid dissolves in water, none of the original acid molecules remain. Nitric acid (HNO_3), found in acid rain, is a strong acid. The complete ionization in a nitric acid solution is represented this way:

$$H_2O(l) + HNO_3(aq) \longrightarrow H_3O^+(aq) + NO_3^-(aq)$$

All of the original HNO_3 molecules react to form hydronium ions and nitrate ions in the water solution.

Conductivity Test Directions

Using a clean beaker for each solution, test the 0.1 M acid solutions for their relative conductivities. Repeat for the 1.0 M basic solutions. Continue in this manner until adding water no longer changes the conductivity of the acetic acid.

If a pH meter is available and used, show students not only how to use the meter but also how to calibrate the instrument with a solution of known pH close to the value to be measured. Indicators may also be used to supplement the pH paper or meter observations.

In a weak acid, only a relatively few dissolved acid molecules ionize to form hydronium ions and anions. Weak acids are only slightly ionized. Nitrous acid (HNO_2), sometimes also found in acid rain, is a weak acid. The partial ionization that occurs within an aqueous nitrous acid solution can be represented as follows:

$$H_2O(l) + HNO_2(aq) \rightleftharpoons H_3O^+(aq) + NO_2^-(aq)$$

Note the double arrow written in the equation. It is used to indicate that the reaction is reversible; that is, it can proceed in both directions. An analogy is useful in illustrating this idea. At a basketball game if the number of people walking out to buy snacks equals the number of people returning to their seats after buying their snacks, the system is described as being in **dynamic equilibrium**—two offsetting processes occur at equal rates, producing a state of balance where no net change is observed. It is important to realize that the *forward and reverse rates*—not the amounts of products and reactants—are equal in a system at equilibrium. In the basketball game analogy, many more people are in the stands watching the game than are at the snack counters. Likewise, at equilibrium, the concentration of HNO_2 is much larger than the concentrations of H_3O^+ and NO_2^-. In both cases, however, a state of dynamic equilibrium is attained.

In addition to nitrous acid, acid rain may contain other weak acids, commonly sulfurous acid (H_2SO_3) and carbonic acid (H_2CO_3). Strong acids, such as sulfuric acid (H_2SO_4), may also be present.

Sodium hydroxide (NaOH) and potassium hydroxide (KOH) are two examples of strong bases. Their solutions are completely composed of metal cations and OH^- ions. Even though the actual OH^- ion concentration may be limited by the solubility of a base in water, such a base is regarded as strong. Magnesium hydroxide, $Mg(OH)_2$, sometimes used as an antacid ingredient, is an example of a sparingly soluble (but strong) base.

A weak base commonly found in the environment is the carbonate anion (CO_3^{2-}), a component of limestone and marble. Why is the carbonate ion basic? The answer is shown by the following equation; water reacts with carbonate ions to form OH^- ions in aqueous solutions.

$$H_2O(l) + CO_3^{2-}(aq) \rightleftharpoons OH^-(aq) + HCO_3^-(aq)$$

C.7 ACIDS, BASES, AND BUFFERS

One early mystery associated with the nature of acid rain was the fact that the pH of some lakes and streams seemed unaffected by acidic precipitation. What protected these bodies of water from large changes in pH?

CD-ROM WWW. Buffer Action

The key to this mystery came when researchers realized that bodies of water suffering from acidification due to acid rain have two features in common. First, they are downwind from a dense concentration of power stations, smelters, or large cities—which produce nitrogen oxides and sulfur oxides. Second, the bodies of water are often surrounded by soils that are unable to neutralize acid carried by the precipitation. If the soil cannot neutralize the acidic precipitation, the lake or stream into which the precipitation drains then becomes acidified.

Demonstration

Dissolve an Alka-Seltzer tablet in 80 mL water in each of four 100-mL beakers on an overhead projector. Add universal indicator solution to each. Reserve the first beaker as a control. Add 1 M HCl dropwise to the second beaker with stirring until a pH of one unit occurred; observe. Then add 1 M NaOH dropwise to the third beaker with stirring until a pH change of one unit has occurred; observe. Finally, add distilled water dropwise to the fourth beaker with stirring; observe. Repeat this procedure in the four beakers of distilled water. Have the students record the observe changes.

MODELING MATTER

STRONG VS. CONCENTRATED

On page 310, students will use two hypothetical acids—one strong (HS) and one weak (HW)—to model the behavior of strong and weak acids in water solution.

The two terms strong and concentrated are often confused by students. Concentration refers to the amount of solute in some specified amount of solution. Strength refers to the percent of molecules that ionize and form ions in solution.

Hydrochloric acid is considered as a strong acid because in aqueous solutions nearly all of its molecules ionize to produce H^+ and Cl^- ions. Acetic acid is considered as a weak acid because only about 1% of the molecules normally ionize.

Measurements will show that despite having the same concentration, the solutions have varying pH values. That is due to the differences in ionization (strength) of the acids and bases.

C.7 ACIDS, BASES, AND BUFFERS

A buffer is a system that resists changes in pH when relatively small amounts of an acid or a base are added to a solution. A buffer must be able to neutralize both acids and bases. The laboratory activity gives students an opportunity to observe the behavior of buffers.

Answers to Questions:

1. a. "Weak" means that there are relatively few ionized particles compared to the number of un-ionized molecules. "Dilute" means that the solution has relatively few solute molecules compared to solvent molecules.

   b. "Strong" acids and bases are highly ionized or dissociated in water solutions. Therefore, there is a high proportion of H^+ ions as compared to un-ionized molecules in strong acid solutions. Strong basic solutions have a high proportion of OH^- and cations in solution, compared to un-ionized molecules. "Concentrated": There are many particles in the solution.
2. See page 314B.
3. The pH of the strong concentrated acid would be less than the pH of the weak concentrated acid. The pH of the strong dilute acid is less than the pH of the weak dilute acid.
4. a. Even at 0.1 mol it would be approximately 10^{22} particles, far too many to draw.

   b. A 0.1 M solution would contain about 6×10^{22} particles per liter.

MODELING MATTER

STRONG VS. CONCENTRATED

In this activity, you will depict concentrated and dilute solutions of two imaginary acids—one strong (HSt) and the other weak (HWe).

In referring to an acid solution, the term "strong" indicates that all (or nearly all) of the acid molecules have reacted with the water to produce hydronium ions and anions. By contrast, "concentrated" indicates that there are a very large number of particles—atoms, molecules, or ions—dissolved in a particular volume of solution.

Figure 38 depicts possible models for strong and weak acids. In a pencil-and-paper activity, you will illustrate the principles you read about in Section C.6. Draw four rectangular boxes, each 10 cm long by 5 cm wide. Each box represents 1.0 L of solution. Use H^+ to represent a hydronium ion. Each symbol will represent 0.1 mol of that species. Do NOT draw water molecules.

In each box, draw between 2 and 20 acid molecules (or appropriate numbers of hydronium ions and anions) to illustrate the following conditions:

- In the first box, depict a concentrated strong acid (HSt).
- In the second box, depict a dilute strong acid (HSt).
- In the third box, depict a concentrated weak acid (HWe).
- In the fourth box, depict a dilute weak acid (HWe).

Questions

1. Explain the difference between the following pairs of words when they are used to describe a solution of an acid or base.
   a. weak, dilute
   b. strong, concentrated
2. Calculate the molar concentration (molarity) of hydronium ions (H^+) in each of your drawings. Remember that every particle you drew represented 0.1 mol of particles.
3. How would the pH values for the four solutions you have modeled compare to one another?
4. a. Why is it easier for a single particle to represent 0.1 mol of particles instead of having to draw each particle in 0.1 mol?
   b. How many total particles are in 0.1 mol of particles? (Refer to page 135 in Unit 2.)
5. In what ways might these models mislead a viewer about the actual situation in each of the four solutions?

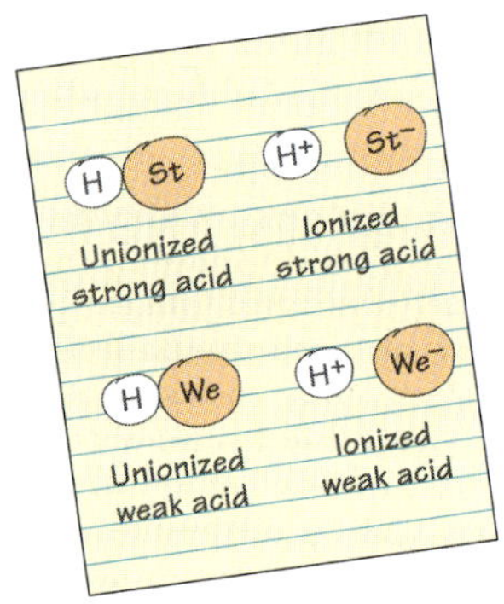

Figure 38 *Suggested models that can be used to represent strong and weak acids in solution.*

5. According to the kinetic theory, particles in solution are in constant motion. Also, since only 1% of the dilute/weak acid is ionized, it is difficult to represent with so few particles in the model. 1 out of 5 is too high and 0/5 looks too low.

Bodies of water not seriously affected by acid rain often benefit from surrounding and underlying rock and soils that can neutralize the acidic precipitation. In particular, the effects of acidic precipitation can be greatly reduced by limestone ($CaCO_3$), which reacts with acids to produce soluble calcium bicarbonate:

$$CaCO_3(s) + H_3O^+ \longrightarrow Ca^{2+}(aq) + HCO_3^-(aq) + H_2O(l)$$

However, the calcium carbonate can only prevent the pH from changing due to added acid. If a base is added, there will be nothing to neutralize it and the pH will increase. To keep the pH of a solution relatively steady—neither increasing or decreasing appreciably if small amounts of base or acid are added—a buffer is needed.

A **buffer** is a substance or combination of substances capable of neutralizing limited quantities of either acid (H_3O^+) or base (OH^-) added to it. Thus, a buffer contains two components—an acid to neutralize added base and a base to neutralize added acid. Although the calcium carbonate does not act as a true buffer—it cannot prevent the addition of base from increasing the pH—the bicarbonate (HCO_3^-) produced does act as a buffer, neutralizing either added acid or base:

$$H_3O^+(aq) + HCO_3^-(aq) \longrightarrow 2\,H_2O(l) + CO_2(g)$$
$$OH^-(aq) + HCO_3^-(aq) \longrightarrow H_2O(l) + CO_3^{2-}(aq)$$

In the next laboratory activity, you will observe the unique behavior of buffers and gain a better understanding of how some bodies of water are able to counter the effects of acid rain.

C.8 BUFFERS

Laboratory Activity

Introduction

In this activity, you will test a buffer solution. In particular, the results of adding acid and base to water, which serves as the control, will be compared to the results of adding acid and base to a buffered solution.

CD-ROM WWW. Lab Video

Procedure

1. Read the entire procedure and construct a data table suitable for collecting all relevant data.
2. Place a 24-well wellplate on a sheet of white paper. The white paper will help you more easily see any color changes during the activity. Add 20 drops distilled water to each of two wells. Add one drop universal indicator to each well. Note and record the colors of the resulting solutions in your data table.
3. Add 20 drops buffer solution to each of two other wells. This buffer is a solution containing 0.1 M sodium hydrogen phosphate (Na_2HPO_4) and 0.1 M sodium dihydrogen phosphate (NaH_2PO_4).
4. Add one drop universal indicator to each well containing buffer solution. Note and record the resulting color.

C.8 LABORATORY ACTIVITY: BUFFERS

The objective of this laboratory activity is to have students compare the results of adding an acid or a base to a buffer solution to that of adding an acid or a base to water.

Time

30 minutes

Materials (for a class of 24 working in pairs)

12 24-well wellplates
12 Beral pipets containing distilled water
12 Beral pipets containing 0.1 M NaOH (sodium hydroxide), 0.4 g NaOH in 100 mL solution
12 Beral pipets containing 0.01 M HCl (hydrochloric acid)
0.08 mL concentrated HCl diluted to 100 mL of water
12 Beral pipets filled with a solution containing 5.36 g $Na_2HPO_4 \cdot 7H_2O$ and 2.76 g $NaH_2PO_4 \cdot H_2O$ in 200 mL water
12 Beral pipets containing universal indicator solution
12 coffee stirrers (or Beral pipet tips)
universal indicator color charts
toothpicks

Safety

Students should wear safety goggles and aprons throughout the laboratory activity.

Add acid to water when diluting hydrochloric acid.

Advance preparation

Label and fill the pipets and place them tip up in the wellplate for distribution.

Pre-lab discussion

Review the function of a buffer as described in Section C.8.

Lab tips

For the best results students should minimize drop size by using Beral pipets drawn out and trimmed to narrow the stem of the pipet.

All references to water in this activity are to distilled (or deionized) water.

Remind students not to share the straws used to blow air into the flask to prepare the hydrogen phosphate buffer system. Used straws should be discarded.

Provide universal indicator color charts or tables so students can estimate the pH values of their solutions. Students should not try to match the intensity of the color, only the color itself.

Universal indicator color changes follow the colors in the rainbow as the pH moves from 1 to 13. The name ROY G. BIV helps keep the color straight: Red, Orange, Yellow, Green, Blue, Indigo, and Violet. A pH of 7 produces a green color.

Expected results

It should take considerably more acid and base than the water to make the buffer solution change pH appreciably.

Answers

1. The buffer solution required much larger amounts of acid or base to change the pH.
2. a. In each case, more acid or base is needed with the hydrogen phosphate buffer. The reason is that the buffer is able to react and neutralize both acids and bases.
3. $HPO_4^{2-}(aq) + H_3O^+(aq) \rightarrow H_2PO_4^-(aq) + H_2O(l)$

Post-lab discussion

Pool the class data so all students gain the benefit of the observations of all groups. Each group should be able to observe that the buffer system, compared with water, strongly resists pH change.

5. Add 5 drops 0.01 M NaOH to one of the wells containing distilled water and universal indicator. Note and record the color produced.

   CAUTION: *The solutions in this experiment are corrosive. If you splash anything on your skin, wash it thoroughly with water and inform your teacher.*
6. Carefully counting each drop, slowly add 0.01 M NaOH to one of the wells containing the buffer solution and universal indicator until the color matches the color in the well from Step 5. Record the number of drops required.
7. Add 5 drops 0.01 M HCl to a second well containing distilled water and universal indicator. Record the color produced.
8. Carefully counting each drop, slowly add 0.01 M HCl to the second well containing the buffer solution and universal indicator until its color matches the color in the well from Step 7. Record the number of drops required.
9. Save the solutions until you have gathered all the information required in the data table.
10. Report your group's data to the class as instructed by your teacher.
11. Dispose of the solutions as directed by your teacher.
12. Wash your hands thoroughly before leaving the laboratory.

Questions

1. What observations suggest that the solution you used is a buffer?
2. a. How many drops of NaOH were needed to bring the buffer solution to the same color (or pH) created by adding 5 drops of NaOH to the distilled water? Explain any difference.

   b. Make the same comparison for the added HCl.
3. The buffer used in this activity included hydrogen phosphate ions (HPO_4^{2-}). Write an equation showing how these ions would prevent the pH of lake water from decreasing if limited quantities of acid (H_3O^+) were added to it.

C.9 AIR QUALITY AND THE LHMP

Making Decisions

In designing the Lunar Habitation Module Prototype (LHMP), it will be helpful to focus on these air-quality questions:

1. List two or three possible needs for combustion that will result in the production of sulfur oxides and nitrogen oxides within the LHMP.
2. For each need listed in Question 1, identify one alternative way to meet the same need while reducing the generation of sulfur oxides or nitrogen oxides.
3. What damage could be caused within the LHMP by the generation of oxides of sulfur and nitrogen?
4. What strategies might be used to remove sulfur oxides and nitrogen oxides within the LHMP?

Questions & Answers CD-ROM WWW.

C.9 LUNAR HABITATION MODULE PROTOTYPE

Making Decisions

1. Sources of SO_x and NO_x within the LHMP include the combustion of fossil fuels and emissions from plants and human activities.
2. Alternatives: Instead of relying on the combustion of fossil fuel for energy, use solar or nuclear power.
3. Damage caused by the generation of oxides of sulfur and nitrogen includes the killing of plants and the deterioration of buildings and structures containing metals, marble, or concrete. There is also damage to lakes or ponds, which contain aquatic organisms.
4. Two possible answers: Locate the LHMP separate from power plants and industrial complexes; use soil that contains limestone that is able to neutralize acids, or find other natural buffers. Use scrubbers to treat combustion sources.

SECTION SUMMARY

Reviewing the Concepts

1. What is the pH range of

a. typical rain water?

b. acid rain?

a. 5.6

b. pH lower than 5.6.

2. Provide an explanation (including a chemical equation) for why rain is naturally acidic.

Carbon dioxide in the atmosphere reacts with water vapor in the atmosphere to form carbonic acid: $CO_2(g) + H_2O(l) \rightarrow H_2CO_3(aq)$

3. List two natural sources of sulfur oxides and nitrogen oxides and two human-generated sources of these gases.

Natural sources of sulfur oxides and nitrogen oxides are volcanic eruptions and lighting bolts. Human-generated sources are the burning of fossil fuels and industrial processes.

4. Through what strategies can humans reduce emissions of sulfur oxides and nitrogen oxides to the atmosphere?

Reduce the burning of fossil fuels and clean up industrial processes.

5. Name the major acidic components in acid rain.

The acidic components of acid rain are nitrogen oxides, sulfur oxides, and carbon dioxide.

6. Write a balanced equation representing what happens when SO_3 is dissolved in water.

$SO_3(g) + H_2O(l) \rightarrow H_2SO_4(aq)$

7. Carbon dioxide is not a significant contributor to acid rain. Why?

Carbon dioxide in the atmosphere reacts with water vapor to form a slightly acidic solution because carbonic acid is a weak acid.

8. Which of these compounds are acids? Which are bases? Which are neutral?

a. LiOH	**base**
b. CH_3COOH	**acid**
c. $C_{12}H_{22}O_{11}$ (table sugar)	**neutral**
d. H_2SO_3	**acid**

9. What is the difference between a strong acid and a concentrated acid?

A strong acid refers to the ability of an acid to ionize in water to produce H^+ or H_3O^+ ions. A concentrated acid refers to the number of particles (ions, molecules, atoms) that are dissolved in a volume of solution.

10. Explain why the compound referred to by the name "hydrogen hydroxide" has neither acidic nor basic properties.

Hydrogen hydroxide is actually water, HOH. It ionizes into small but equal amounts of hydrogen ion and hydroxide ion.

11. What is the difference between a hydrogen ion and a hydronium ion?

The hydronium ion consists of a hydrogen ion bonded to a water molecule. $H^+ + H_2O \rightarrow H_3O^+$

12. A methane molecule (CH_4) contains several hydrogen atoms. How could you determine whether or not CH_4 is an acid?

If the solution has more hydronium ions than hydroxide ions, it would be acidic.

13. Compare the concentration of hydroxide ions and hydrogen ions in pure water.

In pure water, the number of hydrogen ions is equal to the number of hydroxide ions; thus water is a neutral substance.

14. What would happen to the concentration of hydrogen ions in a solution of a strong acid if an equal number of hydroxide ions were added to the solution?

Water, a neutral substance, is formed.

15. A certain solution contains 1×10^{-2} mol H_3O^+ and 1×10^{-12} mol OH^-. Is the solution acidic or basic? Explain.

It would be acidic. It contains more hydronium than hydroxide ions.

16. Why do chemists often express the acidity of a solution in terms of pH instead of hydrogen ion concentration?

The pH values are simply numbers from 1–14 and are easier to understand.

17. Coffee has a pH of 4 and pure water has a pH of 7. How many more times acidic is coffee than water?

coffee, pH 4; pure water, pH 7; coffee is 10^3 or 1000 times more acidic than water.

18. A sample of acid rain has a pH of 3.

a. What is its hydronium ion concentration?

b. What is its hydroxide ion concentration?

a. The pH = 3. The hydronium ion concentration = 0.001 M.

b. The hydroxide ion concentration = 0.00000000001 M

19. Consider these three solutions:

- lemon juice, containing 0.001 M H^+
- stomach fluid, containing 0.1 M H^+
- drain cleaner, NaOH, containing 0.1 M OH^-

a. What is the pH of each solution? See Figure 37.

b. Which is most acidic?

c. Which is most basic?

a. The pH of lemon juice is 3; stomach fluid is 1; and drain cleaner is 13.

b. Stomach fluid is the most acidic.

c. Drain cleaner is the most basic.

20. Someone proposes that the acid rain problem could be fixed by adding large quantities of basic substances into affected lakes. Do you agree? Explain.

In theory it could work, but it would take tons of material and would only last for a limited time.

21. How would a lake surrounded by limestone respond differently to acid rain from one that was surrounded by rock that does not react appreciably with acid rain?

A lake surrounded by limestone can neutralize acid rain because the calcium carbonate of limestone reacts with the acid. A lake surrounded by granite contains no material able to neutralize it.

22. What are the general components of a buffer?

A buffer system can act as both an acid and a base and can neutralize additions of either a base or an acid. The acid in the buffer system provides H^+ to neutralize the added base, and the base in the buffer system neutralizes added acid.

23. Do buffers prevent all changes in pH? Explain your answer.

No, they only resist the change, and they have a limit to their total buffering capacity.

24. Why do some lakes appear unaffected by acid rain?

They may have natural buffering action.

Connecting the Concepts

25. Explain how it might be possible for a strong acid and a weak acid to have the same pH.

The pH depends on the concentration of hydronium ions in solution. Thus it is possible for solutions of a strong acid and a weak acid to have the same number of hydronium ions.

26. Why are shampoos and medicines often buffered?

Shampoos and medicines are buffered because their action is dependent on matching the pH of the body where they are used.

27. How do antacids differ from buffers?

Antacids neutralize an acid. A buffer can neutralize both an acid and a base.

28. What pattern would you expect to find among pH readings taken of precipitation in the immediate vicinity of an acid-rain producing smokestack and at regular intervals downwind and upwind from the site? Sketch a graph of what you would expect to find.

Downwind would have the lower pH, thus the more acid. The pH in the immediate vicinity of the smokestack may be higher indicating the presence of lower acid. Tall stacks may cause the pollution to be deposited miles away from the source.

Extending the Concepts

29. With the permission of a parent or guardian, make an acid-base indicator at home from red cabbage juice and test its properties using common household materials. (Consult your teacher for instructions.)

Take pieces of raw red cabbage and rub them on the surface of index cards. Rub hard enough to transfer some of the color. Use a dropper to add the substances to be tested, and note the color. Test some known acids and bases for comparison.

30. Use the Internet and library to research serious instances of acid rain in the past decade.

Student research will reveal a number of well-reported instances, including a "killer smog" in London in the 1950s.

31. Look at the list of ingredients for several antacid products. Which ingredients are responsible for neutralizing acids?

Many antacids contain $Mg(OH)_2$, a base, as the active ingredient to neutralize the excess stomach acid.

32. Is it possible to have a pH of zero? Explain your answer.

Yes. A pH of zero would have a H^+ concentration of 1 M. An acid with a pH of zero is very acidic.

33. Sometimes the zone between warring or feuding nations is called a "buffer zone." How does that use of the term "buffer" compare to a chemist's use of that word?

In chemistry a buffer maintains a neutral state, despite limited additions of acids or bases. A "buffer zone" between nations helps maintain peace by maintaining a neutral area between opposing forces.

Sample Student Drawings

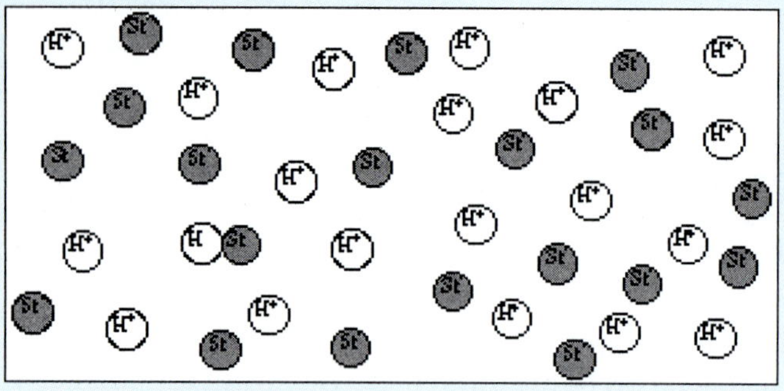

Strong/concentrated

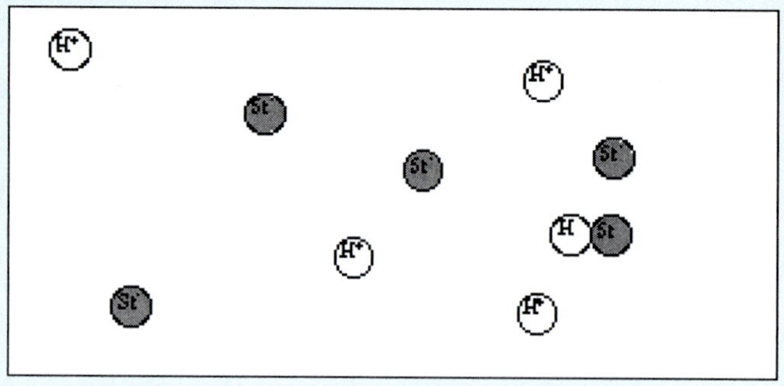

Strong/dilute

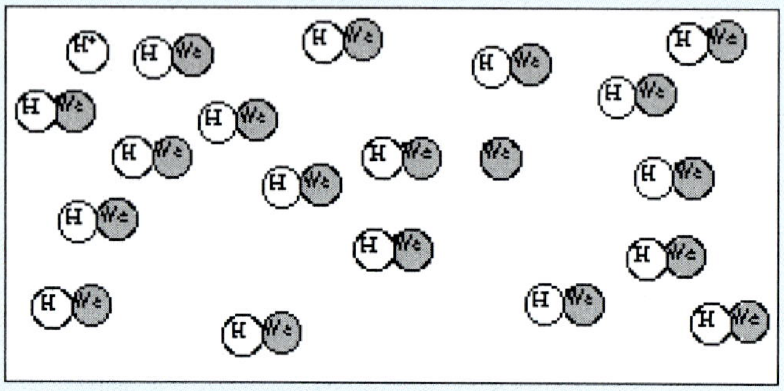

Weak/concentrated

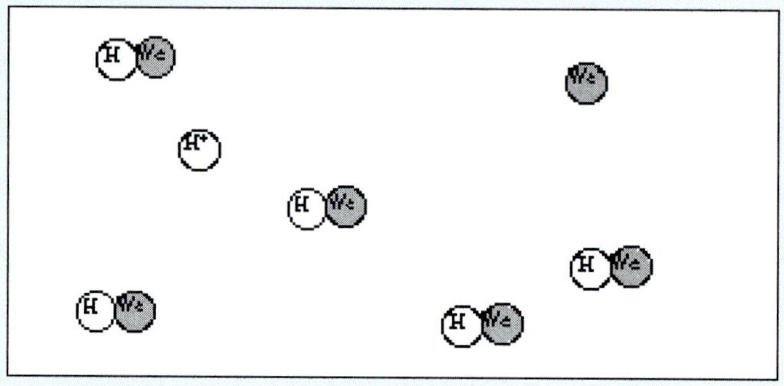

Weak/dilute

2. Calculations—molar concentrations of hydrogen ions for drawings.

Strong/concentrated

$$\frac{19\ H^+\text{particles}}{1.0\ \text{L solution}} \times \frac{0.1\ \text{mol}}{1\ \text{particle}} = 1.9\ \text{M}\ H^+$$

Strong/dilute

$$\frac{4\ H^+\text{particles}}{1.0\ \text{L solution}} \times \frac{0.1\ \text{mol}}{1\ \text{particle}} = 0.4\ \text{M}\ H^+$$

Weak/concentrated

$$\frac{1\ H^+\text{particle}}{1.0\ \text{L solution}} \times \frac{0.1\ \text{mol}}{1\ \text{particle}} = 0.1\ \text{M}\ H^+$$

Weak/dilute

$$\frac{1\ H^+\text{particle}}{1.0\ \text{L solution}} \times \frac{0.1\ \text{mol}}{1\ \text{particle}} = 0.1\ \text{M}\ H^+$$

SECTION D

AIR POLLUTION—SOURCES, EFFECTS, AND SOLUTIONS

Polluted air is common enough in the United States (see Figure 39) that weather reports for some cities include the levels of certain contaminants in the air. Depending on where you live, motor vehicles, power plants, or industries may be key sources of those air pollutants.

CD-ROM WWW. Air Pollution—Sources, Effects, and Solutions

At its "best," contaminated air may smell unpleasant, look unsightly, or block a view of stars at night. But more serious than its unpleasantness, air pollution causes billions of dollars of damage every year. It can corrode buildings and machines. It can stunt the growth of agricultural crops and weaken livestock. It has been associated with diseases such as bronchitis, asthma, emphysema, and lung cancer, adding to hospital costs worldwide. The quest for clean air has become a goal for increasing numbers of individuals, organizations, municipalities, and governments.

D.1 SOURCES OF AIR POLLUTANTS

Many air contaminants are the result of natural processes, such as those that contribute rainwater's natural acidity (Section C). In most cases natural air contamination occurs over wide regions and is seldom noticed. Furthermore, the environment may dilute, transform, or disperse such naturally emitted substances before they accumulate to harmful levels in the air.

By contrast, air pollutants from human activities are usually generated within more localized areas, such as in the vicinity of smokestacks, exhaust pipes, or large populations of humans. When the quantity of an air contaminant overwhelms the ability of natural processes to disperse or dispose of

Figure 39 *Hazy polluted air over this city almost completely hides nearby mountains.*

AIR POLLUTION—SOURCES, EFFECTS, AND SOLUTIONS

Introduction

Section D explores the human impact on air quality by comparing natural and human sources of common pollutants. The impact of human activity on the pollutant source, the control and prevention of pollutants, and the importance of preserving the ozone layer are discussed. The laboratory activity is a demonstration of two methods of controlling air pollution.

D.1 SOURCES OF AIR POLLUTANTS

When the contamination of the air occurs faster than natural processes can dispose of it, air pollution becomes a problem. Primary air pollutants are identified, and the natural sources are compared to the human generated sources. Secondary air pollutants and particulate and synthetic substances as sources of contaminants, are also discussed.

It may be beneficial to point out some of the effects of air pollution. Examples include

- minor nuisance: slight odor and reduced visibility
- property damage: discoloration of buildings, accelerated weathering, corrosion of metals
- damage to crops: lower crop yield, plants killed
- damage to animals: eye and respiratory irritation in humans, dead wildlife

it, air quality can become a serious problem. Many large cities are prone to high concentrations of air pollutants. If air pollutants generated by activity in large cities were evenly spread over the entire nation, they would be much less noticeable and would have substantially lowered concentrations. Unfortunately, weather-related air movement does not result in such even spreading.

Several categories of air contaminants must be considered. They are described below.

Primary air pollutants directly enter the atmosphere. For example, methane gas (CH_4, the simplest hydrocarbon) that enters the atmosphere is a by-product of fossil-fuel use and a component of natural gas. However, methane is also produced naturally in large quantities by anaerobic bacteria as they break down organic matter. Natural sources produce more of almost every major air contaminant than human sources do.

Secondary air pollutants are substances formed in the atmosphere by chemical reactions between primary air pollutants and/or natural components of air. For example, atmospheric sulfur dioxide (SO_2) and oxygen gas react to form sulfur trioxide (SO_3), a secondary air pollutant. Reactions with water in the atmosphere can convert sulfur trioxide to sulfates (SO_4^{2-}) or sulfuric acid (H_2SO_4), a secondary contaminant partly responsible for acid rain.

Particulates (Figure 40) include all solid particles that enter the air from either human activities (such as power plants, waste burning, road building, or mining) or natural processes (such as forest fires, wind erosion, or volcanic eruptions). Common particulates include visible emissions ("smoke") from smokestacks and motor-vehicle tail pipes.

Synthetic substances are present in air solely as a result of human activity—if not for people, they would not be there at all. One example is the fluorine released in the stratosphere when high-energy ultraviolet photons interact with a group of synthetic compounds known as chlorofluorocarbons (CFCs). These compounds will be discussed later in this section. Because no natural sources of stratospheric fluorine are known, the presence of fluorine has provided evidence that CFCs play a key role in stratospheric ozone depletion—a topic that you will consider soon.

Figure 40 *Particulate contaminants found in air may include dust, pollen, and soot, all shown here.*

D.2 IDENTIFYING MAJOR AIR CONTAMINANTS

Making Decisions

Air contaminants are by-products of manufacturing, transportation, and energy production, as well as the result of natural processes. Figure 41 gives a detailed picture of sources of some major air contaminants in the United States. Use this information to answer the questions that follow.

1. Overall, what is the main source of U.S. air contaminants?
2. For which contaminants is one-third or more contributed
   a. by industry?
   b. by transportation?
   c. by fuel combustion for electrical, industrial, home heating, and other so-called "stationary fuel burning" applications?
3. Name one general type of volatile organic compound (VOC) that might be associated with transportation. Explain its source.
4. Based on data in Figure 41, would the conversion of gasoline-fueled vehicles to electrically powered vehicles eliminate transportation as a source of sulfur dioxide (SO_2) emissions? Why?

Selected U.S. Air Pollutants, 1997 (in 10^3 metric tons/yr)

Source	CO	Pb	NO_x	VOCs	PM_{10}	SO_2	Totals
Transportation	60 922	0.475	10 541	6 964	667	1 255	80 349
Fuel Combustion							
Electric Utilities	369	0.058	5 616	46	264	11 893	18 188
Industrial	1 009	0.015	2 973	197	285	3 059	7 523
Other (residential, commercial, institutional)	3 001	0.377	1 160	539	452	739	5 891
Industrial Processes	5 502	2.634	834	8 942	1 161	1 562	18 004
Miscellaneous (fires, agriculture and forestry, fugitive dust)	8 698	—	315	780	—	12	9 805
Totals	79 501	3.559	21 439	17 468	2 829	18 520	

Key:
- CO Carbon monoxide
- Pb Lead
- NO_x Nitrogen oxides
- VOCs Volatile organic compounds
- PM_{10} Particulate matter (<10 μm diameter)
- SO_2 Sulfur dioxide

Source: U.S. EPA, *National Air Quality and Emissions Trends Report*, 1997.

Figure 41 *Sources of U.S. Air Pollution, 1997 (in 10^6 metric tons per year).*

D.2 IDENTIFYING MAJOR AIR POLLUTANTS

Making Decisions

Students analyze data from Figure 41 (comparing quantities of pollutants from human and natural sources).

Students should realize that people contribute to the pollution problem by their lifestyles. This assumption is based on the high consumption of energy and the use of many industrial products.

Answers

1. transportation
2. a. Pb, VOCs, PM_{10}
   b. CO, NO_x, VOCs
   c. NO_x, PM_{10}, SO_2
3. Answers might include gasoline, natural gas, propane, or butane resulting from incomplete combustion.
4. No. Although electrically powered vehicles would reduce transportation sulfur dioxide emissions, the resulting increase in electric utility emissions would be as bad or worse.

D.3 SMOG: HAZARDOUS TO YOUR HEALTH

Photochemical smog refers to polluted air found over cities; it is the result of the interaction of industrial emissions and automotive exhaust with solar radiation. The major chemicals involved in photochemical smog include hydrocarbons, nitrogen oxides, and ozone. The type of smog produced depends on the type of activity, the climate, and the geography of the area. The relationship between a temperature inversion and the severity of photochemical smog are presented along with the key ingredients of smog.

What proportion of total air contamination does human activity produce? Automobile use alone contributes about half the total mass of human-generated air contaminants. But air is also contaminated each time a building is heated or cooled, when fossil-fuel-based electricity is used, or when food and other products are delivered to stores.

D.3 SMOG: HAZARDOUS TO YOUR HEALTH

Sometimes meteorological conditions interact with air contaminants to form a potentially hazardous condition called **smog.** Smog is a combination of smoke and fog: thus its name. In 1952 a deadly smog over London, England, lasted five days and contributed to the deaths of nearly 4000 people. Four years earlier in the coal-mining town of Donora, Pennsylvania, smog-laden air killed 20 people and hospitalized hundreds of others. Similar though less deadly episodes have occurred in other cities.

Because substances in smog can endanger health, their levels in the air are of major public interest; many weather forecasters report an air quality index along with humidity and temperature data. The U.S. Environmental Protection Agency (EPA) has devised the Air Quality Index (AQI) based on concentrations of pollutants that are major contributors to smog in metropolitan areas of more than 350 000 people. Figure 42 underscores the fact that the combined health effects of these pollutants can become quite serious.

Fatality rates in severe smog episodes have been higher than predicted from known hazards of sulfur oxides or particulates alone. According to some researchers, this increase may be due to **synergistic interactions,** or interactions where the combined effect of several substances is greater than the sum of their separate effects alone.

A brownish haze that irritated the eyes, nose, and throat and also damaged crops first appeared in the air in and around Los Angeles, California, in the 1940s. Researchers were puzzled for some time because Los Angeles has no significant industrial or heating activities. However, the city has a large number of motor vehicles and sunshine, and is bordered by mountains on three sides. Although its valley location makes Los Angeles smog-prone, the smog experienced there was much worse than seemed reasonable. There had to be more to this story.

Normally, air at Earth's surface is warmed by solar radiation and by reradiation from surface materials. This warmer, less dense air rises, carrying pollutants with it. Cooler, less polluted air then moves in below. In a **temperature inversion,** a cool air mass is trapped beneath a less-dense warm air mass, often in a valley or over a city (Figure 43, page 320). In Los Angeles, the combination of sunny weather and mountains can produce temperature inversions about 320 days annually. During a temperature inversion, air pollutants cannot escape, so their concentrations may rise to dangerous levels. The production and severity of **photochemical smog**—which can occur even without a temperature inversion—is amplified by that accumulation of pollutants.

Any reaction initiated by light is a photochemical reaction.

U.S. Pollutant Standards Index Air Pollutant Levels							
Air Quality Index (AQI) Range	Air Quality Description	Particulate Matter (<10 μm diameter) (24-hour) $\mu g/m^3$	Sulfur Dioxide (24-hour) ppm	Carbon Monoxide (8-hour) ppm	Ozone (1-hour) ppm	Nitrogen Dioxide (1-hour) ppm	Effects and Suggested Actions
0-50	Good	0-54	0.000-0.034	0.0-4.4	—	Not reported	No significant effects.
51-100	Moderate	55-154	0.035-0.144	4.5-9.4	—	Not reported	No significant effects.
101-150	Unhealthy, especially for sensitive groups	155-254	0.145-0.224	9.5-12.4	0.125-0.164	Not reported	Some damage to materials and plants. Human health not affected unless level continues for many days.
151-200	Unhealthy	255-354	0.225-0.304	12.5-15.4	0.165-0.204	Not reported	Those with lung or heart disease should reduce physical exertion. Healthy persons notice irritations.
201-300	Very unhealthy	355-424	0.305-0.604	15.5-30.4	0.205-0.404	0.65-1.24	High-risk group has more symptoms, and should stay indoors and reduce physical activity. All persons notice lung irritation.
301-400	Hazardous	425-504	0.605-0.804	30.5-40.4	0.405-0.504	1.25-1.64	High-risk group should stay quietly indoors. Others should avoid outdoor activity.
401-500	Hazardous	505-604	0.805-1.004	40.5-50.4	0.505-0.604	1.65-2.04	Normal activity impossible. All should remain indoors with windows and doors closed. Fatal for some in high-risk group.*

*High-risk group includes elderly people, children, and those with heart or lung diseases.
Adapted from: U.S. EPA, *Guidelines for Reporting of Daily Air Quality—Air Quality Index (AQI)*, 1999.

Figure 42 *U.S. Pollutant Standards Index.*

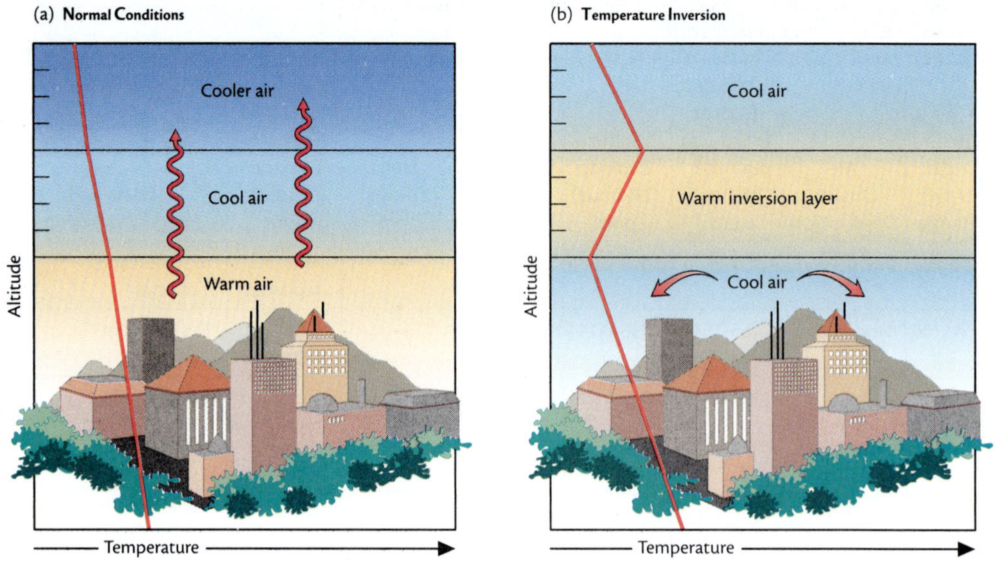

Figure 43 *A temperature inversion. Unlike (a) normal conditions, in a temperature inversion (b) a layer of cool air is trapped close to Earth's surface below a warm layer, preventing contaminants from escaping.*

The simplified equation that follows represents the key ingredients and products of photochemical smog. Hydrocarbons, carbon monoxide, and nitrogen oxides from vehicle exhausts are irradiated by sunlight in the presence of oxygen gas. The resulting reactions produce a potentially dangerous mixture, including other nitrogen oxides, ozone, and irritating organic compounds, as well as carbon dioxide and water vapor:

$$\underset{\text{(auto exhaust)}}{\text{Hydrocarbons}} + \text{Sunlight} + O_2(g) + CO(g) + NO_x(g) \longrightarrow$$

$$O_3(g) + NO_x(g) + \underset{\text{(oxidizing agents and irritants)}}{\text{Organic compounds}} + CO_2(g) + H_2O(g)$$

See Unit 2, page 125, for background on oxidizing agents. Such materials are also called oxidants.

Nitrogen oxides are essential ingredients of photochemical smog. At the high temperatures and pressures of automotive combustion (several hundred degrees Celsius and about 10 atm), nitrogen gas and oxygen gas react in the vehicle's engine cylinders to produce the pollutant nitrogen monoxide (NO):

$$N_2(g) + O_2(g) + \text{Energy} \longrightarrow 2\,NO(g)$$

Exposed to the atmosphere, nitrogen monoxide from the vehicle's exhaust is oxidized to reddish-brown nitrogen dioxide gas (NO_2), visible in polluted urban air:

$$2\,NO(g) + O_2(g) \longrightarrow 2\,NO_2(g)$$

The photochemical smog cycle begins when photons from sunlight—the second essential ingredient—break N-O bonds in NO_2 into NO molecules and highly reactive oxygen (O) atoms, also known as oxygen radicals:

$$NO_2(g) + \text{Sunlight} \longrightarrow NO(g) + O(g)$$

Atomic oxygen then reacts with oxygen molecules (O_2) in the troposphere to produce ozone (O_3), just as it does in the stratosphere:

$$O(g) + O_2(g) \longrightarrow O_3(g)$$

Ozone produced in photochemical smog remains largely in the troposphere.

Two harmful and unpleasant components of photochemical smog have now been identified—NO_2 and O_3. Nitrogen dioxide has a pungent, irritating odor. Even at relatively low concentrations (0.5 ppm), nitrogen dioxide can inhibit plant growth. And at 3–5 ppm this pollutant can cause respiratory distress in humans after one hour of exposure. Ozone is a very powerful oxidizing agent (oxidant). At concentrations as low as 0.1 ppm, ozone can crack rubber, corrode metals, and damage plant and animal tissues.

Recall that the concentration unit "ppm" refers to "parts per million."

Ozone undergoes a complex series of reactions with hydrocarbons—principally volatile organic compounds (VOCs)—the third essential ingredient of photochemical smog. Hydrocarbons escape from gasoline tanks and are also emitted during incomplete combustion of gasoline.

The concentrations of the substances associated with photochemical smog vary over a 24-hour period. In the following activity you will consider factors that may influence such concentration changes.

D.4 VEHICLES AND SMOG
Making Decisions

In this activity, students analyze the graph in Figure 44 to determine why the concentration of certain air pollutants varies over a 24-hour period.

Answers

1. a. Nitrogen oxides and hydrocarbons peak between about 6 a.m. and 9 a.m.

   b. These times correspond to the morning rush hours, when traffic is at a peak.
2. A given pollutant may decrease because it is being physically dispersed, adsorbed onto surfaces, or chemically converted to another substance.
3. Nitrogen monoxide is the source of nitrogen dioxide. Most of the NO_x emissions from automobiles are in the form of nitrogen monoxide. Once released into the atmosphere, nitrogen monoxide is further oxidized to form nitrogen dioxide.
   $2NO(g) + O_2(g) \rightarrow 2NO_2(g)$
4. a. Ozone is at a maximum when NO is at a minimum.

   b. This might suggest that NO is involved in ozone formation

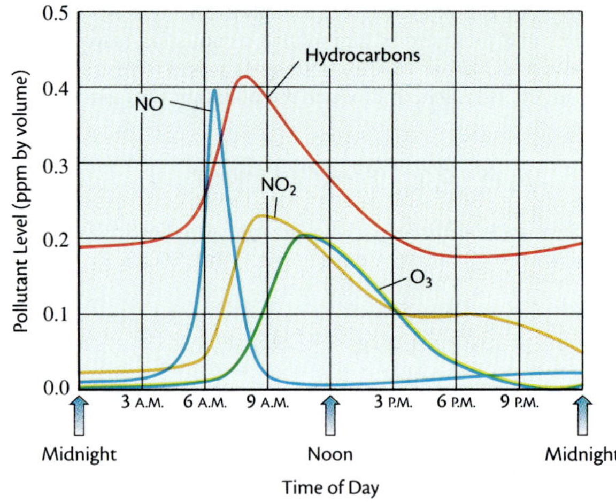

Figure 44 Photochemical smog formation.

D.4 VEHICLES AND SMOG
Making Decisions

Use data in Figure 44 to answer these questions.

1. a. Between what hours do the concentrations of nitrogen oxides and hydrocarbons peak?

   b. Account for this fact in terms of automobile traffic patterns.
2. Give two reasons why a given pollutant may decrease in concentration over a period of several hours.
3. The concentration maximum for NO_2 occurs at the same time as the concentration minimum for NO. Explain this phenomenon.
4. Although ozone is necessary in the stratosphere to protect Earth from excessive ultraviolet light, at Earth's surface it is a major component of photochemical smog.

   a. Using Figure 44, determine which substances or other species are at their minimum concentrations when O_3 is at maximum concentration.

   b. What does this suggest about the production of O_3 in polluted tropospheric air?

Many nations—including the United States—have made considerable progress in smog control. Many cities have cleaner air than they did decades ago. As you read on, you will explore how the reduction and control of air pollutants formerly contributed by motor vehicles and other sources have been addressed.

D.5 POLLUTION CONTROL AND PREVENTION

Several strategic options can limit and prevent air pollution:

- Energy technologies that cause air pollution can be replaced by technologies that do not involve combustion, such as solar power, wind power, or nuclear power.
- Pollution from combustion can be reduced by energy-conservation measures, such as extracting more energy from fuels and thereby burning less of them.
- Potentially harmful substances can be removed from fuel before burning; for example, most sulfur can be removed from coal.
- The combustion process can be modified so the fuel is more completely burned (oxidized).
- Pollutants can be removed after combustion.

All pollution-reduction options involve new costs. When evaluating a pollution prevention or control measure, decision makers must answer three key questions: What will the prevention or control cost? What benefits will it offer? What costs or risks will be involved in *not* using the prevention or control?

Power plants and smelters generate more than half of the particulate matter emitted in the United States. However, notable progress has been made in cleaning up air pollution from these sources and others over the past decade. Since the 1970s, there has been a significant decrease in the amount of most (but not all) air pollutants in the United States. See Figure 45.

Manufacturers have used large-scale techniques to reduce particulates emerging from industrial plants. Several cost-effective methods, such as those described next, are used for controlling particulates and other emissions.

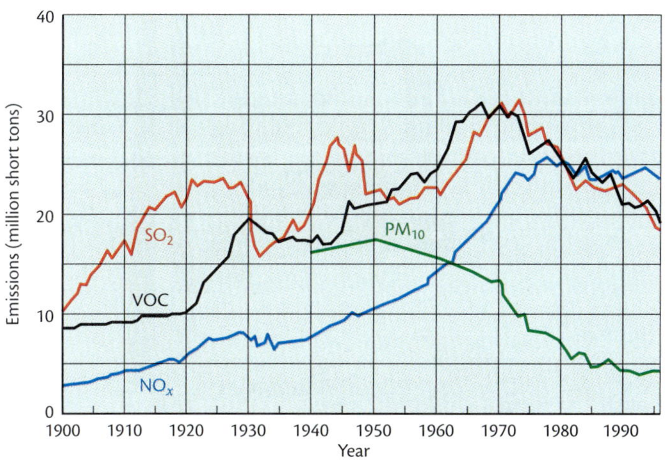

Figure 45 *Air Pollutants in the United States; Changes in Average Concentrations, 1900–1996. One short ton is equal to 2000 pounds.*

D.5 POLLUTION CONTROL AND PREVENTION

The reduction and control of air pollutants contributed by motor vehicles and other sources have made considerable progress in pollution control. Several strategies to control pollution are presented while the question of relative costs and benefits is raised. Descriptions of the most cost-effective methods of controlling emissions are discussed, including mechanical filtering, electrostatic precipitation and scrubbing. This discussion is followed by a related ChemQuandary and a laboratory activity.

Electrostatic precipitation is currently the most important technique for controlling particulate pollutants. Combustion waste products pass through a strong electrical field, where they become electrically charged. The charged particles then collect on plates of opposite charge. This technique can remove 99% (or more) of particulates, leaving only particles with diameters smaller than about one-tenth of a micrometer (0.1 μm, where 1 $\mu m = 10^{-6}$ m). Dust and pollen collectors installed in home ventilation systems are often based on this technique.

Mechanical filtering works much like a vacuum cleaner. Combustion waste products pass through a cleaning room (bag house) where huge filters trap up to 99% of the particles.

Scrubbing is a method that controls particles and sulfur oxides. In the example of wet scrubbing shown in Figure 46, sulfur dioxide gas is removed by using an aqueous solution of calcium hydroxide, $Ca(OH)_2(aq)$, also called limewater. The sulfur dioxide reacts to form solid calcium sulfite, $CaSO_3$:

$$SO_2(g) + Ca(OH)_2(aq) \longrightarrow CaSO_3(s) + H_2O(l)$$

Scrubbers can remove up to 95% of sulfur oxides. Unfortunately, because of higher operating costs resulting from maintenance and disposal expenses—combined with lower net power production—scrubbers add significantly to the cost of electrical power.

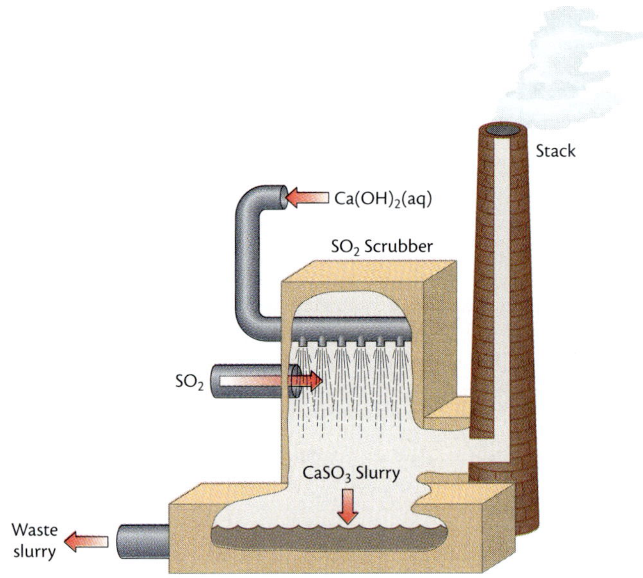

Figure 46 A wet scrubber for removing sulfur dioxide and particulates from products of industrial combustion processes.

324

LIMITING REACTANTS AND SCRUBBERS

ChemQuandary 3

In Section B (page 294) you were introduced to the concept of limiting reactants. Consider the removal of sulfur dioxide in a wet scrubber according to the equation on page 324. In properly designing this scrubber, which reactant should be designated as the limiting reactant? Why? What would be the consequences if the other reactant were the limiting reactant instead?

D.6 CLEANSING AIR

Laboratory Activity

Introduction

Your teacher will demonstrate two control methods for air pollutants—electrostatic precipitation and wet scrubbing. Prepare a sketch of each laboratory setup as part of your note taking for this activity.

CD-ROM WWW. Lab Video

Part 1: Electrostatic Precipitation

1. Observe the generation of smoke and what happens to it. Record your observations.
2. Observe the chemical reaction that takes place on the copper rod. Record your observations.

Part 2: Wet Scrubbing

3. As the reaction proceeds, observe the color of the universal indicator in the liquid and the color of the pH paper in each flask. Compare those colors with color-key information in each case. Record your observations.

Questions

1. Write an equation representing the smoke-generating reaction that occurred between HCl and NH_3 in Part 1.
2. What information did the universal indicator provide about the contents of each flask in Part 2?
3. What information did the pH paper in each flask provide?
4. What was the overall effect of the scrubbing, as shown by the indicators?
5. List the two ways in which the quality of the air in the reaction vessel was changed by wet scrubbing.
6. a. What advantages do electrostatic precipitators have over wet scrubbers?
   b. What are their disadvantages?

D.6 CLEANSING AIR

Laboratory Activity

For instructions and procedures for setting up and performing laboratory activity D.6 Cleansing Air, see pages 4F and 4G of the Planning Guide.

Answers

1. $HCl(g) + NH_3(g) \rightarrow NH_4Cl(s)$
2. Universal indicator gives the pH of the liquid after the "smoke" enters the flasks and dissolves in the water.
3. The pH paper indicates the pH of the gases hovering above the water layer just before being exhausted.
4. The pH of the water solution and gas above the water layer approaches neutrality as the reaction proceeds to the second flask and on to the aspirator. Scrubbing can effectively remove acidic or basic compounds from exhaust gases prior to their emission.
5. Wet scrubbing can remove particulate matter and lower the acidity (or alkalinity) of exhaust gases.
6. a. Electrostatic precipitators yield a solid waste product, whereas wet scrubbers produce a water solution that needs to be concentrated.

   b. Precipitators require expensive electricity to operate and do not remove gases such as sulfur dioxide and nitrogen dioxide that can be removed by wet scrubbers.

LIMITING REACTANTS AND SCRUBBERS

ChemQuandary 3

There should be an excess of $Ca(OH)_2$ so that all the undesirable SO_2 is removed. If calcium hydroxide $Ca(OH)_2$ is used as the limiting reactant, some SO_2 gas may remain when the reaction is complete. The SO_2 should be the limiting reactant.

CHEMISTRY AT WORK

A Lean, Green, Dry-Cleaning Machine

Dr. Sid Chao is the president of Raytheon Technologies, Inc. in El Segundo, California. We talked to him about a special project his company is working on: a new, environmentally friendly way to dry-clean clothes using carbon dioxide.

Q. *Can you provide a little background on your company?*

A. A few years ago, we started doing business as a subsidiary of Hughes Aircraft. Our mission is to find new uses and customers for products and technologies that Hughes has developed for the defense industry. Our specialty is developing products that help protect or preserve the environment—for example, our experimental liquid CO_2 dry-cleaning process. Some people refer to such processes as examples of "green chemistry." To us, they're primarily good products that we believe will sell.

"Our process reuses CO_2 that has already been produced, so we're not adding to existing levels."

Q. *How did you happen to focus on dry-cleaning?*

A. We knew there was a market demand for better, more cost-effective dry-cleaning methods. And we saw an opportunity to introduce a process that was better for the environment.

Our parent company had already developed a product that uses liquid CO_2 at ultra-high pressures to clean the surfaces of metals, glasses, and materials used in defense equipment, while leaving behind certain desirable substances. We theorized that the same basic idea might be the basis of a good alternative dry-cleaning solvent, if we could "scale it down" to work in a typical dry-cleaning shop. We developed a liquid CO_2 system that operates at much lower pressures, and at temperatures that a typical dry-cleaning shop can handle easily and cost effectively.

Q. *How is it better than regular dry-cleaning?*

A. For one thing, it's much better for the environment. Typical dry-cleaning processes are responsible for adding significant levels of halogenated hydrocarbons to the atmosphere. Our process reuses CO_2 that has already been produced, so we're not adding to existing levels. We are also designing a process for recovering the used solvent and "cleaning" it so that it can be used again and again.

326

CHEMISTRY AT WORK

Q. *Will we be seeing these systems soon at our own dry-cleaners?*

A. Right now we're building machines for field testing. Using the results of this testing, we're planning to develop a process that is easy to operate and costs no more to buy or use than what's already on the market. We still have to fine-tune the marketing infrastructure for the product—that is, the manufacturing, delivery, and service systems—and other details. But in the next few years, we hope to have our product available on the market.

The Greening of Cleaning

1. Since 1995 the White House has awarded the Presidential Green Chemistry Challenge (PGCC) Awards. The PGCC Awards are administered by the Environmental Protection Agency and are presented to individuals and organizations for promoting the design of chemical products and processes that prevent pollution and are economically competitive. Research one of the projects that has won a PGCC award within the last five years.
2. One concept in green chemistry is atom economy. It basically is concerned with how many reactant atoms are incorporated into the desired product and how many are wasted. Think back over the past few laboratory activities you have completed in class and evaluate them in terms of their atom economy.
3. Another green chemistry concept is that it is better to prevent waste than to treat it or clean it up after it is formed. How does the process described in this Chemistry at Work exemplify this principle?

D.7 CONTROLLING AUTOMOBILE EMISSIONS

The EPA set emissions standards for automobiles, and the automobile industry has developed several techniques to reduce emissions. Although students may be familiar with the term catalytic converter, most do not understand the chemistry of how it operates. The chemical principle behind its operation leads to an introduction of catalysts and their role in reducing the activation energy in a chemical reaction.

The activation-energy diagram (see Figure 47) should be discussed in some detail to insure students' understanding of the concept.

D.7 CONTROLLING AUTOMOBILE EMISSIONS

The Clean Air Act of 1970 authorized the Environmental Protection Agency to set emissions standards for new automobiles. Maximum allowable limits were set for hydrocarbon, nitrogen oxide, and carbon monoxide emissions. One major contribution toward meeting those standards was the development of the **catalytic converter,** a reaction chamber built into to the exhaust system of motor vehicles.

In the catalytic converter, exhaust gases and outside air pass over several catalysts that help accelerate the conversion of potentially harmful nitrogen oxides to nitrogen gas, carbon monoxide to carbon dioxide, and hydrocarbons to both carbon dioxide and water. Exhaust gases enter the catalytic converter, where, for example, nitrogen oxides and carbon monoxide are removed according to these equations:

$$2\ NO(g) + 2\ CO(g) \xrightarrow{\text{Catalyst}} N_2(g) + 2\ CO_2(g)$$

$$2\ NO(g) + 2\ H_2(g) \xrightarrow{\text{Catalyst}} N_2(g) + 2\ H_2O(g)$$

The hydrogen gas is the result of the reaction of water vapor and unburned hydrocarbon gas from the engine.

As you may recall, a catalyst is a material that speeds up a chemical reaction that would proceed far more slowly without that catalyst. Enzymes that aid in digesting food and in promoting other body functions are organic catalysts. Although the catalyst participates in the chemical reaction, it is not considered a reactant because it remains unchanged when the reaction is completed.

How can a catalyst speed up a reaction and still remain unchanged? Reactions can occur only if molecules collide with sufficient energy and with suitable orientation to disrupt bonds. The minimum energy required for such effective collisions is called the **activation energy.** You can think of the activation energy as an energy barrier that stands between the reactants and the products. See Figure 47. Reactants must have enough energy to get over the barrier before a reaction can occur. The higher the barrier, the fewer the molecules with sufficient energy to get over it, and the slower the reaction proceeds.

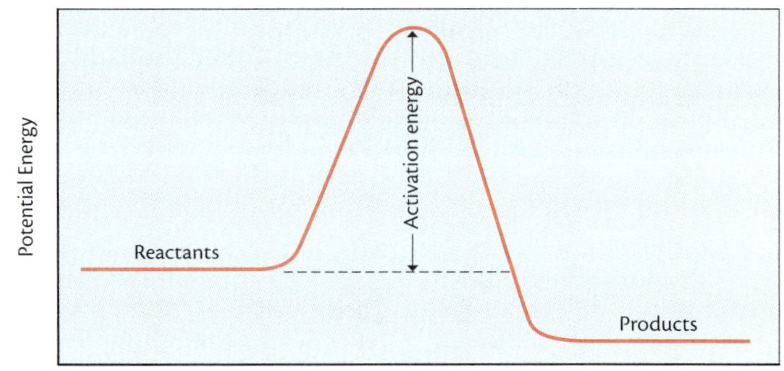

Figure 47 *Activation energy diagram for a typical chemical reaction. The activation energy represents the minimum energy needed to initiate the reaction.*

A catalyst works by providing a different reaction pathway, one with a lower activation energy. In effect, the catalyst lowers the barrier. The result is that more molecules have sufficient energy to react and form products within a given period of time. In automotive catalytic converters, a few grams of platinum, palladium, or rhodium, embedded in a material such as aluminum oxide (Al_2O_3), act as the catalyst.

Clean Air Act Amendments of 1990 set new tailpipe emission standards for U.S. automobiles manufactured after 1992. Those national tailpipe emission levels set achievable goals that significantly helped reduce photochemical smog.

CONTROLLING AIR POLLUTION

ChemQuandary 4

From a practical viewpoint, why might controlling air pollution from all automobiles, recreational vehicles, buses, and trucks in the United States be more difficult than controlling air pollution from U.S. power plants and industries?

D.8 OZONE AND CFCs: A SUCCESS STORY

Although small doses of ultraviolet radiation are necessary for health, too much is dangerous. In fact, if all ultraviolet radiation in sunlight reached Earth's surface, many forms of life would be seriously damaged. Ultraviolet photons, as you learned in Section B, have enough energy to break covalent bonds, even within vital biological structures such as DNA. The resulting chemical changes can cause sunburn and cancer in humans and damage to many biological systems.

CD-ROM WWW. Ozone Formation

Fortunately, Earth has an ultraviolet shield high in the stratosphere. The shield consists of a layer of gaseous ozone (O_3) that absorbs ultraviolet radiation. However, the stratospheric "ozone shield" is a fragile system containing a remarkably low concentration of ozone. If all ozone molecules in the shield were moved to Earth's surface at atmospheric pressure, they would form a layer only 3 mm thick (about the thickness of the cover of this textbook).

CD-ROM WWW. Ozone Depletion

The ozone layer is vitally important to supporting and protecting life on Earth. A National Research Council study suggests that for each 1% decrease in stratospheric ozone, a 2–5% increase in various forms of skin-cancer will occur. Such ozone depletion could also lower yields of some food crops, sharply increase sunburn and eye cataracts, and damage some aquatic plant species.

How does Earth's ozone shield work? As sunlight penetrates the stratosphere, high-energy ultraviolet photons react with oxygen gas molecules, splitting them into individual oxygen atoms:

$$O_2 + \text{high-energy UV photon} \longrightarrow O + O$$

CONTROLLING AIR POLLUTION

ChemQuandary 4

It is easy and cheap to install, maintain, and update control technologies when relatively few sites and relatively few people are involved. However, automobiles, buses, and trucks are located at many sites and are controlled primarily by hundreds of thousands of individuals, which makes it far more difficult to ensure compliance with standards. However, many states require inspection of motor vehicles before licenses are issued to assure compliance with emission standards.

D.8 OZONE AND CFCs: A SUCCESS STORY

The formation of ozone and its decomposition are explained The depletion of the ozone layer by CFCs has been virtually stopped by the use of HCFCs. However, HCFCs were also found to contribute to global warming and the depletion of the ozone layer. The success of the Montreal Protocol requiring a phase out of all CFCs is discussed. Continued research has lead to the development of HCFCs that contain no ozone-depleting chlorine atoms.

These highly reactive oxygen atoms are an example of a class of chemical species called **free radicals.** Radicals contain unstable arrangements and numbers of electrons. Radicals quickly enter into chemical reactions that allow them to attain stable arrangements of electrons. For example, oxygen free radicals in the stratosphere can combine with oxygen molecules to form ozone, as the equation below describes. A third molecule (typically N_2 or O_2, represented by M in the equation) carries away excess energy from the reaction but remains unchanged:

$$O_2 + O + M \longrightarrow O_3 + M$$

Each ozone molecule formed in the stratosphere can absorb an ultraviolet photon with a wavelength of less than 320 nm. This energy absorption, which keeps potentially harmful ultraviolet radiation from reaching Earth's surface, causes the ozone to decompose, producing an oxygen molecule and an oxygen free radical:

$$O_3 + \text{medium-energy UV photon} \longrightarrow O_2 + O$$

These products can then carry on the cycle by replacing ozone in the protective stratospheric layer:

$$O_2 + O + M \longrightarrow O_3 + M$$

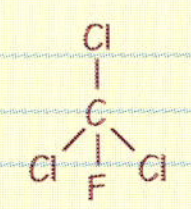

CFC-11

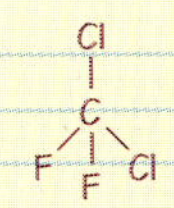

CFC-12

Human activities have already affected this ozone layer. The major culprit has been identified as chlorine atoms from chlorinated hydrocarbon molecules, called **chlorofluorocarbons (CFCs).** The Freons—CCl_3F (CFC-11) and CCl_2F_2 (CFC-12)—are the most familiar CFCs. Chlorofluorocarbons are synthetic substances that were widely used (and still are, in some countries) as propellants in aerosol cans, cooling fluids in air conditioners and refrigerators, and cleaning solvents for computer chips. Developed in the 1930s, the nontoxic and nonflammable CFCs quickly replaced toxic ammonia or sulfur dioxide as coolants of choice in refrigerators and air conditioners. For almost half a century, CFCs were well accepted as refrigerants, propellants, and cleaning solvents.

In the 1970s, two U.S. chemists—F. Sherwood Rowland and Mario Molina—proposed that chlorofluorocarbons (CFCs) pose a threat to the world's stratospheric ozone layer. Since then, extensive research has developed a more complete picture of the threat.

Rowland and Molina received the 1995 Nobel Prize in Chemistry for their pioneering research in stratospheric ozone depletion.

A satellite launched later in the 1970s carried instruments to monitor changes in worldwide stratospheric ozone concentrations. By 1986, data sent from the satellite showed seasonal ozone reduction (formation of an "ozone hole") in the stratosphere above Antarctica. See Figure 48. The satellite data indicated an annual variation in ozone concentration of 1% to 2.5%, well beyond normal variation noted before CFCs. That annual depletion cycle was observed to increase annually over subsequent years. In fact, during September and October of 1998, nearly 70% of the stratospheric ozone disappeared over Antarctica, creating an area of ozone depletion as large as North America. More recent evidence shows some thinning of stratospheric ozone over Arctic regions as well, although not to the same extent as over Antarctica.

330

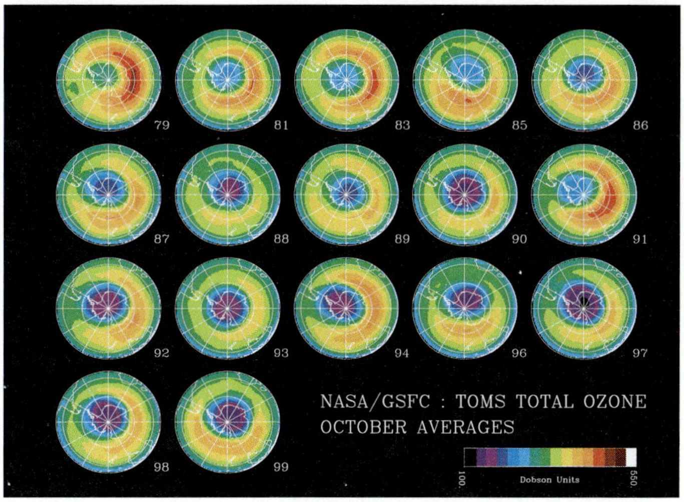

Figure 48 *Average October concentrations of ozone in the stratosphere over Antarctica, 1979 to 1999. Reds and greens indicate high ozone concentrations; blues and purples indicate low ozone concentrations. The lower the Dobson units, the lower the ozone concentration.*

Several bodies of growing evidence gathered since the mid-1980s have converged, linking the Antarctic ozone depletion to chlorine atoms from CFCs, not from natural sources. Although CFCs are highly stable molecules in the troposphere, high-energy ultraviolet photons in the stratosphere split chlorine radicals (Cl) from CFCs by breaking their C-Cl bonds.

The freed chlorine radicals are very reactive and can participate in a series of reactions that destroy ozone by converting it to O_2:

$$\cdot Cl + O_3 \longrightarrow \cdot ClO + O_2$$
$$\cdot ClO + O \longrightarrow \cdot Cl + O_2$$

The net result of these reactions can be summarized this way:

$$O_3 + O \longrightarrow 2\,O_2$$

Notice that every chlorine radical (Cl) that participates in the first reaction can later be regenerated. Thus each chlorine radical acts as a catalyst—participating in not just one, but an average of 100 000 ozone-destroying reactions. In so doing, it speeds up ozone destruction but remains unchanged itself. The net effect is depletion of ozone molecules (O_3) by their conversion into oxygen molecules (O_2), a conversion catalyzed by chlorine radicals released from CFC molecules. See Figure 49 on the next page.

Concerned about the growing threat to the stratospheric ozone layer, representatives of the United States and 55 other countries met in Montreal in 1987 and signed a historic agreement, *The Montreal Protocol on Substances That Deplete the Ozone Layer.* This initial treaty, with amendments in the 1990s, established a timetable to phase out all new CFC production in developed nations by 1996.

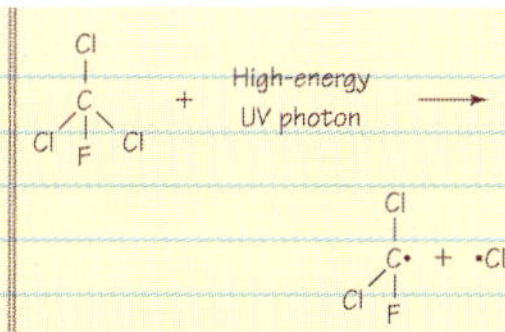

Breaking a C-Cl bond in CFC-11 by a high-energy ultraviolet light photon.

A particular chlorine radical is eventually removed from this stratospheric cycle due to other chemical reactions, such as its combination with methane (CH_4) to produce hydrogen chloride, HCl.

331

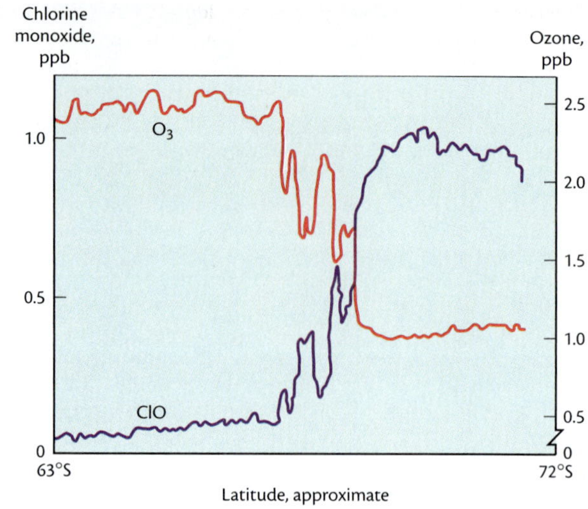

Figure 49 *Stratospheric ozone and chlorine monoxide concentrations over Antarctica.*

The Montreal Protocol, to which over 170 countries are now Party, seems to be working. U.S. production of CFCs was terminated at the close of 1995. Production of CFCs in other developed nations has stopped as well. The amended treaty also includes a multi-million dollar fund to assist other nations phase out CFC production by 2010 by helping them build industries that do not rely on CFCs.

Another outcome of the Montreal Protocol restrictions is that significant chemical research has been accomplished in developing CFC substitutes that will not harm the Earth's protective ozone layer. Chemists have synthesized such substitutes by modifying the molecular structures of CFCs. Air-conditioned vehicles currently manufactured in the United States use CFC substitutes known as **hydrochlorofluorocarbons (HCFCs)** as the coolant. HCFCs replace some of the offending chlorine atoms with atoms of hydrogen and fluorine. The hydrogen atoms allow HCFCs to decompose in the troposphere. For example, CHF_2Cl has only about 5% of the ozone-depleting capacity of CFC-12.

Unfortunately, HCFCs are not the perfect substitute for CFCs. Although some HCFC decomposition occurs in the troposphere, significant numbers of undecomposed HCFC molecules rise into the stratosphere. And like CFCs, HCFCs are greenhouse gases; thus there is some concern that they may contribute to global warming. Because of these potential problems, HCFCs are considered as only transitional replacements for CFCs. Amendments to the Montreal Protocol call for HCFC production to be phased out gradually in coming years.

Chemists continue their research to develop even better CFC substitutes. That research has produced a group of synthetic compounds known as **hydrofluorocarbons, HFCs.** Molecules of HFCs, such as CH_2FCF_3, contain only carbon, fluorine, and hydrogen atoms. They do not contain any ozone-depleting chlorine atoms, and the hydrogen atoms promote HFC decomposition in the lower atmosphere.

As you have seen, human activity has the potential to create large-scale effects on the envelope of gases surrounding the planet. Such effects may influence the world's average temperature by altering the normal atmospheric concentrations of substances such as carbon dioxide. Or they may increase potential exposure of living systems to ultraviolet radiation by thinning the stratospheric ozone layer.

The chemistry of the atmosphere is quite complex and difficult to study. However, increased knowledge of the chemical and physical processes involved, all gained through scientific research, has already helped guide decision making to protect and enhance that vital envelope of gases that encircle Earth. That knowledge can also guide design decisions involved in creating supportive environments for humans in off-Earth locations, such as the work you have already started in designing a suitable LHMP.

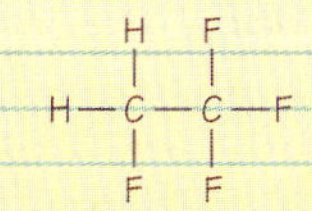

CH_2FCF_3, a typical hydrofluorocarbon, HFC

CD-ROM WWW. **Questions & Answers**

333

SECTION SUMMARY

Reviewing the Concepts

1. What distinguishes a primary air pollutant from a secondary air pollutant?

Primary air pollutants directly enter the atmosphere. Secondary air pollutants are formed in the atmosphere by a chemical reaction between primary air pollutants and/or natural components of air.

2. List some examples of primary and secondary air pollutants.

Primary sources of air pollutants	Secondary sources of air pollutants
Burning fossil fuels	**Water in the atmosphere**
Combustion of natural gas	**Oxygen in the atmosphere**
Sewage and decomposition	

3. What is the major source of each of the following pollutants?

a. CO_2

b. CH_4

c. NO_x

d. H_2S

The primary sources of the following air pollutants are

a. burning of wood and fossil fuels, decomposition.

b. burning fossil fuels and natural gas, decomposition of organic matter.

c. automobile emissions.

d. burning fossil fuels and organic decomposition.

4. What are the ingredients of photochemical smog?

The ingredients of photochemical smog include nitrogen oxides, hydrocarbons, ozone, and carbon monoxide.

5. Are all cities subject to temperature inversions? Explain your answer.

No. Cities subjected to temperature inversions are located in a valley or near mountains with sunny weather.

6. What conditions are shared by cities that suffer from photochemical smog?

Cities that suffer from photochemical smog have sunny weather and are located near mountains.

7. What are some methods by which photochemical smog can be reduced?

Photochemical smog may be reduced by using electrostatic precipitators, mechanical filtering, scrubbing, and reduction of sources.

8. What is the meaning of NO_x?

NO_x indicates the oxides of oxygen (such as NO and NO_2).

9. Why is the general expression NO_x used instead of a particular chemical formula?

NO and NO_2 are generally present together in smog. NO_x is the generic reference to represent both nitrogen oxides.

10. List five ways that air pollutants from motor vehicles and industries can be reduced.

Pollutants from autos and industry may be reduced by using electrostatic precipitators, scrubbing, mechanical filtering, catalytic converters, and reduced use.

11. Describe how a catalyst works to speed up a reaction.

A catalyst speeds up a reaction by lowering the activation energy needed to start a reaction, thus increasing the number of molecules with sufficient energy to react.

12. Consider these three air-pollutant control technologies: scrubbers, catalytic converters, and electrostatic precipitators. Which technique is most effective in combatting each of these classes of air pollutants?

a. sulfur oxides	**scrubbing**
b. particulates	**electrostatic precipitators**
c. hydrocarbons	**catalytic converters**

13. Explain how electrostatic precipitators can reduce air pollution.

Waste products pass through a strong electrical field, becoming charged. The charged particles collect on plates of opposite charge.

14. List several examples of CFCs.

Examples of CFCs include propellants for aerosol cans, refrigerants for air conditioners and refrigerators, and cleaning solvents for computer chips.

15. What properties of CFCs made them useful in industry?

The property of CFCs that made them important for industrial use are they were nontoxic, excellent propellants, good nonpolar solvents and nonflammable.

16. What are some alternatives for the industrial uses of CFCs?

Alternatives to CFCs in industry include HCFCs (hydrochlorofluorocarbons) and HFCs (hydrofluorocarbons).

17. Why are the problems associated with CFCs global rather than national?

Once the CFCs get into the atmosphere they affect the climate worldwide, regardless of where they came from.

18. Explain how chlorine acts as a catalyst in the upper atmosphere.

CFCs contain chlorine atoms, which in the stratosphere are split off by high-energy ultraviolet photons. In a series of reactions these chlorine atoms break down ozone.

Connecting the Concepts

19. Why is the *presence* of ozone a problem at ground level while the *absence* of ozone is a problem in the stratosphere?

The presence of ozone is a problem at ground level because it is a strong oxidant. At low concentrations it can crack rubber, corrode metals, and cause damage to plant and animal tissues. The absence of ozone in the upper atmosphere results in lower yields of some food crops, damage to some aquatic plants, and increased incidents of sunburns, skin cancer, and eye cataracts.

20. Explain the pattern in the data displayed in Figure 49, page 332. If comparable data were taken in the northern hemisphere, would you expect the pattern to be the same?

Stratospheric ozone seems to be decreasing, leading to the "ozone hole." A similar pattern might be expected in the northern hemisphere. In fact, comparable—but less severe—effects have been observed in the northern polar region.

21. Find Internet sites that report the current patterns of ozone distribution in the upper atmosphere. Write a report to summarize the current information.

Internet sites that students might use to begin their research include www.unep.ch/ozone/ or www.ozoneinfo@unep.org. These sites include links to other sites about ozone.

Extending the Concepts

22. Examine the labels of some products that are sprayed during use, such as cleaners, paints, hair sprays, and cooking sprays. Identify the specific substances such products use as propellants.

Student answers will vary.

23. Find the scientific research that motivated the Montreal Protocols.

A good starting point are the sites listed above for Question 21.

Putting it all Together

AIR QUALITY IN THE LHMP

CHECKLIST FOR LHMP PROPOSAL PROGRESS

— **Phase I:** Review the form and requirements for the LHMP proposal from page 247. Teams should discuss the requirements amongst themselves.

— **Phase II:** B. 8—*Planning for Greenhouse Gases.* Have students review the Making Decisions exercise (p. 294), and discuss how the answers can be incorporated in their LHMP proposals.

— **Phase III:** C.9 *Lunar Habitation Module Prototype.* After students complete the Making Decisions exercise (p. 312), discuss how their answers can be incorporated in their LHMP proposals.

— **Phase IV:** *Putting It All Together.* Have each team develop a preliminary proposal, then share it with the other teams. Then have the various teams negotiate the final air-quality management plans.

— **Phase V:** *Putting It All Together.* Have all students participate in some aspect of the presentation of the final plan. Presentations can include writing, oral, and multimedia presentations and displays that include artwork and models.

PUTTING IT ALL TOGETHER

Air Quality in the LHMP

Putting It All Together CD-ROM WWW.

As you have learned throughout the unit, in the atmosphere there are complex sets of interactions among naturally occurring components of the atmosphere, solar radiation, and emissions created by human activity. These interactions must be considered in any large-scale human-support system, including the proposed LHMP.

Because of the complexity of the atmospheric system, air-quality management involves making well-reasoned decisions. The particular choices of materials, processes, or fuels may influence many aspects of the system. What you have learned will help you propose appropriate choices for managing air-quality aspects of transportation, waste, energy, radiation, pressure, smog, and humidity within the LHMP.

1. Each team of students within your class will prepare a response to one issue (such as transportation and waste) identified in the LHMP Air-Quality Request for Proposals. Each team's response will address the following questions:
    - What are the key air-quality issues?
    - What are the major emissions?
    - What primary gases are involved? What are their roles?
    - Which (if any) emissions should be monitored? How should they be monitored?
2. Each team will also develop options for managing emissions and protecting air quality as it relates to that team's aspect of the Request for Proposals.
3. Once the preliminary proposals are complete, representatives from the teams will meet to share their air-quality plans and negotiate an overall air-quality management plan for the LHMP.
4. When a final plan has been negotiated, each team will refine its proposal for consistency with this plan and to ensure uniformity in presentation format across the groups. The final plan will then be presented as a formal response to the Request for Proposals, both in writing and as an oral or multimedia presentation.

LOOKING BACK AND LOOKING AHEAD

This concludes the first four units of *Chemistry in the Community.* Many chemical concepts remain to be explored, such as chemical equilibrium, biochemistry, and nuclear chemistry. The remaining three *ChemCom* units—Industry, Atoms, and Food—also focus on relevant chemistry-related issues but can be studied in any order. Those units all build on the chemistry knowledge you have gained throughout the first four units.

PLANNING GUIDE UNIT 5

Section	Time class	Student Edition Features	Program Resources	Media and Technology
Section A. The Chemistry of Nitrogen **Objectives:** • Fertilizers contain many essential nutrients including nitrogen, phosphorus, and potassium. • The element nitrogen is transformed chemically as it cycles through living systems and the physical environment. • The atoms in nitrogen gas are only accessible and useful to most living things if they are first converted to nitrogen-containing compounds. • The relative tendency of an atom to attract electrons within a chemical bond can be estimated by the electronegativity of that element.	5 Class Periods	**Building Skills** #1. Plant Nutrients #2. Electronegativity and Oxidation State **Laboratory Activity** • A.3 Fertilizer Components • A.5 Phosphates **Modeling Matter** • The Nitrogen Cycle	**Activity Book** • A.3 Fertilizer Components • A.5 Phosphates **Black Line Copy Masters** • Electronegativity Values	**ChemCom CD-ROM** • Unit V Industry—Section A • Laboratory Video **Transparencies** • Products of the Chemical Industry • Fertilizer • Electronegativity Values of Selected Elements **Test Bank**
Section B. Nitrogen and Industry **Objectives:** • When a system is in dynamic equilibrium, the rate of the forward reaction equals and is thus balanced by the rate of the reverse reaction.	5 Class Periods	**Building Skills** #3. Chemical Systems at Equilibrium #4. Explosive Nitrogen **ChemQuandary** #1. The Top Chemicals **Laboratory Activity** B.2 LeChatelier's Principle **Making Decisions** • What Does Riverwood Want?	**Activity Book** • B.2 LeChatelier's Principle • B.7 Making Decisions Supplement **Skill Handbook** Understanding Equations and Stoichiometry	**ChemCom CD-ROM** • Unit IV Industry—Section B • Laboratory Video **Transparency** • Top 25 Chemicals Produced in the United States

PLANNING GUIDE

Section	Time class	Student Edition Features	Program Resources	Media and Technology
• A change in temperature or concentration within a system at equilibrium may cause the equilibrium position to shift to favor formation of more reactants or products. The direction of this change can be predicted by LeChatelier's Principle. • The rate of a particular reaction depends on the temperature, concentration of reactants, and the influence of a catalyst. • Ammonia, which has many important uses, is commonly produced industrially by the Haber-Bosch process. • Initiatives such as Responsible Care, Green Chemistry, and EPA standards stress conservation, safety, and pollution prevention in decisions regarding manufacturing, storing, transporting, and disposing of chemical materials.				

PLANNING GUIDE

Section	Time class	Student Edition Features	Program Resources	Media and Technology
Section C. Metal Processing and Electrochemistry **Objectives:** • An operating voltaic cell, which is based on a spontaneous oxidation-reduction reaction, converts chemical energy to electrical energy. The magnitude of the current generated by a cell depends in part on its electrical potential. • The activity series can be used to predict the direction of electron flow within a particular voltaic cell. • Any oxidation-reduction can be described in terms of two half-reactions. Batteries, based on particular oxidation-reduction reactions and consisting of one or more voltaic cells, provide convenient, portable ways to energize many common electrical devices. • The Hall-Héroult process is an industrially important way to reduce aluminum cations (Al^{3+}) to aluminum metal.	5 Class Periods	**Building Skills** #5. Getting a Charge from Electrochemistry **Laboratory Activity** • C.2 Voltaic Cells **Making Decisions** • C.6 Asset or Liability?	**Activity Book** • C.2 Voltaic Cells • C.3 Building Skills: Supplement: Getting a Charge from Electricity **Skill Handbook** Understanding Equations and Stoichiometry	**ChemCom CD-ROM** • Unit V Industry—Section C • Laboratory Video **Transparencies** • An Alkaline Battery • Discharging and Charging a Car Battery • Electrolysis of Aqueous NaCl • The Hall-Héroult Process for Manufacturing Aluminum • Worldwide Aluminum Production through 1997

PLANNING GUIDE

Section	Time class	Student Edition Features	Program Resources	Media and Technology
• All technologies—new and old—have both positive and negative consequences. Benefit-burden analysis can help guide decisions about selecting and using these technologies. **Putting It All Together** • A Chemical Plant for Riverwood? • Confronting Final Choices			**Black Line Copy Masters** • PIAT Evaluation **Activity Book** PIAT Evaluation	

SUPPLY LIST FOR UNIT 5

Expendable Items and Unusual Materials per 24 students working in pairs

Alphabetical by Item Name			
Quantity	**Unit**	**Item**	**Sections Used**
120	mL	Ammonium molybdate-sulfuric reagent	A.5
5	Grams	Ammonium nitrate (NH_4NO_3)	A.3
80	Grams	Ammonium nitrate (NH_4NO_3) (alternate fertilizer)	A.5
20	Grams	Ascorbic acid	A.5
1.2	Grams	Barium chloride ($BaCl_2$)	A.3
12	Each	Beakers, 100 mL	A.3
36	Each	Beakers, 400 mL	A.5, B.2
40	Grams	Calcium chloride, anhydrous	B.2
2.37	Grams	Cobalt chloride ($CoCl_2 \cdot 6\ H_2O$)	B.2
12	Each	Cobalt-blue glass squares	A.3
20	Grams	Commercial fertilizer	A.5
1.2	Grams	Copper(II) nitrate ($Cu(NO_3)_2 \cdot 3\ H_2O$)	C.2
36	Cm	Copper strips, 0.25 cm wide	C.2
36	Cm	Copper strips, 0.5 cm wide	C.2
36	Cm	Copper strips, 1 cm wide	C.2
10	Liters	Distilled water	A.5
24	Each	Droppers	B.2
24	Each	Electrical wires with alligator clips	C.2
100	mL	Ethanol, 95%	B.2
12	Each	Gas burner	A.3, A.5
12	Each	Graduated cylinders, 10 mL	A.5
12	Each	Graduated cylinders, 50 or 100 mL	A.5
12	Each	Hotplate	A.3, A.5, B.2
6	mL	Hydrochloric acid (2 M HCl)	A.3
15	mL	Hydrochloric acid (6 M HCl)	A.3

SUPPLY LIST

Alphabetical by Item Name (continued)			
Quantity	**Unit**	**Item**	**Sections Used**
1.4	Grams	Iron(II) sulfate ($FeSO_4$)	A.3
5	Grams	Iron(III) nitrate ($Fe(NO_3)_3$)	A.3, C.2
36	Cm	Iron metal strips or small iron nails	C.2
1	Vial	Litmus paper, red	A.3
1.3	Grams	Magnesium nitrate ($Mg(NO_3)_2$)	C.2
36	Cm	Magnesium ribbon	C.2
12	Each	Nichrome or platinum wire loops (10 cm)	A.3
160	Each	Pipets, Beral	A.3, C.2
35	Grams	Potassium chloride (KCl)	C.2
0.5	Grams	Potassium nitrate (KNO_3)	A.3
0.5	Grams	Potassium thiocyanate (KSCN)	A.3
12	Each	Ring stands with iron rings & wire gauze square	A.3, A.5, B.2
4.5	Grams	Silver nitrate ($AgNO_3$)	B.2
4	Grams	Sodium chloride (NaCl)	B.2
6	Grams	Sodium hydroxide (NaOH)	A.3
4.3	Grams	Sodium nitrate ($NaNO_3$)	A.3
1.9	Grams	Sodium phosphate (Na_3PO_4)	A.3
380	Grams	Sodium phosphate (Na_3PO_4) (alternate fertilizer)	A.5
1.6	Grams	Sodium sulfate (Na_2SO_4)	A.3
12	Each	Stirring rods	A.5
24	mL	Sulfuric acid, conc. (6 M H_2SO_4)	A.3
24	Each	Test tube rack	A.5
60	Each	Test tubes (18 × 150 mm)	A.3, A.5
24	Each	Test tubes (18 × 150 mm)	B.2
24	Each	Test tubes (10 × 75 mm)	A.3

SUPPLY LIST

Alphabetical by Item Name (continued)			
Quantity	**Unit**	**Item**	**Sections Used**
1	Box	Toothpicks, wood	A.3
12	Each	Voltmeter	C.2
12	Each	Wellplates (24 wells)	A.3, C.2
1	Package	Wood splints	A.3
1.5	Grams	Zinc nitrate ($Zn(NO_3)_2 \cdot H_2O$)	C.2
36	cm	Zinc strips, 0.25 cm wide	C.2
36	cm	Zinc strips, 0.5 cm wide	C.2
36	cm	Zinc strips, 1.0 cm wide	C.2

SUPPLY LIST

By Section			
Quantity	**Unit**	**Item**	**Sections Used**
0.5	Grams	Potassium nitrate (KNO_3)	A.3
0.5	Grams	Potassium thiocyanate (KSCN)	A.3
1	Box	Toothpicks, wood	A.3
1	Package	Wood splints	A.3
1	Vial	Litmus paper, red	A.3
1.2	Grams	Barium chloride ($BaCl_2$)	A.3
1.4	Grams	Iron(II) sulfate ($FeSO_4$)	A.3
1.6	Grams	Sodium sulfate (Na_2SO_4)	A.3
12	Each	Beakers, 100-mL	A.3
12	Each	Cobalt-blue glass squares	A.3
12	Each	Nichrome or platinum wire loops (10-cm)	A.3
15	mL	Hydrochloric acid (6 M HCl)	A.3
24	Each	Test tubes, small	A.3
24	mL	Sulfuric acid, conc. (6 M H_2SO_4)	A.3
4.3	Grams	Sodium nitrate ($NaNO_3$)	A.3
6	Grams	Sodium hydroxide (NaOH)	A.3
6	mL	Hydrochloric acid (2 M HCl)	A.3
12	Each	Gas burner	A.3, A.5
382	Grams	Sodium phosphate (Na_3PO_4) (alternate fertilizer)	A.3, A.5
85	Grams	Ammonium nitrate (NH_4NO_3) (alternate fertilizer)	A.3, A.5
12	Each	Hotplate	A.3, A.5, B.2
12	Each	Ring stands with iron rings & wire gauze square	A.3, A.5, B.2
60	Each	Test tubes (18 $\times$ 150 mm)	A.3, A.5, B.2
12	Each	Wellplates, 24-wells	A.3, C.2
160	Each	Pipets, Beral	A.3, C.2
5	Grams	Iron(III) nitrate ($Fe(NO_3)_3$)	A.3, C.2

SUPPLY LIST

By Section (continued)			
Quantity	**Unit**	**Item**	**Sections Used**
10	Liters	Distilled water	A.5
12	Each	Graduated cylinders, 10-mL	A.5
12	Each	Graduated cylinders, 50- or 100-mL	A.5
12	Each	Stirring rods	A.5
120	mL	Ammonium molybdate–sulfuric reagent	A.5
20	Grams	Ascorbic acid	A.5
20	Grams	Commercial fertilizer	A.5
24	Each	Test tube rack	A.5
36	Each	Beakers, 400-mL	A.5, B.2
100	mL	Ethanol, 95%	B.2
2.37	Grams	Cobalt chloride ($CoCl_2 \cdot 6H_2O$)	B.2
24	Each	Droppers	B.2
4	Grams	Sodium chloride (NaCl)	B.2
4.5	Grams	Silver nitrate	B.2
40	Grams	Calcium chloride, anhydrous	B.2
1.2	Grams	Copper(II) nitrate ($Cu(NO_3)_2 \cdot 3H_2O$)	C.2
1.3	Grams	Magnesium nitrate ($Mg(NO_3)_2$)	C.2
1.5	Grams	Zinc nitrate ($Zn(NO_3)_2 \cdot 6H_2O$)	C.2
12	Each	Voltmeter	C.2
24	Each	Electrical wires with alligator clips	C.2
35	Grams	Potassium chloride (KCl)	C.2
36	Cm	Copper strips, 0.25-cm wide	C.2
36	Cm	Copper strips, 0.5-cm wide	C.2
36	Cm	Copper strips, 1-cm wide	C.2
36	Cm	Iron metal strips or small iron nails	C.2
36	Cm	Magnesium ribbon	C.2

SUPPLY LIST

By Section (continued)			
Quantity	**Unit**	**Item**	**Sections Used**
36	Cm	Zinc strips, 0.25-cm wide	C.2
36	Cm	Zinc strips, 0.5-cm wide	C.2
36	Cm	Zinc strips, 1.0-cm wide	C.2

UNIT 5 INDUSTRY: APPLYING CHEMICAL REACTIONS

Introduction to Unit

Industrial chemistry encompasses a broad range of manufacturing processes, including the refining of petroleum (Unit 3), the preparation of metals from their ores (Unit 2), and the production of pharmaceuticals. This unit is focused on the industrial processes associated with the production of synthetic inorganic fertilizers and the preparation of aluminum from bauxite ore.

This unit gives students an opportunity to evaluate the operations of the EKS and WYE companies.

Teacher notes

Because it is impossible to demonstrate industrial processes adequately in the classroom, it is strongly recommended that you use the following strategies:

- Obtain videos or films of industrial processes to view with your class. *The World of Chemistry* videotapes and *TV Ontario's Videotapes* are good sources.
- Arrange field trips to industrial plants such as electroplating companies, jewelry makers, electronic firms, and aluminum companies, etc.
- Arrange for classroom visits by knowledgeable individuals such as industrial chemists, electroplating and electrolysis workers, and silversmiths.
- Arrange to have some of your students shadow a professional, or assign students to a mentor for short periods of time.

UNIT 5

HOW does the chemical industry transform elements and compounds into other useful materials?

WHAT chemical principles are involved in converting nitrogen gas to nitrogen-containing compounds?

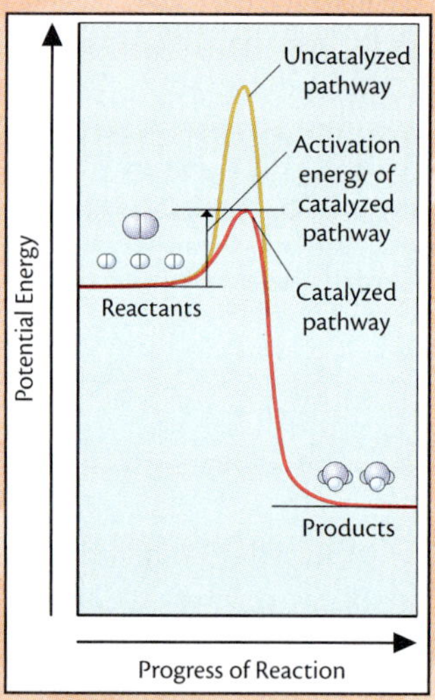

INDUSTRY: APPLYING CHEMICAL REACTIONS

SECTION C **Metal Processing and Electrochemistry**
(page 380)

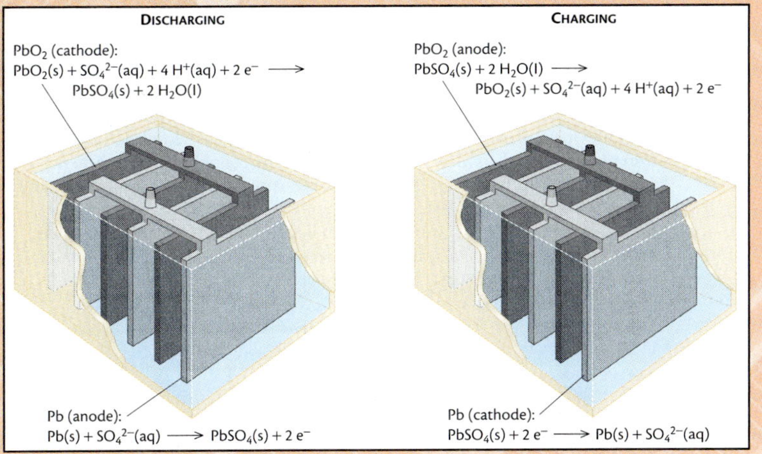

HOW can chemical changes be produced or caused by electrical energy?

The town of Riverwood needs new jobs; two large chemical companies are prepared to provide them. Turn the page to learn about issues Riverwood must address before either company is invited to build a plant in the community.

EKS OR WYE? EITHER MAY SPELL JOBS FOR RIVERWOOD

Commentary by Greta B. Lederman
Riverwood Resident

Several weeks ago, the town council announced that two industrial firms are interested in building chemical manufacturing plants in Riverwood—EKS Nitrogen Products Company and WYE Metals Corporation. Since then, many friends and neighbors have expressed relief at this news.

Many residents either lost their jobs or know someone who did when Riverwood Corporation declared bankruptcy last year. Although some have found new jobs, many have not. Unfortunately, this has contributed to an unacceptable unemployment rate—near 15% for most of the past year.

No one in our community has been immune to the impact of this level of unemployment. Since the Riverwood Corporation closed its doors, several smaller businesses have also closed due to a decline in customers. Some businesses have dealt with their smaller customer base in a different way—by laying off workers.

Can anything be done to boost our local economy? Yes. The answer lies in allowing one of these eager companies to build a chemical plant in our town. New jobs would be created, from chemists and chemical technicians to office personnel. Each company claims that about 200 new employees would need to be hired.

In addition, representatives from both companies have confirmed that they plan to occupy the former Riverwood Corporation site. This is good news, since that vacant property and adjoining parking lot seem only to attract flea-market vendors on weekends and displays of used cars for sale.

Let's consider negatives. Sure, we need jobs—no one denies that. But is a chemical plant the best option for Riverwood? Why not a furniture manufacturer or book warehouse? Aren't chemical plants dangerous?

The reality is that no other companies have expressed an interest in locating here. EKS and WYE are attracted by our well-educated workforce and by access to abundant resources and electrical energy, among other factors. And both are well suited to our community.

Second, both firms have strong safety and environmental records. They are members of the American Chemistry Council's Responsible Care initiative. Both advocate "green" technologies. Reports on the companies' job-safety and pollution-prevention records are available at City Hall.

We have a unique opportunity to choose between two companies that are competing to locate here. Usually it's the reverse—two or more cities competing for one company. Learn more about each company—attend the scheduled citywide discussions and visit their websites.

If you still have doubts, notice how this news has affected your friends and neighbors. A new sense of hope has emerged over recent weeks. Let's turn weeks of hope into years of prosperity for Riverwood.

340

EKS OR WYE? THE PRICE OF NEW JOBS MAY BE TOO HIGH

COMMENTARY BY PAK JIN-WOO
Riverwood Resident

While no one can argue that our Riverwood community needs new jobs, it's foolish to invite either EKS Nitrogen Products Company or WYE Metals Corporation to locate here to manufacture, respectively, ammonia or aluminum, without first considering all the consequences.

First, the promise of 200 new jobs sounds alluring. However, all of these jobs would be located within only one company. What will happen if there's a decline in the market for ammonia or aluminum metal?

Either of the proposed Riverwood plants would be small compared to the EKS and WYE companies' other plants. Thus the Riverwood plant would likely be among the first to be shut down. If that happens, the Riverwood area would quickly be back in the same economic situation it is experiencing now.

Wouldn't it be more prudent to distribute those 200 jobs among several different companies rather than placing all of our eggs in one basket? Let's learn from our recent experience with the city's former largest employer, the now-bankrupt Riverwood Corporation.

Although long-term economic health is important, we must question the potential safety and environmental risks that each of these two chemical plants would pose.

For instance, ammonia is produced at very high pressures and temperatures. Although accidents are uncommon, the potential consequences of an explosion or spill are great. Several illnesses and even deaths of workers have been documented at ammonia plants.

Aluminum production activity also involves risks to its workers. For every 100 employees, aluminum-production illnesses and injuries involve—on average—about 20 workers annually. Which 40 Riverwood residents would be included in this statistic if we allow an aluminum plant to be established here?

Perhaps the greatest long-term concern to all Riverwood residents is the potential for environmental harm. Ammonia-based contamination of the Snake River could surely cause another fish kill, an all-too-recent occurrence in the memories of all Riverwood citizens.

Along the same lines, the modern industrial process involved in manufacturing aluminum metal produces, among other things, toxic carbon monoxide gas. What might happen if containment of this gas were accidently mishandled?

I realize that many of these possible scenarios are unlikely to be realized in practice. However, we must be certain that we are absolutely willing to accept the consequences should something negative happen. The promise of new Riverwood jobs now may be followed by an extremely heavy price later. With what level of risk are you and your friends willing to live?

341

As you can infer from the newspaper commentaries you just read, two companies seek to establish a chemical plant in Riverwood. One of them—EKS Nitrogen Products Company—wishes to establish a plant that manufactures ammonia. The other—WYE Metals Corporation—is interested in building an aluminum-production plant. As both commentators acknowledge, either company would provide at least 200 new job opportunities to the Riverwood community. However, as the commentaries suggest, job creation is far from the only factor under consideration as residents examine their options.

Later, you will help decide whether either EKS or WYE should be invited to build a chemical plant in Riverwood. The chemistry you learn in this unit will prepare you to make informed decisions regarding risks and benefits that might accompany the operation of such a chemical plant. Keep in mind the concerns voiced in the two newspaper commentaries as you learn about the chemistry involved in manufacturing ammonia and aluminum.

342

SECTION A THE CHEMISTRY OF NITROGEN

Throughout *ChemCom* you have learned that chemistry is concerned with the composition and properties of matter, changes in matter, and the energy involved in those changes. In this unit you will explore ways that chemical industries use chemical knowledge and reactions to produce a wide range of material goods and services.

In particular, this unit offers you the opportunity to evaluate the chemical operations of both EKS and WYE. This knowledge will help you later as you participate in a debate on whether a new chemical plant should be located at Riverwood—and, if so, which one should be invited.

CD-ROM WWW. Introduction

A.1 CHEMICAL PROCESSING IN YOUR LIFE

Making Decisions

To build a better sense of how pervasive the products of chemical processing are in everyday life, try to list five items or materials around you that have *not* been manufactured, processed, or altered from their natural form. Start by considering everyday items—clothes, household materials, means of transportation, books, communication devices (such as phones and computers), sports and recreation equipment—whatever you routinely encounter.

Based on your list, answer these questions. Be prepared to share and discuss your answers in class.

1. a. Which items on your list were wrapped, boxed, or shipped in materials that had been manufactured? Explain.
   b. Is the packaging or shipping material essential or simply a convenience? Why?
2. In what ways might each item or material on your list be better or worse than a manufactured, processed, or synthetic alternative? Consider factors such as cost, convenience, availability, and quality.
3. If a product is "100% natural," does that necessarily mean it was not involved in any processing or chemical or physical changes? Why? Support your answer with at least one example.

A.2 CHEMICAL PRODUCTS

The chemical industry's focus is changing natural materials and resources into useful products to meet a wide variety of needs and purposes. New substances are also created as replacements for natural ones—for example, plastics may substitute for wood and metals, or synthetic fibers may replace cotton or wool.

THE CHEMISTRY OF NITROGEN

Introduction

The main focus of Section A is to give students an opportunity to explore methods used by the chemical industry to produce useful products by using fundamental chemical principles.

A.1 CHEMICAL PROCESSING IN YOUR LIFE

Making Decisions

This introductory activity is designed to have students think about the impact of the chemical industry on their lives. Students' answers to questions will vary. Be sure to allow sufficient time for class discussion.

1. a. Unless you grow your own food, spin your own fibers, or purchase items from individuals who do, most purchased items (even those labeled as 100% natural) are packaged in chemically processed paper, cardboard, or plastics.

   b. Packaging is especially important to help maintain product quality—freshness and purity—during shipping and storage.
2. In most cases, synthetic products cost less than their natural alternatives (examples: nylon vs. silk, synthetic vitamins vs. natural). Often natural materials are produced in particular regions, which necessitates more transport for distribution than required by locally produced synthetics. Typically, synthetics are at least of comparable if not superior quality (example: synthetic rubber for natural). Sometimes a blend of natural and synthetic provides the best combination of cost and quality (example: clothing made by blending cotton with synthetic fibers).
3. "100% natural" does not mean that no chemical processing occurred; it only indicates that no synthetic additives are included in the product. In fact, almost all natural products have undergone some chemical processing. Examples include 100% natural cereals and vitamins and all-cotton or all-wool clothing items. This question also provides an opportunity to reemphasize two points: (a) all matter is made up of chemical substances, and (b) assuming it is pure, a given synthetic compound is identical to the natural version of the same compound and often is less expensive.

A.2 CHEMICAL PRODUCTS

The chemical industry changes natural materials into useful products and creates new products to replace natural materials. The importance of fertilizers to the production of food and the appropriate choice of a specific fertilizer are discussed. The EKS Nitrogen Products Company, a fertilizer manufacturer, is the focus of the discussion.

Figure 1 *The chemical industry produces not only substances for consumer use, but also goods ranging from paint to fiberglass.*

Even though the chemical industry is a worldwide, multibillion dollar enterprise that affects everyone's life daily through its products and economic impact, most people are not aware of what happens in the production of new materials in the industry. This fosters an aura of mystery about how chemical industries operate, how they manufacture new products, and what those products contain.

The modern chemical industry employs well over a million people worldwide. Over the past 80 years, it has grown through mergers of smaller companies and the creation of new companies. During that time the industry's focus has expanded from a limited range of basic products to more than 70 000 different products. Hundreds of chemical companies form the third largest manufacturing industry in the United States—only the industries that produce machinery and electrical equipment are larger. Indeed, if the food and petroleum industries are included in the chemical industry category, this category is the largest industry in the world. For example, 75 billion pounds of plastic were manufactured in the United States in 1995; plastics production has had an average annual growth rate of about 12% for over 70 years.

Most chemical products reach the public indirectly, since many of them are used in producing other consumer materials. For instance, the automobile and home-construction industries use enormous supplies of industrial chemicals. Paints and plastics are needed for automobile body parts such as bumpers, dashboard panels, upholstery, and carpeting, and for synthetic rubber in tires and hoses. Home construction involves large quantities of plastics for carpeting and flooring, insulation, siding, window frames, piping, and appliances. Paints, metals, and air-conditioning coolants are also used extensively. The range of products from chemical industries, as shown in Figure 1, represents a wide array of materials.

In Riverwood, the two chemistry-related companies under consideration manufacture products involving nitrogen and aluminum. Among the products of EKS Nitrogen Products Company are nitric acid and ammonia, which are often used in chemical reactions that produce other materials. By contrast, the sheet aluminum produced by WYE Metals Corporation is most often used directly in its manufactured form.

EKS is committed to producing high-quality fertilizer in Riverwood at reasonable cost, using the best available technologies. Fertilizer may sound unappealing as a product, but its manufacture and sale represent a worldwide multimillion dollar business that employs thousands of people and affects the lives of nearly everyone, from farmers and gardeners to food producers and consumers.

How can you decide whether a fertilizer is best for a particular application, such as houseplants, a lawn, or a cornfield? One way is to find out if the fertilizer contains the proper ingredients. Complete fertilizers, such as those produced by EKS, contain the three main elements needed by growing plants—nitrogen, phosphorus, and potassium—as well as trace ions and filler material.

345

A.3 FERTILIZER COMPONENTS

Laboratory Activity

Students use the techniques of analytical chemistry to determine if various nutrients are present in a fertilizer.

A qualitative test is performed for each of six different ions.

Time

50 minutes

Materials (for a class of 24 working in pairs)

12 100-mL beakers
24 red litmus paper strips
12 10-cm nichrome or platinum (Pt) wire loops, each attached to a cork stopper or handle
12 24-well wellplates
12 gas burners
12 18 × 150 mm test tubes (for cleaning nichrome wires)
12 squares of cobalt-blue glass
12 ring stands
12 rings
12 wire gauze squares
60 toothpicks
24 wood splints soaked in 0.5 M KCl (4 g KCl/100 mL solution)
120 Beral pipets
50 mL 6 M HCl for cleaning nichrome wires (25 mL conc. HCl/50 mL solution)
solutions
- 50 mL 6.0 M HCl (25 mL conc. HCl/50 mL solution)
- 50 mL 3.0 M NaOH (60 g NaOH/500 mL solution)
- 50 mL conc. H_2SO_4 in dropper bottles
- 50 mL 0.1 M $FeSO_4$ (1.4 g $FeSO_4 \cdot 7H_2O$/50 mL solution)
- 50 mL 0.1 M $(BaCl_2) \cdot$(1.2 g $BaCl_2 \cdot 2\,H_2O$/50 mL solution)
- 50 mL 0.2 M KSCN (0.5 g KSCN/50 mL solution)
- 24 mL 6 M H_2SO_4

known ion solutions
- 50 mL 0.1 M Na_3PO_4 (1.9 g $Na_3PO_4 \cdot 2\,H_2O$/50 mL solution)
- 50 mL 0.1 M $NaNO_3$ (4.3 g $NaNO_3$/50 mL solution)
- 50 mL 0.1 M Na_2SO_4 (1.6 g $Na_2SO_4 \cdot 10\,H_2O$/50 mL solution)
- 50 mL 0.1 M KNO_3 (0.50 g KNO_3/50 mL solution)
- 50 mL 0.1 M NH_4NO_3 (0.40 g NH_4NO_3/50 mL solution)
- 50 mL 0.1 M Fe $(NO_3)_3$ (2.0 g $Fe(NO_3)_3 \cdot 9\,H_2O$/50 mL solution)

unknown fertilizer solutions in Beral pipets, labeled with a code number. See advanced preparation notes.

Figure 2 *The 4–5–2 values shown on the fertilizer bag indicate 4% nitrogen, 5% P_2O_5, and 2% K_2O.*

Reporting P and K as P_2O_5 and K_2O originated during early research on plant fertilizers, when plants were burned to ash, which was analyzed and the resulting quantities of P_2O_5 and K_2O weighed.

Figure 2 shows the label of a typical commercial fertilizer bag. The sequence of integers on the label indicates the percentages of key ingredients contained in the fertilizer—percent nitrogen, N; phosphorus, P (expressed as percent P_2O_5); and potassium, K (expressed as percent K_2O). The proportion of each component varies according to crop needs. Most lawn grasses need nitrogen; a 20–10–10 fertilizer would be a good choice for them. Phosphorus is especially useful in promoting fruit and vegetable growth, so a 10–30–10 mixture would be preferred over a balanced (10–10–10) composition.

Many plant nutrients provided in fertilizers are in the form of cations and anions. Cations are likely to include potassium (K^+), ammonium (NH_4^+), and iron(II) (Fe^{2+}) or iron(III) (Fe^{3+}); anions include nitrate (NO_3^-), phosphate (PO_4^{3-}), and sulfate (SO_4^{2-}).

Fertilizer Components CD-ROM WWW.

In the following laboratory activity you will use confirming tests to determine whether various nutrients are present in a fertilizer solution.

A.3 FERTILIZER COMPONENTS

Laboratory Activity

Introduction

Lab Video CD-ROM WWW.

You used confirming tests in water testing in Unit 1—see page 35.

In this laboratory activity you will test a fertilizer solution for six particular ions (three anions and three cations). In Part 1 you will perform tests on known solutions of those ions to become familiar with each confirming test. In Part 2 you will decide which ions are present in an unknown fertilizer solution. Read the complete procedure and prepare a suitable data table to record your observations.

Procedure

Part 1. Ion Tests

1. Prepare a warm-water bath for use in Step 6d. Pour about 30 mL of water into a 100-mL beaker. Place the beaker on a hot plate. The water must be warm, but should not boil. Control the heat accordingly.
2. Obtain a Beral pipet set containing each of six known ions—nitrate (NO_3^-), phosphate (PO_4^{3-}), sulfate (SO_4^{2-}), ammonium (NH_4^+), iron(III) (Fe^{3+}), and potassium (K^+). Record the color of each solution.

$BaCl_2$ Tests

Several ions you are studying in this activity can be identified first by their reaction with barium cations (Ba^{2+}) and then their behavior in the presence of acid.

3. a. Place a clean sheet of white paper under a multiple-well wellplate. Write the formulas of the six ions to be tested on the paper, locating each near a separate depression in the wellplate.
   b. Place 2 to 3 drops of each ion solution into its corresponding well.
   c. Test each sample solution individually by adding to it 1 or 2 drops of 0.1 M barium chloride ($BaCl_2$) solution. Use a new toothpick to mix each solution.
   d. Add 3 drops of 6 M hydrochloric acid (HCl) to each of the six wells containing $BaCl_2$ solution. **CAUTION:** *6 M HCl is corrosive. If any splashes on your skin, wash it off thoroughly with water and inform your teacher. Do not inhale HCl fumes.* Record your observations. Light-colored precipitates may be easier to observe if you temporarily remove the white paper beneath the wellplate. Record your observations.
   e. Clean and rinse the wellplate.
4. Dispose of all solutions as instructed by your teacher.

Each time you mix a solution in a well of a wellplate, be sure to use a new toothpick. That prevents contamination of the solutions.

Brown-Ring Test

In the presence of nitrate ions (NO_3^-), mixing iron(II) ions (Fe^{2+}) and sulfuric acid (H_2SO_4) produces a distinctive result. This "brown-ring test" can be used to detect nitrate ions in a solution.

5. a. Place 8 drops of sodium nitrate ($NaNO_3$) solution in a small, clean test tube. Place 8 drops of distilled water in a second small, clean test tube, which will serve as the control.
   b. Add about 1 mL of iron(II) sulfate ($FeSO_4$) reagent to both test tubes. Gently mix each tube.
   c. Have your teacher carefully pour about 1 mL of concentrated sulfuric acid (H_2SO_4) along the inside of each test tube so the acid

Advance preparation

Label Beral pipets and fill them with the test solutions (except for concentrated sulfuric acid) and the six known ions. (If student teams are to use sulfuric acid, it should be supplied in dropper bottles made specifically for concentrated sulfuric acid.) Place the pipets in the wellplate with the points up to avoid spillage.

For the unknowns, you can use any available solutions that contain the appropriate ions. Samples of commercial fertilizers are not recommended for the unknowns because some contain insoluble matter, time-released substances, or ions that interfere with the simple identifying tests suggested in this procedure. Concentrations should be at least 0.1 M. Examples include (1) a solution that contains both 0.1 M NH_4NO_3 and 0.1 M Na_3PO_4 (testing positive for NH_4^+, NO_3^-, and PO_4^{3-}); (2) a solution of 0.1 M $Fe_2(SO_4)_3$ (testing positive for Fe^{3+} and SO_4^{2-}). It is not necessary to prepare a different unknown for each pair of students.

Prepare the unknown solutions and place them in Beral pipets labeled with a code. Instruct students to record the code for their unknown in their laboratory report.

If necessary, clean the used flame test wires yourself with a higher concentration of HCl, such as 12 M, after student use.

Lab tips

- Wellplates must be rinsed with distilled water before the same depression is used a second time during the activity. Remind students to clean the wellplates thoroughly when they finish the laboratory activity.

Safety

Students should wear safety goggles and aprons throughout the laboratory activity.

Concentrated sulfuric acid is extremely hazardous: students should use it only under strict supervision. We recommend that teachers add the concentrated sulfuric acid in Step 3b.

Barium compounds are toxic. Direct student attention to the **CAUTION** statement in the student text.

- To prevent contamination of the solutions, a new toothpick should be used each time a solution is mixed in a well of a wellplate.
- The burner setting should be adjusted to produce a hot, blue flame. A hot platinum or nichrome wire will itself display a color. It is important to record the color that appears before the wire is red hot.
- Remind students that red litmus paper turns blue (or slightly violet) in a base.

Lab discussion

- Stress the importance of keeping a careful record of each solution's location in the wellplate. Show students the techniques of labeling a piece of paper underneath the plate to keep track of their results.

forms a second layer under the unagitated liquid already in the tube. **CAUTION:** *Concentrated H_2SO_4 is a very strong, corrosive acid. If any contacts your skin, immediately wash affected areas with abundant running tap water and inform your teacher.*

d. Allow the two test tubes to stand—without mixing—for 1 to 2 minutes.

e. Observe any change that occurs at the interface between the two liquid layers. Record your observations.

NaOH and Litmus Tests

One or more of the three cations can be identified through observing their characteristic behavior in the presence of a strong base.

6. a. Add 4 drops of each cation test solution to three separate, clean test tubes.

b. Moisten three pieces of red litmus paper with distilled water; place them on a watch glass.

c. Add 10 drops of 3 M sodium hydroxide (NaOH) directly to the solution in one test tube. **CAUTION:** *3 M NaOH is corrosive. If any splashes on your skin, wash it off thoroughly with water and inform your teacher.* Do not allow any NaOH solution to contact the test tube lip or inner wall. Immediately stick one of the three moistened red litmus paper strips from Step 6b onto the upper inside wall of the test tube. The strip must not contact the solution.

d. Warm the test tube gently in the hot water bath for 1 minute. Note and record your observations after waiting about 30 seconds.

e. Repeat Steps 6c and 6d for the other two test tubes.

Flame Test

Many metal ions can be identified by the characteristic color they emit when heated in a burner flame. It is common to use a flame test to identify potassium ions, for example.

7. a. Obtain a platinum or nichrome wire inserted into glass tubing or a cork stopper.

b. Set up and light a burner. Adjust the flame to produce a light blue, steady inner cone and a more luminous, pale blue outer cone.

c. To clean the wire, place about 10 drops of 2 M hydrochloric acid (HCl) in a small test tube. **CAUTION:** *2 M HCl is corrosive. If any splashes on your skin, wash it off thoroughly with water and inform your teacher. Do not inhale HCl fumes.*

d. Dip the wire into the hydrochloric acid; then heat the wire tip in the burner flame. Position the wire in the outer "luminous" part of the flame, not in the center cone. As the wire heats to a bright red, the burner flame may become colored. See Figure 3. The colors are due to metallic cations held on the surface of the wire.

e. Continue dipping the wire into the acid solution and inserting the wire into the flame until there is little or no change in flame color as the wire is heated to redness.

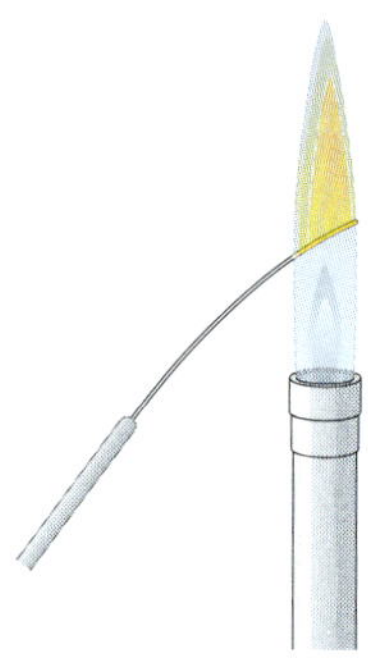

Figure 3 *Conducting a flame test. Metallic ions give off characteristic colors when heated in a flame.*

The lack of a color change in the flame indicates the wire is clean.

Expected Results

Test	Ion Tests					
	NH_4^+	Fe^{3+}	K^+	NO_3^-	PO_4^{3-}	SO_4^{2-}
Color	NC	Brown	NC	NC	NC	NC
$BaCl_2$	NC	NC	NC	NC	Precipitate	Precipitate
$BaCl_2$ & HCl	NC	NC	NC	NC	Precipitate dissolved	Precipitate remained
Brown ring test	X	X	X	Brown ring	X	X
NaOH & litmus	Blue litmus	NC	NC	X	X	X
Flame test			Violet	X	X	X
Flame test & cobalt glass			Pink	X	X	X
KSCN test	NC	Red	NC	NC	NC	NC

NC: no change
X: The test was performed.

f. Place 7 drops of potassium ion solution into a clean well in a wellplate. Dip the cool, cleaned wire into this solution. Then insert the wire into the flame. Note any change in flame color, the color intensity, and the time (in seconds) that the color is visible.

g. Repeat the potassium-ion flame test, this time observing the burner flame through cobalt-blue or didymium glass. Again, note the color, intensity, and duration of the color. Your partner can hold the wire in the flame while you observe through the colored glass. Then change places. Record all observations.

The colored glass may help make the characteristic potassium flame color easier to detect against the burner-flame color.

KSCN Test

In Unit 1 (page 37) you learned that when potassium thiocyanate (KSCN) is added to an aqueous solution containing iron(III) ions (Fe^{3+}), a deep red color appears due to formation of $[FeSCN]^{2+}$ cations. Appearance of this characteristic color confirms the presence of iron(III) in the solution.

8. a. Place 3 drops of iron(III)-containing solution into a well in a wellplate.

b. Add 1 drop of 0.1 M potassium thiocyanate (KSCN) solution to the well. Record your observations.

c. Clean and rinse the wellplate.

Part 2. Tests on Fertilizer Solution

1. Obtain a Beral pipet containing an unknown fertilizer solution. Record the code number of the solution. Your unknown solution contains one of the anions and one of the cations you tested in Part 1. Observe and record the color of the unknown solution.
2. Conduct suitable laboratory tests on the unknown solution until you are confident that you have identified which anion and cation from Part 1 are present. Record all observations and conclusions. Repeat a particular test if you wish to confirm your observations.
3. Dispose of all solutions used in this activity as directed by your teacher.
4. Wash your hands thoroughly before leaving the laboratory.

Questions

1. Name and give the formulas for two compounds that could supply the ions you found in your unknown solution. For example, sodium chloride (NaCl) would supply sodium ions (Na^+) and chloride ions (Cl^-) to a solution, while potassium carbonate (K_2CO_3) would furnish potassium ions (K^+) and carbonate ions (CO_3^{2-}).
2. Describe a test you could complete to verify whether a fertilizer sample contains phosphate ions (PO_4^{3-}).
3. Explain why the kind of information gathered in this laboratory activity about a given fertilizer solution is not enough to allow you to judge whether it is suitable for a particular use.

Answers

1. Student answers will vary, but must be consistent with the supporting data.
2. A white precipitate in the barium chloride test indicates the presence of either phosphate or sulfate. If the precipitate then dissolves in 6 M hydrochloric acid, the test shows that phosphate ion is present.
3. No, these tests provide only qualitative information. Quantitative information is necessary for deciding whether a fertilizer is suitable for a particular use.

Post-lab discussion

Discuss the use of the data chart to determine the ions present in the unknown sample. Emphasize that in this activity, students played the role of an analytical technician, one of the positions that will be available at the new plant.

A.4 FERTILIZERS AND THE NITROGEN CYCLE

The primary purpose of any fertilizer is to add enough nutrients to soil so that plants will have an adequate supply. This section examines the role played by each of the primary plant nutrients (nitrogen, phosphorus, and potassium) included in a typical fertilizer to ensure the production of strong healthy plants. Chemical fertilizers are an important factor in increasing the growth rate and the yield of crops. The use of these fertilizers has been a key factor in enabling the United States to be a major food producer in the world.

Ammonia can also be directly applied to soil as fertilizer.

The Nitrogen Cycle CD-ROM WWW.

A.4 FERTILIZER AND THE NITROGEN CYCLE

Each year EKS manufactures about 3 million tons of ammonia and more than 1.5 million tons of nitric acid. Most of this production is used to manufacture fertilizers sold to farmers and gardeners.

The purpose of all fertilizers is to add enough of all needed nutrients to soil so growing plants have adequate supplies of each. The raw materials that growing crops use are mainly carbon dioxide from the atmosphere and water and nutrients from the soil. Water and nutrients such as phosphate, magnesium, potassium, and nitrate ions are absorbed by plant roots from the soil.

Phosphate becomes part of the energy-storage molecule ATP (adenosine triphosphate), the nucleic acids RNA and DNA, and other compounds. Magnesium ions are a key component of chlorophyll, which is essential for photosynthesis. Potassium ions are found in the fluids and cells of most living things. Without adequate potassium ions, a growing plant's ability to convert carbohydrates from one form to another and to synthesize proteins would be diminished.

Nitrogen is critical in plant growth. Plant cells are largely protein; nitrogen makes up about 16% of the mass of those protein molecules. Although nitrogen gas (N_2) is abundant in the atmosphere, it is so highly unreactive that plants cannot use it directly. However, nitrogen gas can be "fixed"—that is, converted to nitrogen-containing compounds that plants are able to use chemically. Lightning or combustion can "fix" atmospheric nitrogen (see Figure 4) by causing it to combine with other elements, especially hydrogen and oxygen, to form compounds usable by plants. In addition, some plants called legumes, such as clover and alfalfa, have nitrogen-fixing bacteria in their roots.

Figure 4 *Nitrogen in the atmosphere can become fixed in a lightning storm such as this one.*

Figure 5 *Many U.S. farmers regularly apply nitrogen-containing anhydrous ammonia fertilizer to their fields.*

Rather than relying on legumes, scientists are exploring biological methods for making atmospheric nitrogen more available to plants. These include engineering some microorganisms and plants to contain genes that will direct the production of nitrogen-fixing enzymes. This would make it possible for plants or their bacteria to produce their own nitrogen-based fertilizers, just as the bacteria associated with legumes do.

Conversion of N_2 to nitrogen-containing compounds usable by plants is called nitrogen fixation.

When added to soil from decaying matter and from other sources (see Figure 5), ammonia (NH_3) and ammonium ions (NH_4^+) are oxidized to nitrate ions by soil bacteria. Plants first reduce nitrate ions to nitrite ions (NO_2^-), then to ammonia. Plants then use ammonia directly in synthesizing amino acids. Unlike humans or other animals, many plants are able to synthesize all their needed amino acids by using ammonia or nitrate ions as initial nitrogen-containing reactants.

Nitrifying bacteria are in many plants, as well as blue-green algae, some marine algae, and lichens.

When organic matter decays, much of the released nitrogen recycles among plants and animals, and some returns to the atmosphere. Thus, some nitrogen gas removed from the atmosphere through nitrogen fixation eventually cycles back to its origin.

The **nitrogen cycle** consists of these steps:

1. Atmospheric nitrogen (N_2) is converted to ammonia or ammonium ion by nitrogen-fixing bacteria that live in legume root nodules or in soil, or N_2 is converted to nitrogen oxides by lightning.
2. Ammonia and ammonium ions, in turn, are oxidized by various soil bacteria—first to nitrite ions and then to nitrate ions.
3. Plants take in nitrogen from the soil in the form of nitrate ions.
4. The nitrogen then passes along the food chain to animals that feed on these plants and to animals that feed on other animals.
5. When those plants and animals die, bacteria and fungi take up and use some of the nitrogen from plant/animal protein and other nitrogen-containing molecules. The remaining nitrogen is released as ammonium ions and ammonia gas.
6. Denitrifying bacteria convert some ammonia, nitrite, and nitrate back to nitrogen gas, which returns to the atmosphere.

Some nitrogen is able to recycle through the living world without returning to the atmosphere.

MODELING MATTER

THE NITROGEN CYCLE

The nitrogen cycle is compared to the carbon and hydrological cycles. The main idea, however, is that the nitrogen cycle is a natural recycling system that supplies a useable form of nitrogen to plants (nitrogen-fixation).

Students are asked to construct their own visual models of the nitrogen cycle. Answers to the questions will vary based on students' experience.

Hint

After students have revised their nitrogen-cycle models, they could be divided into groups to discuss and decide which visual model by group members best illustrates the nitrogen cycle. Then each group could construct a larger "best model" to display in the classroom and compare with those of other groups.

MODELING MATTER

THE NITROGEN CYCLE

In Unit 1, you learned how the hydrologic cycle can purify water (page 70). You also realized that carbon-containing molecules are transformed as carbon cycles among living and nonliving components on Earth (page 286). For instance, carbon found as CO_2 can be transformed into complex molecules, such as carbohydrates, by plants. Carbon is also held as dissolved carbon dioxide in the oceans and as carbonate rocks in Earth's crust.

In addition to reading about the hydrologic and carbon cycles, you examined figures illustrating those processes. Such visual models help organize related information, allowing interactions and connections to be readily noted and traced.

You have just learned nitrogen cycles among the atmosphere, soil, and organisms. However, a visual model depicting that nitrogen-cycle information has not been presented. In this activity you will create that missing diagram.

Look back at the illustrations found on pages 70 and 286. Notice how the hydrologic and carbon cycles are depicted. Then review Section A.4 in this unit (and earlier textbook material if needed) to guide your completion of these steps:

1. Construct your own diagram of the nitrogen cycle. Follow these general guidelines:
   a. Use arrows to show the direction of flow as nitrogen atoms cycle among the atmosphere, soil, and living organisms.
   b. Include symbols and names for key molecules and ions at each cycle stage.
   c. Use pictures and color as needed to clarify details in your nitrogen-cycle model.
   d. Make your model easy to follow. A classmate should be able to summarize the steps correctly by studying your diagram.
2. Exchange your model with that of a classmate.
3. Select an appropriate starting point on your classmate's diagram and trace nitrogen through its cycle.
4. Repeat Step 3. This time, however, use your classmate's diagram to write a description of the key steps in the cycle. Your written description should be limited to information in your classmate's diagram, even if some features are different from those in your diagram.
5. Exchange diagrams and written descriptions with your classmate. You should now have the nitrogen-cycle diagram you originally drew and your classmate's written description based on it.

Answer these questions:

1. Compare the difficulty you experienced in completing these two tasks:
   a. transforming the book's description of the nitrogen cycle into a diagram
   b. transforming your classmate's diagram of the cycle into a written description
2. How closely does your classmate's written description reflect the actual structure and details of your diagram? Explain.
3. Compare the description your classmate wrote about your diagram with the description in Section A.4.
   a. Compared to the textbook description, did your classmate's description omit or add any details or steps? Explain.
   b. Which description is more detailed? Explain.
4. Based on your classmate's description,
   a. how easy was your diagram to interpret and follow? Explain.
   b. how would you modify your diagram to improve its accuracy or clarity? Explain.
5. Considering your answers to Question 4, make any needed changes to your diagram so it more clearly depicts the nitrogen cycle.

PLANT NUTRIENTS — Building Skills 1

1. Why do some farmers over several growing seasons alternate plantings of legumes and plantings of grain crops?
2. Why is it beneficial to return unused parts of harvested crops to the soil?
3. How might research on new ways to fix nitrogen help lower farmers' operating costs?

In Laboratory Activity A.3 you identified some major ions present in fertilizer. In the next activity you will complete a more detailed study of one of these ions dissolved in water—phosphate (PO_4^{3-}).

A.5 PHOSPHATES — Laboratory Activity

CD-ROM WWW. Lab Video

See page 346 for background regarding key components of commercial fertilizers, including phosphorus.

Introduction

Fertilizers can be evaluated in part by the percentages (by mass) of essential nutrients contained in them. In this activity you will analyze a fertilizer solution to determine the mass and percent of phosphate.

The chemical method you will employ, **colorimetry,** is based on the fact that the intensity of color in a solution is directly related to the concentration of that colored substance. To analyze colorless phosphate ions (PO_4^{3-}) in that manner, you will first convert all phosphate ions to a new, colored ion. Then, to determine the percent phosphate, you will compare the intensity of color in the unknown solution to the intensities of solutions with known phosphate concentrations—that is, to a set of phosphate-based color standards.

If the color intensities in equal volumes of the sample and the standard are the same, then the phosphate concentrations must be identical in the two solutions. Likewise, if the color intensity of the unknown sample is lower than that of the standard, the unknown sample must have the lower of the two phosphate concentrations. One goal of the laboratory procedure is to reduce the unknown phosphate solution concentration sufficiently through dilution so its color intensity will be within the range of the prepared color standards.

Procedure

1. Label five clean test tubes as follows: 10 ppm, 7.5 ppm, 5.0 ppm, 2.5 ppm, and *x* ppm.
2. To prepare a water solution of the unknown fertilizer, place a 0.50-g sample of the solid fertilizer in a 400-mL beaker. Add 250 mL of distilled water and stir until the sample is completely dissolved.

As prepared in Step 2, your fertilizer solution is still much too concentrated to analyze by colorimetry. In other words—milliliter for milliliter—your

PLANT NUTRIENTS — Building Skills 1

These questions highlight the importance of using chemical technologies to work with nature.

Answers

1. Legumes contain nitrogen-fixing bacteria in their roots. If such crops are rotated with plants that extract nitrogen compounds from the soil, the soil's store of nitrogen is replenished.
2. Nonharvested plant parts contain nitrogen, phosphorus, potassium, and other nutrients, originally extracted from the soil by the growing plant. By plowing the plants back into the soil, farmers make these nutrients available for the next season's crops; less fertilizer is needed than would otherwise be the case.
3. If scientists developed plants capable of fixing nitrogen, energy costs would be reduced because less fertilizer would be needed and less tractor fuel would be burned applying the fertilizer to the soil.

A.5 PHOSPHATES — Laboratory Activity

Using colorimetric techniques, students determine the mass and percent of phosphate in a fertilizer solution. The intensity of the color of a solution is related to the concentration of a known colored substance.

Time

50 minutes

Materials (for a class of 24 working in pairs)

- 20 g ascorbic acid
- 20 g granular fertilizer (20-20-20); or others as available; or, as an alternative, a mixture of 80 g NH_4NO_3 and 380 g $Na_3PO_4 \cdot 12\ H_2O$
- 1.0 L 10.0-ppm standard phosphate buffer solution
- 325 mL ammonium molybdate–sulfuric acid reagent
- 10 L distilled water
- 60 test tubes (18 X 150 mm)
- 12 test tube holders
- 12 test tube racks
- 12 ring stands
- 12 iron rings
- 12 wire screens
- 36 400-mL beakers
- 12 10-mL graduated cylinders
- 12 50-mL or 100-mL graduated cylinders
- 12 stirring rods
- 12 gas burners
- 12 balances

solution contains considerably more phosphate ion than can be evaluated by the color standards you will use later. Therefore, it is necessary to prepare a more dilute fertilizer solution for analysis—one that is only 1/50 as concentrated.

One way to accomplish a "dilution" goal of 1/50 concentration would also be highly impractical: Imagine adding distilled water (with stirring) to your 250-mL fertilizer solution until its volume becomes 50 times larger. The phosphate concentration would then be lowered to 1/50 of its initial value. However, the diluted solution volume would become almost thirteen liters (0.250 L × 50)—over three gallons!

Fortunately, there is a more practical approach. As you complete Steps 3 and 4, think about how the objective is being met in a more useful way.

3. Measure out and retain 1/50 of your total fertilizer solution (that is, 5.0 mL of the original 250 mL). Discard the remaining volume of original solution as directed by your teacher.
4. Pour the 5.0-mL sample of solution into a clean 400-mL beaker. Then add enough distilled water to bring the total volume of solution in the beaker to 250 mL. Stir thoroughly.
5. Pour 20 mL of the diluted solution into the test tube labeled "*x* ppm." Discard the rest of the fertilizer solution remaining in the 400-mL beaker as directed by your teacher.
6. Your teacher has already prepared a supply of 10-ppm phosphate ion standard solution. Place 20 mL of that standard solution in the test tube labeled "10 ppm."
7. Given supplies of 10-ppm solution and distilled water, decide what volumes of 10-ppm solution and distilled water should be measured and mixed to prepare 20-mL samples, respectively, of 7.5-ppm, 5.0-ppm, and 2.5-ppm phosphate solutions. Write your plan in your laboratory notebook.
8. Ask your teacher to check your solution-preparation plan. After receiving your teacher's approval, prepare the three solutions. Pour each standard solution into its appropriately labeled test tube.
9. Add 2.0 mL of ammonium molybdate–sulfuric acid reagent to each of the four phosphate standards and also to the unknown solution.
10. Add a few crystals of ascorbic acid (no more than the volume of a pencil eraser tip) to each tube. Stir to dissolve. Rinse and dry the stirring rod after mixing each tube.
11. Place a 400-mL beaker half-full of water on a hot plate. Carefully place your five test tubes into the beaker. Heat the water bath until a blue color develops in the 2.5-ppm solution. Do not boil the water. Turn off the hot plate.
12. Using a test-tube holder, remove the test tubes from the water bath and place them in numerical order in a test-tube rack.
13. Compare the color intensity of the unknown solution ("*x* ppm") with the intensities of the four color-standard solutions. Place the unknown-solution test tube between the two tubes containing standard solutions with the closest-matching color intensities.

Safety

Students should wear safety goggles and aprons throughout the laboratory activity.

Advance preparation

For optimum results, the solutions should be made up a week ahead of time. They will keep indefinitely.

- The standard phosphate solution is prepared by dissolving 0.92 g K_2HPO_4 in a small volume of water and then diluting with water to 1.00 L in a volumetric flask. After this solution is thoroughly dissolved, dilute 20.0 mL to 1.00 L, which produces 1 L of standard (10.0 ppm) phosphate solution. (0.92 g K_2HPO_4 × (95 g PO_4/174 g K_2HPO_4) × (1 g PO_4/1000 g solution in first dilution) × (20 mL first solution/1000 mL final dilute solution) = 10.0 ppm)
- Prepare the ammonium molybdate–sulfuric acid reagent by dissolving 7.5 g of ammonium molybdate, $(NH_4)_2MoO_4$, in 75 mL of distilled water. Set the solution aside to cool in an ice bath.
- Slowly, with mixing, add 125 mL of ice-cold 18 M H_2SO_4 to 125 mL of ice-cold distilled water. **Caution:** *Diluting concentrated sulfuric acid generates considerable heat, and may produce splattering if not added slowly with continual stirring.* Perform this dilution with considerable care. Add the acid to the water, not the water to the acid.
- Slowly add the cold sulfuric acid solution (with stirring) to the ice-cold ammonium molybdate solution. Keep the ammonium molybdate solution in an ice bath during this step.

Pre-lab discussion

Point out the connection between this activity and earlier qualitative work with fertilizer.

Lab tips

The intensity of the color is affected by the length of time the solution is allowed to boil; some consistent method of heating must be agreed upon to ensure comparable results by all teams. An acceptable reference point is the point at which the first bubble of boiling occurs.

If a spectrophotometer is available, data can be collected

14. Estimate the concentration (ppm) of your unknown phosphate solution from the known color standards. For example, if the unknown solution color falls between the 7.5-ppm and 5.0-ppm color standards, you might decide to call it "6 ppm," or "6 g PO_4^{3-} per 10^6 g solution." Record the estimated phosphate ion concentration in ppm.
15. Discard your test-tube solutions as directed by your teacher.
16. Wash your hands thoroughly before leaving the laboratory.

Recall that ppm represents parts per million, or, as here, grams per 10^6 grams.

Calculations

1. Calculate the mass (grams) of phosphate present in your original 250-mL fertilizer solution (or in the original 0.50-g solid fertilizer sample). Note that a multiplication factor of 50 is needed to compensate for the fact that only 1/50 of the original fertilizer sample was used—see Step 3. In other words, the phosphate concentration in the original fertilizer must have been 50 times higher than the value found for the solution in the "*x* ppm" test tube. Thus the calculation becomes

$$\text{Mass of phosphate (in grams)} = \frac{\text{?? g phosphate}}{10^6\ \cancel{\text{g solution}}} \times 250\ \cancel{\text{g solution}} \times 50$$

   In place of "??," substitute the value of the unknown solution concentration (in ppm) you obtained in Step 14. Record your calculated mass of phosphate.

2. Based on your answer to Calculation 1, find and record the percent phosphate (by mass) in your original 0.50-g fertilizer sample.

Questions

1. Name two common household products for which you can estimate relative concentration just by observing their color intensities.
2. Chemists often use an instrument called a colorimeter to determine solution concentrations. A colorimeter measures the quantity of light that passes through an unknown sample and compares it to the quantity of light that passes through a known standard solution. What advantages does a colorimeter offer over the human eye?
3. Explain this statement: "The accuracy of the results of a colorimetric analysis depends, in part, on the care taken in preparing the standards used in the analysis."
4. In Laboratory Activity A.3 (page 346), the tests were qualitative—they were used only to indicate whether particular ions were present. How could an ion test involving formation of a colored solution be modified so it would be quantitative—indicating the concentration of ion present?

and plotted to obtain an absorption curve for the phosphate standards. This curve can then be used to compare sample values with the standards. A wavelength of 610 nm gives maximum absorption.

Expected results

Using 20-20-20 fertilizer the actual value of phosphate as diphosphorus pentaoxide (P_2O_5) is 20%. Student results will range between 17% and 25% phosphate.

Answers

1. Coffee, tea, juices, and powdered drinks are possible answers.
2. The colorimeter is able to provide repeatable, quantitative data.
3. Because colorimetric analysis depends on comparing an unknown to a known series of concentrations, if the standards are not carefully prepared, the data will be erroneous.
4. The test could yield quantitative results if the amount of sample was carefully measured and a series of known solutions was made for comparison.
5. a. Farmers have soils analyzed to determine nutritional deficiencies, which then can be corrected by applying the correct fertilizer formulation.

   b. Fertilizer is expensive. Adding too much increases the costs of producing the crop, damages some crops, and causes environmental damage due to runoff into waterways and aquifers.
6. The percent error of a smaller sample would be greater.

Post-lab discussion

Center the discussion on the questions that emphasize the importance of colorimetric techniques and the importance of quantitative data to farmers.

A.6 FIXING NITROGEN

Nitrogen in the atmosphere exists as $N_2(g)$ ($N{\equiv}N$). The stability of the triple bond in this molecule makes the gas molecules extremely hard to break. The breaking of the triple bond between the atoms of nitrogen in the N_2 molecules and the subsequent formation of ammonia is accomplished through the Haber-Bosch process. It converts atmospheric nitrogen into ammonia molecules (nitrogen fixation). The Haber-Bosch process is used to review the concepts of oxidation and reduction.

DEMONSTRATION IDEAS

Demonstration #1: A Traffic Light Reaction

A flask containing a pale yellow solution is gently swirled. The solution turns red. The flask is shaken, and the solution turns green.

Procedure

Wear safety goggles and an apron throughout the procedure.

1. Place 50 mL of solution A in a 250-mL flask.
2. Add 5–10 mL of solution B. Before mixing with solution A, the solution should be light yellow.
3. Stopper the flask.
4. Gently swirl the flask to produce the red color.
5. Give the flask a quick shake to produce the green color.

Reactions

1. The indicator is reduced by alkaline dextrose, and a yellow color is produced.
2. When the flask is swirled, oxygen gas is dissolved, the indicator is oxidized, and the red color is produced.
3. Shaking the flask introduces even more oxygen and causes further oxidation of the indicator to the green color.
4. Upon standing, the dextrose reduces the indicator back to the yellow color.

Solutions

1. **Solution A:** 3 g dextrose (glucose) and 5 g NaOH in 250 mL water.
2. **Solution B:** 1.0% solution indigo carmine indicator is made up by dissolving 1 g indigo carmine in 100 mL water.

Teaching tips

1. If the red color does not persist, increase the number of drops of indicator.
2. The "traffic light" system has an advantage over other

---

5. a. Why must farmers know the composition of the fertilizers they use?
   b. What risks are involved in applying more of a soil nutrient than is actually needed?
6. In Step 3 you retained 1/50 of the total fertilizer solution and discarded the rest. Consider this seemingly simpler alternative: In Step 2 you could have taken 1/50 of the starting mass of fertilizer ($0.50\ g \times 1/50 = 0.010\ g$) and then omitted Steps 3 and 4 completely. What are possible disadvantages of this shorter procedure?

A.6 FIXING NITROGEN

The impact of using commercial ammonia for fertilizer on agriculture and world food supplies cannot be overemphasized. World ammonia production has increased dramatically over the last 50 years, as farmers have increased their use of fertilizer to meet the food needs of growing populations. The U.S. chemical industry annually produces about 40 billion pounds of ammonia; most of it is dedicated to making fertilizer.

Before modern methods of manufacturing ammonia were developed, nitrogen-containing fertilizers came either from animal waste or from nitrate compounds. Large quantities of such compounds came from Chilean nitrate beds in South America. At the turn of the twentieth century, speculation arose that this nitrate source would be depleted by about 1930, raising fears of an agricultural crisis and the spectre of world famine. However, the manufacture of fertilizers made possible by commercial ammonia production has largely eliminated dependence on all natural nitrate sources.

In seeking ways to "fix" atmospheric nitrogen artificially, scientists in 1780 first combined atmospheric nitrogen and oxygen by exposing them to an electric spark. However, the cost of electricity made this too expensive for any commercial use. A less expensive method, the **Haber-Bosch process,** replaced it. From 1912 to 1916, Fritz Haber and Karl Bosch in Germany developed the technique for making ammonia from hydrogen and nitrogen, according to this equation:

$$N_2(g) + 3\,H_2(g) \rightleftharpoons 2\,NH_3(g)$$

During the combination of hydrogen and nitrogen to form ammonia in the Haber-Bosch process, electrons are transferred from hydrogen to nitrogen. As you know, whenever atoms lose one or more electrons, the process is called oxidation (see page 122). For example, the conversion of metallic sodium atoms into sodium ions (Na^+) is oxidation; electrically neutral sodium atoms are oxidized to +1 sodium ions. You will also recall that the opposite process—gaining electrons—is called reduction. The formation of chloride ions, Cl^-, from electrically neutral chlorine atoms is an example of reduction. Electrons can be transferred to or from particular atoms, molecules, or ions. The products of such oxidation-reduction reactions also include atoms, molecules, or ions.

Regardless of how the electron transfer occurs, the same redox principles apply.

You will also recall that the relative tendency of bonded atoms to attract electrons in compounds can be estimated from an element's electronegativity. Nonmetals typically have higher electronegativities than metals. Electronegativity values for most elements are shown in Figure 6.

For a review of the concept of electronegativity, see page 59.

Consider the key chemical change in the Haber-Bosch process, as depicted with electron-dot formulas and space-filling models:

$$:\mathrm{N}:::\mathrm{N}: \quad + \quad 3\ \mathrm{H}:\mathrm{H} \quad \longrightarrow \quad 2\ :\overset{\displaystyle \mathrm{H}}{\underset{\displaystyle \mathrm{H}}{\ddot{\underset{\cdot\cdot}{\mathrm{N}}}}}:\mathrm{H}$$

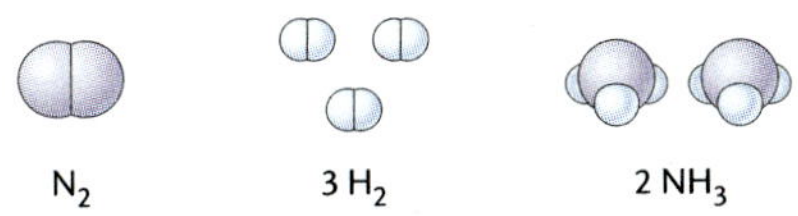

Note that each nitrogen atom in N_2 shares six electrons with another nitrogen atom, resulting in a triple covalent bond. Both nitrogen atoms have equal attraction for their shared electrons. As the reaction progresses, each nitrogen atom becomes covalently bonded to three hydrogen atoms. The bonded nitrogen and hydrogen atoms each share an electron pair—but they do not share equally. Nitrogen atoms are more electronegative than hydrogen atoms. Nitrogen atoms (electronegativity = 3.0) have a greater attraction for the shared electrons than do hydrogen atoms (electronegativity = 2.1). The nitrogen atom in each NH_3 molecule acquires a greater share of hydrogen's electrons; because of this, the nitrogen in ammonia is assigned a **negative oxidation state.** Likewise, each hydrogen atom in ammonia has lost some of its share of bonding electrons in the reaction—it is assigned a **positive oxidation state.**

Increasing Electronegativity →

Increasing Electronegativity ↑

						H 2.1										
Li 1.0	Be 1.5											B 2.0	C 2.5	N 3.0	O 3.5	F 4.0
Na 0.9	Mg 1.2											Al 1.5	Si 1.8	P 2.1	S 2.5	Cl 3.0
K 0.8	Ca 1.0	Sc 1.3	Ti 1.5	V 1.6	Cr 1.6	Mn 1.5	Fe 1.8	Co 1.9	Ni 1.9	Cu 1.9	Zn 1.6	Ga 1.6	Ge 1.8	As 2.0	Se 2.4	Br 2.8
Rb 0.8	Sr 1.0	Y 1.2	Zr 1.4	Nb 1.6	Mo 1.8	Tc 1.9	Ru 2.2	Rh 2.2	Pd 2.2	Ag 1.9	Cd 1.7	In 1.7	Sn 1.8	Sb 1.9	Te 2.1	I 2.5
Cs 0.7	Ba 0.9	Lu 1.2	Hf 1.3	Ta 1.5	W 1.7	Re 1.9	Os 2.2	Ir 2.2	Pt 2.2	Au 2.4	Hg 1.9	Ti 1.8	Pb 1.9	Bi 1.9	Po 2.0	At 2.2
Fr 0.7	Ra 0.9															

Figure 6 *Electronegativity values of selected elements.*

similar reactions: It does not require a magnetic stirrer.

3. Do not exceed the volume of Solution A recommended.

Questions to ask students

1. Is this a redox reaction? If so, what is oxidized and what is reduced?
2. What role does the indicator play?
3. What happens when the flask is swirled? Shaken?
4. Will this reaction eventually stop working if the stopper remains in the flask?

Demonstration #2 The Blue Bottle Reaction

A large flask contains a colorless solution. When shaken vigorously, the solution become blue. After a few seconds, the blue color fades and the solution again becomes colorless. The process can be repeated.

Procedure

Wear safety goggles throughout the procedure.

1. Show your students a stoppered flask containing approximately 300 mL of solution.
2. Give the flask a few vigorous shakes, keeping the stopper secure with your thumb or finger. Notice that the solution turns blue.
3. After a few seconds, the solution becomes colorless again.
4. Ask for suggestions to explain these reactions.
5. The process can be repeated several times. It may be necessary to remove the stopper periodically between shakes.

Reactions

1. The students should deduce that shaking the flask to cause the reaction that in turn causes the blue color is essentially one in which gas + liquid produces blue color and that the clearing of the solution is simply blue-colored substance + colored-substance produces colorless.
2. The actual reaction involves the reduction of methylene blue by an alkaline dextrose solution. Upon shaking, the reduced product is reoxidized to the blue dye. Essentially, the reaction occurs in four steps:

   (1) $O_2(g) + H_2O(l) \rightarrow O_2(aq)$

   (2) methylene blue (colorless) $\xrightarrow{O_2(aq)}$ methylene blue (blue)

   (3) glucose + $OH^- \rightarrow$ glucoside

   (4) glucoside + methylene blue (blue solution) $\rightarrow$ methylene blue + OH^- (colorless solution)

Solution

Prepare the solution for the flask as follows: Add 8 g KOH to 300 mL water. Cool the solution and add 10 g glucose (dextrose). Add a few drops of methylene blue indicator solution or a small amount of the solid indicator (no larger than a match head). Do not add too much.

Additional activity

Have students conduct library and/or electronic research on the Haber-Bosch process to reinforce the difference between industrial and laboratory chemistry.

ELECTRONEGATIVITY AND OXIDATION STATE

Building Skills 2

Positive and negative oxidation states are related to the concept of electronegativity. The oxidation state refers to a hypothetical charge assigned to atoms in a chemical species based on the assumption that all bonding electrons "belong" to the more electronegative atom in each bond. The idea is not that charges are actually formed but that the more electronegative atom has more control over the shared electrons in each covalent bond.

In this activity, students decide whether an element should be assigned a positive or negative oxidation state in a given compound by comparing electronegativities.

Encourage students to look for patterns in the electronegativity values and the arrangement of elements in the Periodic Table. For example, all of the metals have relatively low electronegativities. The nonmetals have relatively greater electronegativities.

The oxidation state of an atom in a particular substance depends on the identity of the neighboring atoms to which it is bonded. Consider a nitrogen molecule. In N_2 each nitrogen atom has a **zero oxidation state.** (This is true of any atom of any element that is not combined chemically with any other element.) That is, the two nitrogen atoms share their bonding electrons equally—there is no separation of electrical charge. However, in NO_2, oxygen (electronegativity = 3.5) attracts bonding electrons more strongly than does nitrogen (electronegativity = 3.0). The nitrogen atom in NO_2 has a lesser share of its bonding electrons than it had in N_2. As a result, the nitrogen atom in NO_2 has a positive oxidation state, while the oxygen atoms have a negative oxidation state.

By contrast, consider what happens when nitrogen (N_2) is reduced in the Haber-Bosch reaction, producing ammonia (NH_3). Since the nitrogen atom in NH_3 has a greater share of its bonding electrons than it had in N_2, the nitrogen atom in ammonia has a negative oxidation state. Thus, depending on the electronegativity of its bonding neighbor, it is possible for nitrogen to have a zero oxidation state, a positive oxidation state, or a negative oxidation state.

The Haber-Bosch process converts difficult-to-use nitrogen molecules from air into ammonia molecules, a form of "fixed" nitrogen. Once nitrogen is chemically combined with another element, it can be readily converted to other nitrogen-containing compounds. For example, under proper conditions, ammonia will react readily with oxygen gas to form nitrogen dioxide:

$$4\ NH_3(g) + 7\ O_2(g) \longrightarrow 4\ NO_2(g) + 6\ H_2O(g)$$

This is an oxidation-reduction reaction. In forming NO_2, the nitrogen atom in ammonia has been oxidized—the nitrogen has lost part of its share of electrons. Why? Because oxygen is more electronegative than nitrogen (N = 3.0; O = 3.5); oxygen attracts bonded electrons more strongly than does nitrogen. So, each oxygen atom has been reduced—each atom has gained more control of electrons than it originally had in O_2.

Electronegativity and Oxidation State CD-ROM WWW.

ELECTRONEGATIVITY AND OXIDATION STATE

Building Skills 2

Oxidation state is a convenient, although arbitrary, way to express the degree of oxidation or reduction of atoms in substances. Each atom in an element or compound can be assigned a numerical oxidation state. The higher (more positive) the oxidation state, the more an atom has been oxidized. The lower (less positive) the oxidation state, the more it has been reduced.

In assigning oxidation states to atoms in binary compounds (compounds composed of two elements), atoms of the element with lower electronegativity are assigned positive oxidation states—corresponding to loss of electrons (an oxidized state). Likewise, atoms of the more electronegative element are assigned negative oxidation state values, corresponding to a gain of electrons (a reduced state).

For example, which element in aluminum oxide (Al_2O_3) has the positive oxidation state? Figure 6 (page 357) indicates that aluminum has an

Answers

1.

Compound	Positive oxidation state	Negative oxidation state
a. NH_3	H	N
b. HCl	H	Cl
c. NaCl	Na	Cl
d. SO_2	S	O
e. OF_2	O	F
f. IF_3	I	F
g. MnO_2	Mn	O

electronegativity of 1.5 and oxygen has 3.5. In a chemical bond, oxygen has greater electron-attracting ability than aluminum. In aluminum oxide, therefore, aluminum is assigned a positive oxidation state.

The formation of aluminum oxide from aluminum and oxygen gas can be depicted this way:

$$4\,Al(s) + 3\,O_2(g) \longrightarrow 2\,Al_2O_3(s)$$

This is an oxidation-reduction reaction in which aluminum metal becomes oxidized. Why? Because the oxidation state of aluminum changes from zero (in the uncombined element) to a positive value in the product, aluminum oxide. By contrast, oxygen gas becomes reduced—its oxidation state has been reduced from zero to a negative value.

1. Consider each of these compounds. Using electronegativity values from Figure 6, decide which element in each compound has a positive oxidation state and which has a negative oxidation state.
   a. Ammonia, NH_3
   b. Hydrogen chloride, HCl
   c. Sodium chloride, NaCl
   d. Sulfur dioxide, SO_2
   e. Oxygen difluoride, OF_2
   f. Iodine trifluoride, IF_3
   g. Manganese dioxide, MnO_2
2. Each of these compounds is composed of a metallic element and a nonmetallic element. Decide which element in each compound has a positive oxidation state and which has negative oxidation state.
   a. Sodium iodide, NaI
   b. Lead(II) fluoride, PbF_2
   c. Lead(II) sulfide, PbS
   d. Potassium oxide, K_2O
   e. Iron(III) chloride, $FeCl_3$
3. Consider your answers to Questions 1 and 2. What conclusions can you draw about the oxidation states of metals and nonmetals in binary compounds?
4. Consider this chemical equation: $Ni + S \longrightarrow NiS$.
   a. Does the equation represent an oxidation-reduction reaction?
   b. If so, identify the element oxidized and the element reduced. If not, explain why.
5. The element iron is part of a system essential to energy transfer within human cells. In that system, Fe^{2+} ions are converted to Fe^{3+} ions. Does that chemical change represent an oxidation or a reduction?
6. Within the nitrogen cycle (see page 351), nitrogen gas (N_2) undergoes chemical reactions in which it is oxidized and reactions in which it is reduced. Identify—by name and formula—a nitrogen-cycle product that forms when nitrogen gas is
   a. oxidized.
   b. reduced.

CD-ROM WWW. Questions & Answers

3. When a metal and a nonmetal form a binary compound, the metal has a positive oxidation state and the nonmetal has a negative oxidation state.
4. a. The reaction is an oxidation-reduction reaction.

   b. Nickel is oxidized because it loses electrons and thus assumes a more positive oxidation state, while sulfur is reduced because it gains electrons and its oxidation state becomes more negative.
5. This is oxidation because it loses electrons and the oxidation state becomes more positive.
6. a. Nitrogen gas is oxidized, for example, to nitrite ions (NO_2^-).

   b. Nitrogen gas is reduced, for example, to ammonia, NH_3, and to ammonium ions, NH_4^+.

2.

Compound	Positive oxidation state	Negative oxidation state
a. NaI	Na	I
b. PbF_2	Pb	F
c. PbS	Pb	S
d. K_2O	K	O
e. $FeCl_3$	Fe	Cl

SECTION SUMMARY

Reviewing the Concepts

1. What would the expression"7–7–7" mean on a fertilizer-bag label?

The expression "7-7-7" refers to the percent of the primary nutrients contained in the fertilizer; thus in this case there is 7% Nitrogen; 7% Phosphate, P_2O_5; and 7% Potassium, K_2O.

2. Describe the role in plant growth of each of the three essential ingredients in a typical fertilizer.

The role of the primary nutrients in a typical fertilizer is as follows: *Phosphate* ions become part of the energy storage molecule ATP and the nucleic acids RNA and DNA. *Potassium* ions activate enzymes and maintain water and ion balance. *Nitrogen* ions are used to synthesize amino acids.

3. Why is the commonly used term "plant food" misleading?

The term "plant food" is misleading because plants do not take in food directly but require nutrients needed to synthesize food.

4. Why do fertilizer-product compositions, such as 7–7–7 and 20–10–10, differ?

The proportion of each nutrient found in fertilizer varies according to the crop's need. Examples: To promote growth, grass needs more nitrogen, but fruits and vegetables need phosphorus.

5. Why do particular plants—not all—depend on external sources of fixed nitrogen?

Some plants such as alfalfa and clover have nitrogen-fixing bacteria in their roots, thus eliminating the need for an external source of nitrogen fixation. These plants are referred to as legumes.

6. Summarize the steps of the nitrogen cycle.

These are the steps in the nitrogen cycle:

a. Atmospheric nitrogen is converted to ammonia or nitrogen oxides.

b. Ammonia or ammonium ions are oxidized to nitrite and nitrate ions.

c. The plant roots absorb nitrate ions from the soil.

d. The nitrogen passes through the food chain to animals.

e. When plants and animals die, some of the nitrogen is assimilated by bacteria and fungi. The remaining nitrogen is released as ammonium or ammonia gas.

f. Some nitrogen is returned to the atmosphere as nitrogen gas.

7. How do plants and animals differ in the ways they obtain nitrogen?

Plants obtain nitrogen directly from ammonia or nitrate ions in the soil. Animals must obtain nitrogen from eating food.

8. How do plants use the nitrogen they absorb?

Plants use the nitrogen they absorb to synthesize amino acids.

9. Given the fact that nitrogen is abundant in the atmosphere, why is it necessary to include it as an ingredient in fertilizers?

Nitrogen in the atmosphere exists as $N_2(g)$. The stability of the triple bond makes the molecule unreactive and therefore unavailable to plants. Lightning and/or some bacterial species are capable of fixing nitrogen in forms that plants can use.

10. What is the role of denitrifying bacteria in the environment?

The role of denitrifying bacteria is to convert ammonia, nitrite, and nitrate back to nitrogen gas.

11. What does "fixing" nitrogen gas mean?

"Fixing nitrogen" gas refers to the process by which atmospheric nitrogen is converted into nitrogen-containing compounds that plants can use.

12. List two nitrogen-containing ions that are useful to plants.

Nitrogen containing ions that are useful to plants are nitrate, NO_3^-, and ammonium, NH_4^+, ions.

13. How could you determine whether fixed nitrogen is present in your soil?

"Fixed" nitrogen exists in soil in the form of nitrate or ammonium ions. Chemical tests can detect the presence of fixed nitrogen. Nitrate ions are essential for the plant to synthesize amino acids and proteins. If there is an insufficient supply, plant growth will be stunted and the general health of the plant reduced.

14. Define electronegativity.

Electronegativity is the relative tendency of bonded atoms to attract electrons in compounds.

15. Describe how electronegativity values change

a. as you move from left to right across a row of the Periodic Table.

b. as you move down a vertical column of the Periodic Table.

a. The electronegativity value increases from left to right across a row of the Periodic Table.

b. The electronegativity value decreases from top to bottom down a vertical column of the Periodic Table.

16. How does the electronegativity of an atom relate to its ability to be oxidized?

Oxidation is a loss of electrons; electronegativity is the tendency of bonded atoms to attract electrons. Therefore, the greater an atom's electronegativity, the less easily it can be oxidized.

17. Arrange the following sets of elements in order of their increasing attraction for electrons within a bond:

a. silicon, sodium, sulfur

b. nitrogen, phosphorus, potassium

c. bromine, fluorine, lithium, potassium

a. sodium, silicon, sulfur

b. potassium, phosphorus, nitrogen

c. potassium, lithium bromine, fluorine

18. In general, how do the electronegativities of metals and nonmetals differ?

Nonmetals have lower electronegativities and tend to be easily oxidized.

19. What is the oxidation state of an atom that is not combined with an atom of another element?

The oxidation state of an uncombined atom is zero.

20. How does the oxidation state of an atom change

a. when it is oxidized?

b. when it is reduced?

a. **The oxidation state of an atom increases when it is oxidized.**

b. **The oxidation state of an atom decreases when it is reduced.**

21. How is it possible for the same element to be oxidized in one reaction and reduced in another?

In oxidation-reduction reactions, one species gains partial control of one or more electrons (is reduced) and the other loses partial control over one or more electrons (is oxidized). The more electronegative element gains control of the electrons. If an atom is the more electronegative element in one reaction, the atom is reduced. If the atom is the less electronegative atom in another reaction, the atom is oxidized.

22. What type of element—metal or nonmetal—is more often found in negative oxidation states when combined with other elements?

Nonmetals are most often found in a negative oxidation state when combined with other elements.

Connecting the Concepts

23. Describe one advantage and one disadvantage of using commercial fertilizer instead of manure to fertilize crops.

One advantage is that commercial fertilizer has a precisely known composition and can be used to make up for deficiencies in the soil. One disadvantage is that it is expensive.

24. A magazine article claims that "oxygen is needed for all oxidation reactions."

a. Do you agree or disagree with that statement?

b. Use your knowledge of chemistry to defend your answer to Question 24a.

a. **Disagree: Elements other than oxygen can attract electrons.**

b. **Oxidation is a loss of electrons. Any atom that gives up electrons during a chemical reaction undergoes oxidation. The electrons can be given up to oxygen or other more electronegative elements.**

25. How does the concept of a limiting reactant apply to the use of fertilizers?

Fertilizers typically contain a mixture of nitrogen, phosphorus, and potassium compounds. If any of these are not present in adequate amounts, plants will not be healthy. By adding these compounds to the soil, plants may be healthier.

26. Why is colorimetry effective in measuring the concentration of only certain kinds of solutions?

Colorimetry is based on comparing the intensity of the color of a solution to that of a known concentration but is not effective in measuring the concentration of a colorless solution.

Extending the Concepts

27. Why do some vegetarians claim that their diets make more economical use of world food resources than the diets of nonvegetarians?

Vegetarians eat plants, which have synthesized their own food. Nonvegetarians eat animals that have eaten plants and/or other animals. Therefore, vegetarians have fewer energy conversions and use food more efficiently.

28. The development of the Haber-Bosch process is believed to have prolonged World War I. Explain why.

The production of ammonia by the Haber-Bosch process provided a convenient source of ammonia from which military ammunition could be made. This method of producing military explosives allowed Germany to continue fighting after its shipping connection to Chilean nitrate deposits were cut off by the British Navy.

29. Research and make a diagram of the basic parts of a colorimeter, which is a simple version of a spectrophotometer.

Basic colorimeters consist of a light-tight sample chamber, a light source, and a photo detector.

30. **a.** If magnesium is a key component of chlorophyll, explain the fact that it is generally not included in commercial fertilizers.

b. List some other substances that are required by plants but are not included in fertilizers.

a. Adequate magnesium is found in the soil for most situations.
b. Other nutrients not listed in fertilizers include iron, carbon, calcium, and manganese.

31. Review the list of ingredients found in a multipurpose vitamin for humans, and compare this to the ingredients found in a typical commercial fertilizer. Suggest reasons for the similarities and differences you find.

The top macronutrients for plants are phosphorous, potassium, and nitrogen. In humans they are calcium, chlorine, magnesium, phosphorous, potassium, sodium, and sulfur. In addition humans consume a dozen or so vitamins such as B12, E, and biotin.

NITROGEN AND INDUSTRY

Introduction

This section examines the synthesis of ammonia from nitrogen and hydrogen gas. This is an example of a reversible reaction used for the commercial production of a useful product.

B.1 INDUSTRIAL NITROGEN FIXATION

The *ChemCom* approach of chemistry on a need-to-know basis is evident in this section. The concepts of chemical kinetics and chemical equilibrium are used to explain how chemists overcame the obstacles encountered in producing ammonia.

LeChatelier's Principle is used to describe how the production of ammonia may be controlled to produce the maximum amount of product at minimum expense.

SECTION B

NITROGEN AND INDUSTRY

Introduction CD-ROM WWW.

As you have learned, many industrial raw materials are extracted from Earth's crust (such as minerals, precious metals, sulfur, petroleum), oceans (for example, magnesium, bromine), and atmosphere. Nitrogen gas and oxygen gas—both obtained by low-temperature distillation from liquefied air—are valuable starting materials in the production of substances such as ammonia and nitric acid. As you will learn soon, production of ammonia, an EKS company product, also depends on understanding implications of reversible reactions and chemical equilibrium.

B.1 INDUSTRIAL NITROGEN FIXATION

Producing ammonia from nitrogen gas and hydrogen gas is a chemical challenge. The first reason, as you learned in Section A, is that molecular nitrogen (N_2) is very stable. This means that nitrogen fixation—the chemical combination of nitrogen gas with other elements—has a substantial activation-energy barrier. As you learned in Unit 4 (page 328), a reaction with a large energy barrier requires that either the reactant particles must have substantial kinetic energy or that a catalyst is found to reduce the energy required to initiate the reaction.

A second reason the reaction of nitrogen gas with hydrogen gas is difficult involves the tendency of some ammonia molecules to decompose back to nitrogen gas and hydrogen gas under the conditions of the ammonia-synthesis reaction. As you learned in Unit 4 (page 309), this kind of reaction—one in which products re-form reactants at the same time that reactants form products—is known as a **reversible reaction.** The double arrows used below indicate that both forward and reverse reactions are occurring simultaneously and at the same rate.

$$N_2(g) + 3\,H_2(g) \rightleftharpoons 2\,NH_3(g)$$

How do chemists and chemical engineers—whether at the EKS Nitrogen Products Company or elsewhere—overcome these obstacles to produce ammonia?

Kinetics: Making More Ammonia in Less Time

The rate or speed at which nitrogen fixation occurs determines the time required for a certain amount of ammonia to form. The **reaction rate** expresses how fast a particular chemical change occurs. In chemistry, the study of reaction rates is often referred to as **kinetics.** High temperatures increase the reaction rate by providing more reacting molecules with sufficient energy to overcome the activation energy barrier. Catalysts, on the other hand, increase the reaction rate by lowering the activation energy required for the reaction to occur. See Figure 7.

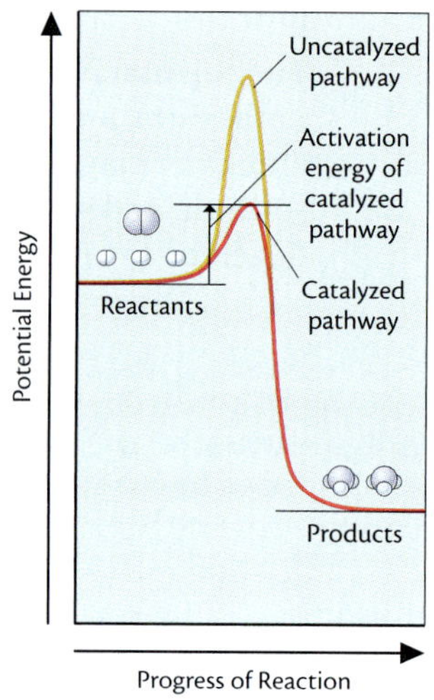

Figure 7 *A catalyst can reduce the size of the potential-energy barrier involved in the ammonia-making reaction.*

Although a higher reaction temperature increases the average kinetic energies of the nitrogen and hydrogen molecules that react to form ammonia, ammonia itself becomes increasingly unstable at higher temperatures. The result is that ammonia decomposes back to nitrogen gas and hydrogen gas. If the reaction takes place at lower temperatures, however, fewer nitrogen and hydrogen molecules have enough energy to overcome the activation energy barrier, thus slowing the net rate of ammonia formation (even though less ammonia decomposes at that lower temperature). What can be done to increase the rate and yield in the production of ammonia?

The major breakthrough that led to workable, profitable ammonia production was the discovery of a suitable catalyst. Catalysts made it possible to produce ammonia at lower temperatures (450–500 °C), thus slowing the rate of ammonia decomposition. Haber and his colleagues spent a great deal of time and energy in the early 1900s systematically searching for good catalysts. Today, the ammonia industry employs a catalytic mixture of iron oxides and aluminum oxide.

Equilibrium: Favoring the Forward Reaction

Any reversible reaction appears to "stop" when the rate at which the product forms becomes equal to the rate at which it reverts back to reactants—that is, when the reactants and products attain dynamic equilibrium. At **equilibrium,** both forward and reverse reactions continue, but there are no further net changes in the amounts of reactants or products. At the point of dynamic equilibrium the two opposing changes are in exact balance, as modeled in Figure 8.

The net amount of ammonia that can be formed from a certain amount of nitrogen gas and hydrogen gas at constant temperature is limited. One way to increase the amount of ammonia produced is to cool and remove the ammonia as soon as it forms, thus preventing it from decomposing back to reactants. If ammonia is continuously removed, the rate of the reverse reaction (the decomposition of ammonia) is decreased dramatically because there is less gaseous ammonia available to decompose. This causes the overall reaction (see page 362) to favor production of more ammonia.

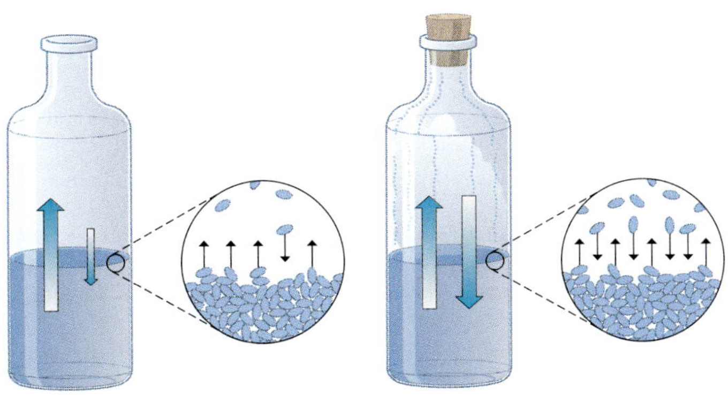

Figure 8 *Liquid water in an open bottle (left) slowly evaporates and escapes from the bottle as water vapor. This happens because the rate of evaporation of water molecules (ovals) is greater than the rate of condensation—the system is not in equilibrium. If the bottle is stoppered (right), no overall change in the liquid level is observed because the evaporation rate is equal to the condensation rate. The system in the stoppered bottle is an example of dynamic equilibrium.*

The optimum conditions for producing ammonia are:

1. maintaining a high concentration of starting materials (nitrogen and hydrogen).
2. removing the product from the reaction as it is formed.
3. strict temperature control. It must be high enough to maintain the forward reaction, but low enough to inhibit the decomposition of the product.
4. using a catalyst to speed up the reaction; the temperature reduction then lowers the energy requirements.
5. controlling pressure. It must be high enough to increase the number of collisions between molecules and therefore encourage the formation of products.

Consider using the video by *TV Ontario* on Chemical Equilibrium—especially Segment 6, The Haber Process, which covers the chemical and historical events mentioned in Section B.

CHEMICAL SYSTEMS AT EQUILIBRIUM

Building Skills 3

Students apply LeChatelier's Principle to chemical reactions.

Answers

1. $N_2(g) + 3\,H_2(g) \rightleftarrows 2\,NH_3(g) + \text{Heat}$

   a. Increasing temperature would favor the reverse reaction.

   b. Temperatures below 400 °C would not be high enough to provide the nitrogen and hydrogen molecules with adequate kinetic energy to react.

   c. The reactants and product of this reaction are all gases and the reaction requires high pressure for optimum conditions.

   d. A combination of low temperature and high pressure would yield the greater amount of ammonia.

2. a. See answer in student text.

   b. $2\,SO_2(g) + O_2(g) \rightleftarrows 2\,SO_3(g) + \text{Heat}$

   i. The concentration of SO_2 or O_2 could be increased.

   ii. The concentration of SO_3 should be decreased by removing the product.

   iii. Cool the system.

   c. $H_2(g) + Cl_2(g) \rightleftarrows 2\,HCl + \text{Heat}$

   i. The amount of H_2 could be increased.

   ii. The amount of HCl should be decreased by removing the product.

   iii. Cool the system.

3. (2b) Decrease the concentration of SO_2 and O_2, increase the concentration of SO_3, or heat the system.

   (2c) Decrease the amount of H_2, increase the concentration of HCl, and raise the temperature of the reaction.

All reversible reactions in closed containers eventually reach equilibrium if conditions such as temperature remain constant.

When an equilibrium system is disturbed to cause one reaction (either forward or reverse) to be favored over the other, chemists have learned that the initial equilibrium has shifted in the direction of the favored reaction. This phenomenon, first described by the French chemist Henri LeChatelier, is commonly referred to as **LeChatelier's Principle.**

The external disturbance imposed on a system at equilibrium, sometimes called a "stress," can be a change in the concentration of a particular reactant or product, a change in the temperature of the system, or—for a system including gases—a change in the total pressure. In the case of the industrial production of ammonia, the removal of ammonia (a change in its concentration in the reaction vessel) results in the initial equilibrium position being shifted toward the right, producing a larger amount of product (ammonia).

Another external disturbance (stress) used to increase ammonia production is to add reactant molecules—nitrogen gas and hydrogen gas—continuously under high pressure. This increases the concentration of reactants, which favors the forward reaction—and thus increases the amount of ammonia that can be formed.

In many cases changing the temperature can also cause an equilibrium system to shift. The direction of that effect depends on whether the forward reaction is exothermic or endothermic. For example, the synthesis of ammonia is exothermic:

$$N_2(g) + 3\,H_2(g) \rightleftarrows 2\,NH_3(g) + \text{Heat}$$

That is, heat (thermal energy) can be considered a product of the forward reaction. Stressing the equilibrium by raising the temperature would favor the reverse reaction—the chemical change that removes heat. The result is that at higher temperatures less ammonia is formed at equilibrium. Remember, though, that the temperature must be high enough to provide the nitrogen and hydrogen molecules with adequate kinetic energy to react. A delicate balance is needed—the temperature must be high enough to produce significant amounts of ammonia, but not so high that the rate of ammonia decomposition is excessively promoted.

Visualizing an Ionic Equilibrium CD-ROM WWW.

CHEMICAL SYSTEMS AT EQUILIBRIUM

Building Skills 3

1. Use the equation for the synthesis of ammonia found above and Figure 9 to help answer these questions:
   a. Based on Figure 9, what generalization can you make about the effect of temperature on the yield of ammonia?
   b. Do you think that generalization would remain valid for temperatures considerably lower than 400 °C? Explain your answer.
   c. Based on Figure 9, what generalization can you make about the effect of total pressure on the yield of ammonia?
   d. What combination of temperature and pressure results in the highest ammonia yield?

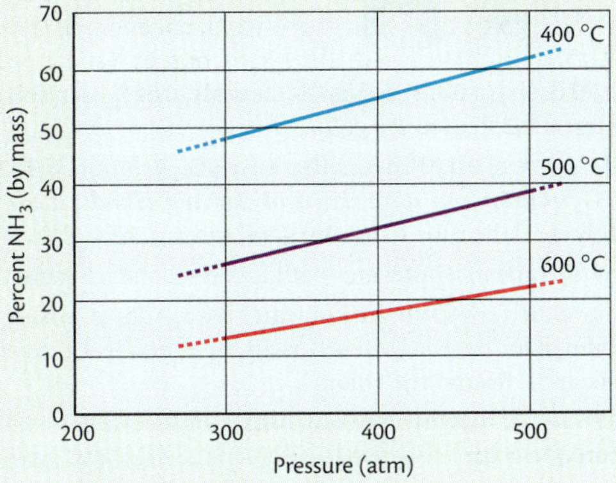

Figure 9 *Ammonia yield in the system $N_2(g) + 3\,H_2(g) \rightleftharpoons 2\,NH_3(g)$ + Heat. The graph depicts the influence of pressure and temperature on the percent NH_3 obtained at equilibrium.*

2. For each equilibrium system below, describe three different changes you could make to favor the forward reaction. Question 2a has been completed as an example.

a. Heat + $2\,NO_2(g) \rightleftharpoons 2\,NO(g) + O_2(g)$

Answer:

i. The system could be heated to a higher temperature.

ii. The concentration of NO_2 could be increased.

iii. The concentration of NO or O_2 could be decreased.

b. $2\,SO_2(g) + O_2(g) \rightleftharpoons 2\,SO_3(g)$ + Heat

c. $H_2(g) + Cl_2(g) \rightleftharpoons 2\,HCl$ + Heat

3. For Questions 2b and 2c, describe three different changes you could make to favor the reverse reaction.

B.2 LeCHATELIER'S PRINCIPLE

Laboratory Activity

CD-ROM WWW. Lab Video

Introduction

In this activity you will take a system at equilibrium and use what you have learned about LeChatelier's Principle to investigate the effect of changes in concentration and temperature on a system at equilibrium. The chemical system you will investigate is described by this equation:

$$\text{Heat} + [Co(H_2O)_6]^{2+}(aq) + 4\,Cl^-(aq) \rightleftharpoons [CoCl_4]^{2-}(aq) + 6\,H_2O(l)$$

A **complex ion** is made up of a single central atom or ion, usually a metal ion, to which other atoms, molecules, or ions are attached. Part of your task will be to determine which complex ion in the chemical equation above—$[CoCl_4]^{2-}$ or $[Co(H_2O)_6]^{2+}$—is blue and which is pink.

Time

30 minutes

Materials (for 24 students working in pairs)

24 test tubes
24 400-mL beakers
12 gas burners or hot plates
12 ring stands
12 iron rings
12 wire gauze
ice (2 lbs)
24 dropper bottles
12 mL 0.1 M $CoCl_2{\cdot}6\,H_2O$
40 g $CaCl_2$ anhydrous
265 mL 0.1 M $AgNO_3$
(4.5 g $AgNO_3$/265 mL)
0.3 g NaCl

Pre-lab discussion

Review the conditions that affect a chemical reaction using LeChatelier's Principle.

Demonstration

Prepare the 0.1 M $CoCl_2$ tincture solution by dissolving 2.37 g $CoCl_2{\cdot}6\,H_2O$ in 100 mL 95% ethanol. Add approximately 40 g anhydrous $CaCl_2$. Swirl until the solution turns a deep blue. Decant the solution into dropper bottles for student use. Slightly more than 40 grams of solute may be needed if the supply of calcium chloride is partially hydrated.

The following demonstration may be used to arouse curiosity before initiating the laboratory activity. Ask students to record their observations as you add several drops of water to a test tube containing the prepared solution; it will turn pink. Then add several pellets of solid $CaCl_{2-}$; the solution will turn blue.

To move work along, set up the hot water baths (on hot plates or based on hot tap water)

B.2 LeCHATELIER'S PRINCIPLE

Laboratory Activity

Students use LeChatelier's Principle to investigate a system at equilibrium.

Note

Student experience with chemical systems at equilibrium could be expanded beyond this activity to include investigations of the iron(III)-thiocyanate system, as well as acid-base indicators.

in advance. Hot water bath temperature is not critical but should be no higher than 60 °C to 70 °C—well short of boiling. Similarly, the ice-water baths should be set up in advance.

An alternate demonstration: Add 7 drops of 12 M HCl to 5 drops of 0.1 M $CoCl_2$ in a small test tube. The resulting solution can be alternately heated in warm water and cooled in an ice bath, demonstrating the effect of temperature on the equilibrium.

CAUTION: *Although the color changes are quite vivid, this procedure poses hazards that arise from working with and neutralizing concentrated HCl.* Due to hazards associated with 12 M HCl, this alternate procedure is more suitable as a teacher demonstration. One test tube might be set up and students allowed to alternately heat and cool it under your supervision.

Answers

1. The reverse reaction is favored by cooling the solution. Cooling favors the exothermic reaction.
2. The forward reaction is favored (it increases the concentration of reactants).
3. AgCl(s) is insoluble; therefore a precipitate is formed.
4. The Ag^+ removed the Cl^- ion from the solution (it removed one of the reactants).
5. The $[CoCl_4]^{2-}$ complex was formed in Step 6 but not in Step 4 because the Ag^+ removed most of the Cl^- from the solution.
6. The $[CoCl_4]^{2-}$, or cobalt salt ion complex, is the blue ion. The $[Co(H_2O)_6]^{2-}$, or hydrated cobalt ion, is the pink ion.

Procedure

1. Add 20 drops of 0.1 M cobalt(II) chloride, $CoCl_2$, solution to a clean, dry test tube. Record the color.
2. Add 7 drops of 0.1 M silver nitrate, $AgNO_3$, solution. **CAUTION:** *$AgNO_3$ solution can stain skin and clothing. Handle it carefully.* Gently swirl the tube to ensure good mixing. Record the color.
3. Heat the tube in a hot water bath for 30 seconds. Record the color.
4. Remove the tube from the hot water bath. Add approximately 0.3 g sodium chloride, NaCl. Gently swirl the tube. Heat the solution for 30 seconds. Record the color.
5. Place the test tube in a beaker containing ice water for 30 seconds. Record the color.
6. Reheat the test tube in the hot water bath. Record the color.
7. Dispose of the mixture in the test tube as directed by your teacher.
8. Wash your hands thoroughly before leaving the laboratory.

Questions

In answering these questions, refer to your observations and to the equilibrium expression that appears in the introduction.

1. Which reaction (forward or reverse) was favored by cooling the solution?
2. Which reaction was favored by adding more chloride ions?
3. What is the identity of the white precipitate that formed in Step 2? (*Hint:* Refer to Unit 1, page 37.)
4. Why did adding $AgNO_3$ solution affect the equilibrium, even though neither Ag^+ ions nor NO_3^- ions appear in the equilibrium equation?
5. Why did the color change after heating in Step 4, but not in Step 3?
6. Which complex ion is pink—$[CoCl_4]^{2-}$ or $[Co(H_2O)_6]^{2+}$? Which complex is blue? Explain how you decided.

Disturbing an Equilibrium

B.3 SYNTHESIS OF AMMONIA: THE HABER-BOSCH PROCESS

Industrial ammonia production involves much more than just allowing nitrogen gas and hydrogen gas to react in the presence of a catalyst. First, of course, the reactants must be obtained. Nitrogen gas is liquefied from air, and—as you will soon learn—hydrogen gas can be obtained chemically from natural gas (mainly methane, CH_4). Consequently, opening an ammonia plant in Riverwood would mean building a natural-gas pipeline.

B.3 SYNTHESIS OF AMMONIA: THE HABER-BOSCH PROCESS

Ammonia is synthesized by using the Haber-Bosch process. Nitrogen from the air is reacted with hydrogen at elevated temperatures (500 °C) and high pressure (150–300 atm) in the presence of iron oxides and Al_2O_3 catalysts. The hydrogen needed for the reaction is produced by the reaction of steam with methane (CH_4); however, the ratio of methane to steam must be controlled carefully to prevent the formation of other carbon compounds. Temperature and pressure are important to this process. The process, however, is not very efficient, since it produces only 20% to 30% of liquid ammonia.

To produce the hydrogen gas needed, natural gas is first treated to remove sulfur-containing compounds; then the methane present is allowed to react with steam:

$$\text{Heat} + CH_4(g) + H_2O(g) \longrightarrow 3\,H_2(g) + CO(g)$$

In modern ammonia plants, this endothermic reaction takes place at 200–600 °C and at pressures of 200–900 atm. The ratio of methane to steam must be controlled carefully to prevent formation of various other carbon compounds.

Carbon monoxide, a product of the hydrogen-generating reaction shown above, is converted to carbon dioxide, accompanied by production of additional hydrogen gas:

$$CO(g) + H_2O(g) \longrightarrow H_2(g) + CO_2(g)$$

All the hydrogen gas produced is separated from carbon dioxide and from any unreacted methane from the CO-producing step.

In the Haber-Bosch process (see Section A.6, page 356), the reactants—hydrogen gas and nitrogen gas—are first compressed to high pressures (150–300 atm). Ammonia forms as the hot gases (at about 500 °C) flow over a catalyst of reduced iron oxide (Fe_3O_4) and aluminum oxide (Al_2O_3). Ammonia gas is removed by converting it—under pressure—to liquid ammonia. Unreacted nitrogen gas and hydrogen gas are recycled, mixed with new supplies of reactants, and passed through the reaction chamber again.

CD-ROM WWW. Nitrogenase—An Enzyme Catalyst

The carbon dioxide can be removed in several ways, including allowing it to react with calcium oxide, CaO (lime), which forms solid calcium carbonate, $CaCO_3$, or by dissolving carbon dioxide gas at high pressures in water.

B.4 NITROGEN'S OTHER FACE

The Haber-Bosch process has provided relatively inexpensive ammonia for use in a variety of applications. For example, ammonia reacts directly with nitric acid to produce ammonium nitrate, a substitute for natural nitrates traditionally used as fertilizers.

$$\underset{\text{Ammonia}}{NH_3(g)} + \underset{\text{Nitric acid}}{HNO_3(aq)} \longrightarrow \underset{\text{Ammonium nitrate}}{NH_4NO_3(aq)}$$

The widespread availability of ammonia and nitrates has changed the course not only of agriculture, but also of warfare. Ammonia has found use as a reactant in the production of explosives, most of which are nitrogen-containing compounds. See Figure 10 on page 368. Development of the Haber-Bosch process provided a convenient source of ammonia from which both fertilizers and military munitions could be made. Production of military explosives in this manner allowed Germany to continue fighting in World War I even after its shipping connections to Chilean nitrate deposits were cut off by the British Navy.

About 80% of ammonia is used in fertilizer; 5% in explosives. What other uses for ammonia might account for the remaining 15%?

Nitrogen-based explosives have nonhostile uses. Air bags in automobiles are one such modern application. An air bag quickly inflates like a big pillow during a collision to reduce injuries to the driver and passengers. The

B.4 NITROGEN'S OTHER FACE

This section discusses the use of nitrogen compounds as explosives, emphasizing the work of Alfred Nobel and his family. Explosions are rapid, exothermic reactions in which the volume of hot, gaseous products is much greater than the volume of the solid or liquid reactants. In explosive reactions containing nitrogen, the nitrogen is released as stable, triple-bonded $N_2(g)$ molecules.

Nitrogen provides an excellent example of the ways in which a particular chemical technology can be used for either constructive or destructive purposes. Technology-based nitrogen fixation may be used to produce fertilizers or explosives.

Additional information

See "Government Industry Efforts Yield Array of Tools to Combat Terrorism," *Chemical & Engineering News* (July 24, 1995), pp. 10–19.

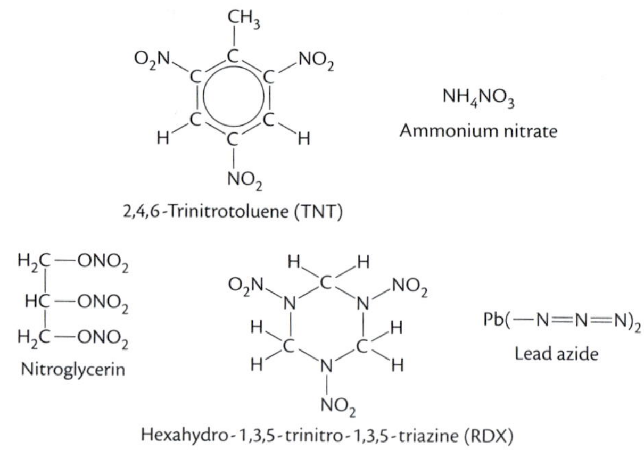

Figure 10 *Formulas of some common explosives*

uninflated air bag assembly contains solid sodium azide, NaN_3. In a collision, sensors initiate a sequence of events that rapidly decompose the sodium azide, forming nitrogen gas:

$$3\ NaN_3(s) \longrightarrow Na_3N(s) + 4\ N_2(g)$$

It takes 60 milliseconds for a 150-L passenger-vehicle air bag to inflate.

The nitrogen gas inflates the driver's air bag fully (to about 50 L) within 50 milliseconds (0.050 seconds) after the collision begins. See Figure 11.

The forces released by explosives also blast road-cuts through solid rock in highway construction. To cut through the stone faces of hills and moun-

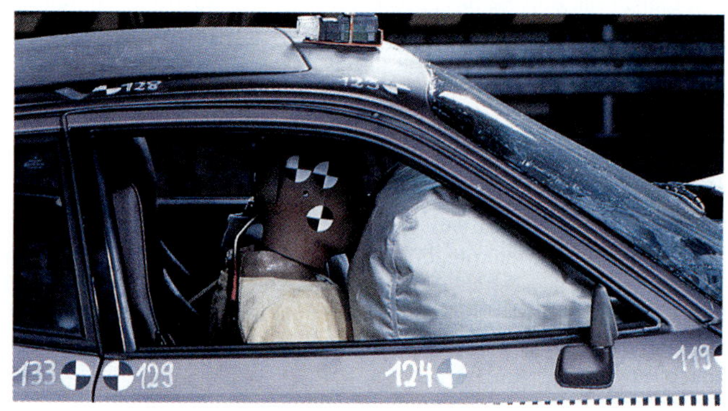

Figure 11 *Automobile air bags deploy using a nitrogen-based explosive.*

tains, road crews drill holes, drop in explosive canisters, and then detonate the explosives.

Explosions in general result from the rapid formation of gaseous products from liquid or solid reactants. Gases produced by the detonation of an explosive such as sodium azide, dynamite, or nitroglycerin occupy more than a thousand times the volume of the original solid or liquid explosive itself.

Many compounds used as explosives involve nitrogen atoms in a positive oxidation state and carbon in a negative oxidation state within the same reactant molecule. This creates conditions for very rapid transfer of electrons from carbon to nitrogen, accompanied by the release of vast quantities of energy. The energy released in this type of explosive reaction is due in part to the formation of N_2, a very stable molecule.

Chemical explosions are rapid, exothermic oxidation-reduction reactions that release large volumes of gas.

The powerful explosive nitroglycerin (see Figure 10) was invented in 1846. However, it was too sensitive to be useful—one never knew when it was going to explode. The Nobel family built a laboratory in Stockholm to explore ways to control this unstable substance. Although the father and four sons were all interested in explosives, Alfred, one of the sons, was the most persistent experimenter.

Carelessness, as well as ignorance of the properties of nitroglycerin, led to many accidental explosions. Alfred's brother Emil was killed in one of them. The city of Stockholm finally insisted that Alfred move his experimenting elsewhere. Determined to continue research to make nitroglycerin less unpredictable and dangerous, Alfred rented a barge and completed further experimentation in the middle of a lake.

He finally discovered that adsorbing the oily nitroglycerin on finely divided material (diatomaceous earth) made nitroglycerin stable enough to be transported and stored. However, it would still explode if activated by a blasting cap. This new, more stable form of nitroglycerin carried a new name—dynamite.

A new era in explosives had begun. At first, dynamite served peaceful uses in mining and in road and tunnel construction. By the late 1800s, however, dynamite also found destructive use in warfare.

Military use of his invention caused Alfred Nobel considerable anguish and finally motivated him to use his fortune to benefit humanity. His will specified that his money be dedicated to annual international prizes for advances in physics, chemistry, physiology and medicine, literature, and peace. (The Swedish parliament later added economics as an award category.) Nobel prizes, first awarded in 1901, are still regarded as the highest honors individuals can receive in these fields. Recent Nobel Laureates in chemistry are listed in Figure 12 (page 370), together with their contributions to chemical science.

369

EXPLOSIVE NITROGEN CHEMISTRY

Building Skills 4

This activity reinforces the information presented in Section B.4.

Answers

1. The reaction begins with 4 mol reactant and produces 29 mol gaseous products. Therefore, 1 mol nitroglycerin produces approximately 7.25 mol gaseous products.
2. 1 mol nitroglycerin produces 7.25 mol gaseous products $\times \frac{22.4\ \text{L}}{\text{mol}} = 160$ L, so the volume increases approximately 1600 times.
3. The volume would suddenly increase by $8 \times 1600 = 13\,000$ times.
4. In addition to the thermal energy released in the reaction, the tremendous increase in volume exerts enormous pressure in the vicinity of the reaction.

Nobel Laureates in Chemistry 1994–1999

Year	Awardees	Contributions
1999	Ahmed H. Zewail, United States, California Institute of Technology	Studied the transition states of chemical reactions using femtosecond spectroscopy.
1998	Walter Kohn, United States, University of California, Santa Barbara John A. Pople, United States, Northwestern University	Developed computational methods in quantum chemistry.
1997	Paul D. Boyer, United States, University of California, Los Angeles John E. Walker, United Kingdom, Cambridge University	Described the enzymatic mechanism underlying the synthesis of adenosine triphosphate (ATP).
	Jens C. Skou, Denmark, Aarhus University	Discovered an ion-transporting enzyme, Na^+, K^+-ATPase.
1996	Robert F. Curl Jr., United States, Rice University Sir Harold W. Kroto, United Kingdom, University of Sussex Richard E. Smalley, United States, Rice University	Discovered fullerenes, a new class of carbon allotropes (such as "buckyballs").
1995	Paul J. Crutzen, Germany, Max Planck Institute Mario J. Molina, United States, Massachusetts Institute of Technology F. Sherwood Rowland, United States, University of California, Irvine	Discovered and evaluated roles of nitrogen oxides and chlorofluorocarbons in stratospheric ozone depletion.
1994	George A. Olah, United States, University of Southern California	Developed classes of "super acids" to use in characterizing unstable chemical species.

Figure 12 *Alfred Nobel's will established an annual award to those who "shall have made the most important chemical discovery or improvement."*

EXPLOSIVE NITROGEN CHEMISTRY

Building Skills 4

The explosion of nitroglycerin is described by this equation:

$$\underset{\text{Nitroglycerin}}{4\ C_3H_5(NO_3)_3(l)} \longrightarrow 12\ CO_2(g) + 6\ N_2(g) + 10\ H_2O(g) + O_2(g) + \text{Heat}$$

1. How many total moles of gaseous products are formed in the explosion of one mole of liquid nitroglycerin?
2. One mole of gas at standard temperature and pressure occupies a volume of 22.4 L. One mole of nitroglycerin occupies approximately 0.1 L. By what factor does the volume increase when one mole of nitroglycerine explodes? (Assume temperature remains constant.)

3. In fact, when nitroglycerin explodes, the rise in temperature causes the gas volume to increase eight times more than the factor you just calculated in Question 1b. Thus, by what combined factor does the total volume suddenly increase during an actual nitroglycerin explosion?
4. How does that information help explain the destructive power of such an explosion?

B.5 FROM RAW MATERIALS TO PRODUCTS

Some chemical reactions you have observed in this chemistry class are essentially the same as reactions used in industry to synthesize chemical products. However, chemical reactions in industry must be scaled up to produce very large quantities of high-quality products at low cost.

Four considerations become crucial in attempting to scale up chemical reactions—engineering, profitability, waste, and safety. Sometimes an industrial reaction is conducted as a batch process—a single "run" of converting reactants to products, such as the reactions you completed to produce esters (Unit 3, page 227). If more product is needed, additional batch runs are completed. Early industrial production of nylon was a batch process.

More commonly and less expensively, industrial reactions are run as a continuous process: reactants flow steadily into the reaction chamber and products continuously flow out. Rate of flow, time, temperature, and catalyst composition must all be carefully controlled to ensure successful production during continuous processes.

Chemical engineers face many challenges in designing manufacturing systems for industry. In your classroom laboratory, the small quantity of thermal energy generated by a test-tube- or well-plate-sized reaction may seem insignificant. However, in exothermic industrial processes where thousands of liters may react in huge vats, the release of large quantities of heat (thermal energy) must be anticipated and carefully managed. Otherwise, reaction temperatures may spiral upward, creating potentially dangerous, costly, and even destructive situations.

The need to be cost effective heightens the challenge for chemical engineers who work in industry. As you learned earlier in the case of ammonia production, few chemical reactions produce 100% of the sought product. Profitability is often enhanced by modifying reaction conditions to increase the amount of desired product formed.

The drive to enhance profitability has made the identification of new or better catalysts a major area of continued research. Catalysts not only increase the rate at which products can be produced (and thus sold), but they almost always reduce the energy demands involved in producing a product. For some products, a difference of even a penny per liter in energy costs can mean the difference between profit or loss.

B.5 FROM RAW MATERIALS TO PRODUCTS

The four factors (engineering, profitability, waste, and safety) involved in scaling up laboratory reactions to industrial size products are discussed. The need to reduce chemical waste in the manufacturing process as mandated by the EPA has led to the reprocessing of waste materials and additional financial benefit for companies.

THE TOP CHEMICALS

ChemQuandary 1

In addition to mass or weight, relative production levels of chemical materials may be compared by listing them by volume, cost, density, molar mass, or the amount (in moles). Each type of ranking would yield a unique order.

B.6 RESPONSIBLE CARE AND GREEN CHEMISTRY

To be able to make informed decisions concerning their community, students must be aware of industry's responsibility to the community. The chemical industry is obligated to produce useful products while causing as little damage as possible to the environment and to public health. On the other hand, citizens should understand industry's requirements and the extent of society's need for its products and services. Therefore, communities and industries can work together to prevent emissions before they occur, rather than cleaning them up after they occur.

More information on the Responsible Care Program can be obtained from:

American Chemistry Council
(Formerly known as Chemical Manufacturers Association)
1300 Wilson Boulevard
Arlington, VA 22209
http:www.cmahq.com

Industry also faces challenges in dealing with unwanted materials that result from chemical processes. Wastes can quickly accumulate when reactions occur on an industrial scale. A major responsibility of the Environmental Protection Agency (EPA) is managing the cleanup of hundreds of chemical waste dumps in the United States. Those dumps are legacies of earlier times when a prevailing attitude was "out of sight, out of mind" and when disposal meant merely releasing unwanted materials directly into the air, into bodies of water, or into the ground.

When the EPA put an end to such waste releases, many chemical industries discovered that with a little additional processing, some previously unwanted materials or products could become valuable commodities. Such former "waste" compounds can often become intermediates in the production of other substances. Thus, instead of contaminating the environment, such wastes-turned-resources offer new sources of income. Additionally, recently developed catalysts and new or modified processes have allowed manufacturers to increase the efficiency of making certain products, while decreasing the amounts of starting materials (reactants).

Chemical manufacturers have learned that pollution prevention pays off. For example, 3M Corporation, a major producer and user of chemical materials, has maintained a pollution prevention program for more than 25 years. During that time, 3M has saved nearly a billion dollars by eliminating more than 1.5 billion pounds of pollutants. 3M's post-2000 goal is to approach zero pollution. Most pollution-prevention suggestions and their implementation came directly from 3M employees.

THE TOP CHEMICALS — ChemQuandary 1

Figure 13 lists the top-produced chemical substances in the United States in 1998. In this list, notice that production is reported in billions of pounds. What other quantities—in addition to mass or weight—could be used to compare the relative production levels of chemical materials? How would the relative rankings be affected—if at all—if each quantity you can suggest were used for comparisons?

B.6 RESPONSIBLE CARE AND GREEN CHEMISTRY

On learning that EKS and WYE were interested in locating in Riverwood, the town council initiated a draft of general criteria that must be met by either company if it wishes to locate in Riverwood. Those criteria are based in part on two national initiatives regarding industry's social and technical responsibilities. The first initiative—Green Chemistry—addresses the nature of the chemistry involved in an industrial process and its effects on humans and the environment. The second initiative—Responsible Care—considers how an industry can safely and productively contribute to a community.

Top 25 Chemicals Produced in the United States in 1998			
Rank	Name	Formula	Billions of Pounds
1	Sulfuric acid	H_2SO_4	95.2
2	Nitrogen	N_2	75.7
3	Oxygen	O_2	57.7
4	Ethene (ethylene)	C_2H_4	51.7
5	Calcium oxide (lime)	CaO	45.0
6	Ammonia	NH_3	39.5
7	Phosphoric acid	H_3PO_4	28.8
8	Propene (propylene)	C_3H_6	28.7
9	Chlorine	Cl_2	25.7
10	Sodium hydroxide	NaOH	23.0
11	Sodium carbonate	Na_2CO_3	21.4
12	1,2-Dichloroethane (ethylene dichloride)	$C_2H_4Cl_2$	19.5
13	Methyl *tert*-butyl ether (MTBE)	$C_5H_{12}O$	18.9
14	Nitric acid	HNO_3	18.7
15	Urea	$(NH_2)_2CO$	17.6
16	Ammonium nitrate	NH_4NO_3	17.2
17	Vinyl chloride	C_2H_3Cl	17.0
18	Benzene	C_6H_6	16.3
19	Ethylbenzene	C_8H_{10}	13.0
20	Methanol	CH_3OH	12.5
21	Styrene	C_8H_8	11.4
22	Terephthalic acid	$C_8H_6O_4$	8.7
23	Formaldehyde	H_2CO	8.6
24	Hydrochloric acid	HCl	8.6
25	Toluene	$C_6H_5CH_3$	8.1

Based on data from *U.S. Chemical Industry Handbook, 1999,* Chemical Manufacturers Association, 1999 (Table 2.17, page 44)

Figure 13 *The U.S. chemical industry produces billions of pounds of products each year.*

373

TRI data are available on CDs and diskettes, as well as on-line and in books.

Implementation of those criteria can be partly monitored by the EPA's Toxics Release Inventory (TRI). The TRI is a state-by-state database of annually reported releases of about 600 different toxic materials into the air, water, or land by more than 21 000 facilities. This major resource is directly accessible by any individuals, such as the citizens of Riverwood, interested in monitoring toxic-materials releases in their local area.

Green Chemistry is an approach that aims to improve industrial chemical products and processes by evaluating every aspect of an industrial approach. The goal is to make the production of chemical products less hazardous to human health and to the environment. This initiative is sometimes termed "Benign by Design." In meeting Green Chemistry objectives, industries also strive to improve their processes by making them more efficient and more profitable.

Principles guiding the Green Chemistry movement include these general points:

- It is better to prevent waste than treat it or clean it up after it is formed.
- Synthetic methods should be designed so that as much as possible of the material used in the process appears in the final product.
- Whenever it is feasible, reactants used and waste generated should be as benign as possible.
- The use of solvents and other materials should be made unnecessary wherever possible and innocuous when not.
- Energy requirements should be recognized for their environmental and economic impacts and should be minimized wherever possible.
- Catalysts should be used whenever appropriate.
- Raw materials should be obtained from renewable resources wherever possible.
- Chemical products should be designed so that if they decompose, resulting products are innocuous.
- Specified substances and their forms used in a chemical process should minimize the potential for chemical accidents, including releases, explosions, and fires.

These principles have encouraged work that has already produced new industrial techniques, such as bleaching paper without the use of chlorine, recycling cellulose-based wastes into fuels, synthesizing drugs and polymers more efficiently, and developing reduced-risk pesticides highly specific for certain pests.

There are 188 members of the American Chemistry Council and 98 Responsible Care Partners and Partner Associations.

Responsible Care, initiated in 1988, is an international program in which chemical manufacturing companies voluntarily agree to public scrutiny and evaluation according to specific criteria. In the United States, member companies and partners of the American Chemistry Council (ACC) pledge to follow these Responsible Care principles:

- Recognize and respond to community concerns.
- Develop chemicals that can be safely made, used, transported, and disposed of.

374

- Make health, safety, and environmental protection priorities in planning products and processes.
- Report information on chemical-related health hazards promptly to officials, workers, and the public and to recommend any protective measures.
- Advise customers on how chemicals can be safely used, transported, and disposed of.
- Operate plants in such a way as to protect the environment and the health and safety of workers and the public.
- Conduct and support research on health, safety, and environmental effects of products, processes, and resulting waste.
- Resolve problems created by past handling and disposal of hazardous materials.
- Participate with government and others to create responsible laws and regulations crafted to safeguard the community, workplace, and environment.
- Offer assistance to others who produce, handle, use, transport, and dispose of chemicals.

As with Green Chemistry, adhering to the Responsible Care pledge has allowed many companies to increase safety for their workers and their surrounding communities, reduce the amount of waste generated, and achieve greater overall efficiency.

The following activity provides an opportunity for you to develop criteria that any chemical industry interested in establishing a plant in Riverwood would have to address.

B.7 WHAT DOES RIVERWOOD WANT?

Making Decisions

The first of several town meetings to discuss the possibility of a chemical plant in Riverwood will be held soon. Representatives from both EKS and WYE and town council members will attend. All interested local citizens are also encouraged to attend.

1. Compare the principles of Green Chemistry with those of the Responsible Care program.
2. a. Develop at least six expectations the town council should ask the companies to meet if they choose to locate in Riverwood.
   b. Classify the six expectations into those that are mandatory and those that are desirable.
3. Related to the expectations you developed in Question 2, list at least two questions that you would ask either company to answer.

CD-ROM WWW. Questions & Answers

B.7 WHAT DOES RIVERWOOD WANT?

Making Decisions

This feature provides a preliminary perspective for the Town Council meeting to be held at the end of the unit. Students should be assigned to work in small groups, possibly in citizen groups similar to those in Unit 1 (see page 388 in the student text for suggestions).

CHEMISTRY AT WORK

Searching for Solutions in Research Chemistry

Someday, when you reach for a pill to relieve an aching muscle or a headache, you might have **Todd Blumenkopf** to thank. Todd, who uses a wheelchair, is a research chemist at Pfizer, Inc. Todd and his laboratory staff are working on medications that reduce the swelling, or inflammation, of joints and muscles.

Todd and his colleagues in the chemical research division are working to conceive of and synthesize new compounds that target the human enzymes and receptors that contribute to inflammation in the body. Todd's team investigates the relationship between a chemical compound's biological activity and its structure. Then they devise ways to prepare and synthesize the compound in the laboratory. Later, each compound is tested *in vitro* (in an artificial environment outside the body) and *in vivo* (in the living body of a plant or animal) to determine whether the compound produces desirable results without unwanted side effects.

The process of developing a new product from an initial idea can take many years.

After a drug has passed the required laboratory testing, researchers collect additional data about the drug to file with the Food and Drug Administration (FDA). When trying to get a new drug approved, companies like Pfizer must prepare and submit large quantities of documents, including laboratory and toxicology data.

As a child, Todd had a strong interest in science. His parents encouraged him to pursue a career in the profession, in part because they felt his disability might be met with less resistance in that field than in others. During his undergraduate years, Todd decided to focus on research as a career. As he progressed in his studies, he chose a career in pharmaceuticals. After receiving his Ph.D. and completing some post-doctoral work, Todd began his career as a research chemist. He now works at Pfizer's pharmaceutical research facilities in Groton, Connecticut.

Todd also works on the American Chemical Society's Committee on Chemists with Disabilities, which sponsors the development of materials and other efforts to improve opportunities in the chemical sciences for individuals with disabilities.

Todd believes the qualities needed for success in chemistry research are curiosity, creativity, and motivation. He also stresses the importance of patience and perseverance, since the process of developing a new product from an initial idea can take many years.

Scientific Methods vs. The Drug Approval Process

1. Use your library or the Web to investigate the major steps necessary to get a new drug approved by the FDA.

376

CHEMISTRY AT WORK

2. How do methods represented in the steps followed in the FDA drug-approval process compare to scientific methods? Explain.
3. The case of the drug thalidomide is widely regarded as providing support for the desirability of implementing a deliberate, thorough process for obtaining approval of a new pharmaceutical product. Research thalidomide's history, and write a brief report on how the FDA process worked in this case.
4. Over the past few years, a powerful new chemical technique called "combinatorial chemistry" has emerged. This technique for making new compounds that may have medicinal properties allows for the creation and testing of hundreds of substances in the time it used to take to make one sample. Use your library or the Web to investigate the techniques used in combinatorial chemistry, and prepare a chart or graphic that helps explain how they work.
5. Students and workers with disabilities face special challenges. Use the information and resources below to report on the obstacles facing disabled students and workers and some of the assistance available to them.

Teaching Chemistry to Students with Disabilities (booklet)
American Chemical Society
Committee on Chemists with Disabilities
To order, please send a request to:
Staff Liaison, Committee on Chemists with Disabilities
American Chemical Society
1155 16th Street, NW
Washington, DC 20036
Tel: 202-872-6070 or 800-227-5558, ext. 6070
Fax: 202-872-6338

American Association for the Advancement of Science (AAAS)
Project on Science, Technology and Disability
Directorate for Education & Human Resources Programs
AAAS Resource Directory of Scientists and Engineers with Disabilities (3rd ed., 1995)
1200 New York Avenue, NW
Washington, DC 20005
Tel/TDD: 202-326-6672
Fax: 202-371-9849

Foundation for Science and Disability
236 Grand Street
Morgantown, WV 26505
Tel: 304-292-4554 or 304-293-5201
Fax: 630-357-0087

When trying to get a new drug approved, companies . . . must prepare and submit large quantities of documents . . .

Job Accommodation Network (JAN)
West Virginia University
918 Chestnut Ridge Road, Suite 1
Morgantown, WV 26506
Tel/TDD: 800-526-7234
Fax: 304-293-5407

AMERICANS WITH DISABILITIES ACT HANDBOOK
Equal Employment Opportunity Commission
1801 L Street, NW
Washington, DC 20507
Tel: 800-669-EEOC (for publications orders), 800-669-4000 (for investigation queries)
Fax: 513-489-8692
TDD: 800-800-3302

377

SECTION SUMMARY

Reviewing the Concepts

1. What is meant by the "rate" of a reaction?

The rate of a reaction expresses how fast a particular chemical change occurs.

2. Why is the term "dynamic equilibrium" used to characterize some chemical reactions?

Dynamic equilibrium occurs when the rate of product formation equals the rate at which the product converts back to the reactants.

3. What is equal about equilibrium?

At equilibrium both forward and reverse reactions continue, but there are no further net changes in the amounts of reactants or products. At the point of dynamic equilibrium, opposing changes are in balance.

4. How can you recognize when chemical equilibrium has been established?

When chemical equilibrium has been established, the reaction appears to stop.

5. How is dynamic equilibrium symbolized in a chemical equation?

In a chemical equation, dynamic equilibrium is symbolized by a double arrow.

6. What is the effect of removing some heat from an exothermic reaction that is at equilibrium?

Removing heat from an exothermic reaction would favor the forward reaction, thus increasing the amount of product formed.

7. Consider this reaction at equilibrium:

$$PCl_5(g) + Heat \rightleftharpoons PCl_3(g) + Cl_2(g)$$

What effect will each of these changes have on the position of that equilibrium system?

a. adding more Cl_2

b. lowering the temperature

c. removing some PCl_3 as it forms

For the system $PCl_5(g) + heat \rightleftharpoons PCl_3(g) + Cl_2(g)$:

a. Addition of more Cl_2; the reaction would shift to the left (reverse reaction).

b. Removal of heat; the reaction would shift to the left (reverse reaction).

c. Removal of some PCl_3 as it forms; the reaction would shift to the right (forward reaction).

8. Consider this reaction at equilibrium:

$$C(s) + H_2O(g) + Heat \rightleftharpoons CO(g) + H_2(g)$$

What effect will each of these changes have on the position of that equilibrium system?

a. lowering the temperature

b. adding water to the equilibrium system

c. adding a catalyst

d. raising the temperature

For the system $C(s) + H_2O(g) + Heat \rightleftharpoons CO(g) + H_2(g)$:

a. Lowering the temperature; the reaction shifts to the left (reverse reaction).

b. Adding water to the equilibrium system; the reaction shifts to the right (forward reaction).

c. Adding a catalyst; there is no shift in the equilibrium position.

d. Raising the temperature; the reaction shifts to the right (forward reaction).

9. Explain why reactions tend to speed up with increased temperature.

According to the kinetic molecular theory, an increase in temperature will cause an increase in the motion of the molecules, thus increasing the probability that the molecules will collide with sufficient energy to react.

10. Explain why reactions tend to speed up with increased concentration.

Increasing the concentration of the reactants will increase the number of molecules, thus increasing the probability that the molecules will collide and react.

11. How does adding a catalyst influence the rate of a reaction?

A catalyst influences the rate of reaction by lowering the activation energy required for the reaction to take place. Thus catalysts increase the rate of reaction.

12. Refrigerated food lasts longer than food left at room temperature. Explain.

Refrigerated food lasts longer because the decrease in temperature causes the rate of reaction to decrease. Refrigeration does not prevent the reaction from taking place. It slows down the rate of the reaction; thus it takes longer for food to spoil.

13. What are some uses for ammonia?

Ammonia is used in the production of fertilizer and explosives.

14. What conditions did Haber and Bosch identify as necessary for the industrial production of ammonia?

Haber and Bosch identified the following conditions as being necessary for the industrial production of ammonia:

- **Nitrogen gas must be liquefied from air.**
- **Hydrogen gas must be obtained from natural gas, mainly methane, CH_4.**
- **Reactant gases must be compressed at pressures of 150–300 atm.**
- **There must be catalysts consisting of reduced iron oxide and aluminum oxide, Al_2O_3.**

15. What are the advantages of synthesizing ammonia instead of obtaining it from animal waste?

Does not rely on limited resources, purity, more reliable, and easier distribution

16. What resources are required to produce ammonia by the Haber-Bosch process?

The resources needed to produce ammonia by the Haber-Bosch process are a high concentration of reactants, removal of products, relatively lower temperatures, use of catalysts, and a pressurized system.

17. The Haber-Bosch process often involves only a 20–30% conversion of reactants to ammonia. Why is it still commercially feasible to produce ammonia by that process?

The Haber-Bosch process is only 20% to 30% efficient, but it is still commercially feasible because unreacted nitrogen and hydrogen gases are recycled; they are mixed with a new supply of reactants and passed through the reaction chamber again, producing more ammonia.

18. List six basic principles of Green Chemistry.

The basic principles of Green Chemistry are as follows:

- **It is better to prevent waste than treat or clean up waste after it is formed.**
- **Synthetic methods should be designed so that as much as possible of the material used in the process appears in the final product.**
- **Whenever it is feasible, reactants used and waste generated should be as benign as possible.**
- **The use of nonaqueous organic solvents and other materials should be made unnecessary wherever possible and innocuous when necessary.**
- **Energy requirements should be recognized for their environmental and economic impacts and should be minimized wherever possible.**
- **Raw materials should be obtained from renewable resources whenever possible.**

19. Consider this old saying: "An ounce of prevention is worth a pound of cure." Explain how this relates to the Green Chemistry initiative.

The quotation is basically saying that it is better to take measures to prevent waste than to try to clean it up after it is produced.

20. Provide an example of how Green Chemistry principles can help a chemical-manufacturing company become more profitable.

Green chemistry focuses on preventing pollution problems before they occur, rather than cleaning them up afterward. By using the principles of green chemistry, many industrial sites have encouraged work that has produced new and improved industrial techniques and products.

Connecting the Concepts

Copy the following graph, which shows the concentrations of the reactant and product in a simple chemical reaction that is conducted at constant temperature. The line connecting A, C, and E represents the reactant, while line B, C, D represents the product.

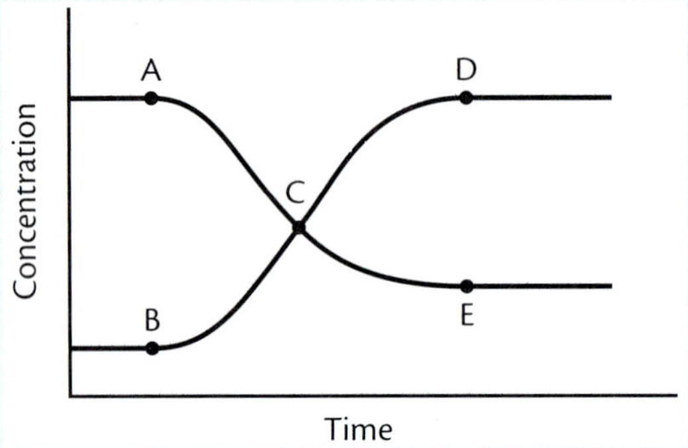

21. What chemical change is represented by curve BD? by curve AE?

Curve BD represents an increase in concentration of the product with time. The curve AE represents a decrease in the concentration of the reactant with time.

22. Redraw the graph to illustrate what would happen if a catalyst were present.

If a catalyst was present, the same concentration changes would take place but in less time.

23. Assume that this reaction is exothermic. Extend the graph to show what would happen if the temperature were increased at time DE.

For an exothermic reaction, an increase in the temperature would favor the reverse reaction. More of E would be produced while there would be less of D.

Extending the Concepts

24. Compare the nitrogen crisis of the 1930s to current concerns about petroleum supplies.

Both cases are very similar. In each case an important resource was only available from a foreign source, which was cut off.

25. Describe some examples, other than chemical reactions, of dynamic equilibria. One example might be that of a juggler who keeps three balls in the air while holding a fourth ball.

Student answers will vary. One example might be birth rate and death rate being equal, keeping the total population constant. Another would be money earned and money spent, while total amount in the bank remains constant.

26. Select two common substances produced by the chemical industry (Figure 13, page 373) and investigate their uses.

Encourage students to choose something familiar, such as paper or nylon.

27. The toxic release inventory (TRI) is available from the EPA. Use the TRI to determine how much your state or metropolitan area has accomplished in reducing emissions from manufacturing facilities.

Have students go to *www.em.doe.gov1* for TRI information.

28. A pressure cooker reduces the amount of time needed to cook foods. Investigate the design of a pressure cooker and explain why it speeds up cooking times.

A pressure cooker allows water to be heated to a higher temperature, speeding up "cooking" reactions.

METAL PROCESSING AND ELECTROCHEMISTRY

Introduction

WYE Metals Corporation is one of two industries competing to locate in Riverwood.

Aluminum, the metal produced by WYE, is reduced from its ore by electrolysis. Students are introduced to electrochemical cells and batteries. The discussion of this section will help students to understand the importance of electrochemistry to the chemical industry and to their lives.

C.1 ELECTROCHEMICAL CHANGES

Aluminum metal is reduced from its ore by electrolysis. The electrical energy for this process comes from hydroelectric or other power-generating plants. Another source of electricity is voltaic cells, although they would not generate enough electricity for aluminum production. A physical description of a voltaic cell is discussed, including the necessary vocabulary.

SECTION C

METAL PROCESSING AND ELECTRO-CHEMISTRY

Introduction CD-ROM WWW.

You have learned how EKS Nitrogen Products Company produces ammonia using the Haber-Bosch process. Such a chemical plant could impact Riverwood both positively and negatively. To help determine whether EKS or WYE should be invited to build in Riverwood, you will now learn more about the WYE Metals Corporation.

In producing sheet aluminum, the WYE corporation specializes in **electrochemistry**—chemical changes that produce or are caused by electrical energy. The following discussions and laboratory activity provide background on principles of electrochemistry. This information will help you understand how WYE's proposed new plant would operate.

C.1 ELECTROCHEMICAL CHANGES

Using the Metal Activity Series CD-ROM WWW.

Bauxite, the primary commercial source of aluminum, is a mixture mainly of aluminum oxides and aluminum hydroxides.

The WYE Metals Corporation proposes to use its extensive experience with electrochemical processes to produce aluminum metal by using oxidation-reduction methods. Since aluminum is the most abundant metallic element in Earth's crust, that might sound like an easy task. However, aluminum in Earth's crust is not present as aluminum metal. Instead it is found within clay soils and as the ore bauxite, in which aluminum ions (Al^{3+}) are strongly bonded to silicon and oxygen atoms. The aluminum in clays is not readily accessible, so most aluminum is extracted from bauxite, its primary ore. This process is accomplished through **electrolysis**—a process in which a chemical reaction is caused by passing an electrical current through a solution of ions (an **electrolyte**). In this case, the flow of electrons is used to reduce Al^{3+} ions to aluminum metal.

Electrolysis requires considerable electrical energy—thus the cost of electricity is a factor in plant location. The hydroelectric plant at the Snake River Dam generates considerably more electricity than is needed to meet the needs of Riverwood and surrounding communities. To encourage WYE Corporation to consider locating in Riverwood, power company officials have offered it large quantities of electrical power at very competitive rates.

In the laboratory activity on metal reactivities in Unit 2 (page 118), you learned that some metals lose electrons (become oxidized) more readily than others—that is, some metals are more chemically active than others. The relative tendencies of metals to release electrons can be summarized in an activity series of the metals. See Figure 14. A metal higher in the activity

Activity Series of Common Metals				
Metal		Products of Metal Reactivity		
Li(s)	$\longrightarrow$	$Li^{+}(aq)$	+	e^{-}
Na(s)	$\longrightarrow$	$Na^{+}(aq)$	+	e^{-}
Mg(s)	$\longrightarrow$	$Mg^{2+}(aq)$	+	$2\ e^{-}$
Al(s)	$\longrightarrow$	$Al^{3+}(aq)$	+	$3\ e^{-}$
Mn(s)	$\longrightarrow$	$Mn^{2+}(aq)$	+	$2\ e^{-}$
Zn(s)	$\longrightarrow$	$Zn^{2+}(aq)$	+	$2\ e^{-}$
Cr(s)	$\longrightarrow$	$Cr^{3+}(aq)$	+	$3\ e^{-}$
Fe(s)	$\longrightarrow$	$Fe^{2+}(aq)$	+	$2\ e^{-}$
Ni(s)	$\longrightarrow$	$Ni^{2+}(aq)$	+	$2\ e^{-}$
Sn(s)	$\longrightarrow$	$Sn^{2+}(aq)$	+	$2\ e^{-}$
Pb(s)	$\longrightarrow$	$Pb^{2+}(aq)$	+	$2\ e^{-}$
Cu(s)	$\longrightarrow$	$Cu^{2+}(aq)$	+	$2\ e^{-}$
Ag(s)	$\longrightarrow$	$Ag^{+}(aq)$	+	e^{-}
Au(s)	$\longrightarrow$	$Au^{3+}(aq)$	+	$3\ e^{-}$

Figure 14 *The higher a metal is in an activity series such as this, the more readily it gives up electrons.*

series will give up electrons more readily than a metal that is lower. For example, aluminum atoms are oxidized (lose electrons) more easily than iron atoms.

The differing tendency of metals to lose electrons allows electrical energy to be generated in an oxidation-reduction reaction. A simple device called a **voltaic cell** can be constructed from two **half-cells** connected in a circuit. See Figure 15 on page 382. Each half-cell contains a metal partially immersed in a solution of ions of that metal—for example, a piece of copper metal immersed in a solution of Cu^{2+} ions and a piece of zinc metal immersed in a solution of Zn^{2+} ions. You know from Unit 2 (Laboratory Activity C.5, page 142) that the reaction between Cu^{2+} and Zn proceeds spontaneously when both reactants are present. By separating the reactants into half-cells, the electrons are forced to flow through a wire and provide electrical energy to an external circuit.

The solutions in the two half-cells are prevented from mixing by a barrier. The two metals, or **electrodes,** in the half-cells are connected by a wire that allows electrons to flow between them, as shown in Figure 15 (page 382). Such a flow of electrons constitutes an **electric current.** To complete the circuit and maintain a balance of electrical charges within the system, dissolved ions must also be allowed to flow between the electrodes. Without

381

C.2 VOLTAIC CELLS

Laboratory Activity

In this laboratory activity, students construct and test various electrochemical cells. The activity applies oxidation-reduction reactions to the concrete, practical production of electrical energy.

Time

50 minute

Materials (for a class of 24 working in pairs)

12 24-well wellplates
48 Beral pipets
50 mL 0.1 M copper(II) nitrate [1.2 g $Cu(NO_3)_2 \cdot 3\ H_2O$/50 mL]
50 mL 0.1 M zinc nitrate [1.5 g $Zn(NO_3)_2 \cdot 6\ H_2O$/50 mL]
50 mL 0.1 M magnesium nitrate [1.3 g $Mg(NO_3)_2 \cdot 6\ H_2O$/50 mL]
50 mL 0.1 M iron(III) nitrate [2.02 g $Fe(NO_3)_3 \cdot 9\ H_2O$/50 mL]
12 wide copper strips (1 cm × 3 cm × 0.1 mm)
2 medium copper strips (0.5 cm × 3 cm × 0.1 mm)
12 narrow copper strips (0.25 cm × 3 cm × 0.1 mm)
12 wide zinc strips (1.0 cm × 3 cm × 0.25 mm)
12 medium zinc strips (0.5 cm × 3 cm × 0.25 mm)
12 narrow zinc strips (0.25 cm × 3 cm × 0.25 mm)
12 3-cm lengths of magnesium ribbon
12 iron metal strips, 0.5 cm × 3 cm (or small iron nails)
12 voltmeters
24 small alligator clips with fine wire leads
36 strips of filter paper, 8 mm × 3 cm
saturated KNO_3 solution (about 35 g/100 mL water)

Safety

Students should wear safety goggles and aprons throughout the laboratory activity.

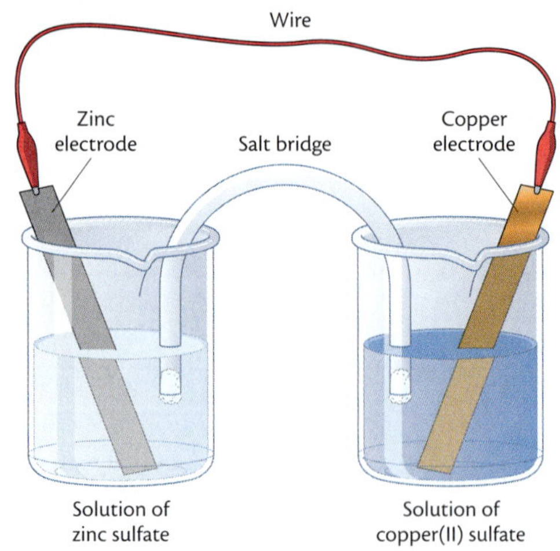

Figure 15 *A voltaic cell consists of two half-cells connected in a circuit through which electricity can flow. The salt bridge—containing a solution of ions such as K^+ and NO_3^-—is in electrical contact with the two half-cell solutions and thus completes the internal circuit.*

this flow of ions, a positive charge would build up in one half-cell, and a negative charge would build up in the other half-cell. That situation would prevent any further flow of electrons in the cell.

When two metals in different positions on the activity series are connected in a voltaic cell, an **electrical potential** is created between the metals. Electrical potential volts (V) is somewhat like water pressure in a pipe. Just as pressure causes water to flow in the pipe, electrical potential is the "push" that drives electrons through the wire connecting the two metals. The greater the difference between the chemical activities of the two metals, the greater the electrical potential generated by the cell.

Constructing a Voltaic Cell — CD-ROM WWW.

A voltaic cell can be prepared by immersing each of two metals in a solution of its ions. The metals are connected by a wire and to a voltmeter to measure the cell's electrical potential. In addition, a pathway is provided for ions to flow from one solution to the other, such as the salt bridge in Figure 15. In the following laboratory activity, filter paper saturated with potassium chloride solution provides that needed dissolved-ion pathway.

Lab Video — CD-ROM WWW.

C.2 VOLTAIC CELLS — Laboratory Activity

Introduction

In this activity you will construct several voltaic cells and measure and compare their electrical potentials. You will also explore factors that may help determine the electrical potential generated by a particular voltaic cell.

Advance preparation

Prepare the solutions listed above. Label and fill Beral pipets with the appropriate solutions. Place them in a wellplate with the points up to avoid spillage. Soak the filter-paper strips in the saturated KNO_3 solution to make the salt bridges. A physics teacher may have the voltmeters to loan to you.

Lab tips

If using analog meters, remind students to avoid allowing the voltmeter needle to deflect in a negative direction for any length of time as this damages the meter.

Also remind them to clean and rinse the wellplates immediately after using. The metals may be collected in an appropriately labeled container.

Procedure

Part I: Constructing a Voltaic Cell

1. Add 1 mL of 0.1 M $Cu(NO_3)_2$ to one well in a wellplate.
2. Add 1 mL of 0.1 M $Zn(NO_3)_2$ to an adjacent well.
3. Add a Cu strip (electrode) to the well containing $Cu(NO_3)_2$ solution.
4. Add a Zn strip (electrode) to the well containing $Zn(NO_3)_2$ solution.
5. Drape a small strip of filter paper saturated with KNO_3 solution between the wells containing the two solutions. Ensure that the filter paper strip is immersed in both solutions. Do not allow the metal strips to touch each other.
6. Obtain a voltmeter and two electrical wires with alligator clips. Attach one end of each wire to a separate voltmeter terminal.
7. Attach one wire from the voltmeter to the copper electrode. Lightly touch the second wire to the zinc electrode. If the needle deflects in the direction of a positive potential or if the digital readout is positive, attach the clip to the zinc metal. If the needle deflects in a negative direction (or the readout value is negative), reverse the clip connections to the electrodes. Figure 16 depicts a completed cell.
8. Record the electrical potential indicated by the voltmeter.

Part II: Measuring Electrical Potentials

9. Using what you learned in Part I about constructing a voltaic cell, construct additional cells, and measure electrical potentials for the voltaic cells composed of all possible pairs of half-cells listed below. Record your data.
   a. 0.1 M $Cu(NO_3)_2$ and Cu strip
   b. 0.1 M $Zn(NO_3)_2$ and Zn strip
   c. 0.1 M $Mg(NO_3)_2$ and Mg strip
   d. 0.1 M $Fe(NO_3)_2$ and Fe "strip" (nail)

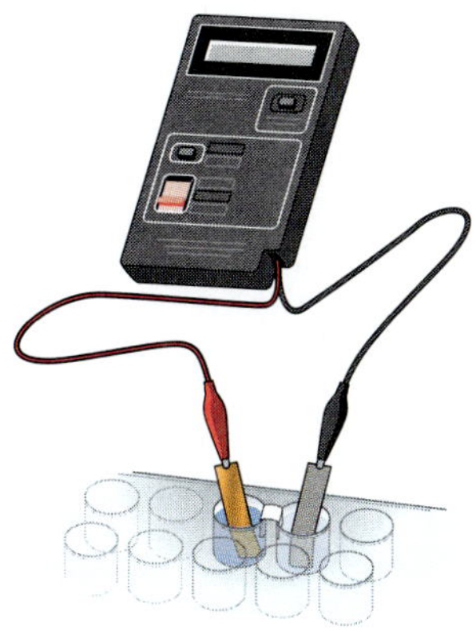

Figure 16 *Setting up an operating voltaic cell.*

Part III: Exploring Electrode Size Effects

10. Based on Parts I and II, design an experiment to investigate whether the size of an electrode influences the total electrical potential generated by a cell and—if so—how that effect can be described. Base your design on copper and zinc electrodes of different widths—0.25 cm (narrow), 0.50 cm (medium), and 1.0 cm (wide).
11. Construct an appropriate data table.
12. Set up and conduct your experiment. Record your data.
13. Dispose of your experimental materials as directed by your teacher.
14. Wash your hands thoroughly before leaving the laboratory.

Expected Results

Negative lead	Positive lead	Volts
medium Cu	Mg	2.7
medium Cu	medium Zn	1.1
medium Cu	Fe	0.8
narrow Cu	Mg	2.7
narrow Cu	narrow Zn	1.1
narrow Cu	Fe	0.8
wide Cu	Mg	2.7
wide Cu	wide Zn	1.1
wide Cu	Fe	0.8

Answers

1. a. Mg/Cu, Zn/Cu, and Fe/Cu

   b. Mg, Zn, and Fe are all above Cu in the activity series, which means that they lose electrons more readily than does Cu. Mg and Cu have the greatest difference in electronegativity, so they generate the greatest cell potential.
2. a. Less: Zn and Cr are closer in the activity series than are Zn and Cu.

   b. More: Zn and Cu are closer together.

   c. Less: Sn and Cu are closer together.
3. Neither increasing nor decreasing the size of the electrode has any observable effect.
4. No. There is not much difference in the tendencies of silver and gold to lose electrons. The battery would also be very expensive.

Post-lab discussion

Ask students to draw a conclusion from their data about the factors affecting electrical potential. Although results obtained by various groups may differ slightly, the rank order should be the same for all. The activity series arranges elements in order of decreasing tendency to lose electrons, so the results reported in the activity are consistent with the elements' arrangement in the activity series. Magnesium loses electrons more readily than does zinc, which in turn loses electrons more readily than iron. At the conclusion of the discussion, be sure that students understand that in an electrochemical cell, electrons flow spontaneously from one metal to another. The greater the electrical potential (potential difference), the greater the voltage of the cell.

Questions

1. a. In Part II, you constructed voltaic cells using copper and three other metals—zinc, magnesium, and iron. List those three copper-based cells in order of decreasing electrical potential (highest cell potential first).

   b. Explain the resulting list in terms of the relative activity of the metals involved.
2. Would the electrical potential generated by cells composed of each of the following pairs of metals be larger or smaller than that of the Zn-Cu cell? See Figure 14 (page 381) for more information.

   a. Zn and Cr

   b. Zn and Ag

   c. Sn and Cu
3. a. How did changing the size of the zinc and copper electrodes affect the measured electrical potential?

   b. Explain your results.
4. a. Would an Ag-Au cell be a commercially feasible voltaic cell?

   b. Why?

C.3 ELECTROCHEMISTRY

In the voltaic cells you constructed in the laboratory, each metal immersed in a solution of its ions represented a half-cell. The activity series predicts that zinc is more likely to be oxidized (lose electrons) than copper. Thus, in the zinc-copper cell you investigated, oxidation (electron loss) occurred in the half-cell with zinc metal immersed in zinc nitrate solution. Reduction (electron gain) took place in the half-cell composed of copper metal in copper(II) nitrate solution. The **half-reactions** (individual electron-transfer steps) for that cell are as shown here:

$$\text{Oxidation:}\quad Zn(s) \longrightarrow Zn^{2+}(aq) + 2\,e^-$$
$$\text{Reduction:}\quad Cu^{2+}(aq) + 2\,e^- \longrightarrow Cu(s)$$

One way to distinguish the electrodes is to note that anode and its associated process (oxidation) both begin with vowels; while cathode and its process (reduction) both start with consonants.

The electrode at which oxidation takes place is the **anode.** Reduction occurs at the **cathode.**

The overall reaction in the zinc-copper voltaic cell is the sum of the two half-reactions, added so that electrical charges are balanced—that is, so electrons lost and gained are equal and therefore cancel.

$$\begin{array}{r} Zn(s) \longrightarrow Zn^{2+}(aq) + \cancel{2\,e^-} \\ Cu^{2+}(aq) + \cancel{2\,e^-} \longrightarrow Cu(s) \\ \hline Zn(s) + Cu^{2+}(aq) \longrightarrow Zn^{2+}(aq) + Cu(s) \end{array}$$

C.3 ELECTROCHEMISTRY

A voltaic cell (a battery) allows the chemical energy of an oxidation reduction reaction to do electrical work. Electrochemical cells are made from various combinations of metals and ions. They are convenient because they can be made portable. Dry cell and alkaline batteries are discussed as oxidation-reduction reactions.

A display of "dissected" batteries is instructive. **CAUTION:** *Do not allow students to handle caustic and corrosive materials from the batteries.*

TV Ontario videos on electrochemistry have good segments about electrochemical

Since a barrier separates the two reactants (Zn and Cu^{2+}) in the cell, the electrons released by zinc must travel through the wire to reach (and reduce) the copper ions.

The greater the difference in reactivity of the two metals in a voltaic cell, the greater the tendency for electron transfer to occur, and the greater the electrical potential of the cell. A zinc-gold voltaic cell, therefore, would generate a larger electrical potential than a zinc-copper cell. (See Figure 14, page 381, to compare the placement of these metals in the activity series.)

GETTING A CHARGE FROM ELECTROCHEMISTRY

Building Skills 5

Each question below addresses voltaic cells, their properties, and the equations used to describe them. Answers to the first question are already completed as an example.

1. Consider a voltaic cell in which lead (Pb), silver (Ag), and solutions of lead(II) nitrate, $Pb(NO_3)_2$, and silver nitrate, $AgNO_3$, are appropriately arranged.

a. Predict the direction of electron flow in the wire connecting the two metals.

b. Write equations for the two half-reactions.

c. Which metal is the anode and which the cathode?

Answers:

1a. Figure 14 (page 381) shows that lead is a more active metal than silver. Therefore, lead will be oxidized, and electrons will flow from lead to silver.

1b. One half-reaction involves forming Pb^{2+} from Pb, as shown in Figure 14. The other half-reaction produces Ag from Ag^+, which can be written by reversing the equation found in Figure 14.

$$Pb \longrightarrow Pb^{2+} + 2\,e^-$$
$$Ag^+ + e^- \longrightarrow Ag$$

1c. In the cell reaction, each Pb atom loses two electrons. Lead metal is therefore oxidized. Thus, lead metal is the anode. Each Ag^+ ion gains one electron. This is a reduction reaction. Since reduction takes place at the cathode, silver metal must be the cathode.

Now try these questions.

2. Predict the direction of electron flow in a voltaic cell made from each pair of metals in solutions of their ions.

a. Al and Sn

b. Pb and Mg

c. Cu and Fe

cells (both lab- and industrial-scale) and related topics treated in this section. See the *ChemCom CD-ROM* for an activity related to voltaic cells.

GETTING A CHARGE FROM ELECTROCHEMISTRY

Building Skills 5

In this activity, students apply their understanding of redox reactions to electrochemical cells. The principle is the same, but some of the reactions are quite complex. Help students to see the similarities without becoming overwhelmed by the details.

Answers

1. Answered in student text.

2. a. $Al \rightarrow Sn$

b. $Mg \rightarrow Pb$

c. $Fe \rightarrow Cu$

3. a. $Sn^{2+}(aq) + 2e^- \rightarrow Sn(s)$ (reduction)

   $Cd(s) \rightarrow Cd^{2+}(aq) + 2e^-$ (oxidation)

   b. Cd loses electrons more readily.

4. Student drawings should resemble those in text.

5. a. Anode = Zn, cathode = Cu

   b. Anode = Al, cathode = Zn

   c. Anode = Mg, cathode = Mn

   d. Anode = Ni, cathode = Au

Maximizing the Cell Potential CD-ROM WWW.

3. A voltaic cell is constructed using tin (Sn) and cadmium (Cd) as the electrodes. The overall equation for the cell reaction is

$$Sn^{2+}(aq) + Cd(s) \longrightarrow Cd^{2+}(aq) + Sn(s)$$

   a. Based on information in that redox equation, write the two half-reactions.
   b. Which metal, Sn or Cd, loses electrons more readily?

4. Sketch a voltaic cell composed of a Ni–$Ni(NO_3)_2$ half-cell and a Cu–$Cu(NO_3)_2$ half-cell.

5. For each cell designated below, identify the anode and the cathode. Assume that appropriate ionic solutions are used in each cell.
   a. Cu-Zn cell
   b. Al-Zn cell
   c. Mg-Mn cell
   d. Au-Ni cell

While zinc–manganese dioxide systems are the most common dry cells, other types include magnesium–manganese dioxide, mercuric oxide–zinc, silver oxide–zinc, and lithium cells.

Voltaic cells are a convenient way to convert chemical energy to electrical energy. The cells can be packaged in small, portable containers. Various combinations of metals and ions can be used to make commercially useful voltaic cells. In ordinary dry cells—often called batteries—zinc is the anode, and a graphite rod surrounded by a water-based paste of manganese dioxide (MnO_2) and graphite serves as the cathode. A mixture of ammonium chloride and zinc chloride in an aqueous paste serves as the electrolyte. In alkaline batteries, commonly used in portable tape and CD players, similar zinc and graphite/MnO_2 electrodes are present, but the electrolyte is an alkaline (basic) aqueous potassium hydroxide (KOH) paste. See Figure 17.

Both zinc–MnO_2 dry cells and alkaline batteries generate an electrical potential of 1.54 V. These redox equations describe the chemical changes involved.

Dry Cell

Oxidation: $Zn(s) \longrightarrow Zn^{2+} + \cancel{2e^-}$

Reduction: $2\,MnO_2(s) + 2\,NH_4^+ + \cancel{2e^-} \longrightarrow 2\,MnO(OH)(s) + 2\,NH_3$

Overall: $Zn(s) + 2\,MnO_2(s) + 2\,NH_4^+ \longrightarrow Zn^{2+} + 2\,MnO(OH)(s) + 2\,NH_3$

Alkaline Battery

Oxidation: $Zn(s) + \cancel{2\,OH^-}(aq) \longrightarrow Zn(OH)_2(s) + \cancel{2e^-}$

Reduction: $2\,MnO_2(s) + H_2O(l) + \cancel{2e^-} \longrightarrow Mn_2O_3(s) + \cancel{2\,OH^-}(aq)$

Overall: $Zn(s) + 2\,MnO_2(s) + H_2O(l) \longrightarrow Zn(OH)_2(s) + Mn_2O_3(s)$

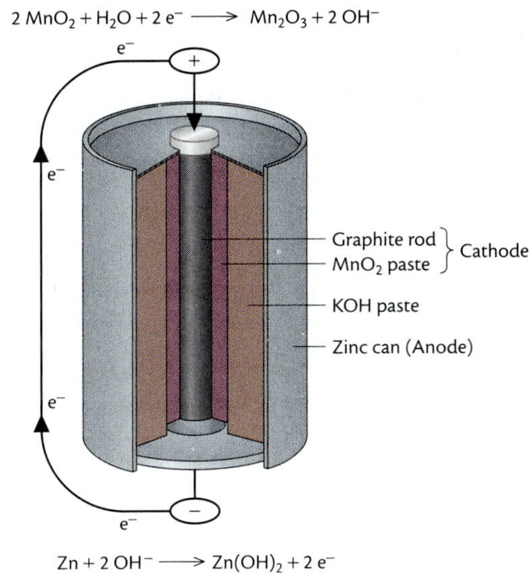

Figure 17 *An alkaline battery.*

BATTERY SIZES

ChemQuandary 2

Each of the batteries shown in Figure 18 generates the same electrical potential—1.5 V. So, why is there any need for 1.5-V batteries larger than the smallest one? Wouldn't it save space, weight, and perhaps even resources to restrict consumer use to AAA batteries?

Figure 18 *Despite their difference in size, each battery generates about 1.5 V, since they are all based on the same redox reaction.*

BATTERY SIZES

ChemQuandary 2

Different applications require batteries of various sizes. A lightweight radio for jogging will be different than a stationary desk clock. Also, some applications may need greater total capacity (battery hours), which is provided by a larger size.

Figure 19 *Cross-section of a Ni-Cad battery.*

Dry cells and alkaline batteries both can be used only as long as all starting materials (reactants) remain. Such batteries, called primary batteries, can no longer be used when the limiting reactant (see Unit 4, page 290) is depleted. That is, primary batteries cannot be recharged (returned to their original state) after use. By contrast, nickel-cadmium (NiCad) and automobile batteries are rechargeable—their systems can be recharged to their original states and thus the batteries can be reused.

The common nickel-cadmium (NiCad) rechargeable battery is based on a nickel-iron battery developed by Thomas Edison in the early 1900s. The NiCad battery anode is cadmium and the cathode is nickel oxide. These electrodes have a rolled design, as shown in Figure 19; this structure increases surface area and subsequently increases available current (number of electrons in motion). When this battery operates, Cd and NiO_2 are converted to $Cd(OH)_2(s)$ and $Ni(OH)_2(s)$. These solid products cling to the electrodes, allowing them to be converted back to reactants when the battery is connected to a recharging circuit. In a recharging circuit, an external electrical potential is used to cause the flow of electrons to move in the opposite direction, thus reversing the reaction.

Lead-acid batteries are found in such diverse equipment as trucks, aircraft, motorcycles, self-starting lawnmowers, and golf carts.

A 12-V automobile battery, known as a lead-acid secondary battery, is composed of a series of six electrochemical cells. As illustrated in Figure 20, each cell consists of uncoated lead (Pb) plates that serve as anodes and lead plates coated with lead dioxide (PbO_2) that serve as cathodes. The electrodes are immersed in a dilute solution of sulfuric acid, H_2SO_4, the electrolyte in this system. When the vehicle's ignition is turned on, an electrical circuit is completed. Metallic lead at the anode is oxidized to Pb^{2+}. The freed electrons travel through the wire to the lead dioxide cathode, reducing PbO_2, which forms additional Pb^{2+}. Lead ions produced at both electrodes then form $PbSO_4$ by reacting with the electrolyte, as shown in the following equations. As the electrons travel from one electrode to the other, they provide energy for the car's electrical systems.

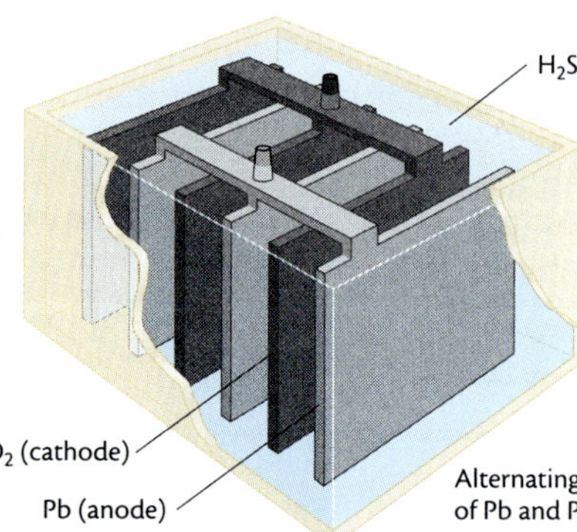

Figure 20 *One cell of a lead storage battery.*

388

The redox reactions in a car battery (while discharging) are

$$\text{Oxidation:} \quad Pb(s) + SO_4^{2-}(aq) \longrightarrow PbSO_4(s) + \cancel{2e^-}$$
$$\text{Reduction:} \quad PbO_2(s) + SO_4^{2-}(aq) + 4\,H^+(aq) + \cancel{2e^-} \longrightarrow PbSO_4(s) + 2\,H_2O(l)$$
$$\text{Overall:} \quad PbO_2(s) + Pb(s) + 4\,H^+(aq) + 2\,SO_4^{2-}(aq) \longrightarrow 2\,PbSO_4(s) + 2\,H_2O(l)$$

If an automobile battery is used too long without being recharged, it "runs down"—that is, the redox reaction slows down and stops. The lead(II) sulfate formed eventually coats the electrodes, which reduces their ability to react and thus produce a current and voltage. In an automobile, recharging is accomplished by an alternator or generator, which converts some mechanical energy from the vehicle's engine into electrical energy and causes electrons to move in the opposite direction through the battery. This reverses the direction of the chemical reactions, thus recharging the battery. See Figure 21.

A "run-down" automobile battery can sometimes be recognized by observing layers of solid white $PbSO_4$ that coat the electrodes.

Lead storage batteries can pose dangers if they are rapidly charged by an outside source of electrical energy. Hydrogen gas, formed by the reduction of H^+ present in the acidic electrolyte, is released at the lead electrode and can be ignited by a spark or flame, causing an explosion. This is one reason why care must be exercised when using jumper cables to start a vehicle with a "dead" battery.

An estimated 70% of U.S. used car batteries are recycled, leading to the recovery of lead metal, sulfuric acid, and battery-casing polymer.

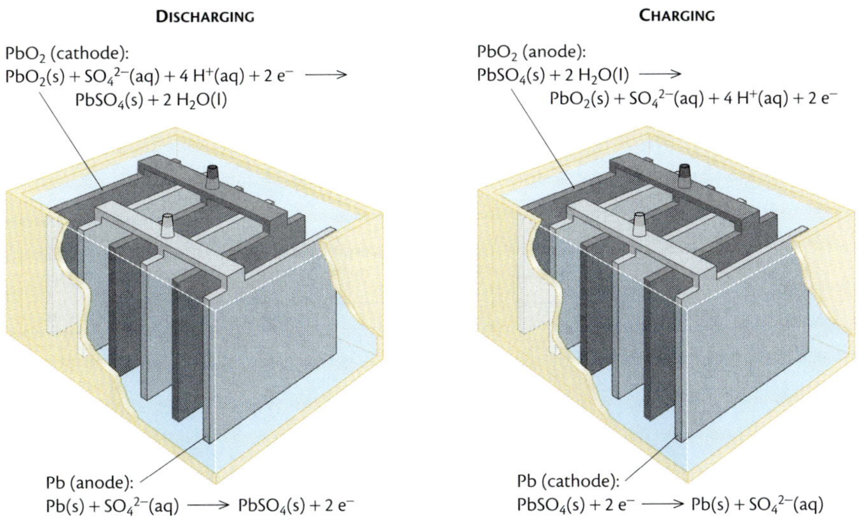

Figure 21 *Discharging and charging a car battery.*

389

C.4 INDUSTRIAL ELECTROCHEMISTRY

This section focuses on the electrolysis of brine (aqueous NaCl), and on the industrial production of aluminum using the Hall-Héroult process. The Hall-Héroult process is responsible for a vast number of aluminum products—and led to locating aluminum plants near electric power generating plants. Emphasize that although the scale of operations is vastly different, the same chemical reactions occur on the laboratory bench as in the industrial plant.

C.4 INDUSTRIAL ELECTROCHEMISTRY

One economically significant application of electrolysis in the chemical industry is the chlor-alkali process. In this process, sodium chloride dissolved in water (a solution known as brine) is electrolyzed, producing chlorine gas, Cl_2, at the anode, and hydrogen gas, H_2, at the cathode (Figure 22). Positive sodium ions remain in solution, along with negative hydroxide ions produced by electrolysis. The hydroxide ions replace the chloride ions lost from the brine as chlorine gas, thus maintaining charge balance, as shown by the following equation:

$$\text{Electrical energy} + 2\,Na^+(aq) + 2\,Cl^-(aq) + 2\,H_2O(l) \longrightarrow 2\,Na^+(aq) + 2\,OH^-(aq) + H_2(g) + Cl_2(g)$$

Electrolysis uses electrical energy to cause chemical changes.

This electrolysis reaction generates three widely produced industrial substances—hydrogen gas, chlorine gas, and sodium hydroxide. More than 13 million tons of chlorine and over 11 million tons of sodium hydroxide are produced in this way annually by the U.S. chemical industry.

The hydrogen gas and chlorine gas produced in the chlor-alkali process can be used without additional purification, because no other gases are released at the anode and cathode. The sodium hydroxide must be purified, since it remains in solution along with unreacted sodium chloride.

In earlier years, the large industrial electrolysis cells for the chlor-alkali process included mercury cathodes. Before the environmental hazards of metallic mercury were established, it was dumped into nearby rivers and lakes during maintenance and cleaning of the cells. Mercury is no longer used in chlor-alkali cells. Responding to the need to replace mercury in electrolytic cells, research chemists and chemical engineers modified the chlor-alkali process considerably. Both electrodes were replaced—mercury with iron for the cathode and graphite with new metals for the anode—and new permeable polymer membranes were developed to separate the anode and cathode compartments.

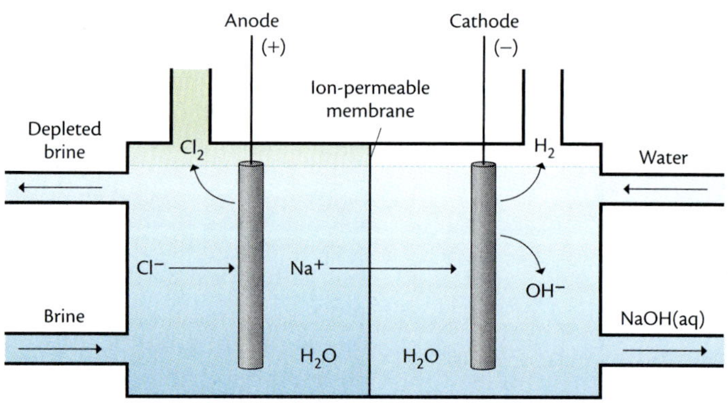

Figure 22 *Electrolysis of aqueous NaCl to produce $H_2(g)$, $Cl_2(g)$, and NaOH.*

From: Kotz, J.C. and Treichel, P.: *Chemistry & Chemical Reactivity,* Saunders College Publishing, Fort Worth, TX 1999.

The industrial production of aluminum—the focus of the WYE Metals Corporation—is another major process that uses electrical energy to produce chemical change. It is very difficult to reduce the Al^{3+} in its ores to the metallic state. Aluminum was first isolated as a metal in the 1820s by an expensive and potentially dangerous process that used highly reactive sodium or potassium metal as the reducing agent. As a result, for more than 60 years, aluminum metal was very expensive, despite the fact that aluminum is the most plentiful metal ion in Earth's crust. Highly prized aluminum metal was even used in jewelry, including the French crown jewels and the Danish crown. In 1884, a 2.8-kg (6-lb.) aluminum cap was installed on top of the Washington Monument in Washington, D.C., as ornamentation and the tip of a lightning rod system. At that time, the aluminum for the cap cost considerably more than the same mass of silver.

The challenge for industrial users was how to produce aluminum at a much lower cost. No common substance gives up electrons readily enough to reduce aluminum cations to aluminum metal. For example, carbon, an excellent reducing agent for metal compounds such as iron oxide or copper sulfide, cannot reduce aluminum compounds.

A young man named Charles Martin Hall solved the problem. In 1886, one year after graduating from Oberlin College (Ohio), Hall devised a method for reducing aluminum using electricity. See Figure 23. Hall's breakthrough was in discovering that aluminum oxide (Al_2O_3), which melts at 2000 °C, dissolved in molten cryolite (Na_3AlF_6) at 950 °C. Thus, he found a relatively easy way to get aluminum ions from aluminum ore into "solution" for electrolysis. Hall's discovery became the basis of a rapidly growing aluminum industry; he founded the Aluminum Company of America (Alcoa) and was a multimillionaire at the time of his death in 1914.

Within two months after Hall developed his process, Paul-Louis Héroult, a young French scientist, independently developed the same process. These two workers, linked through their common discovery, also shared the same birth and death years (1863–1914).

In the Hall-Héroult process, aluminum oxide (bauxite) is dissolved in molten cryolite in a large steel tank lined with carbon. The carbon tank lining is given a negative electrical charge by a source of direct current. This carbon cathode transfers electrons to aluminum ions, reducing them to

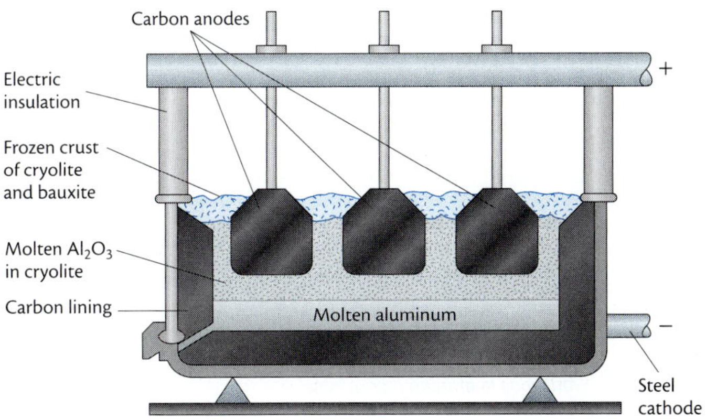

Figure 23 *The Hall-Héroult process for manufacturing aluminum.*

391

molten metal. The molten aluminum sinks to the bottom, where it is periodically drawn off.

The anode, also made of carbon, is oxidized during the reaction. As the tips of the carbon-rod anodes are consumed, the rods are lowered deeper into the molten cryolite bath.

The half-reactions for producing aluminum metal by the Hall-Héroult process are

$$4\ Al^{3+}(melt) + 12\ e^- \longrightarrow 4\ Al(l)$$
$$3\ C(s) + 6\ O^{2-}(melt) \longrightarrow 3\ CO_2(g) + 12\ e^-$$

Ions from cryolite carry the electric current in the molten mixture.

This process had a sudden and long-lasting impact on aluminum production. Earlier, only about 1000 kg of aluminum were produced annually worldwide. In 1884, a pound of aluminum metal sold for about $12, a significant sum in those times. Within three years of their discovery, Hall and Héroult each had commercial operations producing aluminum, and a million tons per year were produced worldwide by the turn of the century. See Figure 24. The price of aluminum plummeted—five years after large-scale production started, aluminum sold for 70 cents a pound. Thus, this process opened doors for later wide-ranging uses of aluminum—from beverage cans and stepladders to foil wrap and aircraft components.

More than 10 billion kilograms (over 22 billion pounds) of aluminum are produced in the United States each year, requiring about $2 000 000 000 worth of electrical energy (about 20 cents per kilogram aluminum). Because aluminum plants use such large quantities of electricity, they are often located near sources of hydroelectric power. Many are located in the Pacific

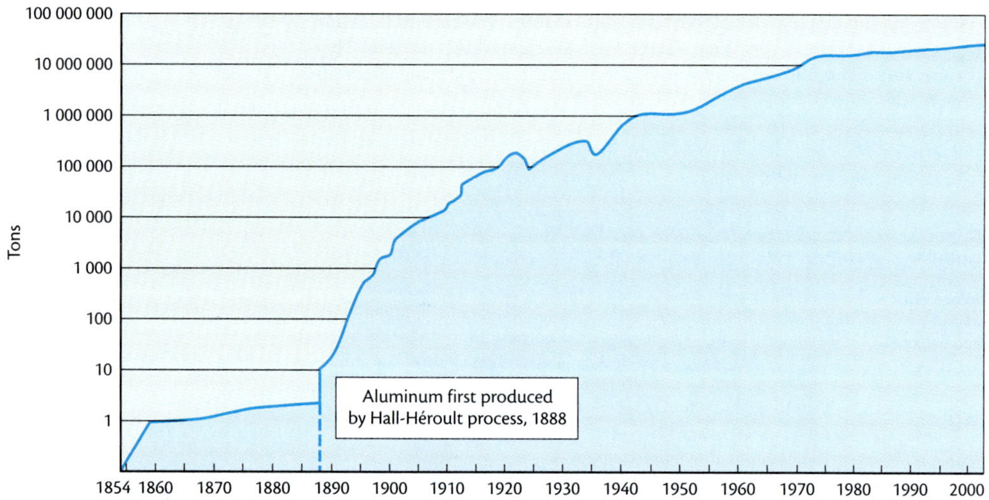

Figure 24 *Worldwide aluminum production.*

392

Northwest of the United States—electricity from hydropower generally costs less than that generated from the combustion of fossil fuels. The WYE Metals Corporation is interested in Riverwood due to availability of hydroelectric power and the town's proximity to aluminum ore sources.

You have now considered the chemical processes by which EKS and WYE companies make their products. To further inform the choices you will make about a possible chemical industry for Riverwood, some additional aspects of the chemical industry will now be examined.

C.5 INDUSTRY AS A SOCIAL PARTNER

As you learned earlier in this unit, industries, governments, and citizens in the United States and elsewhere have recognized their joint responsibility to ensure that chemical products are manufactured with maximum benefit and minimum risk to society. EPA's Design for Environment program, which includes Green Chemistry (pages 372–374), has already influenced the way chemical processing, printing, and dry-cleaning industries do business.

Over the past decade, the chemical industry has begun to develop new synthesis methods, based on using safe starting materials to replace toxic or environmentally unsafe substances. Green Chemistry focuses on preventing environmental pollution directly at the point of manufacturing. In this approach, and in others such as Responsible Care (pages 374–375), the chemical industry works as a social partner to sustain development and international trade without damaging the environment.

New Green Chemistry processes use more environmentally benign reactants than traditional processes and create waste products that do less damage to air and water. For example, chemical processes have been developed that use D-glucose, ordinary table sugar, to replace benzene, a known carcinogen. The D-glucose serves as chemical feedstock to produce reactants that eventually led to nylon and various medicinal drugs. Also, nontoxic food dyes have been demonstrated in some processes to be effective substitutes for catalysts composed of toxic metals such as lead, chromium, or cadmium.

A chemical feedstock is a starting material from which many other materials can be made. For example, crude oil is a chemical feedstock for fuels, plastics, and many other types of substances.

Phosgene, a toxic gas, can be replaced by carbon dioxide in the manufacture of isocyanates. Isocyanates are substances used to make polyurethanes, materials widely used to produce seat cushions, insulation, and contact lenses. In addition, Green Chemistry efforts seek ways to synthesize industrial chemicals using water as a solvent in place of toxic alternatives, whenever possible, and to employ materials that can be recycled and reused, thus significantly reducing waste disposal problems. To succeed commercially, such new processes must also be cost effective.

The chemical industry has the responsibility to make products in ways that are as hazard-free as possible. Industry must deal honestly with the public to ensure that risks and benefits of chemical operations are clearly known. In 1998, the occupational injury and illnesses rate of the U.S. chemical industry was less than half that of U.S. manufacturing industries as a

C.5 INDUSTRY AS A SOCIAL PARTNER

The chemical industry is obligated to produce useful products or services while causing as little damage as possible to the environment and public health. Compliance with worldwide industry standards is monitored by independent outside organizations. An international program (ISO 14 000) is in the process of developing a single set of environmental management system standards that will be accepted worldwide. As responsible citizens, students are expected to know the risks and benefits associated with the chemical industry and to weigh those risks and benefits to make wise decisions.

C.6 ASSET OR LIABILITY?

Making Decisions

The main purpose of this activity is not just to have students make a decision concerning the proposed plants; it is to have students also consider the issues affecting the decision and issues that are affected by it. A number of such issues are outlined in the text. Encourage students to think of others.

Another option for this exercise is based on a grid (large cross) that has "high" and "low" on the vertical axis and "unlikely" and "likely" on the horizontal axis. Students are asked to "plot" each factor on this grid.

whole. Chemical companies also must assure consumers, of course, that their products are safe when used as intended.

Chemical industries must comply with relevant laws and regulations, as well as with voluntary standards set by many manufacturers themselves. Often, compliance with voluntary or mandatory requirements is monitored by independent, outside organizations. A worldwide initiative calling for voluntary quality-management systems has been developed by the International Organization for Standardization (ISO) in Geneva, Switzerland. Representatives from 74 nations, including the United States, have established five standards (called ISO 9000) to guide corporations in setting up quality-management systems.

In addition, governments of various nations around the world have established laws regulating environmental pollution, although the laws frequently lack consistency from country to country. An international program (ISO 14000) is developing environmental standards to help manufacturers and organizations evaluate the impact of their operations, products, and services. The goal of ISO 14000 is to establish internationally accepted environmental management system standards that can help manufacturing companies in any country achieve sustainable development.

No activities or initiatives of the chemical industry or the government can eliminate all risks involved in manufacturing chemical substances, any more than the risks of automobile travel can be completely eliminated. However, knowing the risks, continuing to explore the sources of and alternatives to those risks, and prudent decision making all remain essential.

As users of the industry's chemical products, all consumers share these responsibilities. Having studied some basic concepts about the manufacturing of chemical products, you can use such knowledge to weigh the risks and benefits—to you, the community, and to the environment—of particular decisions.

C.6 ASSET OR LIABILITY?

Making Decisions

If you were a citizen of Riverwood, what would be your view—should a chemical plant be invited to locate in Riverwood? On what basis should such a choice be made?

To clarify such decision-making challenges, it is helpful to consider and evaluate both benefits (positive factors) and burdens (negative factors) associated with an option.

First, you will consider some positive and negative points regarding a chemical plant in Riverwood. Then you will begin to address key questions that confront Riverwood citizens.

Positive Factors

- *The local economy would be improved.* Either plant would employ about 200 local residents. This would add about $8 million to Riverwood's economy annually. In addition, each plant employee would indirectly provide jobs for another four people in local businesses. This is quite desirable, because 15% of Riverwood's labor force of 21 000 is currently unemployed.
- *Farming costs could be lowered.* Fertilizer for farms near Riverwood is now trucked from a fertilizer plant 200 miles away. These transport costs increase farmers' expenses by $15 for each ton of ammonia-based fertilizer. Each year 700 tons of fertilizer are spread on farms in the Riverwood area. Local farmers thus stand to save about $10 000 annually in transportation costs.
- *New local industries could result.* With direct access to aluminum metal locally, several Riverwood entrepreneurs would consider starting a factory to make aluminum window frames and studding for home construction. Without significant transportation costs for raw materials, the proposed company could establish itself in the market by selling its products at lower prices. The projected aluminum-fabrication factory would probably hire 30–45 employees, mainly shop workers and drivers, adding $700 000 annually to the local economy.

  Ammonia is a commercial refrigerant used to produce ice in large quantities. Thus ready access to ammonia supplies could lead to development of a commercial ice-making plant in Riverwood. Such a company might employ 25–35 individuals, principally drivers and some plant workers, which could add $500 000 annually to Riverwood's economy as well as provide ice for the region.
- *Riverwood air quality could be enhanced.* A natural gas transmission company would build a line to deliver natural gas (mainly methane, CH_4) to a Riverwood ammonia plant. Town residents, who have traditionally burned fuel oil in their home furnaces, could convert to natural gas. If all 11 000 Riverwood homes and businesses burned natural gas rather than oil, emissions of sulfur dioxide and particulate matter would decrease significantly.
- *The tax base would be improved.* Although tax incentives are part of the appeal of a Riverwood site, the aluminum or ammonia plant will still contribute to the tax base in Riverwood. This will provide a large increase in revenues for the community.

Negative Factors

- *Injuries and accidents could occur.* Ammonia is manufactured at high pressure and high temperature. Existing as a gas at ordinary temperature and pressure, ammonia is extremely toxic at high concentrations.

An accident on the roadway or at the plant that resulted in the release of large amounts of ammonia would create a health hazard and could injure or kill nearby workers. For each 100 U.S. workers in ammonia-based fertilizer plants, several cases of work-related injury or illness have been reported annually. A few worker-related deaths have also been reported in such plants. A 1994 explosion at a nitrogen fertilizer plant near Sioux City, Iowa, killed four employees and seriously injured two more. The blast, suspected to be caused by explosion of ammonium nitrate at a nearby production facility, tore open a 15 000-ton anhydrous ammonia storage tank, releasing ammonia vapor. Many of the other 24 employees on that shift suffered from ammonia inhalation.

Aluminum metal itself is nontoxic, but its production has its own dangers. High temperatures are used and molten aluminum is drawn from large electrolytic cells. As the carbon anodes burn off in the cryolite bath, toxic carbon monoxide gas is produced and must be handled properly. Injuries are also associated with mining and transporting aluminum ore. Aluminum production illness and injury rates annually are about 20 per 100 full-time employees. Infrequently deaths at such plants have also been reported. However, the injury and accident rates in aluminum and ammonia-based fertilizer production are both lower than the rates in motor vehicle manufacturing, meat packing, and steel foundries.

- *Water quality could be threatened.* If leaks occurred during production, so that ammonia or ammonia-bearing wastewater entered the Snake River, the resulting water solution would become basic—possessing a pH higher than 7—and thus possibly threatening to aquatic life.
- *The aluminum or ammonia markets might decline.* Aluminum is commercially valuable because it is inexpensive, corrosion-resistant, and has low density (low mass for a given volume). The market for aluminum is driven by applications where such properties are important—building trades (insulation, window casings, reinforcement strips, ventilation devices), transportation (sheeting for truck trailers, campers, and mobile homes), and packaging containers. If aluminum substitutes were developed in any of these major areas, demand for aluminum would weaken. These substitutes might come from new alloys or by developing new plastics or ceramics. The Riverwood plant would be in jeopardy if demand for aluminum declined.

The fertilizer industry is among the largest consumers of ammonia. Current fertilizer-intensive agricultural methods have created controversy. In some cases crop yields have declined despite application of increased amounts of synthetic fertilizer. Some farmers have elected to use less synthetic fertilizer, so ammonia demand may decline in coming years. This could hurt Riverwood's future economy, even though the plant would provide short-term economic assistance.

396

Considering Risks and Benefits

The poet William Wordsworth used the phrase, "Weighing the mischief with the promised gain . . ." in assessing technological advancements (in his case, new railroads). Since the dawn of civilization, people have often accepted the risks and burdens of new technologies in order to reap the benefits. Fire, one of civilization's earliest tools, gave people the ability to cook, warm themselves, and forge tools from metals. Yet fire out of control can destroy property and life. Every technology offers its benefits at a price.

Some people oppose new technologies in general, arguing that the new benefits are not worth the added burdens. For example, many argued against the introduction of trains and automobiles. The benefits generated by the chemical industry also come with associated burdens. One must weigh the potential benefits of the technology against its costs, including the harm or threats it poses to individuals, society, and the environment.

One way to identify an "acceptable" new technology or venture is as one that has a relatively low probability of producing harm—that is, delivers benefits that far outweigh the risks. Unfortunately, benefit-burden analysis—weighing what Wordsworth called "mischief" against "promised gain"—is far from an exact science. For instance, some technologies may present high risks immediately, while others may deliver chronic, low-level risks for years or even decades. Many risks are impossible to predict or assess with certainty. Individuals can control some potential risks, but others must be addressed and controlled at regional or national levels. In short, it is quite difficult to compare or "weigh" various risks. Still, decisions such as whether to invite a chemical plant to Riverwood must be made.

In many cases, electing not to make a decision is, in fact, also a decision—one accompanied by its own set of risks and benefits.

1. Based on what you have already learned in this unit, create two lists for the chemical plants (EKS and WYE)—one summarizing benefits and one summarizing burdens/risks.
2. Check your summaries of burdens/risks. Are any of those negative factors completely unacceptable? If so, the plant associated with that risk or burden is probably not a viable option for Riverwood.
3. Within each list, mark the most valuable benefits with a plus symbol and the most serious burdens with a minus symbol.
4. Benefit-burden analysis also considers the likelihood of occurrence of each benefit and burden. A burden that is fairly minor but almost certain to occur might merit more consideration than a burden that is more serious but extremely unlikely. Using a scale from 1 (highly unlikely) to 5 (extremely likely), rate each item on your lists in terms of its likelihood of occurrence.
5. Based on your responses to Questions 3 and 4, select a personal position on the question of building a chemical plant in Riverwood. Discuss your view with your group.
6. As a class, discuss whether Riverwood should allow a chemical plant to be built. Be sure that concerns of students who do not share the opinion of the majority are also heard and considered.

CD-ROM WWW. Questions & Answers

CHEMISTRY AT WORK

A Career That's Good from Start to Finish

Gene had just pulled into the parking lot when his cellular phone rang. It was a "code call"—a serious problem at Ardmore Manufacturing. "Well," he thought to himself, "so much for lunch with Pat."

Gene Kropp is a technical field engineer and salesman for Technic, Inc., in Rhode Island. Gene sells equipment and supplies to electroplating manufacturers across a wide geographic area. Usually his days are fairly predictable: appointments, meetings, trips to suppliers. Gene visits four or five customers, dropping off chemicals and equipment and taking new orders. Gene also visits prospective customers to develop new business.

But sometimes he gets a code call, an emergency call from a customer who is having trouble with Technic's electroplating system. Gene listens to the customer describe the problem. Then, using the process of elimination and a few "tricks of the trade," Gene helps get things running productively again.

Sometimes he gets a code call from a customer who is having trouble. . . .

Electroplating is the process of attaching a metal (such as chromium, silver, or gold) to the surface of an article to protect it and make it more attractive. Gene's knowledge of chemistry in general (and silver and other metals in particular) helps him meet clients' needs and solve their problems.

Gene attended a two-year college and earned a degree in general studies with a concentration in chemistry. Since then, Gene has worked in various positions in the electroplating field, increasing his knowledge through additional coursework and on-the-job training.

The Science of Repair

It is easy to make comparisons between scientific research and equipment repair. In each case there are fundamental problem-solving skills that need to be used.

1. Locate a user's manual for one of your household appliances or devices (such as a refrigerator, VCR, or microwave oven.) See if it has a

Technic, Inc.

Call Log
Date: March 3

To Do:	Call Barbara @ Robinson Electroplating-set up plant visit
	Call Bill Smith, ask about new valve
8:30 am	Robinson Electroplating-meet new plant manager, inspect equipment
	Visit Bill in purchasing; drop off bid on replacing backup pump
10:00	Drop off chemicals at headlamp plant; visit Bill in lab
10:35	CODE CALL from Ardmore/Tom D says pump has quit
10:50	Ordered new pump from L-Way, will pick up on way to Ardmore
	Called service, will have Joe and engineer meet me at Ardmore
11:00	Left for Ardmore
12:30	~~Lunch with Pat~~ Cancelled

398

CHEMISTRY AT WORK

Photograph courtesy of Technic, Inc.

"troubleshooting" section. When you find a user's manual that does have this type of section, review the content and see what similarities you find between the approach in the manual and how you do scientific problem solving in chemistry.

2. Call or visit your local two-year or community college and get a list of training programs for technicians in fields related to science.
3. Look in the employment listings of a newspaper classified-advertising section under "technicians" and see what kinds and numbers of jobs are available.
4. Keeping a notebook is a job requirement for many employees. It may be important for workers to document work and actions for a variety of reasons. Some typical requirements include the following:

Typical Rules for Keeping an Industrial Laboratory Notebook

MATERIALS
- bound notebook
- pen

LEGAL ISSUES
Notebooks are used to support applications for patents and to document all laboratory tests done for a company. To be legally acceptable, laboratory notebooks must be completed and signed by technicians daily. Each page must be
- numbered
- dated
- witnessed by another person

When a worker keeps a laboratory notebook it must be neat, accurate, and written in pen. It is a permanent record of all procedures and data. Others may need to replicate what the worker has done.

Using the process of elimination and a few "tricks of the trade," Gene helps get things running productively again.

How do the requirements listed above for an industrial-type laboratory notebook compare to the type of notebook you may keep for chemistry class?

SECTION SUMMARY

Reviewing the Concepts

1. What is meant by the term "electrochemistry"?

Chemical changes that produce or are produced by electrical energy

2. Explain how chemical reactions can produce electricity.

Chemical reactions can produce electricity through oxidation-reduction reaction. Two half-cells made up of two specific metals are partially immersed in solutions of the respective metal's ions. The half-cells are connected, but a barrier between them prevents the solutions from mixing. The two metals (electrodes) are connected by a wire that allows electrons to flow through them. (Electricity is a flow of electrons.)

3. Does the electrical potential produced by a voltaic cell depend on

a. the mass of the electrodes? Explain.

b. the specific metals used? Explain.

a. The amount of material does not affect the potential difference, although it may make the cell last longer.

b. Yes, electrical potential depends upon the difference between the chemical activities of the two metals used. The greater the difference in the location of the metals on the activity series, the greater the electrical potential.

4. What determines whether an oxidation-reduction reaction will be spontaneous?

Redox reactions tend to be spontaneous when a more active metal reduces a less active metal.

5. Place these three cells in the order of increasing electrical potential. Refer to Figure 14 (page 381) for additional information.

a. tin and silver	**c. Cu and Pb**
b. chromium and silver	**b. Cr and Ag**
c. copper and lead	**a. Sn and Ag**

6. Using Figure 14, select the pair of metals in Question 5 that would generate the largest electrical potential.

Chromium and silver have the greatest distance between metals in the activity series.

7. The following questions refer to this equation, which describes the reactions in a typical automobile lead storage battery:

$$PbO_2(s) + Pb(s) + 4\,H^+(aq) + 2\,SO_4^{2-}(aq) \longrightarrow 2\,PbSO_4(s) + 2\,H_2O(l)$$

a. Does this equation represent the charging or discharging of the battery?

b. Electrical energy or work can be extracted from this reaction. What does that fact imply about the reverse reaction?

c. Why can the condition of a lead storage battery be tested with a hydrometer, a device that measures liquid density? (*Hint:* Sulfuric acid solutions have a greater density than liquid water.)

d. In this reaction identify the species being oxidized and the species being reduced.

e. Under what conditions may a lead storage battery produce hydrogen gas?

a. Discharging

b. The reverse reaction would consume energy (endothermic) and is nonspontaneous.

c. Sulfuric acid is used up in discharging the battery. The greater the density of the sulfuric acid, the less the battery has discharged.

d. The lead in the lead dioxide (spongy lead) is being reduced (gaining additional electrons); metallic lead is being oxidized (losing electrons).

e. When it is being charged from an outside source, hydrogen may be produced as a by-product at the lead dioxide electrode, owing to the reduction of hydrogen ion $H^+(aq)$ to hydrogen gas $H_2(g)$.

8. Use Figure 14 to predict the direction of electron flow in a voltaic cell composed of each of these pairs of metals and their ions:

a.	Al and Cr	**a. Al → Cr**
b.	Mn and Cu	**b. Mn → Cu**
c.	Fe and Ni	**c. Fe → Ni**

9. Write half reactions for each of these oxidation-reduction equations:

a. $Pb(s) + Cu^{2+}(aq) \longrightarrow Pb^{2+}(aq) + Cu(s)$

b. $Cr(s) + 3\,Ag^+(aq) \longrightarrow Cr^{3+}(aq) + 3\,Ag(s)$

a. $Pb(s) \rightarrow Pb^{2+}(aq) + 2e^-$	**Oxidation**
$Cu^{2+}(aq) + 2e^- \rightarrow Cu(s)$	**Reduction**
b. $Cr(s) \rightarrow Cr^{3+}(aq) + 3e^-$	**Oxidation**
$3Ag^+ + 3e^- \rightarrow 3Ag(s)$	**Reduction**

10. Sketch a voltaic cell that involves magnesium metal and chromium metal. Identify the appropriate solutions, the anode, the cathode, and the direction of electron flow.

See Figure 15 for a sample diagram of a voltaic cell.

11. Write the overall equation for making aluminum by the Hall-Héroult process.

$4\,Al^{3+} + 3\,C(s) + 6\,O^{2-} \rightarrow 4\,Al(l) + 3\,CO_2(g)$

12. Why are aluminum-producing plants generally located in areas with inexpensive electrical power?

The most important factor in determining where to locate an aluminum-producing plant is the availability of an inexpensive source of electricity. Large quantities of generated electricity are required to reduce aluminum from its minerals because the Al^{3+} ions have a very weak attraction for electrons from other chemical substances.

13. Why is the reduction of aluminum oxide more difficult to accomplish than the reduction of copper oxide?

Aluminum is found in nature in its $^{+}3$ oxidation state (Al^{3+}) compared to copper (Cu^{+} or Cu^{2+}). Al^{3+} is far more difficult to reduce to its metallic state. The more active the metal, the harder it is to reverse the process and add electrons.

14. How many moles of electrons are needed to reduce enough Al^{3+} ions to produce a 378-g roll of aluminum foil?

$$378\text{ g Al} \times \frac{1\text{ mol Al}}{27\text{ g Al}} \times \frac{3\,e^-}{1\text{ mol Al}^{3+}} = 42\text{ mol electrons}$$

15. Describe the economic and societal impact of the Hall-Héroult process.

The Hall-Héroult process reduced the cost of producing aluminum and significantly increased the amount of aluminum that could be produced. Plentiful and inexpensive aluminum is the building material for thousands of common products.

16. Identify one benefit and one burden associated with each of the these activities or technologies:

a. playing high-school basketball

b. driving an automobile

c. jogging

d. receiving a dental X-ray

e. applying pesticides to garden plants

f. using food preservatives

Activity	Benefit	Burden (possible)
a. playing basketball	exercise, teamwork	injury
b. driving an automobile	getting you where you want to go	cost, accidents
c. jogging	exercise, weight control	injury
d. receiving a dental X-ray	locate cavities	exposure to radiation
e. applying pesticides	kills pests, improves yield of garden plants	damage to environment
f. using food preservatives	food lasts longer	undesirable health side effects

17. List some positive and negative aspects associated with the technologies involved in producing ammonia and aluminum.

	Positive	Negative
Aluminum	can be produced in large quantities	uses a large amount of electrical energy
Ammonia	fertilizers increase the production of food	used in making military munitions (could also be positive)

18. Why do you think chemical industries sometimes have a negative public image?

The chemical industries often have a negative image because most people do not realize what is involved in making a product. Industrial pollution, accidents, and unsightly production plants add to the negative image.

19. Describe the kind of partnership that would be desirable between the chemical industry and society.

The chemical industry and society should be partners in reducing pollution and developing uses of waste materials.

Connecting the Concepts

20. How do processes involved in electrochemical cells differ from those found in electrolysis?

Difference between electrochemical cells and electrolysis:

Electrochemical Cells	Electrolysis
exothermic	endothermic
spontaneous	nonspontaneous
electrons flow from more active reactants to less	electrons flow from less active reactants to more active one

21. Explain why recycling aluminum is more cost effective than processing aluminum ore by the Hall-Héroult method.

Recycling aluminum requires less electrical energy than reducing the metal from its ore. Bauxite reserves are decreasing; therefore, bauxite use should be reduced by recycling aluminum.

22. Unwanted materials from chemical processing plants can often be regarded as "resources out of place." Explain why.

Unwanted materials can become the starting materials for new products.

23. Identify the top four substances produced by the chemical industry in the United States. List two uses of each of those substances.

The top four chemicals produced by the chemical industry are

a. sulfuric acid: manufacturing fertilizer, refining petroleum

b. nitrogen: manufacturing ammonia, liquid nitrogen for cooling

c. oxygen: used in acetylene torches, used in oxygen tents for medical purposes

d. ethene: fuel, raw materials for polyethylene

24. Why are some battery types rechargeable and others not?

Primary batteries (alkaline and dry cells) can be used only as long as the reactants have not been used up. They do not use reversible reactions. A rechargeable battery produces solid products that cling to the electrodes. The process can be reversed by connecting the battery to an external electrical potential, which causes the flow of electrons to move in the opposite direction and recharge the battery.

25. Why do batteries eventually stop operating? Explain this in terms of limiting reactants.

A battery will stop operating when one of the reactants is used up.

Extending the Concepts

26. The reasons behind naming chemical processes are sometimes both controversial and political. Investigate the history of how two processes described in this unit—Haber-Bosch and Hall-Héroult—got their names.

Early reports may give credit to one individual, when in fact two or more researchers may have made simultaneous discoveries.

27. Find out how aluminum metal is welded. What special techniques involved in welding aluminum differ from those used in welding other metals?

When being welded, aluminum must be bathed in a noble gas to keep the aluminum from oxidizing. The shielding gas is generally delivered through the head of the welding torch.

28. Investigate how the printing and dry cleaning industries are changing their operations to become more environmentally sensitive and responsible.

Green Chemistry strives to reduce pollution in these industries by reinventing chemical processes, using safer solvents, reducing waste, and generating less toxic by-products.

29. As a class, identify several major activities and technologies that can be found in your community. Use the risk-assessment table below to assign an appropriate letter to each activity. Be prepared to defend and discuss your decisions. Compare your rankings with those made by others in your class.

Assessment of Possible Risks Involved with an Activity or Technology

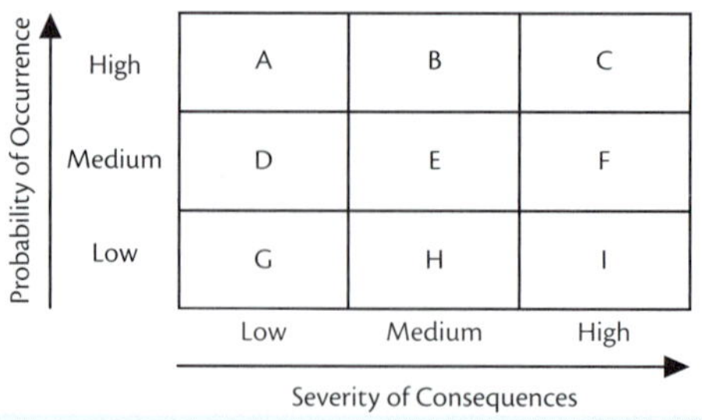

Answers will vary from community to community.

PUTTING IT ALL TOGETHER

A Chemical Plant for Riverwood? Confronting Final Choices

CD-ROM WWW. Putting It All Together

DECISION NEARS: IS CHEMICAL PLANT IN RIVERWOOD'S FUTURE?

BY GARY FRANZEN
Riverwood News Staff Reporter

After months of study and discussion, the Riverwood Town Council is prepared to act on separate proposals from EKS Nitrogen Products Company and WYE Metals Corporation to locate a chemical plant in Riverwood. At tonight's town council special meeting, the council will decide which, if either, plant should be invited to Riverwood. The meeting starts at 7:30 p.m. in Town Hall.

Mayor Cisko said, "I'm very pleased with the turnout we've had for the town meetings held on this issue. I want to emphasize that tonight's council meeting is open to the public. I encourage all community members to attend and express their views about a chemical plant in Riverwood."

At the request of the council, both companies have prepared comprehensive summaries to inform citizens of their plans for a Riverwood plant. These summaries were circulated at the previous town meetings on the subject and have been widely distributed throughout the community. Additional copies can be obtained from the mayor's office.

A CHEMICAL PLANT FOR RIVERWOOD? CONFRONTING FINAL CHOICES

The Town Council Meeting

- Detailed instructions for this activity are in the student text. Because this is the second town meeting activity, students should carry it out with less direction from the teacher. To maximize student participation, ask half of the class to advocate the positive position and the other half to advocate the negative position.
- Allow some class time for groups to organize and assign roles within each group.
- Monitor the discussion so that one person or group does not dominate. About 35 to 45 minutes should be enough to bring out the important points.
- At the end of the discussion, allow time for the town council to meet and discuss options.
- Voting or conducting a referendum provides closure for this activity and for the unit.

TOWN COUNCIL MEETING

Meeting Rules and Penalties for Rule Violations

1. The order of presentations is decided by council members and announced at the start of the meeting.
2. Each group will have a specified time for its presentation. Time cards will notify the speaker of time remaining.
3. If a member of one group interrupts the presentation of another group, the offending group will be penalized 30 seconds for each interruption, to a maximum of one minute. If the group has already made its presentation, it will forfeit its rebuttal time.

Town Council Members: Background Information

Your group is responsible for conducting the meeting in an orderly manner. Be prepared to

1. decide and announce the order of presentations at the meeting. Groups presenting factual information should be heard before groups voicing opinions.
2. decide how the presentation area will be organized: where town council members controlling the meeting will be located, where the presenters will speak, and where the groups and observers will be positioned.
3. explain the meeting rules and the penalties for violating those rules.
4. recognize each group at its assigned presentation time.
5. enforce established presentation time limits by preparing time cards with "one minute," "30 seconds," and "time up" written on them. These cards, placed in the line of sight of the speaker, can serve as warnings.
6. control the rebuttal discussions and open-forum speeches.
7. summarize the options when testimony has been completed.
8. conduct a vote of all town council members.
9. report the results of the vote and future actions mandated by it.

404

EKS Nitrogen Products Company Officials: Background Information

Your company is a multinational organization with a long-term record of safely manufacturing and handling a wide variety of industrial chemicals. EKS helped develop chemical industry standards for limiting environmental pollution and agrees to meet the council's criteria for any new industries. Your company is a member of Responsible Care.

Ammonia production is potentially dangerous because of the high temperatures and pressures used. However, long-standing proper engineering and safety measures make ammonia production safer than many non-chemical manufacturing processes.

WYE Metals Corporation Officials: Background Information

The WYE Corporation is a top U.S. aluminum producer. Your plants are modern, efficient, and safe. Your company has been cited as an industry model for its environmentally sound practices and community outreach. WYE has agreed to abide by the council's criteria for a new industry and is a member of Responsible Care.

Aluminum production relies on procuring adequate raw materials and abundant, inexpensive electrical energy. The Riverwood plant would be similar in design and operation to several other WYE aluminum production facilities known for their safe, efficient operation.

Riverwood Industrial Development Authority Members: Background Information

Because of unemployment as high as 15%, Riverwood needs to attract additional employers. EKS or WYE would provide significant new jobs, which will help revive the local economy. A new chemical manufacturing plant might encourage other industries to locate in Riverwood.

Your organization is involved in efforts to bring additional industries to Riverwood. Your presentation should emphasize the availability in Riverwood of abundant electrical power, excellent railroad and highway access, and favorable tax rates.

405

Riverwood Environmental League Members: Background Information

Your organization is concerned about the potential problems of air and water pollution that might be caused by either EKS or WYE. Your presentation should address how closely either plant would comply with federal, state, and local policies governing the manufacturing of its products.

Although each company has a strong safety record, accidents do occur in chemical manufacturing, some with significant environmental impact. This is a great concern since either plant would be adjacent to the Snake River. Other types of industries might offer fewer potential problems while creating as many jobs as EKS or WYE.

Riverwood Taxpayer Association Members: Background Information

Your organization is concerned about the financial effects of locating a chemical plant in Riverwood. Thus some of the questions to be answered at the town council meeting should be addressed in your presentation. Include these questions:

- What level of new taxes will be needed to upgrade the local water treatment plant to meet growing industrial needs?
- Who will pay for construction and installation of new road signs and perhaps even new roads?
- Will WYE or EKS pay a fair proportion of taxes compared to other Riverwood businesses?

Because your presentation may be influenced by the testimony of other groups, you may find it useful, if possible, to obtain their written briefs before the council meeting.

406

LOOKING BACK

Whether or not you decided to allow a chemical plant to locate in Riverwood, you have learned some valuable chemistry in the process of making that decision. You learned how some key substances such as ammonia are produced and used and how nitrogen cycles among the air, soil, and living organisms. You also learned how principles of electrochemistry can be used not only to harness chemical energy from spontaneous chemical reactions but also to provide energy to cause other desired reactions to occur.

One day you may need to decide whether a chemical plant locates in or near your community. But whether that happens or not, the chemical knowledge gained and skills practiced in weighing risks and benefits will help you deal more effectively with a wide range of future decision-making challenges.

Looking Back

It is worthwhile to have students look back and compare the knowledge they now have with the background they brought to the course at its beginning. In addition to learning about chemistry, they should now have a better idea of how important chemistry and technology are in their world. Students owe it to themselves as responsible citizens to continue weighing risks and benefits to help them deal more effectively with future decision making challenges.

Section	Time class	Student Edition Features	Program Resources	Media and Technology
Section A. The Nature of Atoms **Objectives** • Radiation can be classified as either ionizing or nonionizing radiation. • The constituent atoms of a radioactive substance are unstable and undergo spontaneous changes in nuclear structure. Radioactivity usually results in the emission of alpha, beta, and gamma radiation. • Rutherford's gold-foil experiment resulted in a model of the atom having a tiny, massive, positively charged region at the center of the atom—the nucleus—with negatively charged electrons surrounding the nucleus. • Atoms of an element having different numbers of neutrons are isotopes of that element. Isotopes that are radioactive are called radioisotopes.	6 Periods	**Building Skills** #1. Isotope Notation #2. Molar Mass and Isotope Abundance **ChemQuandary** #1. Scientific Discoveries **Modeling Matter** • Isotopic Pennies **Making Discoveries** • Fact or Fiction?	**Activity Book** • A.4 Building Skills 1 Supplement • A.5 Building Skills 2 Supplement **Black Line Copy Masters** • Your Annual Ionizing Radiation Dose • PIAT Evaluation	**ChemCom CD-ROM** • Unit VI Air—Section A • Laboratory Video **Transparencies** • Simple Cathode Ray Tube • Electromagnetic Spectrum • Rays in a Magnetic Field and Alpha-Particle Scattering Experiment • Results of Alpha-Particle Scattering Experiment • Isotopes

PLANNING GUIDE

Section	Time class	Student Edition Features	Program Resources	Media and Technology
Section B. Nuclear Radiation **Objectives** • Radiation emitted by naturally occurring radioisotopes is the source of background radiation in the environment. • Alpha, beta, and gamma radiation differ in properties, including the extent of shielding required to stop the radiation. Three key factors in minimizing exposure to radiation are shielding, exposure, and proximity. • Ionizing radiation has sufficient energy to break chemical bonds. • The emission of nuclear radiation changes the composition of the nucleus. These changes are represented by nuclear equations. • Ionizing radiation may be detected by its interaction with matter using a variety of detectors such as unexposed photographic film, scintillation counters, solid-state detectors, cloud chambers, or Geiger-Mueller counters.	6 Periods	**Building Skills** #3. Nuclear Balancing Act **ChemQuandary** #2. Radiation Exposure Standards **Making Decisions** • Your Annual Ionizing Radiation Dose • Ensuring Public Safety **Laboratory Activity** • B.4 Alpha, Beta, and Gamma Radiation • B.8 Cloud Chamber	**Activity Book** • B.2 Making Decisions: Supplement: Common Sources of Radiation • B.4 Alpha, Beta, and Gamma Radiation • B.5 Building Skills: Supplement: Nuclear Balancing Act • B.8 Cloud Chambers	**ChemCom CD-ROM** • Unit VI Air—Section B • Laboratory Video **Transparencies** • U.S. Sources of Ionizing Radiation • Distance vs. Source and Intensity of Radiation • Alpha Particle Emission • Changes Resulting from Nuclear Decay • Relative Penetrating Power of Rays • U-238 Decay Series

PLANNING GUIDE

Section	Time class	Student Edition Features	Program Resources	Media and Technology
Section C. Using Radioactivity **Objectives** • Half-life is a measure of the time required for a radioisotope to decay. These essentially unchangeable half-lives vary from element to element. • Radioisotopes can be used as tracers for diagnostic purposes. • Ionizing radiation emitted by some radioisotopes can be used to kill cancerous cells. • The artificial conversion or transmutation of one element to another can be accomplished by bombarding nuclei with subatomic particles or other nuclei.	5 Periods	**Building Skills** #4. Half-Lives #5. Bombardment Reactions **ChemQuandary** #3. Carbon-14 Dating #4. Using Radioisotopes in Medicine #5. Transmutation of Elements **Modeling Matter** • Understanding Half-life **Making Decisions** • Opinions About Radioactivity	**Activity Book** • C.1 Building Skills: Supplement: Half-Lives • C.4 Building Skills: Supplement: Bombardment Reactions	**ChemCom CD-ROM** • Unit VI—Atoms: Section C • Laboratory Video **Transparencies** • Half-Lives of Selected Radioisotopes • Selected Medical Radioisotopes
Section D. Nuclear Energy—Benefits and Burdens **Objectives** • Certain large nuclei, upon bombardment with neutrons, split into two smaller nuclei and several additional neutrons. This process is referred to as nuclear fission.	5 Periods	**Building Skills** • High- and Low-Level Waste Disposal **Modeling Matter** • The Domino Effect **ChemQuandary** #6. Risk Free Travel? **Making Decisions** • The Safest Journey		**ChemCom CD-ROM** • Unit VI—Atoms: Section D **ChemCom Connections Videodisc** • Electrical Energy From Nuclear Fission **Transparencies** • Nuclear Chain Reaction • Nuclear Power Plant • Mined Geologic Disposal Plan

PLANNING GUIDE

Section	Time class	Student Edition Features	Program Resources	Media and Technology
• The strong force holds the nucleus of an atom together. Nuclear reactions result in a small loss of mass, which is converted into large quantities of energy. • A critical mass of fissionable material is required to sustain a chain reaction. • The electricity produced by a nuclear power plant originates from the energy released by fission of U-235 in a controlled chain reaction. • Nuclear fusion results from the combination of two relatively small nuclei into a new, more massive atom. This process, which powers the Sun and other stars, requires extremely high temperature and pressure. • Permanent disposal of nuclear waste poses problems related to the volume, level of radioactivity, and half-lives of radioisotopes present in the waste.				
Putting It All Together • The Truth about Nuclear Chemistry			**Black Line Copy Master** • PIAT Evaluation	

SUPPLY LIST FOR UNIT 6

Expendable Items and Unusual Materials per 24 students working in pairs

Alphabetical by Item Name			
Quantity	Unit	Item	Sections Used
100	mL	Alcohol (methyl, ethyl, or isopropyl)	B.8
12	Set	Alpha, beta, and gamma radiation sets	B.4
12	Each	Balance	MM Section A
12	Each	Bottles, squeeze (for alcohol)	B.8
12	Each	Cloud chamber kit	B.8
12	Each	Container or box (plastic or cardboard)	MM Section C
12	Each	Crucible tongs	B.8
12	Set	Dominoes	MM Section D
10	Pounds	Dry ice	B.8
12	Each	Forceps	B.4, B.8
24	Each	Glass plates (about 5-cm square, 0.1-cm thick)	B.4
1	Box	Gloves, disposable	B.8
1	Package	Graph paper	B.4
12	Each	Index cards, 3″ × 5″ (or cut as necessary)	B.4
24	Each	Lead plates (about 5 cm square, 0.1 cm thick)	B.4
12	Each	Light, high intensity (bright flashlight)	B.8
12	Each	Meter sticks and rulers	B.4
1200	Each	Pennies	MM Section C
72	Each	Pennies, post-1982	MM Section A
72	Each	Pennies, pre-1982	MM Section A
12	Each	Radiation counter	B.4
12	Each	Radiation source (usually part of cloud chamber)	B.8
12	Each	Sealed opaque containers (possibly film canisters)	MM Section A

SUPPLY LIST

Expendable Items and Unusual Materials per 24 Students working in pairs.

By Section			
Quantity	**Unit**	**Item**	**Sections Used**
12	Set	Alpha, beta, and gamma radiation sets	B.4
24	Each	Glass plates (about 5 cm square, 0.1 cm thick)	B.4
1	Package	Graph paper	B.4
12	Each	Index cards, 3" x 5" (or cut as necessary)	B.4
24	Each	Lead plates (about 5-cm square, 0.1-cm thick)	B.4
12	Each	Meter sticks and rulers	B.4
12	Each	Radiation counter	B.4
12	Each	Forceps	B.4, B.8
100	mL	Alcohol (methyl, ethyl, or isopropyl)	B.8
12	Each	Bottles, squeeze (for alcohol)	B.8
12	Each	Cloud chamber kit	B.8
12	Each	Crucible tongs	B.8
10	Pounds	Dry ice	B.8
1	Box	Gloves, disposable	B.8
12	Each	Light, high intensity (bright flashlight)	B.8
12	Each	Radiation source (usually part of cloud chamber)	B.8
12	Each	Balance	MM Section A
72	Each	Pennies, post-1982	MM Section A
72	Each	Pennies, pre-1982	MM Section A
12	Each	Sealed opaque containers (possible film canisters)	MM Section A
12	Each	Container or box (plastic or cardboard).	MM Section C
1200	Each	Pennies	MM Section C
12	Set	Dominoes	MM Section D

UNIT 6
ATOMS: NUCLEAR INTERACTIONS

Introduction to Unit

Atoms: Nuclear Interactions confronts one of the most emotionally charged issues in modern society, the use of nuclear energy. The unit traces the history and development of nuclear energy from the discovery of radioactivity to modern-day reactors and fusion, with strong emphasis on exploring the risks and benefits of nuclear technologies.

The goal here is to provide a firm background on the nature of atoms by providing the basic facts and principles related to nuclear science and technology. It is also important that students use these facts and principles to weigh the potential benefits of the technology against the harm or threats it poses to individuals, society, and the environment.

The PIAT involves students making a presentation to a senior citizens group about nuclear science and technology. The purpose is to enable senior citizens to make informed decisions about nuclear energy.

UNIT 6

WHAT discoveries led to modern understanding of the composition and structure of atoms?

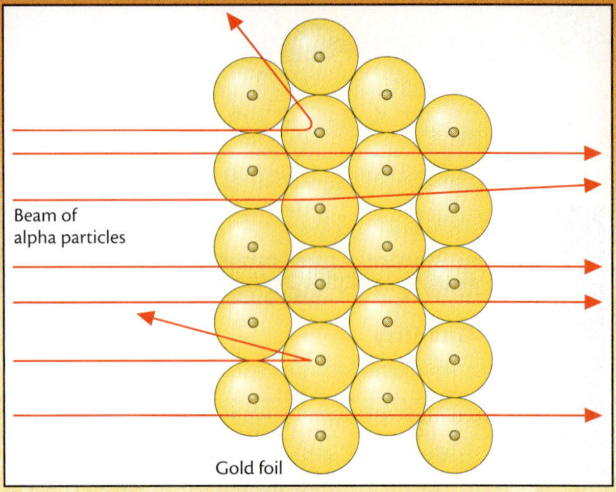

SECTION A **The Nature of Atoms** (page 412)

WHY does exposure to some types of radiation cause health problems in humans?

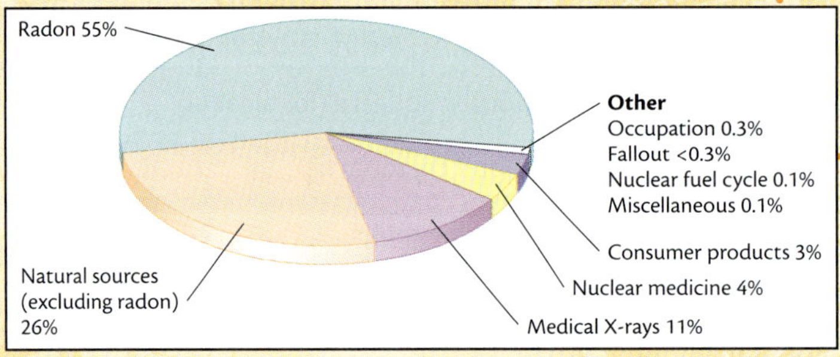

SECTION B **Nuclear Radiation** (page 428)

ATOMS: NUCLEAR INTERACTIONS

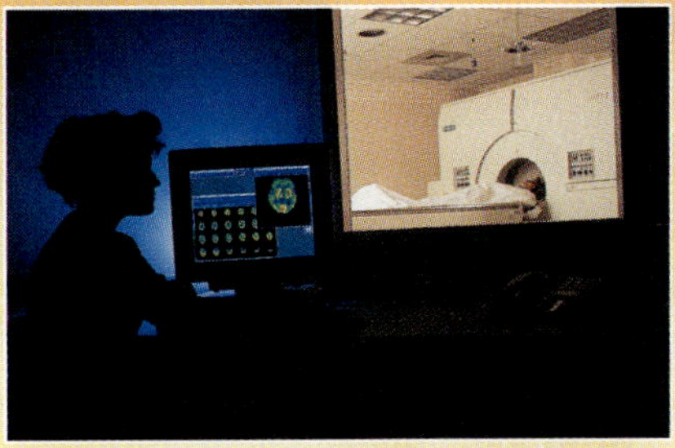

HOW does the rate of radioactive decay influence decisions about using nuclear radiation?

WHAT balance of risks and benefits does nuclear energy production pose?

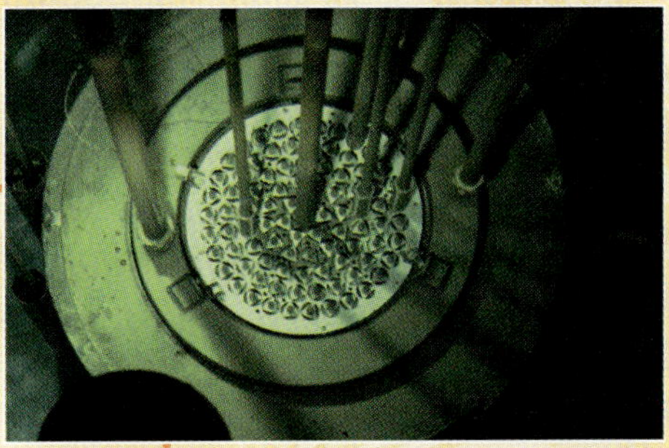

A citizen's group wants to ban all nuclear radiation from Riverwood. Turn the page to learn what you might contribute to community discussions of the group's proposals.

THE NATURE OF ATOMS

Introduction

Students are introduced to the processes associated with changes in the nuclei of atoms, namely, nuclear radiation and nuclear energy. Remind students of the general *ChemCom* theme, that scientific understanding alone will not provide the answers to complex societal problems and technological issues, but it can provide a more rational basis for public debate and action.

From a discussion of the work of Roentgen, Rutherford, and Becquerel, students learn that certain scientific discoveries are accidental and that in searching for the answer to one question, scientists often stumble upon another question that is even more interesting.

Citizens Against Nuclear Technology (CANT)
invite you to an informative seminar on banning the use of all nuclear energy and materials in the Riverwood area.
See you Friday, 7:00 p.m., Town Council Hall

Do you know whether each statement below is true or false? If not, you could be in danger of serious nuclear exposure!

1. Home smoke detectors contain radioactive materials.
2. Radioactive materials and radiation are unnatural—they did not exist on Earth until created by scientists.
3. All radiation causes cancer.
4. Human senses can detect radioactivity.
5. Individuals vary widely in how they are affected by exposure to radiation.
6. Small amounts of matter change to immense quantities of energy in nuclear weapons.
7. Physicians can distinguish cancer caused by radiation exposure from cancer resulting from other causes.
8. Medical X-rays are dangerous.
9. Nuclear power plants create serious hazards to public health and the environment.
10. An improperly managed nuclear power plant can explode like a nuclear weapon.
11. Some nuclear wastes must be stored for centuries to prevent dangerous radioactivity from escaping.
12. New, dangerous elements are being invented every day.
13. Nuclear power plants produce material that could be converted into nuclear weapons.
14. All nuclear medical techniques are highly dangerous.

410

All the chemistry you have learned thus far has involved chemical changes—those changes due to sharing or transferring outer electrons among atoms. You will encounter changes of a much different sort in this unit—those associated with the nuclei of atoms. This unit looks closely at nuclear radiation, radioactivity, and nuclear energy as well as implications of their use and development.

"Nuclei" is the plural of "nucleus."

Nuclear energy is not new. The Sun, like all stars, has always run on nuclear power. What might be regarded as new about nuclear energy, however, is how humans use it. Knowing the nature of the nucleus, scientists and engineers are able to harness the Universe's strongest known force. Nuclear science makes contributions to industry, to biological research, to energy production, and especially to medicine. However, the production and use of nuclear energy, like all other applications of technology, involve burdens and risks as well as potential benefits. This unit will help you evaluate these benefits and risks and make decisions about appropriate roles of nuclear technology in modern society.

In this unit you return to the town of Riverwood. This time a local organization, Citizens Against Nuclear Technology (CANT), has been formed to prevent all uses of nuclear power, the disposal of nuclear waste, food irradiation, and even nuclear medicine, in the Riverwood area. The flyer on page 410 is an example of the organization's communication with Riverwood residents.

Some Riverwood citizens have become concerned about the restrictions proposed by CANT, afraid that such restrictions would hinder access to a full range of medical diagnosis and treatment options. The grandmother of Riverwood High School chemistry teacher Lynn Paulson is among those concerned. She asked Paulson if any chemistry students at the high school would be willing to provide some background about nuclear science and technology to senior citizens. They plan to attend the announced CANT seminar but would like to acquire some background and be ready to ask seminar representatives questions about their proposals. Paulson agreed to help. In this unit you and your classmates will play the part of Riverwood High School chemistry students preparing for their presentation.

Given the questions raised by the flyer, as you advance through this unit, keep track of ideas and applications you want to share in your presentation. Consider issues that would be meaningful to your audience, why such topics would be meaningful, and how the science behind those issues can help clarify them.

411

A.1 THE GREAT DISCOVERY

This section provides a historical background to the discovery of radioactivity, including the often-startling revelation that Becquerel's crucial experiment, the exposure of photographic film by uranium, was something of an accident. However, Becquerel's knowledge and attitude led him to investigate the accident and make an important discovery.

The discovery of radioactivity also demonstrates how scientific inquiries are often interrelated. In this case, Roentgen's X-rays led Becquerel to seek X-rays from fluorescent materials such as uranium. Becquerel's plates were exposed by the uranium salts. Photographic plates are composed of silver halides (chloride, bromide, or iodide), salts that are sensitive to radiant energy. Any radiation more energetic than infrared radiation will cause a photochemical reaction, leading to silver ions being reduced to silver atoms during photographic development. This reduction results in the darkening of a photographic plate. Thus, uranium salts must have given off some type of radiant energy, as hypothesized by Becquerel.

SECTION A

THE NATURE OF ATOMS

Introduction CD-ROM WWW.

Determining and describing the internal structure of atoms ranks among the great scientific accomplishments of the twentieth century. The story of these discoveries will help you understand atomic and nuclear chemistry and how the methods of science have been employed to gain this knowledge.

In this section, you will find out how both well-planned experiments and chance observations led to the modern model of the atom and to the discovery of radioactivity.

A.1 THE GREAT DISCOVERY

The history of modern scientific investigation of the atom began with the study of radiation. Scientists have long been interested in light—because of its essential role in life—and other types of radiation. See Unit 4, page 276. By the close of the nineteenth century, many types of radiation had already been extensively studied.

In 1895, however, a series of observations significantly broadened scientists' understanding of radiation. The German physicist W. K. Roentgen was studying **fluorescence,** the phenomenon in which certain materials emit light when struck by radiant energy such as ultraviolet rays. Roentgen found that certain materials fluoresced when exposed to beams of cathode rays. They were known as **cathode rays** because they were emitted from the cathode when electricity passed through an evacuated glass tube such as the one shown in Figure 1. A few years after Roentgen's work, cathode rays were demonstrated to be beams of electrons.

Fluorescent minerals glow when illuminated by ultraviolet radiation.

Computer monitors and televisions are based upon a modern version of the cathode ray tube.

While working with a cathode ray tube shielded with black cardboard, Roentgen observed an unexpected glow of light coming from a piece of paper across the room. The paper, coated with a fluorescent material, would be expected to glow when exposed to radiation. However, visible radiation could not pass through the black cardboard covering the cathode ray tube and the fluorescent paper was not in the path of electrons from the tube. Roentgen concluded that some other radiation that could pass through the black shielding cardboard had been emitted by the cathode-ray tube. He named the mysterious radiation **X-rays,** using "X" to represent the unknown.

In modern X-ray devices, the X-rays are generated when an electron beam strikes a metal target, often made of tungsten.

Further experiments revealed that these X-rays could penetrate many materials, but could not easily pass through dense materials such as lead and bone. Scientists now know that X-rays are a form of high-energy electromagnetic radiation. Scientists soon realized how useful X-rays could be in medicine. In fact, the first X-ray photograph that Roentgen took was of his wife's hand. Figures 2 and 3 (page 414) show some modern X-ray images.

Roentgen's discovery excited other scientists. Soon many other researchers, including the French physicist Henri Becquerel, began studying

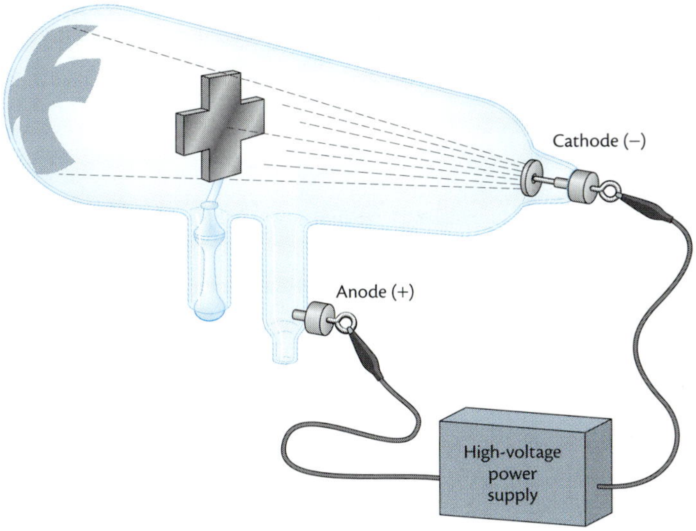

Figure 1 *A cathode-ray tube. The shadow demonstrates that the cathode rays are emitted by the cathode, instead of the anode, and that cathode rays are stopped by metal. The sharpness of the shadow demonstrates that the rays travel in straight lines.*

this new form of radiation. Since X-rays could produce fluorescence, Becquerel wondered if fluorescent minerals might give off X-rays. In 1896, he placed in sunlight some crystals of a fluorescent mineral that happened to contain uranium. He then wrapped a photographic plate in black paper and placed the mineral crystals on top of the wrapped plate. If the mineral did emit X-rays as it fluoresced, the X-rays would penetrate the black paper and the film would darken, even though it was shielded from light.

Following some initial successes, Becquerel was prevented by cloudy weather from completing more experiments. He stored the wrapped photographic plates in a drawer together with the uranium-containing mineral.

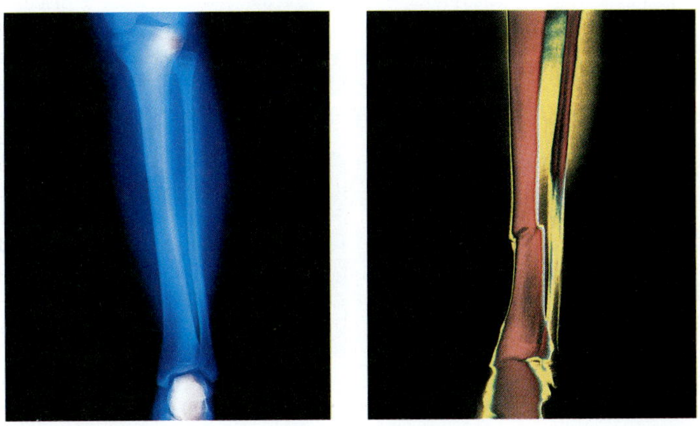

Figure 2 *These X-rays show a healthy human leg (left) and a broken leg (right).*

Some common objects, such as certain types of Frisbees and light sticks, fluoresce, or glow in the dark, when activated. These objects contain compounds that are excited by ultraviolet light and re-emit the absorbed energy as visible light. Diverse materials, such as laundry additives that get clothes "whiter than white" and certain flowers, also illustrate the principle of absorbing ultraviolet light and emitting light of a different energy. A white shirt washed in such detergents "glows" when exposed to ultraviolet light.

TV screens and computer monitors (CRTs) also demonstrate fluorescence. They apply the idea of using a beam of electrons to excite certain substances. As the substances return to a more stable state, energy may be released in the form of X-rays. TV picture tubes and X-ray machines are both CRTs that have been modified to decrease and increase the emission of X-rays, respectively.

Focus attention on the medical benefits of X-rays, using Figures 2 and 3 (pages 413–414) and the students' own experiences.

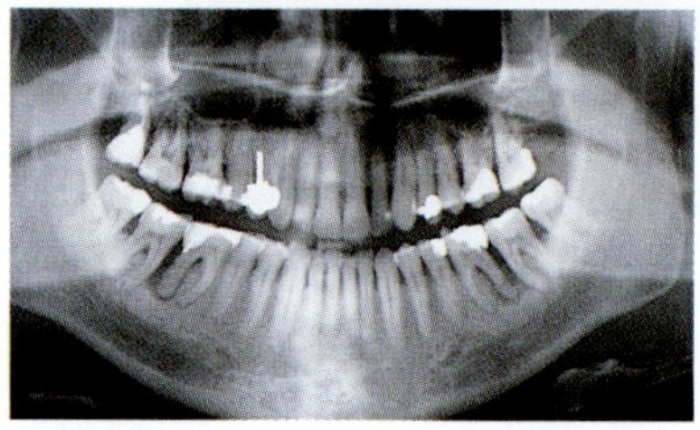

Figure 3 *A dental X-ray photograph of a human jaw. Such X-rays are useful for detecting cavities and other dental problems.*

Figure 4 *Becquerel's investigation. (a) After sitting on a window sill, the radioactive mineral sample was placed on a blank and wrapped photographic plate (b). The radiation exposed the film on the plate (c). Even when the wrapped plate was placed in a drawer and kept from light (d), the mineral sample exposed the film on the plate (e).*

After several days, he decided to develop some of the stored plates on the chance that some mineral fluorescence might have persisted, causing some fogging of the plates. But when Becquerel developed the plates from the drawer, he was astounded. Instead of faint fogging, the plates had been strongly exposed. Figure 4 illustrates the chain of events in Becquerel's investigation.

Fluorescence stops as soon as the external source of radiation (in this case, the Sun) is removed. Thus, a fluorescent mineral in a darkened desk drawer could not cause such a high level of exposure. At that time no satis-

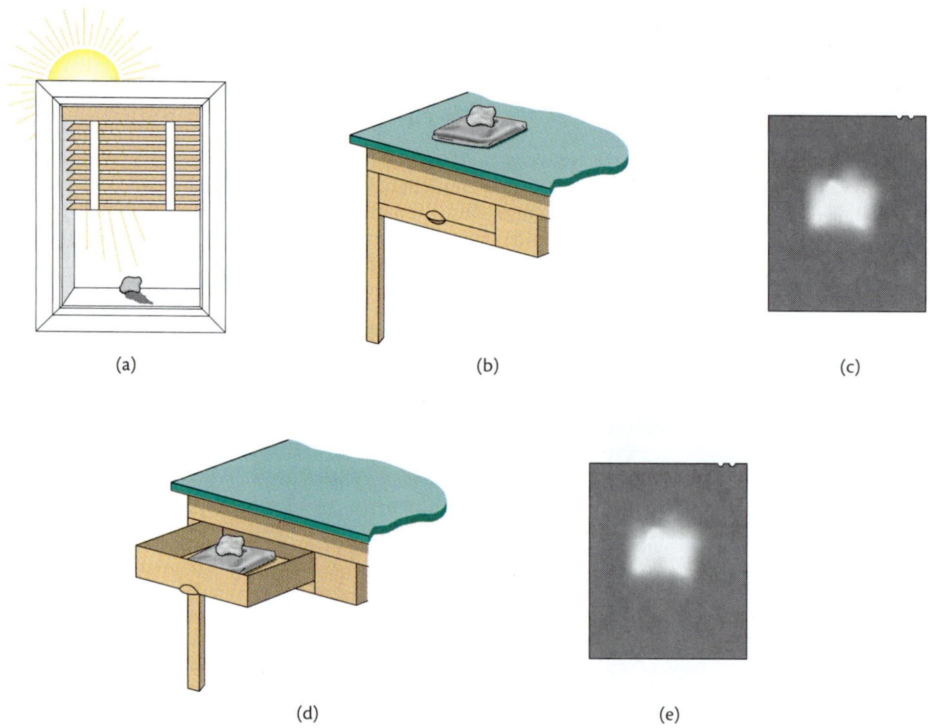

414

factory explanation could be offered for those observations. Becquerel suspected that the rays that exposed the photographic plates in the drawer were more energetic and possessed much greater penetrating power than X-rays. He interrupted his X-ray work to study the mysterious radiation apparently given off by the fluorescent uranium-containing mineral. Although he was unable to explain it, Becquerel had discovered the phenomenon now known as **radioactivity**—the spontaneous emission of nuclear radiation. As scientists know now, radioactivity originates in the nucleus of an atom, making it distinctly different from X-ray production and fluorescence.

SCIENTIFIC DISCOVERIES — ChemQuandary 1

What do the events described below have in common with Becquerel's discovery of radioactivity?

1. Charles Goodyear was experimenting with natural rubber (a sticky material that melts when heated and cracks when cold) when he accidentally allowed a mixture of rubber and sulfur to touch a hot stovetop. He noted that the rubber-sulfur mixture did not melt. Vulcanization, the process that makes rubber more widely useful by modifying its properties, resulted from this observation.
2. Research chemist Roy Plunkett used gaseous tetrafluoroethene ($F_2C{=}CF_2$) from a storage cylinder, but the gas flow stopped long before the cylinder was empty. He cut open the cylinder and discovered a new, white solid. Today this solid is known as Teflon (polytetrafluoroethene).
3. James Schlatter, a research chemist working to produce an anti-ulcer drug, inadvertently got some of the substance on his fingers. When he later licked his fingers to pick up a piece of paper, his fingers tasted very sweet, and he correctly linked the sweetness to the anti-ulcer drug. Instead of finding an anti-ulcer drug, he discovered aspartame (NutraSweet), now used as an artificial sweetener.

Becquerel suggested to Marie Curie, a graduate student working with him, that she attempt to isolate the radioactive component of pitchblende, a uranium ore, as her Ph.D. research. Her preliminary work was successful, and her physicist husband, Pierre Curie, changed his research work to join her on the pitchblende project. Working together, Marie and Pierre Curie discovered that the level of radioactivity in pitchblende was four to five times greater than expected from its known uranium content. The Curies suspected the presence of another radioactive element. After processing more than a thousand kilograms of pitchblende, they isolated tiny amounts (milligrams) of two previously unknown radioactive elements—polonium (Po) and radium (Ra).

Further research revealed that all uranium-containing compounds are radioactive. Interest in radioactivity continued to grow among scientists. Some realized that a better understanding of this radiation might provide clues to the structure of atoms. As you will learn, Ernest Rutherford was one scientist who was particularly successful in penetrating the mysteries of the atom.

SCIENTIFIC DISCOVERIES

ChemQuandary 1

Each discovery was similar to Becquerel's uranium work in that something unexpected happened. In all cases, the observed events were discrepant with current scientific explanations. Becquerel did not expect the plates to be exposed, nor did Goodyear expect sulfur to harden the rubber. Plunkett did not expect to find a polymer of tetrafluoroethane, nor did Schlatter expect to find artificial sweetener when he licked his fingers. The point to be made is that careful observation is needed to reveal deeper scientific principles from seemingly ordinary events or mistakes. Pasteur's statement, "Chance favors the prepared mind," is good advice for all scientists, present and future. These cases may be used for student library research reports.

A.2 NUCLEAR RADIATION

The electromagnetic spectrum is used to compare and contrast ionizing and nonionizing radiation by focusing on the energies of the various types of radiation. Make sure that students are familiar with the differences highlighted in this section.

SUPPLEMENTARY ACTIVITY: THE BLACK BOX

This activity simulates how scientists use indirect evidence to infer properties that cannot be perceived by our senses. The focus of this activity is to have students use reasoning to solve a problem.

Time

30–45 minutes

Materials (for a class of 24 working in pairs)

Twenty-four sealed, coded boxes (cigar or shoe boxes), each containing three unconstrained objects such as small scissors, small rubber ball, small fork, plastic comb, large paper clip, or small key. The use of common laboratory materials such as test tube holders, flame spreaders, or rubber stoppers is also recommended. Commercially available black boxes could also be used.

Before the activity begins, discuss how the study of atomic structure is analogous to a black box. A brief preview of how scientists discovered alpha, beta, and gamma rays is useful here.

Post-activity discussion

Focus attention on how the activity simulated the work of atomic scientists rather than on students who got right answers. Ask students to explain how they arrived at their answers.

A.2 NUCLEAR RADIATION

To many people, anything associated with the word "nuclear" and its companion term "nuclear radiation" is cause for alarm—sometimes even panic. Before you learn more about the origins and effects of nuclear radiation, it will be helpful to understand that there are different types of radiation.

All radiation can be classified as either ionizing or nonionizing. Electromagnetic radiation in the visible and lower-energy regions (see Figure 5) is **nonionizing radiation.** Nonionizing radiation has much lower energy than ionizing radiation. Nonionizing radiation transfers its energy to matter, causing atoms or molecules to vibrate (infrared radiation) or move their electrons to higher energy levels (visible radiation). In some cases, chemical reactions occur as energy from this radiation is transferred to molecules. Reactions caused by microwave ovens in cooking food are one example. Nonionizing radiation is also involved in radio and TV reception and electric light bulbs. Excessive exposure to nonionizing radiation can be harmful. Sunburn, for example, results from overexposure to nonionizing radiation from the Sun. In fact, intense microwave and infrared radiation can cause lethal burns.

Ionizing radiation, which includes all nuclear radiation and other high-energy electromagnetic radiation (short-wavelength ultraviolet radiation, X-rays, and gamma rays), carries more energy and potential for harm than nonionizing radiation. Energy from ionizing radiation can eject electrons from atoms and molecules, forming molecular fragments and ions. These fragments and ions can be highly reactive; if formed within a living system they can cause disruption of normal cellular chemistry. Ionizing radiation can cause serious damage to cells. Thus, exposure to ionizing radiation should—at best—be limited.

Nuclear radiation is a form of ionizing radiation that results from changes in the nuclei of atoms. In chemical reactions the atomic number (number of protons) does not change—an atom of aluminum (13 protons) always remains aluminum, and an iron atom (26 protons) is always iron in chemical changes. However, atoms with unstable nuclei—radioactive atoms—can spontaneously change their identities. Such atoms change spontaneously through emission of nuclear radiation. Usually, radioactive

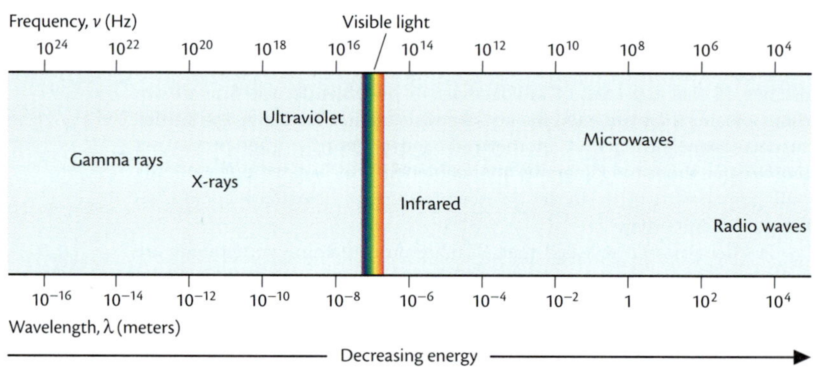

Figure 5 *The electromagnetic spectrum.*

atoms change to produce an atom of a different element (that is, one with a different number of protons) by emitting a particle and energy. This process is referred to as **radioactive decay;** the emitted particles and energy together make up nuclear radiation.

In 1899 Ernest Rutherford showed that nuclear radiation included at least two different types of rays, which he named **alpha rays** and **beta rays.** He placed thin sheets of aluminum in the pathway of radiation from uranium and found that beta rays penetrated more layers of the sheets than did alpha rays. Shortly afterward, another kind of radiation produced by radioactive elements was discovered. This third type of nuclear radiation was called **gamma rays.**

The three types of nuclear radiation were passed through magnetic fields to investigate their electrical properties. Scientists already knew that when charged particles move through a magnetic field they are deflected—their paths are altered—by the magnetic force. They knew that positively charged particles are deflected in one direction, while negatively charged particles are deflected in the opposite direction, and electrically neutral particles and electromagnetic radiation are not deflected by magnetic fields. See Figure 6. Experiments with nuclear radiation in magnetic fields revealed that alpha rays were composed of positively charged particles and beta rays were made up of negatively charged particles. Gamma rays, which were not deflected by a magnetic field, had no electric charge and were later identified as high-energy electromagnetic radiation, similar to X-rays.

Thus, the nature of radioactivity was established. However, as often happens in science, a new discovery toppled an old theory—in this case, that atoms were the smallest, most fundamental units of matter. Once alpha, beta, and gamma rays were identified, scientists became convinced that atoms must be composed of smaller particles.

From the results of another experiment, described below, Rutherford proposed a fundamental model of an atom that remains useful even today. To do so, he developed an ingenious, indirect way to "look" at the structure of atoms.

Alpha (α), beta (β), and gamma (γ) are the first three letters of the Greek alphabet.

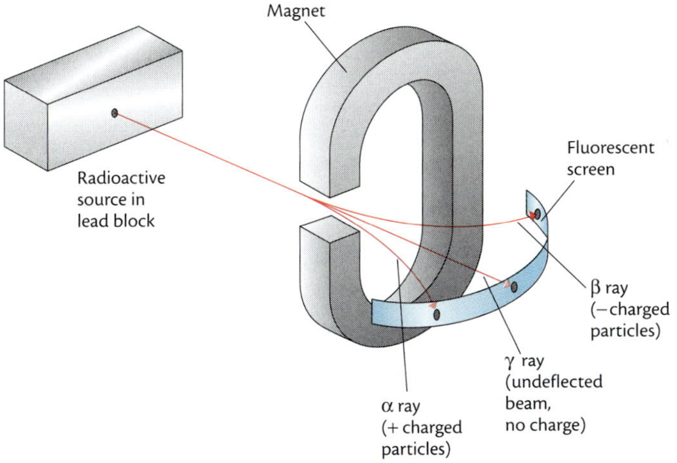

Figure 6 *Behavior of alpha (α), beta (β), and gamma (γ) rays in a magnetic field.*

Questions

1. Which of your senses did you use to collect the data?
2. In what ways does this activity resemble the efforts of scientists in exploring atomic and molecular structures?
3. Name some theories about the nature of the world that are based primarily on indirect evidence.

Answers

1. Hearing.
2. The structure of the atom is like a black box that scientists cannot directly open in order to examine its contents.
3. Indirect evidence is used to formulate explanations (theories) about the natural world whenever the events examined occur beyond the range of our perceptions. The time for events to occur may be too short (kinetic molecular theory) or too long (evolution), or the objects of interest may be too small (atomic theory) or too large (the universe).

Demonstration ideas

An interesting variation on the black box involves suspending a disposable object (key, paper clip, fork, comb, scissors, rubber ball, etc.) inside a sealed shoe box. Cut rigid construction paper strips (white) 50–55 cm long by 7 cm wide to fit through slits in the back of the box, to serve as an imaging screen. Spray paint through a hole in the front of the box to create a "shadow image" of the concealed object. The demonstration reinforces the notion of using a source and a target, and of shooting something at the object in the atomic black box to acquire information about its contents.

A.3 THE GOLD-FOIL EXPERIMENT

This section provides an example of how experimental observations are sometimes unexpected when they are based upon a current model. The discovery of the nucleus was not the purpose of the gold foil experiment. Rutherford was investigating the behavior of alpha particles with metals. The unusual deflections prompted him to propose the existence of a very massive and very small particle—the nucleus. To help students visualize the physical arrangement of the experiment, focus attention on Figure 7. A transparency of the experiment maybe beneficial.

In the late 1800s this was known as the "plum pudding model," because it resembled the distribution of raisins within that traditional English dish.

Alpha rays and beta rays are more commonly referred to as particles since they have measurable mass and do represent forms of electromagnetic radiation.

A.3 THE GOLD-FOIL EXPERIMENT

Prior to Rutherford's research, the arrangement of electrons and positively charged particles in atoms had been explained in several ways. In the most widely accepted model, an atom was viewed as a volume of positive electrical charge, with the negatively charged electrons embedded within, like peanuts in a candy bar.

Working in Rutherford's laboratory in Manchester, England, Hans Geiger and Ernest Marsden focused a beam of alpha particles—the most massive of the three types of nuclear radiation—at a sheet of gold foil 0.00004 cm (about 2000 atoms) thick. Geiger and Marsden used a small zinc sulfide screen to observe the alpha particles after they passed through the gold foil (Figure 7). The zinc sulfide-coated screen emitted a flash of light where each alpha particle struck it. By observing the tiny light flashes at different positions with respect to the gold foil, the researchers were able to deduce the paths of the alpha particles as they interacted with the gold foil.

Rutherford expected the alpha particles to scatter slightly as they were deflected by the gold atoms in the foil, producing a pattern similar to water spraying from a nozzle. He was in for quite a surprise. First, most alpha particles passed straight through the gold foil as if nothing were there (Figure 8). This implied that most of the volume taken up by the gold atoms was essentially empty space. But what surprised Rutherford even more was that a few alpha particles—about one in every 20 000—bounced back toward the source. Rutherford described his astonishment: "It was about as incredible as if you had fired a 15-inch shell at a piece of tissue paper and it came back and hit you."

Whatever repelled these few deflected alpha particles must have been extremely small, because most alpha particles went straight through the foil. But whatever the particles encountered must also have been quite massive (compared to the alpha particles) and most likely possessed a positive electrical charge, since it repelled positively charged alpha particles.

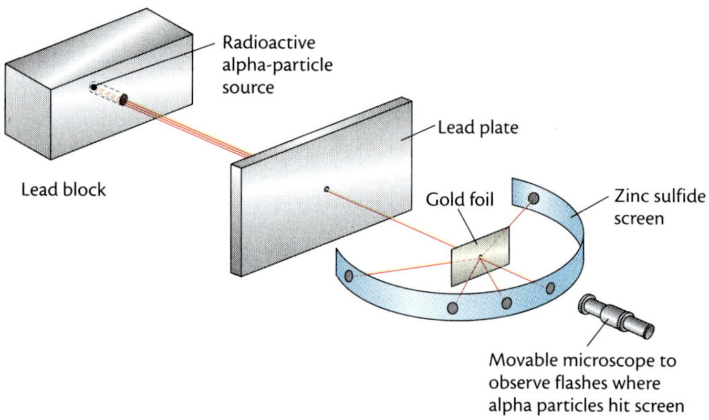

Figure 7 *The alpha-particle scattering experiment.*

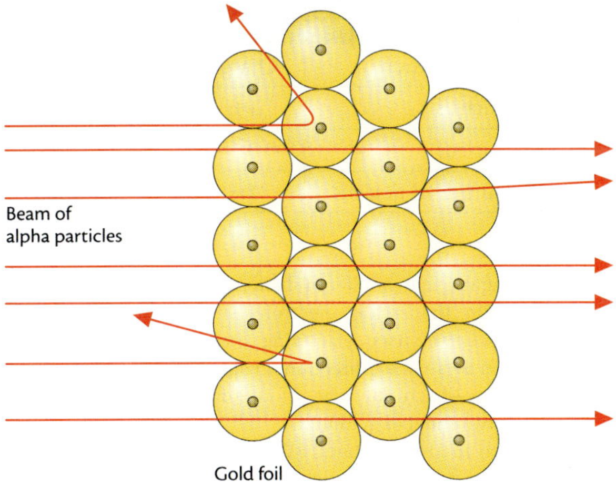

Figure 8 *Result of the alpha-particle scattering experiment. Most alpha particles passed through the foil; a few were deflected, some at very large angles.*

From these results, Rutherford developed the modern model of the nuclear atom. He called the tiny, dense, positively charged region at the center of the atom the nucleus. He envisioned that electrons orbited around the nucleus, somewhat like planets orbit the Sun.

A.4 ARCHITECTURE OF ATOMS

Since Rutherford's time, scientific understanding of atomic structure has expanded and changed in some ways. Rutherford's image of a central, dense nucleus surrounded mostly by empty space is still valid. From subsequent research it is also known that the idea of "orbiting electrons"—although useful—is incorrect. The general regions in which electrons are most likely to be found can be described, but particular movements or locations of electrons cannot.

Further research revealed that the nucleus is composed of **nucleons** (nuclear particles) of two types—neutrons, which are electrically neutral, and protons, possessing a positive charge. These particles—as well as electrons—are known as **subatomic particles.** Both types of nucleons have about the same mass, 1.7×10^{-24} g. Although this mass is incredibly small, it is still much larger than the mass of an electron, 9.1×10^{-28} g. As shown in Figure 9 (page 420), a mole of electrons (6.02×10^{23} electrons) has a mass of about 0.0005 g. The same number of nucleons would have a total mass of about 1 g. In other words, a proton or a neutron is about 2000 times more massive than an electron. Thus nucleons account for nearly all of the mass of every atom and also for nearly all of the total mass of every object you encounter.

The diameter of a typical atom is about 10^{-10} m, but an average nuclear diameter is 10^{-14} m, only one ten-thousandth ($10^{-14}/10^{-10}$) as large as the diameter of the entire atom. Looking at this another way, the nucleus occupies only about one trillionth (10^{-12}) of the volume of an atom. As a

Neutrons are actually very slightly more massive than protons.

Protons—positive electrical charge
Neutrons—no electrical charge
Electrons—negative electrical charge

A.4 ARCHITECTURE OF ATOMS

The components of the nucleus are identified and isotopes are explained in terms of their relationship to mass numbers. Nuclear notation can be used to help identify the number of neutrons in a given isotope. Analogies can help students gain a sense of atomic scale. There are several points to emphasize:

- Atoms are very small. If people were the size of atoms, the entire population of the world (approximately six billion) would fit on the head of a pin.
- Most of the volume of an atom is essentially empty space. If the nucleus were the size of a billiard ball, the average electron would be located about 1.6 km away.
- Protons and neutrons account for most of the mass of the atom because their masses are approximately 2000 times greater than that of electrons.
- Encourage students to come up with other appropriate analogies. A brief review of the structure of atoms vs. ions may be beneficial at this point.

Figure 9 *The proton, neutron, and electron are subatomic particles.*

Three Components of an Atom			
Particle	**Location**	**Charge**	**Molar Mass (g/mol)**
Proton	Nucleus	1+	1
Neutron	Nucleus	0	1
Electron	Outside of nucleus	1−	0.0005

comparison, imagine that a billiard ball represents the diameter of an atom's nucleus. On that scale, electrons surrounding this billiard-ball nucleus would occupy space extending more than one-half kilometer (about a half-mile) away in all directions. That model explains why most of the alpha particles—each the size of a helium nucleus—passed right through Rutherford's sheet of gold foil.

In most Periodic Tables, the atomic number is written above the symbol of the element.

As you learned in Unit 2 (page 101), each atom of an element has the same number of protons in its nucleus, and each element has a unique number of protons. This number, called the atomic number, identifies the element. For example, since each carbon atom nucleus contains six protons, carbon has atomic number 6.

However, all atoms of one element do not necessarily have the same number of neutrons in their nuclei. Atoms of the same element having different numbers of neutrons are called isotopes of that element. Naturally occurring carbon atoms, each containing six protons, may have six, seven, or even eight neutrons. The composition of these three carbon isotopes is summarized in Figure 10.

Isotopes are distinguished by their different mass numbers. The mass number, as you learned in Unit 2 (page 102), represents the total number of nucleons (protons and neutrons) in an atom. The three carbon isotopes in Figure 10 have mass numbers of 12, 13, and 14. To specify a particular isotope, the atomic number and the mass number are written in front of the element's symbol in a particular way. For example, an isotope of strontium (Sr) with an atomic number of 38 and a mass number of 90 is written this way:

$$^{90}_{38}\mathrm{Sr}$$

A 2+ ion of strontium-90 would be shown as

$$^{90}_{38}\mathrm{Sr}^{2+}$$

Three Carbon Isotopes				
Name	**Total Protons (Atomic Number)**	**Total Neutrons**	**Mass Number**	**Total Electrons**
Carbon-12	6	6	12	6
Carbon-13	6	7	13	6
Carbon-14	6	8	14	6

Figure 10 *Isotopes of an element have the same number of protons, but different numbers of neutrons.*

420

Another way to identify a particular isotope is to write the name or symbol of the element followed by a hyphen and the mass number. For example, the isotope depicted above is called strontium-90, or Sr-90. The symbols, names, and nuclear composition of some isotopes are summarized in Figure 11.

The atomic number is a lower-left subscript; the mass number is an upper-left superscript.

ISOTOPE NOTATION

Building Skills 1

Suppose you know that one product of a certain nuclear reaction is an isotope containing 85 protons and 120 neutrons. It therefore has a mass number of 205 (85 protons + 120 neutrons = 205 nucleons). What is the symbol of this element?

$$^{205}_{85}?$$

Consulting the Periodic Table, you can see that astatine (At) is element 85.

$$^{205}_{85}At$$

1. Prepare a summary chart similar to Figure 11 for the following six isotopes. (Consult the Periodic Table for any needed information.)
   a. $^{12}_{?}C$
   b. $^{14}_{7}?$
   c. $^{16}_{?}O$
   d. $^{24}_{12}?^{2+}$
   e. $^{202}_{?}Hg$
   f. $^{238}_{92}?$
2. Using the Periodic Table as a source of information, what general relationship do you note between total number of protons and total number of neutrons
   a. for atoms of lighter elements, with atomic numbers less than 20?
   b. for atoms of heavier elements, with atomic numbers greater than 50?

Some Common Isotopes

Symbol	Name	Total Protons (Atomic Number)	Total Neutrons	Mass Number	Total Electrons
$^{7}_{3}Li$	Lithium-7	3	4	7	3
$^{67}_{31}Ga$	Gallium-67	31	36	67	31
$^{201}_{81}Tl$	Thallium-201	81	120	201	81
$^{208}_{82}Pb$	Lead-208	82	126	208	82
$^{208}_{82}Pb^{2+}$	Lead-208, 2+ ion	82	126	208	80

Figure 11 *Examine the names and symbols associated with these isotopes.*

ISOTOPE NOTATION

Building Skills 1

This activity provides practice in using nuclear chemistry notation and illustrates that atoms with higher atomic numbers have a larger neutron/proton ratio than do the atoms of elements with lower atomic numbers.

Answers

1. See bottom.
2. a. The lighter elements exhibit roughly a one-to-one ratio between the number of neutrons and protons.

   b. Heavier elements such as mercury (1.5 : 1 ratio) and uranium (1.6 : 1 ratio) exhibit higher neutron to proton ratios. Generally, the more massive the atom, the greater the neuton/proton ratio.

Symbol	Name	Total Protons (Atomic Number)	Total Neutrons	Mass Number	Total Electrons
a. C	Carbon	6	6	12	6
b. N	Nitrogen	7	7	14	7
c. O	Oxygen	8	8	16	8
d. Mg^{2+}	Magnesium	12	12	24	10
e. Hg	Mercury	80	122	202	80
f. U	Uranium	92	146	238	92

MODELING MATTER

ISOTOPIC PENNIES

This activity simulates the determination of the isotopic composition of an element. The simulation involves determining the number of pre- or post-1982 pennies in a sealed container of 10 pennies without opening the container.

Materials (for a class of 24 working in pairs)

12 sealed opaque containers (coded) with different 10 coin mixtures of pre-1982 and post-1982 pennies (35-mm canisters make ideal containers)
72 pre-1982 pennies
72 post-1982 pennies
Balance

Advance preparation

1. Provide the mass of the container for students to use in their calculations. Fill the containers, using various combinations of pennies. Use a coding system to identify the containers.
2. Avoid using 1982 pennies, since some have the pre-1982 composition and others are similar to post-1982 pennies. If you want all the pennies to look equally "new," soak them briefly in vinegar then rinse and dry.
3. One possible challenge with this activity is finding containers with consistent mass. Film canisters vary by over one gram. Envelopes may be used, although they are not as sturdy as the film canisters.
4. Put the pennies in pre-weighed containers. Or use a Seal-a-Meal machine to seal the pennies into little bags, after squashing all the pennies together so students cannot see the dates on the individual coins.

MODELING MATTER

ISOTOPIC PENNIES

You learned earlier (Unit 2, page 96) that pre-1982 and post-1982 pennies have different compositions. As you might suspect, they also have different masses. In this activity, a mixture of pre- and post-1982 pennies will model or represent the atoms of a naturally occurring mixture of two isotopes of the imaginary element "coinium." Using the pennies, you will simulate one way scientists determine the relative amounts of different isotopes present in a sample of an element.

You will be given a sealed container that holds 10 pennies, a mixture of pre-1982 and post-1982 pennies. Your container might hold any particular atomic mixture of the two "isotopes." Your task is to determine the isotopic composition of the element coinium without opening the container.

PROCEDURE

1. Your teacher will give you a pre-1982 penny, a post-1982 penny, and a sealed container with a mixture of 10 pre- and post-1982 pennies, and will tell you the mass of the empty container. Record that information and the code number of your sealed container.
2. Determine the isotopic composition of the element "coinium." That is, find the percent pre-1982 and percent post-1982 pennies in your container.

QUESTIONS

1. Describe the procedure you used to find the percent composition of "coinium."
2. What property of the element "coinium" is different in its pre- and post-1982 forms?
3. a. In what ways is the penny mixture a good analogy or model for actual element isotopes?
   b. In what ways is the analogy misleading or incorrect?
4. Name at least one other familiar item that could serve as a model for isotopes.

A.5 ISOTOPES IN NATURE

Most elements in nature are mixtures of isotopes. Some isotopes of an element may be radioactive, while others are not. All isotopes of an element behave virtually the same chemically—they have the same electron arrangement and only differ slightly in mass. Thus if one considers only chemical changes, the existence of isotopes is not particularly important. The atomic and molar masses of an element, as shown on the Periodic Table, represent averages based on the relative natural abundances of isotopes of that element.

5. To minimize differences in individual pennies, students may weigh 10 pre-1982 pennies and 10 post-82 pennies and use the average masses in their calculations.
6. To provide a conceptual bridge between the simulation and "real" isotopes, point out again that atomic-weight values for various elements are weighted averages of the masses of each element's isotopes.
7. Students in a technical chemistry class were very successful in using a non-math "technical" approach to the calculations. They gathered systematic mass data and determined the ratio of old vs. new pennies from the data. They weighed

MOLAR MASS AND ISOTOPE ABUNDANCE

Building Skills 2

To calculate the molar mass of an element, it is useful to use the concept of a weighted average. You can calculate the weighted average mass of the coins in the element "coinium" from your modeling matter activity on page 422.

Suppose you found that the composition of your mixture was 0.4 (40%) pre-1982 pennies and 0.6 (60%) post-1982 pennies. These decimal fractions, together with the mass of each type of penny (the two isotopes of "coinium"), can be used to calculate the average mass of a penny:

Notice that the decimal fractions must add up to 1. Why?

Average mass of a penny = (Decimal fraction of pre-1982 pennies) × (Mass of pre-1982 penny) + (Decimal fraction of post-1982 pennies) × (Mass of post-1982 penny)

1. Calculate the average mass of a penny in your coinium mixture as described above.
2. Calculate the average mass of a penny in your mixture another way: Divide the total mass of your entire penny sample by 10.
3. Compare the average masses you calculated in Questions 1 and 2. These results should convince you that either calculation leads to the same result. If not, consult your teacher.

Now consider an actual isotopic mixture. Naturally occurring copper, Cu, consists of 69.1% copper-63 and 30.9% copper-65. The molar masses of the two isotopes are

Copper-63	62.93 g/mol
Copper-65	64.93 g/mol

Using these data, calculate the average molar mass of naturally occurring copper. The equation for finding average molar masses is the same as the earlier equation you used for "coinium":

Molar mass = (Fraction of isotope 1) × (Molar mass of isotope 1) + (Fraction of isotope 2) × (Molar mass of isotope 2) + . . . (for each isotope involved)

Because there are two naturally occurring isotopes for copper, the average molar mass of copper is found this way:

Molar mass of copper = (0.691)(62.93 g/mol) + (0.309)(64.93 g/mol) = 63.5 g/mol

This is the value for copper given in the Periodic Table, rounded to the nearest 0.1 unit.

all of the combinations (one new, nine old; eight new, two old; and so on), then compared their mass with this table of known masses.

Answers

1. Total mass of pennies = (Total pre-1982 pennies) × (Mass of one pre-1982 penny) − (Total post-1982 pennies) × (Mass of one post-1982 penny)

   The penny calculation can be explained this way:

   There are 10 pennies in the container. So, if (p) represents the total number of pre-1982 pennies, then $(10 - p)$ = number of post-1982 pennies. Further, the mass of all pre-1982 pennies equals the number of pre-1982 pennies (p) multiplied by the mass of one pre-1982 penny. Likewise, the total mass of the post-1982 pennies equals the number of post-1982 pennies $(10 - p)$ times the mass of one post-1982 penny. That is: Total mass of pennies = (p) × (Mass of pre-1982 penny) − (10 − p) × (Mass of post-1982 penny).
2. The pre-1982 and post-1982 pennies differ in mass.
3. a. The penny mixture represents different isotopes of the fictitious element "coinium." The isotopes look and "behave" the same, yet they have different masses. Similarly, isotopes of the same element behave the same chemically, yet have different masses.

   b. This analogy is limited, in that

   (1) There are only two "isotopes," while actual elements often have more than two.

   (2) In the case of real isotopes, one isotope is sometimes more useful than another. This is especially true in the case of isotopes used in nuclear applications.

   (3) One cannot weigh individual atoms using a balance.
4. Eggs, car tires, soft drinks, or any class of objects that come in different masses, but look alike and serve the same function, might be used.

A.5 ISOTOPES IN NATURE

Most elements in nature occur as mixtures of isotopes. Although the various isotopes exhibit the same chemical properties, their masses differ because of the varying numbers of neutrons.

MOLAR MASS AND ISOTOPE ABUNDANCE

Building Skills 2

This activity is designed to provide students with practice in calculating the average molar mass of an element from isotopic molar masses, and in determining the percent abundance of isotopes from an element's molar mass and isotopic molar masses.

Answers

1. Student data. Students should use the formula given in the text.
2. Student data.
3. The answers to 1 and 2 should be the same.
4. a. 238. The final mass should be closest to the most prevalent isotope.

   b. Average mass = $(238.1 \times 0.9928) + (235.0 \times 0.0071) + (234.0 \times 0.000054) = 238.1$

4. Naturally occurring uranium (U) is a mixture of three isotopes:

Isotope	Molar Mass (g/mol)	% Natural Abundance
Uranium-238	238.1	99.28%
Uranium-235	235.0	0.71%
Uranium-234	234.0	0.0054%

a. Do you expect the molar mass of naturally occurring uranium to be closest to 238, 235, or 234? Why?
b. Calculate the molar mass of naturally occurring uranium.

As noted earlier, no stable isotopes have yet been found for elements with atomic numbers of 83 or greater.

Marie Curie originally thought that only heavy elements were radioactive. It is true that naturally occurring radioisotopes (radioactive isotopes) are more common among the heavy elements. In fact, all naturally occurring isotopes of elements with atomic numbers greater than 83 (bismuth) are radioactive. However, many natural radioactive isotopes are also found among lighter elements. Modern technology has made it possible to create a radioactive isotope of any element. Figure 12 lists some naturally occurring radioisotopes and their isotopic abundances.

Some Natural Radioisotopes

Name	Symbol	Relative Isotopic Abundance (%)
Hydrogen-3	${}^{3}_{1}H$	0.00013
Carbon-14	${}^{14}_{6}C$	Trace
Potassium-40	${}^{40}_{19}K$	0.0012
Rubidium-87	${}^{87}_{37}Rb$	27.8
Indium-115	${}^{115}_{49}In$	95.8
Lanthanum-138	${}^{138}_{57}La$	0.089
Neodymium-144	${}^{144}_{60}Nd$	23.9
Samarium-147	${}^{147}_{62}Sm$	15.1
Lutetium-176	${}^{176}_{71}Lu$	2.60
Rhenium-187	${}^{187}_{75}Re$	62.9
Platinum-190	${}^{190}_{78}Pt$	0.012
Thorium-232	${}^{232}_{90}Th$	100
Uranium-235	${}^{235}_{92}U$	0.72
Uranium-238	${}^{238}_{92}U$	99.28

Figure 12 *The percent abundance indicates the proportion of that element's atoms that consist of the specified isotope.*

A.6 FACT OR FICTION?

Making Decisions

Look again at the statements on the flyer depicted at the start of this unit (page 410). Answer the following questions about that flyer, which will help you start preparing your presentation for Riverwood senior citizens.

1. Which particular statements on the flyer can you now conclusively identify as either true or false?
   a. List two pieces of evidence that helped you make your decision about each statement you identified.
   b. List two public concerns about each statement you identified that you plan to address in your senior-citizen presentation.
2. a. Choose one statement from the flyer that you understand more completely now, but are still unable to confirm or deny.
   b. What else do you need to know before you can make a decision about that statement?
3. How helpful do you think it will be to discuss the history of some of the discoveries you studied in this section when you talk to the senior citizens? Explain your answer.
4. a. Which new terms introduced in this section will you define and explain in your presentation?
   b. Select two terms from your answer to Question 4a. Prepare a definition or explanation of each that you think your audience will be able to understand. Also consider using examples or "real-world" applications.

CD-ROM WWW. Questions & Answers

The history of science is full of discoveries that build on earlier discoveries. The discovery of radioactivity was such an event, beginning with the work of Roentgen, Becquerel, and Rutherford, and leading in turn to new knowledge of atomic structure. As you turn to considering some current applications of nuclear radiation, think about the history and scientific reasoning that made these technologies possible.

A.6 FACT OR FICTION?

Making Decisions

Students' answers will help them in preparing the presentation for the Riverwood senior citizens.

1. The following statements are false:

   "Radioactive materials and radiation are unnatural—they did not exist on Earth until created by scientists." Radioactive materials exist in nature. The first nuclear scientists extracted the nuclear material from tons of pitchblende.

   "All radiation causes cancer." Radiation only leads to cancer when the radiation is of sufficient energy and the exposure limits are reached. All life is exposed to natural background radiation all the time.

   "Human senses can detect radioactivity." Human senses cannot detect alpha and beta particles or gamma waves. Special sensors are needed to detect and measure radiation.

2–4. Assist students as they clarify their ideas about the topic and make plans for their presentation.

SECTION SUMMARY

Reviewing the Concepts

1. Distinguish between ionizing and nonionizing radiation.

Nonionizing radiation includes electromagnetic radiation in the visible and lower energy region of the spectrum. It transfers energy to matter, which causes the atoms or molecules to vibrate or move their electrons to a higher energy level. Ionizing radiation is of a higher energy that includes short wavelength ultraviolet radiation, X-rays, and gamma rays. Ionizing radiation can eject electrons from atoms and molecules, thus forming molecular fragments and ions. These fragments are highly reactive if formed within living systems. They can cause disruptions of normal cellular chemistry.

2. A local politician proposes to ban all sources of radiation in the community. Explain why this proposal has little merit or chance of success.

The proposal would have little chance of success because "all radiation" includes the entire electromagnetic spectrum, both ionizing and nonionizing. It is impossible to ban all sources of radiation, as that would mean banning radio waves, microwave equipment, visible light, and infrared heaters.

3. Describe the relationship between the energy associated with particular radiation and its classification as either ionizing or nonionizing.

Ionizing radiation has high energy while nonionizing radiation has low energy.

4. Classify each of the following as ionizing or nonionizing radiation.

a. visible light	**a. nonionizing**
b. X-rays	**b. ionizing**
c. ultraviolet radiation	**c. ionizing**
d. gamma rays	**d. ionizing**
e. radio waves	**e. nonionizing**

5. Give two examples of ionizing and two examples of nonionizing radiation.

Examples of nonionizing radiation include visible light, radio and TV reception, and electric light bulbs. Examples of ionizing radiation include cosmic rays and X-rays.

6. What is a spontaneous nuclear reaction?

A natural change of an element to produce a different element by emitting a nuclear particle and energy

7. Describe the sequence of events that led to Becquerel's discovery of radioactivity.

Becquerel wrapped a photographic plate in black paper, then placed it next to a fluorescent mineral. He then stored the covered photographic plate and mineral in a drawer. Because the skies were cloudy for several days, the film was left in the drawer with the mineral. Unexpectedly, the film showed faint fogging when developed.

8. How can the identity of a radioactive element be affected when it undergoes spontaneous nuclear change?

The element changes into a different element that has a different number of protons in the nucleus.

9. How were some of the differences among alpha, beta, and gamma radiation determined?

Alpha, beta, and gamma particles were passed through a magnetic field. The alpha particles or positive charges were deflected in one direction, the beta particles or negative charges were deflected in the opposite direction, but the gamma rays, which are neutral, were not deflected by the magnetic field.

10. What change in the statement of the atomic theory was required due to the discovery of radioactivity?

Atoms are not the smallest part of matter, but atoms are composed of smaller particles.

11. Make drawings to show the concept of an atom before and after Rutherford's gold-foil experiment.

Student drawings representing the atom before the gold-foil experiment should show the electrons and protons evenly distributed throughout the atom, like blueberries in blueberry muffins. Drawings of models after the experiment should show a small, dense nucleus and electrons around the nucleus, but most of the volume is empty space.

12. What characteristic of alpha particles made them desirable as the probing beam in the Rutherford gold-foil experiment?

Alpha particles are relatively massive and they have an electrical charge, both of which help them be detected.

13. Why was Rutherford astonished to find that alpha particles bounced back from the gold foil?

Rutherford was surprised because he did not believe there was anything in the atoms massive enough to bounce the particles back.

14. What was the basic structure of the atom that Rutherford proposed?

Rutherford's atom had a massive, positively charged region at the center, the nucleus, with the electrons orbiting around the nucleus.

15. How did Rutherford determine that atomic nuclei are positively charged?

In Rutherford's gold-foil experiment, most of the alpha particles passed straight through or had small deflections, but a few bounced back toward the source at large angles. Only a positively charged nucleus would repel a positively charged alpha particle.

16. How does carbon-12 differ from carbon-13 and carbon-14?

Carbon-12 differs from carbon-13 and carbon-14 in the number of neutrons in the nucleus.

17. Fill in the missing information, letters *a* through *p*, in this table:

Symbol	Number of protons	Number of neutrons	Mass number
$^{2}_{1}H$	*a*	1	*b*
$^{37}_{c}Cl$	*d*	*e*	*f*
$^{g}_{h}Tc$	43	56	*i*
$^{137}_{j}Cs$	*k*	*l*	*m*
$^{n}_{o}Ag$	47	60	*p*

Symbol	Number of Protons	Number of Neutrons	Mass Number
	a = 1		b = 2
c = 17	d = 17	e = 20	f = 37
g = 99 h = 43			i = 99
j = 55	k = 55	l = 82	m = 137
n = 107 o = 47			p = 107

18. Carbon-12 is a stable isotope, while carbon-14 is radioactive. What structural difference in these two atoms is responsible for this difference?

The number of neutrons in the nucleus; thus they have different mass numbers.

19. Neon (Ne) is composed of three isotopes with the molar masses and relative abundances shown here: Ne-20 (19.99 g/mol), 90.51%; Ne-21 (20.99 g/mol), 0.27%; and Ne-22 (21.99 g/mol), 9.22%.

a. Based on these data, should neon have an atomic weight closer to 20, 21, or 22? Why?

b. Calculate the atomic weight of neon.

a. Ne would have an atomic mass closer to 20. This is because the most prevalent isotope of neon is Ne-20, which makes up 90.51% of neon found in nature.

b. Atomic weight = (fraction, isotope 1) × (molar mass, isotope 1) + (fraction, isotope 2) × (molar mass, isotope 2) + (fraction, isotope 3) × (molar mass, isotope 3)

Atomic weight of neon = (19.99 g/mol) × (0.9051) + (20.99 g/mol) × (0.0027) + (21.99 g/mol) × (0.0922) = 18.09 + 0.0567 + 2.0275 = 20.17 g/mol

Connecting the Concepts

20. Describe the fundamental difference between nuclear reactions and chemical reactions.

In a chemical reaction the changes are due to the arrangement of the outer electrons. In a nuclear reaction, the changes take place in the nucleus.

21. Why are some isotopes radioactive and others not?

The lighter elements exhibit roughly a one-to-one ratio between the number of neutrons and protons. These elements are more likely to be stable. Heavier elements with proton-to-neutron ratios of 1.5 to 1 or higher are more likely to be radioactive.

22. Describe how the atomic theory can be used to illustrate the way in which one scientific discovery can build upon previous scientific knowledge and experiments.

In the case of atomic structure, Bequerel's observations led to questions about the nature of atomic structure, which led to further reasearch by Rutherford and others. Science is a social, cooperative process.

23. Explain how the history of the atomic theory demonstrates the importance of indirect evidence.

Atoms are too small to observe directly; therefore, indirect evidence is used to formulate theories, explanations, and so on, about the atom.

24. Why is it possible to get a suntan from ultraviolet radiation but not from radio waves?

Radio waves are nonionizing while ultraviolet rays can be ionizing. Sunburns results from overexposure to ionizing radiation from the Sun.

Extending the Concepts

25. Investigate the properties of gold to find out why Rutherford chose that metal as the material for his alpha-particle target.

Gold is a very dense but not very active metal. It is extremely malleable; it can be hammered into a sheet only a few atoms thick.

26. Assume that Rutherford had used beta particles instead of alpha particles in his gold-foil experiment. What result would you predict?

The beta particles are too lightweight. The kind of scattering Rutherford observed would never have occurred with beta particles.

27. The neutron was discovered decades after the discovery of the proton and the electron. What made the discovery of the neutron a more difficult challenge?

The neutron is neutral; it cannot be deflected by an opposite-charged particle.

NUCLEAR RADIATION

Introduction

This section introduces the nature and properties of the three most common types of nuclear radiation—alpha, beta, and gamma rays. The amount of radiation considered safe for humans is discussed in terms of background radiation and exposure standards. Some of the consequences of exposure to radiation are presented, leading to a discussion of the devices used to detect and measure the intensity of radiation. Laboratory activities give students an opportunity to use these devices.

B.1 EXPOSURE TO IONIZING RADIATION

Humans are exposed regularly to low-level natural background radiation due to the spontaneous decay of radioactive elements from natural sources in the environment. Exposure to additional radiation is the result of radiation generated from human activities, such as medical X-rays or industrial activities. The SI units used to measure radiation exposure are discussed (gray and sievert) and compared to the more traditional absorbed dose (rad) and its effect on living organisms (rem). Although the level of background radiation from natural and human sources is well below the limits set by the U.S. government as being reasonably safe, it is best to avoid any unnecessary exposure to ionizing radiation.

SECTION B NUCLEAR RADIATION

Introduction CD-ROM WWW.

Although most of the isotopes you encounter in daily life are not radioactive, of the nearly 2000 known isotopes, the total number of radioactive isotopes far exceeds the number of nonradioactive (stable) isotopes. The radiation emitted by naturally occurring radioisotopes provides all individuals with a constant but small exposure to radioactivity. This radiation comes from building materials (such as brick and stone) in your school and home, from air, land, and sea, from the food you eat—even from within your own body. Since nuclear radiation cannot be detected by human senses, various devices have been developed to detect it and to measure its intensity.

B.1 EXPOSURE TO IONIZING RADIATION

When radioactive isotopes spontaneously decay, they usually give off alpha, beta, and/or gamma radiation. The type and intensity of radiation emitted helps determine the possible medical and industrial applications of particular radioisotopes. In addition, each of these three types of nuclear radiation poses distinct hazards to human health.

A relatively constant level of natural radioactivity, called **background radiation,** is always present around you. Everyone receives background radiation at low levels from natural sources and also from sources related to human activity. Thus there is no way to escape at least some exposure to ionizing radiation.

Natural sources of background ionizing radiation include the following:

- Exceedingly high-energy particles that bombard Earth from outer space.
- Radioisotopes in rocks, soil, and groundwater—uranium (U-238 and U-235), thorium (Th-232), and their **decay products,** which are radioactive isotopes formed when uranium and thorium radioactively decay.
- Radioisotopes in the atmosphere—radon (Rn-222) and its decay products, including polonium (Po-210).
- Naturally occurring radioisotopes found in foods and the environment, such as carbon-14 and potassium-40.

*The number of radioisotopes released to the environment from fossil-fuel power plants is **greater** than that released from nuclear power plants.*

Technology has also created additional sources of background radiation, such as:

- Residual radioactive fallout from above-ground nuclear-weapon testing.
- Increased exposure to radiation during high-altitude airplane flights.
- Radioisotopes released into the environment from both fossil fuel and nuclear electrical power generation and other nuclear technologies.

- Release of radioisotopes through the disturbance and use of rocks in mining and in the making of cement, concrete, and sheet rock.

Because of its effect on living tissue, the amount of ionizing radiation to which people are exposed over time must be monitored. The **gray** (Gy) is the SI unit used to measure the quantity of ionizing radiation delivered to a particular sample, typically human tissue. An absorbed dose of one gray is defined as one joule of energy delivered per kilogram of body tissue.

Not all forms of ionizing radiation, however, produce the same effect on living organisms. Alpha radiation will cause more harm than the same quantity of gamma radiation. The **sievert** (Sv) is the SI unit that expresses the ability of radiation—regardless of type or activity—to cause ionization in human tissue. Any exposure to radiation that produces the same detrimental effects as one gray of gamma rays is said to represent one sievert of exposure. It is usually most convenient to express exposure in sieverts, since this unit allows for direct comparisons across different types of ionizing radiation.

While the SI units for radiation exposure are the gray and the sievert, in the United States two other units have traditionally been used—the rad and the rem. The **rad** (like the gray) measures the asorbed dose of radiation, while the **rem** measures the ionizing effect on living organisms (like the sievert). Both the rad and the rem are one-hundredth of their corresponding SI units. Even though a rem is only one-hundredth as large as a sievert, it is still much larger than typical exposures. Normal human exposures are so small that dosage is expressed in millirem (mrem), where 1 mrem = 0.001 rem. One millirem of any type of radiation produces essentially the same biological effects, whether the radiation is composed of alpha particles, beta particles, or gamma rays.

Some of the ionizing radiation you encounter comes from sources within your own body, such as those depicted in Figure 13. On average, people living in the United States receive above 360 mrem per person each year; about 300 mrem (82%) comes from natural sources. Figure 14 shows the percent that comes from each source.

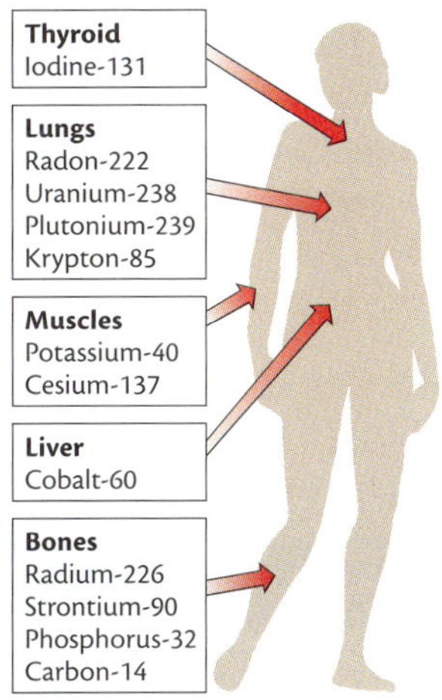

Figure 13 *All living things contain some radioactive isotopes, including these, which are found in specific parts of the human body.*

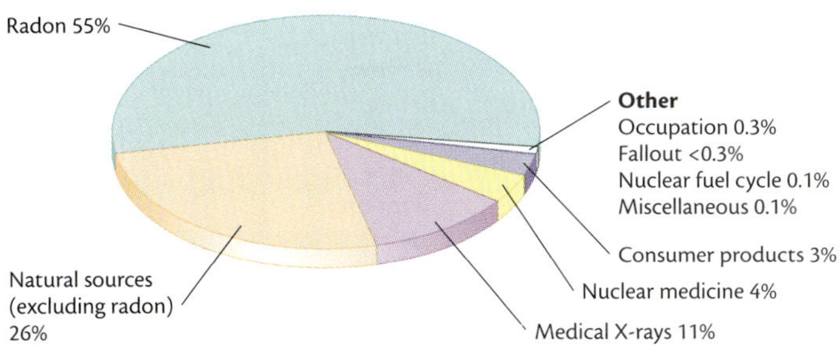

Figure 14 *Sources of ionizing radiation in the United States.*

RADIATION EXPOSURE STANDARDS

ChemQuandary 2

Certain individuals who have more rapidly dividing cells, such as children and pregnant women, are more at risk from a given dose level than the general population. On the other hand, for some people needing the benefits of radiation treatment, higher doses may be worth the risks. Also, for practical (including economic) reasons, occupationally exposed individuals (such as medical X-ray technicians and workers handling radioactive pharmaceuticals, defense program workers, and civilian workers in nuclear power plants) are allowed to receive higher, though apparently still safe, doses.

B.2 YOUR ANNUAL IONIZING-RADIATION DOSE

Making Decisions

The purpose of this activity is to raise student awareness of the sources and quantities of radiation in their everyday environment. Students are asked to estimate their annual radiation dose in mrem, analyze the sources, and then consider ways to reduce the quantity of their exposure.

*The units rad and rem, originally proposed by H. M. Parker, are abbreviations for **r**oentgen **a**bsorbed **d**ose (rad) and for **r**oentgen **e**quivalent **m**an (rem). 1 rad = 10^{-2} Gy; 1 rem = 10^{-2} Sv.*

It is clearly in your best interest to avoid unnecessary ionizing radiation exposure. What ionizing radiation level is considered reasonably safe? The U.S. government's background radiation limit for the general public is 500 mrem (0.5 rem) per year for any individual. The U.S. average exposure value of 360 mrem falls below this. The established U.S. limit for an individual's annual maximum safe exposure in the workplace is 5000 mrem (5 rem).

RADIATION EXPOSURE STANDARDS

ChemQuandary 2

Why might radiation exposure-level standards for some individuals be different from those for the general public? Why are standards different for those who are occupationally exposed?

B.2 YOUR ANNUAL IONIZING-RADIATION DOSE

Making Decisions

On a sheet of paper, list the numbers and letters of each category in the following table. Then fill in the blanks in the right column with appropriate quantities. Add all those quantities to estimate your annual radiation dose.

Common Sources of Radiation				Your Annual Dose (mrem)
1. Where You Live				
a. Cosmic Radiation (from outer space)				
Your exposure depends on elevation. These are annual doses.				
sea level	26 mrem	4000–5000 ft	47 mrem	
0–1000 ft	28 mrem	5000–6000 ft	52 mrem	
1000–2000 ft	31 mrem	6000–7000 ft	66 mrem	
2000–3000 ft	35 mrem	7000–8000 ft	79 mrem	
3000–4000 ft	41 mrem	8000–9000 ft	96 mrem	
				____ mrem
b. Terrestrial Radiation (from the ground)				
If you live in a state bordering the Gulf or Atlantic coasts			16 mrem	
If you live in AZ, CO, NM, or UT			63 mrem	
If you live anywhere else in the continental United States			30 mrem	
				____ mrem
c. House Construction				
If you live in a stone, adobe, brick, or concrete building			7 mrem	
				____ mrem
d. Power Plants				
If you live within 50 miles of a nuclear power plant			0.009 mrem	
If you live within 50 miles of a coal-fired power plant			0.03 mrem	
				____ mrem

Common Sources of Radiation		Your Annual Dose (mrem)
2. Food, Water, Air		
Internal Radiation (based on average values)		
From food (carbon-14 and potassium-40) and from water (radon dissolved in water)	40 mrem	____ mrem
From air (radon)	200 mrem	____ mrem
3. How You Live		
Weapons test fallout*	1 mrem	____ mrem
Travel by jet aircraft (per hour in air)	0.5 mrem	____ mrem
If you have porcelain crowns or false teeth	0.07 mrem	____ mrem
If you wear a luminous wristwatch	0.06 mrem	____ mrem
If you go through airport security (each time)	0.002 mrem	____ mrem
If you watch TV*	1 mrem	____ mrem
If you use a video display (computer screen)*	1 mrem	____ mrem
If you live in a dwelling with a smoke detector	0.008 mrem	____ mrem
If you use a gas camping lantern with an old mantle	0.2 mrem	____ mrem
If you wear a plutonium-powered pacemaker	100 mrem	____ mrem
4. Medical Uses (radiation dose per procedure)		
X-rays: Extremity (arm, hand, foot, or leg)	1 mrem	____ mrem
Dental	1 mrem	____ mrem
Chest	6 mrem	____ mrem
Pelvis/hip	65 mrem	____ mrem
Skull/neck	20 mrem	____ mrem
Barium enema	405 mrem	____ mrem
Upper GI	245 mrem	____ mrem
CT scan (head and body)	110 mrem	____ mrem
Nuclear medicine (e.g., thyroid scan)	14 mrem	____ mrem
Your Estimated Annual Radiation Dose		____ **mrem**

*The value is less than 1 mrem, but adding that value would be reasonable.
Adapted from "Estimate your personal annual radiation dose," American Nuclear Society, 2000.

Questions

1. a. Compare your annual radiation dose to the U.S. limit of 500 mrem.
   b. Compare it to average background radiation (360 mrem).
2. a. Why is it useful to monitor how many X-rays you receive annually?
   b. When choosing a place to live, what factors might decrease your annual ionizing-radiation dose?
3. a. What are some lifestyle changes that could reduce a person's exposure to ionizing radiation?
   b. Would you want to make those changes? Explain.

Answers

1. a. Individual student answers.

   b. Individual student answers, comparing their values with the standard 360 mrem/year.
2. a. Avoiding unnecessary X-rays will help reduce your annual risk due to radiation. This can be accomplished by keeping records of medical and dental X-rays, inquiring about the necessity of each X-ray, and being sure that technicians use lead to shield any of your body parts not being examined.

   b. Factors include living at lower altitudes, living in nonmasonry structures, living away from nuclear power plants, and living near the Gulf Coast or the Atlantic Ocean. Point out to students that even after making all these changes, the reduction in exposure would still be very small.
3. a. Lifestyle changes that would reduce a person's exposure to ionizing radiation include flying less (which may involve switching to hazardous automobile transportation), moving to lower elevations, switching to a wood home, or never watching TV or using a computer.

   b. Individual student answers. In many cases students would not want to make these changes because the activity may be necessary or the changes will cause them to alter their lifestyles.

B.3 IONIZING RADIATION—HOW MUCH IS SAFE?

The damage caused by ionizing radiation to human tissue depends upon the amount of radiation received and the energy of the radiation. The biological effects of large doses of ionizing radiation on human tissues include changes to DNA, which may result in birth defects, cancer, and other diseases. One problem associated with assessing radiation damage is that low-level radiation does not produce unique effects that can be distinguished from the results of other injuries or illnesses; therefore, the results are inconclusive. Also, it often takes years for symptoms to appear.

B.3 IONIZING RADIATION—HOW MUCH IS SAFE?

Each type of ionizing radiation can cause damage to human tissues in different ways. The two main factors that determine tissue damage are radiation density (the number of ionizations within a given volume) and dose (the quantity of radiation received).

Gamma rays and X-rays are ionizing forms of electromagnetic radiation that penetrate deeply into the body. Ionizing radiation causes damage to tissues by breaking bonds in molecules and thus tearing the molecules apart. At low levels of ionizing radiation, only a few molecules are damaged. In most low-dose cases, a body's systems can repair the damage. As the dose (or quantity of ionizing radiation) received by tissues increases, so does the number of molecules affected by the radiation. Generally, damage to proteins and nucleic acids is of greatest concern due to their role in body structures and functions. Proteins form much of the body's soft tissue structure and make up enzymes, molecules that control the rates of cellular chemical reactions. If a large number of protein molecules are destroyed within a small region, too few functioning molecules may remain to enable the body to heal itself in a reasonable time.

See Unit 7 for more information on proteins and enzymes.

Nucleic acids in DNA can be damaged by radiation. Minor damage causes **mutations**—changes in the structure of DNA that may result in the production of altered proteins. Often mutations will kill the cell in which they occur. If the cell is a sperm or ovum, a mutation may lead to a birth defect in offspring. Some mutations can lead to cancer, a disease in which cell growth and metabolism are out of control. When the DNA in many body cells is severely damaged, new proteins cannot be synthesized to replace damaged ones and the organism or person dies.

DNA molecules control cell reproduction and synthesis of cellular proteins.

Figure 15 lists factors that determine the extent of biological damage from radiation. Figure 16 summarizes the biological effects of large dosages of ionizing radiation. Note that values in the table in Figure 16 are so large they are reported in rems, not millirems.

Becquerel observed a red spot on his chest after carrying a radium sample in his breast pocket.

Large ionizing radiation doses can have drastic effects on humans. Conclusive evidence that such high radiation doses produce increased cancer rates has been gathered from uranium miners, nuclear-accident victims, and atomic-bomb survivors of World War II. Some of the first cases of exposure to large doses of ionizing radiation occurred among workers who used radium compounds to paint numbers on watch dials that would glow in the dark. The workers used their tongues to smooth the tips of their paint brushes, and unknowingly swallowed small amounts of radioactive radium compounds. These workers often began to lose hair and became quite weak. Sometimes this exposure eventually led to death.

The radium dial markings glowed in the dark. Modern glow-in-the-dark watches do not contain radium.

Leukemia, a rapidly developing cancer of the white blood cells, is commonly associated with exposure to high doses of ionizing radiation. Exposure to such radiation often encourages other forms of cancer, as well as anemia, heart-related problems, and cataracts (opaque spots on the lens of the eye). In general, each 1-mrem dose received increases an individual's risk of dying from cancer by one chance in 4 million.

Biological Damage from Radiation	
Factor	**Effect**
Dose	Most scientists assume that an increase in radiation dose produces a proportional increase in risk.
Exposure time	The more a given dose is spread out over time, the less harm it does.
Area exposed	The larger the body area exposed to a given radiation dose, the greater the damage.
Tissue type	Rapidly dividing cells, such as blood cells and sex cells, are more susceptible to radiation damage than are slowly dividing or non-dividing cells, such as nerve cells. Fetuses and children are more susceptible to radiation damage than are adults.

Figure 15 *These four factors determine the actual effects of particular radiation exposure.*

Radiation Effects	
Dose (rem)	**Effect**
0–25	No immediate observable effects.
25–50	Small decreases in white blood cell count, causing lowered resistance to infections.
50–100	Marked decrease in white blood cell count. Development of lesions.
100–200	Radiation sickness—nausea, vomiting, hair loss. Blood cells die.
200–300	Hemorrhaging, ulcers, deaths.
300–500	Acute radiation sickness. Fifty percent die within a few weeks.
>700	One hundred percent die.

Figure 16 *You can see how the consequences of radiation exposure change as dose increases.*

B.4 ALPHA, BETA, AND GAMMA RADIATION

Laboratory Activity

This activity is designed to acquaint students with the characteristics of emissions from radioactive materials. In Part 1, students compare the penetrating power of alpha, beta, and gamma radiation. In Part 2 they determine the effect of distance on radiation intensity, and in Part 3 they compare the radiation-shielding effects of lead and glass. As they account for background radiation in their observations, students also discover that their everyday environments contain natural radiation.

Only a few detectors are sensitive to all three types of ionizing radiation. Alpha radiation is commonly the hardest to detect. Thus, depending on the equipment available, you may have difficulty demonstrating alpha-particle detection. If equipment is in short supply, this activity can be presented as a teacher-led demonstration. Because the activity covers a number of important concepts, however, students should at least observe a demonstration.

Note: The American Nuclear Society has a special program to distribute working Geiger-Mueller counters to schools. A local college or university physics department may have Geiger counters available for loan.

Figure 17 *A person using a Geiger-Muller counter to measure radiation dose.*

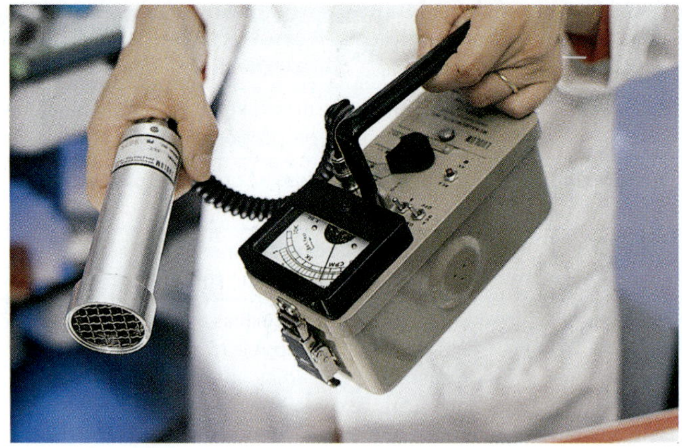

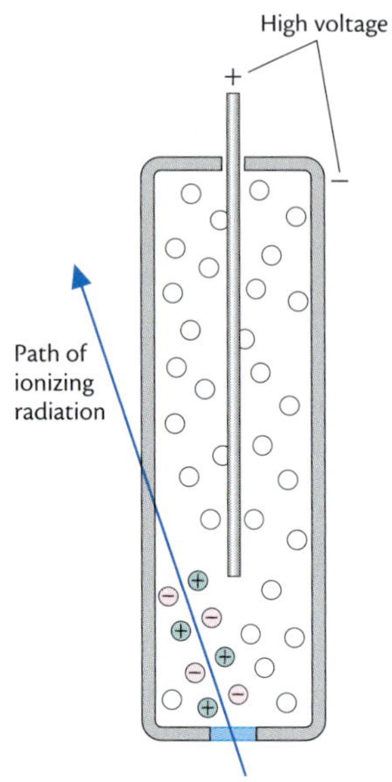

Figure 18 *The detector tube of a Geiger-Muller counter. As ionizing radiation passes through the detector tube, ions form in the gas inside. The positive ions are attracted to the negatively charged outer wall of the tube, while the negative ions are attracted to the positively charged center. These movements of electrically charged particles constitute a pulse of electrical current. Each pulse of current is detected and counted.*

Considerable controversy continues regarding whether very low doses of ionizing radiation, such as those from natural sources, can cause cancer. Most data on cancer have come from human exposure to high doses of such radiation, with mathematical extrapolation back to much lower doses. Few studies have directly linked low radiation doses with cancer development. Although most scientists agree that natural levels of ionizing radiation are safe for most people, some authorities argue that any increase in such radiation above natural levels increases the probability of developing cancer.

B.4 ALPHA, BETA, AND GAMMA RADIATION

Laboratory Activity

Introduction

One early device used to detect radioactivity was the **Geiger-Mueller counter.** It is still often used today to detect radiation. See Figure 17. As shown in Figure 18, when particles from a radioactive source strike the detector, electrical signals are produced. In this activity or demonstration, you will use a Geiger-Mueller or other detector to compare the penetrating abilities of different types of ionizing radiation through cardboard, glass, and lead.

When ionizing radiation enters the counter's detecting tube (or probe), an electronic signal is produced. Most radiation counters register these electronic signals as both audible clicks and a meter reading. The intensity of the radiation is indicated by the number of electronic signals or counts detected per minute (cpm).

Time

50–90 minutes

Materials (for a class of 24 students)

12 Geiger-Mueller counters (or equivalent)
Alpha, beta, gamma radiation source set
12 meter sticks
12 lead plates (5 cm × 5 cm × 0.1 cm)
24 glass plates (5 cm × 5 cm × 0.3 cm)
12 pieces of cardboard (5 cm × 5 cm × 0.1 cm)
Vernier calipers (optional) for measuring the exact thickness of plates
12 sheets of graph paper
12 forceps or crucible tongs (for handling isotopes)
12 pairs disposable safety gloves

Advance preparation

Check the background radiation before students begin the activity. If student readings are too variable, have them take more readings.

An initial reading of background radiation must be taken to establish a baseline before readings are taken from a known radioactive source. The background count is then subtracted from each reading taken from the radioactive source.

With proper handling, the radioactive materials in this activity pose no danger to you. Nuclear materials are strictly regulated by state and federal laws. The radioactive sources you will use emit only very small quantities of radiation; using them requires no special license. Nevertheless, you should handle all radioactive samples with great care, including using rubber or plastic gloves. Do not allow the radiation counter to come in direct contact with the radioactive material. Check your hands with a radiation monitor before you leave the laboratory.

CD-ROM WWW. Lab Video

Procedure

Part 1: Penetrating Ability

1. Prepare a data table with four data columns and three data rows for recording the number of counts detected per minute. The heading of each column should represent the type of shielding used, while each row should be labeled with the type of radiation being used.
2. Set up the apparatus shown in Figure 19. There should be room between the source and the detector for several sheets of glass or metal.
3. Turn on the counter; allow it to warm up for at least 3 minutes. Determine the intensity of the background radiation by counting clicks for one minute in the absence of any specific radioactive sources. Identify and record this background radiation in counts per minute (cpm) below your data table.
4. Put on protective gloves. Using forceps, place a gamma radiation source on the ruler at a point where it produces a high reading on the meter (Figure 19). Observe the meter for 30 seconds and

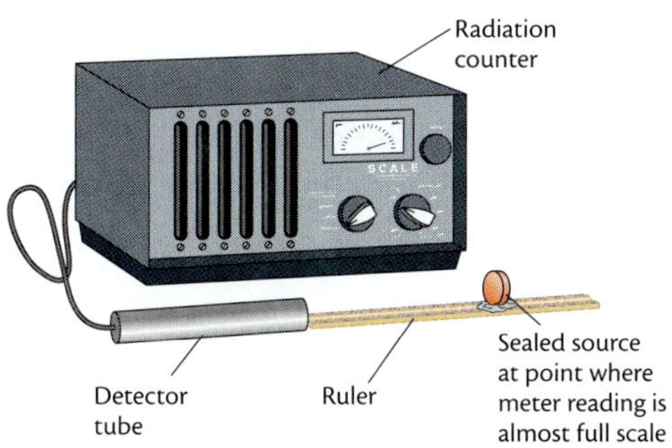

Figure 19 *Initial laboratory set-up.*

Lab tips

- Cover lab tables with butcher paper for nuclear labs.
- Radioactive sources sealed in plastic discs are available from most suppliers of school science materials.
- The thickness of the lead plate is easily determined by measuring the plate on a centimeter scale. Most plates available for this purpose are 1 mm thick.
- Because even a few centimeters of air will block alpha particles, the thick glass on the probe of the counter will not allow alpha particles to be detected. You can order a very thin glass plate or totally remove the glass plate on the end of the counter probe so that students can discover how a few centimeters of air act as a barrier to alpha particles.
- It may be important to adjust the laboratory instruction to the apparatus available. Not all detectors will measure all forms of radiation. Since the radioactive sources used will probably be weak, the amount of time needed may need to be extended from 30 seconds to several minutes. If possible, it is better to count clicks rather than just an average meter reading. The use of gloves in this activity is not to shield the students from radiation, but to protect the sample from chemical contamination and possible subsequent ingestion. With the sealed commercial sources, there is little risk of this type of contamination.
- Great care must be taken to ensure that the radioactive materials you and your students will be handling constitute no danger.

Safety precautions

- Always wear rubber or plastic gloves and use tongs when handling radioactive materials.
- Do not bring food into the room when working with radioactive materials (or at any other time). Food can be easily contaminated during handling and its presence could result in the ingestion of radioactive materials.
- Be sure that no extraneous radioactive materials come in contact with your radiation counter. The counter can become contaminated with radiation and record higher than normal background readings.

- Collect all materials used in the activity in an appropriate storage container and check for radiation before disposal.
- Check students' hands with a radiation monitor before students leave the laboratory. If the monitor reading is high, hands should be rewashed.

Part 1 Answers

1. Lead greatly reduces the level of alpha, beta, and gamma radiation. Cardboard greatly reduces the alpha radiation. Glass reduces the alpha and beta radiation.
2. Gamma radiation is the most penetrating; alpha radiation is the least penetrating. Gamma radiation is without mass and is classified as electromagnetic radiation. Alpha particles are the most massive and can be retarded easily by collisions with other particles.
3. a. Lead

   b. Cardboard
4. Density and thickness.

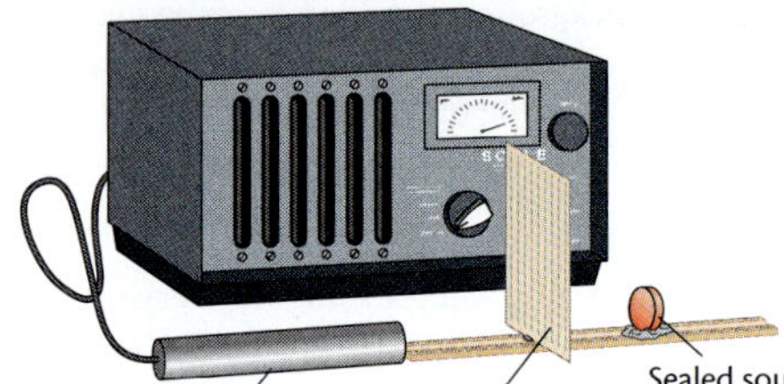

Figure 20 *Laboratory set-up showing placement of shield.*

estimate the average cpm detected during this period. Record that cpm value. Then subtract the background reading from that value and record the corrected result.

5. Without moving the radiation source, place a piece of cardboard (an index card) between the detector and the source, as shown in Figure 20.
6. Again observe the meter for 30 seconds. Record the average reading. Then correct that reading for background radiation and record the corrected result.
7. Repeat Steps 5 and 6 replacing the cardboard with a glass or plastic plate.
8. Repeat Steps 5 and 6 replacing the cardboard with a lead plate.
9. Repeat Steps 4 through 8 using a beta-particle source.
10. Repeat Steps 4 through 8 using an alpha-particle source.

Part 1 Questions

1. Analyze your results from Steps 5 through 10. Which shielding materials were effective in reducing the intensity of each type of radiation?
2. How do the three types of radiation you tested compare in their penetrating ability?
3. Of the shielding materials tested, which do you conclude is

   a. the most effective in blocking radiation?
   b. the least effective in blocking radiation?
4. What properties of a material appear to affect its radiation-shielding ability?

Part 2: Effect of Distance on Intensity

11. Prepare a data table for recording your Part 2 observations.
12. Place a radioactive source designated by your teacher at the point on the ruler that produces nearly a full-scale reading (usually a distance of about 5 cm).

13. Measure the average reading over 30 seconds. Determine the average counts per minute and record that value. Then correct this reading by subtracting the background value. Record your corrected value (cpm) in the data table.
14. Move the source so its distance from the detector is doubled.
15. Again measure the average reading over 30 seconds. Record that initial value and the corrected value (cpm).
16. Move the source twice more, so the original distance is first tripled, then quadrupled, recording the initial and corrected reading after each move. (For example, if you started at 2 cm, you would take readings at 2, 4, 6, and 8 cm.)
17. Prepare a graph, plotting the corrected cpm on the y axis and the distance from source to detector (in cm) on the x axis.

Part 2 Questions

Analyze the graph you prepared in Step 17.

5. By what factor did the intensity of radiation (measured in counts per minute) change when the initial distance was doubled?
6. Did this same factor apply when the distance was doubled again?
7. State the mathematical relationship between distance and intensity, using that factor and Figure 21.

A factor is a number by which an original value is multiplied to become a new value.

Part 3: Shielding Effects

18. Prepare a data table containing two columns—one for glass and one for lead. The table should include three rows, corresponding to the number of sheets of material. Remember that you are starting with zero sheets.

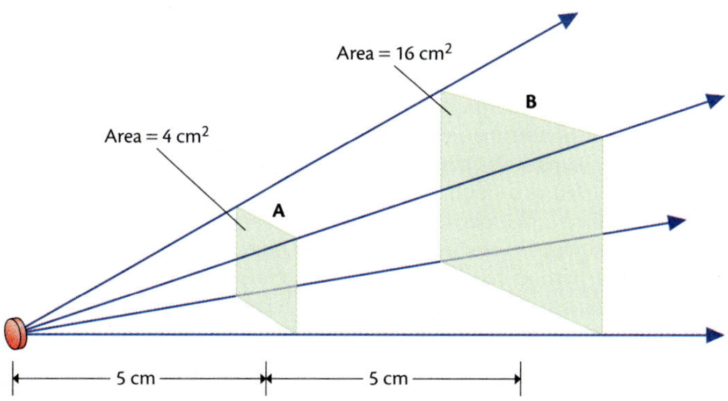

Figure 21 *The relationship between distance from the source and intensity of radiation. Intensity is counts-per-minute within a given area. Note how the same quantity of radiation spreads over a larger area as distance from the source increases.*

Part 2 Answers

5. Decreased by a factor of four.
6. No, now it is approximately one-sixteenth the original.
7. The intensity varies inversely with the square of the distance ($I = k/d^2$).

Part 3 Answers

8. a. Doubling the shield made little difference.

   b. Doubling the shield reduced the intensity of the radiation.

9. a. Lead is most effective at blocking radiation, including X-rays.

   b. Lead is very dense and can stop high-energy radiation.

10. a. Externally, gamma radiation is the most dangerous from a source outside the body.

    b. It has the greatest penetrating power and the greatest energy per unit; thus it can cause the most cellular and molecular disruption. Alpha radiation poses a serious problem if it is taken into the body through air passages or swallowing. Although it has less penetrating power, it has greater ionizing potential.

Post-lab discussion

Focus on the data collected in the activity and the answers to the questions.

B.5 NATURAL RADIOACTIVE DECAY

The characteristics of alpha, beta, and gamma radiation are explained. Given that radioactive decay is a dynamic process occurring below the limits of human vision, the use of computer and film simulations can be effective aids to enhance student understanding.

19. Using forceps, place a source designated by your teacher on the ruler at a distance that produces nearly a full-scale reading.
20. Take an average reading over 30 seconds. Determine the counts per minute, correcting for background radiation. Record your corrected value on the table.
21. Place one glass plate between the source and the detector. Do not change the distance between the detector and the source. Take an average reading over 30 seconds. Then determine and record the corrected counts per minute.
22. Place a second glass plate between the source and the detector. Take an average reading over 30 seconds. Determine and record the corrected counts per minute.
23. Repeat Steps 21 and 22, using lead sheets rather than glass plates.
24. Wash your hands thoroughly before leaving the laboratory.

Part 3 Questions

8. How effective was doubling the shield thickness in blocking the radiation intensity
   a. for glass?
   b. for lead?
9. When you have a dental X-ray, your body is shielded with a special blanket.
   a. What material would be a good choice for this blanket?
   b. Why?
10. a. Which type of nuclear radiation from a source outside the body is likely to be most dangerous to living organisms?
    b. Why?

You have found that the three kinds of nuclear radiation differ greatly in their penetrating ability. Why is this so? What are alpha and beta particles? Where do they originate? Those questions will be addressed next.

B.5 NATURAL RADIOACTIVE DECAY

An alpha particle is composed of two protons and two neutrons; it is identical in composition to the nucleus of a helium-4 atom. Since they have no electrons, alpha particles have a double positive electrical charge. They are often symbolized as a doubly charged helium-4 atom, ${}^{4}_{2}He^{2+}$. Alpha radiation (also called alpha particle emission) is commonly given off by many radioactive isotopes of elements with atomic numbers greater than 83.

An alpha particle has five to fifty times more energy and is over 7000 times more massive than a beta particle. The large mass of alpha particles makes them easy to stop outside the body. However, once inside the body,

In addition, students are introduced to the concept of balanced nuclear equations. In the materials unit, students learned that mass is conserved in chemical reactions. Emphasize that mass numbers are conserved in nuclear reactions, but the identity of elements may change. After working through examples of nuclear equations, draw attention to Figure 23 (page 440), which summarizes radioactive decay and the nuclear notation used.

the alpha particle's electrical charge and energy can cause great damage to tissues. This damage occurs over very short distances (about 0.025 mm).

Since alpha particles are very powerful tissue-damaging agents once inside the body, alpha emitters in air, food, or water are particularly dangerous to human life. Fortunately, outside the body, alpha particles are easily blocked. As you noted in Laboratory Activity B.4, alpha particles are stopped by a few centimeters of air.

Because of their relatively large mass, slower velocities, and large (+2) charge, alpha particles lose most of their energy within a small distance.

Figure 22 illustrates a radium-226 nucleus emitting an alpha particle. During this process, the radium nucleus loses two protons, so its atomic number drops from 88 to 86, forming an isotope of a different element, radon. In addition to the two protons, the radium-226 loses two neutrons, so its mass number drops by 4 (2 protons and 2 neutrons are lost), producing radon-222. The decay process can be represented by this nuclear equation:

$$\underset{\text{Radium-226}}{{}^{226}_{88}\text{Ra}} \longrightarrow \underset{\text{Alpha particle}}{{}^{4}_{2}\text{He}} + \underset{\text{Radon-222}}{{}^{222}_{86}\text{Rn}}$$

Ionic charges, such as the 2+ of the alpha particle, are usually not included in nuclear reaction equations.

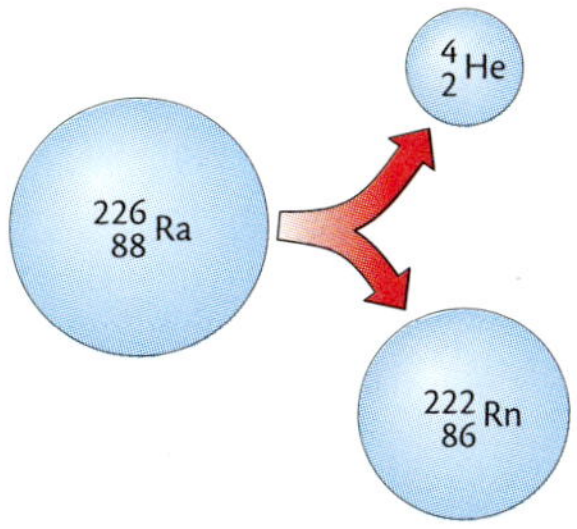

Figure 22 *Alpha particle emission from radium-226. The mass number decreases by four (2p + 2n) and the atomic number decreases by two (2p).*

Atoms of two elements—helium and radon—have been formed from one atom of radium. Note that atoms are not necessarily conserved in nuclear reactions, as they are in chemical reactions. Atoms of different elements can appear on the two sides of a nuclear equation. Total mass numbers and atomic numbers, however, are conserved. In the equation above, the sum of mass numbers of reactants equals that of products (226 = 4 + 222). Also, the sum of atomic numbers of reactants equals that of products (88 = 2 + 86). Both relations hold true for all nuclear reactions.

Beta particles are fast-moving electrons emitted from the nucleus during radioactive decay. Since they are so much lighter than alpha particles and travel at very high velocities, beta particles have much greater penetrating ability than alpha particles. On the other hand, beta particles are not as damaging to living tissue.

During beta decay, a neutron in a nucleus decays to a proton and an electron. The proton remains in the nucleus, but an electron is ejected at high speed. The high-speed electron emitted from the nucleus is a beta particle. This equation describes the process:

$$\underset{\text{Neutron}}{{}^{1}_{0}\text{n}} \longrightarrow \underset{\text{Proton}}{{}^{1}_{1}\text{p}} + \underset{\text{Beta particle (electron)}}{{}^{0}_{-1}\text{e}}$$

In the equation note that *n*, *p*, and *e* are used as symbols, respectively, for a neutron, a proton, and a beta particle. It is also acceptable to symbolize an electron (beta particle) by the Greek letter beta: $^{0}_{-1}\beta$.

Because of its negligible mass and negative electrical charge, a beta particle is assigned a mass number of 0 and an "atomic number" (nuclear charge) of −1. The overall result of beta emission is that a neutron has been converted to a proton.

The equation below shows the nucleus of lead-210 undergoing beta decay—the nucleus loses one neutron but gains one proton. Thus the mass number remains unchanged at 210, but the atomic number increases from 82 to 83. The new nucleus formed is that of bismuth-210:

$$\underset{\text{Lead-210}}{{}^{210}_{82}\text{Pb}} \longrightarrow \underset{\text{Bismuth-210}}{{}^{210}_{83}\text{Bi}} + \underset{\text{Beta particle (electron)}}{{}^{0}_{-1}\text{e}}$$

439

Figure 23 *Summary of the results of alpha, beta, and gamma radioactive decay.*

Changes Resulting from Nuclear Decay				
Type	Symbol	Change in Atomic Number	Change in Neutrons	Change in Mass Number
Alpha	${}^{4}_{2}He$	Decreased by 2	Decreased by 2	Decreased by 4
Beta	${}^{0}_{-1}e$	Increased by 1	Decreased by 1	No change
Gamma	${}^{0}_{0}\gamma$	No change	No change	No change

Once again, the sum of all mass numbers remains constant during this nuclear reaction (210 on each side of the equation). The sum of atomic numbers (the nuclear charge) remains constant as well (82 on each side).

Alpha and beta decay (emission) often leave nuclei in an energetically excited state. This type of excited state is described as **metastable** and is designated by the symbol "m". For example, ${}^{99m}Tc$ represents a technetium isotope in a metastable, excited state. Energy from isotopes in such excited states is released as gamma rays—high-energy electromagnetic radiation having as much or more energy than X-rays. Because gamma rays have neither mass nor charge, their release does not change the mass balance or charge balance in a nuclear equation. Figure 23 summarizes general information about natural radioactive decay.

Of these three forms of nuclear radiation, gamma rays are the most penetrating. Under most circumstances, however, they cause the least amount of damage to tissue over comparable distances. Here is the reason: Tissue damage is related to the extent of ionization created by the radiation, expressed as the number of ionizations within each unit of tissue. Alpha particles cause considerable damage over short range, but protecting against them is easy. Beta and gamma radiation do less damage over longer range, but it is more difficult to protect against them. The relative penetrating power of the three types of nuclear radiation are shown in Figure 24.

New isotopes produced by radioactive decay may also be radioactive, and therefore undergo further nuclear decay. Uranium (U) and thorium (Th) are the "parents" (reactants) in three natural decay series that begin

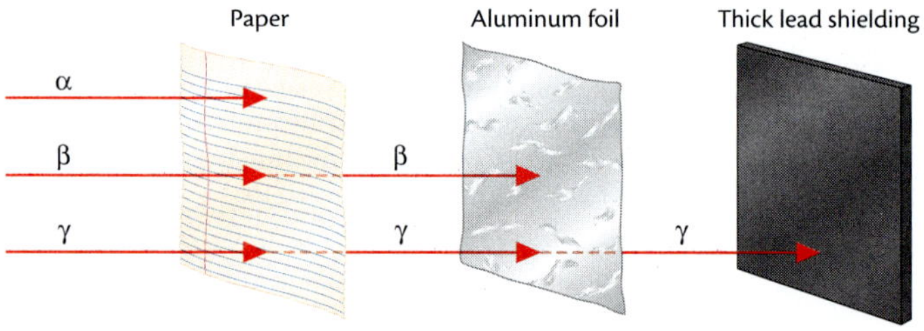

Figure 24 *Relative penetrating powers of alpha (α), beta (β), and gamma (γ) radiation. Gamma rays are the most penetrating, alpha particles the least.*

440

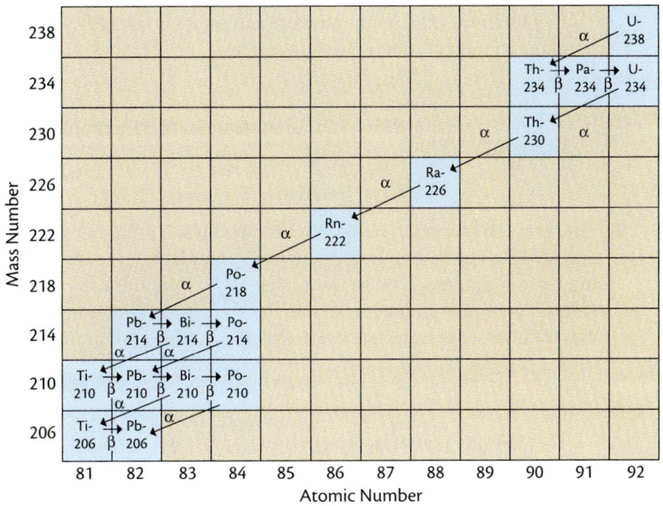

Figure 25 *The uranium-238 decay series. Diagonal lines show alpha decay, horizontal lines show beta decay. Here is how to read the chart: Locate radon-222 (Rn-222). The arrow to the left shows that this isotope decays to polonium-218 by alpha (α) emission. This nuclear equation applies:*

$$^{222}_{86}Rn \longrightarrow {}^{218}_{84}Po + {}^{4}_{2}He$$

with U-238, U-235, and Th-232, respectively. Each decay series ends with the formation of a stable isotope of lead (Pb). The decay series starting with uranium-238 contains 14 steps, as shown in Figure 25.

NUCLEAR BALANCING ACT

Building Skills 3

The key to balancing nuclear equations is recognizing that both atomic numbers and mass numbers are conserved. Use information in Figure 23 to complete the following exercise.

Cobalt-60 is a common source of ionizing radiation for medical therapy. Complete this equation for the beta decay of cobalt-60:

$$^{60}_{27}Co \longrightarrow {}^{0}_{-1}e + \textbf{?}$$

Beta emission causes no change in mass number, as noted in Figure 23. Therefore, the new isotope will also have mass number 60. Thus the unknown product can be written as $^{60}\textbf{?}$. Because the atomic number increases by one during beta emission, the new isotope will have atomic number 28, one more than cobalt. The periodic table indicates that the element with atomic number 28 is Ni, nickel. The final equation is

$$^{60}_{27}Co \longrightarrow {}^{0}_{-1}e + {}^{60}_{28}Ni$$

1. Write the appropriate symbol for the type of radiation given off in each of these reactions:
   a. The decay process shown below allows archaeologists to date the remains of ancient biological materials. Living organisms take in carbon-14 and maintain a relatively constant amount of it over

NUCLEAR BALANCING ACT

Building Skills 3

This activity provides students with practice in balancing nuclear equations. Explain that during beta decay, a neutron is transformed into a proton and an electron. Since the new proton remains in the nucleus, the atomic number increases by 1 while the mass number remains the same. The mass number remains constant because the number of neutrons is decreased by 1 and the number of protons is increased by the consequence is the formation of a different element. Work through several equations in class to ensure students' understanding.

Answers

1. a. $^{0}_{-1}e$
   b. $^{4}_{2}He$
2. a. $^{228}_{88}Ra$
   b. $^{228}_{89}Ac$
   c. $^{0}_{-1}e$
   d. $^{228}_{90}Th$
   e. $^{4}_{2}He$

B.6 RADON

The radon issue is particularly interesting to explore with students for the following reasons:

- It is a relatively recent problem.
- It is a case where nature, not humans, is the source of the pollution.
- It illustrates how solving one problem (high home energy bills) can sometimes help create a new one.
- It provides a good context for discussing risk-benefit analysis and the concept of comparative risks.

their lifetime. After death, no more carbon-14 is taken in, so the amount gradually decreases due to decay:

$$^{14}_{6}C \longrightarrow ^{14}_{7}N + \mathbf{?}$$

b. This decay process takes place in some types of household smoke detectors:

$$^{241}_{95}Am \longrightarrow ^{237}_{93}Np + \mathbf{?}$$

2. Thorium (Th) occurs in nature as three isotopes: Th-232, Th-230, and Th-228. The first of these is the most abundant. The radiation intensity of thorium is quite low. In fact, its compounds can be used without great danger if kept outside the body. Thorium oxide (ThO_2) was widely used for gas lights in mantles in Europe and America during the gas-lighting era and more recently in outdoor camping lanterns, producing the surface that glows brilliant white as the fuel burns.

Th-232 is the parent isotope of the third natural decay series. This series and the U-238 series are believed to be responsible for much of the thermal energy generated inside Earth. (The thermal contribution from the U-235 series is negligible, since the natural abundance of U-235 is quite low.)

Complete these equations, which represent the first five steps in the Th-232 decay series. Identify the missing items A, B, C, D, and E. Each code letter represents a particular isotope or a type of radioactive emission. For example, in the first equation, Th-232 decays by emitting alpha radiation to form "A." What is A?

An alpha particle can be symbolized either as $^{4}_{2}He$ or $^{4}_{2}\alpha$.

a. $^{232}_{90}Th \longrightarrow ^{4}_{2}\alpha + A$

b. $^{228}_{88}Ra \longrightarrow ^{0}_{-1}e + B$

c. $^{228}_{89}Ac \longrightarrow C + ^{228}_{90}Th$

d. $D \longrightarrow ^{4}_{2}He + ^{224}_{88}Ra$

e. $^{224}_{88}Ra \longrightarrow E + ^{220}_{86}Rn$

B.6 RADON

The gaseous element radon (Rn)—the most massive of the noble gases—has always been a component of Earth's atmosphere. It is a radioactive decay product of uranium. In the 1980s, unusually high concentrations of radioactive radon gas in a relatively small number of U.S. homes became a public health concern.

You can locate this radioactive decay product as Rn-222 in the decay series chart shown in Figure 25.

Radon is produced as uranium-238 radioactively decays in the soil and in building materials. Some radon produced in the soil dissolves in groundwater. Houses may have cracks in their foundations and basement floors that permit radon gas to seep into them. In older homes, outdoor air enters through doors, windows, and the gaps around them and thus dilutes the radon or removes it from the house. However, to conserve energy, new homes are built more air-tight than older homes. In a tightly sealed house, radon gas does not have as many opportunities to escape. Since indoor air cannot mix freely with outdoor air, radon concentrations may reach high levels. As a result, radon poses a problem in some areas because of changes in how homes are built and used. Remedies for high radon levels in homes include increased ventilation, sealing cracks in floors, and removing radon from groundwater. Relatively inexpensive radon test kits are available for home use.

The most serious danger of radon gas results from reactions that occur after it is inhaled. Radon radioactively decays to produce, in succession, radioactive isotopes of polonium (Po), bismuth (Bi), and lead (Pb). When radon gas is inhaled, it enters the body and is transformed, through radioactive decay, into toxic heavy-metal ions that cannot be exhaled as gases. These radioactive heavy metal ions emit potentially damaging alpha particles within the body.

Estimates indicate that about 6% of homes in the United States have radon levels higher than the exposure level recommended by the EPA. It is estimated that 10–14% of U.S. deaths from lung cancer annually are due to the effects of radon gas. These figures, although sobering, should be kept in perspective, however. About 80% of all U.S. lung-cancer deaths annually are attributed to cigarette smoking.

B.7 RADIATION DETECTORS

The only way to detect radioactive decay is to observe the results of radiation interacting with matter. In the Geiger-Mueller tube, argon gas is ionized by radiation that enters the tube. The ionized gas conducts an electric current, and an electric signal is generated as the ions and components of radiation pass through the detector.

You learned earlier in this unit that emissions from radioactive materials will expose photographic film. Ionizing radiation promotes the same chemical reactions in photographic film that are triggered by visible light. Workers who handle radioisotopes wear film badges or other detection devices to measure their exposure. The more ionizing radiation they encounter, the greater extent of photographic film exposure in the badge. If they receive a dose in excess of federal limits, they are temporarily reassigned to jobs that minimize their exposure to ionizing radiation.

Devices called **scintillation counters** detect ionizing radiation using light emitted by atoms of a solid that are excited by ionizing radiation. This is similar to the way that alpha particles were detected in the gold foil experiment (page xx). In modern scintillation counters, the flashes of light are detected electronically rather than by eye. The scintillation counter probe pictured in Figure 26 has a sodium iodide (NaI) detector, which emits light when ionizing radiation strikes it.

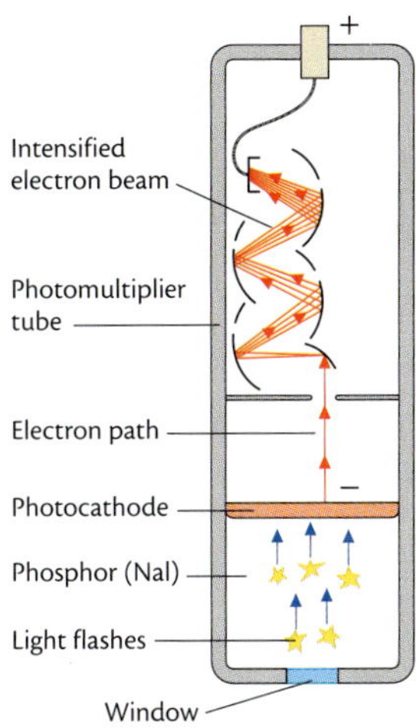

Figure 26 *A scintillation counter probe. Ionizing radiation causes flashes of light (scintillations) in the detector (NaI crystal). Each light flash is converted to an electron pulse that is increased many times as it moves through the photomultiplier tube.*

B.7 RADIATION DETECTORS

The types of radiation detectors—ionization, scintillation, and solid state detectors—are discussed. A Geiger counter is a kind of ionization counter that can detect alpha, beta, and gamma particles. Scintillation counters emit light when radiation strikes them, then a photomultiplier tube counts the flashes of light. Solid state detectors monitor changes in the movement of electrons when they are exposed to ionizing radiation and then count them electronically.

B.8 CLOUD CHAMBERS LABORATORY ACTIVITY

In this activity, students construct a simple device for indirect observation of ionizing radiation. The objective is to use a simple chamber supersaturated with 2-propanol (isopropyl alcohol) vapor and observe the condensation trails left by the radiation source.

Time

45 minutes

Materials (for a class of 24 working in pairs)

12 cloud chamber kits (see note below)
100 mL of 2-propanol
1 lb dry ice (thin slabs)
6–12 high-intensity light sources (20 W or greater)
12 squeeze wash bottles for 2-propanol
24 pairs of disposable gloves
12 crucible tongs

Advance preparation

You will need to obtain cloud chambers and dry ice. (Cloud chamber kits, including the radiation source, may be purchased. Sometimes dry ice can be purchased from ice cream stores or supermarkets. Check the Yellow Pages to find a local source.) Place the radioactive source in the cloud chamber with forceps. Do not allow students to handle it.

Solid-state detectors monitor changes in the movement of electrons through silicon and other semiconductors as they are exposed to ionizing radiation. These detectors are often used for detecting and measuring radioactivity in research laboratory settings.

Ionizing radiation can also be detected in a cloud chamber. You will investigate this detection method in the following activity.

B.8 CLOUD CHAMBERS — Laboratory Activity

Introduction

Lab Video CD-ROM WWW.

A **cloud chamber** is a container filled with air that is saturated with water or other vapor, similar to air on a humid day. If cooled, the air inside the container becomes supersaturated. This is an unstable condition (see page 46). If a radioactive source is placed near a cloud chamber filled with supersaturated air, the radiation will ionize gas molecules as it passes through the chamber. Vapor condenses on the ions formed, leaving a white trail behind each passing radioactive emission and revealing the path of the particle or ray. Figure 27 is a photograph taken of particle tracks under such conditions.

Cloud chamber trails resemble the "vapor" trails from high-flying aircraft.

The temperature of dry ice, solid carbon dioxide, is −78 °C.

The cloud chamber you will use consists of a small plastic container and a felt band moistened with 2-propanol (isopropyl alcohol). The alcohol evaporates faster than water and saturates air more readily. The cloud chamber will be chilled with dry ice to promote supersaturation and cloud formation.

Procedure

1. Fully moisten the felt band inside the cloud chamber with alcohol. Also place a small quantity of alcohol on the container bottom.

Figure 27 *A cloud chamber, showing several particle trails.*

Lab tips

Students need only to dampen the felt band and base disk with alcohol, avoiding direct contact with the radiation source. Place the lid on the chamber and warm it with your hands before placing it on dry ice. You may need to adjust the height of the radioactive source inside the chamber in order to get a result.

If available, a small piece of old lantern mantle (newer ones do not work) can be used as a source. The dry ice should be crushed unless you have a smooth block. The objective is to get good contact between the dry ice and the bottom of the cloud chamber. The light source should be directed at an oblique angle rather than straight down.

Practice is necessary for this activity. Rubbing the outside of the chamber with a silk cloth to promote a charged atmosphere may also help.

See the *ChemCom CD-ROM* for video footage of a cloud chamber in action.

2. Using gloves and forceps, quickly place the radioactive source in the chamber. Replace the lid.
3. To cool the chamber, place it on a flat surface of crushed dry ice. Ensure that the chamber remains level.
4. Leave the chamber on the dry ice for three to five minutes.
5. Dim or turn off the room lights. Focus the light source at an oblique angle (not straight down) through the container so that the chamber base is illuminated. If you do not observe any vapor trails, try shining the light through the side of the container.
6. Observe the air in the chamber near the radioactive source. Record your observations.

Questions

1. What differences—if any—did you observe among the tracks?
2. Which type of radiation do you think would make the most visible tracks? Why?
3. What is the purpose of the dry ice?

B.9 ENSURING PUBLIC SAFETY — Making Decisions

Consider again the statements found on the flyer (page 410) and the report you are going to present to the senior citizens. Within a small group of classmates, select two of the four proposals below. Discuss their appropriateness as possible suggestions for protecting the public from radiation dangers. Why might the flyer encourage public groups to make such proposals?

1. Since all radiation is bad, government entities should ban all exposure to all forms of radiation.
2. Although there are several different types of ionizing radiation (alpha, beta, and gamma), government standards for protecting the public against possible dangers should treat all radiation exposures identically.
3. Because nuclear radiation can be harmful, doctors should be required by law to inform patients of any medical procedure that involves nuclear radiation. Patients can then reject the treatment if they have personal concerns about radiation exposure.
4. The government should provide a waste disposal site for permanently storing all wastes that have ever been identified as radioactive.

In this section you have studied exposure to ionizing radiation, the origins of ionizing radiation, and some possible consequences of exposure. Keep these ideas and concerns in mind as you explore some uses of nuclear radiation in the next section.

CD-ROM WWW. Questions & Answers

Answers

1. The length, direction, and intensity of the tracks will differ.
2. Alpha particles, owing to their large mass and size and their relatively slow speed.
3. The dry ice reduces the temperature of the gases inside the closed chamber. This allows the vaporized alcohol to supersaturate the air. The supersaturated gaseous mixture will condense to a liquid with the slightest disturbance, such as an ion speeding through the chamber.

Post-lab discussion

Discuss the conditions that support the direct observation of ion vapor trails (supersaturation) and the importance of cloud chambers, even in current research.

B.9 ENSURING PUBLIC SAFETY

Making Decisions

Students discuss proposals in small groups. Student discussions will vary. Some typical comments may include:

1. All radiation is not bad, and even if it were, it is impossible to ban it by passing legislation. Radiation is all around, some natural and some a consequence of human activity. It comes in many forms and intensities. Depending on the nature of the radiation, its properties range from harmless to deadly.
2. All radiation is not the same, and it would be wasteful to treat all radiation as if it were. Some has little penetrating power, so it would be a waste to use a lot of shielding to block it. Other sources have extremely short half-lives. The radiation may be expended by the time protection is in place.
3. Patients must always weigh risks and benefits. Nearly all medical treatment involves some risk, from adverse drug reactions to complications from surgical procedures.
4. Just because something is radioactive does not mean it should go to a specialized waste site. If that were true, many of our homes and household fixtures would end up in such a dump.

The flyer has some half-truths and generalizations that can lead to inappropriate calls for action.

SECTION SUMMARY

Reviewing the Concepts

1. List five sources of background radiation.

Sources of background radiation include building materials, air, land, food, and so on—even the human body.

2. Why does background radiation vary from one locale to another?

Variations in altitude, soil type, mineral deposits, and other natural factors. Fallout and contamination may play a role in some areas.

3. Is it possible or desirable to eliminate background radiation? Explain.

It is not possible or desirable to eliminate background radiation because background radiation is everywhere. It is in our soil, water, air, food, cosmic rays, and so on.

4. What is the minimum shielding needed for protection against exposure to each type of radiation?

The types of shielding that is needed for protection against exposure to

Alpha	**A few centimeters of air, cardboard**
Beta	**Aluminum foil**
Gamma	**Lead**

5. Explain why materials that emit alpha radiation are not very dangerous if they are outside the body but are very dangerous if they are taken internally.

Alpha particles have little penetrating power. If the source is outside of the body, the skin blocks their entry. However, if the source is internal, the alpha particles can travel short distances and cause damage to surrounding tissue. If an alpha emitter is ingested or inhaled and if its chemical behavior causes it to be concentrated by the body, more severe damage can result.

6. Why is different shielding appropriate for different kinds of radiation?

Different kinds of radiation have different penetrating power; therefore, each requires different kinds and amounts of shielding.

7. Radioactive wastes are generally stored deep in the earth. Explain how this serves to keep populations safe from excessive exposure to radiation.

Radioactive materials decay with half-lives that cannot be changed by any known means. We must be able to isolate most of our nuclear wastes for thousands of years from both human (terrorism and war) and natural (water seepage and geological shifts) intrusions. The shielding materials must block beta and gamma radiation and withstand thermal, mechanical, and chemical stress.

8. If you could live anywhere in the United States and live in any kind of house you wanted, what decisions would you make to minimize exposure to ionizing radiation? Explain your choices.

To minimize exposure to ionizing radiation one should live in one of the states along the Gulf or Atlantic coast in a wood, glass, or metal building. Also, live at lower altitudes. Living at lower altitudes blocks more incoming cosmic radiation. The Gulf and Atlantic coasts have lower natural levels. Wood and glass have lower levels than masonry and brick, due to the elements they contain.

9. Why is the breaking of chemical bonds in living cells by ionizing radiation potentially detrimental?

The amount of damage caused by ionizing radiation to human tissues is determined by the level of radiation and the number of ionizations within a given volume. If a large number of protein and DNA molecules are destroyed within a small region, the body may not be able to heal itself. Damaged DNA may lead to cancer.

10. Which biological molecules are of the greatest concern in terms of the effects of ionizing radiation?

DNA molecules are of greatest concern because when DNA in the cells is damaged, the body cannot synthesize new proteins to replace the damaged ones, or cancer may develop and the organism or person may die.

11. How does ionizing radiation lead to an increase in mutation rates?

Ionizing radiation may cause damage to the nucleic acids in DNA, which results in altered nucleotide sequences, which in turn can increase mutation rates.

12. Why does ionizing radiation break some bonds but not others?

Low levels of ionizing radiation break few bonds. As radiation scatters, only some of the energy is absorbed by chemical bonds. Bonds only break if they absorb enough radiation.

13. Write out the following equations and predict the unknown products, lettered *a* through *d*:

a. ${}^{6}_{3}\text{Li} + {}^{1}_{0}\text{n} \longrightarrow {}^{4}_{2}\text{He} + a$

b. ${}^{42}_{19}\text{K} \longrightarrow {}^{0}_{-1}\text{e} + {}^{42}_{20}\text{Ca}$

c. ${}^{235}_{92}\text{U} \longrightarrow {}^{4}_{2}\text{He} + {}^{231}_{90}\text{Th}$

d. ${}^{1}_{0}\text{n} + {}^{235}_{92}\text{U} \longrightarrow {}^{142}_{56}\text{Ba} + {}^{91}_{36}\text{Kr} + {}^{31}_{0\text{n}}$

a. $\mathbf{{}^{6}_{3}Li + {}^{1}_{0}n \rightarrow {}^{4}_{2}He + {}^{3}_{1}H}$

b. $\mathbf{{}^{39}_{19}K \rightarrow {}^{0}_{-1}e + {}^{39}_{20}Ca}$

c. $\mathbf{{}^{238}_{92}U \rightarrow {}^{4}_{2}He + {}^{234}_{90}Th}$

d. $\mathbf{{}^{1}_{0}n + {}^{235}_{92}U \rightarrow {}^{140}_{56}Ba + {}^{93}_{36}Kr + 3\,{}^{1}_{0}n}$

14. Describe how each of the following instruments detect the presence of radiation:

a. Geiger-Mueller tube

b. scintillation counter

c. solid-state detector

d. cloud chamber

The radiation strikes the silver salts, causing a chemical reaction and color change in the photographic film.

a. Geiger-Mueller tube—Gas is ionized by radiation entering the tube, which produces an electrical signal as the component passes through the probe.

b. Scintillation counter—Detects light emitted by excited atoms of the detector.

c. Solid-state detector—Detects movement of electrons through a semiconductor.

d. Cloud chamber—Ionizes air as it passes through a chamber. The vapor condenses on the ion, leaving a "cloud" indicating the active emission.

15. Why do radiation detectors register counts even though no apparent source of radioactivity is in the area?

Because of background radiation in the environment.

16. In what way is a white trail observed in a cloud chamber similar to the trail sometimes seen behind an aircraft in the upper atmosphere?

A cloud chamber and the trail seen behind an aircraft in the upper atmosphere are both formed by condensation of vapor in the area.

17. Would you expect alpha, beta, and/or gamma rays to produce the same kinds of trails in a cloud chamber? Explain your answer.

Alpha particles will produce more visible tracks due to their larger mass and size, and to their relatively slow speed. Gamma rays lack mass and do not cause a track to form.

Connecting the Concepts

18. Are the effects of shielding and distance the same for both ionizing and nonionizing radiation? Explain your answer.

In general, yes. They both vary inversely with the square of the distance. They are stopped by various amounts of shielding, but nonionizing radiation is lower energy and can be stopped with less shielding.

19. A student sets up a cloud chamber and sees no white trails. What are some possible explanations for this result?

The student may not see track formation in a cloud chamber if:

a. The temperature is not low enough to condense the air.

b. The air is not saturated. (Must be supersaturated.)

c. The source giving off beta or gamma particles is inactive.

d. The angle of the light is not at the correct angle. (Mist is at an oblique angle.)

20. Two Geiger-Mueller detectors register different counts for the same source. What factors might account for the difference?

The amount of shielding of the source, sensitivity of the counters.

21. A heavy apron is provided for a patient receiving a dental X-ray.

a. What element is most probably is used in the apron?

b. What is the function of the apron?

c. Why does the dentist or hygienist leave the room while the X-ray machine is turned on?

a. The element most likely used is lead.

b. The function of the apron is to block harmful radiation.

c. To avoid being exposed or becoming overexposed to ionizing radiation.

Extending the Concepts

22. In terms of radiation, how is a sunburn different from a suntan? How do sunscreens work to prevent both?

Suntan is the body's defense mechanism. It is used to minimize the influx of damaging UV radiation. Sunburn is usually associated with UV radiation in that it is energetic enough to cause disruption of some molecular bonds within skin cells. If DNA molecules are disrupted and mutations occur, skin cancer may result. Sunscreens block the harmful UV radiation.

23. In the 1940s and 1950s, shoe stores generally let their customers check the fit of shoes by X-raying their feet. Why was that practice discontinued?

X-rays are high-energy electromagnetic radiation (ionizing). Unnecessary exposure should be avoided. Scientists were not aware of all the dangers of radiation at the time.

24. Radon is an inert noble gas. As such, it is relatively harmless to living things. Explain why its presence in homes constitutes a health hazard for occupants.

Radon is radioactive. It undergoes natural decay by alpha emission. The main health risk is cancer. The risk of developing cancer increases as the level of radon concentration and the length of exposure (time) increases. The products of the decay of radon (polonium, bismuth, and lead) are solid particles that can become trapped in the lungs, where they may continue to do damage even after the original source of radon is removed.

25. A newly discovered element with an extremely short half-life is detected by analyzing its decay products. Explain how scientists can work backwards to identify the original element.

Many of the newer elements are extremely unstable. They decay quickly after formation, into one or more "daughter" nuclei, eventually stopping at a stable isotope. By tracing this chain of decay products backward, scientists can discover the identity of the new element.

26. Two students live next door to each other. One receives three times more annual radiation than the other. Explain how that could be possible.

Determining factors might include the type of material from which each house is built (wood, glass, concrete, brick, and so on), the design of the house (built on slab, basement, amount and kind of insulation), and differences in lifestyles (frequency of air travel, extent of TV watching, extent of video-game playing, number and type of medical and dental treatments).

USING RADIOACTIVITY

Introduction

The half-life of a radioactive sample is important in determining how useful or harmful the isotope may be. This section examines the half-life of a radioisotope and the use of half-life in carbon-14 dating. The medical benefits of radioactive tracers are discussed and the section concludes with the transmutation of elements and artificial radioactivity.

C.1 HALF-LIFE: A RADIOACTIVE CLOCK

Half-life may be measured by determining the time required for one-half of the original atoms to decay. Students may have heard of the term *half-life* but may not thoroughly understand it. Allow time for a brief discussion to correct misconceptions. Common misconceptions include (a) that one-half of the material changes in one half-life and the other half in the next half life; and (b) that the atoms disappear rather than being changed into a decay product. Keep the math simple. Avoid obscuring the concept by using difficult mathematical concepts.

SECTION C USING RADIOACTIVITY

Introduction CD-ROM WWW.

Scientists use the term *decay* to refer to the process of emitting ionizing radiation during radioactivity because the nucleus usually changes its identity.

Different radioactive isotopes decay and emit ionizing radiation at different rates. Scientists have devised convenient ways to measure, analyze, and report how rapidly (or slowly) various radioisotopes decay. In this section you will learn about radioactive decay rates, a factor that helps determine how useful or hazardous a particular radioisotope may be.

C.1 HALF-LIFE: A RADIOACTIVE CLOCK

How long does it take for a sample of radioactive material to decay? There is no simple answer to this question. However, understanding how radioactive materials decay allows scientists to predict how long a radioisotope used in medical diagnosis or treatment will remain active within the body, to plan how long hazardous nuclear wastes must be stored, and to estimate the ages of ancient organisms, civilizations, or rocks.

In 1904 Ernest Rutherford proposed the concept of half-life to describe the random process of radioactive decay. **Half-life** expresses the time within which any particular radioactive atom has a 50:50 chance of undergoing radioactive decay. Since one gram of any element contains well over 10^{21} atoms, it is easy to see that typical sample sizes involve very large numbers of atoms.

Half-life and radioactive decay rates have different meanings. Half-life is constant for a particular radioisotope, while the radioactive decay rate is the number of atoms decaying within a certain period of time. Decay rate gradually decreases as the number of radioactive atoms becomes smaller and smaller over time.

Statistics related to very large numbers provide two other useful ways to interpret half-life. The first—and perhaps most common—interpretation is that one half-life is the time it will take for one-half of the total atoms originally present in a sample to decay. The second interpretation is that at the end of one half-life, the rate of radioactive decay in a sample will decrease to one-half its original value. Figure 28 lists the half-lives of several radioisotopes.

Although it is not possible to predict when any individual radioactive atom will decay, each way of thinking about half-life is equally valid when working with the amounts of matter (numbers of atoms) found in typical chemical samples. Consider carbon-14, a naturally occurring radioactive isotope whose relative abundance remains reasonably constant.

Half-Lives of Selected Radioisotopes		
Hydrogen-3	${}^{3}_{1}H \longrightarrow {}^{3}_{2}He + {}^{0}_{-1}e$	12.3 y
Carbon-14	${}^{14}_{6}C \longrightarrow {}^{14}_{7}N + {}^{0}_{-1}e$	5.73×10^3 y
Phosphorus-32	${}^{32}_{15}P \longrightarrow {}^{32}_{16}S + {}^{0}_{-1}e$	14.3 d
Potassium-40	${}^{40}_{19}K \longrightarrow {}^{40}_{20}Ca + {}^{0}_{-1}e$	1.28×10^9 y
Radon-222	${}^{222}_{86}Rn \longrightarrow {}^{218}_{84}Po + {}^{4}_{2}He$	3.82 d

Figure 28 *Decay equations and half-lives for five radioactive isotopes.*

ADDITIONAL INFORMATION

Radioactive half-life and the rate of radioactive decay are different concepts. The *rate of radioactive decay* is an extensive property of matter (like mass) that depends on the quality of material present. It expresses the number of atoms decaying over a specified time—such as counts per minute (CPM). One metric ton of U-238 has a decay rate one million times greater than that of one gram of U-238 because the larger sample has a million times more atoms.

The decay of carbon-14 is represented by this equation:

$$^{14}_{6}C \longrightarrow ^{14}_{7}N + ^{0}_{-1}e$$

All living matter contains carbon—and therefore a small amount of radioactive C-14. Thus all living organisms emit a small but constant level of radioactivity. As C-14 atoms decay in living matter they are constantly being replaced by new carbon atoms. However, no new carbon atoms are taken in by an organism after death. The radioactive C-14 atoms continue to decay within the organism. Thus the longer an organism has been dead, the fewer C-14 atoms it contains.

Figure 28 indicates that carbon-14 has a half-life of 5730 years. Thus, if 50.0 billion atoms of carbon-14 are contained in an organism at the time of death, after 5730 years half of those atoms (that is, 25.0 billion atoms) will have decayed, leaving 25.0 billion atoms of carbon-14. During the next 5730 years, one-half of the remaining 25.0 billion atoms will decay, leaving 12.5 billion atoms of carbon-14. Therefore, if a sample of matter from a previously-living organism contains only one-fourth the number of C-14 atoms found in living organisms, the age of the sample can be predicted as about 11 460 years—that is, two half-lives must have passed since the organism died.

In 1991 a frozen human body was found high up in the Alps, in the Similaun Glacier on the Italian-Austrian border. At first, this discovery was investigated as a suspected murder. However, on closer inspection, it was found to be the body of a Neolithic man. Also preserved were several of his belongings. He quickly became identified in the media as the Iceman.

To determine how long the Iceman had been locked in his frozen grave, scientists performed carbon-14 dating. Scientists measured the rate at which carbon-14 was radioactively decaying in his body and belongings by observing beta particle emissions. They found this rate to be slightly more than 50% of the rate found in living organisms today. This implied that the number of carbon-14 atoms remaining in the Iceman's body was slightly greater than 50% of that found in living humans. Since nearly half of the Iceman's original carbon-14 atoms must have decayed, nearly one half-life of C-14 must have passed since he had died. Therefore, it was estimated that the Iceman had died about 5300 years ago.

Scientists have determined that the relative abundance of C-14 has remained reasonably constant over thousands of years. That is because C-14 is constantly produced in the upper atmosphere by cosmic rays interacting with nitrogen atoms, while—at the same time—C-14 atoms undergo radioactive decay.

Since the rate of decay is directly related to the number of C-14 atoms present, scientists usually measure the rate of radioactive decay rather than directly measuring the number of C-14 atoms.

CARBON-14 DATING

ChemQuandary 3

It is not possible to determine the age of everything using carbon-14 dating. What kinds of materials might be good candidates for carbon-14 dating? What are some materials that could not be dated using carbon-14? Why do you think carbon-14 dating has a practical limit of about 50 thousand years?

Each radioisotope has a specific half-life. Half-lives can be as short as a fraction of a second or as long as several billion years. For example, the half-life of polonium-212 is 3×10^{-7} seconds, while that of uranium-238 is

CARBON-14 DATING

ChemQuandary 3

The age of once-living organic materials may be determined using carbon-14 dating. Metals or concrete cannot be dated using carbon-14 dating. After 50 000 years, carbon-14 would have gone through 8.7 half lives. Therefore the amount of carbon-14 remaining is so small, it will not yield an accurate reading. This method is so accurate because the rate of decay is exceptionally steady. It is not affected by heat, pressure, concentration, or any other external factors.

Radioactive half-life is an intensive property of matter (like density). It does not depend on the quantity of material present. Both a metric ton of U-238 and a single gram of U-238 have the same half-life. To prevent student confusion, you can avoid associating half-life values with decay rate. Instead, clearly define half-life and refer to that concept in that language.

UNDERSTANDING HALF-LIFE

MODELING MATTER

This activity models the exponential decay curve of radioactive samples through several half-lives. The "decaying" atoms are pennies, and decay is represented when a coin flips from heads to tails. The simulation makes the point that because atomic decay is a statistical matter, we can never be sure which particular atom will survive and which will decay at any given time.

The activity should take about 30 minutes to complete. Each student group will need 100 pennies, and a plastic freezer container with lid or a cardboard box. Consider using M&Ms as an alternative to pennies. Use the "printed" side and the "non-printed" side of each candy piece as its distinguishing trait.

The graphs made by the students represent a radioactive decay curve. Make sure students understand the random, statistical nature of radioactive decay. One cannot know when a given atom will decay but can predict the decay behavior for a large number of atoms. The use of various computer and video simulations can be helpful. If a computer-graphing program is available, have students use a computer to generate graphs.

MODELING MATTER

UNDERSTANDING HALF-LIFE

In this activity you will investigate the relationship between the passage of time and how many radioactive nuclei (heads-up pennies) decay.

Assume each heads-up penny represents an atom of the radioactive isotope headsium. Its decay produces a tails-up penny—the isotope tailsium.

You will be given 100 pennies and a box. Placing all pennies heads up will represent the starting sample of headsium. Each shake of the closed box will represent one half-life. During this time a certain number of headsium nuclei will decay to produce tailsium (that is, some pennies will flip over).

PROCEDURE

1. Prepare a data table for recording total remaining undecayed headsium and decayed tailsium atoms after each of four half-lives. Include initial values for 0 half-lives.
2. Place the 100 pennies heads up in the box.
3. Close the box and shake it vigorously.
4. Open the box. Remove all atoms that decayed into tailsium (that is, remove all coins that turned over). Record the number of undecayed (headsium) and decayed (tailsium) atoms after this first half-life.
5. Repeat Steps 3 and 4 three more times. At that point you will have simulated four half-lives. Remember to record your results.
6. Follow your teacher's instructions to obtain pooled class data for total undecayed headsium atoms (coins that did not turn over) remaining after each half-life.
7. Using your own data and class-pooled data, prepare a graph by plotting the number of half-lives on the *x* axis and the number of undecayed atoms remaining after each half-life on the *y* axis. Plot and label two graph lines—one for your own data and one for pooled class data.

QUESTIONS

1. a. Describe the appearance of your two graph lines. Are they straight or curved?
   b. Which set of data—yours or pooled class data—provides a more convincing demonstration of half-life? Why?
2. About how many headsium nuclei would remain after three half-lives if the initial sample had 600 headsium atoms?
3. If 190 headsium nuclei remain from an original sample of 3000 headsium nuclei, about how many half-lives must have passed?
4. Describe one similarity and one difference between your model based on pennies and actual radioactive decay. (*Hint:* Why was it advisable to pool class data?)
5. How could you modify this "decay" model to demonstrate that different isotopes have different half-lives?
6. a. How many half-lives would it take for one mole (6.02×10^{23} atoms) of a radioactive isotope to decay to 6.25% (0.376×10^{23} atoms) of the original number of atoms?
   b. Is it likely that any of the original radioactive atoms would still remain
      i. after 10 half-lives?
      ii. after 100 half-lives?
7. a. In this simulation, can you predict when a particular penny will "decay?"
   b. If you could follow the fate of an individual atom in a sample of radioactive material, could you predict when it would decay? Why or why not?
8. What other methods could be used to model the concept of half-life?

Answers

1. a. The graph slopes downward as a curved line.
   b. The pooled class data should produce a smoother curve and be more reliable, representing a larger number of decayed atoms.

2.

Number of half lives	Headsium nuclei remaining
0	600
1	about 300
2	about 150
3	about 75

3.

Number of half lives	Headsium nuclei remaining
0	3000
1	about 1500
2	about 750
3	about 375
4	about 187

4.5 billion years. Thus in one year it would be likely that all atoms in a small sample of polonium-212 would have decayed, while well over 99% of the original uranium-238 atoms would still be present.

After 10 half-lives, only about 1/1000th or 0.1% of the original radioisotope atoms are still left to decay. (You can verify that statement with your own calculations.) That means that the radioactivity intensity of the isotope has dropped to 0.1% of its initial level. This reduced level is often considered safe because it is roughly approaching the level of normal background radiation.

Since there is no way to change the rate of radioactive decay significantly for a particular isotope, radioactive waste disposal (or storage) can pose a challenging problem, especially for radioisotopes with very long half-lives. You will examine that issue later in this unit.

The graph you constructed in the Modeling Matter activity applies to the decay of any radioactive isotope, with one important difference: half-life is specific to particular isotopes. The following exercise will give you practice applying the concept of half-life.

HALF-LIVES

Building Skills 4

1. Suppose you were given $1000 and told you could spend one-half in the first year, one-half of the balance in the second year, and so on. (One year thus corresponds to one half-life in this analogy.)
   a. If you spent the maximum allowed each year, at the end of which year would you have $31.25 left?
   b. How much would be left after 10 half-lives (that is, 10 years)?
2. Potassium is a necessary nutrient for all living things and is the seventh most abundant element on the surface of Earth, making up about 1.5% of Earth's crust. About 0.01% of natural potassium atoms are the radioisotope potassium-40. K-40 has a half-life of nearly 1.3 billion (1.3×10^9) years.
   a. How much of the K-40 present at Earth's formation remained after one half-life?
   b. Assuming the Earth is about 4.5 billion years old, roughly how many times more potassium-40 was present when Earth formed than is present now?
3. Strontium-90 is one of many radioactive products generated by nuclear-weapon explosions. This isotope is especially dangerous if it enters the food supply. Strontium behaves chemically like calcium; the two elements belong to the same chemical family. Thus, rather than passing through the body, radioactive strontium-90 is incorporated into calcium-based material such as bone. In 1963 the United States, the former Soviet Union, and several other countries

5. To model the half-lives of different isotopes, use an object with more than two sides, such as dice or sugar cubes.
6. a. Four half-lives
   b. (i) Yes (ii) Yes
7. a. No
   b. A half-life value only predicts how long it will take for half of a sample of atoms to decay, not which ones.
8. If you have 1000 automobiles and half break down every 10 years, the half-life of the automobile is 10 years. Students will suggest other ways.

HALF-LIVES

Building Skills 4

This activity provides practice in calculating half-lives.

Answers

1. a. Fifth year (first year $500, second year $250, third year $125, fourth year $62.50, fifth year $31.25)
   b. $0.98 (sixth year $15.62, seventh year $7.81, eighth year $3.91, ninth year $1.95, tenth year $0.98)
2. a. One-half of the natural K-40 atoms.
   b. 3.46 times more when Earth was formed than is present now.

4. The simulation is similar to the actual radioactive decay, in that the graph results model the half-life relationship; that is, during any half-life there is a fifty-fifty chance an atom will decay. This simulation, however, involves only 100 pennies, and the half-life of any radioisotope is based upon a much larger number of particles. The decay of nuclei is a chance occurrence, as observed in this simulation. The averaging of several results from the simulation helps adjust for this fact. Also, identification and separation of radioactive atoms are not as simple as this simulation might suggest.

3. a. (i) Zero—1963 One—1992 Two—2021 Three—2050 Four—2079 Five—2108

(ii) 1963–100% 1992–50% 2021–25% 2050–12.5% 2079–6.25% 2108–3.13%

b. (i) Values can be read from graph. The value for 2000 is around 40%.

(ii) Approximately 3%.

C.2 RADIOISOTOPES IN MEDICINE

The rapidly expanding use of radioisotopes in medicine is beneficial. They serve as both diagnostic tools and a means for therapeutic treatment of cancer. Radioisotope tracers usually have short half-lives. They are used to track cellular abnormalities, locate damaged areas, and diagnose some forms of cancer. Remind students that excessive exposure to ionizing radiation should be avoided when possible. Focus students' attention on the medical radioisotopes in Figure 30 (page 455).

signed a nuclear test-ban treaty that ended most above-ground weapons testing. Some strontium-90 released in previous above-ground nuclear weapon testing still remains in the environment.

a. Sr-90 has a half-life of 28.8 years. Track the decay of Sr-90 atoms that were present in the atmosphere in 1963 by following these instructions:
   i. Using 1963 as year zero, when 100% of released Sr-90 was present, identify the years that represent one, two, three, four, and five half-lives.
   ii. Calculate the percent of the original 1963 Sr-90 radioactivity present at the end of each half-life.
b. Prepare a graph, plotting the percent of the original 1963 Sr-90 radioactivity level on the *y* axis and the years 1963 to 2110 on the *x* axis. Connect the data points with a smooth curve.
   i. What percent of Sr-90 formed in 1963 still remains this year?
   ii. What percent will remain in the year 2100?

C.2 RADIOISOTOPES IN MEDICINE

During the four decades following discovery of X-rays and radioactivity, the public was fascinated with these scientific developments. In fact, some patent medicines advertised that they contained "radium," and healthful effects were claimed for the substance. Unfortunately, some of those medicines actually contained some radioactive materials and were quite hazardous. Indiscriminate exposure to excessive ionizing radiation should always be avoided.

Careful use of ionizing radiation and radioisotopes, however, can be quite effective in medicine. Their uses can be classified as either diagnostic or therapeutic. Diagnostic use helps doctors understand what is happening inside the body, while therapeutic use involves treating a disease or medical condition.

An example of a common diagnostic application is the use of radioisotope-tracer studies, based on detecting an isotope in particular parts of the body. Knowing that certain elements collect in specific parts of the body—for instance, calcium in bones and teeth—physicians can investigate a given part of the body by using an appropriate radioisotope as a tracer. In a tracer study, radioisotopes with short half-lives are placed in a patient's body for diagnostic purposes. Such radioisotopes can identify cellular abnormalities, highlight damaged areas, and help physicians select appropriate therapy.

Tracers have properties that make them ideally suited to this task. First, radioisotopes behave the same chemically as stable isotopes of the element. To investigate a part of the body, a solution of an appropriate tracer isotope is supplied to the body, or a biologically active compound—synthesized to contain a radioactive tracer element—is fed to or injected into the patient.

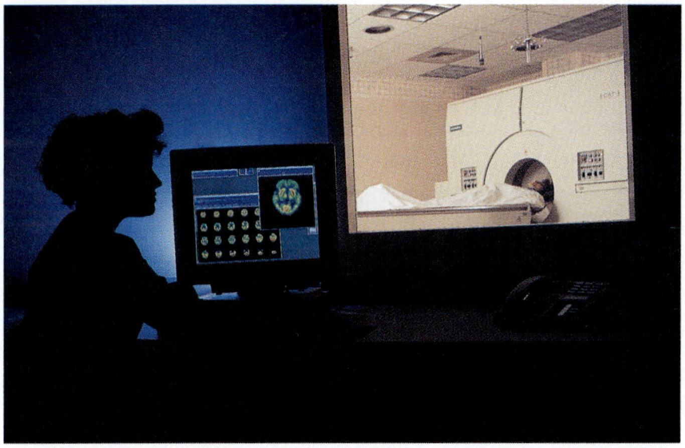

Figure 29 *Injection of a radioisotope makes it possible to produce a noninvasive image of the abdomen.*

See Figure 29. A nuclear radiation detection system then allows the physician to track the location of this tracer throughout the body.

A specific example of a radioisotope tracer is iodine-123. This radioisotope is used to diagnose problems of the thyroid gland, which is located in the neck. A patient drinks a tracer solution containing sodium iodide (NaI), in which some of the iodide ions are the radioactive isotope I-123. The physician, using a radiation detection system, monitors the rate at which the tracer is taken up by the thyroid. A healthy thyroid will incorporate a known amount of iodine. An overactive or underactive thyroid will take up, respectively, more or less iodine. The physician then compares the measured rate of I-123 uptake in the patient to the normal rate for an individual of the same age, gender, and weight, then takes appropriate therapeutic action.

Technetium-99m (Tc-99m), a synthetic radioisotope, is the most widely used diagnostic radioisotope in medicine. It has replaced exploratory surgery as a way to locate tumors in the brain, thyroid, and kidneys. Tumors are areas of runaway cell growth, and the technetium concentrates where cell growth is fastest. A bank of radiation detectors around the patient's body can pinpoint the Tc-99m at the tumor's precise location.

The *m* in Tc-99m indicates that the isotope is metastable—it readily changes to a more stable form of the same isotope, releasing gamma rays in the process.

Therapeutic radioisotopes use the energy carried by the radiation to destroy living tissue. In some cancer treatments, the diseased area is exposed to ionizing radiation to kill cancerous cells. For thyroid cancer, the patient receives a concentrated internal dose of radioiodine, which collects and destroys the cancerous portion of the thyroid gland. In other cancer treatments, an external beam of ionizing radiation (from cobalt-60) may be directed at the cancerous spot. Such irradiation treatments must be administered with great care—high radiation doses can also damage or kill normal cells.

453

USING RADIOISOTOPES IN MEDICINE

ChemQuandary 4

The rate of decay of any radioactive isotope is constant and not affected by outside factors. Using the radiation given off does not "use up" more radiation from the sample, just as an increase in the number of people warming themselves in front of a campfire does not "use up" the firewood any faster.

C.3 EMERGING NUCLEAR MEDICINE TECHNOLOGIES

Modern technology and the use of computers have led to the development and use of Positron Emission Topography (PET Scan) and Magnetic Resonance Imaging (MRI) to diagnose medical conditions. Emphasize to students that MRI does not depend on ionizing radiation, but on radio waves of very low frequencies, produced when a strong magnetic field interacts with atomic nuclei. MRI involves no known health risks.

USING RADIOISOTOPES IN MEDICINE

ChemQuandary 4

Cobalt-60 is used as a source of ionizing radiation in cancer treatment. Rapidly dividing cancer cells are killed when exposed to cobalt's radiation. Cobalt-60 has a half-life of about five years. However, no matter how frequently a particular Co-60 sample might be used in cancer treatment, the time interval before it must be replaced does not change. Why?

Radiosodium is usually administered as a NaCl solution.

Other medical applications include the use of radiosodium (Na-24) to search for circulatory system abnormalities and radioxenon (Xe-133) to help search for lung embolisms (blood clots) and abnormalities. The table in Figure 30 outlines other medical uses for radioisotopes.

You have learned that the ionizing radiation like that observed in a cloud chamber (page 444) is emitted from an unstable, radioactive isotope that eventually changes to a stable, nonradioactive isotope. Do you think it is possible to reverse the process, converting a stable isotope into an unstable, radioactive isotope? Think about it. The question will be addressed later in this section.

C.3 EMERGING NUCLEAR MEDICINE TECHNOLOGIES

Nearly all aspects of modern life have been touched by uses of computers during the last two decades. Two nuclear medicine technologies have emerged during this time that rely heavily on the use of computers to make sense of the large quantities of data obtained. These technologies are Positron Emission Topography and Magnetic Resonance Imaging. Both techniques would have been regarded as science fiction 50 years ago.

Positron Emission Topography (PET) scans are based on a very unusual form of radioactive decay involving a few particular radioisotopes. While most radioisotopes emit alpha, beta, and/or gamma radiation, a few isotopes emit radiation in the form of **positrons.** Positrons originate in the nucleus and have the same mass as beta particles (electrons). However, positrons are different from electrons in two fundamental ways. Positrons have a positive electrical charge, while electrons are negatively charged. Second, positrons are not made of matter as it is commonly understood—positrons are composed of antimatter. When a positron encounters an electron, both particles are annihilated (destroyed) and produce two gamma rays, which fly off in opposite directions. Positron Emission Topography detects these gamma-ray pairs and, with the help of computers, determines where they were formed. By observing a large number of such events, a computer-generated image gradually emerges.

Selected Medical Radioisotopes

Radioisotopes	Half-Life	Use
Used as tracers		
Technetium-99m	6.01 h	Measure cardiac output; locate strokes and brain and bone tumors.
Gallium-67	78.3 h	Diagnosis of Hodgkin's disease
Iron-59	44.5 d	Determine rate of red blood cell formation (these contain iron); anemia assessment
Chromium-51	27.7 d	Determine blood volume and lifespan of red blood cells
Hydrogen-3 (tritium)	12.3 y	Determine volume of body's water; assess vitamin D usage in body
Thallium-201	72.9 h	Cardiac assessment
Iodine-123	13.3 h	Thyroid function diagnosis
Used for irradiation therapy		
Cesium-137	30.1 y	Treat shallow tumors (external source)
Phosphorus-32	14.3 d	Treat leukemia, a bone cancer affecting white blood cells (internal source)
Iodine-131	8.0 d	Treat thyroid cancer (external source)
Cobalt-60	5.3 y	Treat shallow tumors (external source)
Yttrium-90	64.1 h	Treat pituitary gland cancer internally with ceramic beads

Figure 30 *How various radioactive isotopes are used in medicine.*

The radioisotope tracer that emits positrons in PET scans is attached to a sugar molecule. The movement of each "tagged" sugar molecule as it progresses through the body can be accurately determined. Since cancers grow at a faster rate than normal tissues, cancerous tissue concentrates more of the sugar. PET technology differs from other nuclear medicine technologies in that it is able to detect and display metabolic activity. Thus, it is useful

455

C.4 ARTIFICIAL RADIOACTIVITY

A brief history of the development of nuclear chemistry is given, beginning with Rutherford's experiments with alpha particles and transmutation, which had been the alchemists' goal. The phosphorus-30 produced by the Joliot-Curies has a half-life of 3.5 minutes. The transuranic elements have been produced by bombarding heavy nuclei with particles such as helium nuclei. Recently, a few atoms of elements 110 and 111 have been produced by bombarding bismuth-209 nuclei with relatively heavy particles such as nickel-64. *Suggestion:* Plan a field trip to an MRI lab or have a technician visit the class to talk about MRI and/or PET Scans.

not only for medical diagnosis but can also be used to investigate brain functioning without invasive surgery.

Magnetic Resonance Imaging (MRI) does not employ ionizing radiation but relies on special properties of protons in the nuclei of atoms. MRI is an application of a laboratory process known as nuclear magnetic resonance (NMR), developed shortly after World War II. It is a noninvasive technique that can be used to identify atoms within a sample without altering and affecting the sample. It took considerable time to adapt NMR technology to medical uses, since the patient must be placed inside a very large electromagnet. This large-scale technique requires considerable calculations to produce images from the vast quantity of collected data.

A major benefit of MRI is that it does not depend on ionizing radiation being absorbed by the body. MRI imaging can "see through" bone and can produce useful images of soft tissues. Unlike most other nuclear-medicine technologies, MRI uses radio waves of very low energies and involves no known health risks. Although this procedure has very low risk, some patients were hesitant to undergo the procedure when it was called by its original name, nuclear magnetic resonance, due to fear evoked by the term *nuclear.* This unfounded fear sparked a name change to magnetic resonance imaging.

The time required to produce usable images has been decreasing steadily with advances in the technology for PET scans and MRI. In the mid-1990s, 30 to 90 minutes were required for an exam, while in the most modern machines this time has dropped to less than one minute.

C.4 ARTIFICIAL RADIOACTIVITY

In 1919 Ernest Rutherford enclosed nitrogen gas in a glass tube and bombarded the sample with alpha particles. After analyzing the gas remaining in the tube, he found that some nitrogen had been converted to an isotope of oxygen, according to this equation:

$$\underset{\text{Helium-4 (Alpha particle)}}{{}^{4}_{2}\text{He}} + \underset{\text{Nitrogen-14}}{{}^{14}_{7}\text{N}} \longrightarrow \underset{\text{Oxygen-17}}{{}^{17}_{8}\text{O}} + \underset{\text{Hydrogen-1}}{{}^{1}_{1}\text{H}}$$

Rutherford had produced the first synthetic or artificial **transmutation** of an element—that is, the first documented conversion of one element to another. He continued this work but was limited by the moderate energies of alpha particles then available. By 1930, particle accelerators were developed that could produce highly energetic particles needed for additional bombardment reactions. Using these higher-energy particles, scientists were able to create many other synthetic nuclei, some of which were radioactive.

The first artificial radioactive isotope was produced in 1934 by French physicists Irène and Frédéric Joliot-Curie (the daughter and son-in-law of

Pierre and Marie Curie). They bombarded aluminum with alpha particles, producing radioactive phosphorus-30:

$$^{27}_{13}\text{Al} + {}^{4}_{2}\text{He} \longrightarrow {}^{30}_{15}\text{P} + {}^{1}_{0}\text{n}$$

Since then, many transformations of one element to another element have been accomplished. In addition, new radioactive isotopes of various elements have been synthesized. Many diagnostic radioisotopes noted in Figure 30 are synthetic. Tc-99m, for example, is both a synthetic element and a radioisotope.

Most synthetic radioisotopes are produced by bombarding elements with neutrons, resulting in their capture by target nuclei. This requires less energy than many other bombardment reactions, since neutrons have no electrical charge and thus are not repelled by the positive charge of the nucleus. Such reactions produce radioactive nuclei that tend to give off beta particles.

The examples below show the formation of two synthetic radioactive isotopes often used as medical tracers, calcium-45 from calcium-44 and radioactive iron-59 from iron-58.

$$^{44}_{20}\text{Ca} + {}^{1}_{0}\text{n} \longrightarrow {}^{45}_{20}\text{Ca}$$

$$^{58}_{26}\text{Fe} + {}^{1}_{0}\text{n} \longrightarrow {}^{59}_{26}\text{Fe}$$

BOMBARDMENT REACTIONS

Building Skills 5

Nuclear-bombardment reactions generally involve four particles:

- **Target nucleus**—the stable isotope that is bombarded.
- **Projectile particle** (bullet)—the particle fired at the target nucleus.
- **Product nucleus**—the isotope produced in the reaction.
- **Ejected particle**—the lighter nucleus or particle emitted from the reaction.

For example, consider the Joliot-Curies' production of the first synthetic radioactive isotope, phosphorus-30. The four types of particles involved are identified below.

$^{27}_{13}\text{Al}$	+	$^{4}_{2}\text{He}$	$\longrightarrow$	$^{30}_{15}\text{P}$	+	$^{1}_{0}\text{n}$
Aluminum-27		Alpha particle		Phosphorus-30		Neutron
Target nucleus		Projectile particle		Product nucleus		Ejected particle

Nobelium (No) can be produced by bombarding curium (Cm) with projectile nuclei of a light element. That element can be identified by completing the following equation:

$$^{246}_{96}\text{Cm} + \textbf{?} \longrightarrow {}^{254}_{102}\text{No} + 4\,{}^{1}_{0}\text{n}$$

More than one ejected particle may be released. See the Cm-246 example below.

BOMBARDMENT REACTIONS

Building Skills 5

The four types of particles involved in nuclear bombardment reactions are defined and used in writing equations for bombardment reactions.

	Target nucleus		Projectile particle		Product nucleus		Ejected particle
1.	$^{59}_{27}\text{Co}$	+	$^{1}_{0}\text{n}$	$\rightarrow$	$^{60}_{27}\text{Co}$		
2.	$^{96}_{42}\text{Mo}$	+	$^{2}_{1}\text{H}$	$\rightarrow$	$^{97}_{43}\text{Tc}$	+	$^{1}_{0}\text{n}$
3.	$^{58}_{26}\text{Fe}$	+	$^{209}_{83}\text{Bi}$	$\rightarrow$	$^{266}_{109}\text{Mt}$	+	$^{1}_{0}\text{n}$

TRANSMUTATION OF ELEMENTS

ChemQuandary 5

Lead and iron could be changed into gold; however, the method would probably be far too expensive to be profitable due to the huge equipment and energy costs required to make the conversion. Reactions of the type used to produce elements 109–114 could possibly be used. A possible reaction might be $^{1}_{0}n + ^{207}_{82}Pb \rightarrow ^{197}_{79}Au + ^{7}_{3}Li + 4^{1}_{0}n$.

C.5 EXTENDING THE PERIODIC TABLE

The controversy over the naming of new elements highlights the complex and political world of modern-day science. The processes involved in this type of research are very expensive and complicated. In addition, there is a great deal of communication among scientists through publication in research journals and attendance at scientific meetings. Research proceeds more like the activity of a corporation than like a lone scientist toiling at a lab bench. In the naming of the new elements, national pride was quite evident in the opposing claims.

As you learned earlier, completing nuclear equations involves balancing atomic and mass numbers.

Because the sum of the product atomic numbers is 102, the projectile must have the atomic number 6, making the sum of reactant atomic numbers also 102. Therefore, the projectile must be a carbon atom. The total mass numbers of products is 258, indicating that the projectile must have been carbon-12 (246 + 12 = 258). The completed equation is

$^{246}_{96}Cm$	+	$^{12}_{6}C$	$\longrightarrow$	$^{254}_{102}No$	+	$4\,^{1}_{0}n$
Target nucleus		Projectile particle		Product nucleus		Ejected particle

Complete the following equations by supplying the missing numbers or symbols. Name each particle. Then identify the target nucleus, projectile particle, product nucleus, and ejected particle.

1. $^{59}_{27}? + ^{?}_{?}n \longrightarrow ^{60}_{?}?$ (Most medically useful isotopes today are produced by bombarding stable isotopes with neutrons. This process converts the original nuclei to a radioactive form of the same element.)
2. $^{96}_{42}? + ^{?}_{?}H \longrightarrow ^{97}_{43}? + ^{1}_{0}?$ (Until its synthesis in 1937, technetium, Tc, existed only as an unfilled gap in the periodic table; all its isotopes are radioactive. Any technetium originally on Earth has decayed. Technetium, the first element artifically produced, is now used extensively in industry and medicine. For example, each year millions of bone scans are obtained using technetium.)
3. $^{58}_{?}? + ^{209}_{?}Bi \longrightarrow ^{?}_{109}Mt + ^{1}_{0}?$ (In 1992 the GSI research group in Darmstadt, Germany, created element 109 by bombardment of Bi-209 nuclei. The name meitnerium, symbol Mt, has been chosen for this new element to honor Lise Meitner, an Austrian physicist who first proposed the concept of nuclear fission. See page 464.)

Not only does the ability to transform one element into another provide new and powerful technological capabilities, it also has changed how elements are viewed.

TRANSMUTATION OF ELEMENTS

ChemQuandary 5

Ancient alchemists searched in vain for ways to transform lead or iron into gold (the transmutation of one element into another). Has such transmutation now become a reality? From what you know about nuclear reactions, do you think that lead, iron, or even mercury atoms could be changed to gold? If so, try writing equations for the possible reactions.

C.5 EXTENDING THE PERIODIC TABLE

Since 1940, 23 **transuranium** elements—elements with atomic numbers greater than the atomic number of uranium, 92—have been added to the Periodic Table. These elements have been synthesized in nuclear reactions.

From 1940 to 1961, Glenn Seaborg and coworkers at the University of California at Berkeley discovered ten new elements with atomic numbers 94 to 103—a prodigious feat. None of those ten elements occur naturally. All are made by high-energy bombardment of heavy nuclei with various particles. For example, alpha-particle bombardment of plutonium-239 produced curium-242:

$$^{239}_{94}\text{Pu} + ^{4}_{2}\text{He} \longrightarrow ^{242}_{96}\text{Cm} + ^{1}_{0}\text{n}$$

Bombarding Pu-239 with neutrons yielded americium-241, a radioisotope now used in home smoke detectors:

$$^{239}_{94}\text{Pu} + 2\,^{1}_{0}\text{n} \longrightarrow ^{241}_{95}\text{Am} + ^{0}_{-1}\text{e}$$

Seaborg's work was internationally recognized through the 1951 Nobel prize in chemistry. Albert Ghioso, a colleague of Seaborg, has led the way in producing a number of new elements beyond lawrencium (element 103). One such element is 106, produced by bombarding a californium-249 target with a beam of oxygen-18 nuclei, producing an isotope of element 106. To honor Seaborg's pioneering work, element 106 has been formally named seaborgium (Sg). Glenn Seaborg, shown in Figure 31 with colleague Darleane Hoffman, can properly be called the father of the modern Periodic Table.

Traditionally the discoverer of an element selects its name. For example, when Marie Curie first discovered element number 84 she named it polonium (Po) in honor of Poland, her home country. However, several scientific laboratories claimed to have synthesized elements with atomic numbers greater than 92. For example, discovery of elements 104 and 105 were claimed by both Soviet and U.S. laboratories. The Soviets proposed the names kurchatovium (Ku) and dubnium (Db), while U.S. scientists proposed the names rutherfordium (Ru) and hahnium (Ha). The International

Scientific research—like other human endeavors—can involve strong personalities, competition, and controversy.

Figure 31 *Darleane Hoffman and Glenn Seaborg.*

459

C.6 OPINIONS ABOUT RADIOACTIVITY

Making Decisions

Making decisions is another activity in helping students prepare for the PIAT at the end of this unit. Have students compare their responses with that of other classmates. Possible answers might include:

1. Isotopes are all around us; they occur naturally; some are radioactive while others are not. Sources of isotopes include groundwater, air, rocks, sand, and stars; and even our own bodies contain isotopes.
2. It is impossible to live near anything that is not radioactive. Radioactive isotopes are a part of building materials, medical and dental X-rays, television sets, and air travel.
3. The production of new elements has expanded our understanding of the atomic nucleus, and these new elements have basically replaced the need for exploratory surgery in many medical diagnoses.
4. Carbon-14 dating is used to determine the age of the Egyptian mummies. Carbon-14 decays naturally with a half-life of 5730 years. As long as plants or animals are alive, the amount of carbon-14 present in the organism is the same as the carbon-14 in the atmosphere. Since the organism does not exchange carbon with the atmosphere when it dies, the carbon-14 lost to radiation is not replaced.
5. Radioisotopes can kill cancerous cells and treat tumors. These radioisotopes are carefully selected for the area of the body to be treated and administered with care to prevent exposing healthy cells to ionizing radiation.

Union of Pure and Applied Chemistry (IUPAC) examines claims for element discovery and later confirming evidence before recommending official names for the elements. In 1997 the IUPAC formalized the name rutherfordium for element 104 and dubnium for element 105.

At present, the IUPAC recognizes official names for the first 109 elements. Although claims for the discovery of six additional elements have been filed, the IUPAC has not yet recommended their official names. Such unnamed elements are temporarily identified by Latin prefixes indicating their atomic numbers. For example, element 110 is temporarily called ununnilium (un = 1, un = 1, nil = 0), symbolized Uun, until the IUPAC makes a decision on the original discovery of that element.

Few transuranium elements have significant industrial or medical uses at present. Production of these elements has expanded our understanding of the atomic nucleus. With the synthesis and identification of elements beyond atomic number 92, the Periodic Table has expanded to fill out the **actinide series** as well as nearly all of Period 7. At the start of this new millennium, only three of the first 118 elements still remained undiscovered—elements 113, 115, and 117.

C.6 OPINIONS ABOUT RADIOACTIVITY

Making Decisions

Many older people tend to associate nuclear technologies only with the use of atomic bombs during World War II and tensions associated with the Cold-War era of the 1950s and beyond. Some of these people are likely to be in the audience when you speak to the Riverwood senior citizens. The following opinions might be expressed by such community members. Decide how you would respond to each opinion, using knowledge you have gained in this section and in other parts of the unit.

1. "I'm against the presence or use of any isotopes in Riverwood. They're too dangerous."
2. "I'm opposed to living near anything that is radioactive."
3. "I don't know why scientists keep trying to make new elements—all the new ones are radioactive."
4. "I don't know how radioactivity can be used to date those old Egyptian mummies. No one has been around that long."
5. "I don't understand how they can treat cancer with radiation. I thought radiation caused cancer."

Questions & Answers CD-ROM WWW.

Join with a classmate and share your responses to each listed opinion. How are your responses similar? How are they different?

CHEMISTRY AT WORK

High-Tech Soldiers in the War Against Cancer

Beverly Buck works in radiation oncology, the field of medicine that uses radiation to treat patients who have tumors.

Beverly is a radiation therapist at the Joint Center for Radiation Therapy in Boston. She also serves as education and development coordinator, making sure patients are properly treated and overseeing the systems that monitor the radiation equipment. Beverly and other radiation therapists at the center deliver the actual dosages of radiation prescribed by each patient's radiation oncologist (cancer physician).

Radiation therapy is one weapon among several in the fight against cancer, and is often used in combination with chemotherapy. Radiation therapy is associated with risks; high-energy radiation can genetically alter or kill normal body cells, but it can also kill cancer cells. With any cancer treatment, the patient, the patient's family, and the oncologist must carefully weigh the potential benefits and risks.

While radiation therapists must understand the science and technology behind radiation oncology, it's also critical that they focus on each patient. Most cancer patients are going through major crises in their lives, so radiation therapists try to recognize and address patient fears and emotions.

To become a radiation therapist, a person must go through an educational program that involves at least two years of study. Students learn about nuclear chemistry, anatomy, physiology, radiation physics, radiobiology, radiation safety, pathology, oncology, and patient-care methods.

Activity

Chances are that near you there is a health facility that employs radiation therapists.

1. If you were interested in applying for a job opening there, what type of training would you need? Where could you get such training?
2. Would you need any type of state license or certification? If so, what is involved in getting licensed or certified?
3. What other skills or abilities do you think would be useful for the job?

Radiation therapy is one weapon among several in the fight against cancer. . . .

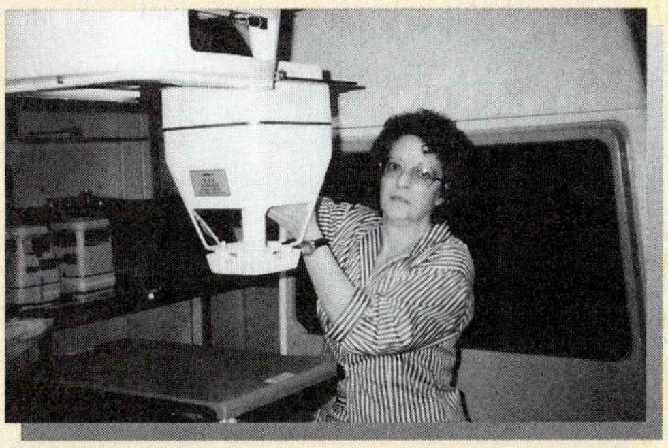

Photograph courtesy of Beverly Buck

461

SECTION SUMMARY

Reviewing the Concepts

1. How can the concept of half-life be used to determine the age of a substance?

The concept of half-life may be used to determine the age of organic substances by using carbon-14 dating. Carbon-14 decays naturally with a half-life of 5730 years. As long as plants or animals are alive, the amount of carbon-14 present in the organism is the same as the carbon-14 in the atmosphere. Since the organism does not exchange carbon with the atmosphere when it dies, the carbon-14 lost to radiation is not replaced. Therefore, the fraction of carbon-14 to carbon-12 can be used to determine the age of the organism.

2. A student wrote this statement on a homework assignment: "After one half-life, half of the mass of a material is gone." Do you agree or disagree? Explain.

Disagree: One half-life refers to the time required for one-half of the total atoms originally present in a sample to decay. The decayed atoms do not disappear but are just changed.

3. The half-life of carbon-12 is 5730 years. The half-life of astatine-209 is 5.4 hours. Provide rough estimates of the percent of the original radioisotopes that would be left

a. after 24 hours.

b. after 100 years.

a. Carbon-12: 100% after 24 hours. Astatine-209: About 4% after 24 hours.

b. Carbon-12: About 98% left after 100 years. Astatine-209: Virtually none left after 100 years.

4. Scientists originally were uncertain whether the oxygen gas produced during photosynthesis came from CO_2, from H_2O, or from both. How could radioisotope tracers be used to settle that question?

Radioactive isotopes of oxygen should be introduced into the CO_2 or the H_2O source, then use a radiation detection system to trace the radioactive molecules through the plant.

5. Give one example of how a radioactive tracer can be used for diagnostic purposes.

A patient is injected or fed a radioactive sample of a biological compound that is active in a particular organ. The rate at which the tracer is taken up by that organ is measured with a radiation detection system and compared to the uptake in the same organ of a healthy individual.

6. Why is a radioisotope of iodine used for detecting thyroid problems?

A healthy thyroid is known to incorporate a certain amount of iodine. When a patient is given a solution containing a radioactive iodine tracer (I-123), the rate of the uptake of the radioactive iodine is compared to that of a healthy thyroid.

7. What makes a radioisotope suitable or unsuitable for diagnostic use?

A radioisotope used for diagnostic purposes must have a short half-life and must be conducive to the particular organ it is being used to diagnose. It must also emit a suitable level of radiation.

8. Medical personnel are selecting a radioisotope for diagnostic use. Why is each of the following considerations important in selecting the most suitable radioisotope?

a. half-life

b. mode of decay

c. chemical properties of the element

a. Half-life—Must have short half-life, decay rapidly, and leave the body of the patient quickly.

b. Mode of decay—Since alpha particles can cause extensive damage inside the body, the tracer should not decay by alpha emission.

c. Chemical properties must behave the same as chemically stable isotopes of the element.

9. How are cancer cells different from normal cells?

Cancer cells divide faster than normal cells. They are also undifferentiated.

10. How can ionizing radiation be used to treat cancer?

Ionizing radiation kills the cancer cells when directed at the specific spot of the cancer. Caution: Ionizing radiation can also damage normal cells.

11. What could happen if the source of ionizing radiation used to treat cancerous growth were

a. too weak?

b. too strong?

a. Too weak—Does not kill all of the cancer cells, which may continue to divide and allow the cancer to spread.

b. Too strong—Can cause damage to surrounding normal cells.

12. Radioactive sodium chloride is appropriate for diagnosing circulatory problems, but radioactive xenon is helpful in searching for lung problems. Explain why.

Sodium ions are normally found dissolved in blood, while the gas xenon is normally found in the lungs.

13. Explain why an externally applied alpha emitter is an ineffective treatment for tumors deep within the body.

Alpha particles have little penetrating power. If the source is outside of the body, the skin blocks their entry.

14. European alchemists of medieval times sought to change lead into gold. Although this feat may now be possible with modern technology, the transmutation of lead to gold is still impractical. Explain why.

The process is too expensive.

15. Complete each of the following transmutation equations:

a. _____ $\longrightarrow {}^{4}_{2}\alpha + {}^{259}_{104}\text{Rf}$ — **a. Sg-263**

b. ${}^{238}_{92}\text{U} \longrightarrow$ _____ $+ {}^{234}_{90}\text{Th}$ — **b. α**

c. ${}^{210}_{83}\text{Bi} \longrightarrow {}^{0}_{-1}\beta +$ _____ — **c. Po-209**

d. ${}^{95}_{40}\text{Zr} \longrightarrow$ _____ $+ {}^{95}_{41}\text{Nb}$ — **d. β**

e. _____ $\longrightarrow {}^{4}_{2}\alpha + {}^{214}_{80}\text{Hg}$ — **e. Pb-218**

Extending the Concepts

16. Would carbon dating be useful to determine the age of dinosaur remains? Explain.

No. Carbon-14 would not be useful to determine the age of dinosaur remains. Dinosaurs lived too long ago, outside the range of accuracy of C-14 dating.

17. The half-life of potassium-42 is 12.4 hours. How much of the original isotope of a 750-g sample is left after 62.0 hours?

Time	**Mass remaining**
12.4 h	**375 g**
24.8 h	**187.5 g**
37.2 h	**93.75 g**
49.6 h	**46.875 g**
62.0 h	**23.44 g**

18. Explain why the disposal of nuclear waste is a challenging issue in many nations, using the concepts of radioactive decay, half-life, and radiation shielding.

The half-lives of radioactive materials are unalterable by any known means. We must be able to isolate most of our nuclear wastes for thousands of years from both human (terrorism and war) and natural (water seepage and geological shifts) intrusion. The shielding material must block beta and gamma radiation and withstand thermal, mechanical, and chemical stress.

Connecting the Concepts

19. Scientists doing carbon-14 dating generally do not go beyond seven half-lives.

a. Explain why that is so.

b. Given that information, what is the maximum number of years a substance can be dated using carbon-14 dating?

a. After seven half-lives, the amount of the original sample remaining will be less than one percent.

b. The half-life of carbon-14 is 5730 years. Therefore, seven half-lives would represent 40 110 years.

20. A patient is injected with a radioisotope tracer sample registering 10 000 cpm (counts per minute) of radioactivity. Later, 6 mL of blood is drawn, and the sample shows an activity of 10 cpm. What is the total blood volume of the patient?

10 000 cpm for 6 mL of blood: 10 cpm for 6 liters of total blood volume (assuming that the source of radioactivity did not decay significantly during the short time before the blood was drawn).

21. Scientists have yet to find a way to change the half-life of any radioactive element. What problems does this fact raise for the storage of radioactive waste?

It means the material must be stored for at least seven half-lives.

NUCLEAR ENERGY—BENEFITS AND BURDENS

Introduction

Students are introduced to the huge quantity of energy that is locked inside the atom. These forces are released via chain reactions in nuclear power plants, in nuclear weapons, and on the Sun and other stars. Major components of a nuclear power plant are pictured and discussed. The risks and benefits of nuclear reactions are explored. The problem of disposing of high and low levels of nuclear waste is presented for consideration.

D.1 UNLEASHING NUCLEAR FORCES

Students are introduced to the strong force that holds the protons and neutrons together in the nucleus. This force is unleashed by bombarding the nucleus of a fissionable isotope. If a sufficient amount of the isotope is present (critical mass), a chain reaction results. Energy changes in nuclear reactions are compared and contrasted with those of chemical reactions. You may want to emphasize that a small quantity of mass, if converted into energy, yields an enormous quantity of energy, according to the equation $E = mc^2$.

SECTION D

NUCLEAR ENERGY—BENEFITS AND BURDENS

Introduction CD-ROM WWW.

In the 1930s a bombardment reaction involving uranium unlocked a new energy source and led to the development of both nuclear power and nuclear weapons. This marked the start of the nuclear age. How did scientists first unleash the enormous energy of the atom, and how has atomic energy been harnessed for both useful and destructive purposes?

D.1 UNLEASHING NUCLEAR FORCES

Shortly before the start of World War II, German scientists Otto Hahn and Fritz Strassman bombarded uranium with neutrons in the hope of creating a more massive nucleus. Much to their surprise, they found that one reaction product was atoms of barium—with only about half the atomic mass of the target uranium atoms.

Lise Meitner, a Jew, had fled to Sweden when Nazis assumed control of Germany and Austria.

The first to understand what had happened in the procedure was the Austrian physicist Lise Meitner, then living in Sweden, who had earlier worked with Strassman and Hahn. Meitner, working in collaboration with her nephew Otto Frisch, suggested that neutron bombardment had split the uranium atom into two parts of nearly equal size. Other scientists quickly verified Meitner's explanation.

Hahn and Strassman had actually triggered an array of related reactions. One of the reactions that produced barium is this:

$$\underset{\text{Neutron}}{{}^{1}_{0}\text{n}} + \underset{\text{Uranium-235}}{{}^{235}_{92}\text{U}} \longrightarrow \underset{\text{Barium-140}}{{}^{140}_{56}\text{Ba}} + \underset{\text{Krypton-93}}{{}^{93}_{36}\text{Kr}} + \underset{\text{Neutrons}}{3\,{}^{1}_{0}\text{n}} + \text{Energy}$$

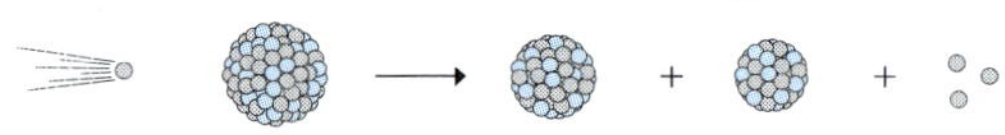

Splitting an atom into two smaller atoms as shown above is known as **nuclear fission.** Scientists soon found that the uranium-235 nucleus can fission (split) into numerous combinations of atoms. Usually the uranium did not split into two equal halves but into one element accounting for about 60% of uranium's mass (such as barium) and another element equivalent to about 40% of the uranium's mass (such as krypton). Here is another example of a nuclear fission reaction involving uranium-235:

$$\underset{\text{Neutron}}{{}^{1}_{0}\text{n}} + \underset{\text{Uranium-235}}{{}^{235}_{92}\text{U}} \longrightarrow \underset{\text{Xenon-143}}{{}^{143}_{54}\text{Xe}} + \underset{\text{Strontium-90}}{{}^{90}_{38}\text{Sr}} + \underset{\text{Neutrons}}{3\,{}^{1}_{0}\text{n}} + \text{Energy}$$

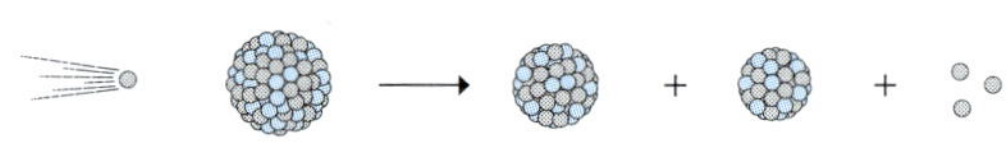

The nuclear fission of heavy atoms such as uranium releases a huge quantity of energy—gram for gram, at least a million times more than is produced in any chemical reaction. This is what makes nuclear explosions so devastating and nuclear energy so powerful.

Why does a nuclear reaction release so much more energy than a chemical reaction? Recall what you know about chemical reactions, such as burning petroleum. Chemical reactions involve breaking chemical bonds in reactants and making new chemical bonds to form products. When bonds are stronger in products than in reactants, energy is released—often as thermal energy (heat). Thus chemical energy is converted into thermal energy. No overall energy loss or gain occurs. Similarly, mass is conserved in a chemical reaction. The rearrangement of electrons and atoms accounts for chemical changes, but the nucleus of each atom and thus its identity, remains intact throughout all chemical reactions. As a result, the number of atoms of each element remains unchanged—the atoms simply become rearranged. Balanced chemical equations illustrate this conservation of atoms and mass.

Not all nuclei are fissionable. Uranium-235 is the only naturally occurring isotope that undergoes fission with lower-energy (thermal) neutrons. However, many synthetic nuclei—in particular uranium-233, plutonium-239, and californium-252—also fission under neutron bombardment.

Nuclear reactions are also based on conserving energy and mass. However, during nuclear fission, very small quantities of mass are converted into appreciable quantities of energy. Where does this energy originate?

The source of nuclear energy lies in the force that holds protons and neutrons together in the nucleus. This force, called the **strong force,** is fundamentally different from—and a thousand times stronger than—the electrical forces that hold atoms and ions together in chemical bonds. The strong force operates over very short distances, extending only across an atom's nucleus.

It would take the fissioning of about a kilogram of uranium-235 to cause one gram of mass to be converted into energy.

Just as chemical bonds in products of exothermic reactions are stronger than those in the reactants involved, forces holding nuclear particles together in the two atomic nuclei produced during nuclear fission are stronger than those in the nucleus of the atom that was split. A small loss of mass results from the process of forming two new nuclei. This mass is converted into large quantities of energy.

How much mass and energy is involved? The mass loss is very small, often less than 0.1% of the total mass of the fissioning atom. Even so, the conversion of these small quantities of mass into energy accounts for the vast power of nuclear energy. Albert Einstein's famous equation relates mass and energy: $E = mc^2$. This equation indicates that the energy released (E) equals the mass lost (m) multiplied by the speed of light (a very large number) squared (c^2). If one gram of matter were fully converted to energy, the energy released would equal that produced by burning 700 000 gallons of high-octane gasoline!

One gram of matter (1×10^{-3} kg) times the speed of light (3×10^8 m/s) squared yields 9×10^{13} J of energy.

Such nuclear energy release has been harnessed by humans to generate electricity and to create atomic weapons. However, the fission of one atom does not produce energy sufficient for practical use. How are fission reactions sustained to involve much larger quantities of atoms?

Note from the equations on page 464 that another product of nuclear fission is the emission of two or three neutrons. It is these emitted neutrons that can sustain the fission reaction by serving as reactants to split additional fissionable nuclei, which produce additional neutrons, which can split additional fissionable nuclei, and so on. The result is a **chain reaction.**

Recall, however, that most of an atom is empty space. The probability that a neutron from a fission reaction will hit and split another fissionable nucleus depends on the amount of fissionable material available. Unless a certain **critical mass** (minimum quantity) of fissionable material is present, the neutrons are not likely to encounter enough fissionable nuclei to sustain the reaction.

However, if a critical mass of fissionable material is present, a chain reaction can occur as depicted in Figure 32. Recognition that such large-scale nuclear reactions were possible and could be employed in military weapons came shortly after the first fission reactions were explained in 1939. Germany and the United States soon initiated projects to build "atomic bombs" during World War II. In 1945, two such bombs were dropped by U.S. aircraft on Hiroshima and Nagasaki in Japan, leading rapidly to the end of World War II.

More recently the energy produced by nuclear fission chain reactions has been used to generate electrical power. However, the rate of fission is carefully monitored and controlled for such uses. Nuclear power plants have been designed to utilize the enormous energy produced by nuclear fission reactions while minimizing risks of an uncontrolled chain reaction. You will soon learn more about those design features.

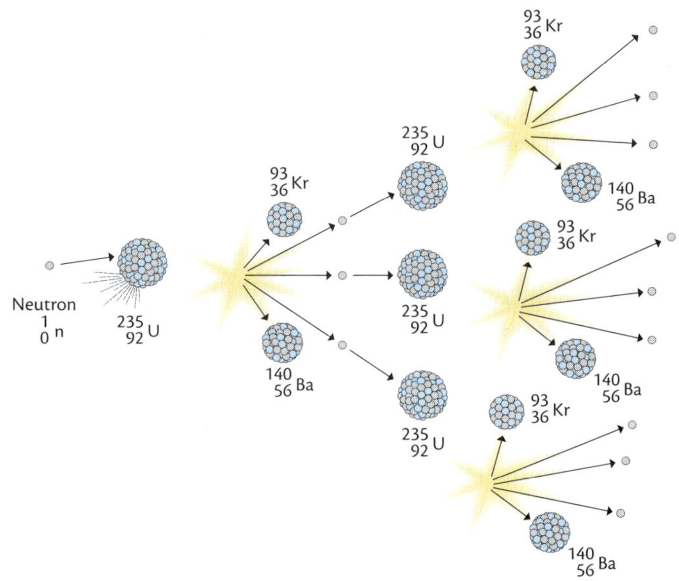

Figure 32 *A nuclear chain reaction. The reaction is initiated (left) by a neutron colliding with a uranium-235 nucleus. The reaction continues and increases, in a chainlike fashion, as long as emitted neutrons encounter and split the nuclei of other fissionable atoms.*

MODELING MATTER

THE DOMINO EFFECT

This activity demonstrates the concept of expanding versus limited chain reactions. Each student group should be provided with a standard set of 28 dominoes. You will need two to three sets of dominoes for a demonstration.

Answers

1. a. No answer required

   b. students' drawings

   c. Each domino that fell caused two additional dominoes to fall.

   d. Each neutron produced two additional neutrons that can bombard two additional U-235 atoms and two more

MODELING MATTER

THE DOMINO EFFECT

Chain reactions that sustain nuclear fission reactions are used in applications such as electrical power generation and atomic weapons. In this activity, dominoes will model some aspects of a chain reaction. Each domino that falls over will represent an atom that has been split during fission. Figure 33 shows one way dominoes can be set up so that causing one domino to fall causes all dominoes to fall. Since you will model a specific fission reaction, your models will not match Figure 33.

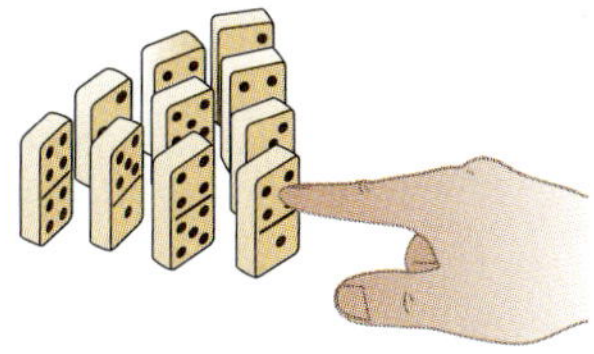

Figure 33 *Dominoes are useful in modeling chain reactions.*

1. Over 100 pathways are known for the fission of a uranium-235 nucleus. One produces tellurium-137 and zirconium-97:

   $${}^{1}_{0}n + {}^{235}_{92}U \rightarrow {}^{137}_{52}Te + {}^{97}_{40}Zr + 2\,{}^{1}_{0}n + \text{Energy}$$

   As the equation shows, two neutrons are released from splitting one U-235 atom.

   a. Set up all the dominoes supplied by your teacher so that each falling domino will cause two more erect dominoes to fall.
   b. Draw a sketch of your setup.
   c. Push over the first erect domino and record what happens.
   d. Explain how this models the release of neutrons during the fission of U-235 as shown in the equation above.
   e. What aspects of the U-235 fission equation are not represented in your model?
2. You learned that a critical mass of fissionable atoms is needed to create the conditions for a chain reaction.
   a. Set up the dominoes as in Question 1a.
   b. Assume that only dominoes (atoms) with seven total dots are fissionable.
   c. Carefully remove all dominoes from your setup that do not have seven total dots.
   d. Draw a sketch of your new setup.
   e. Push over the first erect domino and record what happens.
   f. In what way does this model help clarify the idea of critical mass?
   g. How would you ensure that you had a critical mass of "fissionable" dominoes in a particular setup?
3. Suppose you want to control the number of neutrons allowed to sustain fission.
   a. Set up the dominoes as in Question 1.
   b. Devise a plan so that only half of the dominoes will fall down after the first one is pushed. You should not remove any of the dominoes that are already set up.
   c. Describe and sketch your strategy.
   d. Carry out your plan and record what happens.
   e. What stopped the dominoes from falling?
   f. Use the domino model to propose a way to control fission chain reactions.
4. a. Which domino arrangement (that in Question 1 or in Question 3) is a better model of an atomic-weapon explosion? Explain.
   b. Which domino arrangement (that in Question 1 or in Question 3) is a better model of fission in a nuclear power plant? Explain.
5. Propose another way to model a nuclear chain reaction.
   a. Explain in what ways your model illustrates features of a nuclear chain reaction.
   b. Explain the limitations of your model.

neutrons. This produces an expanding (out of control) chain reaction because the number of falling dominoes increases with each row.

e. The new atoms formed during the chain reaction are not represented in the model. This model is linear; nuclear fission is three-dimensional. Little energy is released from the dominoes.

2. a. no answer required

   b. no answer required

   c. no answer required

   d. students' drawings

   e. A few dominoes may fall, but the "reaction" stops quickly when a domino is missing.

   f. If there are no adjacent atoms, neutrons will not find a target and the reaction will stop.

   g. Dominoes have to be close enough so that one strikes another.

3. a. (No answer required)

   b. Each domino is arranged so that it strikes only one other domino.

   c. Student drawings will vary.

   d. Half the dominoes fall.

   e. One domino was struck by a falling domino. One of the impacts from Question 1 setup was prevented.

   f. Control rods are inserted to absorb some of the neutrons and prevent them from striking other uranium atoms.

4. a. The arrangement of the dominoes in Question 1 models the explosion of a nuclear weapon, in which a tremendous amount of energy is released in a very short time.

   b. The arrangement of the dominoes in Question 3 models a nuclear power plant. In this case the energy is released over a longer period of time and is kept under control.

5. One model might be to spread news via a "calling tree" vs. one-way calling. If one person is not at home, the "chain reaction" stops for that section. Other modeling alternatives include loaded mousetraps and Ping-Pong balls in a closed container.

   a. This simulation is like a nuclear chain reaction in that the action proceeds spontaneously at ever-increasing rates.

   b. The simulation is limited in that it does not produce new substances or produce significant quantities of energy.

D.2 NUCLEAR POWER PLANTS

This section deals with the basic components and operation of nuclear power plants. The functions of plants are explained, with diagrams and text. Several different types of reactors are in use: pressurized water, boiling-water, high-temperature gas-cooled, and breeder reactors. The text explains pressurized and boiling-water reactors.

D.2 NUCLEAR POWER PLANTS

The first nuclear reactors were designed and built during World War II. Since the end of that war, many nuclear reactors have been built for the purpose of generating electricity. In the year 2000, 104 commercial nuclear reactors like that shown in Figure 34 generated approximately one-fifth of the electricity in the United States. Figure 35 shows the locations of these reactors. On a global scale, an estimated 426 nuclear reactors in over 30 nations produce about one-sixth of the world's electricity.

Most power plants generate electricity by burning fossil fuels to boil water and produce steam. A nuclear power plant operates in much the same way. However, unlike coal-, oil-, and natural gas-powered generators that use the heat of fossil-fuel combustion to boil water, nuclear power plants use the thermal energy released during nuclear fission reactions to heat water and produce steam. The steam resulting from either process spins the turbines of giant generators, producing electrical energy.

The essential parts of a nuclear power plant, diagrammed in Figure 36, include the fuel rods, control rods, moderator, generator, and cooling system.

Fuel Rods

Coal-fired power plants consume thousands of tons of coal daily. In contrast, the fuel for a nuclear reactor occupies a fraction of the volume needed for coal, and is replenished only about once each year. Nuclear reactor fuel is in the form of uranium dioxide pellets about the size and shape of short pieces of chalk. The energy contained in one uranium fuel pellet is the equivalent of the energy contained in 1 ton of coal or 126 gallons of petroleum. As many as 10 million fuel pellets may be used in one nuclear power plant. These pellets are arranged in long, narrow steel cylinders—the fuel rods. The fission chain reaction takes place inside these rods.

Figure 34 *A nuclear power plant. Notice the reactor (containment) building with its domed roof.*

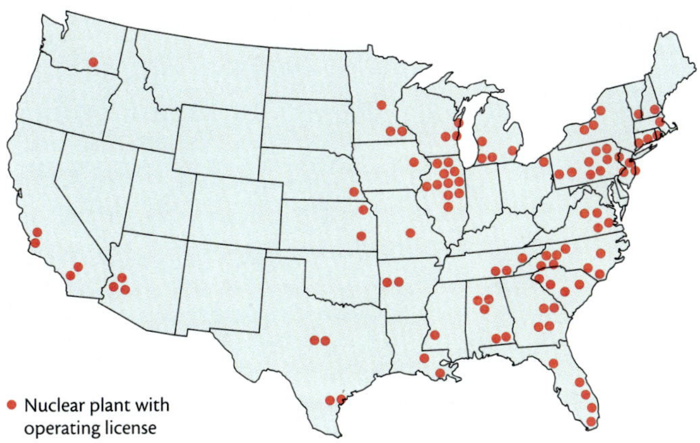

Figure 35 *Locations of all licensed U.S. nuclear power plants.*

Most of the uranium dioxide (UO_2) in fuel pellets contains the non-fissionable uranium-238 isotope. Only 0.7% of natural uranium is U-235, the fissionable isotope. In a nuclear reactor, this quantity of U-235 has been "enriched" to about 3%, still only a small fraction of the total material. It is enough to sustain a chain reaction, but far from enough to cause a nuclear explosion.

Weapons-grade uranium usually contains 90% or more of the fissionable U-235 isotope.

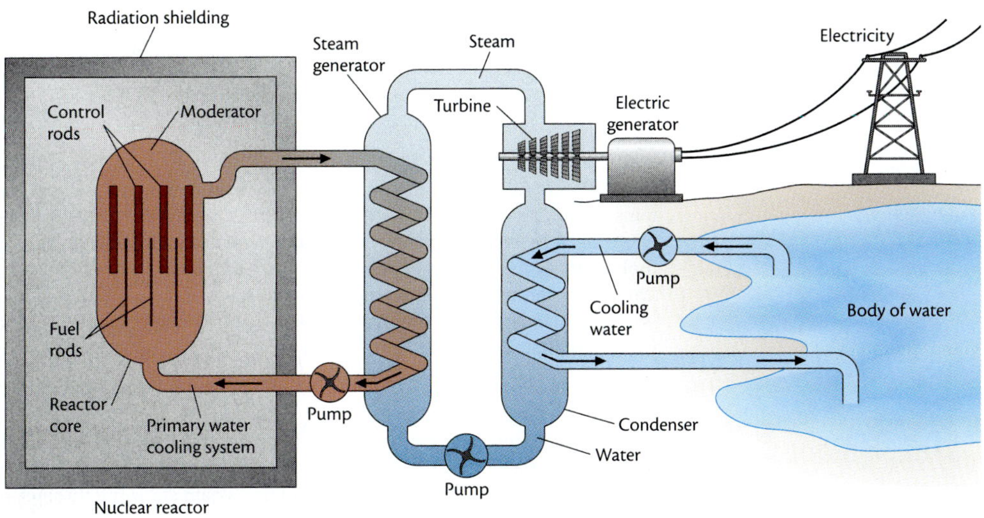

Figure 36 *The parts of a nuclear power plant. Note the flow of water between the reactor and the steam generator and between the steam generator and the outside body of water. The water flow is important to thermal regulation in the power plant.*

Figure 37 *A nuclear reactor core.*

Control Rods

The nuclei of certain elements, such as boron or cadmium, can absorb neutrons very efficiently. Such materials are placed in control rods, which regulate the number of neutrons available for causing fission. The rate of the nuclear chain reaction is controlled by moving the control rods up or down (see Figure 37) between the fuel rods. The reaction can be terminated altogether by dropping the control rods all the way down between the fuel rods and absorbing nearly all of the neutrons released when the U-235 fissions.

Moderator

In addition to fuel rods and control rods, the core of a nuclear reactor contains a moderator, which slows down high-speed neutrons. This allows the neutrons to be more efficiently absorbed by fuel rods, which enhances the probability of fission. Heavy water (with the formula D_2O, where D is the symbol for the hydrogen-2 isotope), regular water (termed light water by nuclear engineers), and graphite (carbon) are the three most common moderator materials.

Generator

In commercial nuclear reactors, the fuel and control rods are usually surrounded by a system of circulating water. In simpler reactors, the heat from the fuel rods boils this water, and the resulting steam spins the turbines of

470

the electrical generator. In another type of reactor, this water is superheated under pressure and does not boil. Instead, it circulates through a heat exchanger, where it boils the water in a second cooling loop. This is the type of reactor shown in Figure 36.

Cooling System

Steam that has moved past the turbines then travels through pipes where it is cooled by water drawn from a nearby body of water. The cooled steam condenses to liquid water and recirculates inside the generator. So much thermal energy is generated that some must also be released into the air. The largest and most prominent features of most nuclear power plants are tall, gracefully tapered concrete cylinders—the cooling towers—where excess thermal energy is released. Some people mistakenly assume that the cooling tower is the nuclear reactor.

The white plumes seen rising from such cooling towers are composed of condensed steam, not smoke.

A nuclear reactor is designed to prevent the escape of radioactive material, should some malfunction cause radioactive material—including cooling water—to be released within the reactor itself. The core of a nuclear reactor is surrounded by concrete walls two to four meters thick. Further protection is provided by enclosing the reactor in a building with thick walls of steel-reinforced concrete, designed to withstand a chemical explosion or an earthquake. The reactor building is also capped with a domed roof that can withstand significant internal pressure.

The well-known nuclear accident at one of the four reactors at the Chernobyl power plant in the Ukraine in 1988 occurred because too many control rods were withdrawn from the reactor and were not replaced fast enough. Thus, there was little control of the fission process and considerable steam was produced. The resulting explosion was not nuclear but was due to the build-up of steam and to chemical reactions triggered by the high-temperature steam. Unfortunately, the plant had been built without a surrounding concrete containment building, so a large quantity of nuclear material from the reactor was released directly into the environment.

In the Chernobyl accident, one chemical reaction that caused trouble was high-temperature steam reacting with carbon from the moderator to produce CO and H_2. When the H_2 mixed with air, it became explosive.

Nuclear fission is not the only way to release nuclear energy. You will next learn about the kind of nuclear reaction that fuels the stars (and, indirectly, all living matter).

D.3 NUCLEAR FUSION

In addition to releasing energy by splitting massive nuclei (fission), large quantities of nuclear energy can be generated by fusing—or combining—small nuclei. **Nuclear fusion** involves forcing two relatively small nuclei to combine into a new, heavier atom. As with fission, the energy released by nuclear fusion can be enormous, again due to the conversion of mass into energy. Gram for gram, nuclear fusion liberates even more energy than nuclear fission—between three and ten times more.

D.3 NUCLEAR FUSION

This section explains nuclear fusion and introduces positrons. The tremendous energy potential of fusion plants, as well as difficulties with their use, is explained. The promise of generating electricity via nuclear fusion is that fusion reactors will produce neither the troublesome air pollutants of fossil fuel plants, nor the long-lived radioactive waste resulting from fission reactors. Mid-1970s forecasts of full-scale fusion power plants by the year 2020 are now considered unrealistically optimistic by most experts. Hopeful announcements of "cold fusion" in the 1990s proved to be unfounded.

CHEMISTRY AT WORK

Nuclear Power in Action: Submarines

With the possible exception of the Space Shuttle, nothing has as much technology crammed into as small a space as a nuclear-powered submarine. Nevertheless, you can be fully trained to operate the nuclear propulsion plant on a submarine by the time you are 19—that is, if you pass Navy Lieutenant **Mike Arnold's** class. After six months of intensive training on land in Charleston, South Carolina, Lieutenant Arnold's students are ready for a challenging career under the oceans.

What does a submariner do?

Submarines must generate their own electricity and propel themselves underwater for months at a time.

Before becoming a trainer at the Nuclear Power Training Unit, Navy Lieutenant Mike Arnold spent three years as a junior officer on the nuclear submarine *USS L. Mendel Rivers*. There he directed the operation of the nuclear propulsion plant, "drove" through some of the most beautiful regions of the oceans, and, along with his crewmates, carried out defense missions.

Each submariner has a distinct duty called a "watch." Each watch is so distinct and essential to the mission of the submarine that the entire crew on duty operates as a single entity. Individuals work in six-hour shifts, around-the-clock, so that no watch is ever left unattended.

Most ships, land vehicles, and aircraft have combustion engines. Why do submarines use nuclear power?

Submarines must generate their own electricity and propel themselves underwater for months at a time. The key to the role of nuclear power, however, is the word *underwater*. Burning fuel—combustion—requires oxygen. Under water there is no way to provide enough usable oxygen to fuel lengthy trips. That is why controlled nuclear reactions are used to power the submarine. Some of this energy is also used to generate electricity to operate the equipment and make life under water possible for the crew. For example, oxygen gas and fresh water are made from seawater using energy generated in nuclear reactions.

472

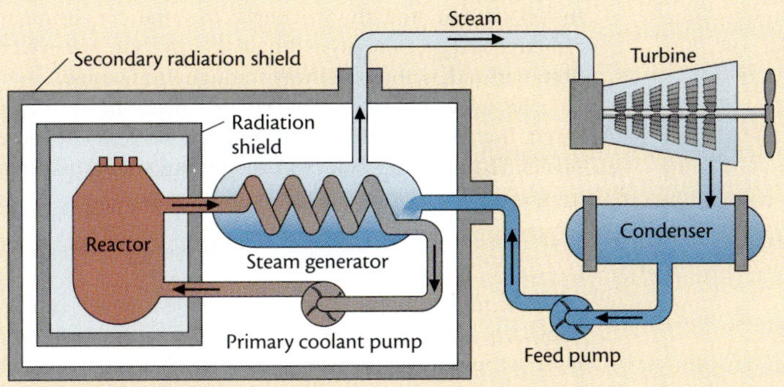

How does the nuclear propulsion plant on board a submarine work?

On a submarine, the turbines that generate electricity and the ones that turn the propeller are spun by steam. The energy used to make this steam comes from the controlled fission of uranium atoms in the core of the nuclear reactor. The design of the nuclear reactor is classified information, but the principles are not.

The heat generated by the fission reaction is transferred to water in a highly pressurized loop of piping called the "primary loop." This water enters a steam generator, where it transfers heat to water running through a separate loop of piping, the "secondary loop." The water in the primary loop goes back to the reactor. The water in the secondary loop turns to steam and heads for the turbines. Here the energy from the steam is converted to electricity and to mechanical energy. The steam then goes through a condenser, where seawater is used to cool the steam to its liquid state, the liquid then heads back to the steam generator, and the cycle begins again.

Questions to Explore

1. Look up the dimensions of a submarine, and find out how many people share the living space during a typical voyage.
2. The world's first nuclear-powered submarine, the *USS Nautilus*, was launched in January of 1955. However, submarines were used in the Civil War, and even Alexander the Great's soldiers are said to have used them. Search the Web to learn how submarines were powered before the advent of nuclear propulsion.
3. Since the end of the Cold-War era, some scientists have been able to use Navy submarines for their underwater research. Find out what they have been studying in the ocean depths by searching www.pbs.org/wgbh/nova/subsecrets/inside.html. While you are at this site, take a virtual tour of a submarine.

Nuclear fusion powers the Sun and other stars. Scientists believe that the Sun formed when a huge quantity of interstellar gas, mostly hydrogen, condensed under the force of gravity. As the volume of gas decreased, its temperature increased to about 15 million degrees Celsius, and hydrogen atoms began fusing into helium. The very high temperature was necessary because the nuclei that fused together were all positively charged and tended to repel one another. The high temperature gave each nucleus considerable kinetic energy, which helped overcome the repulsions. Once fusion was initiated, the Sun began to shine, converting nuclear energy into radiant energy. Scientists estimate that the Sun, believed to be about 4.5 billion years old, is about halfway through its life cycle.

The nuclear fusion reactions that occur in the Sun are rather complicated, but the end result is the conversion of hydrogen nuclei into helium nuclei. That overall result can be summarized by this equation:

$$\underset{\text{Hydrogen-1}}{4\,{}^{1}_{1}\text{H}} \longrightarrow \underset{\text{Helium-4}}{{}^{4}_{2}\text{He}} + \underset{\text{Positrons}}{2\,{}^{0}_{+1}\text{e}}$$

How much energy does such a nuclear fusion reaction produce? In the above equation, a comparison of the total mass of reactants to the total mass of products reveals that 0.0069005 g is lost when one gram of hydrogen-1 atoms are fused to produce one gram of helium-4. Using Einstein's equation, $E = mc^2$, the energy released through the fusion of one gram (one mole) of hydrogen atoms is found to be 6.2×10^8 kJ. Here is one way to put that very large quantity of energy in perspective: The nuclear energy released from the fusion of one gram of hydrogen-1 equals the thermal energy released by burning nearly 5000 gallons of gasoline or 20 tons of coal.

It would take the fusion of about 150 g of hydrogen-1 to convert one gram of mass into energy.

Nuclear fusion reactions have been used in the design of powerful weapons. The hydrogen bomb—also known as a thermonuclear device—is based on a fusion reaction that uses the heat from a small atomic (fission) bomb to initiate the fusion.

Can the energy of nuclear fusion be harnessed for beneficial purposes, such as producing electricity? That remains to be seen. Scientists have spent more than five decades pursuing this possibility. Many schemes have been tried, but none have yet been successful. The major difficulties have been maintaining the high temperatures necessary for fusion while at the same time containing the reactants and fused nuclei. Figure 38 shows one of these fusion experiments.

The challenge of maintaining the incredibly high temperatures (at least 100 million degrees) needed to sustain controlled nuclear fusion has been difficult to meet. The nuclear fusion facility at Princeton University has achieved temperatures sufficient to initiate fusion, but so far the quantity of energy put into the process is more than the energy released.

If scientists finally succeed in controlling nuclear fusion in the laboratory, there is still no guarantee that fusion reactions will become a practical source of energy. Low-mass isotopes to fuel such reactors are plentiful and inexpensive, but confinement of the reaction could be very costly. Further-

474

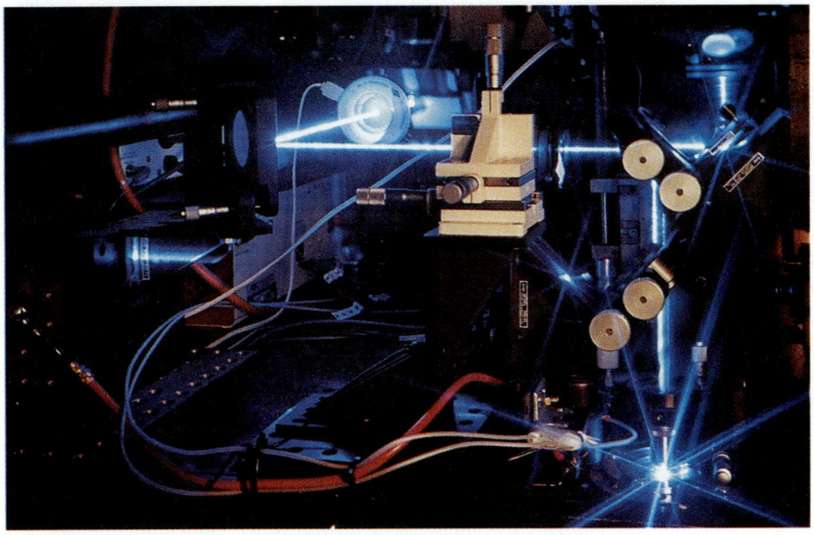

Figure 38 *A deuterium-tritium fuel pellet in the path of a laser, used for laser fusion.*

more, although the fusion reaction itself produces less radioactive waste than nuclear fission, capturing the positrons and shielding the heat of the reaction could generate just as much radioactive waste as is produced by fission power plants.

In splitting and fusing atoms, the nuclear energy that fuels the universe has been unleashed. Much good has come of it, but scientific, social, and ethical questions must also be answered. Along with great benefits, there are also great risks. How much risk is worth the potential benefit? The next activity will provide you with experience in conducting a risk-benefit analysis by analyzing the risks and benefits associated with a common activity—traveling to see a friend.

In many cases, electing not to make a decision is, in fact, an actual decision—one with its own risks and benefits.

D.4 THE SAFEST JOURNEY

Making Decisions

Suppose you want to visit a friend who lives 500 miles (800 km) away, using the safest means of transportation. Insurance companies publish reliable statistics on the safety of different methods of travel. Using Figure 39 on page 476, answer these questions:

1. Assume there is a direct relationship between distance traveled and chance of accidental death. (That is, assume that doubling distance would double your risk of accidental death.)
   a. What is the risk factor value for traveling 500 miles by each mode of travel in the table? For example, Figure 39 shows that the risk

D.4 THE SAFEST JOURNEY

Making Decisions

This activity is designed to point out that risks and benefits enter into every aspect of our lives. How we deal with risks may be a function of information coming from the news media and shaping our perception. For example, even though traveling by air is much safer then by auto or bicycle, it is still considered more dangerous by most people. Controversy arises when the benefits do not seem to outweigh the risks, or when society (government) makes decisions in opposition to the wishes of certain individuals or subgroups.

Answers

1. a. Bicycle: $(500) \times (0.000001/10) = 0.00005 = 5 \times 10^{-5}$

   Auto: $(500) \times (0.000001/100) = 0.000005 = 5 \times 10^{-6}$

   Train: $(500) \times (0.000001/120) = 0.0000042 = 4.2 \times 10^{-6}$

   Bus: $(500) \times (0.000001/500) = 0.000001 = 1.0 \times 10^{-6}$

   Scheduled airline: $(500) \times (0.000001/1900) = 0.00000026 = 2.6 \times 10^{-7}$

   b. Scheduled airline travel is the safest.

   c. Bicycling is the least safe.

2. Students may have the false impression that air travel is the most dangerous form of transportation, owing to media attention given to air disasters. Rarely do we read about non-local bicycle fatalities. Some people, because of their familiarity with driving, have the false impression that automobiles are the safest form of transportation.

   a. Traveling by bicycle is the cheapest; autos, trains, and buses are faster than the bicycle; and transportation by scheduled airlines is the fastest and most worry-free in terms of responsibility for guiding the vehicle but it is expensive.

   b. Bicycle travel is the most risky, due to collisions with cars or road hazards. Cyclists are relatively unprotected, especially if no helmet is worn. Automobile crashes are common; risk of injury goes up if no seat belts are worn, the driver has been drinking, or bad weather conditions exist. Plane crashes are infrequent but most passengers die when they occur. There are few ways to protect oneself in an air disaster.

c. Increased time, cost, personal responsibility, and increased risk of accidents would be seen by most as negative factors. In this case the cost saved in traveling 500 miles by bicycle probably does not seem worth the risk of death and the extended travel time. Other factors, such as the desire for personal mobility and luggage capacity provided by the auto, may influence decision making.

d. Some roads are much more dangerous than others. Some airlines have poor safety records. Travel conditions can increase the risk of travel.

3. These statistics will probably not hold true for 25 years, given an increase in the population and increased automobile transportation and air travel. On the other hand, new technologies, such as proposed "smart highways," might make transportation safer or even provide totally new options.

4. Cost (including meals and overnight expenses), time, responsibility for vehicle, space, payload (luggage, load), freedom to stop vehicle or detour, exercise, pleasure, lack of worry regarding mechanical breakdown.

5. a. (Individual answers)

b. A risk-benefit analysis could never yield the same result for every individual because each person places a different priority and value on each mode of transportation. Sometimes the importance of a trip is worth considerable risk.

Figure 39 *Risk of travel.*

Risk of Travel	
Mode of Travel	**Distance (miles) at which one person in a million will suffer accidental death***
Bicycle	10
Automobile	100
Train	120
Bus	500
Scheduled Airline	1900

*On average, chance of death is increased by 0.000001.

factor (chance of accidental death) for biking increases by 0.000001 for each 10 miles. Therefore, the bike-riding risk factor in visiting your friend would be

500 miles × 0.000001 risk factor/10 miles = 0.00005 risk factor

b. Which is the safest mode of travel (the one with the smallest risk factor)?

c. Which is the least safe?

2. The results obtained in Question 1 might surprise many people. Why?
   a. List some benefits associated with each mode of transportation.
   b. List some risks associated with each mode of transportation.
   c. In your view, do the benefits of riskier ways to travel outweigh their increased risks? Explain your reasoning.
   d. List some situations in which assumptions made in Question 1 would not be valid.
3. Do you think these same statistics will apply 25 years from now? Explain.
4. What factor(s), beyond the risk to personal safety, would you include in a risk-benefit analysis before you decided how to travel?
5. a. Which mode of travel would you choose? Why?
   b. Would someone else's similar risk-benefit analysis always lead to the same decision as yours? Why or why not? (Putting it another way, is there one "best" way to travel?)

RISK-FREE TRAVEL? ChemQuandary 6

Is there any way to travel to visit your friend that would be completely risk-free? Would it be safer not to visit your friend at all?

RISK-FREE TRAVEL? ChemQuandary 6

There is no risk-free way to visit a friend. Even deciding not to visit a friend may pose risks. A conscious decision not to take a risk may put one in a position of greater risk because of events unknown at the time of the decision. For example, if you decide to stay home, a tree might fall on your house. The idea here is to minimize total risk by taking a more knowledgeable approach toward known risks. Students may be surprised to learn that

D.5 NUCLEAR WASTE: PANDORA'S BOX

Imagine this situation. You live in a home that was once clean and comfortable, but you have a major problem: You cannot throw away your garbage. The city forbids this because it has not decided what to do with the garbage. Your family has compacted and wrapped garbage as well as possible for about 40 years. You are running out of room. Some bundles leak and are creating a health hazard. What can be done?

The U.S. nuclear power industry, the nuclear weapons industry, and medical and research facilities share a similar problem. Spent (used) nuclear fuel and radioactive waste products have been accumulating for about 40 years. Some of these materials are still highly radioactive while other materials—even initially—display only low levels of radioactivity. It is uncertain where these materials, regardless of radioactivity levels, will be permanently stored—or how soon.

Nuclear wastes can be broadly classified into two categories: high-level and low-level. **High-level nuclear wastes** are either (1) the products of nuclear fission, such as those generated in a nuclear reactor; or (2) transuranics, products formed when neutrons are absorbed by the original U-235 fuel. For example, plutonium-239 is a transuranic material. **Low-level nuclear wastes** have much lower levels of radioactivity. These wastes include used nuclear laboratory protective clothing, diagnostic radioisotopes, and air filters from nuclear power plants. Figure 40 illustrates the composition of radioactive wastes in the United States.

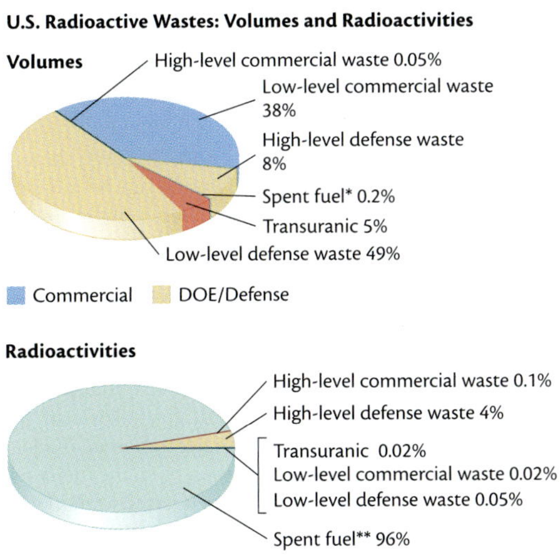

*Including spacing between fuel assembly rods.
**DOE/Defense program to reprocess spent fuel is not shown.

Figure 40 *Volumes and radioactivities of radioactive wastes*

D.5 NUCLEAR WASTE: PANDORA'S BOX

This section introduces students to a highly controversial and pressing issue. The nature of high-level nuclear waste disposal is presented and clarified along with the proposed solution to the problem. Even if all high-level nuclear waste generation were to end tomorrow, it would still be an issue requiring continued attention. Encourage students to watch for newspaper articles pertaining to this issue. Given the nature of radioactive decay (particularly the concept of half-life), the problem will not go away in the near future.

traffic fatalities per passenger mile (30 deaths per 108 passenger miles on horses) were approximately 30 times higher in the pre-automobile days than today, or roughly equivalent to the fatality rate for motorcycles. There is no such thing as a risk-free ride.

Disabuse students of the misconception of "gamblers' logic." Just because the risk of accident may be one per thousand miles, it does not mean that after traveling a thousand miles you are "due" for an accident. The risk is the same for each mile.

HIGH- AND LOW-LEVEL WASTE DISPOSAL

Building Skills 6

1. a. By volume: 38% + 49% = 87%

   b. 8% + 5% = 13%
2. Defense
3. Commercial and defense
4. Evaluate student answers according to how well they defend their answers with evidence. High-level wastes constitute a small fraction of the volume, but the greater part of the radioactivity. It would make sense to deal with it first. On the other hand, the great volume of low-level waste is expensive to store and requires a great deal of attention, even though the levels are low.

Plutonium-239 could be extracted from spent fuel rods of commercial nuclear reactors. However, such extraction is illegal in the United States, because extracted plutonium-239 could be used to produce nuclear weapons.

Figure 40 emphasizes distinguishing waste volume from radioactivity level when considering nuclear wastes; high volume does not mean high radioactivity. For example, Figure 40 shows that defense efforts produce the largest volume of radioactive wastes (57.1%) but this waste accounts for less than 0.1% of the total radioactivity of nuclear wastes. On the other hand, spent fuel from nuclear reactors occupies less than 1% of the volume of all radioactive wastes, yet accounts for almost 96% of the total radioactivity. Moreover, high-level nuclear waste species also can have extended half-lives, some as long as thousands of years.

HIGH- AND LOW-LEVEL WASTE DISPOSAL

Building Skills 6

Use Figure 40 to answer the following questions.

1. Approximately what percent of nuclear waste is
   a. low-level waste?
   b. high-level, including transuranic, waste?
2. What source accounts for the greatest volume of high-level nuclear waste?
3. What two sources account for most low-level nuclear waste?
4. Should high-level or low-level nuclear wastes receive greater attention? Explain your answer.

Because high-level and low-level nuclear wastes have different characteristics, they are disposed of differently. Low-level wastes can be put into sealed containers and buried in lined trenches 20 feet deep. Low-level military nuclear waste is disposed of at federal sites maintained by the Department of Energy. Since 1993, each state is responsible for disposal of its own low-level commercial nuclear wastes. Groups of states in several regions have formed compacts; a disposal site in one state will be used for disposal of low-level wastes from all compact members.

High-level nuclear waste disposal requires a much different approach. Spent nuclear fuels present the greatest challenge. To get a sense of the amount of high-level radioactive waste, consider that a commercial nuclear reactor (power plant) typically produces about 30 tons of spent fuel each year. This means that roughly 3100 tons of waste are generated annually by the 104 commercial nuclear power plants operating in the United States. Approximately 22 000 tons of spent fuel are in storage in 34 states. Eventually, the components of the nuclear power plant reactors themselves will become nuclear waste. By 2000, 64 of the 104 U.S. reactors, most licensed for 40 years, were more than 20 years old.

The U.S. nuclear weapons program has contributed about 20 million liters of additional stored waste. The volume of military waste is much greater than the amount of commercial waste. This waste is in the form of a sludge—the waste product of extracting plutonium from spent fuel rods of military reactors. (Nuclear weapons are created from the plutonium). As radioisotopes in the waste decay, they emit radiation and thermal energy. In fact, without cooling, such wastes can become hot enough to boil. This makes military waste containment extremely challenging. Because of nuclear disarmament treaties, by 2003 the United States will need to remove from nuclear weapons and dispose of as much as 410 tons of highly enriched uranium and about 50 tons of plutonium.

Up to a third of a nuclear reactor's fuel rods must be replaced annually. This is necessary because the uranium-235 fuel becomes depleted as fissions occur and because accumulated fission products interfere with the fission process. The spent fuel rods are highly radioactive, with some isotopes continuing to decay for many thousands of years. Figure 41 lists the half-lives of a few isotopes produced by nuclear fission. All of these and many more are found in spent fuel rods.

Some Isotopes in Spent Fuel Rods

Isotope	Half-Life
Plutonium-239	24 110 y
Strontium-90	28.8 y
Barium-140	12.8 y

Figure 41 *Spent fuel rods contain the radioactive isotopes produced by nuclear fission.*

By federal law, nuclear reactor waste must be stored on site, usually in nuclear waste storage tanks, until a permanent repository is created. Available storage space on site is limited, however. Final storage of high-level radioactive waste is the responsibility of the federal government, but permanent long-term disposal sites for high-level wastes have not yet been opened. Legal negotiations, congressional debates, and developmental work are in process to develop such a site at Yucca Mountain, Nevada. If this site is viewed as suitable, high-level nuclear wastes will be stored deep underground there.

The method of long-term radioactive waste disposal favored by the U.S. government (and by many other nations) is mined geologic disposal, illustrated in Figure 42 on page 480. The radioactive waste would be buried at least a kilometer below Earth's surface in vaults that would presumably remain undisturbed permanently.

To prepare the waste for burial, it would first be allowed to cool for several decades by putting the spent fuel rods in very large tanks of water. Over time, many of the radioisotopes would decay, lowering the level of radioactivity to a point at which the materials could be handled safely. Such on-site nuclear waste storage would be only a temporary measure; the tanks would require too much maintenance to be safe for longer time intervals.

This cooling process for wastes has already been implemented at each nuclear reactor site.

Next, cooled radioactive wastes would be locked inside packages engineered to be leak-proof. The currently favored method is to transform the wastes by **vitrification** into a type of ceramic. Although the encased material would still be highly radioactive, the waste would be much less likely to leak or leach into the environment because of the glasslike "envelope" surrounding it. The vitrified radioactive wastes would be sealed in containers made of glass, stainless steel, or concrete. The Savannah River vitrification plant in South Carolina is the world's largest facility of its kind. Vitrification has been used in France for over a decade, where over 75% of that nation's electrical energy is produced by nuclear reactors.

479

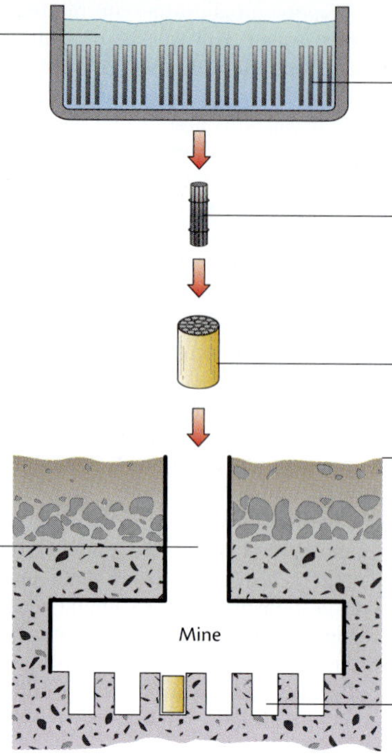

Figure 42 *The mined geologic plan for disposing of nuclear wastes.*

Nowhere in the world, however, has nuclear waste been permanently buried. The challenge is to find technically, politically, and socially acceptable sites. In Japan, the government has even considered deep-ocean burial.

Some geologic sites formerly assumed to be stable enough for radioactive waste disposal were later discovered to be unsafe. For example, some plutonium was buried at Maxey Flats, Kentucky, in a rocky formation that geologists believed would remain stable for thousands of years. Within a decade, however, some buried plutonium had moved dozens of meters away.

Questions & Answers CD-ROM WWW.

What are current plans to resolve our long-term nuclear waste disposal problems? By U.S. law, Congress has selected two sites for permanent radioactive waste disposal from options provided by the Department of Energy. These sites, located in regions of presumed geologic stability, are the Waste Isolation Pilot Plant (WIPP) near Carlsbad, New Mexico, and at Yucca Mountain, Nevada, an extinct volcanic ridge 100 miles northwest of Las Vegas. Both proposals have generated considerable debate and controversy about both the site locations and the means of transportation of radioactive wastes to them. Legal, environmental, and development concerns will continue to be addressed before final decisions are made regarding these sites. It will probably be at least the year 2010 before a suitable site is ready to receive high-level nuclear wastes for long-term storage.

480

SECTION SUMMARY

Reviewing the Concepts

1. Name three isotopes that can undergo neutron-induced fission.

Uranium-235 **Plutonium-239**

Uranium-233 **Californium-252**

2. Explain the difference between nuclear fission and nuclear fusion.

Nuclear fission is the splitting of radioisotopes, by neutron bombardment, into lighter isotopes. Nuclear fusion is the combining of lighter isotopes to form heavier isotopes. In effect, fission is the breakdown of larger nuclei into smaller pieces. Fusion is the joining together of small nuclei to make a larger one.

3. Write a balanced nuclear equation for the fission of U-235 by a neutron, producing Br-87, La-146, and several neutrons.

$^{235}_{92}\text{U} + ^{1}_{0}\text{n} \rightarrow ^{87}_{35}\text{Br} + ^{146}_{57}\text{La} + 3^{1}_{0}\text{n}$

4. State Einstein's mass-energy relationship, and give the meaning of each symbol.

$E = mc^2$; where E = Energy, m = mass, and c = speed of light in a vacuum.

5. How are the conservation of mass and the conservation of energy related in nuclear reactions?

The small quantity of mass loss in the process of forming two nuclei is converted into a large quantity of energy as indicated by Einstein's equation $E = mc^2$. 1 gram of matter yields 9×10^{13} J of energy. The amount of combined matter and energy is constant.

6. Name the force that holds subatomic particles together, and describe its characteristics.

Strong force—stronger than the electrical force that holds the atoms and ions together in chemical bonds. The strong force operates over very short distances.

7. If nuclear fusion is so effective for powering stars, why isn't it practical for generating electricity in power plants?

Currently in research reactors, the quantity of energy put into the process is greater than the energy released.

8. Describe the requirements needed to sustain a fission reaction.

To sustain a fission reaction there must be a minimum amount of fissionable material (critical mass), and an ample supply of neutrons.

9. Why is it impossible for a nuclear power plant to become a nuclear bomb?

Nuclear reactor fuel pellets are only 3% uranium-235. Weapon-grade uranium is enriched to 90% uranium-235, enough to cause a chain reaction leading to a nuclear explosion. The uranium used in nuclear power plants is sufficient to allow the chain reaction to occur, but not enough to cause a nuclear explosion.

10. How can nuclear power plants release so much energy from so little fuel?

Because of chain reaction: The reaction will continue,

releasing large amounts of energy, since as long as an emitted neutron encounters a fissionable atom, the reaction will continue without an outside energy source. Matter is converted to energy in nuclear fission.

11. Explain why the word *chain* is a good descriptor of a chain reaction process.

A chain reaction is a self-propagating reaction in which the products of one reaction initiate the next. When fissionable material such as uranium-235 is bombarded with neutrons, its atoms are split into lighter isotopes, producing more neutrons. These neutrons split other uranium-235 nuclei, thus setting off a chain reaction, just like one link in a chain pulls the next link.

12. Why is it necessary to have a critical mass of fissionable material before a chain reaction can begin?

A chain reaction will occur when the fissionable mass is near enough or large enough to capture neutrons creating the chain reaction. The specific mass in this volume is termed the critical mass.

13. Why is U-235 used in nuclear power plants instead of smaller, simpler radioisotopes?

U-235 is the only abundant naturally occurring radioisotope that can undergo fission.

14. Explain how it is possible for both of these statements to be regarded as true:

a. Nuclear fusion has not been used as an energy source on Earth.

b. Nuclear fusion is Earth's main energy source.

a. Nuclear fusion in the laboratory is problematic and is not yet available for use as an economically sound source of energy.

b. Because energy from the Sun drives the physical and chemical processes of the Earth, and because this energy is derived from nuclear fusion reactions, it may be said that nuclear fusion is already the "number one" energy source on Earth.

15. Why are high pressures and temperatures needed to produce fusion reactions?

A very high temperature is needed to produce fusion because the nuclei that fuse together are all positively charged and tend to repel one another. The high temperature is needed to give each nucleus sufficient kinetic energy to overcome the repulsions. A high pressure is needed to get the nuclei together.

16. How does the process of nuclear fusion help to explain the origin of all chemical elements?

The elements are formed by the fusing together of smaller nuclei into larger nuclei in the middles of stars and supernovae.

17. How does nuclear fusion compare to a reaction involving two hydrogen atoms combining to form a hydrogen molecule, H_2?

Nuclear fusion is the fusing of two hydrogen nuclei under high temperature and pressure, producing a helium atom. Two hydrogen atoms combine to form hydrogen gas molecules by a chemical reaction at atmospheric conditions.

18. Why must nuclear waste be stored rather than just returning it to its place of origin?

Nuclear waste is still highly radioactive. The process of mining and processing nuclear fuel concentrates it far beyond what it was in nature. It would be difficult to disperse it so widely again.

19. What are current methods for disposing of high-level nuclear waste?

High-level nuclear waste must be (1) allowed to cool for thirty years by storing in large tanks of water on site. (2) Cooled radioactive waste must be locked inside leak-proof packages and transformed into a type of ceramic by vitrification. (3) Waste must be buried deep underground in a stable location.

20. A simple way to minimize the need for long-term storage of radioactive waste might seem to be just to speed up all of radioactive decay rates involved. Why won't that plan work?

There is no way to change the rate of radioactive decay.

21. Burning is sometimes regarded as a good way to dispose of certain types of extremely toxic material. Why would burning be an unacceptable plan for destroying nuclear waste?

Matter cannot be created nor destroyed; thus, burning does not destroy atoms, the material would still be radioactive.

Connecting the Concepts

22. Fusion reactions that produce elements with atomic numbers up to iron ($^{56}_{26}Fe$) can take place in the core of an ordinary star. However, elements with higher atomic numbers generally result from violent stellar explosions. Explain the difference.

Fusion reactions that produce elements with atomic numbers up to 26 (iron) are the results of the fusing together of small nuclei. The higher atomic numbers generally require forces even greater than those found in normal stars.

23. What are some factors that complicate the cleanup of nuclear waste that has been abandoned or improperly stored?

The continuous decay of the radioactive waste and the contamination of groundwater make it dangerous even to enter the contaminated area. Also, it may be very difficult to concentrate the waste for proper handling.

Extending the Concepts

24. Research two of the storage sites for high-level nuclear waste that have been proposed or used. List one advantage and one disadvantage of each site.

Yucca Mountain, Nevada Waste Isolation Pilot Plant near Carlsbad, New Mexico. Each site has the advantage of being remote and geologically stable. Each site has the disadvantage that it is impossible to predict what the area will be like in thousands of years. Each site has popular opposition to the dumps. No one wants a nuclear waste facility in his or her backyard.

25. Explain the process that might be involved in reducing the existing volume of high-level nuclear wastes.

To reduce existing volumes of high-level nuclear waste it may be possible to concentrate the waste by removing inert material in the high-level waste.

26. When confronted with the problems surrounding nuclear waste disposal, students sometimes propose putting the waste in a rocket and shooting it to the Sun. What are some problems with that plan?

The high cost and the possibility of a launch accident that will spread radioactive waste over a large area.

27. Nuclear fusion is a potentially desirable technology. What would be some advantages of nuclear fusion over nuclear fission for producing electricity?

With fusion there would be no need to worry about running out of fuel; fusion produces a stable, nonradioactive and nonreactive product, helium, which would eliminate concerns about pollution and storing nuclear waste.

28. In the late twentieth century, a group of scientists announced they had discovered a new process termed "cold fusion." Their claim was later discredited, but the goal of producing sustained nuclear fusion at room temperature remains desirable. Why?

"Cold fusion" would make it unlikely that we would have to look elsewhere to find fuels for the future. It would provide an endless source of nonpolluting energy at low cost.

PUTTING IT ALL TOGETHER

The Truth about Nuclear Chemistry—Communicating Scientific and Technological Knowledge

Throughout your chemistry studies this year, including your work in this unit, you have used scientific ideas to evaluate published claims and to draw your own conclusions. To put this unit all together, your final task is to prepare and present what you have learned to help some Riverwood residents draw their own conclusions about nuclear concerns.

CD-ROM WWW. **Putting It All Together**

As you already know, your audience will be senior citizens at a local community center. They are concerned about the impact of regulations proposed by the Citizens Against Nuclear Technology on their access to a full range of medical care and are alarmed by some of the organization's claims about nuclear waste disposal. They have asked that your class address the scientific aspects of the statements made in the flyer from CANT so that they will be able to evaluate the group's message more fully when they attend the informational meeting.

Each group of students will be responsible for responding to one or more of the statements. In preparing your presentation, it will be essential to coordinate with groups that are working on statements involving similar scientific concepts.

For each statement, the presentation should include the following:

- Determination of whether the statement is true, false, or partially true.
- An explanation of the science or technology upon which the statement is based. For instance, for statements about nuclear power plants, you should explain how such plants work; for general statements about radiation, an introduction to the types of radiation would be appropriate.
- A replacement statement that more accurately and even-handedly addresses the same topic/issue.

Design the presentation to be understandable to an adult with at least a high school education, which may have been completed many years before. When possible, the presentation should include visual aids and everyday examples and applications. At the end of the presentation, the audience should be able to make informed decisions regarding nuclear policy in their community.

THE TRUTH ABOUT NUCLEAR CHEMISTRY

The ultimate goal of *ChemCom* is for students to be able to apply the science knowledge and skills they gain to their everyday lives. This PIAT is a good example of this process. Hopefully, students will use this exercise to create a presentation that will reflect their knowledge and inform the senior citizens.

Consider actually taking your class to a senior citizen center and have students make their presentations. Such encounters are often meaningful for both groups—the students and the senior citizens.

The grading guide found in the Appendix may be helpful in evaluating students' work.

PLANNING GUIDE UNIT 7

Section	Time class	Student Edition Features	Program Resources	Media and Technology
SECTION A. Food as Energy **Objectives:** • The technique of calorimetry can be used to determine the quantity of energy contained in a particular food. • Energy flows from the Sun to plants, then in turn to herbivores, to carnivores, and decomposers. As the energy flows as indicated, most is used to maintain cellular functions. • The energy contained in food is released when cellular reactions break the bonds of complex organic molecules, forming lower-energy molecules. The released energy is temporarily stored in phosphate bonds of ATP and is used to power various cellular functions. • The energy required by a human to complete a physical activity depends on the particular activity, time involved, and weight of the person engaged in the activity.		**Building Skills** #1. Energy in Action #2. You Gain Some, You Lose Some **Laboratory Activity** • A.3 Energy Contained in a Snack **ChemQuandary** #1. How Does Your Garden Grow? **Making Decisions** • A.2 Diet and the Food Pyramid • A.7 Energy Intake	**Activity Book** • A.2 Making Decisions • A.3 Energy Contained in a Snack **Black Line Copy Masters** • Sample Food Diaries	**ChemCom CD-ROM** • Unit 7: Section A • Questions & Answers • Laboratory Video **ChemCom Connections** (Videodisk) • Chapter 21. Energy from Food **Transparencies** • Food Guide Pyramid • Energy Use in a Caterpillar • Energy Expended during Various Activities

PLANNING GUIDE

Section	Time class	Student Edition Features	Program Resources	Media and Technology
SECTION B. Energy Storage and Use **Objectives** • Fat and carbohydrate molecules are composed of carbon, hydrogen, and oxygen. Differences in structure between fat and carbohydrate molecules account for their different properties and energy content. • Carbohydrates include simple sugars such as glucose and also substances involving chemical combinations of two or more simple-sugars units. • A triglyceride (a typical fat molecule) is formed by a condensation reaction between three fatty acid molecules and a glycerol molecule. • Fats may be saturated (all C–C bonds) or unsaturated (some C–C bonds). Unsaturated fats can become more saturated through hydrogenation. • *Cis* and *trans* isomers are distinguished by how two functional groups are positioned on either side of a C=C double bond.		**Building Skills** #3. Calories from Fat #4. Fats in the Diet #5. Limiting Reactants in Chemical Reactions **Making Decisions** • Analyzing Fats and Carbohydrates **ChemQuandary** #2. Fat-Free Food? **Modeling Matter** Limiting Reactants	**Activity Book** B.5 Making Decisions **Demonstration** B.4 Limiting Reactants	**ChemCom CD-ROM** • Unit 7: Section B • Questions & Answers **ChemCom Connections (Videodisk)** #21. Food and Nutrition #19. Growing Crops (Limiting Reactants) **Transparencies** • Structural Formula of Glucose • Formation of Sucrose and Structural Formulas of Starch and Cellulose • Common Carbohydrates • *Cis* and *Trans* Fatty Acids

PLANNING GUIDE

Section	Time class	Student Edition Features	Program Resources	Media and Technology
• Chemical reactions involve substances interacting in certain fixed ratios. The limiting reactant determines the quantity of product that can be produced.				
SECTION C. Proteins, Enzymes, and Chemistry **Objectives** • Proteins, the major structural components of living tissue, perform many cellular functions. • Amino acids—small molecules that contain at least one amino group and at least one carboxylic acid group—are chemically combined to form protein polymers. The human body is unable to synthesize all the types of amino acids it needs; essential amino acids must be obtained directly from foods. • Some protein molecules function as enzymes—biological catalysts that speed up cellular reactions.		**Building Skills** #6. Daily Protein Requirements **Laboratory Activity** • C.3 Enzymes • C.5 Amylase Tests **Making Decisions** • Protein Contents **ChemQuandary** #3. A Problem to Chew On! **Modeling Matter** • Molecular Structure of Proteins	**Activity Book** • C.3 Enzymes • C.5 Amylase Tests • C.6 Making Decisions	**ChemCom CD-ROM** • Unit 7: Section C • Questions & Answers • Laboratory Video **Transparencies** • Types and Functions of Major Proteins in the Body • RDAs for Protein • Enzyme-Substrate Interaction

PLANNING GUIDE

Section	Time class	Student Edition Features	Program Resources	Media and Technology
SECTION D. Other Substances in Foods **Objectives** • Vitamins, organic molecules necessary for basic life functions, are classified as fat-soluble or water-soluble. • A titration can be used to determine the amount or concentration of solute in a particular solution. • Minerals are elements that are essential to life. • Food additives are used to increase the nutritive value, storage life, visual appeal, or ease of production of foods. • Chromatography is a useful technique for separating and identifying the components of a solution.		**Building Skills** #7. Vitamins in the Diet #8. Minerals in the Diet **Laboratory Activity** • D.2 Vitamin C • D.5 Food Coloring Analysis **Making Decisions** • Food Additive Survey • Analyzing Vitamins and Minerals **ChemQuandary** #4. Nitrite Additives	**Activity Book** • D.2 Vitamin C • D.5 Food Coloring Activity • D.9 Making Decisions Supplement • Unit Making Decisions Supplement: What Kind of Food Am I? • Unit Making Decisions Supplement: Using Food Label Information • PIAT Supplement: Food Diary Project	**ChemCom CD-ROM** • Unit 7: Section D • Questions & Answers • Laboratory Video **Transparencies** • RDAs for Selected Vitamins • Vitamins by Category and Vitamin Content of Selected Vegetables • Minerals • Food Additives • RDAs for Minerals • Paper Chromatography with R_f Values
Putting It All Together.			**Black Line Copy Masters** • PIAT Evaluation	

SUPPLY LIST FOR UNIT 7

Expendable Items and Unusual Materials per 24 Students Working in Pairs

Alphabetical by Item Name			
Quantity	Unit	Item	Sections Used
2.3	Grams	Amylase	C.5
1	Each	Apple	C.3
1	Each	Apple (Demo)	D.6
12	Each	Balance	A.2
8	Each	Balloons, 4-inch diameter (Demo)	B.4
300	mL	Benedict's solution	C.5
12.5	Grams	Boric acid, anhydrous	C.5
12	Each	Bottles, 8 oz. gas connecting	D.5
60	mL	Buffer solution, pH 10	C.5
60	mL	Buffer solution, pH 4	C.5
60	mL	Buffer solution, pH 6	C.5
60	mL	Buffer solution, pH 8	C.5
60	mL	Buffer solution, pH 2	C.5
1	Lb	Candy, M&Ms	D.5
1	Lb	Candy, Skittles	D.5
1	Pkg.	Chromatography paper, 1 inch wide	D.5
10.5	Grams	Citric acid, hydrate	C.5
17.3	Grams	Copper(II) sulfate pentahydrate ($CuSO_4 \cdot 5\ H_2O$)	C.5
8	Each	Culture tubes (Demo)	B.4
1	Each	Food coloring, blue	D.5

SUPPLY LIST

Alphabetical by Item Name (continued)			
Quantity	Unit	Item	Sections Used
1	Each	Food coloring, green	D.5
1	Each	Food coloring, red	D.5
1	Each	Food coloring, yellow	D.5
12	Each	Gas burners	C.5
12	Each	Graduated cylinders, 10 mL	C.5, D.2
50	mL	Hydrochloric acid, 3 M (Demo) (HCl)	B.4
12	mL	Hydrogen peroxide, 3% (H_2O_2)	C.3
5	Lbs	Ice	A.2
12	Each	Index cards	A.2
1	Pkg	Legos (Demo)	C.1
0.5	Lb	Liver, raw	C.3
55	Cm	Magnesium ribbon (Demo)	B.4
1	Box	Matches, long stem	A.2
72	Each	Micropipets	D.5, D.2
1	Each	Molecular model kit (Demo)	C.1
12	Each	Paper clips	A.2
1	Gram	Potassium iodate (KIO_3)	D.2
1	Gram	Potassium iodide (KI)	D.2
1	Each	Potato, white or baking	C.3
1	Each	Refrigerator	C.5

SUPPLY LIST

Alphabetical by Item Name (continued)			
Quantity	**Unit**	**Item**	**Sections Used**
12	Each	Ring stand	A.3, C.5
24	Each	Ring clamps	A.3, C.5
12	Each	Rubber stoppers to fit bottles	D.5
173	Grams	Sodium citrate dihydrate ($Na_3C_6H_5O_7$•2 H_2O)	C.5
2.5	Grams	Sodium sulfite (Demo) (Na_2SO_3)	D.6
12	Each	Soft drink cans with pull tabs	A.2
150	mL	Starch solution, 0.1% (0.15 g in 150 mL water)	C.5
12	Each	Stirring rods	A.3
60	Each	Stoppers, rubber to fit test tubes	C.5
12	Each	Stoppers, cork—size 20	A.2
1	Liter	Sulfuric acid, 0.05 M (H_2SO_4)	D.5
12	Each	Test tube holders	C.5
84	Each	Test tubes (16 $\times$ 125 mm)	C.3, C.5
12	Each	Thermometers	A.3
1	Box	Toothpicks	D.5, D.2
100	Grams	Sodium carbonate monohydrate (Na_2CO_3•H_2O)	C.5
38	Grams	Trisodium phosphate dodecahydrate (Na_3PO_4•12 H_2O)	C.5
1	Each	Vitamin C tablets (100 mg)	D.2
12	Each	Wellplates, 24 wells	D.5
12	Each	Wire gauze	C.5
1	Pkg	Yeast cake	C.3

SUPPLY LIST

By Section			
Quantity	**Unit**	**Item**	**Sections Used**
12	Each	Balance	A.2
5	Lb	Ice	A.2
12	Each	Index cards	A.2
1	Box	Matches, long stem	A.2
12	Each	Paper clips	A.2
12	Each	Soft drink cans with pull tabs	A.2
12	Each	Stoppers, cork, size 20	A.2
12	Each	Stirring rods	A.3
12	Each	Thermometers	A.3
24	Each	Ring clamps	A.3, C.5
12	Each	Ring stand	A.3, C.5
8	Each	Balloons, 4-in diameter (Demo)	B.4
8	Each	Culture tubes (Demo)	B.4
50	mL	Hydrochloric acid, 3 M (HCl) (Demo)	B.4
55	Cm	Magnesium ribbon (Demo)	B.4
1	Pkg	Legos (Demo)	C.1
1	Each	Molecular model kit (Demo)	C.1
1	Each	Apple	C.3
12	mL	Hydrogen peroxide, 3% (H_2O_2)	C.3
1	Each	Potato, white or baking	C.3
0.5	Lb	Liver, raw	C.3
24	Each	Test tubes (16 × 125 mm)	C.3
1	Pkg	Yeast cake	C.3
2.3	Grams	Amylase	C.5
300	mL	Benedict's solution	C.5
12.5	Grams	Boric acid, anhydrous	C.5
12	Each	Gas burners	C.5

SUPPLY LIST

By Section (continued)			
Quantity	**Unit**	**Item**	**Sections Used**
60	mL	Buffer solution, pH 2	C.5
60	mL	Buffer solution, pH 4	C.5
60	mL	Buffer solution, pH 6	C.5
60	mL	Buffer solution, pH 8	C.5
60	mL	Buffer solution, pH 10	C.5
10.5	Grams	Citric acid, hydrate	C.5
17.3	Grams	Copper(II) sulfate pentahydrate ($CuSO_4{\cdot}5H_2O$)	C.5
12	Each	Graduated cylinders, 10-mL	C.5
1	Each	Refrigerator	C.5
173	Grams	Sodium citrate dihydrate ($NaC_6H_5O_7{\cdot}2\ H_2O$)	C.5
60	Each	Stoppers, rubber, to fit test tubes	C.5
12	Each	Test tube holders	C.5
60	Each	Test tubes (16×125 mm)	C.5
100	Grams	Sodium carbonate monohydrate ($Na_2CO_3{\cdot}H_2O$)	C.5
38	Grams	Trisodium phosphate dodecahydrate ($Na_3PO_4{\cdot}12\ H_2O$)	C.5
12	Each	Wire gauze	C.5
150	mL	Starch solution, 0.1% (0.15 g in 150 mL water)	C.5
12	Each	Graduated cylinder, 10-mL	D.2
1	Gram	Potassium iodate (KIO_3)	D.2
1	Gram	Potassium iodide (KI)	D.2
1	Each	Vitamin C tablets (100 mg)	D.2
12	Each	Bottles, 8 oz. gas connecting	D.5
1	Lb	Candy, M&Ms	D.5
1	Lb	Candy, Skittles	D.5
1	Pkg.	Chromatography paper, 1-in wide	D.5
1	Each	Food coloring, blue	D.5
1	Each	Food coloring, green	D.5

SUPPLY LIST

By Section (continued)

Quantity	Unit	Item	Sections Used
1	Each	Food coloring, red	D.5
1	Each	Food coloring, yellow	D.5
12	Each	Rubber stoppers to fit bottles	D.5
12	Each	Wellplates, 24-well	D.5
1	Liter	Sulfuric acid, 0.05 M (H_2SO_4)	D.5
1	Box	Toothpicks	D.5, D.2
72	Each	Micropipets	D.5, D.2
1	Each	Apple (Demo)	D.6
2.5	Grams	Sodium sulfite (Na_2SO_3) (Demo)	D.6

UNIT 7 FOOD: MATTER AND ENERGY FOR LIFE

Introduction to Unit

The proper amounts and kinds of food we eat are important for both building body tissue and for getting energy. As the subtitle of the unit suggests, the main focus of this unit is the chemistry of foods. Students will review a three-day inventory of the food they consume and analyze it as to its fat, carbohydrate, protein, vitamin, and mineral content. Students investigate how the food they eat provides the energy and nutrients they need to live and to grow. Students should realize that eating a variety of foods increases the likelihood of an adequate balance of all required nutrients. This unit includes laboratory activities to maintain students' interest and enthusiasm.

UNIT 7

HOW is energy transferred, stored, and released by molecules that make up food?

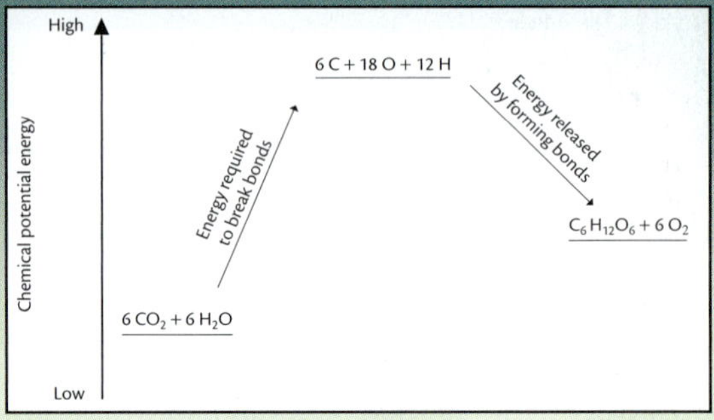

WHAT chemical roles do carbohydrates and fats play in nutrition?

FOOD: MATTER AND ENERGY FOR LIFE

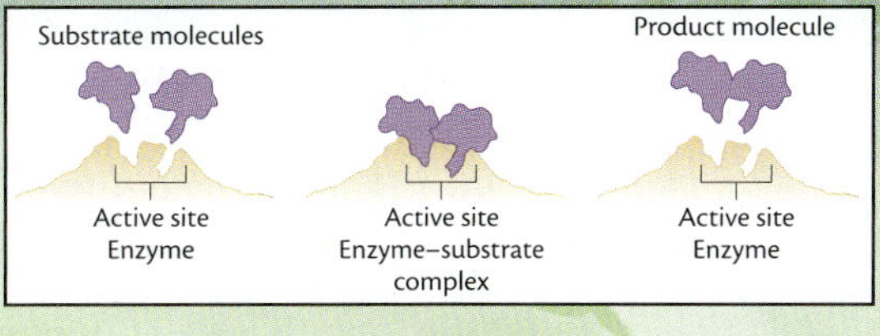

WHY are protein molecules essential to human well-being?

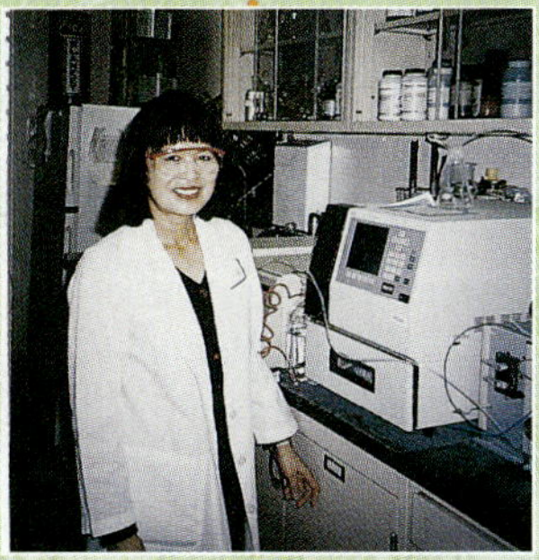

HOW can the chemical composition of foods be analyzed in the chemistry laboratory?

You can use chemistry to explain and evaluate the components of foods that people decide to eat. Turn the page to begin exploring the elements and compounds contained in foods—and the differences they can make in everyone's lives.

BRUCE LEE
CHINESE RESTAURANT
ACAPULCO
TASTE OF INDIA
El Norte
Hot Wok
south east asian cookbook
Sunset International Vegetarian Cook Book
Wolfgang Puck's Modern French Cooking for the American Kitchen
LINDA McCARTNEY ON TOUR
DIANA KENNEDY
THE CUISINES OF MEXICO
SALSA BOOK

486

No matter where you live in the world, how old you are, or what you enjoy doing, you have at least one thing in common with every other person—the need to eat. You may feel hungry enough to eat a large meal, may be "hanging out" with friends eating snacks, or may consume a piece of cake to celebrate a birthday. Regardless of the reasons, eating comes down to this—food provides the energy and nutrients you need to live and grow.

The photographs on the opposite page show various foods, places to eat, cookbooks, and people eating. Why are there so many different kinds of foods? Why are there so many restaurants and cookbooks from which to choose? Why do restaurants—from Mexican and Chinese to Ethiopian and Italian—serve such different foods?

Modern transportation and mobility is one reason why such a wide variety of foods are available in today's restaurants and grocery stores. Before long-distance transportation, people were limited to foods that were grown and prepared locally. The types of foods that were available were tied to the characteristics of the region of the world in which they were grown—such things as climate, weather, soil conditions, and water supply. In centuries past, transportation could be difficult, but people still moved from place to place. But there was no way to ship home-grown foods over long distances. Newcomers' diets were restricted to the foods available in their new towns and cities.

Today, long-distance transportation makes all the difference. In most places a wide variety of foods is available. In the United States you can eat foods that were once only commonly available in such places as Greece, India, China, Japan, Peru, Brazil, and Ghana. Since all these foods can provide most of the energy and nutrients a human body needs, for many people the choice of food becomes a matter of taste and custom.

Lifestyle and dietary needs also influence what you eat. If you decide to lose or gain weight, or if you have diabetes or follow a vegetarian diet, the food you eat will reflect those conditions and decisions. Many different circumstances lead to many different diets.

In this unit, you will learn about the chemistry of foods. You will learn how energy contained in food is stored and released and how substances in food promote growth and repair in the body. You will also investigate the chemistry and the nutritional roles of fats, carbohydrates, proteins, and vitamins and minerals. Based on instructions provided by your teacher, you will focus on analyzing a three-day food-intake inventory, representing all the food an individual consumes in that time interval. By the close of this unit, you will have analyzed that inventory in several ways and will be ready to report the results of your analyses.

You will first learn about the energy stored in and delivered by foods. Read on to see how the foods in your three-day inventory rate in terms of their variety and energy content.

487

A.1 THE FOOD PYRAMID

The Food Pyramid is introduced as a guide for planning food choices that will ensure a healthful diet. The recommendation for a healthful diet is based on a pyramid relationship, with a large proportion of bread, cereal grains, and pasta at the bottom and decreasing amounts of fruits, vegetables, meats, and dairy products. Fats, oils, and sweets represent a very small area at the top, since they are required in lesser amounts. Use the questions raised in the student text as a basis for class discussion.

A.2 DIET AND THE FOOD PYRAMID

Making Decisions

Young people spend a lot of time thinking about what they will eat next. In keeping with the overall theme of *ChemCom* (to make chemistry relevant), students are asked to study a three-day inventory of food consumption, and analyze it for the proportion of each food group included. Three options are given, with the least recommended first. You may give the students a sample inventory that represents a typical diet. Or, you may have students create a typical or "ideal" diet for a teenager. Finally, you may want to have students keep track of their own diets. There is potential for student embarrassment from overeating or other eating disorders, or economically disadvantaged students may have inadequate diets, so sample diaries have been provided. Even students with every dietary advantage may make poor choices, so there is great educational advantage to having students consider their own diet habits. We suggest you do not ask students to share their personal data with other students. Wait until students have analyzed their diets before collecting any class data or individual results.

Sample three-day food diaries

Unit 7 will become more meaningful and motivating to your students if they analyze their own three-day food diaries through their study of this material. If you implement that option, students should be asked to keep records of every food item and beverage they consume over a three-day period. Quantities such as number, volume, or mass should be included, as specified in the text.

SECTION A

FOOD AS ENERGY

Introduction (CD-ROM, WWW.)

Where does the energy required for walking, running, playing soccer—even breathing—come from? The answer is easy—from the food you eat. As you begin to analyze your three-day food inventory, you will gain the chemistry background needed to understand where that energy comes from, how food energy is stored and used, and how decisions about eating can affect body weight.

A.1 THE FOOD PYRAMID

The Energy Value of Food (CD-ROM, WWW.)

Messages abound about eating the right kinds of foods in the proper amounts. But what does "right kinds" mean? What are suitable amounts?

In 1991 the U.S. Department of Agriculture (USDA) approved the Food Guide Pyramid shown in Figure 1 to help people make good dietary choices. The Pyramid is composed of five food groups, plus a sixth category that includes fats, oils, and sweets. Although it is important to eat foods from all five groups, the shape of the Pyramid also conveys information about how frequently you should consume various classes of foods in a healthy diet. The USDA recommends that most of your daily choices come from the bottom section (6 to 11 servings); 3 to 5 servings should come from each of the middle groups, and only a few (2 to 3 servings) from the top section. For good health, fats, oils, and sweets at the very top of the Pyramid should represent only a very small portion of a diet. Although these substances provide energy, they contribute little else to a healthful diet.

Some foods listed in the three-day inventory you will analyze are probably depicted in Figure 1, but others—such as tortillas, tofu, and collard greens—are not. Where should you place foods not illustrated in the Food Guide Pyramid? How does the food inventory you will analyze compare to recommendations of the Food Guide Pyramid? Find out in the next activity.

A.2 DIET AND THE FOOD PYRAMID

Making Decisions

Your three-day food inventory analysis begins with the "big picture." That is, before analyzing the foods in terms of particular features such as energy, fat, carbohydrate, protein, mineral, or vitamin content, you will verify what proportions of each food group are included.

In particular, you will compare the three-day food inventory to the recommended servings indicated in the Food Guide Pyramid. Your teacher may provide you with the three-day inventory. However, if you are instructed to analyze some other food inventory (your own or one you devise), follow these guidelines to construct the list:

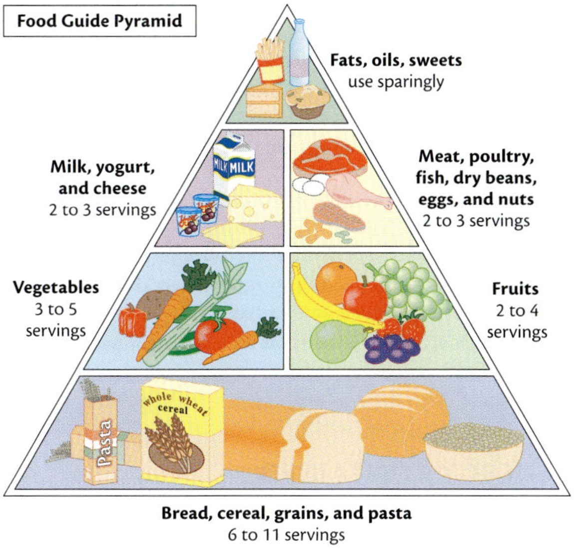

Figure 1 *The Food Guide Pyramid illustrates components of a healthful diet.*

- Include each food item, snack, beverage, and dietary supplement eaten during each of the three days.
- Express the quantity of each item by estimating total number (slices or units, such as the number of eggs or bananas), total mass (grams, ounces, or pounds), or total volume (pints, milliliters, tablespoons, or cups). When possible, use food labels to make your estimates.

For each item in the three-day inventory, indicate to which Food Guide Pyramid group(s) it belongs. Also indicate whether an item belongs in the category of fats, oils, and sweets. You may need to do some research on food items about which you are unsure. Then complete these steps:

1. Construct a data table so that the total number of servings within each Food Guide Pyramid group can be easily recorded. Include columns for Days 1, 2, and 3, and a column for the daily average over the three days.
2. Record in your table the number of servings in each Food Guide Pyramid group for each day. Then calculate and enter the average number of daily servings for each group over the three days.
3. Using daily average values, for which Food Guide Pyramid groups does the food inventory
   a. meet the recommendations?
   b. exceed the recommendations?
   c. fall below the recommendations?

However, if you prefer that your students use less personalized food-diary information, two sample three-day food inventories are provided in the appendix and as Black Line Copy Masters. You can use these, or you can follow their general format to create other three-day food diaries that may better represent typical food choices among your students.

You may wish to give each student a blank data table that meets the report requirements described below.

This activity reveals information about personal nutrition that may surprise students. Allow class time for comparison and group analysis of the charts. Students typically have difficulty estimating serving size. You may want to do a class exercise to have students determine the volume of their fist to serve as a handy comparison for determining serving size. Have students fill a large beaker with water. Place the beaker in a shallow pan. Then have students make a fist and insert it into the beaker. The water will overflow into the pan. Students should measure the volume of overflowed water, which is equal to the volume of their fist. It may be best to measure the volume in cups, as that is a typical measure of serving size in nutrition tables.

Some foods may fit into several Food Guide Pyramid groups. For example, a pepperoni pizza has components that fit into categories of bread, dairy, vegetables, and meat. Students should analyze such "combination" foods carefully. Encourage students to make the best match possible when analyzing their foods. It is not always possible to find the precise description of the food they have eaten.

Answers

1. No answer required.
2. Individual answers.
3. Individual answers.
4. The Food Pyramid is intended to be followed over a longer period, not just for a single day. The balance of foods will vary in the short term.
5. Individual answers.
6. It is better to analyze food consumption over a three-day period to ensure that sufficient data is collected to accurately reflect the amount of nutrients ingested daily.
7. Even over a three-day period, students may not consume nutrients that are typical of their regular diet.

A.3 ENERGY CONTAINED IN A SNACK

Laboratory Activity

Students use the procedures from Unit 3 as a guide in writing the procedure for measuring the energy contained in a snack food. Check student procedures and setup before allowing them to proceed.

Time

50 minutes

Materials (for a class of 24 working in pairs)

(Materials based on similar procedures used in Unit 3.)

12 12-oz empty soft drink cans (with pull tabs)
12 ring stands & clamps
12 thermometers
balances
5 lbs ice
matches
12 paper clips
12 corks, size 20

Safety

Students should wear safety goggles and aprons throughout the laboratory activity. Students should use nonmercury thermometers with careful handling, especially when stirring the water. Caution students to handle the can carefully to avoid cuts from the metal.

A variety of snacks may be tested for their caloric content. It is recommended that two or more groups test the same item so that results can be compared. Some recommended snacks are potato chips, corn chips, and small marshmallows. Potato chips seem to work well, as do small marshmallows. Large chips and marshmallows are not recommended. Listed items are high in sugar/fat/oil content and burn readily. **CAUTION:** *Although peanuts, almonds, cashews, or other nuts work well, some people can experience allergic reactions to nuts, and burning them in this activity is potentially hazardous to those individuals.*

Advance Preparation

If tap water is warmer than about 10 °C below room temperature, prepare a large quantity of chilled water. Remind students not to transfer ice into the can.

Pre-lab discussion

Students should review the Combustion laboratory activity in the Petroleum Unit, particularly the physical setup and the sources of experimental error.

---

4. Why do you think the Pyramid lists ranges of recommended servings—such as 6 to 11 servings in the bread, cereal, rice, and pasta group—rather than a particular number of servings?
5. Complete these steps for food items that were difficult to assign to particular Food Guide Pyramid groups:
   a. List the item and the Pyramid Food Guide group(s) to which you assigned it.
   b. Explain why it was difficult to assign this food item to a group or groups.
   c. Explain how you decided to which group(s) it belonged.
6. Is it better to analyze food consumption over three consecutive days or during one particular day? Explain.
7. Why might a food analysis based on a three-day average be misleading?

A.3 ENERGY CONTAINED IN A SNACK

Laboratory Activity

Introduction

Lab Video CD-ROM WWW.

Figure 2 shows part of a food label from a popular snack. Note the number of grams per serving (30); then note the value "160 Calories." This indicates the quantity of energy in one serving. How is this energy determined? How much energy does 160 Calories actually represent?

In Unit 3 you constructed a simple calorimeter from a soft drink can. See page 204.

The quantity of energy contained in a particular food can be determined by burning a known amount of it under controlled conditions and carefully measuring how much thermal energy is released. You performed a similar activity in Unit 3 when you measured the heat of combustion of a candle. This procedure is called **calorimetry;** the measuring device is called a **calorimeter.**

Figure 2 *Part of the food label from a popular snack. Note the number of Calories in one serving.*

Procedure

In this laboratory activity you will determine the energy contained in a snack-food item. Use the candle-burning procedure (page 203) and information included in the Calculations section below to guide you as you design a procedure for measuring the energy contained in a particular snack. Before performing the laboratory activity, complete these steps:

- Write a complete procedure for determining the energy contained in this snack.
- Draw a sketch of your laboratory setup.
- Construct a data table to record all necessary measurements and observations.
- Have your teacher check and approve your procedure and data table.

After your teacher approves your procedure, you may start the laboratory activity. Complete two trials.

Calculations

If the mass of water and its temperature change are known, the thermal energy needed for that temperature change can be calculated. In this laboratory activity, the energy required to heat the water equals the energy given off by the burning snack item (assuming that no heat is lost in the process).

That quantity of energy can be calculated from the specific heat capacity of water and the mass and temperature change of the water sample. Recall from Unit 3 that the specific heat capacity of water is about 4.2 J/(g·°C). This means that it takes about 4.2 J to raise the temperature of a one-gram sample of water by 1 °C. So, if you know the mass of water that was heated as well as its temperature change, you can calculate the thermal energy absorbed by the water. If necessary, refer to page 204 to remind yourself how to complete this calculation.

Use data you collected in this activity to complete these calculations:

1. Determine the mass (in grams) of water heated.
2. Calculate the overall temperature change in the water.
3. Calculate the total energy (in joules) used to heat the water.
4. The food Calorie, Cal (with an uppercase C), reported on food labels is a much larger energy unit than the joule. One Calorie equals 4184 J, which—for convenience in this case—can be rounded off to 4200 J.
   a. Calculate the total Calories used to heat the water.
   b. How many Calories were released by the burning snack item?
5. a. Calculate the energy released, expressed as Calories per gram of snack item burned, for each trial.
   b. Calculate an average "Cal/g of snack item" value.
6. Use data from the snack-food package to calculate the "Cal/g" value for the brand of snack you used.
7. Calculate the percent difference between the label value and your experimental average value:

$$\% \text{ difference} = \frac{(\text{Experimental value} - \text{Label value})}{\text{Label value}} \times 100\%$$

8. What design errors in your laboratory activity might account for the difference between your Cal/g value and the label value?
9. How could your laboratory setup and procedure be improved to decrease the possibility of design errors?

The more exact specific heat capacity of liquid water is 4.184 J/(g·°C).

The calorie, another common energy unit, is about one-fourth as large as a joule (1 calorie = 4.184 J). One Calorie—used to report energy contained in foods—equals one kilocalorie (1000 cal).

A.4 ENERGY FLOW—FROM THE SUN TO YOU

The snack item you just burned contained enough energy to raise the temperature of a sample of water by several degrees Celsius. Where did that energy come from?

The answer is simple: All food energy originates from sunlight. Through photosynthesis, green plants capture and use solar energy to make large

Calculations (sample data only, actual student calculations will vary)

1. $97.90 \text{ mL} \times \frac{1 \text{ g}}{1 \text{ mL}} = 97.90 \text{ g}$
2. $25.0 - 2.0\ °C = 23.0\ °C$
3. $97.90 \text{ g} \times \frac{4.2 \text{ J}}{\text{g} \cdot °C} \times 23.0\ °C =$ 9500 J
4. a. $9500 \text{ J} \times \frac{1 \text{ Cal}}{4200 \text{ J}} = 2.3 \text{ Cal}$
   b. At least 2.3 Cal. Some of the energy released by the burning snack was lost to the surroundings.
5. a. Using the sample data: $\frac{2.3 \text{ Cal}}{0.52 \text{ g}} = 4.4 \text{ Cal per g}$
   b. Answers for the average value will vary (look for a similar calculation for each trial).
6. Answers will vary, typically about 2 Cal.
7. Answers will vary, typically 20–50%.

8–9. The "design errors" in Questions 8 and 9 can be corrected by changing or improving the design of the experimental apparatus or procedure (for instance, modifying the calorimeter so that more thermal energy from the burning item is transferred to the water). Design errors are distinct from other sorts of "human error", such as "I dropped the chip," which arise mainly from poor laboratory technique.

Lab tips

Students can also account for the heat absorbed by the aluminum can. To do this, students should find the mass of the aluminum can before adding water. The heat released by burning the food equals the heat absorbed by the water and the can. (Specific heat capacity of Al = 0.902 J/(g•°C).)

Sample data and calculations

Mass of snack burned	0.52 g
Mass of water in can	97.90 g
Initial temp. of water	2.0 °C
Final temp. of water	25.0 °C

The label on the snack food is used for calculating the Cal/serving and the percent difference.

Post-lab discussion

The calculations may give students difficulty due to the several units of energy used. Walk them through the stepwise calculations described in the student text. Since students are already familiar with using specific heat capacity in units of

(J/g · °C), they are first asked to find energy content in joules (see Unit 2). However, food energy in the United States is reported in Calories (with an uppercase C). This quantity is much larger than a joule: 1 Calorie = 4180 J = 1000 cal.

No more than two significant figures should be reported in the results. Student results from combustion activities carried out like this one often lack precision. Regardless, the technique still gives students a good concept of calorimetry and the energy content of foods.

One extension is to have students compare this activity to the one in the Petroleum Unit and see how it is different.

A.4 ENERGY FLOW—FROM THE SUN TO YOU

Sunlight is the source of energy stored in food by photosynthesis. The overall process of photosynthesis is endothermic since it requires more energy to break the bonds between carbon dioxide and water molecules than is released when glucose and oxygen molecules form. The original energy from the Sun is reduced as it is transferred from one organism to another through the food chain.

A brief review of endothermic and exothermic reactions (from Units 3 and 4) may be beneficial at this point. Have students trace the energy use in a caterpillar. See Figure 4 (page 493).

Recall that sunlight is also the source of the energy stored in petroleum and coal. See page 195.

molecules from smaller, simpler ones. Recall from Unit 4 (page 286) that green plants use solar energy to convert water and carbon dioxide into carbohydrates and oxygen gas. Although a variety of carbohydrates are produced, an equation for photosynthesis usually depicts the production of glucose:

$$\underset{\text{Carbon dioxide}}{6\,CO_2} + \underset{\text{Water}}{6\,H_2O} + \text{Solar energy} \longrightarrow \underset{\text{Glucose}}{C_6H_{12}O_6} + \underset{\text{Oxygen}}{6\,O_2}$$

For the reaction depicted above to occur, bonds between carbon and oxygen atoms in carbon dioxide molecules and between oxygen and hydrogen atoms in water molecules must be broken. The atoms then must recombine to form glucose and oxygen molecules. Recall from Unit 3 that breaking bonds always requires energy, while bond formation releases energy. In the case of photosynthesis, the bonds in carbon dioxide and water molecules require more energy to break than is released when chemical bonds in glucose and oxygen molecules form. The energy needed to drive this endothermic reaction—as the equation indicates—comes from the Sun. An energy diagram in Figure 3 shows the energy involved in this process.

For a discussion of the energy involved in chemical reactions, see pages 195–197.

In this way, radiant energy from the Sun is converted to chemical energy stored within bonds of carbohydrate molecules. Humans and animals release this chemical energy when they consume and metabolize plants. In effect, metabolism reverses the photosynthesis process—converting glucose and other carbohydrates back into strongly bonded molecules of carbon dioxide and water. Energy originally contained in sunlight continues to flow through ecosystems as carnivores consume animals that ate plants. Eventually, when plants and animals die, they decay, and organisms

A diagram similar to that in Figure 3, showing the energy involved in the combustion of methane, appears on page 197.

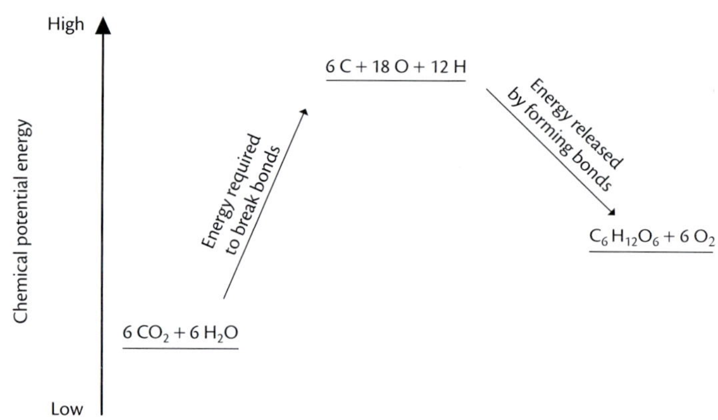

Figure 3 *Photosynthesis. Breaking bonds within carbon dioxide and water molecules requires energy. Energy is released when atoms combine to form glucose and oxygen molecules. The first step (bond breaking) requires more energy than the second step (bond formation) releases, so the overall process of photosynthesis is endothermic.*

that aid in decomposition use the remaining stored-up energy. Thus energy flows from the Sun to photosynthetic plants to herbivores, and then to carnivores and decomposers.

Energy from the Sun becomes more dispersed and less available as it is transferred from organism to organism. For example, as energy flows from one organism to another, some energy is dispersed into the environment as heat. Over half the energy contained in the food that herbivores and carnivores consume and absorb is used to digest food molecules. You may be surprised to learn that a small fraction (only about 10–15%) of food energy consumed by organisms is used for growth—that is, for converting smaller chemical molecules to larger ones that become part of the animal's structure. Only the energy stored within the molecules making up the structure of a plant or animal is available to an organism that consumes the first one. This decline in useful energy continues as energy flows farther from its original source—the Sun. Figure 4 depicts this decline in available energy.

HOW DOES YOUR GARDEN GROW?

ChemQuandary 1

Think about the old nursery rhyme, "Mary, Mary, quite contrary, how does your garden grow?" From a chemical viewpoint, Mary should consider this more scientific question: "As garden plants grow, from where does the material come that causes their mass to increase?" Even though it does not rhyme, how should Mary answer that second question? Why?

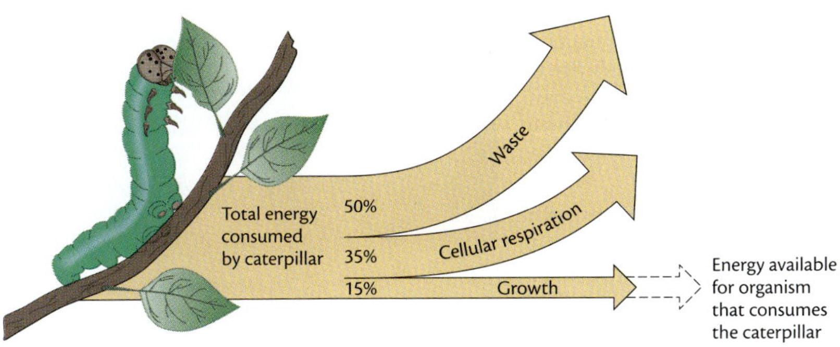

Figure 4 *Energy use in a caterpillar.*

HOW DOES YOUR GARDEN GROW?

ChemQuandary 1

As plants grow, energy from the Sun (the primary energy source) is stored as chemical energy in green plants. A common misconception among students is that most of the materials plants use for growth comes from the soil, instead of carbon dioxide in air. Review the historical Val Helmont experiment from the 18th century. He weighed a quantity of soil, planted a tree in the soil, and let it grow until it weighed some 200 pounds. When he re-weighed the soil its mass had not changed substantially.

Additional ChemQuandary

How does energy flow through ecosystems? As an example, ask students how the number of tiger lilies on Earth relates to the number of tigers on Earth. There are millions of tiger lilies, but only a few tigers. This is because the energy pyramid (similar to the Food Pyramid) to support one tiger is so broad.

CHEMISTRY AT WORK

Food for Thought

The next time you chew gum, drink a canned beverage, or eat a bowl of your favorite breakfast cereal, think about corn. Gum, soda, cereal, ketchup, cookies, pastries, bread, and many other processed foods and drinks are sweetened or colored with products made from corn. Everything you eat and drink has connections to a farm, even if those ties aren't readily apparent.

Sue Adams and her husband John work year-round to run a 1100 acre farm in Atlanta, Illinois. To protect the environment, they use integrated pest management and practice a technique called *no-till farming*—planting new crops directly in the residue of previously harvested crops. This method saves on labor and fuel. More importantly, it reduces the loss of valuable topsoil and moisture, dramatically reducing the need for materials such as fertilizers, pesticides, and herbicides.

> Sue and her husband practice the precise, high-tech agriculture being used on many farms today.

In the spring, Sue and John plant half of the fields in corn and the other half in soybeans. As soon as the crops begin to grow, they scout the fields for evidence of damage by pests and, where necessary, apply special crop protection products that decompose quickly in sunlight. Just like a chemist in a laboratory, Sue carefully monitors and measures how these products are used to ensure maximum protection for the crops and the environment.

Ever Think of a Satellite as a Farm Tool?

Sue and her husband also practice the precise, high-tech agriculture being used on many farms today. One example is their use of Global Positioning System (GPS) technology. GPS employs computers, monitoring units, and radio signals to send, receive, and process information from satellites to study objects on Earth's surface. On the farm, Sue uses GPS technology to plot with great precision which areas of her fields require additional fertilizers or other treatments. She even has a specially equipped all-terrain vehicle that can record and access information while out in the field.

Using GPS data, Sue and John are able to apply fertilizer only to those areas of the farm that need it, minimizing the amount of fertilizer they introduce to the environment. Such increased precision has also reduced the costs of fertilizing (materials, equipment, fuel, and labor) by as much as $10 an acre.

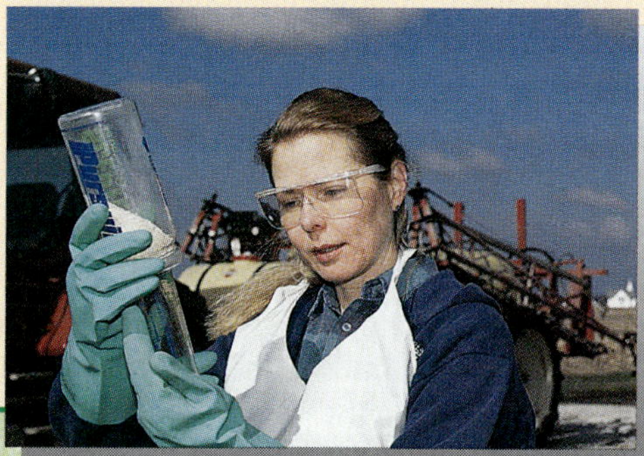

494

Earth and Its Food Supply: What's the Forecast?

In the 1700s, an English economist named Thomas Malthus published his *Essay on the Principle of Population.* The widely read pamphlet argued that unchecked population increases geometrically (1, 2, 4, 8, 16 . . .) while food supplies increase arithmetically (1, 2, 3, 4, 5 . . .). Malthus's theory suggested that one day the world might be unable to feed itself. Indeed, world population has grown significantly. In 1800, world population stood at about 1 billion; in 1999, it reached the 6 billion mark.

In small discussion groups, talk about and answer the following questions.

1. Malthus's prediction has not become reality, at least for a majority of the world's inhabitants. Was there any part of his thinking that was flawed, especially in regard to human population growth? Explain.
2. The world population by the year 2020 will probably be almost 8 billion people. Why do you suppose there is renewed interest in Malthus's theory?
3. One of the problems with increased population is the total land required for human residence versus that needed for cropland. What new techniques might farmers, in collaboration with municipal planners, develop to allow for the space needed by both humans and agriculture?
4. No-till farming is described as having a number of advantages. Why do you suppose it has taken until now for farmers to use no-till techniques instead of the traditional technique of plowing fields?
5. In what ways do the Adamses' farming techniques reflect the principles of Green Chemistry, that is, the movement to prevent environmental problems before they occur?
6. What are the direct advantages to the Adamses of using a "greener," high-tech approach to farming? What are the advantages to society of using these techniques?

Malthus's theory suggested that one day the world might be unable to feed itself.

495

A.5 ENERGY RELEASE AND STORAGE

Cellular respiration, previously discussed in the Air Unit (Unit 4, Section B.5), is now extended to include the concept of energy change, especially in the various types of compounds in food. The discussion of ATP and ADP should tie together previous knowledge of biology. If students have not taken biology, an introduction to cellular respiration may be needed.

A.5 ENERGY RELEASE AND STORAGE

You will learn more about carbohydrates and fats in Section B.

Some energy in the food you eat is used soon after you digest it. The rest of the energy may be stored for later needs. These reserves of body energy are found mainly in fats and—to a much smaller extent—in carbohydrates. Some foods rich in carbohydrates and fats are depicted in Figure 5.

Whether your body uses energy from food recently ingested or from stored fat, the release of energy from these molecules depends on a series of chemical reactions inside your cells. In **cellular respiration,** plants and animals use oxygen to break down complex organic molecules into carbon dioxide and water molecules. The energy required to break bonds in the reactant molecules is less than the energy released when the bonds in carbon dioxide and water form. Thus energy is released in cellular respiration. You can think of that exothermic process as the reverse of the process shown in Figure 3 (page 492). The reaction is similar to burning fossil fuels, except it occurs at much lower temperatures in a series of linked steps under very controlled conditions. The overall process is

Recall that "burning" or "combining with oxygen" was the origin of the term "oxidation." See page 122. In that process, as in all oxidation reactions, one or more electrons are lost by the oxidized species.

$$\text{Organic compounds} + \text{Oxygen} \longrightarrow \text{Carbon dioxide} + \text{Water} + \text{Energy}$$

The released energy is used by specialized structures within each cell to carry out a variety of tasks such as energy production, transport of molecules, waste disposal, storage of genetic material, and synthesis of new molecules. That energy ultimately allows you to walk, talk, run, work, and think—as well as providing energy to power your heart, lungs, and brain.

Carbohydrates, fats, and proteins can all be processed as fuel for your body. However, the chemical process of cellular respiration can be clarified by examining the oxidation of glucose, a carbohydrate. That chemical

Figure 5 *Some foods that contain carbohydrates (left), and some foods rich in fats (right).*

change is described by this overall equation, based on the oxidation of one mole of glucose:

$$C_6H_{12}O_6(aq) + 6\,O_2(g) \longrightarrow 6\,CO_2(g) + 6\,H_2O(l) + 686 \text{ Calories}$$

This equation actually summarizes a sequence of more than 20 integrated chemical reactions that are catalyzed by more than 20 different enzymes. Molecule by molecule, glucose (shown in Figure 6) is oxidized within cells throughout your body.

This equation is the reverse of the equation for photosynthesis. See page 492.

The energy needed to perform individual cellular functions is much less than the energy released by "burning" glucose molecules. Between its release from these energy-rich molecules and its later use in the cell, energy is stored in biomolecules of ATP (adenosine triphosphate), shown in Figure 7. In that energy-storage reaction, one mole of ADP (adenosine diphosphate) uses 7.3 Calories of energy to add a phosphate group, forming ATP and water. The 7.3 Calories of energy involved in the reaction are stored in the phosphate bonds of one mole of newly formed ATP.

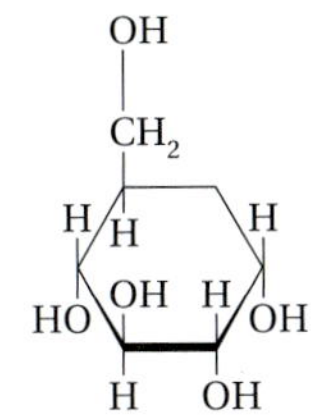

Glucose

Figure 6 *A glucose molecule.*

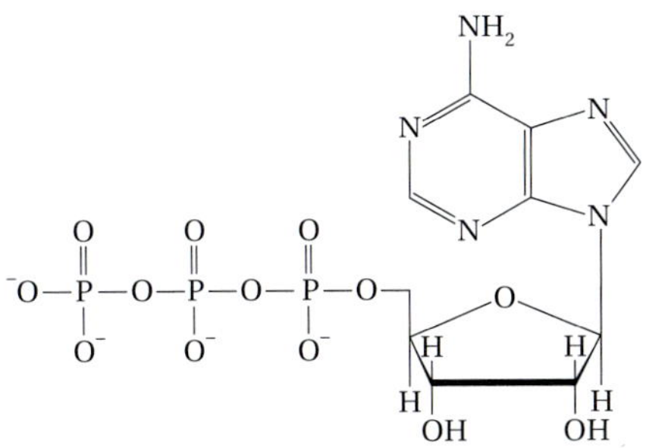

ATP

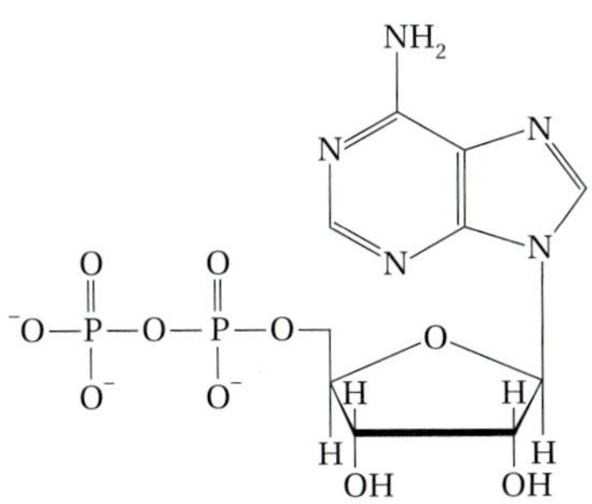

ADP

Figure 7 *Ionic forms of ATP and ADP.*

497

$$7.3 \text{ Cal} + \text{ADP(aq)} + \underset{\text{Hydrogen phosphate ion}}{\text{HPO}_4^{2-}\text{(aq)}} \longrightarrow \text{H}_2\text{O(l)} + \text{ATP(aq)}$$

When that reaction is reversed, each mole of ATP releases 7.3 Calories. This conveniently small quantity of energy—compared to the 686 Calories produced when one mole of glucose is oxidized—can be used to power different steps in cellular reactions. Some steps require less energy than that supplied by one ATP molecule, while other steps need the energy released by several ATP molecules.

Oxidation of each mole of glucose produces enough energy to add 38 moles of ATP to short-term energy storage in a cell. Each day, your body stores and later releases energy from at least 6.02×10^{25} molecules (100 mol) of ATP.

Energy Equivalents for Selected Activities

	Energy Expended (Cal/min) for Individuals of Differing Body Mass			
Activity	**46 kg (100 lb)**	**55 kg (120 lb)**	**68 kg (150 lb)**	**82 kg (180 lb)**
Basketball	6	8	10	12
Bicycling, moderate	6	7	8	10
Football	6	8	10	12
Hockey	6	8	10	12
Martial arts	8	10	12	14
Running, 8 min/mile	10	12	15	18
Sitting	1	1	1	2
Skateboarding	4	5	6	7
Sleeping	1	1	1	1
Soccer	6	7	8	10
Step aerobics	6	7	8	10
Swimming	5	6	7	9
Tennis	6	7	8	10
Volleyball	2	3	4	4
Walking, 17 min/mile	3	4	5	6
Weight lifting	2	3	4	4
Wrestling	5	6	7	9

Figure 8 *Energy expended in various activities.*

498

ENERGY IN ACTION

Building Skills 1

Figure 8 lists the total energy expended, on average, during various activities. Use that information to answer the following questions. The first answer has been worked out as an example.

1. Figure 8 indicates that a 55-kg (120-lb) person burns 5 Cal/min while skateboarding.
   a. That energy is obtained from ATP. If each mole of ATP yields 7.3 Cal, how many moles of ATP will provide the energy needed for a 55-kg person to skateboard for one hour?
   b. If each mole of glucose oxidized produces 38 mol ATP, how many moles of glucose would be needed to provide the energy for that one hour of skateboarding?
   c. What mass of glucose does that represent?

First, find moles of ATP needed for one hour of skateboarding:

$$\frac{5\text{ Cal}}{1\text{ }\cancel{\text{min}}} \times \frac{60\text{ }\cancel{\text{min}}}{1\text{ h}} = 300\text{ Cal/h}$$

$$\frac{300\text{ }\cancel{\text{Cal}}}{1\text{ h}} \times \frac{1\text{ mol ATP}}{7.3\text{ }\cancel{\text{Cal}}} = 41\text{ mol ATP/h}$$

Because 38 mol ATP can be obtained from each mole of glucose, the skateboarder's need for 41 mol ATP can be met with a little more than one mole of glucose:

$$41\text{ }\cancel{\text{mol ATP}} \times \frac{1\text{ mol glucose}}{38\text{ }\cancel{\text{mol ATP}}} = 1.1\text{ mol glucose}$$

What mass of glucose does that represent? One mole of glucose ($C_6H_{12}O_6$) has a mass of 180 g. Because the skateboarder needs more than one mole of glucose, that mass must be greater than 180 g:

$$1.1\text{ }\cancel{\text{mol glucose}} \times \frac{180\text{ g glucose}}{1\text{ }\cancel{\text{mol glucose}}} = 200\text{ g glucose}$$

Now try these:

2. Assume your body produces approximately 100 mol ATP daily.
   a. How many moles of glucose are needed to produce this much ATP?
   b. What mass of glucose does that represent?
3. One minute of muscle activity requires about 0.0010 mol ATP for each gram of muscle mass. How many moles of glucose must be oxidized to energize a pound (454 g) of muscle to dribble a basketball for one minute?
4. a. How many hours do you typically sleep each night?
   b. How many Calories do you use during one night of sleep? (See Figure 8.)
   c. How many moles of ATP does this require?
   d. What kinds of activity require energy while you are sleeping?

More than 200 g glucose would actually be needed since the energy-transfer steps are not 100% efficient.

ENERGY IN ACTION

Building Skills 1

Students use Figure 8 (page 498) to calculate the amount of energy expended during various activities. The calculations continue to move students away from the use of formulas and toward quantitative reasoning. Work through the first problem with students before asking them to solve the remaining problems.

Answers

1. Worked out in student's text.
2. a. $100\text{ mol ATP} \times \frac{1\text{ mol glucose}}{38\text{ mol ATP}} =$ 2.63 mol glucose
   b. $2.63\text{ mol glucose} \times \frac{180\text{ g glucose}}{1\text{ mol glucose}} =$ 473.4 g glucose = approximately 500 g
3. $\frac{0.0010\text{ mol ATP}}{\text{g of muscle}} \times \frac{1\text{ mol glucose}}{38\text{ mol ATP}} \times \frac{454\text{ g muscle}}{1\text{ lb}} =$ 0.012 mol glucose
4. a. Individual answers; 6–10 hours typically.
   b. Answers will vary; sample calculation assumes 8 hours of sleep: $8\text{ hr} \times \frac{60\text{ min}}{1\text{ hr}} \times \frac{1\text{ Cal}}{1\text{ min}} = 480\text{ Cal}$
   c. $480\text{ Cal} \times \frac{1\text{ mol ATP}}{7.3\text{ Cal}} =$ 66 moles ATP
   d. Breathing, movement while sleeping, etc.

A.6 ENERGY IN—ENERGY OUT

People follow various diets for different reasons. Some people may want to lose weight while others may be trying to gain. Still others may put little thought into what they eat—instead, they eat for convenience and pleasure.

Some foods are necessary for the nutrients they deliver, regardless of their energy value. By contrast, other foods provide useful energy but are lacking in nutrients. This latter category, which includes nondiet soft drinks, is sometimes described as furnishing "empty Calories."

If you want to lose weight, you must consume less energy than you expend. If your goal is to gain weight, you need to do just the opposite. If losing weight or gaining muscle mass is desired, then regular physical activity is needed in addition to the particular diet being followed. In the next activity you will explore the interplay between personal activity and diet.

YOU GAIN SOME, YOU LOSE SOME

Building Skills 2

Look again at Figure 8 (page 498). Energy expended depends on the duration of the activity and the weight of the person. For example, during one hour of soccer playing, a 55-kg (120-lb) person expends 420 Cal, whereas a 68-kg (150-lb) person would expend 480 Cal in that same task.

Use information in Figure 8 to answer these questions.

1. Consider eating an ice-cream sundae. Assume that two scoops of your favorite ice cream contain 250 Cal; the chocolate topping adds 125 Cal more.
   a. Assume that your regular diet (without that ice-cream sundae) just maintains your current body weight. If you eat the ice-cream sundae and wish to "burn off" those extra Calories,
      i. how many hours would you need to lift weights?
      ii. how far would you have to walk at 3.5 mph?
      iii. for how many hours would you need to swim?
   b. One pound of weight gain is equivalent to 2700 Cal. If you choose not to exercise, how much weight will you gain from the sundae?
   c. Now assume that you consume a similar sundae once a week for 16 weeks. If you do not exercise to burn off the added Calories, how much weight will you gain?
2. Question 1 implied that eating an ice-cream sundae will cause weight gain unless additional exercise is completed.
   a. Can you think of a plan that would allow you to consume the sundae, do no additional exercise, and still not gain weight?
   b. Explain your answer.
   c. What concerns would you have with this plan?

A.6 ENERGY IN—ENERGY OUT

Students read that people diet for more than one reason. When students compare the total energy consumed with the total energy expended during physical activity, they find one of three results: maintenance of body weight, weight gain, or weight loss.

YOU GAIN SOME, YOU LOSE SOME

Building Skills 2

Students find that the total energy expended during physical activity depends on the duration of the activity and the weight of the person.

Answers

1. Total calories = 250 Cal + 125 Cal = 375 Cal

   a. Assuming a 150-pound person:

   (i) $375 \text{ Cal} \times \dfrac{1 \text{ min}}{4 \text{ Cal}} \times \dfrac{1 \text{ hr}}{60 \text{ min}} = 1.6 \text{ hr}$

   (ii) $375 \text{ Cal} \times \dfrac{1 \text{ min}}{5 \text{ Cal}} \times \dfrac{1 \text{ hr}}{60 \text{ min}} = 1.25 \text{ hrs}$

   (iii) $375 \text{ Cal} \times \dfrac{1 \text{ min}}{7 \text{ Cal}} \times \dfrac{1 \text{ hr}}{60 \text{ min}} = 0.89 \text{ hr}$

   b. $375 \text{ Cal} \times \dfrac{1 \text{ lb}}{2700 \text{ Cal}} = 0.14 \text{ lb}$

   c. $\dfrac{0.14 \text{ lb}}{\text{sundae}} \times \dfrac{1 \text{ sundae}}{\text{week}} \times 16 \text{ wks} = 2.24 \text{ lbs}$

2. a & b. You could lower the intake of some other high-energy foods in your regular diet to compensate for the energy in the ice cream.

   c. Some students may suggest maladaptive methods, such as skipping meals, drinking lots of water, or regurgitating the food.

3. a. $6 \text{ Glasses} \times \dfrac{250 \text{ g}}{\text{glass}} \times \dfrac{4.2 \text{ J}}{\text{g}\cdot{}^\circ\text{C}} \times (37\ {}^\circ\text{C} - 0\ {}^\circ\text{C}) = 230\,000 \text{ J}$

   b. $230\,000 \text{ J} \times \dfrac{1 \text{ Cal}}{4200 \text{ J}} = 55 \text{ Cal}$

   c. $240 \text{ Cal} \times \dfrac{6 \text{ glasses}}{55 \text{ Cal}} = 26 \text{ glasses}$

   d. No, this is not an efficient way to diet. One would need to consume very large quantities of water, although health-food advocates suggest this technique to assist in a diet program. It is considered important for general health to drink eight glasses of water daily.

   e. Answers will vary.

3. Suppose you drank six glasses (250 g each) of ice water (0 °C) on a hot summer day.
   a. Assume your body temperature is 37 °C. How many joules of thermal energy would your body use in heating that ice water to body temperature? Recall that the specific heat capacity of liquid water is about 4.2 J/(g·°C).
   b. How many Calories is this? Recall that 1 Calorie equals about 4200 J.
   c. A serving of French fries contains 240 Cal. How many glasses of ice water would you need to drink to "burn off" the Calories in one serving of French fries?
   d. Does drinking large quantities of ice water seem like a reasonable way to lose weight? Explain.
   e. Select your favorite activity from Figure 8. How many minutes would you need to engage in that activity to burn off the Calories from that serving of French fries?

A.7 ENERGY INTAKE

Making Decisions

You have already evaluated your three-day food inventory in terms of the Food Guide Pyramid. Based on chemical background you have gained, you can now find the total energy contained in that food inventory.

1. Using appropriate resources suggested by your teacher, determine how much food energy (Calories) is contained by each item on the three-day inventory. Record these values.
2. Calculate the total Calories for each of the three days.
3. Calculate the average Calories per day.
4. Using your results from Section A.2, Making Decisions (pages 489–490), how many Calories per day (on average) are supplied by each group in the Food Guide Pyramid?
5. a. Based only on Calories taken in and expended, decide whether the three-day food inventory you are evaluating would be appropriate for each of the following individuals.
      i. A 140-pound female wants to lose 10 pounds without exercising.
      ii. A 150-pound male, who plays basketball an hour each day, wants to gain 10 pounds.
      iii. A 120-pound female wants to maintain her weight. She walks to and from school 15 minutes each way.
   b. Explain each answer in Question 5a.

Use Figure 8 to help guide your decisions in Question 5.

CD-ROM WWW. Questions & Answers

ADDITIONAL

ChemQuandary

In general, why is a weight-loss plan based only on exercise (without controlling caloric intake) not likely to be adequate for anyone who is more than a little overweight?

Answer

Physical activity typically accounts for only about 30% of the body's energy output; the other 70% is to maintain basal metabolism. Even the most vigorous physical activity burns off only about 1000 Cal/h, and 1 lb of body fat is equivalent to about 2700 Cal. Thus, if one wanted to lose 1–2 lb/wk (the maximum recommended without medical supervision) using exercise alone, one would need to set aside 4–6 h/wk for vigorous physical exercise. Such a plan would probably be unhealthful or even impossible for a very overweight person. A balanced combination of low-calorie diet and exercise would be more effective in both the short and long term.

A.7 ENERGY INTAKE

Making Decisions

Students evaluate their three-day food inventory for energy content by:

- Calculating average and total daily Calorie intake.
- Calculating total Calories supplied by each food group.
- Determining if weight would be gained or lost.

Provide students with a nutrition guide such as "Nutritive Values of Foods," U.S. Department of Agriculture. Many cookbooks have guides, or you can purchase inexpensive guides at many grocery stores.

There are several Web sites that have online diet analysis tools, such as:

www.usda.gov/cnpp

The USDA Center for Nutrition Policy and Promotion offers *Dietary Guidelines for Americans*, 2000, 5th Edition, and also offers the *Interactive Healthy Eating Index*, a great tool for analyzing diets.

www.ag.uiuc.edu/~food-lab/nat

This site hosts the USDA Food Composition Data online, or it can be downloaded or can be purchased on CD for $17 (October 2000 prices).

www.nal.usda.gov:80/fnic/etext/000020.html

The Food and Nutrition Information Center found at this site is home to a number of databases for calculating the composition of foods.

SECTION SUMMARY

Reviewing the Concepts

1. Why is it important to know accurately the mass of water used in a calorimeter?

The quantity of energy depends on the change in temperature, mass, and specific heat capacity of water.

2. What does the "specific heat capacity" of water mean?

Specific heat capacity is the quantity of heat needed to raise the temperature of 1 gram of water 1 °C.

3. In what way is a Calorie different than a calorie?

One (1) Calorie is equal to 1000 calories or one (1) kilocalorie.

4. **a.** How much thermal energy is required to increase the temperature of a 110-g sample of water by 10 °C?

b. Suppose that a 2.0-g sample of food was burned in a calorimeter to provide the thermal energy involved in Question 4a.

i. How many cal/g were present in the food sample?

ii. How many Cal/g were present?

a. $110 \text{ g } H_2O \times \frac{1 \text{ cal}}{\text{g}\cdot°\text{C}} \times 10 \text{ °C} = 1100 \text{ calories}$

b. (i) $\frac{1100 \text{ cal}}{2.0 \text{ grams}} = 550 \text{ cal/g}$

(ii) $550 \text{ cal} \times \frac{1 \text{ Cal}}{1000 \text{ cal}} = .55 \text{ Cal/g}$

5. Explain how solar energy "flows" through an ecosystem.

Solar energy is stored by plants, then is transferred through the ecosystem to herbivores, to carnivores, and to decomposers. At each stage, some of the energy is lost to the environment as heat.

6. During photosynthesis, what energy conversion takes place as carbohydrates are formed?

Radiant energy from the Sun is converted to chemical energy that is stored in the carbohydrates.

7. Does all solar energy captured by green plants during photosynthesis continue to flow through an ecosystem from one consumer to the next? Explain.

No. Energy from the Sun is dispersed as it transfers from one organism to the next. It is dispersed as heat energy and/or into less useful substances.

8. What is meant by an "energy rich" molecule?

Energy rich molecules are molecules that store large quantities of energy. Fats and carbohydrates are classified as energy rich molecules.

9. A sign in a small, local restaurant states, "When you eat a pound of fish, you're eating ten pounds of flies." Aside from its questionable value as food-marketing strategy, how accurate is the message? Why?

Only 10–15% of energy is used for growth when consumed by an organism. A fish must eat 10 pounds of flies to make one pound of fish.

10. Compare the process of photosynthesis to the process of respiration.

Photosynthesis and respiration are reversed processes. In photosynthesis, carbon dioxide + water + energy → glucose + oxygen: whereas, in respiration, glucose + oxygen → carbon dioxide + water + energy.

11. Why must the energy released from food go through so many complicated reactions in order to effectively power cells?

Energy is used in energy production, transport of molecules, waste disposal, storage of generic material, and the synthesis of new molecules. Each one of these processes requires a different set of complicated reactions. If the energy was released more simply, as in combustion, the uncontrolled heat would damage the cell.

12. In a science fiction story, a villain creates a pill containing an enzyme that rapidly converts all of the victim's ATP to ADP. What would be the result of swallowing such a pill

a. on a cellular level?

b. to the victim in general?

a. Each mole of ATP molecules would release 7.3 Calories of energy as it is changed into ADP.

b. The victim will "burn up" or at the very least run out of energy.

13. Considering the level of activity required to expend food energy, would you rather increase your exercise level or reduce your food intake if you want to lose some weight? Explain.

Both may be needed. To lose weight a person must consume less energy than he/she expends.

14. Why do people require energy even while they are sitting perfectly still?

Energy is required to carry on body functions, such as breathing, eating, digesting food, synthesizing new molecules, and disposing of waste materials.

15. Explain why two people performing exactly the same exercise may not burn the same number of Calories.

The calories expended during various activities depend on the duration of the activity, the weight of the person doing the exercise, and exercise technique.

16. Explain why low-Calorie food is sometimes described as being "lite" or "light," even though Calories are not a unit of mass.

Low-Calorie food contains less energy than comparable food products, which may lead to weight loss. The connection of "counting Calories" to weight loss reinforces misconceptions among many consumers regarding the difference between mass and energy.

Connecting the Concepts

17. A student argues that eating a certain mass of chocolate or eating the same mass of apples will result in the same gain in body weight. Explain why this is not correct.

Apples and chocolate belong to different food groups. Gram for gram they will be converted into a different number of Calories resulting in a different extent of weight gain.

18. Figure 3 (page 492) shows an energy diagram for the process of photosynthesis.

a. Construct a similar energy diagram for the process of cellular respiration.

b. Compare the energy diagrams for those two processes. What are the similarities and differences?

a. The diagram would be similar, with everything reversed.

b. The same chemicals are involved, but respiration is the reverse process of photosynthesis. Respiration is exothermic while photosynthesis is endothermic.

19. The term respiration is sometimes used to mean "breathing." What is the relationship, if any, between cellular respiration and breathing?

Breathing is part of cellular respiration; it supplies oxygen and gets rid of carbon dioxide. Other than that the processes are distinct from each other.

20. **a.** In what sense does world hunger involve an "energy crisis"?

b. In what sense is it a "national resource crisis"?

a. Providing "fuel" to power humans is similar to providing fuel for vehicles or homes.

b. Most hunger experts agree that world hunger is a resource issue governed by political and economic factors, rather than inadequate food supplies worldwide.

Extending the Concepts

21. Why isn't it possible just to eat pure ATP and eliminate some of the steps in metabolism?

Animal metabolism relies on getting basic energy requirements from food. Additionally, not all food is immediately converted into ATP; some is used for other functions, or may even be fully processed and utilized.

22. Investigate the design of a commercial calorimeter. Make a sketch of the essential parts and explain the calorimeter's operation.

Commercial calorimeters have better insulation, more completely capturing the heat given off from food. They also have technology to completely burn food that doesn't burn well on its own, and more precisely measure the temperature changes.

23. For hibernating animals, the consumption and storage of fat is critical to survival. Investigate a particular animal species to find out how it stores optimum quantities of fat for hibernation.

Students will find any number of animals that have sufficiently low metabolism and high fat stores that allow them to hibernate for long periods of time. Students should be able to find the amount of fat some animals need for a hibernation season and calculate the amount of energy involved.

24. A student decides to lose some weight by not wearing a coat in cold winter weather. What knowledge of food energy might have inspired this idea? Does the plan have merit? Explain.

The body must burn calories to keep warm. To lose weight, a person must burn more calories than they take in. The weight loss possible by such a scheme is small relative to the discomfort it would produce. An interesting calculation is for students to compute the energy loss relative to the progress towards hypothermia for given situations. Hypothermia charts are available in many survival, marine, or outdoor texts, or Web sites.

ENERGY STORAGE AND USE

Introduction

This section examines the basic structures and properties of fats and carbohydrates, including the difference between saturated and unsaturated fats and the concept of *cis*- and *trans*- isomers. The concept of limiting reactants and how it applies to reactions in living systems is introduced.

B.1 CARBOHYDRATES: ONE WAY TO COMBINE C, H, AND O

Carbohydrate structure brings students back to photosynthesis as the energy-storing process in foods. Different types of saccharides, including polysaccharides, are presented, along with some nutritional advice.

Simple sugars can exist in a linear chain form, or they can link with themselves to make a ring form, which is more stable. The two forms have the same molecular formula. These simple sugars serve as monomers to form polymers called complex carbohydrates. The simple sugar subunits can be bonded together in two different ways. The difference in bonding between starch and cellulose is small but critical. In digestible starch, both rings are attached to the connecting oxygen (this is similar to an ether functional group) so that the oxygen is below both rings (this is called the alpha linkage). In cellulose, the oxygen is above one ring and below the other (the beta linkage). Human enzymes can break down the bonding in the starch molecule, but only certain organisms have enzymes capable of digesting cellulose.

SECTION B

ENERGY STORAGE AND USE

Introduction — CD-ROM WWW.

You are now prepared to explore how the chemical energy contained in foods, studied in Section A, is stored and consumed. Keep in mind the three-day food inventory you are evaluating, and think about how this new knowledge may apply to its analysis.

Carbohydrates: Structure and Function — CD-ROM WWW.

B.1 CARBOHYDRATES: ONE WAY TO COMBINE C, H, AND O

Sugars, starch, and cellulose are all carbohydrates.

All **carbohydrate** compounds are composed of atoms of carbon, hydrogen, and oxygen. For example, glucose, the key energy-releasing carbohydrate in biological systems, has the formula $C_6H_{12}O_6$. When such formulas were first established, chemists noted a 2:1 relationship of hydrogen atoms to oxygen atoms in carbohydrates, the same as in water. They were tempted to write the glucose formula as $C_6(H_2O)_6$—implying a chemical combination of carbon with water. They even invented the term "carbo-hydrates" (water-containing carbon substances) to apply to glucose and related compounds. Although it was later established that no water molecules are present in carbohydrates, the name has persisted. In fact, like water, carbohydrate molecules have O—H bonds in their structures.

Carbohydrates may be either simple sugars, such as glucose, or chemical combinations of two or more simple sugar molecules. Simple sugars, called **monosaccharides,** are usually composed of molecules containing five or six carbon atoms. Glucose (like most other monosaccharides) exists principally in a ring form but can also occur in a chain form, as shown in Figure 9. Do both forms have the same molecular formula?

Sugar molecules composed of two monosaccharide units bonded together are called **disaccharides.** They are formed by a condensation reaction between two monosaccharides. Table sugar, sucrose ($C_{12}H_{22}O_{11}$), is a disaccharide formed by the combined ring forms of glucose and fructose, as

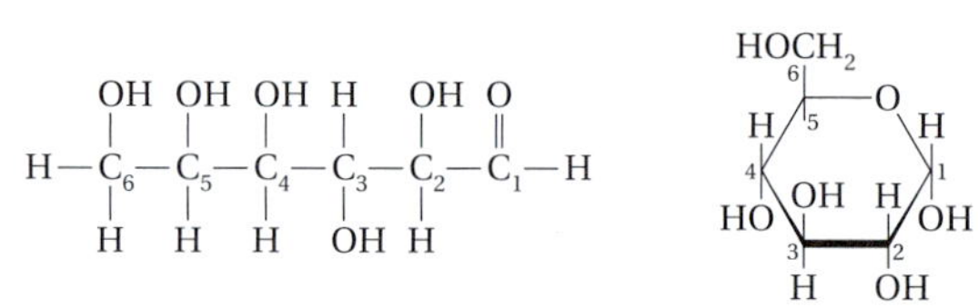

Figure 9 *The structural formula for glucose.*

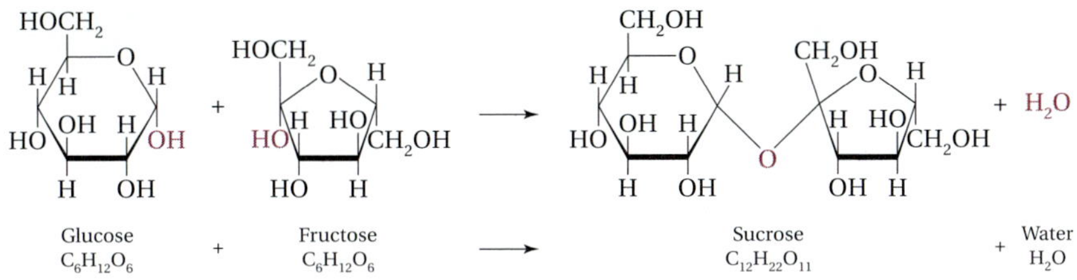

Figure 10 *The formation of sucrose. The two red —OH groups react, with elimination of one H_2O molecule.*

depicted in Figure 10. As their structures suggest, monosaccharide and disaccharide molecules contain polar —OH groups that contribute to the high solubility of sugars in water, a polar solvent.

The same type of condensation reaction that forms disaccharides can also cause many simple sugar units to join to form polymers from these monosaccharides. Such polymers, not surprisingly, are called **polysaccharides** (Figure 11). For example, starch, a major component of grains and many vegetables, is a polysaccharide composed of glucose units. Cellulose, the fibrous or woody material of plants and trees, is another polysaccharide formed from glucose. The major types of carbohydrates are summarized in Figure 12 on page 506.

Condensation reactions were first highlighted in Unit 3, pages 226–227, in ester-formation reactions and also in the formation of condensation polymers.

All polysaccharides are polymers of monosaccharides.

Sugars, starch, and fats are the major energy-delivering substances in human diets. An average U.S. citizen consumes more than 65 kg (145 lb) of sugar each year in beverages, breads, and cakes and as a sweetener. Each gram of a carbohydrate provides 4 Calories of food energy.

Nutritionists recommend that about 60% of dietary Calories should come from carbohydrates. Most people obtain carbohydrates by eating grains, often consumed as rice, corn bread, wheat tortillas, bread, and pasta. People in the United States tend to obtain more carbohydrates, on average, from wheat-based breads, potatoes, and sugar-laden snacks and soft drinks than is common in many other parts of the world. Fruits and vegetables also supply carbohydrates. Meats contain a small amount of the carbohydrate known as glycogen, the form in which animals store glucose.

1 g carbohydrate = 4 Cal energy.

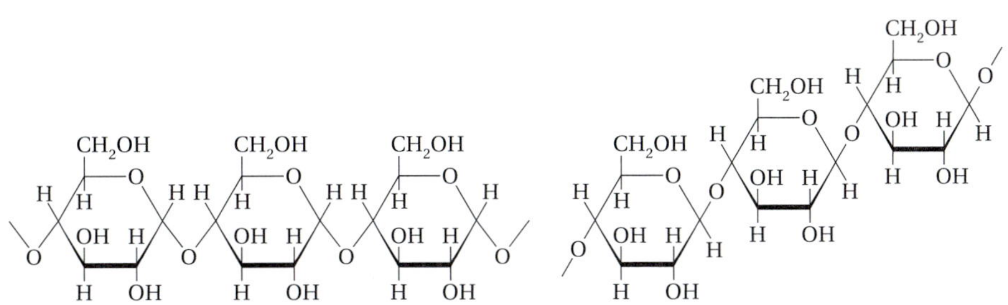

Figure 11 *Structural formulas for starch and cellulose, two polysaccharides.*

B.2 FATS: ANOTHER WAY TO COMBINE C, H, AND O

This section distinguishes between the definition of a fat in common language and the definition of fat as used by scientists. Avoid getting involved at this time in discussion about how good or bad they are, as this is developed more in the next sections.

Figure 12 *The composition of common carbohydrates.*

Carbohydrates			
Classification and Examples	**Composition**	**Formula**	**Common Name or Source**
Monosaccharides		$C_6H_{12}O_6$	
Glucose	—		Blood sugar
Fructose	—		Fruit sugar
Galactose	—		—
Disaccharides	Monosaccharides	$C_{12}H_{22}O_{11}$	
Sucrose	Fructose + glucose		Cane sugar
Lactose	Galactose + glucose		Milk sugar
Maltose	Glucose + glucose		Germinating seeds
Polysaccharides	Glucose polymers	—	
Starch			Plants
Glycogen			Animals
Cellulose			Plant fibers

B.2 FATS: ANOTHER WAY TO COMBINE C, H, AND O

Fats: Structure and Function CD-ROM WWW.

Unlike *carbohydrate* and *protein*, the word *fat* has acquired its own general (and somewhat negative) meaning. Many "fat free" products are featured in advertisements. However, from a chemical point of view, fats are just another major category of biomolecules with special characteristics and functions, like carbohydrates and proteins.

Fats, a significant part of a normal human diet, are present in meat, fish, poultry, salad dressings and oils, dairy products, nuts, and grains. When more food is consumed than is needed to satisfy energy requirements, much of the excess food energy is stored in the body as fat molecules. When food intake is not large enough to supply the body's energy needs, that stored fat is "burned" into energy to make up the difference.

When a substance is burned, it reacts with oxygen gas to form carbon dioxide, water, and thermal energy. When fats, carbohydrates, and protein are "burned" in the body, they react with oxygen gas and are converted into carbon dioxide, water, and energy. These reactions, as you learned in Section A, provide the energy that fuels your body.

Fats are composed of carbon, hydrogen, and oxygen—the same three elements that make up carbohydrates. Fats, however, contain fewer oxygen atoms and more carbon and hydrogen atoms. Thus, fats have a greater number of carbon-hydrogen bonds and fewer carbon-oxygen (and oxygen-hydrogen) bonds than do carbohydrates. You can confirm this by comparing the structure of glyceryl tripalmitate (a typical fat) shown in Figure 13 to that of glucose or starch (Figures 9 and 11, respectively).

Because they have more carbon-hydrogen bonds than carbon-oxygen bonds, fats contain more stored energy per gram than do carbohydrates. In fact, a gram of fat contains over twice the energy stored in a gram of carbohydrate—1 g fat is equivalent to 9 Calories, compared to 4 Calories for each gram of carbohydrate. Consequently, you must run more than twice as far or exercise twice as long to "work off" a given mass of fat as you must to "work off" the same mass of carbohydrate.

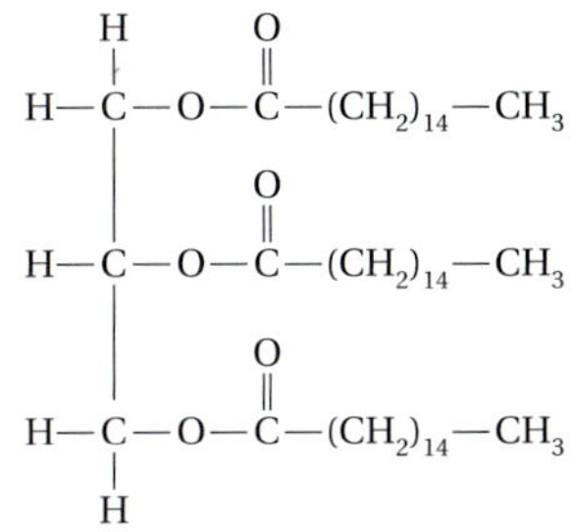

Figure 13 *The structural formula of glyceryl tripalmitate, a typical fat.*

For example, a glazed doughnut contains 11.6 g fat. At 9 Cal/g fat, the doughnut has 11.6 g fat × 9 Cal/g fat, or 104 Calories of energy from fat. It is not surprising that your body uses fat to store excess food energy efficiently.

Generally, fat molecules are nonpolar and only sparingly soluble in water. As you can see from Figure 13, fats have long hydrocarbon-like portions, which make them unlikely to dissolve in a polar solvent such as water. Because of their low water solubility and high energy-storing capacity, fat molecules, unlike carbohydrates, have chemical properties that are similar to those of hydrocarbons.

The chemical properties of fat molecules are due, in part, to the fatty acid groups in the molecule. **Fatty acids** are a class of organic compounds made up of a long hydrocarbon chain with a carboxylic acid group (—COOH) at one end. Two fatty acids are shown in Figure 14.

A typical fat molecule—a **triglyceride**—is a combination of a three-carbon alcohol molecule called **glycerol** and three fatty acid molecules.

In general, the terms "fat" and "triglyceride" are used interchangeably.

1 Glycerol molecule + 3 Fatty acid molecules ⟶
1 Triglyceride (fat) molecule + 3 Water molecules

The formation of a triglyceride is shown in Figure 15 on page 508. Each fatty acid molecule forms an ester linkage when it reacts with an —OH group of glycerol, also producing one water molecule for each ester linkage formed. The main product of this condensation reaction is a fat molecule containing three ester groups. Although the three fatty acids in Figure 15 are identical, fats can contain two or three different fatty acids.

The reaction that produces a fat molecule is similar to the reaction you observed in Unit 3 that produced the ester methyl salicylate (page 227).

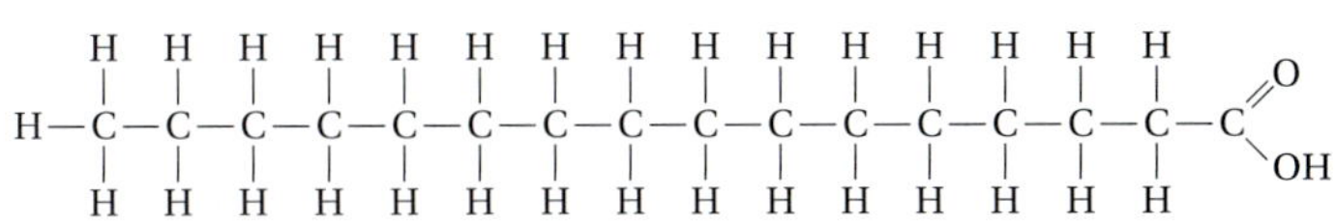

Palmitic acid, a saturated fatty acid

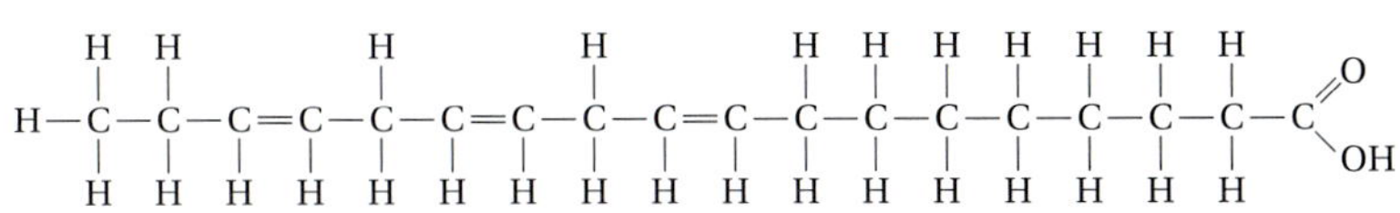

Linolenic acid, a polyunsaturated fatty acid

Figure 14 *Typical fatty acids.*

507

B.3 SATURATED AND UNSATURATED FATS

The section continues by defining saturated fats and polyunsaturated fats. *Cis-* and *trans-* isomerism, hydrogenation, and questions about nutritional safety are discussed. High fat consumption is one factor in obesity. You may want to discuss the popularity of fast foods, which are high in fat, among U.S. high school students. You may also want to discuss the large average intake of sugar in the United States. Both of these are significant contributors to obesity.

Hint: Ball and stick models may be useful in explaining *Cis-* and *trans-*fatty acids.

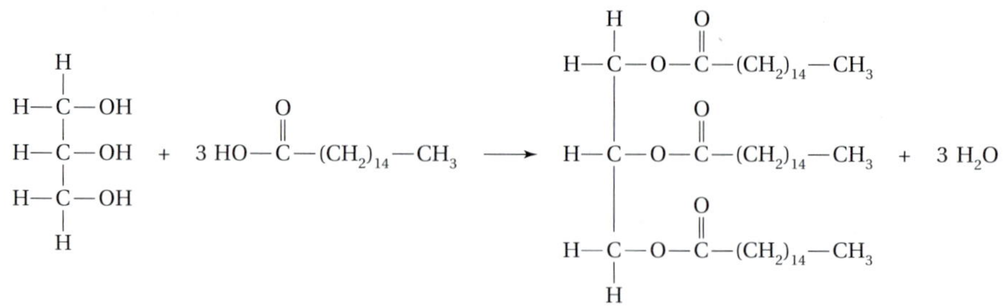

Glycerol — Palmitic acid — Glyceryl tripalmitate (a typical fat) — Water

Figure 15 *Formation of a typical fat, a triglyceride.*

B.3 SATURATED AND UNSATURATED FATS

Recall from Unit 3 that hydrocarbons can be saturated (containing only single carbon-carbon bonds) or unsaturated (containing double or triple carbon-carbon bonds). Likewise, hydrocarbon chains in fatty acids are either saturated or unsaturated. Look again at the fatty acids in Figure 14. Can you identify each as either saturated or unsaturated? Fats containing saturated fatty acids are called **saturated fats;** those containing some unsaturated fatty acids are known as **unsaturated fats.** A **monounsaturated** fat contains just one carbon-carbon double bond in its fatty acid components. A **polyunsaturated fat** contains two or more C=C double bonds in the fatty acid portion of a triglyceride molecule. Based on these definitions, is the fat depicted in Figure 13 saturated, monounsaturated, or polyunsaturated?

Oils are fats that are liquid at room temperature.

In general, higher levels of unsaturation are associated with oils having lower melting points. Triglycerides in butter and other animal fats are nearly all saturated and form solids at room temperature. However, fats from plant sources commonly are polyunsaturated. At room temperature these polyunsaturated fats are liquids (oils), such as safflower oil (13% monounsaturated; 78% polyunsaturated), corn oil (25% monounsaturated; 62% polyunsaturated), and olive oil (77% monounsaturated; 9% polyunsaturated). The levels of saturation in various fats are shown in Figure 16.

Addition reactions were first discussed in Unit 3, pages 216–217.

Due to their C=C double bonds, unsaturated fats undergo addition reactions, but saturated fats cannot. Thus these two types of fats participate differently in body chemistry. Unsaturated fat molecules are much more chemically reactive. Polyunsaturated fats have become newsworthy; increasing evidence suggests that saturated fats may contribute more to health problems than some unsaturated fats. Saturated fats, as well as other factors, are associated with formation of arterial plaque, deposits of fatty

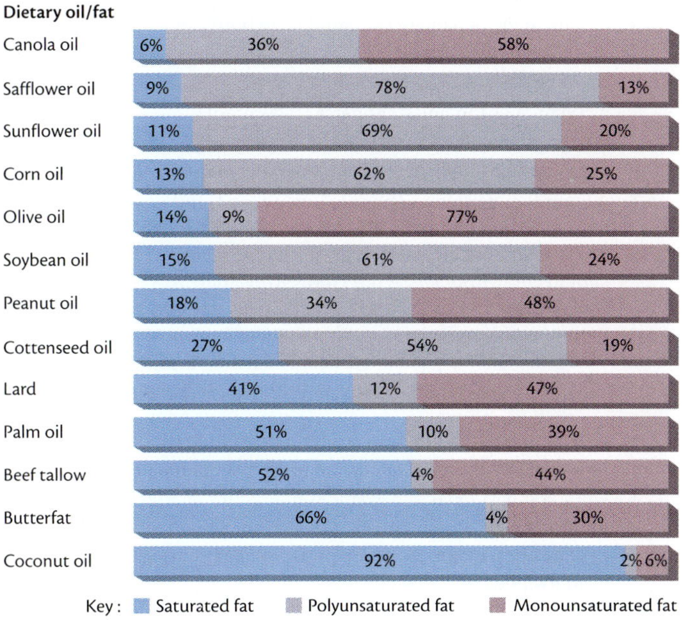

Source: *Food Technology*, April 1989.

Figure 16 *The percent of polyunsaturated fat in fats from plant and animal sources.*

material in blood vessel walls. The result is a condition known as "hardening of the arteries," or atherosclerosis, which is a particular threat to coronary (heart) arteries and arteries leading to the brain. If coronary arteries are blocked, a heart attack can result, damaging the heart muscle. If arteries leading to the brain are blocked, a stroke may result, killing brain cells and disabling various body functions.

CALORIES FROM FAT

Building Skills 3

The label on a package of butter provides the following nutritional data for one serving, which is defined as 1 tablespoon, or 14 g:

Total fat	10.9 g
Polyunsaturated fat	0.4 g
Saturated fat	7.2 g
Calories	100

You can calculate the percent of polyunsaturated fats contained in butterfat by dividing the mass of polyunsaturated fat (0.4 g) by the mass of total fat (10.9 g), giving 3.7%. Thus, approximately 4% of butter's total fat is polyunsaturated.

CALORIES FROM FATS

Building Skills 3

The calculations in this activity point out that nearly all the Calories in both butter and margarine are Calories from fat.

Answers

1. $\frac{7.2 \text{ g}}{10.9 \text{ g}} \times 100\% =$
   66% saturated fat
2. a. There must be some monounsaturated fat in butter.

   b. $10.9 \text{ g} - (0.4 \text{ g} + 7.2 \text{ g}) = 3.3 \text{ g}$
3. $\frac{10.9 \text{ g fat}}{14 \text{ g butter}} \times 100\% = 78\%$
4. a. $10.9 \text{ g} \times \frac{9 \text{ Cal}}{1 \text{ g}} = 98.1 \text{ Cal}$

   b. $98.1 \text{ Cal} \times \frac{100\%}{100 \text{ Cal}} = 98\%$
5. a. $10 \text{ g} \times \frac{9 \text{ Cal}}{1 \text{ g}} = 90 \text{ Cal} = 100\%$

   b. The percent of Calories from fat is about the same. Fat in margarine is generally more unsaturated, and thus may be somewhat more healthful.

1. Calculate the percent of saturated fat in butterfat.
2. The percent values for saturated and polyunsaturated fat do not add up to 100%.
   a. What does the "missing" percent represent?
   b. How many grams of fat does this represent?
3. Calculate the total percent fat in a serving of butter.
4. A common consideration when buying foods is the percent of Calories derived from fat and other nutrients. Health professionals recommend that no more than 30% of Calories in a human diet should come from fats. A direct way to evaluate this is to compare the total number of Calories from fat to the total Calories delivered in that serving of food:

$$\frac{\text{Calories from fat}}{\text{Total Calories}} \times 100\% = \text{Percent of Calories from fat}$$

   a. Determine the total Calories from fat in 1 tablespoon of butter.
   b. Calculate the percent of total Calories from fat in butter.
5. One serving (14 g) of margarine contains 10 g fat and 90 Calories.
   a. Determine the percent of Calories from fat in this margarine.
   b. How does the percent of Calories from fat in margarine compare to the percent in butter?

Recall that 1 g fat = 9 Calories.

A process called partial **hydrogenation** adds hydrogen atoms to some C=C bonds in vegetable oil triglyceride molecules. By reacting with hydrogen, a C=C double bond is converted to a C—C single bond. Since the number of unsaturated sites (C=C) is decreased, the partially hydrogenated product becomes more saturated, and the original liquid oil becomes semisolid. Such partially hydrogenated fats are used in margarines, puddings, and shortening.

The reaction between hydrogen and C=C double bonds in a polyunsaturated fat is typical of how C=C bonds in most alkenes react. Generally, C=C bonds in alkenes can be converted to C—C single bonds when they react with a variety of substances, including hydrogen.

You learned in Unit 3 (page 221) that a double bond in an alkene prevents adjacent carbon atoms in the double bond from rotating around the bond axis. The double-bonded carbon atoms align as shown in Figure 17. This inflexible arrangement between carbon atoms creates the possibility of *cis-trans* isomerism.

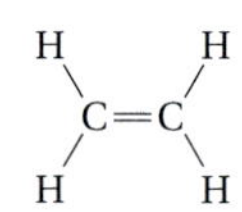

Ethene

Figure 17 *The structural formula for ethene.*

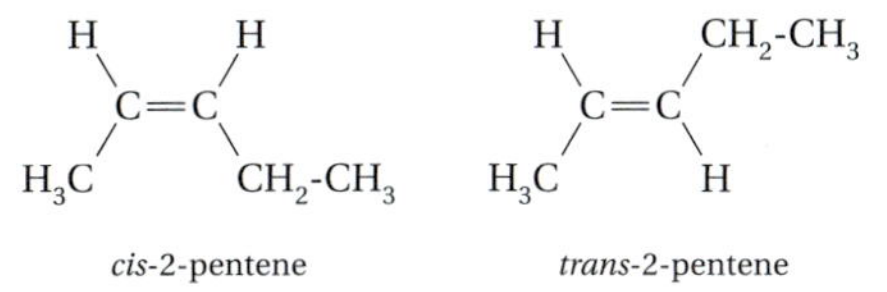

Figure 18 *A comparison of* cis-*2-pentene and* trans-*2-pentene. Note the positions of the functional groups.*

In ***cis-trans*** **isomerism,** two identical functional groups can be in one of two different molecular positions. Both groups can be on the "same side" of the double bond, or they can be across the double bond from the other, as illustrated in Figure 18.

In the case of 2-pentene, the hydrogen atom on each of the double-bonded carbon atoms can be in two different arrangements. Each arrangement represents a different compound, even though the compounds share the same molecular formula! In one case, the molecule has the two hydrogen atoms on the same side of the double bond plane. This arrangement is known as the *cis* (same side) isomer, and the compound is called *cis*-2-pentene. The other compound has the two hydrogen atoms positioned diagonally across the double-bond plane. This arrangement is the *trans* (across) isomer; the compound is known as *trans*-2-pentene.

Cis-trans isomerism, described above for an alkene, is also possible in unsaturated fats, since fatty acid chains often contain C=C bonds. Unsaturated fatty acids in foods that have not been hydrogenated typically have their double bonds arranged in the *cis* arrangement. See Figure 19. During hydrogenation, some of those double bonds rearrange to the *trans* form. Although questions have been recently raised about the nutritional safety of *trans*-fatty acids, studies conducted to address these concerns have not yet produced conclusive answers.

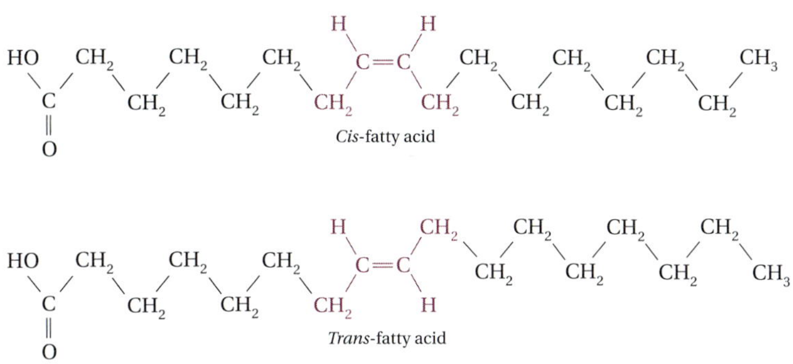

Figure 19
Cis- *and* trans- *fatty acids.*

511

FATS IN THE DIET

Building Skills 4

In this Building Skills students compare the fat contents of ice cream and frozen yogurt.

Answers

1. Ice cream: $\frac{11\text{ g} + 8\text{ g}}{115\text{ g serving}} \times 100\% = 16.5\%$

   Frozen yogurt: $\frac{2.5\text{ g} + 3.5\text{ g}}{112\text{ g serving}} \times 100\% = 5.36\%$

2. a. Yogurt meets 30% Calories from fat guideline, ice cream does not.

   Ice cream: $\frac{19\text{ g}}{310\text{ Cal}} \times \frac{9\text{ Cal}}{1\text{ g}} \times 100\% = 55\%$

   Yogurt: $\frac{6\text{ g}}{200\text{ Cal}} \times \frac{9\text{ Cal}}{1\text{ g}} \times 100\% = 27\%$

   b. No, neither meets the 10% saturated fat guideline.

   Ice cream: $11\text{ g} \times \frac{9\text{ Cal}}{1\text{ g}} \times \frac{100\%}{310\text{ Cal}} = 31.9\%$

   Yogurt: $2.5\text{ g} \times \frac{9\text{ Cal}}{1\text{ g}} \times \frac{100\%}{200\text{ Cal}} = 11.25\%$

FAT-FREE FOOD?

ChemQuandary 2

According to the International Ice Cream Association, classic ice cream must contain at least 10% milkfat by mass. Low-fat ice cream must contain no more than 4.2% fat and nonfat ice cream must have less than 1% fat. Most nonfat ice creams reviewed list no-fat content.

If food energy is not immediately used by the body, it passes through as waste or is stored as fat. Just because a food is fat-free does not mean it lacks calories from carbohydrates or proteins.

Currently, most U.S. citizens obtain about 40% of their total food Calories from fats—well above the 30% recommended by The National Research Council, American Cancer Society, and American Heart Association. Further, these health professionals (and the U.S. Surgeon General) say, saturated fats should contribute less than 10% of total food Calories. The Food Guide Pyramid (page 489) represents these goals for more healthful diets.

High fat consumption is a factor in several "modern" health problems, including obesity and atherosclerosis. Most dietary fat consumed in the United States comes from processed meat, poultry, fish, and dairy products. "Fast foods" and deep-fried foods—such as hamburgers, French fries, fried chicken, and many snack foods—add even more dietary fat. In addition, when your intake of food energy is higher than what you burn off through physical activity, your body converts excess proteins and carbohydrates into fat for storage.

FATS IN THE DIET

Building Skills 4

Your favorite ice-cream flavor just became available in frozen yogurt. The package promotes the yogurt as a less-fatty alternative to ice cream. Here are the nutritional data:

	Serving	Calories	Saturated Fat	Unsaturated Fat
Ice Cream	115 g	310	11 g	8 g
Frozen Yogurt	112 g	200	2.5 g	3.5 g

1. Calculate and compare the total percent fat (by mass) in each of the two desserts.
2. Another way to compare these desserts is to examine the percent of Calories from fat in each.
   a. Does either the ice cream or the frozen yogurt meet the guideline of 30% or less Calories from fat?
   b. Does either meet the guideline of 10% or less Calories from saturated fat?

Recall that 1 g of fat contains 9 Cal of food energy.

FAT-FREE FOOD?

ChemQuandary 2

You just compared ice cream to frozen yogurt. Another frozen dessert option you might consider is a "fat-free" ice cream. Consider the label carefully. Is "fat-free" ice cream really fat free? How do you know? What will happen to the food energy provided by "fat-free" ice cream if it is not immediately used by the body? Do you think it is wise to base a diet entirely on fat-free foods? Why or why not?

B.4 LIMITING REACTANTS

Biochemical reactions in your body convert the fats, carbohydrates, and—under extreme conditions—proteins in the food you eat into energy. These biochemical reactions, like all chemical reactions, require the presence of a complete set of reactants to produce the desired product. Furthermore, the amount of product produced by a chemical reaction depends upon the amounts of reactants present.

To understand how the amounts of reactants determine how much product can be made by a chemical reaction, think about baking a cake. Consider this cake recipe:

2 cups flour	$1\frac{1}{2}$ tablespoons baking powder
2 eggs	1 cup water
1 cup sugar	1/3 cup oil

The proper combination of these quantities (and some thermal energy) will produce one cake. Suppose you have 14 cups flour, 4 eggs, 9 cups sugar, 15 tablespoons baking powder, 10 cups water, and 3–1/3 cups oil. How many cakes can be baked?

Well, 14 cups of flour is enough for 7 cakes (2 cups flour per cake). And there is enough sugar for 9 cakes (1 cup sugar per cake). The supplies of baking powder, water, and oil are sufficient for 10 cakes (confirm this with the recipe). Yet it is not possible to make 10, 9, or even 7 cakes with the available ingredients.

Why? Because only 4 eggs are available—enough just for 2 cakes. The supply of eggs limits the number of cakes that can be made. The excess quantities of other ingredients (flour, sugar, baking powder, water, and oil) remain unused. If more cakes are desired, more eggs (along with required amounts of the other ingredients) will be needed.

In chemical terminology, the eggs in this cake-making example represent the limiting reactant (or limiting reagent). The limiting reactant is the starting material (reactant) that is used up entirely when a chemical reaction occurs. This material limits how much (or how little) product can be formed.

Limiting reactants were first introduced in Unit 4, page 288.

In chemical reactions, just as in recipes, materials react in certain fixed ratios. These relative amounts are indicated in chemical equations by the coefficients. Consider the equation for the oxidation ("burning") of glucose. First, the equation can be interpreted in terms of one glucose molecule:

$$\underset{\text{1 Glucose molecule}}{C_6H_{12}O_6} + \underset{\text{6 Oxygen molecules}}{6\,O_2} \longrightarrow \underset{\text{6 Carbon dioxide molecules}}{6\,CO_2} + \underset{\text{6 Water molecules}}{6\,H_2O} + \text{Energy}$$

Suppose you have available 5 glucose molecules to react with 60 oxygen molecules. Which substance will be the limiting reactant in this reaction? From the equation, you can see that one glucose molecule reacts with 6 oxygen molecules. That means that the 5 glucose molecules require 30 oxygen molecules to react completely. To use up all of the 60 oxygen molecules,

B.4 LIMITING REACTANTS

The concept of limiting reactants (first introduced in Unit 4) is presented in the form of a recipe that should be familiar to all of the students. The concept is applied to balanced chemical equations and then to nutrients and living systems.

In light of possible end-of-year time limitations, student needs, or your own chemistry content preferences, this material on limiting reactants—which extends ideas developed in Unit 4—may be omitted. Omission of Section B.4 will not adversely impact student understanding of material in remaining sections of this or other non-sequential units.

MODELING MATTER

LIMITING REACTANTS

This activity gives students practice in predicting the amounts of products obtained when one reactant is limited. Although none of the systems is "chemical," the calculations correspond closely to those required in chemical systems. This early practice is deliberately limited to concrete, macroscopic systems, so that the central idea will be easier to understand. If students have difficulty with the concept, suggest that they calculate how much product is possible for each reactant and compare the results.

Answers

1. a. Flour is the limiting reactant.

   b. seven cakes.

   c. 12 eggs, 2 c sugar, 4.5 T baking soda, 3 c water, 1 c oil.
2. a. Napkins are used up first.

   b. napkins

   c. 19 lunch boxes are possible.
3. Student answers. Products, such as a simple machine, birdhouse, or first aid kit would work well.

DEMONSTRATION

This demonstration provides a chemical example of the limiting reactant concept, involving easily observed chemical changes. Use a shield during this demonstration. Take care that no flames are used when the balloons are emptied.

Materials

8 culture tubes
8 round balloons (inflated diameter 4 inches)
50 mL 3 M hydrochloric acid
55 cm magnesium ribbon

MODELING MATTER

LIMITING REACTANTS

1. Consider the cake-making example that was just discussed. This time, assume you have 26 eggs and the quantities of the other ingredients specified previously.
   a. Which ingredient now limits the total number of cakes you can make? (In other words, what is the new limiting reactant?)
   b. How many cakes can be made if the other ingredients are present in the same quantities as in the original example?
   c. When the limiting reactant is used up, how much of each other ingredient will be left over?
2. A restaurant prepares carry-out lunches. Each completed lunch includes 1 sandwich, 3 cookies, 2 paper napkins, 1 carton of milk, and 1 box. The current inventory is 60 sandwiches, 102 cookies, 38 napkins, 41 cartons of milk, and 66 boxes.
   a. As carryout lunches are prepared, which item will be used up first?
   b. Which item is the limiting reactant?
   c. How many complete carryout lunches can be assembled from this inventory?
3. Now it is your turn to write a limiting reactant problem. Think about a limiting reactant problem that you encounter in your daily life. First, decide what item you are trying to create. Then, write down all of the necessary "reactants" you need. Finally, identify the limiting "reactant" and how many of the other "reactants" will be left over.

however, requires 10 glucose molecules. Which of those two scenarios is actually possible in this reaction? In other words, which reactant—oxygen or glucose—will be used completely in this reaction?

First, consider glucose molecules. The 5 glucose molecules require 30 oxygen molecules, as noted above. Since 60 oxygen molecules are available, all of the glucose can react, with some oxygen "left over." On the other hand, consider the oxygen molecules. To react completely, the 60 oxygen molecules require 10 glucose molecules—but only 5 glucose molecules are available in the reaction. Thus, through either line of reasoning, glucose must be the limiting reactant. Since the reaction stops once the 5 glucose molecules react with 30 oxygen molecules, 30 oxygen molecules will remain unreacted ("excess") at the end of the reaction. Additional glucose would be needed for the reaction to continue.

The idea of limiting reactants applies equally well to living systems. A shortage of a key nutrient or reactant can severely affect the growth or health of plants and animals. In many biochemical processes, a product from one reaction becomes a reactant for other reactions. If a reaction stops due to a shortage of one substance (the limiting reactant), all reactions following that step will also shut down.

Demonstration tips

Fill the tubes with the listed volumes of hydrochloric acid, clearly label them by number, and place them in test tube racks. Fold each piece of magnesium ribbon so it will fit inside an uninflated balloon.

Part A: Constant amount of HCl, variable amount of Mg

Tube 1: 5 mL 3 M HCl + 2.5 cm Mg ribbon (3 M HCl: 12.5 mL conc. HCl/50 mL solution)
Tube 2: 5 mL 3 M HCl + 5.0 cm Mg ribbon
Tube 3: 5 mL 3 M HCl + 10.0 cm Mg ribbon
Tube 4: 5 mL 3 M HCl + 15.0 cm Mg ribbon

Fortunately, in some cases, alternate reaction pathways are available. If the body's glucose supply is depleted, for example, glucose metabolism cannot occur. One backup system oxidizes stored body fat in place of glucose. More drastically—under starvation conditions—structural protein is broken down and used for energy. If dietary glucose becomes available later, glucose metabolism reactions start up again.

Alternate reaction pathways are not a permanent solution, however. If intake of a particular nutrient is consistently below what the body requires, that nutrient may become a limiting reactant in vital biochemical processes. The results can easily affect personal health.

Producing glucose from protein is much less energy efficient than producing glucose from carbohydrates. However, the human body will use some of its protein to synthesize glucose if necessary.

LIMITING REACTANTS IN CHEMICAL REACTIONS

Building Skills 5

A chemical equation can be interpreted not only in terms of molecules, as discussed in Section B.4, but also in terms of moles and grams (as illustrated on page 137). The glucose-oxidation reaction can be rewritten in terms of moles of all reactants and products:

$C_6H_{12}O_6$	+	$6\ O_2$	$\longrightarrow$	$6\ CO_2$	+	$6\ H_2O$	+ Energy
1 mol Glucose molecules		6 mol Oxygen molecules		6 mol Carbon dioxide molecules		6 mol Water molecules	

Then molar masses can be used to convert those molar amounts to grams, as follows:

$C_6H_{12}O_6$	+	$6\ O_2$	$\longrightarrow$	$6\ CO_2$	+	$6\ H_2O$	+ Energy
1 mol		6 mol		6 mol		6 mol	
		(6 × 32 g) =		(6 × 44 g) =		(6 × 18 g) =	
180 g		**192 g**		**264 g**		**108 g**	

Note that the sum of reactants (180 g + 192 g) equals the sum of products (264 g + 108 g). Why must that be so?

If only 90 g (0.5 mol) glucose is oxidized (half the mass above), only 96 g (3 mol) O_2 is needed (half of 192 g). As illustrated in the following example, these relationships can also be used to identify the limiting reactant.

1. How much CO_2 can be produced if 100 g glucose is allowed to react with 100 g O_2? Which is the limiting reactant, oxygen or glucose?

   First, consider how much CO_2 could be produced by the glucose.

$$100\ \cancel{\text{g glucose}} \times \frac{1\ \cancel{\text{mol glucose}}}{180\ \cancel{\text{g glucose}}} \times \frac{6\ \text{mol CO}_2}{1\ \cancel{\text{mol glucose}}} = 3.3\ \text{mol CO}_2$$

   Next, determine how much CO_2 could be produced by the oxygen gas:

$$100\ \cancel{\text{g O}_2} \times \frac{1\ \cancel{\text{mol O}_2}}{32\ \cancel{\text{g O}_2}} \times \frac{6\ \text{mol CO}_2}{6\ \cancel{\text{mol O}_2}} = 3.1\ \text{mol CO}_2$$

   Finally, identify the limiting reactant. In this example, the limiting reactant is O_2. Why? Because it produces a smaller amount of CO_2 than does the glucose (3.1 mol CO_2 vs. 3.3 mol CO_2). When the 100-g sample of oxygen gas is gone, the reaction will stop.

the air out of the balloon as you attach it. Then stand the tube upright, so that the magnesium ribbon falls into the acid. Note the relative volume of hydrogen gas generated (as indicated by the size of the balloon) and how much of the original sample of magnesium ribbon is consumed. Repeat for each tube in the set.

Finally, display and compare all the tube-balloon systems. Invite students to interpret the results in terms of the limiting-reactant concept. Ask students (after they have observed all tubes carefully) which reactant apparently limited the production of hydrogen in the Part A tubes (1–4). Then invite them to analyze the Part B tubes in a similar fashion. Also, help students sense how this demonstration helps to clarify the limiting-nutrient ideas developed in the unit.

LIMITING REACTANTS IN CHEMICAL REACTIONS

Building Skills 5

This extension of the previous activity applies the concept of a limiting reactant to moles and molecules.

Answers

1. Worked out in student text.

Part B: Constant amount of Mg, variable amount of HCl

Tube 5: 2 mL 3 M HCl + 5.0 cm Mg ribbon
Tube 6: 4 mL 3 M HCl + 5.0 cm Mg ribbon
Tube 7: 6 mL 3 M HCl + 5.0 cm Mg ribbon
Tube 8: 10 mL 3 M HCl + 5.0 cm Mg ribbon

Work with one tube at a time. (Practice this procedure before doing it with students.) The manipulations are not difficult, just a bit awkward. Hold the tube at an angle to prevent the ribbon from prematurely falling into the acid. Carefully stretch the neck of the balloon containing the magnesium strip over the lip of the culture tube. Try to keep all of

2. a. Lactose is the limiting reactant.

   b. 1.5 g lactose × $\frac{1 \text{ mol lactose}}{342 \text{ g lactose}}$ × $\frac{1 \text{ mol glucose}}{1 \text{ mol lactose}}$ = 0.0044 mol glucose

   c. 0.0044 mol glucose × $\frac{180 \text{ g glucose}}{1 \text{ mol glucose}}$ = 0.79 g glucose

   d. water

   e. The 0.0044 moles of glucose requires the same number of moles of water. 0.0044 mol × 18 g/mol = approximately 0.1 g water used. 9.9g water remain.

B.5 ANALYZING FATS AND CARBOHYDRATES

Making Decisions

Students will analyze the main sources of energy found in the contents of their diet from Section A. They will

a. determine the mass of carbohydrate and the percent of Calories from carbohydrate.

b. determine the total mass of fats (saturated and unsaturated).

c. calculate the average mass of fats.

d. calculate average daily energy supplied by fat.

e. identify ways to reduce fats in the diet.

Use the reference materials identified earlier to assist students with these calculations. Student calculations can be verified using the online diet-analysis tools listed for Section A.7 on page 501.

If one of the supplied food diaries is used, student answers can be compared to the food diary analysis in the appendix.

You can calculate how much glucose would remain unreacted when the reaction stops. Here is one way to determine this:

First, find how much glucose reacts with the O_2:

$$100 \text{ g } O_2 \times \frac{1 \text{ mol } O_2}{32 \text{ g } O_2} \times \frac{1 \text{ mol glucose}}{6 \text{ mol } O_2} \times \frac{180 \text{ g glucose}}{1 \text{ mol glucose}} = 94 \text{ g glucose}$$

Then, find how much glucose is left over:

(100 g glucose initially) − (94 g glucose reacted) = 6 g glucose left over.

Enzymes are biological catalysts. You will learn more about roles and properties of enzymes on pages 525–530.

Even though glucose and galactose have the same formula—$C_6H_{12}O_6$—they have different molecular structures. Thus they are two different monosaccharides.

2. Lactose, the sugar found in milk, is a disaccharide. In the human body, the enzyme lactase converts a lactose molecule into a molecule of each of two monosaccharides—glucose and galactose.

$$\underset{\substack{\text{Lactose}\\342\text{ g}}}{C_{12}H_{22}O_{11}} + \underset{\substack{\text{Water}\\18\text{ g}}}{H_2O} \xrightarrow{\text{lactase}} \underset{\substack{\text{Glucose}\\180\text{ g}}}{C_6H_{12}O_6} + \underset{\substack{\text{Galactose}\\180\text{ g}}}{C_6H_{12}O_6}$$

Imagine that you are investigating how lactase hydrolyzes lactose. During one trial, you add lactase to 1.5 g lactose and 10.0 g water.

a. Identify the limiting reactant in this trial.
b. How many moles of glucose will be produced in this reaction?
c. How many grams of glucose will be produced?
d. After the reaction stops, which reactant will be left over (that is, which will be present in excess)?
e. How much of that reactant will remain unreacted?

B.5 ANALYZING FATS AND CARBOHYDRATES

Making Decisions

In Section A, you analyzed the energy content of a diet. Using information in Section B, you can now evaluate the main sources of this energy.

1. Using appropriate resources, determine the mass of carbohydrates (in grams) contained in each item on the three-day food inventory. Then, calculate the percent of Calories provided by carbohydrates. Record both sets of data.
2. Using appropriate resources, determine the mass of fat (in grams) contained in each item on the three-day food inventory. If possible, identify the total mass of saturated and unsaturated fat in each item. Record these data.
3. Calculate the average mass of fat supplied each day.
4. a. Calculate the average daily energy (Calories) supplied by the fat.

   b. Based on that value, is food energy from fat less than 30% of the total food energy supplied?
5. Identify possible ways to reduce the quantity of fat in the food inventory you are analyzing.

Questions & Answers CD-ROM WWW.

SECTION SUMMARY

Reviewing the Concepts

1. Where did the name "carbohydrate" come from?

The name carbohydrate refers to molecules containing carbon plus hydrogen and oxygen in the same proportion as water.

2. Compare the structure of a molecule of fat to that of a molecule of carbohydrate.

Fats are long chain hydrocarbons with a carboxyl acid group (—COOH) on one end. Fats have more C—H bonds than C—O or O—H bonds.

3. What is the chemical explanation for why fats contain more food energy per gram than carbohydrates?

Fats contain more C—H bonds than carbohydrates, thus fats contain more stored energy per gram.

4. Why is the term "burning fat" sometimes used, even though no actual burning is involved?

The term burning fat refers to the energy given off by fats. The reaction is similar to burning, but at a much more controlled rate.

5. List two chemical characteristics shared

a. by all fats.

b. by all carbohydrates.

a. All fats are long chain hydrocarbons with a carboxylic acid on one end. Fats are nonpolar.

b. All carbohydrates have multiple alcohol groups and are polar.

6. Name two common monosaccharides and two common disaccharides.

Monosaccharides: glucose, fructose, or galactose.

Disaccharides: sucrose, lactose, or maltose.

7. Using a shaded oval to represent one molecule of a monosaccharide, sketch a model of a monosaccharide, a disaccharide, and a polysaccharide.

Monosaccharides should be a single oval. Disaccharides should show two ovals joined together. Polysaccharides should show many ovals joined together, perhaps with brackets to show they can be many units long.

8. What kind of chemical reaction links monosaccharides together to make more complex carbohydrates?

A condensation reaction or dehydration synthesis

9. What are fatty acids? In what ways are these substances similar to or different from inorganic acids?

Fatty acids are long chain hydrocarbons with a carboxylic acid (—COOH) on one end. Inorganic acids are generally smaller molecules, composed of a nonmetal atom, sometimes oxygen and always hydrogen. Both ionize to produce protons. Fatty acids are generally weak acids.

10. Describe the reaction that chemically combines three fatty acid molecules to a molecule of glycerol.

Three fatty acids + glycerol joined by an ester linkage make a fat molecule.

11. Fats are examples of "triglycerides." Why does that name provide an appropriate description of all fats?

The combination of glycerol plus three fatty acids makes up a triglyceride.

12. With what is a "saturated fat" actually saturated?

A saturated fat is saturated with hydrogen.

13. How does the degree of unsaturation affect the properties of a fat?

The greater the degree of unsaturation, the lower the melting point. Unsaturated fats are generaly liquid at room temperature.

14. Some margarines are advertised as "partially hydrogenated."

a. What property does the product gain as a result of partial hydrogenation?

b. Why do you think a manufacturer might decide against completely hydrogenating the margarine?

a. Partially hydrogenated fats become more saturated and become more solid.

b. Hydrogenated fats would completely saturate the margarine. Saturated products are not good for the body.

15. What are some natural sources of

a. saturated fats?

b. unsaturated fats?

a. Natural sources of saturated fats include butter and animal fats.

b. Natural sources of unsaturated fats include plant and fish oils.

16. Draw *cis* and *trans* isomers for 2-pentene, which has the formula $CH_3—CH_2—CH=CH—CH_3$.

Student drawings should show a five-carbon chain with an ethyl group and a methyl group angled on the same side relative to the double bond. The double bond is between carbons 2 and 3. The other illustration is similar, but with the ethyl group and methyl groups on opposite sides of the double bond.

17. Draw a fat molecule containing both *cis* and *trans* fatty acids. (A sketch using zig-zag lines to represent the fatty acids is sufficient.)

The bend chain should show two double bonds. One should have adjacent bonds on the same side as the double bond, the other double bond should show adjacent bonds on opposite sides.

18. Why do *cis* and *trans* isomers occur only around a C=C double bond and not around a C—C single bond?

The double bond prevents the adjacent carbon atoms in the double bond from rotating around each other.

19. Is it always possible to use up all the reactants available for a chemical reaction? Explain.

The limiting reactant determines the amount of the reactants that will be used up during the reaction.

20. What is a "limiting reactant"?

The limiting reactant is the starting material that is used up first in a chemical reaction.

21. Why are coefficients in a chemical equation customarily written as integers, not decimals or fractions?

Coefficients represent whole number atoms and molecules.

22. Consider the reaction of hydrogen and oxygen to produce water:

$$2\,H_2(g) + O_2(g) \longrightarrow 2\,H_2O(l)$$

a. If 4.0 mol hydrogen gas is combined with 4.0 mol oxygen gas, how much water can be produced?

b. What is the limiting reactant in that reaction?

a. 4 moles of water can be produced.

b. Hydrogen is the limiting reactant.

23. The reaction used to inflate an air bag in a motor vechicle is

$$2\,NaN_3(s) \longrightarrow 2\,Na(s) + 3\,N_2(g)$$

If 112 g NaN_3 reacts according to that equation, what mass of $N_2(g)$ will be produced?

72 grams N_2

24. One step in an early soap-making method was to allow wood ashes from the fireplace (containing potash, K_2O) to react with water, producing a strongly basic solution. The reaction involved is

$$K_2O(s) + H_2O(l) \longrightarrow 2\,KOH(aq).$$

A 5.4-g sample of K_2O reacts with 9.0 g water.

a. What is the limiting reactant in that reaction?

b. How much KOH can be produced?

a. 5.4 g of K_2O is 0.057 moles, and 9.0 g of water is 0.50 moles, so water is in excess and K_2O is the limiting reactant.

b. $$5.4\text{ g } K_2O \times \frac{1\text{ mol } K_2O}{94\text{ g } K_2O} \times \frac{2\text{ mol KOH}}{1\text{ mol } K_2O} \times \frac{56\text{ g KOH}}{1\text{ mol KOH}} =$$

6.4 g KOH produced

Connecting the Concepts

25. Explain how a relatively small number of subunits can create so many different kinds of fats and carbohydrates.

Different combinations create different molecules, like the letters of the alphabet.

26. How do *cis-trans* isomers compare to the other types of isomers you have learned about? (Recall that isomers were discussed in Unit 3.)

Like other isomers, they have the same molecular formula but different three-dimensional arrangements.

27. Polysaccharides are examples of naturally occurring polymers. How are they alike and how are they different from synthetic polymers?

They are alike in that they are made from repeating subunits. They differ in that they are made in nature. The types of subunits also differ.

28. How do the chemical properties of fats compare to alkanes, alkenes, and other hydrocarbons?

They are similar, but the carboxylic acid group allows fats to undergo unique reactions.

29. How are chemical equations like recipes used for cooking? How are they different?

Both show the proportions of ingredients, but chemical reactions only allow whole number ratios.

30. In thinking about the reaction HCl + NaOH $\longrightarrow$ NaCl + H_2O, one student concluded that one gram of HCl should react completely with one gram of NaOH. Explain why the student's reasoning is incorrect.

1 mol HCl = 36 g. 1 mol NaOH = 40 g. Therefore, 1 g HCl will react with 0.9 g NaOH. The coefficients refer to the number of molecules or moles that will react, not the number of grams.

31. Consider the reaction described in Question 22. Specify a pair of starting masses for hydrogen and oxygen that would allow both of those reactants to be completely used up in that reaction.

4 g H_2 and 32 g O_2, or other similar mass ratios

Extending the Concepts

32. Look at the ingredients listed on several food packages in your home. Identify the names of any specific carbohydrates or fats that you recognize.

Student answers will vary. Carbohydrate names tend to end in *-ose*.

33. Consumption of *trans* fatty acids may possibly be associated with health risks, although research has not been conclusive on this topic. Investigate and report on research that supports that claim and research that dismisses it.

Much of the research on the role of fats in the diet has opposing views.

34. Fat-like products such as Olestra are being designed and marketed as replacements for fat in some foods. Investigate the chemistry of these fat substitutes, and relate that information to any health claims that are made.

These fat replacement products are designed so that the body cannot metabolize them.

35. Assume that you currently consume 3000 Cal of food energy daily and want to lose 30 lb. of body fat over the next two months (60 days).

a. If you decide to lose that weight only through dieting (no extra exercise), how many Calories, on average, would you need to omit from your diet each day?

b. How many food Calories would you still be allowed to consume daily?

c. Is that a sensible way to lose weight? Why?

a. $\frac{\textbf{30 lbs}}{\textbf{60 days}} \times \frac{\textbf{2700 Cal}}{\textbf{1 lb}} = \textbf{1350 Cal/day}$

b. 3000 Cal − 1350 Cal = 1650 Cal

c. This is not a sensible way to lose weight. A large reduction in Calories should not be undertaken without doctor's orders. A more gradual scheme of weight loss is healthier.

PROTEINS, ENZYMES, AND CHEMISTRY

Introduction

Section C focuses on proteins as builder molecules and as enzymes that control chemical reactions in living organisms. The importance of enzymes in controlling the rate at which biochemical reactions occurs is discussed.

C.1 PROTEINS—FUNCTION AND STRUCTURE

This section introduces proteins and protein building blocks and amino acids, and reviews the condensation reaction that produces the peptide bond. The structures of amino acids and their linkage into polypeptides may need some clarification. Formation of the peptide bond may be illustrated with a set of children's pop beads, with male and female ends. Point out that amino acids always combine so that male (carboxyl) reacts with female (amino); the product still has an amino group and a carboxyl group free to react. Also point out that there are two ways to unite the amino acids, producing two different peptides. Small plumbing parts or pen caps may be used to illustrate the same concept.

SECTION C

PROTEINS, ENZYMES, AND CHEMISTRY

Introduction CD-ROM WWW.

So far in this unit you have explored how the energy that helps to keep you alive is delivered by carbohydrates and fats that you eat. The chemical bonds storing that energy can be broken, and needed energy is released in the formation of new, more strongly bonded molecules. In this section, you will investigate the roles that proteins fill in maintaining your well being, such as controlling release of the energy stored in foods and providing the structure and support to maintain your cells and organs—in short, in helping sustain life itself.

Constructing a Protein CD-ROM WWW.

C.1 PROTEINS—FUNCTION AND STRUCTURE

The word "protein" comes from the Greek word *proteios*, which means "of prime importance."

Whenever you look at a living creature, most of what you see is protein—skin, hair, feathers, eyeballs, fingernails, and claws. **Protein** molecules are the major structural components of living tissue. Inside your body, tissues such as muscles, cartilage, tendons, and ligaments are also composed of protein.

At the cellular level, proteins make it possible to transport materials into and out of cells. Your immune system depends on the ability of protein molecules to identify foreign substances. The rates of many chemical reactions that your cells require would be too slow were it not for special proteins called **enzymes**—biological catalysts present in all cells. In fact, your body contains tens of thousands of different proteins. Figure 20 lists just a few major roles that proteins play in the human body. How is it possible for one type of molecule to have so many different uses?

Proteins are polymers built from smaller molecules called **amino acids.** As Figure 21 shows, amino acid molecules contain carbon, hydrogen, oxygen, and nitrogen; a few—like cysteine—also contain sulfur. Just as monosaccharide molecules serve as building blocks for more complex carbohydrates, 20 different amino acids serve as the structural units of all proteins.

All amino acids have similar structural features, but each is unique, as shown in Figure 21. Two functional groups, the amino group ($-NH_2$) and the carboxylic acid group ($-COOH$), are found in every amino acid molecule.

The combination of two amino acid molecules with loss of one water molecule, as illustrated in Figure 22, is a typical condensation reaction. The bonds that link amino acids together are **peptide bonds.** Like starch, nylon, and polyester (see page 226), proteins are condensation polymers. Proteins are composed of chains of amino acids that vary in length, from ten to several hundred or several thousand amino acids. Just as the 26 letters of the alphabet combine in different ways to form hundreds of thousands of words, the 20 amino acids can combine in a nearly infinite number of ways to form different proteins to meet many different needs of a living organism.

Proteins in the Body		
Type	**Function**	**Examples**
Structural proteins		
Muscle	Contraction, movement	Myosin
Connective tissue	Support, protection	Collagen, keratin
Chromosomal proteins	Part of chromosome structure	Histones
Membranes	Control of influx and outflow, communication	Pore proteins, receptors
Transport proteins	Carriers of gases and other substances	Hemoglobin
Regulatory proteins		
Fluid balance	Maintenance of pH, water, and salt content of body fluid	Serum albumin
Enzymes	Control of metabolism	Proteases
Hormones	Regulation of body functions	Insulin
Protective proteins	Antibodies	Gamma globulin

Figure 20 *Major groups of proteins in the human body.*

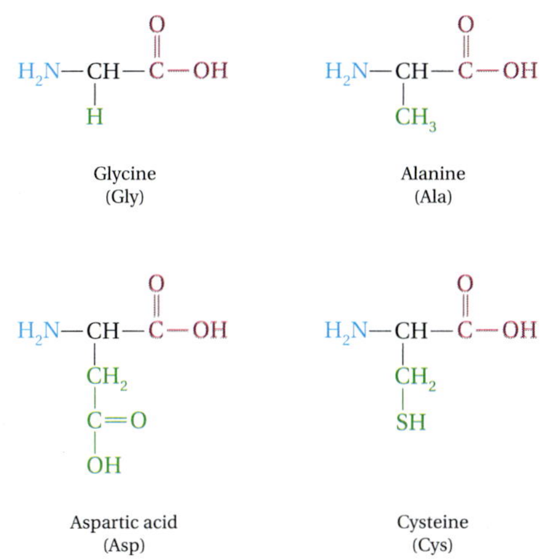

Figure 21 *Structural formulas of glycine, alanine, aspartic acid, and cysteine. The amino groups of each are highlighted in blue; the carboxylic acid groups are in red; and the green highlights the unique side groups that distinguish amino acids from one another.*

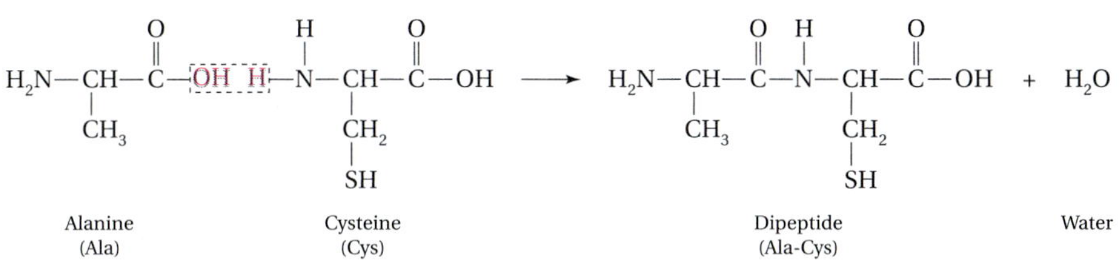

Figure 22 *Formation of a dipeptide from two amino acids (Ala and Cys).*

521

MODELING MATTER

MOLECULAR STRUCTURE OF PROTEINS

Students examine and practice writing structural formulas for amino acids and proteins. Consideration of how many formulas can be written with only a few amino acids helps them realize how a staggering number of different protein structures can be based on a limited number of amino acids.

If molecular model kits are available, you may wish to use them with your students during this exercise. Also, an animation of peptide bond formation would be useful for students to view either on the *ChemCom CD-ROM* or on the Web site. It may also be useful to use Legos, regardless of the size or color. The "out-y" end could represent the carboxylic acid end of an amino acid, while the "in-y" end could represent the amino group. If Legos of different colors (or different sizes) are used to represent the different amino acids, then students can put the Legos together in different ways to convince themselves that the structure of A-B is different from B-A.

MODELING MATTER

MOLECULAR STRUCTURE OF PROTEINS

Protein molecules differ from one another in the number and types of amino acids they contain and in the sequence in which the amino acids are bonded. However, the way in which amino acids bond to one another to form these long polymers is the same for every protein molecule. You will learn how amino acids bond and will investigate—using only three or four amino acids—how the sequence of amino acids affects a protein's structure.

1. Draw structural formulas for glycine and alanine on a sheet of paper. See Figure 21.
   a. Circle and identify the functional groups in each molecule.
   b. How are the two molecules alike?
   c. How do the two molecules differ?
2. Proteins are polymers of amino acids. Examine the equations in Figures 22 and 23 to see how a pair of amino acids join. Notice that the amino group on one amino acid is bonded to the carboxylic acid group on another. This linkage—the peptide bond—is shown here:

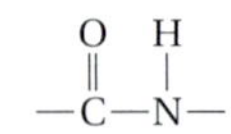

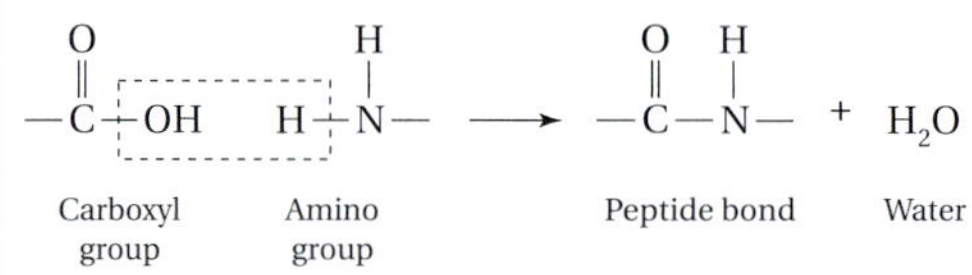

Figure 23 *A carboxylic acid group combining with an amine group.*

Because two amino acid units join to form a peptide bond, the product is called a **dipeptide.**

An amino acid contains at least one amino group and one carboxylic acid group; thus each amino acid can form a peptide bond at either or both ends. Using Figures 22 and 23 as models, complete these steps:

3. Using structural formulas, write the equation for the reaction between two glycine molecules to form a dipeptide. Circle the peptide bond in the dipeptide product.
4. Using structural formulas, write equations for possible reactions between a glycine molecule and an alanine molecule.

C.2 PROTEIN IN YOUR DIET

When foods containing proteins reach your stomach and small intestine, peptide bonds between the amino acids are broken. The separated amino acids then travel through the intestinal walls to the bloodstream, to the liver, then to the rest of the body. Individual cells can use the amino acids as building units for new proteins to meet the body's needs.

If you eat more protein than your body requires—or if your body needs to use protein because carbohydrates and fats are in short supply—amino acids are metabolized in the liver. There nitrogen atoms are removed and converted into urea (Figure 24), which is excreted through the kidneys in urine. Remaining amino acid molecules are either converted to glucose and oxidized—releasing 4 Cal of energy per gram—or stored as fat.

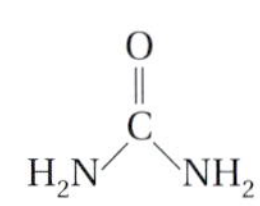

Urea

Figure 24 *The structural formula for urea.*

The human body can normally synthesize adequate supplies of 11 of the 20 required amino acids. The other nine, called **essential amino acids,** must be obtained from protein in the diet. If an essential amino acid is in

Answers

1. a.

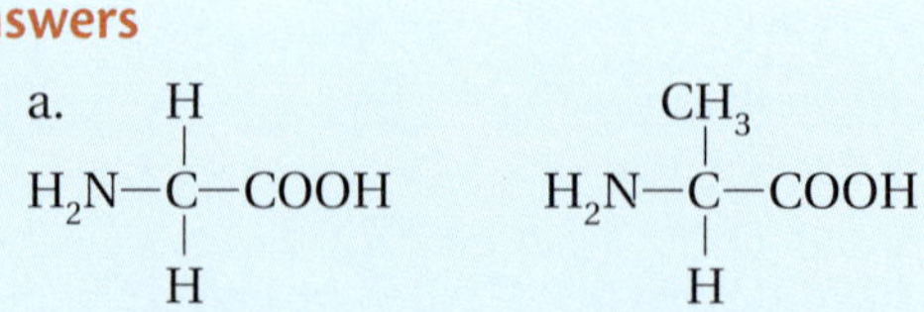

Glycine Alanine

The H_2N group is the amino group and the COOH is the carboxyl group.

b. Functional groups: both molecules contain the amino group (NH_2) and the carboxyl or carboxylic acid group (COOH).

c. Alanine has a methyl (CH_3) group in place of the hydrogen atom on glycine.

5. Examine structural formulas of the dipeptide products identified in Question 4. Note that each dipeptide still contains a reactive amino group and a reactive carboxylic acid group. That means these dipeptides could react further with other amino acids, forming more peptide linkages.

6. Assuming that you have supplies of three different amino acids—A, B, and C—and that each type of amino acid can be used only once, how many different **tripeptides** (three amino acids linked together) can be formed? Write down the possible combinations.

7. Keeping in mind that the carboxylic acid group of the first amino acid forms a peptide bond with the amino group of the second amino acid, explain how the amino acid sequence A-B-C is different from C-B-A.

8. Writing the sequence of amino acids using their chemical names is time consuming and tedious for most protein molecules. As a result, chemists have devised three-letter (and even one-letter) abbreviations for each of the 20 amino acids. Thus a tripeptide with a sequence of glycine–aspartic acid–cysteine is abbreviated Gly-Asp-Cys. Using the three-letter abbreviation system, write all other possible sequences for these three amino acids.

9. How many tetrapeptides (four amino acids joined together) could be formed from four different amino acids? Write the possible tetrapeptides using the three-letter abbreviation system for the four amino acids shown in Figure 21.

10. Given that the cells in your body, using 20 different amino acids, build proteins that can be as long as 10 000 amino acids, would you estimate that the theoretical number of unique proteins your cells could produce is in the hundreds, thousands, or millions-plus? Explain your answer.

11. Why would a living organism benefit from the ability to synthesize so many different proteins? What potential problems do you see?

short supply in the diet, that amino acid can become a limiting reactant in building any protein containing that amino acid. When this happens, the only way the body can make that protein is by destroying one of its own proteins that contains the essential amino acid.

Any dietary source of protein that contains adequate amounts of all essential amino acids represents a source of **complete protein.** Most sources of animal protein contain all nine essential amino acids in needed quantities. Sources of plant protein and some sources of animal protein are incomplete—that is, they do not contain adequate amounts of all nine essential amino acids.

Although no single plant can provide adequate amounts of all essential amino acids, certain combinations of plants can. These combinations of foods, which are said to contain **complementary proteins,** form a part of diets in various parts of the world. See Figure 25 on page 524.

Since your body cannot store amino acids, a balanced protein diet is required daily. The recommended level of protein intake is 10% of total daily

The essential amino acids	
Isoleucine	Phenylalanine
Leucine	Threonine
Lysine	Tryptophan
Methionine	Valine
Histidine (for infants)	

Sources of animal protein include meat, poultry, and seafood.

2. No answer required.

3. 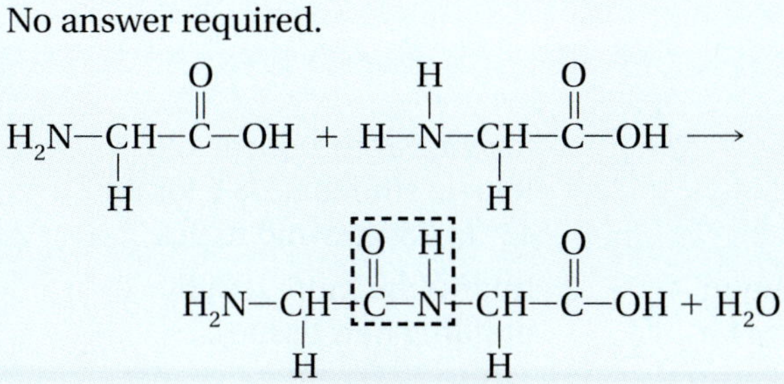

4. 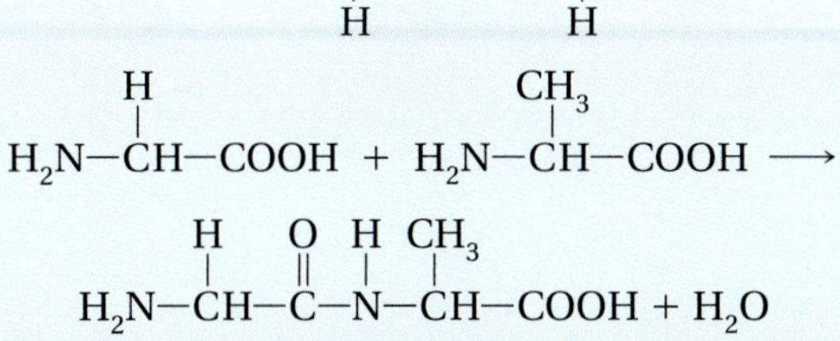

5. No answer required.

6. ABC, BAC, CAB, ACB, BCA, CBA—six combinations.

7. In A-B-C the "A" subunit may bond its carboxyl end. In CBA it would bond its amino end.

8. Cys-Asp-Gly, Cys-Gly-Asp, Asp-Gly-Cys, Gly-Cys-Asp, Asp-Cys-Gly

9. 24 tetrapeptides (4×6): These can be represented as follows.

ABCD	BACD
ABDC	BADC
ACBD	BCAD
ACDB	BCDA
ADBC	BDAC
ADCB	BDCA
CABD	DABC
CADB	DACB
CBAD	DBAC
CBDA	DBCA
CDAB	DCAB
CDBA	DCBA

10. Assuming you could use each amino acid more than once, you could have up to 1 000 020 combinations.

11. With such a vast number of arrangements, unique "genetic codings" would be possible at the molecular level. This allows for a vast array of building materials and enzymes. One problem is that any error in the amino acid sequence can drastically change the characteristics of the protein, causing serious effects.

C.2 PROTEIN IN YOUR DIET

Since the body cannot store excess protein, it is important that we take in a continual supply of these biomolecules for body-maintenance "building" and "rebuilding" purposes. Here students examine and apply the Recommended Daily Dietary Allowances (RDAs) for protein, as summarized in Figure 26 (page 525).

Figure 25 *Complementary proteins, in combination, provide adequate amounts of essential amino acids when eaten together in a single day.*

Complementary Proteins	
Foods	**Country**
Corn tortillas and dried beans	Mexico
Rice and black-eyed peas	United States
Peanut butter and bread	United States
Rice and bean curd	China and Japan
Rice and lentils	India
Wheat pasta and cheese	Italy

Complementary proteins do not need to be combined at each meal—just during the day.

Calories. Too much protein is as harmful as too little. Consumption of excess protein causes stress on the liver and kidneys, the organs that metabolize amino acids. Excessive protein also increases excretion of calcium ions (Ca^{2+}) that function in nerve transmission and in bone and teeth structures. An overly protein-rich diet can even cause dehydration, since more fluids are needed for urinary excretion of urea. This problem is particularly relevant to athletes.

How much protein is actually needed? In 1989 the FDA issued revised Recommended Dietary Allowances (U.S. RDAs) for all required nutrients, including protein. The RDAs for protein shown in Figure 26—as well as for other nutrients—depend on a person's age, gender, and level of physical activity. Specified RDAs are based on actual median heights and weights for a U.S. population of designated age and sex. The use of these values does not imply that those height-to-weight ratios are ideal.

Food labels currently list nutritional information in terms of either a quantity (such as grams) or a percent of Daily Values, as shown in Figure 27. Daily Values are calculated based on a daily 2000 Calorie intake and must be adjusted for an individual's level of physical activity and the energy demands related to such exercise.

A 0–0.5-year-old infant requires 1.0 g protein for each pound of body mass (13 g protein/5.9 kg mass = 2.2 g/kg). By contrast, a 30-year-old female requires only 0.81 g protein per kilogram (50 g/62 kg = 0.81 g/kg).

DAILY PROTEIN REQUIREMENTS — Building Skills 6

Use information from Figure 26 to answer these questions:

1. What mass of protein should you consume each day, on average, according to the Daily Values in Figure 26? (For active individuals, the recommended value increases to 1 g protein per kilogram of body mass daily.)
2. Figure 26 suggests that, for each kilogram of body mass, infants actually require more protein than adults. Why should protein values per kilogram of body mass be highest for infants and become progressively lower as a person ages?

DAILY PROTEIN REQUIREMENTS Building Skills 6

Answers

1. Student answers will depend on their age and gender: 44 g for the 15- to 18-year-old female students; 58 g for the 15- to 18-year-old males.
2. Infants are more rapidly building new tissues.

RDAs for Protein			
Age (yr) or Condition	**Median Weight (lb)**	**Median Height (in)**	**RDA (g)**
Infants			
0–0.5	13	24	13
0.5–1	20	28	14
Males			
11–14	99	62	45
15–18	145	58	58
19–24	160	69	58
25–50	174	69	63
51 +	170	68	63
Females			
11–14	101	62	46
15–18	120	64	44
19–24	128	65	46
25–50	138	64	50
51 +	143	63	50
Pregnant			60
Nursing			
First 6 mo.			65
Second 6 mo.			62

Source: Food and Nutrition Board, National Academy of Sciences—National Research Council. Recommended Dietary Allowances, Revised 1989.

Figure 26 *USDA protein RDAs by age, gender, height, and weight. Like all RDAs, dietary allowances for proteins are average recommendations, meant to provide general guidance. In making dietary choices, an individual needs to consider overall Calorie intake, activity level, genetic makeup, and inherited health factors.*

3. Consider a 57-kg (125-lb), 37-year-old female of median height.
   a. Would this individual's RDA for protein be higher or lower than 50 g?
   b. Why?
4. a. What food do babies consume that meets most of their relatively high protein needs?
   b. Can you find any evidence in Figure 26 to support your answer?

C.3 ENZYMES

Laboratory Activity

Introduction

Think of an incident where you had to respond quickly, such as catching an object that has unexpectedly fallen or getting ready for school when you are late. How do your cells get the energy they need to respond quickly? You have already learned that food is "burned" to meet the body's continuous

Figure 27 *A nutrition label highlighting protein content.*

3. a. probably lower than 50 g.
   b. With lower body weight, protein needs are less.
4. a. mother's milk.
   b. Nursing (lactating) women require higher supplies of protein to produce breast milk.

C.3 ENZYMES

Laboratory Activity

This activity illustrates how enzymes function to speed up certain reactions. In the absence of a catalyst and sunlight, the decomposition of hydrogen peroxide proceeds slowly at room temperature. In this laboratory activity the rates of the catalyzed and uncatalyzed reactions are compared. Raw liver, apple, potato, and yeast are used as sources of enzymes. The decomposition reaction is summarized by the equation $2\ H_2O_2(aq) \rightarrow 2\ H_2O(l) + O_2(g)$.

Time

30 minutes

Materials (for a class of 24 working in pairs)

- 120 mL of 3% hydrogen peroxide (H_2O_2) in brown bottle
- 24 test tubes (16 × 125 mm)
- 6 small pieces of fresh raw liver
- 6 small pieces of boiled liver
- 6 small pieces of fresh apple
- 6 small pieces of boiled apple
- 6 small pieces of fresh potato
- 6 small pieces of boiled potato
- 6 small pieces of fresh (baker's) yeast cake
- 6 small pieces of boiled yeast cake

Safety

Students should wear safety goggles and aprons throughout the laboratory activity.

Advance preparation

Cut the required number of pieces of the fresh samples of potato, apple, liver, and yeast. The pieces must be small enough to fit into a 16 × 125 mm test tube. Boil a potato, an apple, a piece of liver, and a cake of yeast. Cool and cut into pieces.

Pre-lab discussion

Ask students why hydrogen peroxide solutions purchased in drug stores are in brown bottles. As the bottle labels indicate, hydrogen peroxide decomposes with age and exposure to light, heat, or dust.

Lab tips

Provide containers for student disposal of raw and cooked samples, which may then be put into the garbage disposal in the home economics department or cafeteria.

Answers

1. More bubbles of gas are given off, which means that the catalyst speeds up the reaction.
2. The boiled materials do not contain enzymes that catalyze the reaction.
3. a. Answers should be the same.

   b. All of the materials catalyze the decomposition of hydrogen peroxide. Some may be better catalysts than others.
4. If commercial hydrogen peroxide did not contain preservatives, it would all decompose before it could be transported and sold.
5. Enzymes in the blood catalyze the decomposition of the hydrogen peroxide.

energy needs. The rate of "burning" can be turned from low to high, literally within a heartbeat.

Lab Video

What is the secret behind this impressive performance? It lies with biological catalysts called enzymes. **Enzymes**—which are typically proteins—are able to speed up specific reactions without undergoing any lasting change themselves. Before exploring how enzymes fulfill their roles, in the following laboratory activity you will investigate the rate of an enzyme-catalyzed reaction. In particular, you will investigate how the enzyme catalase affects the rate of decomposition of hydrogen peroxide to water and oxygen gas:

$$2\ H_2O_2(aq) \xrightarrow{\text{catalase}} 2\ H_2O(l) + O_2(g)$$

You will be assigned a particular material—apple, potato, or liver—to examine for the presence of the enzyme catalase. You will test a fresh piece of the material and one that has been boiled, to see whether either material catalyzes the decomposition of hydrogen peroxide.

Read the procedure. Prepare a data table with appropriately labeled columns for your data.

Procedure

1. Obtain two pieces of your assigned food sample—one piece that is fresh and one that has been boiled.
2. Label two 16 mm × 125 mm test tubes—one "fresh," the other "boiled."
3. Add 5 mL 3% hydrogen peroxide, H_2O_2, solution to each test tube. Observe whether any bubbles form—that is, whether a gas is produced. (Why is that a sign of hydrogen peroxide decomposition?)
4. Add a portion of the fresh material to the appropriate test tube. Insert a stopper containing a segment of glass tubing into the mouth of the test tube, and arrange the tubing as shown in Figure 28. Be sure the end of the glass tubing is submerged in the beaker of water.
5. For three minutes, record the estimated number of bubbles formed in the test tube each minute.
6. Add a portion of boiled material to the second test tube, and repeat Steps 4 and 5.
7. Discard and dispose of the materials and solutions as directed by your teacher.
8. Wash your hands thoroughly before you leave the laboratory.

Questions

1. What can you conclude from the observations of the test with the fresh material?
2. What can you conclude from the observations of the test with the boiled material?

Post-lab discussion

Students are probably most familiar with hydrogen peroxide as a topical antiseptic. But how does it work? Many people hold the misconception that bubbling of a wound soaked in hydrogen peroxide is a sign of bacterial infection. But in fact the opposite is almost true. Hydrogen peroxide is a strong oxidizing agent that kills cells. Almost all living organisms contain catalase enzyme that decomposes the hydrogen peroxide before it can do damage. Bacteria and other simple single-celled organisms lack this enzyme. That is why hydrogen peroxide is a good medicine. It kills the "bad" cells, but leaves the "good" cells unharmed. This kind of selectivity is very important in medicines and antibacterial agents.

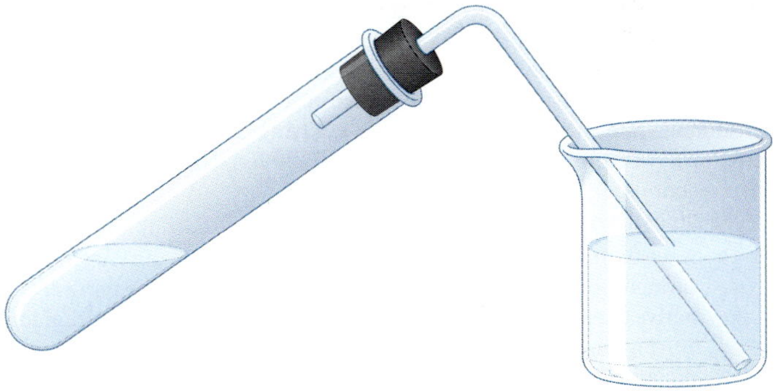

Figure 28 *Setup for laboratory activity.*

3. Compare your experimental data with those of other class members who used
   a. the same material.
   b. a different material.
4. Why does commercial hydrogen peroxide contain preservatives?
5. Your friend tells you that when hydrogen peroxide is added to a cut, foaming shows that infection is present. Explain to her the chemical reason for the foaming.

If the same amount of original hydrogen peroxide solution were left undisturbed without adding any catalase-containing material, it would take days for the same level of hydrogen peroxide decomposition to occur. How do enzymes work so well at catalyzing reactions? Read on.

C.4 HOW ENZYMES WORK

CD-ROM WWW. **Amylase: A Biological Protein Catalyst**

The speed of an enzyme-catalyzed reaction is hard to comprehend. In a single second, one molecule of amylase enzyme assists in converting 18 000 glucose molecules from starch to release needed energy.

$$\underset{\text{(portion of starch molecule)}}{\text{-glucose-glucose-glucose-}} + H_2O \xrightarrow{\text{amylase}} \underset{\text{(remaining portion of starch molecule)}}{\text{-glucose-glucose-}} + \text{glucose}$$

How do enzymes (or catalysts in general) work? Chemical reactions can occur when two reactant atoms or molecules meet with the proper energy and orientation. When the reactant molecules are large, it is especially important that the appropriate reactive groups on the molecules come together in the correct orientation. If molecules collide randomly, this is very unlikely to happen; thus, the reaction rate will be very slow.

The names of most enzymes include the suffix -ase. As with any catalyst, the name of the enzyme is included above the arrow in the equation representing the reaction. This shows that although the enzyme is involved in the reaction, it is not consumed.

C.4 HOW ENZYMES WORK

This section presents information on the mechanism of the interaction between enzymes and substrates. It explains how enzymes are selective, catalyzing only certain reactions. The function of coenzymes is briefly described. For example, alcohol dehydrogenase speeds up the conversion of alcohol to acetaldehyde, which is responsible for many of the adverse effects of alcohol consumption. You may wish to discuss the dangers associated with drinking.

Catalysts make reactions occur more quickly by orienting reactant molecules properly. Reactions in living cells often involve very large molecules reacting at only a single site on the molecule, so catalysts become essential. Without them, many reactions would not occur at all, or at least not at a reasonable rate at normal temperatures. Living systems manufacture a large variety of different enzymes (catalysts), tailored to assist almost all cellular reactions.

In general, enzymes function this way:

- A reactant molecule—known as the **substrate**—and the enzyme are brought together. The substrate molecule fits into the enzyme at an **active site,** where its key functional groups are properly positioned. See Figure 29. The active site of the enzyme determines what kind of reaction an enzyme is able to catalyze. The three-dimensional structure of the enzyme—how the amino acids in the protein polymer arrange themselves and interact—forms a "lock" into which only certain "keys" can fit. In other words, only substrates with the right kind of structure will fit the active site of a particular enzyme.
- The enzyme then interacts with the substrate molecule(s), weakening critical bonds and making the reaction more energetically favorable.
- The substrate is changed into the product by the breaking of weakened bonds and the forming of stronger ones. One or more products then depart from the enzyme surface, freeing the enzyme to interact with other substrate molecules. Each enzyme molecule can participate in numerous reactions without any change to its structure. Because of this regeneration, or "recycling," enzymes are required in much smaller amounts than the reactant substrates.

The discussion of automobile catalytic converters on pages 328–329 described how catalysts function—the reaction is speeded up because the activation energy barrier is lowered.

Each enzyme is as selective as it is fast. It catalyzes only certain reactions—even though substrates for many other reactions are also available. How does an enzyme "know" what to do? For example, why does amylase cleave a starch molecule into thousands of glucose molecules instead of decomposing hydrogen peroxide molecules into water and oxygen gas?

The answer to these questions is the same: The active site of a given enzyme has a three-dimensional shape that only allows certain properly shaped molecules or functional groups to occupy it. Some enzymes are much more selective than others. Thus the enzyme catalyzes only one specific class or type of reaction. Because of the lock-and-key interaction of an enzyme with its substrate(s), cells have precise control over which reactions take place.

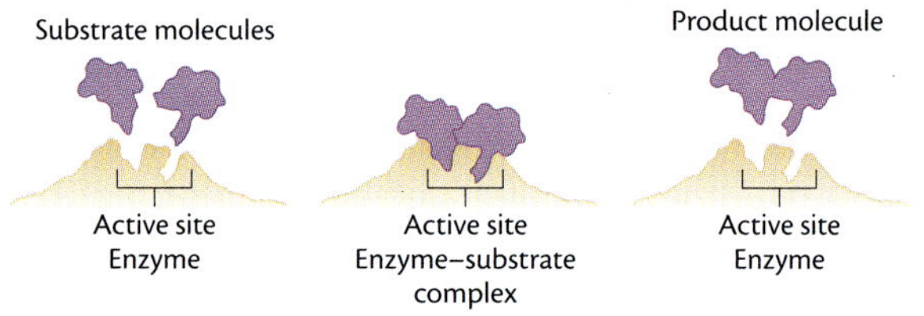

Figure 29 *A model of the way enzyme and substrate molecules interact.*

If a person consumes alcohol, the ethanol (ethyl alcohol, C_2H_5OH) is acted on by the enzyme alcohol dehydrogenase (ADH). Dehydrogenases are enzymes that facilitate the removal of hydrogen atoms (an oxidation process) from a substrate. ADH catalyzes the conversion of ethanol (C_2H_5OH) to acetaldehyde (C_2H_4O). It is believed that acetaldehyde in high concentration in the brain and other tissues is responsible for the effects of a "hangover," or of continued alcohol abuse. Ethanol that is not converted in the stomach enters the bloodstream and ultimately reaches the liver, where ADH converts it to acetaldehyde:

$$CH_3-CH_2-OH \xrightarrow[\text{dehydrogenase}]{\text{alcohol}} CH_3-\overset{\overset{\displaystyle O}{\|}}{C}-H$$

A PROBLEM TO CHEW ON! ChemQuandary 3

Chew an unsalted cracker for a minute or two before swallowing. Describe any changes in taste you detect. Explain what caused any changes in taste.

To meet their energy needs, your body's cells synthesize enzymes that catalyze reactions to convert large molecules into many smaller molecules, releasing energy in the process. The bonds in each of those smaller molecules contain chemical energy that your cells can release through other enzyme-catalyzed reactions. In the next activity, you will investigate the performance of one of many enzymes involved in this process.

C.5 AMYLASE TESTS — Laboratory Activity

Introduction

CD-ROM WWW. Lab Video

In this activity, you will consider how temperature and pH affect the performance of the enzyme amylase. Amylase in saliva breaks down starch molecules into individual glucose units. Glucose reacts with Benedict's reagent to produce a yellow-to-orange precipitate. The color and amount of precipitate formed is a direct indication of the concentration of glucose generated by the catalyst.

Your laboratory team will explore the performance either of amylase at room temperature or of amylase that has been cooled to a lower temperature. Read the procedure below. Prepare a data table with appropriately labeled columns for data to be collected by each group member.

Materials (for a class of 24 working in pairs)

60 test tubes (16 × 125 mm)
60 stoppers for test tubes
12 10-mL graduated cylinders
12 laboratory burners
12 250-mL beakers
boiling chips
12 ring stands
12 rings
12 wire gauze
12 test tube holders
150 mL 0.1% starch suspension (0.15 g in 150 mL water; refrigerate)
300 mL Benedict's solution (see instructions below)
150 mL 0.5% amylase solution (2.3 g amylase in 150 mL solution)
60 mL buffer solution, pH 2
60 mL buffer solution, pH 4
60 mL buffer solution, pH 7
60 mL buffer solution, pH 8
60 mL buffer solution, pH 10
refrigerator

Safety

Students should wear safety goggles and aprons throughout the laboratory activity. The buffer solutions of pH 2 and 4, and of 8 and 10, should be treated as acidic and alkaline solutions, respectively. Caution students to be careful with boiling water and handling hot rings and clamps. Boiling chips should never be added to a boiling or very hot liquid.

Advance preparation

Buffer solutions:
Capsules or envelopes of dry powders can be purchased to make buffer solutions by simply adding the required volume of distilled water. If this is impossible, all the buffer solutions can be made from the following two solutions.

Solution A: Dissolve 12.4 g anhydrous boric acid (H_3BO_3)

A PROBLEM TO CHEW ON! ChemQuandary 3

Students should be able to detect a sweeter taste after prolonged chewing. Amylase, the enzyme in saliva, converts starch into glucose, which tastes sweeter than starch.

C.5 AMYLASE TESTS Laboratory Activity

Students investigate the effects of temperature and pH on the decomposition of starch by amylase. The activities for Day 1 will take 30 minutes; Day 2 requires 45 minutes.

and 10.5 g citric acid hydrate ($C_6H_8O_7 \cdot H_2O$) in enough water to make 1.00 L of solution.

Solution B: Dissolve 38.0 g trisodium phosphate dodecahydrate ($Na_3PO_4 \cdot 12\ H_2O$) in enough water to make 1.00 L of solution.

To make up 200 mL of each buffer, measure the volumes of each solution as specified in the chart below.

Buffer Solutions

pH	mL solution A	mL solution B
2	195	5
4	155	45
7	99	101
8	85	115
10	54	146

Benedict's solution:
Prepared Benedict's solution for qualitative determinations can be purchased. If this is not possible, use the following method of preparation.

Dissolve 100 g sodium carbonate monohydrate ($Na_2CO_3 \cdot H_2O$) and 173 g of sodium citrate dihydrate ($Na_3C_6H_5O_7 \cdot 2H_2O$) in 800 mL distilled water. Warm the mixture slightly to speed up the dissolving process.

Dissolve 17.3 g of copper(II) sulfate pentahydrate ($CuSO_4 \cdot 5H_2O$) in 100 mL distilled water. Slowly pour this solution into the first solution. Stir constantly, cool, and add enough water to make 1.00 L.

To make the starch suspension, dissolve the powdered starch in a small amount of cold water to make a slurry. Then add this to the remainder of the water that has been heated to near boiling. Remove from heat before adding the slurry. Cool before use.

Procedure

Day 1. Preparing the Samples

1. Label each of five test tubes (near the top) with the temperature your group is assigned to investigate, and with pH values of 2, 4, 7, 8, and 10. Also mark each label so that on Day 2 your set of test tubes can be distinguished from those of other groups.
2. Using solutions provided by your teacher, add 5 mL of pH 2, 4, 7, 8, or 10 solutions to the appropriate tubes.
3. Add 2.5 mL of starch suspension to each tube.
4. Add 2.5 mL of 0.5% amylase solution to each tube.
5. Insert a stopper into each tube. Hold the stopper in place with your thumb or finger, and shake the tube well for several seconds.
6. Leave the tubes that are to remain at room temperature in the laboratory overnight as directed by your teacher.
7. Give your teacher the tubes to be refrigerated.
8. Wash your hands thoroughly before leaving the laboratory.

Day 2: Evaluating the Results

9. Prepare a hot-water bath by adding about 100 mL of tap water to a 250-mL beaker. Add a boiling chip. Heat the beaker on a hotplate.
10. Place 5 mL of Benedict's reagent in each tube. Replace each stopper, being careful not to mix them up. Securing the stopper in place with a thumb or finger, shake each tube well for several seconds.
11. Ensure that the tubes are clearly labeled. Remove the stoppers, and place the tubes into the water bath.
12. Heat the test tubes in the hot-water bath until the solution in at least one tube has turned yellow or orange. Then continue heating for 2–3 minutes more.
13. Use a test tube holder to remove the tubes from the water bath. Arrange them in order of increasing pH.
14. Observe and record the color of the contents of each tube.
15. Share your data with your classmates as directed by your teacher.
16. Wash your hands thoroughly before leaving the laboratory.

Questions

1. a. At which temperature did the enzyme perform most effectively?
   b. At which pH?
2. Make a general statement about the effects of temperature and of pH on an enzyme-catalyzed reaction.

Color of Benedict's Solution

Enzyme	pH 2	4	7	8	10
Amylase (room temperature)	blue	blue	orange	orange	blue

As important as enzymes are, remember that your cells synthesize many other kinds of proteins to keep you alive. The amino acids used to synthesize those proteins are best obtained through a diet providing appropriate and adequate amounts of protein. You are now ready to determine whether a particular diet meets those protein needs.

C.6 PROTEIN CONTENT

Making Decisions

You have analyzed your three-day food inventory in terms of energy it provides and the fat and carbohydrate molecules it delivers. Now consider whether that food inventory meets recommended amounts of a key building block of living material—protein.

Use your three-day food inventory to answer these questions. Refer to Figure 26 (page 525) if necessary.

1. What average total mass of protein (in grams) is provided daily?
2. What other information would you need to know about a person who follows this food-intake pattern to evaluate the appropriateness of the protein supplied by these foods?

CD-ROM WWW. Questions & Answers

Answers

1. b. The amylase is most effective at pH 7 and 8.
2. Enzymes work best at body temperature. If the temperature is too low, very little happens; if too high, the enzymes are denatured, as was shown in the previous laboratory activity. Enzymes work best when at a pH similar to that in the body tissue where they are found.

Post-lab discussion

Pool the results from different groups. The amylase is normally found in saliva in the relatively neutral pH of the mouth. It is interesting that because of the pH, digestion of starch stops when swallowed material reaches the stomach, and does not continue until enzymes in the intestine continue breaking down the starch. The advice about chewing one's food thoroughly has an element of truth. Chewing helps mix amylase with starch to begin digestion.

C.6 PROTEIN CONTENT

Making Decisions

Students will analyze their three-day inventory for proteins, the key building block of living materials. Again, provide students with the reference materials cited earlier to assist them with their analysis.

Lab tips

If storage of so many test tubes is a problem, consider dividing the class into groups of four rather than the usual pairs. Remind students to label their test tubes clearly on Day 1, so that they can be identified on Day 2. The starch solution must be freshly prepared and stored in the refrigerator to retard spoilage.

Expected results

In the presence of simple sugars, Benedict's solution produces a yellow, orange, or reddish precipitate of copper(I) oxide when the reagent is heated. The intensity of the color indicates the amount of sugar in the solution.

SECTION SUMMARY

Reviewing the Concepts

1. Name three tissues in your body for which protein is the main structural component.

Skin, hair, nails, muscles, etc.

2. The name "protein" comes from a Greek word meaning "of prime importance." Why is this name appropriate for proteins?

Proteins are used for building the body, and in controlling all chemical reactions of any importance to the body.

3. What are some cellular functions for which proteins are particularly important?

Proteins are used to transport materials in and out of the cell, for contraction, for support, and to control metabolism.

4. Name five food items that are made primarily of protein.

meats, fish, beans, nuts, egg white, and gelatin

5. How does the relatively small number of different amino acids account for the vast variety of proteins found in nature?

Proteins are polymers built up from 20 different amino acid monomers. They can differ from each other in terms of their length (number of monomer units) and the type and sequencing of specific amino acids. Just as 26 letters can form thousands of words, 20 different amino acids can build an immense variety of proteins.

6. What is the range for the total number of amino acids in various protein molecules?

The total number of amino acids in proteins ranges from a few amino acids to more than 10 000 units.

7. From where does the name "amino acid" come?

A carboxylic acid on one end and an amino (NH_2) group on the other.

8. Explain the meaning of each of these terms:

a. complete proteins

b. essential amino acids

c. complementary proteins

a. Complete proteins contain adequate amounts of all essential amino acids.

b. Essential amino acids cannot be synthesized by the body. The nine essential amino acids must be obtained from proteins in the diet.

c. Complementary proteins are a combination of foods that provide all essential amino acids.

9. a. How are enzymes like other catalysts?

b. How are enzymes different from other catalysts?

a. Enzymes are like other catalysts because they speed up a chemical reaction without being used up in the reaction.

b. A specific enzyme can catalyze only one type or class of reaction. Their function depends on temperature and pH.

10. What would be the effect of halting all enzyme activity in the body?

The rate of all chemical reactions would slow down to the point that life would not be possible.

11. Explain the effect of excessively high temperatures on the function of an enzyme.

Enzymes are most effective at body temperature. Excessively high temperatures can cause the enzyme to be denatured, rendering it inactive.

12. Describe how enzymes work to speed up chemical reactions.

Enzymes work by binding with the substrate, making it easier to make or break bonds.

13. a. Explain the interaction of an active site and a substrate in terms of the analogy of a lock and key.

b. Identify at least one limitation in the lock-and-key analogy for enzyme activity.

a. Only the right substrate will fit into the active site and be acted upon, just as only one key will open a lock.

b. With molecules, the substrate changes, but in the analogy the key doesn't. Also, enzymes work on thousands of molecules.

Connecting the Concepts

14. a. How are proteins similar to carbohydrates and fats?

b. How are they different?

a. Proteins, like carbohydrates and fats are made up of carbon, hydrogen, and oxygen.

b. The properties are different and in addition to carbon, hydrogen, and oxygen, proteins also contain nitrogen.

15. If a beefsteak is left on the barbecue too long, it turns black. What does that observation suggest about the chemical composition of the protein in the meat?

It indicates that it contains the element carbon.

16. Using simple block outlines, sketch how an enzyme might act on three fatty acid molecules and one glycerol molecule to build a fat molecule.

The illustration should show three long molecules hooked to one shorter molecule. The enzyme should be shown to fit around the molecules, like a baseball glove, making it easier for the molecules to bond together.

17. The genetic code found in DNA apparently carries blueprints for making all proteins found in the body. Explain why DNA helps to determine the body's physical development and functioning.

By coding for structural proteins it determines physical appearance and properties. By specifying enzymes it controls which chemical reactions will occur.

Extending the Concepts

18. Although many plants contain high levels of protein, vegetarians must be more concerned than non-vegetarians about including adequate protein in their diets.

a. Explain why.

b. How can vegetarians ensure that they obtain all the building blocks needed to build required proteins?

a. Of the 20 different amino acids needed by the human body, eight cannot be internally synthesized and must be received by consumption. Their absence will cause key reactions to be limited even if other necessary "ingredients" are in excess. Most animal sources of protein are complete in that they contain all eight essential amino acids.

b. If one consumes only plant-derived food, one must be sure to eat a variety of foods that "complement" each other's deficiencies so that all the amino acids are present in the diet.

19. The phrase "form follows function" is particularly applicable to enzymes. Explain how form and function are closely related in enzyme molecules.

As illustrated by the lock-and-key analogy, the shape of an enzyme controls how it interacts with substrate molecules.

20. What are some health effects of insufficient protein in a human diet?

Poor growth, bone development, slow maturation.

21. What chemical characteristics make proteins more suitable as structural material than carbohydrates or fats?

The great variety of proteins that can be formed from 20 amino acids makes them ideal for their many functions in living organisms.

OTHER SUBSTANCES IN FOODS

Introduction

In Section D we discuss the roles of vitamins, minerals, and other food additives. Students learn why vitamins and minerals are essential and the importance of RDAs. In the laboratory they determine the quantity of vitamin C by titration and they analyze food colors by paper chromatography. Finally, students briefly examine the benefits and risks of some common food additives.

D.1 VITAMINS

RDAs for selected vitamins are examined in this section, as well as the health effects of consuming too much or too little of particular vitamins.

Another very important food for seafarers trying to avoid scurvy was sauerkraut. It is high in vitamin C and since it is fermented, it lasts for long periods of time at sea. This was a distinct advantage over citrus fruits.

SECTION D

OTHER SUBSTANCES IN FOODS

Introduction CD-ROM WWW.

When speaking in nutritional terms, "minerals" refers to elements (including calcium, sodium, selenium, and zinc) that are necessary to sustain life. However, to geologists, geochemists, and mining engineers, "minerals" are naturally occurring solid compounds found in the crust of Earth, such as calcite ($CaCO_3$), quartz (SiO_2), and hematite (Fe_2O_3). See page 113.

Proteins, carbohydrates, and fats are the major building blocks and fuel molecules of life, but alone they cannot sustain life. Even though your body requires them only in tiny amounts, vitamins and minerals also play vital roles. Small, but absolutely essential, quantities of vitamins and minerals are supplied by the foods you eat or by appropriate dietary supplements. What do vitamins and minerals do in the body that is so important?

D.1 VITAMINS

Biomolecules are organic compounds produced by chemical reactions associated with living systems.

By definition, **vitamins** are biomolecules necessary for growth, reproduction, health, and life. Each vitamin is required in only a tiny amount. The total quantity of all vitamins required daily by an adult is only about 0.2 g—"a little goes a long way" with vitamins. How much is "enough" of each vitamin? That depends on age and gender. See Figure 30.

Vitamins perform very specialized tasks. Vitamin D, for example, helps move calcium ions from your intestines into the bloodstream. Without vitamin D, much of the calcium you ingest would not be used by the body. Some vitamins function as **coenzymes,** organic molecules that interact with enzymes and enhance their activity. For example, the B-vitamin group acts as coenzymes in releasing energy from food molecules. Figure 31 illustrates how a coenzyme functions.

Long before the term vitamin was introduced early in the last century, people had discovered that small quantities of certain substances were necessary for health. For example, scurvy, a condition characterized by symptoms of swollen joints, bleeding gums, and tender skin, was once common among sailors. As early as the 1500s, scurvy was considered a symptom of food deficiency. After the mid-1700s, seafarers loaded citrus fruits on board before long voyages and ate it during the voyage to prevent scurvy. It is now known that the disorder is caused by lack of vitamin C, which is present in citrus fruits. About a dozen different vitamins have been identified. Figure 32 (page 537) illustrates how vitamins support human life.

Vitamins and Solubility CD-ROM WWW.

Recall that "like dissolves like."

Vitamins are classified as fat-soluble or water-soluble (see Figure 32). Water-soluble vitamins have polar functional groups that allow them to pass directly into the bloodstream. Due in part to their solubility, water-soluble vitamins are not stored in the body—they must be ingested regularly. Some water-soluble vitamins, including the B vitamins and vitamin C, are destroyed by cooking.

RDAs for Selected Vitamins

Gender and Age	A (RE)	D (μg)	C (mg)	B_1 (mg)	B_2 (mg)	Niacin (mg)	B_{12} (μg)	K (μg)
Males								
11–14	1000	10	50	1.3	1.5	17	2.0	45
15–18	1000	10	60	1.5	1.8	20	2.0	65
19–24	1000	10	60	1.5	1.7	19	2.0	70
25–50	1000	10	60	1.5	1.7	19	2.0	80
51 +	1000	10	60	1.2	1.4	15	2.0	80
Females								
11–14	800	10	50	1.1	1.3	15	2.0	45
15–18	800	10	60	1.1	1.3	15	2.0	55
19–24	800	10	60	1.1	1.3	15	2.0	60
25–50	800	5	60	1.1	1.3	15	2.0	65
51 +	800	5	60	1.0	1.2	13	2.0	65

Source: Food and Nutrition Board, National Academy of Sciences—National Research Council, Recommended Dietary Allowances, Revised 1989.

Figure 30 *USDA vitamin RDAs by gender and age.*

1 μg = microgram = 10^{-6} g
1 RE = retinol equivalent
1RE = 1 μg retinol

Your body absorbs fat-soluble vitamins into the blood from the intestine with the assistance of fats in the food you eat. Since the nonpolar structures of fat-soluble vitamins allow them to be stored in body fat, it is not necessary to consume fat-soluble vitamins daily. In fact, because fat-soluble vitamins accumulate within the body, they can build up to toxic levels if taken in excessively large quantities.

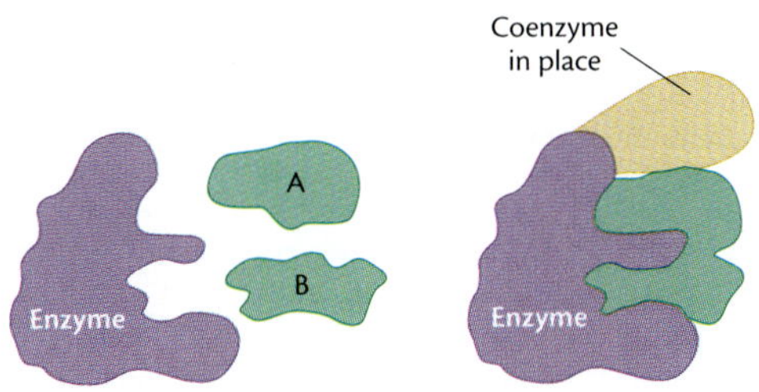

Figure 31 *A vitamin serving as a coenzyme.*

535

VITAMINS IN THE DIET

Building Skills 7

This activity provides students with practice in interpreting data from tables and applying it to their own situations.

Answers

1. a. B_{12} (cobalamin) deficiency results in pernicious anemia and exhaustion; D (calciferol) deficiency results in rickets.

   b. By taking vitamin supplements.

For comparison, one cup of orange juice contains 0.223 mg vitamin B_1 and 124 mg vitamin C.

VITAMINS IN THE DIET — Building Skills 7

1. Carefully planned vegetarian diets are nutritionally balanced. Some individuals following a vegan diet choose not to consume any animal products, including eggs and milk. Such individuals must ensure that they obtain the recommended daily allowances of two particular vitamins.
   a. Use Figure 32 to identify these two vitamins, and briefly describe the effect of their absence in the diet.
   b. How might individuals following a vegan diet avoid this problem?
2. Complete the following table about yourself, using data from Figures 30 and 33.

Vegetable (one-cup serving)	Your RDA		Total Servings to Supply Your RDA	
	B_1	C	B_1	C
Green peas				
Broccoli				

   a. Would any of your table entries change if you were a member of the opposite sex? If so, which entry or entries?
   b. Based on your completed table, why do you think variety in a person's diet is essential?
   c. Why might vitamin deficiencies pose problems even if people receive adequate supplies of food Calories?
3. Nutritionists recommend eating fresh fruit rather than canned fruit and raw or steamed vegetables instead of canned or boiled vegetables.
   a. What does food freshness have to do with vitamins?
   b. To what kinds of vitamins do you think nutritionists might be referring?
   c. Is their advice sound? Explain.

To follow a nutritionally balanced diet, you need to know, among other things, which foods deliver adequate amounts of vitamins your body requires. Although nutritionists are still learning about the kinds and amounts of nutrients needed for best health, it is a good idea to build a diet around a variety of foods. In the following laboratory activity you will find out how to determine the vitamin C content of common foods.

2.

Vegetable (1-c serving)	RDA		Number of Servings to Supply RDA	
	B1	C	B1	C
Green peas				
Males	1.5	60	3.88	1.03
Females	1.1	60	2.84	1.03
Broccoli				
Males	1.5	60	25.9	0.73
Females	1.1	60	19.0	0.73

Vitamins		
Vitamin (Name)	**Main Sources**	**Deficiency Condition**
Water-soluble		
B_1 (Thiamin)	Liver, milk, pasta, bread, wheat germ, lima beans, nuts	Beriberi: nausea, severe exhaustion, paralysis
B_2 (Riboflavin)	Red meat, milk, eggs, pasta, bread, beans, dark green vegetables, peas, mushrooms	Severe skin problems
Niacin	Red meat, poultry, enriched or whole grains, beans, peas	Pellagra: weak muscles, no appetite, diarrhea, skin blotches
B_6 (Pyridoxine)	Muscle meats, liver, poultry, fish, whole grains	Depression, nausea, vomiting
B_{12} (Cobalamin)	Red meat, liver, kidneys, fish, eggs, milk	Pernicious anemia, exhaustion
Folic acid	Kidneys, liver, leafy green vegetables, wheat germ, peas, beans	Anemia
Pantothenic acid	Plants, animals	Anemia
Biotin	Kidneys, liver, egg yolk, yeast, nuts	Dermatitis
C (Ascorbic acid)	Citrus fruits, melon, tomatoes, green peppers, strawberries	Scurvy: tender skin; weak, bleeding gums; swollen joints
Fat-soluble		
A (Retinol)	Liver, eggs, butter, cheese, dark green and deep orange vegetables	Inflamed eye membranes, night blindness, scaling of skin, faulty teeth and bones
D (Calciferol)	Fish-liver oils, fortified milk	Rickets: soft bones
E (Tocopherol)	Liver, wheat germ, whole-grain cereals, margarine, vegetable oil, leafy green vegetables	Breakage of red blood cells in premature infants, oxidation of membranes
K (Menaquinone)	Liver, cabbage, potatoes, peas, leafy green vegetables	Hemorrhage in newborns; anemia

Figure 32 *Vitamins by category, showing sources and deficiency conditions.*

Vitamins B_1 and C in Common Vegetables		
	Vitamin (in mg)	
Vegetable (one-cup serving)	**B_1 (thiamin)**	**C (ascorbic acid)**
Green peas	0.387	58.4
Lima beans	0.238	17
Broccoli	0.058	82
Potatoes	0.15	30

Figure 33 *Vitamin content of selected vegetables.*

3. a. As food ages the vitamins decompose.

b. Water-soluble vitamins are more easily lost in canning and boiling.

c. Long cooking can dissolve and often destroy these vitamins from vegetables. Raw or lightly steamed vegetables retain maximum concentrations of vitamins.

a. yes; vitamin B_1

b. Since there is no perfect food, variety in the diet assures that the proper kinds of vitamins are consumed. Green peas are a better source of B_1 compared to broccoli, but are not as good a source of vitamin C.

c. Even if Calorie intake is adequate, vitamin deficiencies from the lack of variety in foods will limit body-building and burning processes.

CHEMISTRY AT WORK

Keeping Food Supplies Safe for Consumers

NEWS BRIEFS

SUFFOLK

19TH HORSE DIES: A horse from Suffolk has died and is suspected to be the 19th known victim of "moldy corn poisoning" in Virginia. The horse's owners, who were not identified, could not recall where they purchased the corn.

State inspectors widen area for corn feed testing

Associated Press

RICHMOND — Inspectors from the Virginia Department of Agriculture expanded their search Thursday for a toxin in corn feed that has killed 18 horses.

Inspectors began taking samples of from manufacturers and dealers who mix their own feed in the area south of Interstate 64 and east of Interstate 95. The samples will be tested for Fumonisin B1, a mycotoxin that causes leukoence-phalomalacia, also called molfy corn poisoning.

These actual newspaper headlines illustrate the deadly effects that certain microscopic toxins can have. Here, the consumers of the food were horses, but similar toxins can endanger human consumers as well.

Mary Trucksess is a chemist working to understand and eradicate food toxins. She is a research chemist and chief of the Bioanalytical Chemistry Branch in the Division of Natural Food Products, Office of Plant and Dairy Food and Beverages at the U.S. Food and Drug Administration. Mary and her fellow chemists at the FDA help ensure the quality of food and drugs (prescription and nonprescription) to protect people from harmful contaminants. Currently Mary is investigating fumonisins—toxins, found mainly in corn, that are caused by a common soil fungus. Fumonisins can cause horses to develop a fatal disease of the brain. The FDA is investigating these toxins to make certain that they do not pose a threat to humans.

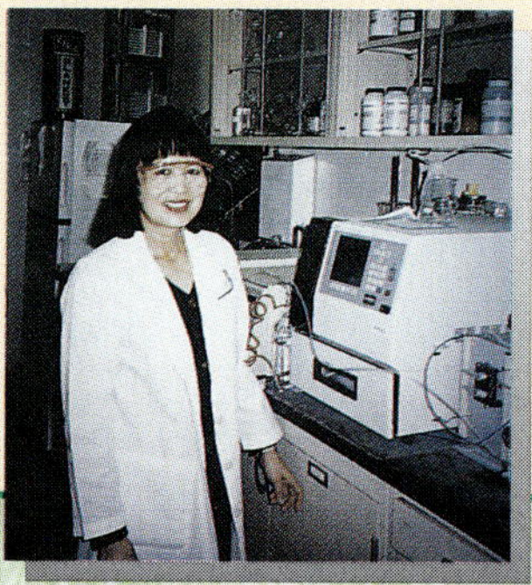

538

Mary's office analyzes other foods, such as peanuts and peanut butter, for aflatoxins. Aflatoxins form when certain molds (*Aspergillus flavus* or *Aspergillus parasiticus*) begin to grow on peanuts. If not controlled, these and other toxins could threaten the health of consumers.

Mary also manages and directs other chemists as they conduct research. The chemists in her branch often discuss with her any problems they encounter in their research. She reviews progress reports from the chemists and recommends different approaches as needed.

For several years, Mary—who holds undergraduate and doctoral degrees in chemistry—has supervised the work of high school honor students who have been selected to intern at the FDA. Mary develops specific projects for the students and trains them in laboratory procedures and data interpretation. Mary has also been a coordinator for Project SEED. This summer program for economically disadvantaged high school students offers them the opportunity to work with a mentor in a chemistry laboratory. For information about this program, visit www.acs.org/education/student/projectseed.html.

Currently Mary is investigating fumonisins—toxins found mainly in corn.

For Research and Practice

1. When preparing food, consumers can unknowingly create conditions in which bacteria thrive.
   a. What should you do to keep your food safe?
   b. What temperature range is the "danger zone" for food, and why is it called that?
2. CNN reported that "A survey by the Centers for Disease Control and Prevention shows that men between the ages of 18 and 29 violate food safety practices more than any other group." Members of this group are also more likely to experience food poisoning than those of any other group. Design an ad campaign that will target men between 18 and 29 and has the goal of teaching them about food safety.

539

D.2 VITAMIN C

Laboratory Activity

Students use microscale techniques and calculations to determine the amounts of vitamin C in several beverages. The use of Beral pipets and wellplates enables teams of students to analyze several different samples in 50 minutes.

Time

50 minutes

Materials (for a class of 24 working in pairs)

12 10-mL graduated cylinders
12 Beral pipets
12 Beral pipets containing iodine solution
12 Beral pipets containing 1% starch suspension
vitamin C solution, 0.100 g/100 mL water
Beral pipets containing beverage samples
toothpicks for stirring

Safety

Students should wear safety goggles and aprons throughout the laboratory activity.

Advance preparation

Iodine solution: Dissolve 1.0 g of potassium iodide and 0.053 g potassium iodate in 25 mL water. Add 6 mL of 3 M sulfuric acid. Add water to make the total volume 100 mL. Fill 12 pipets shortly before class time and provide a capped bottle for refills. The iodine pipets cannot be cleaned, so use the same pipets for every class. The reaction in the iodine solution:

$IO^3 + 5I- + 6H^+ \rightarrow$
$3 I_2 + 3 H_2O$.

Starch suspension: Dissolve 1 teaspoon of liquid laundry starch in 100 mL hot (not boiling) water. Fill 12 pipets and provide a capped bottle for refills.

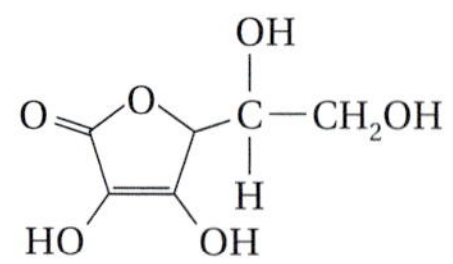

Ascorbic acid

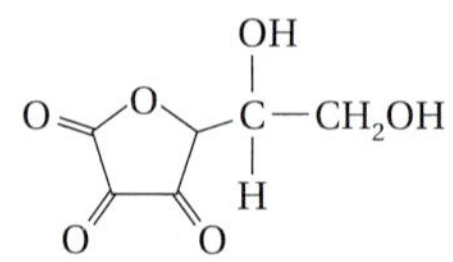

Dehydroascorbic acid

D.2 VITAMIN C

Laboratory Activity

Introduction

Vitamin C, also known as ascorbic acid, is a water-soluble vitamin. It is also among the least stable vitamins. It reacts readily with oxygen gas, and exposure to light and heat can cause it to decompose. In this laboratory activity you will investigate how much vitamin C is contained in a variety of popular beverages, including fruit juices, milk, and soft drinks.

This laboratory procedure is based on chemical properties of ascorbic acid (vitamin C) and iodine. A solution of iodine (I_2) oxidizes ascorbic acid to form the colorless products dehydroascorbic acid, hydrogen ions, and iodide ions:

$$\underset{\text{Iodine}}{I_2} + \underset{\substack{\text{Ascorbic acid}\\ \text{(vitamin C)}}}{C_6H_8O_6} \longrightarrow \underset{\substack{\text{Dehydroascorbic}\\ \text{acid}}}{C_6H_6O_6} + \underset{\substack{\text{Hydrogen}\\ \text{ion}}}{2\,H^+} + \underset{\substack{\text{Iodide}\\ \text{ion}}}{2\,I^-}$$

You will perform a **titration,** a common procedure used to determine the concentrations of substances in solutions. A known amount of one reactant will be added slowly from a Beral pipet to another reactant in a wellplate until just enough has been added for complete reaction. Completion of the reaction—the **endpoint**—is noted by a color change or other highly visible change. Knowing the chemical equation for the reaction involved, you can then calculate the unknown amount of the second reactant from the known amount of the first reactant.

The titration materials are illustrated in Figure 34. Your teacher will demonstrate correct titration procedures.

The endpoint in this titration is the first sign of permanent blue-black color in the beverage-containing well of the wellplate. This color is the result of the reaction of excess iodine with starch. First, starch solution is added to the beverage to be tested. Next, a standardized iodine solution is added drop by drop from a Beral pipet. As long as ascorbic acid is present, the iodine is quickly converted to the colorless iodide ion, and no blue-black iodine-starch product is observed. As soon as all available ascorbic acid has been oxidized to dehydroascorbic acid, the next drop of iodine solution that is added reacts with the starch, producing the blue-black color.

Ascorbic acid is thus the limiting reactant in this titration.

First, you will complete a titration involving a known vitamin C solution. This will allow you to determine the mass of ascorbic acid that can react with one drop of iodine solution. You can then calculate the mass (mg) of vitamin C present in a 25-drop sample of each beverage. This information will allow you to rank the beverages you test in terms of the level of vitamin C found in each sample.

Before beginning the activity, prepare a data table. Label five vertical columns: "Beverage," "Drops vitamin C solution," "Drops I_2 solution," "mg vitamin C per 25 drops of beverage," and "Rank." Leave room in the table to also record the number of drops of vitamin C in 1.0 mL (Step 1) and the number of drops of iodine needed to reach the endpoint (Step 4). Create a horizontal row in your table for each beverage sample to be analyzed.

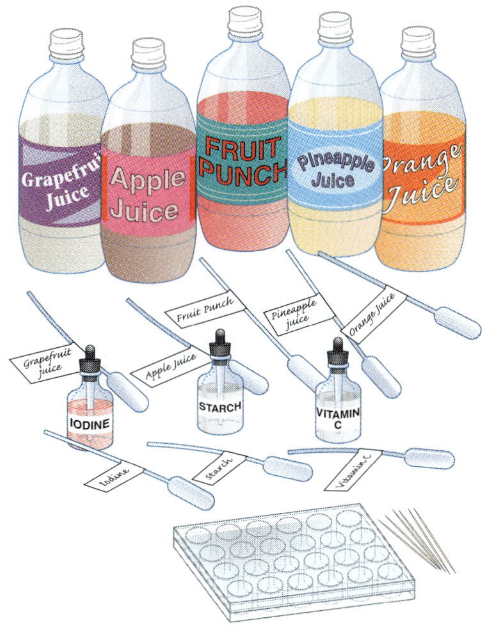

Figure 34 *Titration equipment and supplies for this laboratory activity.*

Procedure

Part 1. Finding the Iodine Solution Concentration

1. Fill a Beral pipet with vitamin C solution provided by your teacher. Then determine how many drops of vitamin C solution delivered by that pipet represent a volume of 1.0 mL.
2. Fill a second Beral pipet with iodine solution. Determine the number of drops from that pipet that represent a 1.0-mL volume, just as you did in Step 1.
3. Add 25 drops vitamin C solution to a well of a clean 24-well wellplate. The vitamin C concentration is 1.0 mg vitamin C/mL.
4. Add one drop of starch solution to the well.
5. Place a piece of white paper underneath the wellplate—this will help you detect the color. Add iodine solution one drop at a time to the well containing the starch/vitamin C mixture. After each addition of iodine solution, gently stir the resulting mixture with a toothpick. Count the drops of iodine solution added to reach the endpoint (the first "permanent" blue-black color). Continue adding iodine

Vitamin C solution: Crush a 100-mg vitamin C tablet and dissolve in water. Dilute to 100 mL. (This is much more than needed, but it is difficult to work with smaller quantities of the tablet.) The solution is somewhat unstable and should be used immediately.

Beverage samples: Choose five or more beverages to test (non-fortified with vitamin C). The color of dark juices and sodas obscures the endpoint of the reaction, so they are best avoided. If colored juices are used, suggest that all five samples titrated be of the same color of Tang, orange juice, Hi-C, and so on. Remove any pulp from juices by filtering through cheesecloth before placing in pipets.

Have students determine the number of drops/mL for each of the beverages tested. There may be some variation in the number of drops/mL for each of the beverages. Discuss how this may affect student rankings in Step 12.

Beverages vary considerably in vitamin C content. Ensure that those selected are not fortified with vitamin C. The natural vitamin C content of some beverages is listed below.

Vitamin C content	Beverage
Very high	Pineapple or orange juice
High	White grape juice, grapefruit juice, or Hi-C
Medium	V-8 juice
Low	Apple juice, milk, or Tang
Very low or none	Gatorade or Sprite (or 7-Up)

Pre-lab discussion

Students have probably heard of vitamin C both as a component of various juices and soft drinks and as a cold preventive. Discuss the types of foods that are high in vitamin C. Students will probably be surprised to find that significant amounts of vitamin C are found in broccoli and green and red peppers. (See *ChemMatters*, April 1995.)

Demonstrate the use of the pipet and wellplate in titration. When titrating with a Beral pipet, the angle at which the pipet is held should be the same so that the drop size stays the same. Otherwise the results will be greatly affected. The pipet should be held as vertical as possible to get consistent drop size. Emphasize the importance of holding the pipet at the same angle each time it is used.

solution drop by drop while stirring, until the solution in the well remains blue-black for 20 seconds. If the color fades before 20 seconds have elapsed, add another drop of iodine solution. Record the number of drops of iodine needed to reach the endpoint.

6. You know that the concentration of the vitamin C solution used is 1.0 mg/mL. How many milligrams of vitamin C react with one drop of iodine solution? In this procedure, 25 drops of vitamin C solution reacted with the iodine solution. How many drops of iodine solution did it take to react with that much vitamin C solution? How do drops of vitamin C solution relate to milligrams of vitamin C? Use that information to calculate how many milligrams of vitamin C react with one drop of iodine solution:

$$\frac{25\ \cancel{\text{drops}}}{\cancel{\text{vitamin C}}} \times \underset{\text{(Step 1)}}{\frac{1\ \cancel{\text{mL vitamin C}}}{\text{Total}\ \cancel{\text{drops vitamin C}}}} \times \frac{1\ \text{mg vitamin C}}{1\ \cancel{\text{mL vitamin C}}} \times \underset{\text{(Step 4)}}{\frac{1}{\text{Total drops I}_2}} = \_\_\_\_\ \text{mg vitamin C/drop I}_2$$

For example, suppose you found in Step 1 that 30 drops vitamin C solution equaled 1.0 mL, and that it took 22 drops iodine solution to reach the endpoint in Step 5. These data lead to the calculated result that 0.038 mg vitamin C reacts with 1 drop iodine solution. (Try the calculation to confirm that result.)

Part 2. Analyzing Beverages for Vitamin C

7. Using a Beral pipet, add 25 drops of a beverage to a clean well of a wellplate.
8. Add one drop of starch solution to the well.
9. With a Beral pipet, add iodine solution one drop at a time to the well containing the beverage/starch mixture. After each addition, stir the resulting mixture with a toothpick. Count the total drops of iodine solution that are added to reach the endpoint (a "permanent" blue-black color). (Note: Colored beverages may not produce a true blue-black endpoint color. For example, red beverages may appear purple at the endpoint.)
10. Continue adding iodine solution drop by drop, while stirring, until the solution in the well remains blue-black for 20 seconds. If the color fades before 20 seconds have elapsed, add another drop of iodine solution. Record the total drops of iodine solution needed to reach the endpoint.
11. Repeat Steps 7 through 10 for each beverage you are assigned to analyze.
12. Use data from Step 10 and the results of the calculation in Step 6 to calculate the milligrams of vitamin C contained in 25 drops of each assigned beverage.

13. Rank the beverages from the highest level to the lowest level of vitamin C.
14. Wash your hands thoroughly with soap and water before leaving the laboratory.

Questions

1. Among the beverages tested, in your opinion, were any vitamin C levels
   a. unexpectedly low? If so, explain.
   b. unexpectedly high? If so, explain.
2. Imagine that while you are performing your titration, you add too much iodine solution and miss the true endpoint. Will this error increase or decrease the calculated milligrams of vitamin C in the sample? Explain.

Now you are ready to consider trace nutrients known as minerals. What are they, and what do they do in the body?

D.3 MINERALS: AN ESSENTIAL PART OF DIETS

CD-ROM WWW. Minerals in the Body

Never taste any materials found in the chemistry laboratory.

Minerals are essential life-supporting materials. Some are quite common; others are likely to be found in large quantities only on research-laboratory shelves.

Some minerals become part of the body's structural material. Others help enzymes do their jobs. Still others help maintain the health of heart, bones, and teeth. The thyroid gland, for example, uses only a miniscule quantity of iodine (only millionths of a gram daily) to produce the hormone thyroxine. The rapidly growing field of bioinorganic chemistry explores how minerals function within living systems.

Of the more than 100 known elements, only 22 are believed essential to human life. For convenience, essential minerals are divided into **macrominerals,** or major minerals, and **trace minerals.** As the name suggests, your body contains rather large quantities—at least five grams—of each of the seven macrominerals. Trace minerals are present in relatively small quantities, fewer than five grams each in an average adult. However, in the diet trace minerals are just as essential as macrominerals. Any essential mineral—macro or trace—can become a limiting reactant if it is not present in sufficient quantity.

The essential minerals and their dietary sources, functions, and deficiency conditions are listed in Figure 35 on page 544. In addition to these, several other minerals—including but not limited to arsenic (As), cadmium (Cd), and tin (Sn)—are known to be needed by animals. These and perhaps

Though student data may not exhibit great precision, results should allow them to rank-order the beverages quite reliably in terms of vitamin C levels.

Relate the laboratory activity to the text materials by discussing how much of each drink would be necessary to acquire the recommended RDA for vitamin C.

Extension

If time permits, some teams can investigate the amount of vitamin C in other foods. Use a blender to obtain the juice. Another interesting investigation is to see what heating or microwaving does to the vitamin C content.

D.3 MINERALS: AN ESSENTIAL PART OF DIETS

This section emphasizes that minerals are also essential to health, as cofactors in biochemical catalysis and as component parts of the body's structure. Minerals are classified as either macrominerals or trace minerals. Although only small quantities of trace minerals are needed, they can act as limiting reagents. Health problems occur with either too much or too little of any needed mineral.

Answers

1. a., b. Answers will vary. Some may be surprised that Gatorade does not contain vitamin C.
2. Increase; it will make it appear as though there is more vitamin C in the beverage than there actually was.

Post-lab discussion

Students may have some difficulty with calculations. Depending on their mathematical ability, they may need help calculating the conversion factor. Working through a sample calculation in Step 6 may be necessary.

Mineral	Source	Deficiency Condition
Macrominerals		
Calcium (Ca)	Canned fish, milk, dairy products	Rickets in children; osteomalacia and osteoporosis in adults
Chlorine (Cl)	Meats, salt-processed foods, table salt	—
Magnesium (Mg)	Seafoods, cereal grains, nuts, dark green vegetables, cocoa	Heart failure due to spasms
Phosphorus (P)	Animal proteins	—
Potassium (K)	Orange juice, bananas, dried fruits, potatoes	Poor nerve function, irregular heartbeat, sudden death during fasting
Sodium (Na)	Meats, salt-processed foods, table salt	Headache, weakness, thirst, poor memory, appetite loss
Sulfur (S)	Proteins	—
Trace minerals		
Chromium (Cr)	Liver, animal and plant tissue	Loss of insulin efficiency with age
Cobalt (Co)	Liver, animal proteins	Anemia
Copper (Cu)	Liver, kidney, egg yolk, whole grains	—
Fluorine (F)	Seafoods, fluoridated drinking water	Dental decay
Iodine (I)	Seafoods, iodized salts	Goiter
Iron (Fe)	Liver, meats, green leafy vegetables, whole grains	Anemia; tiredness and apathy
Manganese (Mn)	Liver, kidney, wheat germ, legumes, nuts, tea	Weight loss, dermatitis
Molybdenum (Mo)	Liver, kidney, whole grains, legumes, leafy vegetables	—
Nickel (Ni)	Seafoods, grains, seeds, beans, vegetables	Cirrhosis of liver, kidney failure, stress
Selenium (Se)	Liver, organ meats, grains, vegetables	Kashan disease (a heart disease found in China)
Zinc (Zn)	Liver, shellfish, meats, wheat germ, legumes	Anemia, stunted growth

Figure 35 *Minerals by category, showing sources and deficiency conditions.*

other trace minerals may be essential to human life as well. You may be surprised to learn that the widely known poison arsenic might be an essential mineral. In fact, many substances beneficial in low doses are toxic in higher doses. Figure 36 summarizes recommended daily doses for several macrominerals and trace minerals.

544

RDAs for Selected Minerals						
Gender and Age	Calcium (mg)	Phosphorus (mg)	Magnesium (mg)	Iron (mg)	Zinc (mg)	Iodine (μg)
Males						
11–14	1200	1200	270	12	15	150
15–18	1200	1200	400	12	15	150
19–24	1200	1200	350	10	15	150
25 +	800	800	350	10	15	150
Females						
11–14	1200	1200	280	15	12	150
15–18	1200	1200	300	15	12	150
19–24	1200	1200	280	15	12	150
25–50	800	800	280	15	12	150
51 +	800	800	280	10	12	150

Source: Food and Nutrition Board, National Academy of Sciences—National Research Council, Recommended Dietary Allowances, Revised 1989.

Figure 36 *USDA mineral RDAs by gender and age.*

MINERALS IN THE DIET — Building Skills 8

Use the values reported in Figure 36 to answer the following questions.

1. A slice of whole-wheat bread contains 0.8 mg iron.
   a. How many slices of whole-wheat bread supply your daily iron allowance?
   b. Predict the health consequences of not consuming an adequate amount of iron.
2. One cup of whole milk contains 288 mg calcium. How much milk would you need to drink each day to reach your daily allowance for that mineral?
3. One medium pancake contains about 27 mg calcium and 0.4 mg iron.
   a. Does a pancake provide a greater percent of your RDA for calcium or for iron?
   b. Explain your answer.
4. a. What total mass of calcium or phosphorus do you need each year?
   b. Why are those values so much higher than the RDA values for other listed essential minerals? (*Hint:* Consider how calcium and phosphorus are used in the body.)
   c. List several good sources of calcium and of phosphorus.
   d. Predict the health consequences of a deficiency of each of those minerals.
   e. Would any particular age group or gender be especially affected by those consequences?

MINERALS IN THE DIET
Building Skills 9

Students use the RDAs to examine the variety of ways to obtain the required minerals.

Answers

1. a. $12 \text{ mg Fe} \times \frac{1 \text{ slice}}{0.8 \text{ mg Fe}} =$ 15 slices (male)
   $15 \text{ mg Fe} \times \frac{1 \text{ slice}}{0.8 \text{ mg Fe}} =$ 19 slices (female)
   b. Anemia, tiredness, apathy.
2. $1200 \text{ mg Ca} \times \frac{1 \text{ c milk}}{288 \text{ mg Ca}} =$ 4.2 c
3. a. Iron.
   b. The pancake provides 2.25% of the RDA calcium, while 3.33% of iron for males (2.67% for females) is provided. Calculations are
   $\frac{27 \text{ mg Ca}}{1200 \text{ mg Ca}} \times 100\% =$ 2.25% of RDA Calcium
   $\frac{0.4 \text{ mg Fe}}{12 \text{ mg Fe}} \times 100\% =$ 3.3% of RDA Iron (males)
   $\frac{0.4 \text{ mg Fe}}{15 \text{ mg Fe}} \times 100\% =$ 2.7% of RDA Iron (females)
4. a. $\frac{1200 \text{ mg}}{\text{day}} \times \frac{365 \text{ days}}{\text{year}} \times \frac{1 \text{ g}}{1000 \text{ mg}} = 440 \text{ g per year}$
   b. Bones and teeth contain large amounts.
   c. Ca: milk, dairy products; P: animal protein
   d. Ca: rickets, osteomalacia, and osteoporosis; P: no symptoms of deficiencies listed
   e. yes; infants and women

5. a. to prevent goiter.
   b. seafood

"Iodized salt" refers to such products.

5. Most table salt includes a small amount of potassium iodide (KI) that has been added to the main ingredient, sodium chloride (NaCl).
   a. Why do you think KI is added to the table salt?
   b. If you decide not to add iodized salt to your food, what other kinds of food could you use as sources of iodine?

Food Additives

Additive Type	Purpose	Examples
Anticaking agents	Keep foods free-flowing	Sodium ferrocyanide
Antioxidants	Prevent fat rancidity	BHA and BHT
Bleaches	Whiten foods (flour, cheese); hasten cheese maturing	Sulfur dioxide, SO_2
Coloring agents	Increase visual appeal	Carotene (natural yellow color); synthetic dyes
Emulsifiers	Improve texture, smoothness; stabilize oil-water mixtures	Cellulose gums, dextrins
Flavoring agents	Add or enhance flavor	Salt, monosodium glutamate (MSG), spices
Humectants	Retain moisture	Glycerin
Leavening agents	Give foods light texture	Baking powder, baking soda
Nutrients	Improve nutritive value	Vitamins, minerals
Preservatives and antimycotic agents (growth inhibitors)	Prevent spoilage, microbial growth	Propionic acid, sorbic acid, benzoic acid, salt
Sweeteners	Impart sweet taste	Sugar (sucrose), dextrin, fructose, aspartame, sorbitol, mannitol

Figure 37 *Types, purposes, and examples of food additives.*

D.4 FOOD ADDITIVES

Small amounts of vitamins and minerals occur naturally in food. Some foods—especially processed foods such as packaged snacks or frozen entrees—also contain small amounts of **food additives**. Those substances are added during processing to increase the nutritive value of foods or to enhance their storage life, visual appeal, or ease of production. A typical food label might provide the following information:

> Sugar, bleached flour (enriched with niacin, iron, thiamine, and riboflavin), semisweet chocolate, animal and/or vegetable shortening, dextrose, wheat starch, monocalcium phosphate, baking soda, egg white, modified corn starch, salt, nonfat milk, cellulose gum, soy lecithin, xanthan gum, mono- and diglycerides, BHA, BHT.

Can you guess the identity of the food product with this label?

That's quite a collection of ingredients! You probably recognize the major ingredients such as sugar, flour, shortening, and baking soda, and some additives such as vitamins (thiamine, riboflavin) and minerals (iron and monocalcium phosphate). But you probably do not recognize the food additives xanthan gum (an emulsifier—an agent that helps produce uniform, non-separating water-oil mixtures) or BHT (butylated hydroxytoluene, a preservative).

Food additives have been used since ancient times. For example, salt has been used for centuries to preserve foods, and spices helped disguise the flavor of food that was no longer fresh. To make foods easier and less expensive to distribute and store, most manufacturers, especially those of processed foods, rely on food-preservation additives. The table in Figure 37 summarizes the major food additive categories. The structural formulas of several common additives are shown in Figure 38.

Color and taste additives are often used to increase the attractiveness of foods and in an attempt to increase sales. In the following laboratory activity you will analyze several commercial food dyes.

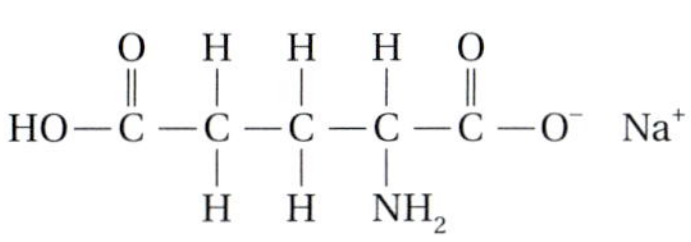

Monosodium glutamate (MSG)

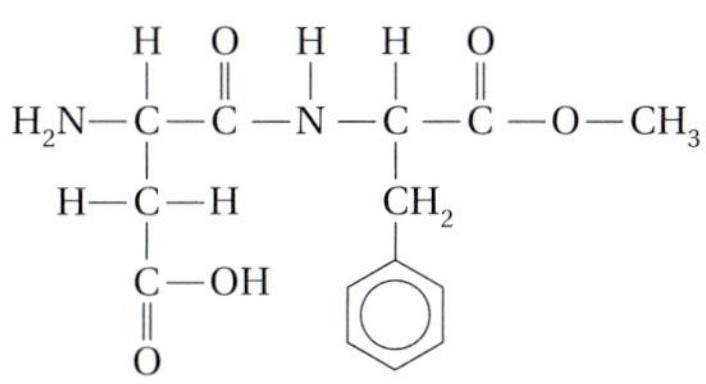

Aspartame

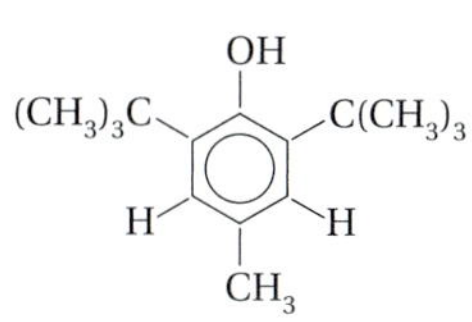

Butylated hydroxytoluene (BHT)

Figure 38 *The structures of MSG, aspartame, and BHT, a preservative.*

D.4 FOOD ADDITIVES

This section briefly surveys the benefits (and some associated risks) involved in modern use of food additives.

D.5 FOOD COLORING ANALYSIS

Laboratory Activity

Students use paper chromatography to separate dyes in candies and compare them to dyes in food coloring. They use the data on the chromatography to calculate R_f values.

Time

50 minutes

Materials (for a class of 24 working in pairs)

12 8-oz gas-collecting bottles
12 rubber stoppers to fit gas bottles
12 24-well wellplates
36 micropipets
chromatography paper, 1-inch-wide strips
toothpicks
1 package M&Ms candies
1 package Skittles candies
red food coloring
blue food coloring
yellow food coloring
green food coloring

Safety

Students should wear safety goggles and aprons throughout the laboratory activity. If possible, handle the ammonia in a hood.

Advance preparation

Cut 12 strips of chromatography paper, each 15 cm long.

D.5 FOOD COLORING ANALYSIS

Laboratory Activity

Introduction

Lab Video CD-ROM WWW.

Many candies contain artificial coloring agents to enhance the visual appeal of the product. Colorless candies would be quite dull! In this laboratory activity, you will analyze the food dyes in two commercial candies and compare them with the dyes in food coloring.

Food Coloring Analysis CD-ROM WWW.

You will separate and identify the food dyes with the help of **paper chromatography.** This technique uses a solvent (the mobile phase) and paper (the stationary phase). Paper chromatography is based on relative differences in attraction between the dye molecules and the solvent and the dye molecules and the paper. Dye molecules that are more attracted to paper will more readily leave the solvent and separate out onto the paper, producing a colored spot on the paper. Dye molecules less attracted to the paper (and more attracted to the solvent) will separate out onto the paper later. Individual areas for different dye molecules will appear on the paper.

To analyze the results of this chromatography experiment, you will calculate the **R_f value** for each spot. The R_f value, as shown in Figure 39, is a ratio comparing the distance each dye has traveled up the paper with the distance the solvent has moved up the paper.

$$R_f = \frac{\text{distance to a spot}}{\text{distance moved by solvent}}$$

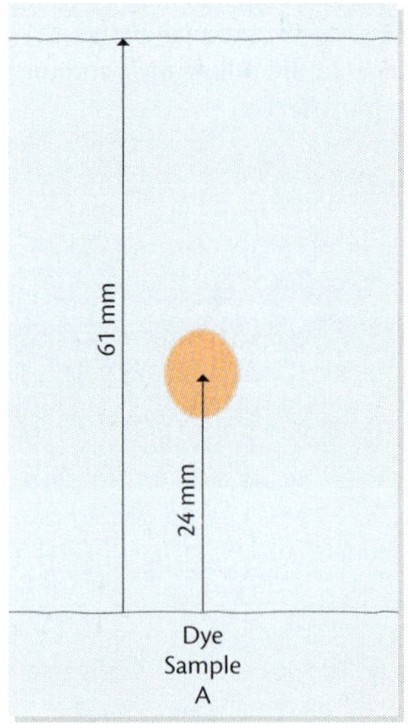

Calculating R_f

$$R_f = \frac{24 \text{ mm}}{61 \text{ mm}} = 0.39$$

Figure 39 *Paper chromatography with R_f values.*

Before beginning the activity, prepare a data table that contains this information: The table should have six vertical columns labeled "Sample," "Initial color of sample," "Dyes observed," "Distance to dye spot (cm)," "Distance solvent moved (cm)," and "R_f value." The table should have three rows—one for each candy sample and one for the food coloring sample.

Procedure

1. Obtain one piece each of two different commercial candies from your teacher, as well as a food-coloring sample. The food coloring and candies should all be the same color.
2. Put each candy into a separate well of a wellplate. Note which candy is in each well. Add 5 to10 drops of water to each well. Stir the mixture in each well with a separate toothpick until the color completely dissolves from the candy. Add 3 to 4 drops of the food coloring sample to a third well in the wellplate. Observe and record the initial color of each sample.
3. Obtain a strip of chromatography paper, handling it only by its edges. With a pencil (not pen), draw one horizontal line 2 cm from the bottom of the paper and another horizontal line 3 cm from the top. Label the strip as shown in Figure 40.
4. Next, you will place a spot of each dye on the bottom line—the spots should not be large. To do this, use a separate toothpick for each of the three samples. Place a drop of the first candy's color solution with a toothpick as indicated in Figure 40. Allow the drop of solution to sit until the spot it makes stops spreading out on the paper. Then apply a second drop of the same sample on top of the spot. Using the same two-drop technique, apply drops of the second candy's color solution and the food-coloring sample to appropriate places on the pencil line, as indicated in Figure 40.
5. Obtain a chromatography chamber. Make a mark on the outside of the chamber that is approximately 1 cm from the bottom. Pour solvent into the chromatography vessel to the 1-cm mark.
6. Lower the spotted chromatography paper into the chromatography vessel until the bottom of the paper rests evenly in the solvent (water). Be sure the spots remain above the solvent surface. Place a lid on top of the chromatography chamber.
7. Allow the solvent to rise past the spots and up the paper until the solvent is close to the penciled line, about 3 cm from the top of the paper. Then remove the paper from the vessel and, using a pencil, mark the farthest point of travel by the solvent. Allow the paper to air-dry overnight.
8. Record the colors observed for the dye sample and candy solutions.
9. Measure the distance (in cm) from the initial pencil line where you placed the spots to the center of each dye spot. Record these distances in your data table.

Figure 40 *Spotting a chromatogram on chromatography paper.*

Pre-lab discussion

A food product may contain a number of artificial dyes solely for the sake of appearance. Paper chromatography is a way to separate these dyes. The various dyes in a mixture respond differently to the paper and the solvent. Some are held more firmly to the paper than others. It is difficult to compare results unless all conditions are carefully controlled. Therefore, a standard is generally run with an unknown, and the R_f values are compared.

Lab tip

This activity is not difficult for students. The most likely error is allowing the spots to be submerged in the solvent. If this happens, the dye merely dissolves in the solvent, and the spots will not appear on the chromatography paper.

Expected results

The R_f values are given for red candy and food coloring. Do not be concerned if student values are slightly different. The spots will be in the same order, but the actual distances may differ. Many things affect the results: temperature of the room, small changes in concentration of the solvent, and even the paper used.

Candy	Dye Color	Distance Dye Spot Moved	Distance Solvent Front Moved	R_f Value
M&Ms	red component	30 mm	68 mm	0.44
	red component	25 mm	68 mm	0.37
Skittles	yellow component	37 mm	68 mm	0.54
Food Coloring	pink component	22 mm	68 mm	0.32
	red component	45 mm	68 mm	0.66

Answers

1. The dyes in the ink will also be separated by paper chromatography.
2. Answers will vary with student results.
3. Answers will vary with student results. The dye with the smallest R_f value.
4. Answers will vary with student results. The dye with the largest R_f value.
5. a. Answers will vary.

   b. Any dye that shows spots with the same R_f value as others is assumed to be the same. However, it could be that more than one dye has the same R_f value.
6. a. b. c. Answers will vary with student results.

Post-lab discussion

The post-lab discussion can focus on two topics: the use of dyes as food additives and the technique of paper chromatography. The answers to the questions can be used as guidelines in discussing the technique of paper chromatography. Get student opinions on the appropriateness of using dyes to improve the appearance of foods. Looking at labels will reveal a somewhat surprising list of foods that contain dyes.

10. Measure the total distance (in cm) that the solvent moved.
11. Calculate the R_f value of each dye spot you observed in your samples. Record these values.

Questions

1. Why were you instructed to use a pencil rather than a pen to mark lines on the paper?
2. Which sample solution, if any, created single spots rather than several spots?
3. Which dye in the sample solutions had the greatest attraction for the paper?
4. Which dye in the sample solutions had the greatest attraction for the solvent?
5. a. Based on your data, do any of the three samples have the same dye(s) in them?
   b. What evidence did you use to answer Question 5a?
6. The candy and food-color packages list each dye they contain. Compare this information with your experimental results.
   a. What similarities did you find?
   b. What differences did you discover?
   c. If differences are found, what are possible reasons for them?

D.6 REGULATING ADDITIVES

Both processed and unprocessed foods may contain contaminants that are not deliberately added to them, such as pesticides, mold, antibiotics used to treat animals, insect parts, food packaging materials, or dirt. Food purchased in grocery stores and restaurants is assumed to be completely safe to eat. For the most part, that is true. Nonetheless, in recent years some food additives and contaminants have been identified as or suspected of posing hazards to human health.

Food quality in the United States is regulated by law. The basis of the law is the Federal Food, Drug and Cosmetic Act of 1938, through which Congress authorized the Food and Drug Administration (FDA) to monitor the safety, purity, and wholesomeness of food. This act has been amended in response to concerns over pesticide residues, artificial colors and food dyes, and potential **carcinogens** (cancer-causing agents) or **mutagens** (agents that cause mutations—or changes—in DNA) in foods. Manufacturers must complete tests and provide extensive information on the safety of

D.6 REGULATING ADDITIVES

In this section we discuss the laws that enable monitoring food safety and purity in the United States. Discussion of the Delaney Clause, which requires prospective food additives to be tested on laboratory animals, leads to the ChemQuandary on weighing risks and benefits of using nitrites to preserve meats. In 1996 the part of the Delaney Clause relating to pesticide any proposed food product or additive. A new food product must be approved by the FDA before it is put on the market.

According to the amended Federal Food, Drug and Cosmetic Act, ingredients that were known not to be hazardous and had been used for a long time prior to the act were exempted from testing. These substances, instead of being legally defined as additives, constitute the "Generally Recognized as Safe" (GRAS) list. The GRAS list, periodically reviewed in light of new findings, includes items such as salt, sugar, vinegar, vitamins (such as vitamin C and riboflavin, a B vitamin), and some minerals.

In accord with the Delaney Clause (added to the act in 1958), every proposed food additive must be tested on laboratory animals (usually mice). The Delaney Clause specifies that "no additive shall be deemed to be safe if it is found to induce cancer when it is ingested by man or animal." Thus, approval of a proposed additive is denied if it causes cancer in the animals.

Since 1958, there have been great advances in science and technology. Improvements in chemical analysis techniques permit scientists to measure smaller amounts of potentially harmful substances than ever before. As a result, scientists can now study contaminants that have always been present in food but were previously undetectable.

That new information has resulted in a greater understanding of potential risks associated with exposure to particular food additives. People now recognize that amounts of additives comparable to those that cause cancer in test animals are often vastly greater than would ever be encountered in a human diet. In light of this, The Food Quality and Protection Act was passed in 1996. This legislation, which replaces the Delaney Clause, states that food additives that present a "negligible risk" may be used.

Yet many concerns about food additives remain. Sodium nitrite, $NaNO_2$, for instance, is a color stabilizer and spoilage inhibitor used in many cured meats, such as hot dogs and bologna. Nitrites are particularly effective in inhibiting the growth of the bacterium *Clostridium botulinum*, which produces botulin toxin, the cause of botulism, an often fatal disease. Sodium nitrite, however, may be a carcinogen. In the stomach, nitrites are converted to nitrous acid:

$$\underset{\text{Sodium nitrite}}{NaNO_2(aq)} + \underset{\text{Hydrochloric acid}}{HCl\,(aq)} \longrightarrow \underset{\text{Nitrous acid}}{HNO_2(aq)} + \underset{\text{Sodium chloride}}{NaCl(aq)}$$

Nitrous acid can then react with compounds formed during protein digestion, producing nitrosoamines that are known to be potent carcinogens. An example of this reaction is shown below. The concentration of carcinogenic compounds produced, however, is generally low.

$$HNO_2 + R\text{—}\underset{\underset{R}{|}}{N}\text{—}H \longrightarrow R\text{—}\underset{\underset{R}{|}}{N}\text{—}N{=}O + H_2O$$

Testing a new food additive usually requires millions of dollars and years of research.

Consider discussing the true story "Eleven Blue Men," a tale of unintentional poisoning in a boarding house. Sodium nitrite was inadvertently substituted for sodium chloride, and men developed methemoglobinemia, a condition that results in reduced oxygen transport in the blood. The story can be found in *The Medical Detectives* by B. Roulche, Truman Talley Books/Plume, New York, 1991.

Consider discussing how the danger associated with exposure to carcinogens is related to the amount, concentration, and duration of the exposure.

Demonstration

Prepare 100 mL of approximately 0.2 M sodium sulfite (2.5 g Na_2SO_3/100 mL). Dip a fresh slice of apple into the sulfite bath for one minute. Remove it and place it on a towel near an untreated "control" apple slice. Continue with your lesson, and view the slices later in the class period.

The treated slice will continue to look fresh long after the control slice begins to darken. Ask students which apple slice they would be more likely to choose at a salad bar (assuming they were unaware of the sulfite treatment). Ask students if they think the use of this additive is justified. Point out that a small fraction of the population is sensitive to sulfites and suffer dire health effects if they are exposed to them.

residues was repealed. It was replaced with a new standard of a reasonable certainty of no harm. The artificial sweetener aspartame and its relationship to phenylketonuria are also discussed.

It may be beneficial to emphasize the difference between $NaNO_3$—sodium nitrate—and $NaNO_2$—sodium nitrite.

NITRITE ADDITIVES

ChemQuandary 4

This ChemQuandary probes the benefits and risks of using nitrites as food additives. Here are some questions that students might ask:

1. How conclusive is the evidence that nitrosamines cause cancer in laboratory animals and in humans?
2. How deadly is the botulin toxin? Are there antidotes for it?
3. How often does botulism occur and how often does nitrite cause cancer?
4. What are some possible substitutes for sodium nitrite as a preservative?
5. What is the incidence of *Clostridium* growth without the use of nitrites?

D.7 ARTIFICIAL SWEETENERS

Sugar substitutes have gained popularity in many foods and beverages. This section opens the concern that these additives may also pose health hazards.

Demonstration

Float unopened cans of regular and diet soda in an aquarium or other transparent container. The regular soda sinks, the diet soda floats. Why? Certainly not because of fewer calories. The amount of sweetener added to diet sodas is much less, lowering its density. This helps illustrate the incredible difference in sweetness between natural and artificial sweeteners.

NITRITE ADDITIVES — ChemQuandary 4

To help decide whether nitrites should be used to preserve meats, consider the following benefits and risks:

	Using Nitrites	Eliminating Nitrites
Benefit	Minimizes botulin toxin formation	Removes risk of possible carcinogens
Risk	Increases possibility of carcinogen formation in body	Increases risk of botulin toxin formation

More information and greater knowledge can help guide the best decision. If the choice of using or eliminating nitrites were up to you, what questions would you want answered first?

Molecular Design of an Artificial Sweetener (CD-ROM, WWW.)

D.7 ARTIFICIAL SWEETENERS

Saccharin now appears on the GRAS list.

Since many people believe they should reduce the amount of sugar in their diets, low-Calorie sweeteners are widely used. Saccharin (available under the trade name Sweet 'N Low) was the first sugar substitute used extensively in the United States. There has been some controversy over the use of saccharin, because early experiments conducted in rats suggested that massive amounts of saccharin may cause cancer. Further investigation indicates that the link between saccharin and cancer is very weak. Now, it is generally believed that saccharin is safe for human consumption. Today, you can find saccharin in many candies and gum, baked goods, jellies, and jams.

The sugar-substitute aspartame (trade names NutraSweet and Equal) is an ingredient in over 160 diet beverages and 3000 other food products. Aspartame is a chemical combination of two natural amino acids, aspartic acid and phenylalanine, neither of which, by itself, is sweet. One gram of aspartame contains roughly the same food energy as one gram of table sugar (4 Cal), but aspartame tastes 200 times sweeter. Since much smaller quantities of aspartame are needed to sweeten a product, aspartame is used as a "low-Calorie" alternative to sugar—one that is also safe for people with diabetes. Annually, thousands of tons of aspartame are used in the United States to sweeten soft drinks and cold foods. One may use aspartame in baked goods such as cakes and cookies. Many recipes have been specifically designed to use aspartame, while others use a combination of aspartame and sugar. However, aspartame cannot be used directly for cooking because it decomposes at cooking temperatures.

Although no serious warning regarding aspartame has been issued for the general population, aspartame poses a health hazard to phenylketonurics, individuals who cannot properly metabolize phenylalanine. An FDA-required warning "Phenylketonurics: Contains Phenylalanine" on the label of foods with aspartame points out the potential risk (see Figure 41).

Figure 41 *The warning to phenylketonurics found on a can of diet drink containing aspartame.*

D.8 FOOD ADDITIVE SURVEY

Making Decisions

Being aware of what you eat is a wise habit for everyone. In the following activity, you will investigate food labels and consider the additives involved.

1. Collect the labels from five or six packaged foods. Select no more than two samples of the same type of food—for example, no more than two breakfast cereals or two canned soups. Bring your set of labels to school.
2. From the ingredient listings on the packages, select three additives that have been used in the foods.
3. Complete a summary table with the following format:
   a. List the three additives in a vertical column along the left side of the page.
   b. Make four vertical columns to the right of the additives with the following headings: "Food Product Where Found," "Purpose of Additive" (if known), "Chemical Formula," and "Other Information."
4. Use the table in Figure 37 (page 546) to review the purposes of particular food additives. Then answer this question for each food in your summary table: Why do you think the particular additives should be included in that particular food?
5. What alternatives to food additives can you propose to prevent food spoilage?

Individuals with specific medical conditions may need to avoid certain foods and food additives. For instance, people with diabetes must restrict their intake of sugar. Persons with high blood pressure (hypertension) must avoid excess salt. Some persons are allergic to certain substances. If such restrictions apply to you, you must always read food labels. Sometimes a new ingredient will be used in a food you have used safely in the past; this new ingredient may put you at risk of an adverse reaction.

D.9 ANALYZING VITAMINS AND MINERALS

Making Decisions

In previous sections, you analyzed your three-day food inventory in terms of the energy it provides from fats and carbohydrates as well as how much protein it provides. Now consider whether the food selections provide adequate quantities of vitamins and minerals.

1. Your teacher will specify two vitamins and two minerals for you to study.
2. Using references provided by your teacher, analyze the vitamin and mineral content of each item of food on your three-day inventory for its content of the selected vitamins and minerals. Record these values.

D.8 FOOD ADDITIVE SURVEY

Making Decisions

The purpose of this activity is to make students more aware of food additives found in products that they consume every day. Students survey five or six packaged foods and select six additives to describe and investigate.

This activity will take about 30 minutes to complete. As an alternative procedure, have each student investigate only one or two food additives. Then, you can ask students to share and discuss their findings with the class.

You may find that this activity runs more smoothly if students are asked to make their summary table at home rather than trying to bring labels to school.

Emphasize the risks and benefits of using food additives in the various products the students have selected.

D.9 ANALYZING VITAMINS AND MINERALS

Making Decisions

Students analyze their diets to determine whether the food selections provide adequate quantities of vitamins and minerals. Limiting the analysis to two vitamins (A and C are recommended) and two minerals (calcium and iron are recommended) will help keep the task to a manageable size.

Materials

Students should have diaries or logbooks to record their information. They also should have access to research materials: library, Internet, reference books. Provide students with a bibliographical form they can use when they report on their research. If you want the final reports prepared by typewriter, word processor, or computer, instruct your students about the format.

Procedure

1. Have each student keep a detailed diary/logbook over a three-day period and/or provide students with sample diaries. Direct them to record the date, time, and quantity for each and any type of food consumed. Have students follow steps 2–7 and the instructions alongside page 559.
2. Students should research the mass, energy, and chemical content of foods.
3. Students should research the recommended daily allowance (RDA) of food content, taking into account any personal deviations required due to health or medical conditions.
4. Students should research recommended daily energy requirements.
5. Students should research methods for computing energy expenditures and make estimates to compute personal expenditure.
6. Students should research the structure, sources, and health aspects of vitamins and minerals. You may wish to assign two of each to students to report on.

3. **Data Analysis**—the body of the report. In this section you bring together the detailed data you have gathered concerning energy, fats, carbohydrates, protein, and vitamins and minerals. Organize and present the information in a way that will make sense to your readers. (Consider organizing it in chronological order, cause-and-effect, or another reasoning pattern.) Include graphs and tables as appropriate.

   In your data analysis you may want to address questions such as these:

   - What is the average daily Calorie intake? How reasonable is this value? Does the average accurately represent each day in the inventory, or does the Calorie intake vary widely from day to day?
   - What percent of Calories come from carbohydrates, fats, and protein? Are these percentages in line with USDA recommendations?
   - What percent of fats are saturated and unsaturated? Do these levels meet USDA recommendations?
   - How does the inventory compare to Food Guide Pyramid guidelines?
   - Which vitamins and minerals are present? Which are lacking?

4. **Conclusions**—summing up your findings and recommendations based on analysis of the three-day inventory. In this section, concentrate on the "big issues" and use details to support your ideas. Issues you may want to address include the following:

   - What parts of the inventory meet USDA recommendations? Given these recommendations, is this inventory well balanced?
   - Do any parts of the inventory exceed or fall below USDA recommendations? How serious are these discrepancies? Recommend changes that could be made to address both excesses and deficiencies in the inventory.
   - Suppose a person follows the diet in the inventory but does not want to gain or lose weight. What level of activity would guarantee such a result? Recommend an exercise plan that balances the Calorie intake of this three-day inventory.

DRAWING YOUR OWN CONCLUSIONS

People often have reasons for choosing the foods they eat. Given what you have learned from this inventory, answer these questions for yourself. Write the answers in your laboratory notebook.

1. What reasons might a person have for eating the foods in this inventory?
2. What type of person might benefit from following a diet represented by this inventory?
3. What problems, if any, might arise if a person strictly followed a diet represented by this inventory?
4. Is this particular food inventory appealing to you? Explain.

LOOKING BACK

This unit focused on the chemistry of something you encounter every day—food. You should now be able to explain how the food you eat provides energy for daily living and structural components for growth. You can attach deeper chemical meaning to words such as "carbohydrate" and can better evaluate the consequences associated with eating certain foods. The next time you hear the phrase "You are what you eat," smile and explain that you know the chemistry behind it!

Analysis and Conclusions

Explain to students the correct procedures for analyzing their information:

1. Compare each computed average daily measurement to the recommended values.
2. Identify areas in which the diet was satisfactory and areas that need improvement. Be sure that students understand that their task is not to discuss the diet or consumed foods but rather concentrate on the content of the food.

Summary

Have students summarize their findings:

1. Require students to evaluate how and where their personal diets and energy profiles can be improved. Have them include thoughts about influences that their families have on their diet.
2. Make comments on the value of this project and how it might be improved.

Results

Explain to students procedures for presenting their information:

1. Sum the total quantities of mass, Calories, and energy provided by the proteins, carbohydrates, fats and the mass of the vitamins and minerals consumed over the test period.
2. Compute the average daily energy balance and quantities of proteins, carbohydrates, fats, vitamins, and minerals over the test period.
3. Make a data table that displays the recommended daily requirements next to the computed values. Students may present their findings as bar graphs to display the comparisons visually.

APPENDIX

CONTENTS

Unit 1: Teacher's Background Information for Data Analysis Determining the Cause of the Fish Kill

Before the Riverwood Town Council meeting, select the scenario the class will investigate. Assign students to one of nine groups, and supply them with the descriptions below for the scenario you have selected. Each group will analyze a set of data, then present it to the rest of the groups. At the start the data assigned to each group is arbitrary; by the end of the sharing session students will have information on all the data. In the text are sets of four questions that all groups should answer, based on the data they are assigned. After all the data has been shared, each group will draw a conclusion on what the data seems to indicate as the cause of the fish kill.

ASSIGNMENTS FOR INITIAL DATA ANALYSIS

Group	Data Provided
Scientists	Water temperature and dissolved oxygen
Engineers	Rainfall
Power company	Water flow
Agriculture group	Pesticides
Mining company	Heavy metals
Sanitation	Nitrate and phosphate levels
Chamber of commerce	pH
Taxpayers	Organic carbon
Town Council	PCBs

Encourage students to look for anomalies in the data. Anything that significantly alters the trend of the past two summers should be considered. Let students know that this summer is the only one with a fish kill. During the previous two summers there were no incidents of fish kill.

STUDENT BACKGROUND FOR TOWN COUNCIL MEETING

The background information that follows supports four possible scenarios for the fish kill (a fifth scenario—inappropriate pH of the river—is not supported by the data):

- *Scenario #1:* Data-Molecular Substances in the River: Pesticide Runoff or Nonpoint Source
- *Scenario #2:* Data—Too Low Oxygen Levels
- *Scenario #3:* Data—Inappropriate Heavy Metal Ions in the River
- *Scenario #4:* Data—Too Much Dissolved Oxygen: Gas Bubble Trauma

To bring some realism to the town meeting, a plausible set of background information supports each scenario.

SCENARIO SUMMARY

#1: Molecular Substances

- Heavy rainfall
- Detectable levels of pesticide in water or fish
- Evidence of pesticide spill

#2: Low Oxygen Level

- Heavy rainfall or large release from dam
- High water temperatures
- Low dissolved oxygen levels
- High nitrate and phosphate levels
- High organic carbon

#3: Heavy Metal or Mercury Poisoning

- Heavy rainfall or large release from dam
- High levels of mercury or heavy metals in water
- Water having a low pH (acidic)

#4: Gas Bubble Trauma

- Large release from dam
- High, supersaturated dissolved oxygen levels
- Lower water temperatures

The text lists four causes for fish kills, but emphasize to students that in real life there also are other causes of fish kills. The scenarios for this activity present typical causes but should not be considered all-inclusive. After distributing the data, ensure that the table labels identifying the scenario are removed or masked out before distribution. **Students should not know the cause of the fish kill before the Town Council Meeting. One of their tasks is to discover the potential cause.**

To supply more background for the discussion, you may choose to share part or all of the following information.

THE HISTORY OF THE DATA

Since the early 1900s various groups have measured the Snake River watershed in a number of different ways. As a result, although some of the measurements and methods have changed, there is an excellent record of data for the past few years.

Joseph Fisker of the County Sanitation Commission has measured the river's pH, dissolved oxygen, and water temperature. For the past few years he has also measured water entering the utility water system daily. Using a portable meter (routinely calibrated), Fisker measures the pH—the acidity or basicity of the water. Using a special probe and working under the bridge near Riverwood Hospital, he samples dissolved oxygen at a depth of one-half meter. Twice a month he verifies the readings with a chemical test. Knowing the water temperature allows Fisker to calculate the saturation of oxygen in the water.

Every day the power company monitors water flow on the Snake River. The data is reported monthly in cubic feet per second. In the same fashion, the company also measures and reports the amount of rainfall in inches. The data is collaborated with measurements taken by the National Stream Water-Quality Monitoring Networks, part of the U.S. Geological Survey.

The Aurgent Mining Company, which once operated several zinc and silver mines in the area, still measures several metal ions in the river. Greta Black, Aurgent's supervisor of environmental programs, has a 20-year record of data for lead, cadmium, mercury, arsenic, and several other metals. The readings are taken each Tuesday near the Riverwood Hospital. Black has provided data for the past two years. She also took daily measurements for the week following the discovery of the fish kill.

Finally, each week an environmental consulting firm, under contract to the Snake River Agricultural Cooperative measures the organic carbon, nitrates, phosphates, pesticides, and other dissolved molecular substances in the river. Don Walker, vice-president of the cooperative, has provided data on these measurements for the past two years. The amount of organic carbon indicates organic matter that has a biological origin, such as dead plant matter or animal wastes. Nitrates and phosphates (along with nitrites, ammonia and other dissolved ionic substances) generally get into the water from fertilizers, animal wastes, or treated wastewater from cities. Measuring pesticides and other dissolved molecular substances is more difficult; detecting these substances involves specialized tests.

Each set of data in this activity is based on actual fish kills. The data should point clearly to the cause. If students cannot come up with an answer, you may want to selectively share the following background. Or you could share all scenarios and ask students to choose which one best fits their data. Use all the data available to bring this matter to a close.

SCENARIO #1—MOLECULAR SUBSTANCES IN THE RIVER: NONPOINT SOURCE POISONING

Background

In a **nonpoint source poisoning,** contaminants get into the watershed from a new location each time. This is in contrast to **point source** pollution, which always has the same origin, such as a factory or a mine. An example of a nonpoint source would be a homeowner who has a bottle of pesticide that he or she no longer wants. Rather than disposing of the pesticide in a safe manner, the homeowner pours it down a storm drain. Rainfall then washes the pesticide to a tributary. Because the poison is initially very concentrated, it can cause many fish to die as soon as the poison hits the river. But as the pesticide travels downstream, it becomes more and more diluted until it soon is below detectable limits. As a result, detecting nonpoint sources of pollution is very difficult. By the time the problem becomes apparent, fish are dying and any chemical evidence of the original pesticide has probably been washed away long ago. There may or may not be any evidence of the original pesticide.

Many pesticides are complex mixtures, which makes them even more difficult to recognize after they have been partially metabolized by organisms in the water. An analogy is that fish are often like canaries in a coal mine, which can indicate a hazardous situation before humans. A slug of toxic pesticide can do considerable damage and then quickly dissipate through dilution and chemical decomposition.

Recognizing a nonpoint source fish kill can be difficult. A heavy rainfall, particularly after a period of dry weather, can contribute to the problem. Toxic molecular substances can accumulate in storm drains or in the ground for a long period of time before a heavy rainfall washes them downstream all at once. By the time the evidence is gathered, some pesticides may be detected in the water or the dead fish, but levels may be very low. All other data regarding the water in the river are likely to be normal.

The nonpoint scenario was chosen because in real life situations it is all too often the cause of fish kills. It is uncommon that the cause is agricultural runoff or errant pesticides spraying by farmers. In the past two decades regulations controlling the application of agricultural chemicals have increased substantially, and those who apply them to large tracts of land do so very appropriately. However, homeowners have to follow very few rules when using garden chemicals. And, unfortunately, homeowners are often less knowledgeable about the hazards posed by household chemicals. It is hoped that students who go through the nonpoint scenario will learn some lessons about individual responsibility in preventing environmental damage.

Scientists: Background Information

After the water samples taken following the fish kill were found to have higher than usual values for Biological Oxygen Demand and Dissolved Organic Carbon, preserved samples were analyzed using gas chromatography. These analyses found high levels of some pesticides. Upon consultation with Don Walker of the Agricultural Cooperative, you learn that the day preceding the fish kill was one of very few clear days in several weeks, and some cooperative members took that opportunity to spray their crops for pests.

Engineers: Background Information

Contaminated runoff from agricultural land is an ongoing concern for Riverwood. Under normal weather conditions, the pesticides dissipate or break down before they reach the Snake River. Rainy seasons greatly increase the potential for high levels of pesticide in the runoff. Consider types of pesticides and methods of application in formulating solutions to this problem.

You were responsible for a decision to make a scheduled heavy release of water from the dam just before the fish kill occurred. This was a routine release, done to make room in the reservoir for impending rains.

Power Company Officials: Background Information

Top company officials approved the water release from the dam, a standard practice given the time of year and the weather forecast. The power company staff monitors water flow and rain fall regularly. It also maintains its own meteorologist on staff to help with weather predictions. Officials are considering a general electricity rate hike, a decision that is unrelated to the fish kill.

Agricultural Cooperative Members: Background Information

Pesticides could be responsible for the Snake River fish kill. It has not been possible to trace the source of the contamination because all the evidence has been washed down stream and diluted. However, you need to do your best to demonstrate that it was not your fault. All members of your cooperative are licensed to use agrochemicals and trained in the appropriate means of applying them. There has not been a fish kill in recent years, even though over that period of time the Coop members have used the same pesticides and other chemicals. Applying chemicals to crops is expensive and time-consuming, so it is not in your best interest to allow the chemicals to be washed away. Finally, many agricultural leaders are environmentalists by nature. They know that the best way to ensure prospering farmland is to care for all land properly. Your presentation should include measures that your members are already taking to keep agrochemicals out of natural waters and proposed improvements that could prevent recurring fish kills.

Mining Company Representatives: Background Information

There is no evidence of any toxic organic substances such as PCBs or pesticides in any of your holding ponds or drainage areas. A recent inventory of substances in the soil and holding ponds amply documented this fact. Even though there is no direct evidence linking your company to the fish kill, it is always difficult for you to convince the public that your company is not the source of the harm. You are considering a new advertising campaign to help convince people that mining is beneficial to them personally.

County Sanitation Commission Members: Background Information

You are concerned that the source of the pesticides may have been a homeowner who disposed a toxic, possibly banned substance, by dumping it in a storm drain, the sewer. For the past few years you have dealt with improper disposal of household chemicals. For the past five years you have sponsored a marketing campaign that urges people to dispose of household and garden chemicals properly. You recently began offering a summer program so that homeowners can bring unwanted household and garden chemicals to a special recycling center. Nonetheless, you know that heavy rain could wash the pesticide through the sewage system and into the river.

Chamber of Commerce Members: Background Information

You are concerned that the negative publicity surrounding the fish kill is very bad for the community image. If the annual fishing tournament is tainted with the prospect of poison fish, it could be the end of the tournament. It is in your best interest to get this settled as quickly, fairly—and quietly—as possible.

Riverwood Taxpayer Association Members: Background Information

Tax revenue is going to suffer as a result of this accident. It is very important to you that the responsible individual is identified and made to pay for the damages. The city cannot really afford the tax revenue it is going to lose. If this matter is not resolved quickly, there could also be implications for next year's fishing tournament.

SCENARIO #2—INAPPROPRIATE OXYGEN/OXYGEN LEVEL TOO LOW

Background

During very hot weather, the level of dissolved oxygen (DO) drops in water. This is because as temperature increases, the solubility of gases in water decreases. Depleted oxygen levels can be a particular problem in calm water, such as ponds. High temperatures may not be sufficient by themselves to cause a fish kill, but when coupled with other events, they can be a significant contributing cause. For example, decaying organic matter uses a great deal of oxygen, and when combined with the effects of warm weather, fish may die in great numbers. An algae bloom can do the same thing; the algae can consume more oxygen than they produce, causing fish to die from lack of oxygen.

High rainfall can also contribute to decreased oxygen levels. The rainfall can wash large amounts of fertilizers or manure into the watershed. This condition would be reflected in unusually high levels of nitrates and phosphates in the water. Higher water temperature would compound the low level of DO.

Dissolved oxygen is increased by aeration (mixing air into the water) or photosynthesis. Sources of aeration are bubbling brooks, rapid rivers, spillways, dams, or any other forms of turbulence. Photosynthesis by the plants in the water can increase the DO during the daylight hours, even though the plants are also using it for their own respiration. At night, the plants continue to consume oxygen, which can result in an overall drop in DO.

As a result, low oxygen levels could result from a combination of high water temperature and high demand for dissolved oxygen. A large release of water from the dam or heavy rain can be contributing factors. A large release of water from the dam could stir up sediments on the bottom of the reservoir or the river, and these sediments contain lots of organic matter. If events like these were the cause of the fish kill, the amount of total organic carbon would likely be at high levels. A large amount of rain could do the same thing. It also could cause fertilizer runoff that might promote an algal bloom.

Scientists: Background Information

In putting together the cause of the fish kill, it is important that all the data be taken into consideration. Often the complex picture of what went on during the incident can only become apparent when the data is considered as a whole.

Engineers: Background Information

Organic material can increase dramatically in rivers and streams in at least two ways. One is when an algal bloom occurs. If the water contains a high concentration of nutrients, such as nitrate and phosphate fertilizers, the algae can grow quickly, and when they die, their decomposing structures add to the total amount of organic carbon in the water. A second way that organic carbon can increase is when undecomposed plant and animal material, which can lay undisturbed underneath mats of inert mud or soil, is flushed off the riverbed by strong rain or water flow. When strong water flow flushes the material out, it begins decomposing, a process that can consume relatively large amounts of oxygen.

Power Company Officials: Background Information

The month appears to be fairly normal in terms of the operation of the dam and the city electrical system. Temperatures were very hot on the days preceding the fish kill and also the days that followed, causing an increase in electricity demand for residential air-conditioning. Water temperature of the river was also high.

Agricultural Cooperative Representatives: Background Information

The Coop contracts with an environmental consulting firm to monitor the levels of phosphate, nitrates, pesticides, PCBs, and organic carbon in the river. Near the time of the fish kill some of these levels were high. Although fertilizer use can increase the levels of phosphates and nitrates in waterways, it is hardly in the best interest of the farmers if the fertilizers wash away during rainstorms. After all, fertilizers are a major expense in growing crops. However, such incidents all around the country are well documented to have occurred each year. Another source of phosphates is home wastewater. Phosphate-based cleaners were major contributors of this substance until they were banned in most communities.

Mining Company Representatives: Background Information

There is no evidence of any toxic organic substances such as PCBs or pesticides in any of your holding ponds or drainage areas. A recent inventory of substances in the soil and holding ponds amply documented this fact. Even though there is no direct evidence linking your company to the fish kill, it is always difficult for you to convince the public that your company is not the source of the harm. You are considering a new advertising campaign to help convince people that mining is beneficial to them personally.

County Sanitation Commission Members: Background Information

City officials regularly monitor the amount of dissolved oxygen, water temperature, and pH of the river. All of these measurements have been done for a number of years, and the data is very reliable. The Sanitation Commission staff has even had technicians win awards for their knowledge and techniques. The amount of dissolved oxygen in a body of water is determined by complex factors. Be prepared to explain some of the factors that can determine the amount of DO in the water.

Chamber of Commerce Members: Background Information

You are concerned that the negative publicity surrounding the fish kill is very bad for the community image. If the annual fishing tournament is tainted with the prospect of poison fish, it could be the end of the tournament. It is in your best interest to get this settled as quickly, fairly—and quietly—as possible.

Riverwood Taxpayer Association Members: Background Information

Tax revenue is going to suffer as a result of this accident. It is very important to you that the responsible individual is identified and made to pay for the damages. The city cannot really afford the tax revenue it is going to lose. If this matter is not resolved quickly, there could also be implications for next year's fishing tournament.

SCENARIO #3—HEAVY METAL OR MERCURY POISONING—SEDIMENT CONTAMINATION

Background

Mercury (and other heavy metal contaminants) can come from many different sources. Often mercury is used during mining operations dedicated to recovering other metals, such as lead, gold, and zinc. It can also come from oceans, volcanoes, and the weathering of rocks. Human activities, such as burning fossil fuels and manufacturing consumer products, can also add mercury to rivers and streams. Heavy metals are not typically found in solution in rivers and streams. More typically, they are suspended in the water after a heavy rain or run-off stirs up the sediments at the bottom of the river.

During normal growth plants absorb mercury from the soil and air. In some cases, plants concentrate mercury to small droplets. Some bacteria convert inorganic mercury to methyl mercury compounds. This constant cycling of mercury from one form to another has gone on for eons without any recognizable toxic effect on the world's food supply. In areas that have had chronic pollution over the years, however, it is possible to find high concentrations of mercury compounds. Even after the source of the contamination has been stopped, river and lake sediments can still have high levels of the compounds. Heavy metal concentration is a considerable problem for the Environmental Protection Agency and others. Containment ponds and dumps at mining sites can be monitored and controlled because regulations dictate how those sites should be managed. But it is difficult to legislate or regulate a river, where the contaminants may have rested quietly for dozens of years.

Typically, toxic problems due to mercury are more chronic than acute. Low levels of mercury can accumulate in aquatic plants. However, the concentration is biomagnified—that is, concentration levels accumulate as the mercury moves through the food chain. If this is the case, fish death would likely occur more gradually. Levels of toxic materials would have to be very high to cause a fish kill.

Two conditions acting together could cause a lethal level of heavy metals: First, there would have to be high levels of heavy metals in river sediment layers. Then the layers would have to be mixed and released by heavy rain or by a large release of water from the reservoir. Tests for heavy metals would show high concentrations in the water. Acidic (low pH) water would add to this problem because acids increase the solubility of heavy metals in water.

Scientists: Background Information

Heavy metal contamination was more common fifty years ago than it is today. Now the dangers of heavy metals have been well documented, and as a result their use is heavily regulated in almost every area. Still, many regions remain heavily contaminated due to the continuing presence in the environment of heavy metals released years ago.

Problems associated with mercury, lead, and the other heavy metals are generally more chronic than acute. Because these elements tend to interfere with metabolic processes, fish tend to die off relatively slowly. Because heavy metals bioaccumulate (become more concentrated as they move up the food chain), it may take considerable time for their ill effects to show. Although elemental mercury is not terribly poisonous, it does have grievous long-term effects. Compounds such as methylmercury are much more toxic. Arsenic is even more acutely poisonous.

Engineers: Background Information

The mining waste storage ponds have been in place for several decades. The mining company has a good record with the EPA for submitting reports and keeping metal levels within permit requirements. However, during a particularly rainy season at least one fish kill occurred for which no cause was ever discovered. Your presentation should include some permanent disposal options for the waste that resulted from the mining of zinc and silver at this site.

Power Company Officials: Background Information

Operations have been normal. In early fall there was an unusual water release due to a long-range forecast of heavy storms. The release was followed closely by amounts of heavy rainfall. Demand for electricity has been typical for the time of year.

Agricultural Cooperative Representatives: Background Information

All the members of your cooperative are anxious to show they were not responsible for the fish kill. There has been tension in the community as family farms are closed or abandoned in favor of larger or corporate-run farms. Together with low crop and livestock prices and increased regulation controlling their farming practices, coop farmers have little patience for inaccurate claims against them. At the same time, they would like to receive credit for the positive steps they have taken to establish safe farming practices.

Mining Company Representatives: Background Information

Years ago a contaminated discharge from your company's mining waste pond was responsible for a Snake River fish kill. Although your company has a good environmental record, this incident brought the company very bad publicity, a situation you should address in your presentation. You should also address some of the permanent solutions to disposal of mining waste, including a discussion of their financial feasibility.

County Sanitation Commission Members: Background Information

County sanitation leaders have sponsored a continuing campaign to remove toxic materials from households and have encouraged citizens to dispose of these materials correctly by hosting "Riverwood Clean-up Day." During this event, residents have free access to the municipal landfill, yard waste recycling yard, and a special station for recycling or disposal of household chemicals. Here, citizens can properly dispose of problematic materials such as automobile batteries, outdated lawn and garden chemicals, and household cleaners.

Chamber of Commerce Members: Background Information

You are concerned that the negative publicity surrounding the fish kill is very bad for the community image. If the annual fishing tournament is tainted with the prospect of poison fish, it could be the end of the tournament. It is in your best interest to get this settled as quickly, fairly—and quietly—as possible.

Riverwood Taxpayer Association Members: Background Information

Tax revenue is going to suffer as a result of this accident. It is very important to you that the responsible individual is identified and made to pay for the damages. The city cannot really afford the tax revenue it is going to lose. If this matter is not resolved quickly, there could also be implications for next year's fishing tournament.

SCENARIO #4—GAS BUBBLE TRAUMA

Background

Gas bubble trauma is caused by the presence of gas bubbles in the gills of fish. Supersaturation of water with oxygen and nitrogen may occur near dams and hydroelectric projects when released water forms "froth," trapping large quantities of air. Water at the dam's base may contain oxygen and nitrogen dissolved up to 130% of saturation. Even 90 km (50-60 miles) downstream, gas supersaturation may be as high as 111% of saturation. The Environmental Protection Agency limit for combined oxygen and nitrogen supersaturation in rivers is 110%. Especially designed spillways that provide gradual water release from dams may substantially lessen or even prevent such gas supersaturation.

Scientists: Background Information

Evidence for gas bubble trauma would be a large release of water from the dam, supersaturated dissolved oxygen levels, and evidence of bubbles in dissected fish gills. However, because some gas bubbles disappear rapidly after the fish die, detecting the condition requires prompt dissection and analysis.

Engineers: Background Information

If engineers know the spillway's physical structure and the volume of water that will be released, they can predict whether large quantities of air will be trapped when water is released from a dam spillway.

The U.S. Corps of Engineers has investigated whether operational or structural changes in dams might reduce the chance of gas supersaturation in released water. Their main goal has been to find ways to reduce the water volume released from spillways and to prevent released water from plunging deeply beneath the river's surface. When water plunges to great depths, the increased pressure can force greater amounts of air to dissolve in the water (see Figure 27, p. 62, in the student edition).

Three specific suggestions might help Riverwood. The first is to enlarge the reservoir located upstream from the dam. This would provide greater water storage capacity during times of heavy rain and runoff. Water could thus be released from the spillway in smaller quantities, which would decrease the formation of high levels of gas supersaturation.

The second suggestion is to start a major fish-collecting operation downstream from the dam. Collected fish would be trucked around the dam and released at upstream locations where dissolved oxygen levels support fish survival.

The third suggestion is to install deflectors on the spillway's downstream side. Spillway deflectors prevent released water from plunging to great depths. Deflectors release the water at the river surface, reducing the chance of large quantities of air dissolving into the river.

Power Company Representatives: Background Information

You are responsible for running the dam and have done so for a considerable period of time. It appears likely that the fish kill was caused by the water release from the dam. However, at the time it was licensed the dam was built to government specifications, and no mandatory modifications have been ordered.

Still, as a public utility it is in your best interest to address a rapid solution to this problem. Your presentation should address if—and how—you plan to avoid this situation in the future and how this proposal will be financed.

Agricultural Cooperative Representatives: Background Information

All the members of your cooperative are anxious to show they were not responsible for the fish kill. There has been tension in the community as family farms are closed or abandoned, in favor of larger or corporate-run farms. Together with low crop and livestock prices and increased regulation controlling their farming practices, Coop farmers have little patience for inaccurate claims against them. At the same time, they would like to receive credit for the positive steps they have taken to establish safe farming practices.

Mining Company Representatives: Background Information

Mining company officials can document a long history of compliance with local, state, and federal regulations governing the mining operation. One of the continuing problems for the mine officials has been the ever changing nature of the regulations. In the 1930s and 1940s there was little in the way of guidance on how the mine should operate. Since then, ever-stricter and more complex rules have been put in place. The mining company has made every effort to meet these changing requirements. However, some problems have occurred because of practices that were legal in the past.

County Sanitation Commission Members: Background Information

County sanitation leaders have sponsored a continuing campaign to remove toxic materials from households, and, by hosting "Riverwood Clean-up Day," have encouraged citizens to dispose of these materials correctly. During this event, residents have free access to the municipal landfill, yard waste recycling yard, and a special station for recycling or disposal of household chemicals. Here, citizens can properly dispose of problematic materials such as automobile batteries, outdated lawn and garden chemicals, and household cleaners.

Chamber of Commerce Members: Background Information

You are concerned that the negative publicity surrounding the fish kill is very bad for the community image. If the annual fishing tournament is tainted with the prospect of poison fish, it could be the end of the tournament. It is in your best interest to get this settled as quickly, fairly—and quietly—as possible.

Riverwood Taxpayer Association Members: Background Information

Tax revenue is going to suffer as a result of this accident. It is very important to you that the responsible individual is identified and made to pay for the damages. The city cannot really afford the tax revenue it is going to lose. If this matter is not resolved quickly, there could also be implications for next year's fishing tournament.

Unit 1: EPA Drinking Water Standards

Substance	Maximum Contamination Level (MCL)
pH	6.5–8.5
Total carbon	No total level set
Total lead	.015 mg/L
Total cadmium	.005 mg/L
Total mercury	.002 mg/L
Total arsenic	.05 mg/L
Nitrate (measured as nitrogen)	10 mg/L
Orthophosphate	No total level set
Pesticides	See specific standards
PCBs	.0005 mg/L

Unit 1: Scenario 1: September's Data

Month/Day	Snake River Water Flow (ft^3 per s)	Water Temperature (°C)	Dissolved Oxygen (ppm)	Rainfall (mm)	pH	Organic Carbon (µg/L)	Lead (µg/L)	Cadmium (µg/L)
Sep-1	29 100	11.0	11.10	18.00	8.4	1.300	29.000	2.000
Sep-2	29 050	11.0	11.10	14.00	8.4	1.300	29.000	2.000
Sep-3	29 050	11.0	11.10	0.00	8.4	1.300	29.000	2.000
Sep-4	28 990	11.0	11.10	0.00	8.4	1.300	29.000	2.000
Sep-5	28 980	11.0	11.10	0.00	8.4	1.300	29.000	2.000
Sep-6	29 005	11.0	11.10	0.00	8.4	1.300	29.000	2.000
Sep-7	28 970	11.0	11.10	0.00	8.4	1.300	29.000	2.000

(continued)

Scenario 1: September's Data (continued)

Mercury (µg/L)	Arsenic (µg/L)	Nitrate (ppm)	Ortho-phosphate (mg/L)	Pesticide (mg/L)	PCBs (µg/L)
0.000	2.000	0.130	0.010	0.08	0.15
0.000	2.000	0.130	0.010	0.08	0.15
0.000	2.000	0.130	0.010	0.05	0.15
0.000	2.000	0.130	0.010	0.02	0.15
0.000	2.000	0.130	0.010	0.01	0.15
0.000	2.000	0.130	0.010	0.00	0.15
0.000	2.000	0.130	0.010	0.00	0.15

Unit 1: Scenario 1: Last Two Years' Data

Month	Snake River Water Flow (ft^3 per s)	Water Temp. (°C)	Dissolved Oxygen (ppm)	Rainfall (mm)	pH	Organic Carbon (μg/L)
May	15 450	8.1	10.40	29.55	8.2	4.40
June	16 520	9.5	9.80	33.15	8.6	3.70
July	15 120	13.0	9.20	19.55	8.1	3.30
Aug	9582	13.5	9.20	17.50	7.8	2.70
Sept	8126	14.0	9.20	15.02	7.5	3.30
Oct	5068	10.5	10.60	14.55	7.3	4.50
Nov	3781	6.5	11.00	17.25	7.5	2.00
Dec	4403	4.0	11.00	14.06	7.5	0.46
Jan	4248	2.5	12.40	16.91	7.5	3.00
Feb	5091	1.0	12.50	12.66	8.1	1.20
March	7693	5.0	12.21	13.31	8.3	0.89
April	7849	7.0	10.60	17.20	8.0	1.50
May	13 440	8.0	11.21	32.58	8.1	3.20
June	16 903	9.0	9.80	32.44	8.6	4.00
July	13 410	12.0	9.20	20.58	8.4	5.00
Aug	9231	13.0	9.20	18.17	8.3	3.50
Sept	6403	14.0	10.01	21.32	8.3	2.50
Oct	3639	12.5	10.60	15.56	7.9	1.10
Nov	2853	5.5	11.00	18.50	7.5	1.30
Dec	2702	0.5	11.00	19.10	8.0	1.60
Jan	2637	2.5	12.70	16.91	7.9	1.20
Feb	2683	3.5	12.50	12.66	8.0	0.80
March	3440	5.0	11.00	13.31	8.1	1.30
April	6422	7.0	10.60	17.20	8.0	2.00
May	13 440	8.0	10.40	30.11	7.9	2.10
June	17 130	11.0	10.00	33.48	8.4	1.30
July	16 120	13.0	9.20	20.58	8.4	0.80
Aug	15 540	14.5	9.50	60.10	8.4	0.95
Sept	7103	13.9	9.99	20.58	8.4	1.05
Oct	3619	12.4	10.55	15.5	7.8	1.20

(continued)

Unit 1: Scenario 1: Last Two Years' Data (continued)

Lead (µg/L)	Cadmium (µg/L)	Mercury (µg/L)	Arsenic (µg/L)	Nitrate (ppm)	Ortho-phosphate (mg/L)	Pesticide (mg/L)	PCBs (µg/L)
29.0	0.0	0.1	2.0	0.120	0.010	0.00	0.010
10.0	0.0	0.1	2.0	0.130	0.010	0.00	0.005
10.0	1.0	0.1	2.0	0.130	0.015	0.00	0.000
4.0	1.0	0.1	2.0	0.125	0.010	0.00	0.005
4.0	1.0	0.1	2.0	0.120	0.010	0.00	0.010
9.0	1.0	0.5	3.0	0.130	0.010	0.00	0.000
9.0	1.0	0.1	2.0	0.140	0.010	0.00	0.000
10.0	1.8	0.1	2.0	0.150	0.010	0.00	0.000
5.0	1.8	0.5	2.0	0.200	0.010	0.00	0.000
7.0	0.5	0.1	2.0	0.220	0.010	0.00	0.000
28.0	1.5	0.1	2.0	0.190	0.010	0.00	0.005
30.0	1.8	0.1	2.0	0.160	0.010	0.00	0.010
24.0	1.0	0.5	2.0	0.120	0.010	0.00	0.030
24.0	1.8	0.5	3.0	0.130	0.010	0.00	0.020
24.0	1.8	0.1	2.0	0.130	0.015	0.00	0.030
20.0	1.8	0.1	2.0	0.125	0.010	0.00	0.010
16.0	1.8	0.1	2.0	0.120	0.010	0.00	0.010
9.0	1.8	0.1	2.0	0.130	0.010	0.00	0.000
9.0	1.8	0.1	3.0	0.140	0.010	0.00	0.000
10.0	1.8	0.1	2.0	0.150	0.010	0.00	0.000
7.0	1.8	0.2	3.0	0.200	0.010	0.00	0.000
11.0	1.0	0.1	2.0	0.195	0.010	0.00	0.000
9.0	1.8	0.1	2.0	0.190	0.010	0.00	0.010
13.0	1.8	0.1	2.0	0.160	0.010	0.00	0.008
30.0	1.8	0.2	2.0	0.140	0.010	0.00	0.005
29.0	1.8	0.0	2.0	0.130	0.010	0.00	0.010
10.0	1.8	0.1	2.0	0.130	0.010	0.00	0.010
29.0	1.8	0.0	2.0	0.130	0.010	0.15	0.010
10.0	1.8	0.1	2.0	0.130	0.010	0.11	0.010
9.0	1.8	0.1	2.0	0.130	0.010	0.00	0.010

Unit 1: Scenario 2: September's Data

Month/Day	Snake River Water Flow (ft^3 per s)	Water Temperature (°C)	Dissolved Oxygen (ppm)	Rainfall (mm)	pH	Organic Carbon (µg/L)	Lead (µg/L)	Cadmium (µg/L)
Sep-1	41 550	28.4	8.00	18.00	8.4	14.000	29.000	2.000
Sep-2	41 600	28.3	8.00	14.00	8.4	13.500	29.000	2.000
Sep-3	40 890	28.0	8.00	0.00	8.4	13.500	29.000	2.000
Sep-4	36 054	27.5	8.00	0.00	8.4	13.500	29.000	2.000
Sep-5	32 050	26.9	8.00	0.00	8.4	13.500	29.000	2.000
Sep-6	31 500	24.7	8.00	0.00	8.4	13.500	29.000	2.000
Sep-7	28 950	23.0	8.00	0.00	8.4	12.500	29.000	2.000

(continued)

Scenario 2: September's Data (continued)

Mercury (µg/L)	Arsenic (µg/L)	Nitrate (ppm)	Ortho-phosphate (mg/L)	Pesticide (mg/L)	PCBs (µg/L)
0.000	2.000	27.000	0.100	0	0.15
0.000	2.000	28.000	0.100	0	0.15
0.000	2.000	26.000	0.100	0	0.15
0.000	2.000	26.500	0.100	0	0.15
0.000	2.000	23.000	0.100	0	0.15
0.000	2.000	21.500	0.100	0	0.15
0.000	2.000	14.000	0.100	0	0.15

Unit 1: Scenario 2: Last Two Years' Data

Month	Snake River Water Flow (ft^3 per s)	Water Temp. (°C)	Dissolved Oxygen (ppm)	Rainfall (mm)	pH	Organic Carbon (µg/L)
May	15 450	8.1	10.40	29.55	8.2	4.40
June	16 520	9.5	9.80	33.15	8.6	3.70
July	15 120	13.0	9.20	19.55	8.1	3.30
Aug	9582	13.5	9.20	17.50	7.8	2.70
Sept	8126	14.0	9.20	15.02	7.5	3.30
Oct	5068	10.5	10.60	14.55	7.3	4.50
Nov	3781	6.5	11.00	17.25	7.5	2.00
Dec	4403	4.0	11.00	14.06	7.5	0.46
Jan	4248	2.5	12.40	16.91	7.5	3.00
Feb	5091	1.0	12.50	12.66	8.1	1.20
March	7693	5.0	12.21	13.31	8.3	0.89
April	7849	7.0	10.60	17.20	8.0	1.50
May	13 440	8.0	11.21	32.58	8.1	3.20
June	16 903	9.0	9.80	32.44	8.6	4.00
July	13 410	12.0	9.20	20.58	8.4	5.00
Aug	9231	13.0	9.20	18.17	8.3	3.50
Sept	6403	14.0	10.01	21.32	8.3	2.50
Oct	3639	12.5	10.60	15.56	7.9	1.10
Nov	2853	5.5	11.00	18.50	7.5	1.30
Dec	2702	0.5	11.00	19.10	8.0	1.60
Jan	2637	2.5	12.70	16.91	7.9	1.20
Feb	2683	3.5	12.50	12.66	8.0	0.80
March	3440	5.0	11.00	13.31	8.1	1.30
April	6422	7.0	10.60	17.20	8.0	2.00
May	13 440	8.0	10.40	30.11	7.9	2.10
June	17 130	11.0	10.00	33.48	8.4	1.30
July	13 410	13.0	9.20	20.58	8.4	0.80
Aug	9231	14.5	9.50	18.17	8.3	3.50
Sept	7501	27.1	8.00	60.10	8.4	10.00
Oct	3619	12.4	10.55	15.5	7.8	1.20

(continued)

Scenario 2: Last Two Years' Data (continued)

Lead (μg/L)	Cadmium (μg/L)	Mercury (μg/L)	Arsenic (μg/L)	Nitrate (ppm)	Ortho-phosphate (mg/L)	Pesticide (mg/L)	PCBs (μg/L)
29.0	0.0	0.1	2.0	0.120	0.010	0.000	0.010
10.0	0.0	0.1	2.0	0.130	0.010	0.000	0.005
10.0	1.0	0.1	2.0	0.130	0.015	0.000	0.000
4.0	1.0	0.1	2.0	0.125	0.010	0.000	0.005
4.0	1.0	0.1	2.0	0.120	0.010	0.000	0.010
9.0	1.0	0.5	3.0	0.130	0.010	0.000	0.000
9.0	1.0	0.1	2.0	0.140	0.010	0.000	0.000
10.0	1.8	0.1	2.0	0.150	0.010	0.000	0.000
5.0	1.8	0.5	2.0	0.200	0.010	0.000	0.000
7.0	0.5	0.1	2.0	0.220	0.010	0.000	0.000
28.0	1.5	0.1	2.0	0.190	0.010	0.000	0.005
30.0	1.8	0.1	2.0	0.160	0.010	0.000	0.010
24.0	1.0	0.5	2.0	0.120	0.010	0.000	0.030
24.0	1.8	0.5	3.0	0.130	0.010	0.000	0.020
24.0	1.8	0.1	2.0	0.130	0.015	0.000	0.030
20.0	1.8	0.1	2.0	0.125	0.010	0.000	0.010
16.0	1.8	0.1	2.0	0.120	0.010	0.000	0.010
9.0	1.8	0.1	2.0	0.130	0.010	0.000	0.000
9.0	1.8	0.1	3.0	0.140	0.010	0.000	0.000
10.0	1.8	0.1	2.0	0.150	0.010	0.000	0.000
7.0	1.8	0.2	3.0	0.200	0.010	0.000	0.000
11.0	1.0	0.1	2.0	0.195	0.010	0.000	0.000
9.0	1.8	0.1	2.0	0.190	0.010	0.000	0.010
13.0	1.8	0.1	2.0	0.160	0.010	0.000	0.008
30.0	1.8	0.2	2.0	0.140	0.010	0.000	0.005
29.0	1.8	0.0	2.0	0.130	0.010	0.000	0.010
10.0	1.8	0.1	2.0	0.130	0.010	0.000	0.010
4.0	1.8	0.1	2.0	0.125	0.010	0.000	0.010
5.0	1.8	0.0	2.0	20.000	0.100	0.000	0.010
9.0	1.8	0.1	2.0	0.130	0.010	0.000	0.010

Unit 1: Scenario 3: September's Data

Month/Day	Snake River Water Flow (ft^3 per s)	Water Temperature (°C)	Dissolved Oxygen (ppm)	Rainfall (mm)	pH	Organic Carbon (μg/L)	Lead (μg/L)	Cadmium (μg/L)
Sep-1	29 100	11.0	11.10	8.00	1.3	1.300	40.000	4.000
Sep-2	29 050	11.0	11.10	4.00	1.3	1.300	40.000	4.000
Sep-3	29 050	11.0	11.10	0.00	1.3	1.300	40.000	4.000
Sep-4	28 990	11.0	11.10	0.00	1.3	1.300	40.000	4.000
Sep-5	28 980	11.0	11.10	0.00	1.3	1.300	40.000	3.000
Sep-6	29 005	11.0	11.10	0.00	1.3	1.300	38.000	2.500
Sep-7	28 970	11.0	11.10	0.00	1.3	1.300	33.000	2.300

(continued)

Scenario 3: September's Data (continued)

Mercury (µg/L)	Arsenic (µg/L)	Nitrate (ppm)	Ortho-phosphate (mg/L)	Pesticide (mg/L)	PCBs (µg/L)
1.000	4.000	1.300	0.010	0	0.15
1.000	4.000	1.300	0.010	0	0.15
0.800	4.000	1.300	0.010	0	0.15
0.750	4.000	1.300	0.010	0	0.15
0.400	3.000	1.300	0.010	0	0.15
0.180	2.500	1.300	0.010	0	0.15
0.150	2.000	1.300	0.010	0	0.15

Unit 1: Scenario 3: Last Two Years' Data

Month	Snake River Water Flow (ft^3 per s)	Water Temp. (°C)	Dissolved Oxygen (ppm)	Rainfall (mm)	pH	Organic Carbon (μg/L)
May	15 450	8.1	10.40	29.55	8.2	4.40
June	16 520	9.5	9.80	33.15	8.6	3.70
July	15 120	13.0	9.20	19.55	8.1	3.30
Aug	9582	13.5	9.20	17.50	7.8	2.70
Sept	8126	14.0	9.20	15.02	7.5	3.30
Oct	5068	10.5	10.60	14.55	7.3	4.50
Nov	3781	6.5	11.00	17.25	7.5	2.00
Dec	4403	4.0	11.00	14.06	7.5	0.46
Jan	4248	2.5	12.40	16.91	7.5	3.00
Feb	5091	1.0	12.50	12.66	8.1	1.20
March	7693	5.0	12.21	13.31	8.3	0.89
April	7849	7.0	10.60	17.20	8.0	1.50
May	13 440	8.0	11.21	32.58	8.1	3.20
June	16 903	9.0	9.80	32.44	8.6	4.00
July	13 410	12.0	9.20	20.58	8.4	5.00
Aug	9231	13.0	9.20	18.17	8.3	3.50
Sept	6403	14.0	10.01	21.32	8.3	2.50
Oct	3639	12.5	10.60	15.56	7.9	1.10
Nov	2853	5.5	11.00	18.50	7.5	1.30
Dec	2702	0.5	11.00	19.10	8.0	1.60
Jan	2637	2.5	12.70	16.91	7.9	1.20
Feb	2683	3.5	12.50	12.66	8.0	0.80
March	3440	5.0	11.00	13.31	8.1	1.30
April	6422	7.0	10.60	17.20	8.0	2.00
May	13 440	8.0	10.40	30.11	7.9	2.10
June	17 130	11.0	10.00	33.48	8.4	1.30
July	13 410	13.0	9.20	20.58	8.4	0.80
Aug	17 130	14.5	9.50	60.10	8.4	1.30
Sept	13 410	13.9	9.99	20.58	8.4	0.80
Oct	3619	12.4	10.55	15.5	7.8	1.20

(continued)

Scenario 3: Last Two Years' Data (continued)

Lead (µg/L)	Cadmium (µg/L)	Mercury (µg/L)	Arsenic (µg/L)	Nitrate (ppm)	Ortho-phosphate (mg/L)	Pesticide (mg/L)	PCBs (µg/L)
29.0	0.0	0.1	2.0	0.120	0.010	0.000	0.010
10.0	0.0	0.1	2.0	0.130	0.010	0.000	0.005
10.0	1.0	0.1	2.0	0.130	0.015	0.000	0.000
4.0	1.0	0.1	2.0	0.125	0.010	0.000	0.005
4.0	1.0	0.1	2.0	0.120	0.010	0.000	0.010
9.0	1.0	0.5	3.0	0.130	0.010	0.000	0.000
9.0	1.0	0.1	2.0	0.140	0.010	0.000	0.000
10.0	1.8	0.1	2.0	0.150	0.010	0.000	0.000
5.0	1.8	0.5	2.0	0.200	0.010	0.000	0.000
7.0	0.5	0.1	2.0	0.220	0.010	0.000	0.000
28.0	1.5	0.1	2.0	0.190	0.010	0.000	0.005
30.0	1.8	0.1	2.0	0.160	0.010	0.000	0.010
24.0	1.0	0.5	2.0	0.120	0.010	0.000	0.030
24.0	1.8	0.5	3.0	0.130	0.010	0.000	0.020
24.0	1.8	0.1	2.0	0.130	0.015	0.000	0.030
20.0	1.8	0.1	2.0	0.125	0.010	0.000	0.010
16.0	1.8	0.1	2.0	0.120	0.010	0.000	0.010
9.0	1.8	0.1	2.0	0.130	0.010	0.000	0.000
9.0	1.8	0.1	3.0	0.140	0.010	0.000	0.000
10.0	1.8	0.1	2.0	0.150	0.010	0.000	0.000
7.0	1.8	0.2	3.0	0.200	0.010	0.000	0.000
11.0	1.0	0.1	2.0	0.195	0.010	0.000	0.000
9.0	1.8	0.1	2.0	0.190	0.010	0.000	0.010
13.0	1.8	0.1	2.0	0.160	0.010	0.000	0.008
30.0	1.8	0.2	2.0	0.140	0.010	0.000	0.005
29.0	1.8	0.0	2.0	0.130	0.010	0.000	0.010
10.0	1.8	0.1	2.0	0.130	0.010	0.000	0.010
40.0	4.0	1.0	5.0	0.130	0.010	0.010	0.000
10.0	2.0	0.1	2.5	0.130	0.010	0.010	0.000
9.0	1.8	0.1	2.0	0.130	0.010	0.000	0.010

Unit 1: Scenario 4: September's Data

Month/Day	Snake River Water Flow (ft^3 per s)	Water Temperature (°C)	Dissolved Oxygen (ppm)	Rainfall (mm)	pH	Organic Carbon (µg/L)	Lead (µg/L)	Cadmium (µg/L)
Sep-1	41 550	11.0	12.80	8.00	8.4	1.300	29.000	2.000
Sep-2	41 600	11.0	12.70	4.00	8.4	1.300	29.000	2.000
Sep-3	40 890	11.0	12.65	0.00	8.4	1.300	29.000	2.000
Sep-4	36 054	11.0	11.80	0.00	8.4	1.300	29.000	2.000
Sep-5	32 050	11.0	11.10	0.00	8.4	1.300	29.000	2.000
Sep-6	31 500	11.0	11.10	0.00	8.4	1.300	29.000	2.000
Sep-7	28 950	11.0	11.10	0.00	8.4	1.300	29.000	2.000

(continued)

Scenario 4: September's Data (continued)

Mercury (µg/L)	Arsenic (µg/L)	Nitrate (ppm)	Ortho-phosphate (mg/L)	Pesticide (mg/L)	PCBs (µg/L)
0.000	2.000	0.130	0.010	0	0.15
0.000	2.000	0.130	0.010	0	0.15
0.000	2.000	0.130	0.010	0	0.15
0.000	2.000	0.130	0.010	0	0.15
0.000	2.000	0.130	0.010	0	0.15
0.000	2.000	0.130	0.010	0	0.15
0.000	2.000	0.130	0.010	0	0.15

Unit 1: Scenario 4: Last Two Years' Data

Month	Snake River Water Flow (ft^3 per s)	Water Temp. (°C)	Dissolved Oxygen (ppm)	Rainfall (mm)	pH	Organic Carbon (μg/L)
May	15 450	8.1	10.40	29.55	8.2	4.40
June	16 520	9.5	9.80	33.15	8.6	3.70
July	15 120	13.0	9.20	19.55	8.1	3.30
Aug	9582	13.5	9.20	17.50	7.8	2.70
Sept	8126	14.0	9.20	15.02	7.5	3.30
Oct	5068	10.5	10.60	14.55	7.3	4.50
Nov	3781	6.5	11.00	17.25	7.5	2.00
Dec	4403	4.0	11.00	14.06	7.5	0.46
Jan	4248	2.5	12.40	16.91	7.5	3.00
Feb	5091	1.0	12.50	12.66	8.1	1.20
March	7693	5.0	12.21	13.31	8.3	0.89
April	7849	7.0	10.60	17.20	8.0	1.50
May	13 440	8.0	11.21	32.58	8.1	3.20
June	16 903	9.0	9.80	32.44	8.6	4.00
July	13 410	12.0	9.20	20.58	8.4	5.00
Aug	9231	13.0	9.20	18.17	8.3	3.50
Sept	6403	14.0	10.01	21.32	8.3	2.50
Oct	3639	12.5	10.60	15.56	7.9	1.10
Nov	2853	5.5	11.00	18.50	7.5	1.30
Dec	2702	0.5	11.00	19.10	8.0	1.60
Jan	2637	2.5	12.70	16.91	7.9	1.20
Feb	2683	3.5	12.50	12.66	8.0	0.80
March	3440	5.0	11.00	13.31	8.1	1.30
April	6422	7.0	10.60	17.20	8.0	2.00
May	13 440	8.0	10.40	30.11	7.9	2.10
June	17 130	11.0	10.00	33.48	8.4	1.30
July	13 410	13.0	9.20	20.58	8.4	0.80
Aug	35 000	11.0	12.40	25.20	8.4	1.30
Sept	13 410	13.0	9.20	20.58	8.4	0.80
Oct	3619	12.4	10.55	15.5	7.8	1.20

(continued)

Scenario 4: Last Two Years' Data (continued)

Lead (µg/L)	Cadmium (µg/L)	Mercury (µg/L)	Arsenic (µg/L)	Nitrate (ppm)	Ortho-phosphate (mg/L)	Pesticide (mg/L)	PCBs (µg/L)
29.0	0.0	0.1	2.0	0.120	0.010	0.000	0.010
10.0	0.0	0.1	2.0	0.130	0.010	0.000	0.005
10.0	1.0	0.1	2.0	0.130	0.015	0.000	0.000
4.0	1.0	0.1	2.0	0.125	0.010	0.000	0.005
4.0	1.0	0.1	2.0	0.120	0.010	0.000	0.010
9.0	1.0	0.5	3.0	0.130	0.010	0.000	0.000
9.0	1.0	0.1	2.0	0.140	0.010	0.000	0.000
10.0	1.8	0.1	2.0	0.150	0.010	0.000	0.000
5.0	1.8	0.5	2.0	0.200	0.010	0.000	0.000
7.0	0.5	0.1	2.0	0.220	0.010	0.000	0.000
28.0	1.5	0.1	2.0	0.190	0.010	0.000	0.005
30.0	1.8	0.1	2.0	0.160	0.010	0.000	0.010
24.0	1.0	0.5	2.0	0.120	0.010	0.000	0.030
24.0	1.8	0.5	3.0	0.130	0.010	0.000	0.020
24.0	1.8	0.1	2.0	0.130	0.015	0.000	0.030
20.0	1.8	0.1	2.0	0.125	0.010	0.000	0.010
16.0	1.8	0.1	2.0	0.120	0.010	0.000	0.010
9.0	1.8	0.1	2.0	0.130	0.010	0.000	0.000
9.0	1.8	0.1	3.0	0.140	0.010	0.000	0.000
10.0	1.8	0.1	2.0	0.150	0.010	0.000	0.000
7.0	1.8	0.2	3.0	0.200	0.010	0.000	0.000
11.0	1.0	0.1	2.0	0.195	0.010	0.000	0.000
9.0	1.8	0.1	2.0	0.190	0.010	0.000	0.010
13.0	1.8	0.1	2.0	0.160	0.010	0.000	0.008
30.0	1.8	0.2	2.0	0.140	0.010	0.000	0.005
29.0	1.8	0.0	2.0	0.130	0.010	0.000	0.010
10.0	1.8	0.1	2.0	0.130	0.010	0.000	0.010
9.0	1.8	0.0	2.0	0.130	0.010	0.010	0.000
10.0	1.8	0.1	2.0	0.130	0.010	0.010	0.000
9.0	1.8	0.1	2.0	0.130	0.010	0.000	0.010

Unit 1: PIAT: Riverwood Town Council Meeting Evaluation

ChemCom's ultimate goal is for students to be able to apply the science knowledge they gain to their everyday lives. Explain the rubric to your students, and instruct them to use the specifications as guides as they develop their project. Students may also use the rubrics to evaluate other groups' presentations. The points indicated for each specification are only suggestions; you can modify them to fit your goals.

Special Interest Group Name ______________________________

Team members: ______________________________

Evaluator's Name:		
1. Organization: Team members were organized and well prepared. Assignments and forms were submitted on time.	Poor Great 0 2 4 6 8 10	**Score**
2. Group Arguments: Students presented well thought-out arguments appropriate to their special interest group and based on the data. Arguments were based on facts, not personal attacks or sensationalism.	Poor Great 0 2 4 6 8 10	
3. Questions for Other Groups: Questions were well conceived and designed to clarify the issue. Alternate questions were also prepared.	Poor Great 0 2 4 6 8 10	
4. Quality of Rebuttals: Teams were prepared for questions and provided effective answers based on facts and data.	Poor Great 0 2 4 6 8 10	
5. Application of Science Concepts: Students applied knowledge gained in the unit as they interpreted data and presented arguments. Answers to questions reflected appropriate application of science concepts.	Poor Great 0 2 4 6 8 10	
6. Visual Aids: Students made appropriate use of audio-visual aids and presentation materials.	Poor Great 0 2 4 6 8 10	
7. Group Involvement: All team members contributed to the presentation.	Poor Great 0 2 4 6 8 10	
8. Presentation: Students effectively adopted their roles and presented information appropriate to their special interest group. Students used appropriate costumes and props to identify their group.	Poor Great 0 2 4 6 8 10	
9. Article or Letter to Editor: Writing effectively summarized the Town Meeting and evaluated the decision from the perspective of the special interest group. The article or letter displayed appropriate clarity, grammar, and style.	Poor Great 0 2 4 6 8 10	
10. Science, Technology, and Society: Students made appropriate connections between science and technology issues and the social issues in this exercise.	Poor Great 0 2 4 6 8 10	
Total (100 points possible)		

Unit 2: Element Data Cards

Atomic mass	1
Melting point (°C)	–259
Boiling point (°C)	–252
O in oxide	0.5
Cl in chloride	1

Atomic mass	40
Melting point (°C)	–189
Boiling point (°C)	–186
O in oxide	—
Cl in chloride	—

Atomic mass	40
Melting point (°C)	851
Boiling point (°C)	1487
O in oxide	1
Cl in chloride	2

Atomic mass	19
Melting point (°C)	–218
Boiling point (°C)	–188
O in oxide	0.5
Cl in chloride	1

Atomic mass	7
Melting point (°C)	179
Boiling point (°C)	1327
O in oxide	0.5
Cl in chloride	1

Atomic mass	31
Melting point (°C)	44
Boiling point (°C)	280
O in oxide	2.5
Cl in chloride	3

Atomic mass	36
Melting point (°C)	–101
Boiling point (°C)	–34
O in oxide	0.5
Cl in chloride	1

Atomic mass	32
Melting point (°C)	119
Boiling point (°C)	445
O in oxide	1
Cl in chloride	2

Atomic mass	9
Melting point (°C)	1287
Boiling point (°C)	2507
O in oxide	1
Cl in chloride	2

Atomic mass	23
Melting point (°C)	97
Boiling point (°C)	889
O in oxide	0.5
Cl in chloride	1

Atomic mass	14
Melting point (°C)	–210
Boiling point (°C)	–196
O in oxide	2.5
Cl in chloride	3

Atomic mass	20
Melting point (°C)	–248
Boiling point (°C)	–246
O in oxide	—
Cl in chloride	—

Atomic mass	39
Melting point (°C)	64
Boiling point (°C)	757
O in oxide	0.5
Cl in chloride	1

Atomic mass	10
Melting point (°C)	2037
Boiling point (°C)	2527
O in oxide	1.5
Cl in chloride	3

Atomic mass	27
Melting point (°C)	659
Boiling point (°C)	2327
O in oxide	1.5
Cl in chloride	3

Atomic mass	12
Melting point (°C)	3470
Boiling point (°C)	4347
O in oxide	2
Cl in chloride	4

Atomic mass	28
Melting point (°C)	1407
Boiling point (°C)	2677
O in oxide	2
Cl in chloride	4

Atomic mass	16
Melting point (°C)	–219
Boiling point (°C)	–183
O in oxide	1
Cl in chloride	2

Atomic mass	4
Melting point (°C)	–272
Boiling point (°C)	–269
O in oxide	—
Cl in chloride	—

Atomic mass	24
Melting point (°C)	650
Boiling point (°C)	1117
O in oxide	1
Cl in chloride	2

Unit 2: PIAT: Making Money Evaluation

ChemCom's ultimate goal is for students to be able to apply the science knowledge they gain to their everyday lives. Explain the rubric to your students, and instruct them to use the specifications as guides as they develop their project. Students may also use the rubrics to evaluate other groups' presentations. The points indicated for each specification are only suggestions; you can modify them to fit your goals.

	Comments	Score
Coin Model Must contain detailed drawing of proposed coin. All models should be 5x actual size. Specifications for actual size and mass given. Model rendered in excellent fashion.		(20)
Materials Design Describe material(s) chosen for coin. Rationale for choices well reasoned and explained. Anti-counterfeiting measures detailed.		(20)
Material Source Data Source of raw materials for coin is listed. Complete explanation of processing or refining presented. Chemistry of process is clearly presented.		(20)
Life Cycle Analysis Estimates of life expectancy of coin are provided. Provisions for reuse or recycling of coin built into design. Sage materials used for coin. Toxic or hazardous materials not part of production process.		(20)
Cost Analysis Cost of production is analyzed and reasonable. Total cost includes location of resources, processing, production, and distribution. Cost of production will allow for profit to mint and does not exceed face value of the coin.		(20)
Total (100 points possible)		

Unit 3: PIAT: Vehicle Ad Evaluation

ChemCom's ultimate goal is for students to be able to apply the science knowledge they gain to their everyday lives. Explain the rubric to your students, and instruct them to use the specifications as guides as they develop their project. Students may also use the rubrics to evaluate other groups' presentations. The points indicated for each specification are only suggestions; you can modify them to fit your goals.

Specification	Scoring Guide	Points Earned (10 pt scale)	Multiplier (to 100 points or percent scale)	Score Earned
Time	**1:** A clear, concise message was conveyed within the "air time" limit. **0:** Standards were not met.		x 10	
Scientific claims	**3:** All vehicle claims were accurate and explained, the vehicle was compared to petroleum vehicles, and fuel efficiency and emission projections were proposed for the design vehicle. **2:** One of the claims was not covered. **1:** Two of the claims were not covered satisfactorily. **0:** None of the claims were adequately covered.		x 10	
Comfort, design, & safety features	**3:** The message presented a vehicle name, model, and special features to enhance consumer appeal and maintain safety parameters. **2:** One of the designed features was not covered satisfactorily. **1:** Two of the designed features were not covered satisfactorily. **0:** None of the design standards were adequately covered.		x 10	
Presentation	**3:** The message was organized, visually stimulating, and motivating, while the script was interesting, clear, concise, and smoothly presented. **2:** The message lacked one preparation feature. **1:** The message lacked two preparation features. **0:** None of the presentation standards were met.		x 10	
	Total (100 points possible)			

Unit 4: PIAT: Air Quality in the Lunar Habitation Module Proposal

ChemCom's ultimate goal is for students to be able to apply the science knowledge they gain to their everyday lives. Explain the rubric to your students, and instruct them to use the specifications as guides as they develop their project. Students may also use the rubrics to evaluate other groups' presentations. The points indicated for each specification are only suggestions; you can modify them to fit your goals.

Specification	Scoring Guide	Points Earned	Multiplier (to 100 points or percent scale)	Score Earned
Presentation	**4:** Message was very organized, visually stimulating, motivating, and smoothly presented, showing outstanding preparation. **3:** Message was organized with an above average presentation showing good preparation. **2:** Message showed average group preparation and presentation. **1:** Message was poorly organized and lacking in preparation. **0:** None of the presentation standards were met.		x 5	
Identification of key air-quality issues and emissions	**3:** The proposal identified at least 5 issues and emissions. **2:** The proposal identified at least 3 issues and emissions. **1:** The proposal identified at least 1 issue and no emission. **0:** The proposal identified no issue and emission.		x 5	
Description of roles of the primary gases involved	**3:** The proposal identified at least 3 roles. **2:** The proposal identified at least 2 roles. **1:** The proposal identified at least 1 role. **0:** None of the criteria were met.		x 5	

(continued)

Unit 4 PIAT (continued)

Monitoring schemes	**3:** The proposal identified at least 2 emission problems and had a scheme that showed innovation, initiative, and a knowledge of the scheme's goals. **2:** The proposal identified at least 2 emission problems and had a scheme that showed good group effort. **1:** The proposal identified 1 emission and had a scheme, which was not well thought out or not feasible in nature. **0:** None of the criteria were met.		x 5	
Recommended and alternative management options	**3:** The options were well researched and showed depth of knowledge. **2:** The options were researched and showed good effort by the group. **1:** The options were not well thought out and showed a lack of preparation. **0:** None of the criteria were met.		x 5	
Formal written report	**4:** All requirements for the proposal were addressed, and the report included a scientific style of writing and appropriate graphics. **3:** All requirements for the proposal were addressed, and the report included a scientific style of writing. **2:** All requirements for the proposal were addressed, but the report was either poorly organized or had a number of grammatical errors. **1:** Not all of the requirements for the proposal were addressed; the report was incomplete. **0:** None of the criteria were met.		x 5	
	Total		x 5	

Unit 5: PIAT: Riverwood Town Council Meeting Evaluation

ChemCom's ultimate goal is for students to be able to apply the science knowledge they gain to their everyday lives. Explain the rubric to your students, and instruct them to use the specifications as guides as they develop their project. Students may also use the rubrics to evaluate other groups' presentations. The points indicated for each specification are only suggestions; you can modify them to fit your goals.

		Score
Group Arguments and Questions: From paper submitted by group, accuracy of facts, complete presentation of position.	Poor Great 0 4 8 12 16 20	
Presentation and Organization: Presentation was organized; students got into their roles and modeled the situation well; all students were actively involved.	Poor Great 0 4 8 12 16 20	
Quality of Rebuttals: Responses to statements given during the meeting.	Poor Great 0 4 8 12 16 20	
Creativity: Uniqueness of presentation.	Poor Great 0 4 8 12 16 20	
Visual Aids and Name Tags: Uniqueness of design, props, costumes, posters, transparencies, etc.	Poor Great 0 2 4 6 8 10	
Audience Participation: Involvement of group members in classroom presentations.	Poor Great 0 2 4 6 8 10	
Total (100 points possible)		

Unit 6: PIAT: Truth About Nuclear Chemistry Evaluation

ChemCom's ultimate goal is for students to be able to apply the science knowledge they gain to their everyday lives. Explain the rubric to your students, and instruct them to use the specifications as guides as they develop their project. Students may also use the rubrics to evaluate other groups' presentations. The points indicated for each specification are only suggestions; you can modify them to fit your goals.

COMMUNICATING SCIENTIFIC AND TECHNOLOGICAL KNOWLEDGE

Specification	Comments	Score
Presentation was clear and understandable; presenter(s) demonstrated a professional posture and maintained visual contact with the audience.		(20)
Determinations of the statements made in the flyer were stated as true, false, or partially true.		(20)
An explanation was provided for each determination, based on applicable science or technology.		(20)
Replacement statements to the flyer accurately addressed the issue, were well written and produced. (Superior work would include a replacement flyer with new statements.)		(10)
Visual aids, such as graphs, power point presentations, transparencies, charts, or graphics, were used well.		(10)
Everyday examples and applications were used to clarify the issues to the audience. (Superior presentations made the everyday examples appropriate for the senior-citizen audience.)		(10)
Presenter(s) responded to questions about the presentation in an informed and professional manner.		(10)
Total (100 points possible)		

Unit 7: Sample Three-Day Food Diaries

Unit 7 will become more meaningful and motivating to your students if they analyze their own three-day food diaries during their study of this material. If you implement that option, students should be asked to keep records of every food item and beverage they consume over a three-day period. Stress that students must include quantities such as number, volume, or mass, as specified in the text.

However, if you prefer that your students use less personalized food-diary information, you may use the two sample three-day food inventories provided below. Or you may follow their general format to create other three-day food diaries that may better represent typical food choices among your students.

Sample Food Diary 1

Day 1

Breakfast
2 waffles
  2 tsp margarine
  1 Tbsp syrup
6 oz grapefruit juice
3 strips bacon

Snack
2 granola bars
10 oz water

Lunch
1 cheeseburger
Medium french fries
12 oz regular cola

Dinner
3 oz grilled chicken
1 cup white rice
1/2 cup mixed vegetables
1 tsp margarine
1/2 cup canned peaches (light syrup)
10 oz water

Snack
1/2 cup vanilla ice cream
2 oz strawberries

Day 2

Breakfast
3-egg omelet
  1/4 cup green pepper
  1/4 cup mushrooms
  1 oz ham
1/2 cup hash browns
1 nectarine
8 oz orange juice

Lunch
2 beef tacos with lettuce, cheese, tomato
12 oz iced tea
2 tsp sugar
1 slice apple pie

Snack
1 medium pear

Dinner
1 cup spaghetti
  1/2 cup spaghetti sauce
  2 tsp parmesan cheese
1/2 cup green beans
2 tsp olive oil
1 slice Italian bread
1 tsp margarine
1/2 tsp garlic powder
1/2 cup fruit cocktail
14 oz water

Sample Food Diary 1 (continued)

Day 3

Breakfast

1 bagel
  2 Tbsp light cream cheese
6 oz orange juice

Snack

1/2 cup chocolate pudding
6 graham crackers

Lunch

Turkey sandwich
  2 slices whole wheat bread
  2 oz sliced turkey
  1 slice American cheese
  1 Tbsp mayonnaise
  1 leaf lettuce
  1/3 tomato, sliced
8 oz skim milk
1/2 cup coleslaw
2 chocolate chip cookies

Snack

1 medium apple
2 Tbsp peanut butter
6 saltine crackers

Dinner

3 oz T-bone steak
1/2 cup mashed potatoes
1/4 cup gravy
1/2 cup broccoli
2 Tbsp Ranch dressing
12 oz water
1 slice angel food cake
1/2 cup raspberries
2 Tbsp whipped cream
6 oz skim milk

Snack

1 cup cocoa made with skim milk

Sample Food Diary 2 (Vegetarian)

Day 1

Breakfast
1 cup oatmeal
 2 Tbsp raisins
 1 medium banana
8 oz 1% milk

Lunch
Peanut butter-&-jelly sandwich
 2 slices whole-wheat bread
 2 Tbsp peanut butter
 2 tsp strawberry jam
1 orange
1/2 cup carrot sticks
8 oz 1% milk

Snack
1 oz mozzarella cheese
5 whole-wheat crackers
12 oz water

Dinner
Bean burrito
 1 flour tortilla
 1 cup black beans
 1/4 cup shredded cheddar cheese
 1/4 avocado sliced
1/3 cup salsa
1 Tbsp sour cream
2 Tbsp Ranch dressing
 1/4 head lettuce
 1/2 tomato
 1/4 cup cucumbers
16 oz water
3 cups popcorn
 2 tsp margarine
8 oz diet cola

Day 2

Breakfast
1 1/2 oz shredded wheat
1/2 cup 1% milk
1 orange
6 oz grapefruit juice

Snack
1/2 cup cottage cheese
1/2 cup pineapple

Lunch
1 1/2 cups vegetarian vegetable soup
6 saltine crackers
1 oz cheddar cheese
1 piece corn bread
 1 tsp margarine
 1 tsp honey
8 oz regular cola

Snack
1 oz almonds
5 dried apricots
10 oz water

Dinner
1 cup brown rice
Stir-fry dish
 4 oz tofu
 2 tsp soy sauce
 1/2 cup broccoli
 1/4 cup onion
 2 tsp peanut oil
 2 Tbsp cashews
1 medium apple
12 oz water

Day 3

Breakfast
8 oz fruit-flavored yogurt
2 slices whole wheat toast
  2 tsp margarine
  1 Tbsp jelly
2 eggs, scrambled
8 oz cranberry juice

Snack
1 oat bran muffin
8 oz 1% milk

Lunch
6 oz cheese ravioli
  1/2 cup spaghetti sauce
  2 tsp parmesan cheese
1 cup cantaloupe, cubed
12 oz iced tea
1 tsp sugar

Snack
20 grapes
1 whole wheat pita bread
  1/4 cup hummus

Dinner
2 slices vegetarian pizza
  2 sliced mushrooms
  1/4 sliced green pepper
  2 sliced black olives
  1 Tbsp chopped onion
1 cup spinach salad
  1 Tbsp Italian dressing
12 oz lemonade

Analyses of Food Diaries 1 and 2

Neither sample diary should be viewed as representing either recommended or idealized food-intake records. Instead, both diaries depict what might be found in actual student food inventories—contrasting patterns of consumption of particular nutrients. Given the short, three-day sampling of foods consumed, such variable results are not unexpected. What follow are analyses of Sample Food Diaries A and B. It is important to note that these analyses involve only a small subset of all needed nutrients; broader conclusions about the overall quality of a particular diet cannot be made from such data.

The data reported here are adapted from analyses completed with Nutritional Analysis Tool (NAT) software developed at the University of Illinois, found at this Website: *http://www.nat.uiuc.edu*. Other Web-based analyses report nutrient values for a total food inventory, but software at the University of Illinois site also provides nutrient values for each individual food within the inventory. The site allows students to input all food consumed for one day. The analysis provides total nutrients for all foods as well as recommended amounts. A list of nutrients for each food can also be generated. In addition to the nutrients listed here, the site can provide analyses of fiber, potassium, sodium, phosphorus, thiamin, riboflavin, niacin, percent water, and cholesterol.

Before making your first diary-analysis assignment to students, we recommend that you visit the NAT website and gain experience with that software tool yourself. Then you can decide what level of background "user support" you wish to provide for your students, or—more fundamentally—whether you will elect to have your students use NAT or alternative nutrition-analysis resources on the Web or in print.

If NAT is used, we recommend that students enter all their food items for a particular day and specify an analysis for all nutrients. They can then print the results for each day and thus refer to those daily summary printouts as they complete their analyses for particular nutrients. Such a day-by-day procedure prevents students from needing to re-enter their food-diary information each time they analyze for a new nutrient.

Note that NAT expresses quantities of Vitamin A in IU (International Units) rather than Retinol Equivalents (RE)—the Vitamin-A quantity used in the text. To align the two quantities, you may wish to let students know that 5 RE = 1 IU. Thus, for example, 1000 RE is the same quantity of Vitamin A as 5000 IU.

Unit 7: Sample Diary 1: Detailed Analysis

	Cal	Carbo-hydrate (g)	Fat (g)		Protein (g)	Vitamins		Minerals (mg)	
			sat	unsat		A (IU)	C (mg)	Ca	Fe
			Day 1						
Breakfast									
2 waffles	174	27	1	4	4	801	0	153	3.0
2 tsp margarine	68	0	2	6	0	311	0	3	0
1 tbsp syrup	57	15	N/A	N/A	0	0	0	0	0
6 oz grapefruit juice	76	18	0	0	1	17	62	15	0.2
3 strips bacon	109	0	3	6	6	0	6	2	0.3
Snack									
2 granola bars	200	35	1	0	4	14	0	0	1.1
10 oz water	0	0	0	0	0	0	0	6	0
Lunch									
1 cheeseburger	295	27	6	6	16	462	2	111	2.5
medium French fries	450	57	4	N/A	6	100	18	20	1.1
12 oz regular cola	140	35	0	0	0	0	0	10	0
Dinner									
3 oz grilled chicken	147	0	1	2	26	25	0	13	0.9
1 cup white rice	205	45	0	0	4	0	0	16	1.9
1/2 cup mixed vegetables	54	12	0	0	3	3892	3	23	0.7
1 tsp margarine	34	0	1	3	0	155	0	1	0
1/2 cup canned peaches (light syrup)	68	18	0	0	1	444	3	4	0.5
10 oz water	0	0	0	0	0	0	0	6	0

Sample Diary 1 Detailed Analysis (continued)

Snack									
1/2 cup vanilla ice cream	133	16	5	2	2	271	0	85	0
2 oz strawberries	17	4	0	0	0	15	32	8	0.2
Totals Day 1	**2227**	**309**	**24**	**29**	**73**	**6507**	**126**	**476**	**12.4**
Day 2									
Breakfast									
Three-egg omlette	269	2	6	12	18	1157	0	74	2.1
1/4 cup green pepper	7	2	0	0	0	158	22	2	0.1
1/4 cup mushrooms	4	1	0	0	0	0	1	1	0.2
1 oz ham	52	1	1	2	5	0	8	2	0.3
1/2 cup hash browns	119	6	4	6	2	0	4	6	0.6
1 nectarine	67	16	N/A	N/A	1	1001	7	7	0.3
8 oz orange juice	100	23	0	0	2	177	75	23	0.5
Lunch									
2 beef tacos with lettuce, cheese, tomato	350	25	8	0	19	1080	9	163	1.7
12 oz iced tea	3	1	0	0	0	0	0	7	0
2 tsp sugar	31	8	N/A	N/A	0	0	0	0	0
1 slice apple pie	411	58	5	14	4	90	3	11	1.7

Sample Diary 1 Detailed Analysis (continued)

Snack									
1 medium pear	98	25	0	0	1	33	7	18	0.5
Dinner									
1 cup spaghetti	230	44	0	0	11	0	0	14	1.0
1/2 cup spaghetti sauce	136	20	1	5	2	1528	14	35	0.9
2 tsp Parmesan cheese	15	0	1	0	1	23	0	46	0
1/2 cup green beans	14	3	0	0	1	237	3	18	0.6
2 tsp olive oil	80	0	1	8	0	0	0	0	0
1 slice Italian bread	80	15	0	1	3	0	0	23	0.9
1 tsp margarine	34	0	1	3	0	156	0	1	0
1/2 tsp garlic powder	5	1	N/A	N/A	0	0	N/A	1	0
1/2 cup fruit cocktail	72	19	0	0	1	262	2	8	0.4
14 oz water	0	0	0	0	0	0	0	8	0
Totals Day 2	**2178**	**270**	**28**	**51**	**71**	**5902**	**155**	**468**	**11.8**
Day 3									
Breakfast									
1 bagel	195	38	0	1	7	0	0	13	2.6
2 tbsp light cream cheese	70	2	4	N/A	3	400	0	40	0
6 oz orange juice	75	17	0	0	1	133	56	17	0.3

Sample Diary 1 Detailed Analysis (continued)

Snack									
1/2 cup chocolate pudding	161	31	2	N/A	5	201	0	152	1.1
6 graham crackers	89	16	1	1	1	0	0	5	0.8
Lunch									
turkey sandwich									
2 slices whole-wheat bread	123	23	0	1	5	0	0	36	1.7
2 oz sliced turkey	62	0	0	0	13	0	0	4	0.2
1 slice American cheese	50	2	2	N/A	5	300	0	150	0
1 tbsp mayonnaise	57	4	1	4	0	32	0	2	0
1 leaf lettuce	2	0	0	0	0	190	2	7	0.1
1/3 tomato, sliced	8	2	0	0	0	230	7	2	0.2
8 oz skim milk	79	11	0	0	7	463	2	279	0
1/2 cup cole slaw	44	8	0	1	1	406	21	29	0.4
2 chocolate-chip cookies	156	19	3	6	2	204	0	12	0.8
Snack									
1 medium apple	81	21	0	0	0	73	8	10	0.3
2 tbsp peanut butter	188	7	3	12	8	0	0	11	0.5
6 saltine crackers	78	13	0	2	2	0	0	21	1.0

Sample Diary 1 Detailed Analysis (continued)

Dinner									
3 oz T-bone steak	129	0	7	8	21	0	0	7	2.3
1/2 cup mashed potatoes	111	18	1	3	2	177	6	27	0.3
1/4 cup gravy	31	3	1	1	2	0	0	4	0.4
1/2 cup broccoli	12	2	0	0	1	1320	41	21	0.4
2 tbsp Ranch dressing	150	1	3	N/A	0	0	0	0	0
12 oz water	0	0	0	0	0	0	0	7	0
1 slice angel-food cake	129	29	0	0	3	0	0	42	0.1
1/2 cup raspberries	30	7	0	0	1	80	15	14	0.4
2 tbsp whipped cream	15	1	1	0	0	55	0	6	0
6 oz skim milk	60	8	0	0	6	347	2	209	0
Snack									
1 cup cocoa made with skim milk	218	26	6	3	9	318	3	298	0.8
Totals Day 3	**2403**	**309**	**35**	**43**	**105**	**4929**	**163**	**1425**	**14.7**

Unit 7: Sample Diary 2 (Vegetarian): Detailed Analysis

	Cal	Carbo-hydrate (g)	Fat (g)		Protein (g)	Vitamins		Minerals (mg)	
			sat	unsat		A (IU)	C (mg)	Ca	Fe
Day 1									
Breakfast									
1 cup oatmeal with	145	25	0	2	6	37	N/A	19	1.6
2 tbsp raisins	54	14	0	0	1	1	1	9	0.4
1 medium banana	105	27	0	0	1	92	10	7	0.3
8 oz 1% milk	127	12	3	1	9	460	2	324	0.2
Lunch									
peanut butter-jelly sandwich									
2 slices whole-wheat bread	123	23	0	1	5	0	0	36	1.7
2 tbsp peanut butter	188	7	3	12	8	0	0	13	0.6
2 tsp strawberry jam	30	7	0	N/A	0	0	2	0	0
1 oz potato chips	152	15	3	6	2	0	9	7	0.5
1 orange	62	15	0	0	1	269	70	52	0.1
8 oz 1% milk	127	12	3	1	9	460	2	324	0.2
Snack									
1 oz mozzarella cheese	79	1	3	2	8	178	0	207	0.1
5 whole-wheat crackers	89	14	1	2	2	0	0	10	0.6
12 oz water	0	0	0	0	0	0	0	7	0

Sample Diary 2 Detailed Analysis (continued)

Dinner									
1 bean burrito									
1 flour tortilla	114	19	0	2	3	0	0	44	1.2
1 cup black beans	227	41	0	0	15	10	0	46	3.6
1/4 cup shredded cheddar cheese	114	0	6	3	7	299	0	204	0.2
1/4 avocado, sliced	81	4	1	5	1	308	4	6	0.5
1/3 cup salsa	39	5	0	0	0	264	0	0	0
1 tbsp sour cream	26	1	2	1	0	95	0	14	0
1 cup tossed salad									
2 tbsp Ranch dressing	134	5	3	9	0	21	0	3	0.1
1/4 head lettuce	18	3	0	0	1	445	5	26	0.7
1/2 tomato	13	3	0	0	1	383	12	3	0.3
1/4 cup cucumbers	3	1	0	0	0	56	1	4	0.1
16 oz water	0	0	0	0	0	0	0	9	0
Snack									
3 cups popcorn with 2 tsp margarine	105	12	2	N/A	3	0	0	0	0
8 oz diet cola	2	0	0	0	0	0	0	9	0
Totals Day 1	**2157**	**266**	**30**	**47**	**84**	**3378**	**118**	**1383**	**13.0**

Sample Diary 2 Detailed Analysis (continued)

Day 2									
Breakfast									
1 1/2 oz shredded wheat	153	34	N/A	N/A	5	0	0	16	1.8
1/2 cup 1% milk	69	7	2	1	5	250	1	176	0.1
1 orange	62	15	0	0	1	269	70	52	0.1
6 oz grapefruit juice	76	18	0	0	1	17	62	15	0.2
Snack									
1/2 cup cottage cheese	102	4	1	1	15	79	0	78	0.2
1/2 cup pineapple	38	10	0	0	0	18	12	5	0.3
Lunch									
1 1/2 cups vegetarian vegetable soup	108	18	0	1	3	4508	2	33	1.8
6 saltine crackers	78	13	0	2	2	0	0	21	1.0
1 oz cheddar cheese	114	0	6	3	7	300	0	204	0.2
1 piece corn bread	188	29	2	4	4	123	0	44	1.1
1 tsp margarine	34	0	1	3	0	156	0	1	0
1 tsp honey	21	6	N/A	N/A	0	0	0	0	0
8 oz regular cola	93	24	0	N/A	0	0	0	7	0
Snack									
1 oz almonds	166	5	1	13	6	0	0	70	1.0
5 dried apricots	83	22	0	0	1	2534	1	16	1.7
10 oz water	0	0	0	0	0	0	0	6	0

Sample Diary 2 Detailed Analysis (continued)

Dinner									
1 cup brown rice	220	46	0	1	5	0	0	20	1.0
Stir-fry dish									
4 oz tofu	164	5	1	8	18	188	0	232	11.9
2 tsp soy sauce	6	1	0	0	1	0	0	2	0.2
1/2 cup broccoli	12	2	0	0	1	678	41	21	0.4
1/4 cup onion	15	3	0	0	0	0	3	8	0.1
2 tsp peanut oil	80	0	2	7	0	0	0	0	0
2 tbsp cashews	98	6	2	6	3	0	0	8	1.0
1 medium apple	81	21	0	0	0	73	8	10	0.3
12 oz water	0	0	0	0	0	0	0	7	0
Totals Day 2	**2061**	**289**	**18**	**50**	**78**	**9193**	**200**	**1052**	**24.4**
				Day 3					
Breakfast									
8 oz fruit-flavored yogurt	225	42	2	1	9	111	1	313	0.2
2 slices whole-wheat toast	139	26	1	2	5	0	0	41	1.9
2 tsp margarine	68	0	1	6	0	313	0	3	0
1 tbsp jelly	48	13	0	0	0	2	2	4	0.1
2 eggs, scrambled	199	3	4	8	13	818	0	85	1.4
8 oz cranberry juice	129	33	0	N/A	0	9	80	7	0.5

Sample Diary 3 Detailed Analysis (continued)

Snack									
1 oat-bran muffin	154	28	1	2	4	10	0	36	2.4
8 oz 1% milk	127	12	3	1	9	460	2	324	0.2
Lunch									
6 oz cheese ravioli	541	62	14	N/A	22	619	0	387	2.8
1/2 cup spaghetti sauce	136	20	1	5	2	1528	14	35	0.9
2 tsp Parmesan cheese	15	0	1	0	1	23	0	46	0
1 cup cantaloupe, cubed	56	13	N/A	N/A	1	5158	68	18	0.3
12 oz iced tea	3	1	0	0	0	0	0	7	0
1 tsp sugar	15	4	N/A	N/A	0	0	0	0	0
Snack									
20 grapes	30	8	0	0	0	48	2	7	0.1
1 whole-wheat pita bread	170	35	0	1	6	0	0	10	1.9
1/4 cup hummus	105	12	1	4	3	15	5	31	1.0
Dinner									
2 slices cheese pizza	281	41	3	3	15	765	3	233	1.1
2 sliced mushrooms	9	2	0	0	1	0	1	2	0.4
1/4 sliced green pepper	5	1	0	0	0	117	17	2	0.1
2 sliced black olives	10	1	0	1	0	35	0	8	0.3
1 tbsp chopped onion	4	1	0	0	0	0	1	2	0
1 cup spinach salad	12	2	0	0	2	3760	16	55	1.5
1 tbsp Italian dressing	55	1	1	N/A	0	0	0	0	0
12 oz lemonade	136	36	0	0	0	71	13	10	0.7
Totals Day 3	**2672**	**397**	**33**	**34**	**93**	**13 862**	**225**	**1666**	**17.8**

Unit 7: Sample Diary 1: Three-Day Summary Analysis

			Day 1			Day 2			Day 3			Three-Day Summary		
	RDA (Ages 15–18)			Percent of RDA			Percent of RDA			Percent of RDA			Overall Percent of RDA	
Dietary Component	Male (%)	Female (%)	Daily Total	Male (%)	Female (%)	Daily Total	Male (%)	Female (%)	Daily Total	Male (%)	Female (%)	Daily Average	Male (%)	Female (%)
Calories	3000	2200	**2227**	74.2	101	**2178**	72.6	99.0	**2403**	80.1	109	**2269**	75.6	103
Carbohydrate (g)	—	—	**309**	—	—	**270**	—	—	**309**	—	—	**296**	—	—
Saturated fat (g)	33.33	24.44	**24**	72.0	98.2	**28**	84.0	115	**35**	105	143	**29**	87.0	119
Unsaturated fat (g)	66.66	48.88	**29**	43.5	59.3	**51**	76.5	104	**43**	64.5	88.0	**41**	61.5	83.9
Protein (g)	59	44	**73**	124	166	**71**	120	161	**105**	178	239	**83**	141	189
Vitamin A (IU)	5000	4000	**6507**	130	163	**5902**	118	148	**4929**	98.6	123	**5779**	116	144
Vitamin C (mg)	60	60	**126**	210	210	**155**	258	258	**163**	272	272	**148**	247	247
Calcium (mg)	1200	1200	**476**	39.7	39.7	**468**	39.0	39.0	**1425**	119	119	**790**	65.8	65.8
Iron (mg)	12	15	**12.4**	103	82.7	**11.8**	98.3	78.7	**14.7**	123	98.0	**13.0**	108	86.7

Unit 7: Sample Diary 2: Three-Day Summary Analysis

			Day 1			Day 2			Day 3			Three-Day Summary		
	RDA (Ages 15–18)			Percent of RDA			Percent of RDA			Percent of RDA			Overall Percent of RDA	
Dietary Component	Male (%)	Female (%)	Daily Total	Male (%)	Female (%)	Daily Total	Male (%)	Female (%)	Daily Total	Male (%)	Female (%)	Daily Average	Male (%)	Female (%)
Calories	3000	2200	**2157**	71.9	98.0	**2061**	68.7	93.7	**2672**	89.0	121	**2297**	76.6	104
Carbohydrate (g)	—	—	**266**	—	—	**289**	—	—	**397**	—	—	**317**	—	—
Saturated fat (g)	33.33	24.44	**30**	90.0	123	**18**	54.0	73.6	**33**	99.0	135	**27**	81.0	110
Unsaturated fat (g)	66.66	48.88	**47**	70.5	96.1	**50**	75.0	102	**34**	51.0	69.6	**44**	66.0	90.0
Protein (g)	59	44	**84**	142	191	**78**	132	177	**93**	158	211	**85**	144	193
Vitamin A (IU)	5000	4000	**3378**	67.6	84.4	**9193**	184	230	**13 862**	277	347	**8811**	176	220
Vitamin C (mg)	60	60	**118**	197	197	**200**	325xx	325xx	**225**	375	375	**181**	302	302
Calcium (mg)	1200	1200	**1383**	115	115	**1052**	88.1xx	88.1xx	**1666**	139	139	**1367**	114	114
Iron (mg)	12	15	**13.0**	108	86.7	**24.4**	203	163	**17.8**	148	119	**18.4**	153	123

Unit 7: PIAT: Food Inventory Evaluation

ChemCom's ultimate goal is for students to be able to apply the science knowledge they gain to their everyday lives. Explain the rubric to your students, and instruct them to use the specifications as guides as they develop their project. Students may also use the rubrics to evaluate other groups' presentations. The points indicated for each specification are only suggestions; you can modify them to fit your goals.

Item	Points Possible	Self-Evaluation	Instructor Evaluation
Detailed diary	20		
Research of food content	10		
Research of RDA	10		
Research of vitamins	10		
Research of minerals	10		
Research of human energy	10		
Summary of diet and average daily computation	5		
Graph	10		
Analysis and conclusion	5		
Summary	5		
Bibliography (format)	5		
Total	**100**		

GLOSSARY

A

acid rain
fog, sleet, snow, or rain with a pH lower than about 5.6 due to dissolved gases such as SO_2, SO_3, and NO_2

actinide series
elements with atomic numbers 89 through 103

activation energy
minimum energy required for successful collision of reactant particles in a chemical reaction

active site
location on an enzyme where a substrate molecule is positioned for a reaction

activity series
ranking of elements in order of chemical reactivity

addition polymer
polymer formed by repeated addition reactions at double or triple bonds within particular monomer units

addition reaction
reaction at the double or triple bond within an organic molecule

adsorb
take up or hold molecules or particles to the surface of a material

aeration
mixing of air into a liquid, as in water flowing over a dam

alcohol
nonaromatic organic compound containing one or more —OH groups

alkali metal
highly reactive metal belonging to the first group of the Periodic Table

alkaline
a basic solution; containing an excess of hydroxide ions (OH^-)

alkane
hydrocarbon containing only single covalent bonds

alkene
hydrocarbon containing a double covalent bond

alkyne
hydrocarbon containing a triple covalent bond

allotropes
two or more forms of an element in the same state that have distinctly different physical and/or chemical properties

alloy
solid solution consisting of atoms of different metals

alpha ray
positively charged particles emitted during decay of some radioactive elements; helium nuclei

amino acid
organic molecule containing a carboxylic acid group and an amine group; used as a protein building block

anion
negatively charged ion

anode
electrode in an electrochemical cell at which oxidation occurs

aqueous solution
solution in which water is the solvent

aquifer
structure of porous rock, sand, or gravel that holds water beneath the surface of Earth

aromatic compound
ringlike compound such as benzene that can be represented as having alternating double and single bonds between carbon atoms

atomic mass
mass of a particular atom of an element

atomic number
number of protons in an atom; distinguishes atoms of different elements

atomic weight
average mass of an atom of an element as found in nature

atom
smallest particle possessing the properties of an element

average value
See mean value

B

background radiation
relatively constant level of natural radioactivity that is always present

bacterial action
conversion by bacteria of organic substances into simpler compounds

barometer
device that measures atmospheric pressure

base unit
metric unit that expresses a fundamental physical quantity (such as length or mass)

beta ray
negatively charged particles emitted during decay of some radioactive elements; high-speed electrons

biomolecule
large organic molecules found in living systems

branched polymer
polymer formed by reactions that create numerous side chains rather than linear chains

branched-chain alkane
alkane in which at least one carbon atom is bonded to three or four other carbon atoms

brittleness
a material's tendency to shatter under pressure

buffer
substance or combination of dissolved substances capable of resisting changes in pH when limited quantities of either acid or base are added

C

calorimeter
device for determining the heat of reaction or other thermal properties

calorimetry
technique for determining the heat of reaction or related properties such as the energy value of a food

carbohydrate
compounds composed of carbon, hydrogen, and oxygen, commonly with a 2:1 ratio of hydrogen to oxygen atoms

carbon chain
carbon atoms chemically linked to one another, forming a stringlike molecular sequence

carbon cycle
movement of carbon atoms within Earth's ecosystems, from carbon storage as plant and animal matter, through release as carbon dioxide due to cellular respiration, combustion, and decay, to reacquisition by plants

carboxylic acid
organic compound containing the —COOH group

carcinogens
materials known to cause cancer

catalyst
substance that speeds up a chemical reaction but is itself unchanged

catalytic converter
reaction chamber in an auto exhaust system designed to accelerate conversion of potentially harmful exhaust gases to nitrogen gas, carbon dioxide, and water vapor

cathode
electrode in an electrochemical cell at which reduction occurs

cathode ray
beam of electrons emitted from a cathode when electricity is passed through an evacuated tube

cation
positively charged ion

cellular respiration
reactions of oxygen with complex organic molecules in plants and animals to produce carbon dioxide, water, and energy

ceramics
materials made by heating or "firing" clay or components of certain rocks

chain reaction
in nuclear fission, reaction that is sustained because it produces enough neutrons to collide with and split additional fissionable nuclei

chemical bond
force that holds atoms or ions together within a substance

chemical bonding
formation of chemical bonds

chemical change
change in matter resulting in formation of one or more new substances

chemical equation
symbolic expression that summarizes a chemical change, such as
$2\ H_2(g) + O_2(g) \longrightarrow 2\ H_2O(g)$

chemical formula
symbolic expression that represents the elements present in a substance, together with subscripts indicating the relative numbers of atoms of each element

chemical property
a property that can only be observed or measured by changing the chemical identity of the sample of matter

chemical reaction
formation of new substances from reactants; involves the breaking and forming of chemical bonds

chemical species
See species

chemical symbol
abbreviation of the name of a chemical element, such as N for nitrogen or Fe for iron

chlorination
adding chlorine to a water supply to kill harmful organisms

chlorofluorocarbons (CFCs)
synthetic substances previously used as aerosol propellants, cooling fluids, and cleaning solvents that led to ozone destruction through production of chlorine radicals

***cis-trans* isomerism**
two substances differing only in the molecular positions of groups either on the "same side" of a double bond or across the double bond from each other

cloud chamber
container filled with supersaturated air that, when cooled and exposed to ionizing radiation, produces visible trails of condensation, tracing the paths of radioactive emissions

coating
treatment in which a material is physically or chemically attached to a product's surface, typically for protection

coefficients
numbers in a chemical equation indicating the relative numbers of units of each species involved in the reaction

coenzyme
organic molecule that interacts with an enzyme to facilitate or enhance its activity

colloid
mixture containing solid particles that are small enough to remain suspended

colorimetry
chemical analysis method that uses color intensity to determine solution concentration

combustion
burning

complementary proteins
multiple protein sources that provide adequate amounts of all essential amino acids when consumed together

complete protein
protein source for humans that contains adequate amounts of all essential amino acids

complex ion
single central atom or ion, usually a metal ion, to which other atoms, molecules, or ions are attached

compound
substance composed of two or more elements in fixed proportions that cannot be broken down into simpler substances by physical means

concentration
See solution concentration

condensation
conversion of a substance from the gaseous to liquid or solid state

condensation polymer
polymer formed by repeated condensation reactions of one or more monomers

condensation reaction
chemical combination of two organic molecules, accompanied by loss of water or other small molecule

conductor
material that allows electricity (or heat) to flow through it

confirming test
laboratory test that can confirm the presence of a particular chemical species

control
in an experiment, a trial that duplicates all conditions except for the variable being investigated

cracking
process in which hydrocarbon molecules from petroleum are converted to smaller molecules, using thermal energy and a catalyst

critical mass
minimum mass of fissionable material needed to sustain a nuclear chain reaction

cycloalkane
saturated hydrocarbon containing carbon atoms joined in a ring

D

data
objective pieces of information, such as information gathered in an experiment

decay product
isotopes formed from radioactive decay

density
the mass per unit volume of a given material

derived unit
metric unit formed by mathematically combining two or more base units

diatomic molecule
molecule made up of two atoms

dipeptide
molecule composed of two amino acids joined by a peptide bond

disaccharide
sugar molecule composed of two monosaccharide units bonded together through a condensation reaction

dissolve
the process of a solute interacting with a solvent to form a solution

distillate
condensed products of distillation

distillation
method of separating liquid substances, based on differences in their boiling points

doping
adding impurities to a semiconductor to modify its electrical conductivity

dot structure
See electron-dot formula

double covalent bond
bond in which four electrons are shared between two adjacent atoms

dynamic equilibrium
state of balance when two offsetting (opposite) processes occur at equal rates

E

ejected particle
lighter nucleus or particle emitted in a nuclear-bombardment reaction

electric current
flow of electrons, as through the wire connecting the electrodes in a voltaic cell

electrical conductivity
ability to conduct an electric current

electrical potential
expresses the tendency for electrical charge to move through an electrochemical cell (based on an element's relative tendency to lose electrons when in contact with a solution of its ions); measured in volts (V)

electrochemistry
chemical changes that produce or are caused by electrical energy

electrode
strip of metal or other conductor serving as a contact between an ionic solution and the external circuit in an electrochemical cell

electrolysis
process in which a chemical reaction is caused by passing an electrical current through an ionic solution

electrolyte
ionic solution; dissolved ions allow passage of electrical charge through the solution

electromagnetic radiation
radiation ranging from low-energy radio waves to high-energy X-rays and gamma rays; includes visible light

electromagnetic spectrum
waves comprising the full range of electromagnetic radiation frequencies; *see* electromagnetic radiation

electrometallurgy
use of electrical energy to process metals or their ores

electron
particle possessing negative electrical charge; found within atoms

electron-dot formula
formula of a substance or ion in which dots represent the outer electrons in each atom

electron-dot structure
See electron-dot formula

electronegativity
an expression of the tendency of an atom to attract shared electrons within a chemical bond

electroplating
deposition of a thin layer of metal on a surface by an electrical process involving oxidation-reduction

electrostatic precipitation
pollution-control method in which combustion waste products are electrically charged, then collected on plates of opposite electric charge

element
fundamental chemical substance from which all other substances are made

endothermic
process requiring the addition of energy

endpoint
point of completion of a reaction, usually in reference to a titration

energy level
expression of the relative potential energy possessed by electrons located within an atom

enzymes
biological catalysts present in all cells

equilibrium
point in a reversible reaction where the rate of products forming from reactants is equal to the rate of reactants forming from products

essential amino acids
nine amino acids not synthesized in adequate quantities by the human body; must be obtained from protein in the diet

ester
organic compound containing the —COOR group, where R represents any stable arrangement of carbon and hydrogen atoms

evaporation
change of a substance from liquid to gaseous state

exothermic
process involving the release of energy

extrapolation
estimating a value beyond the known range

F

family (Periodic Table)
See group

fat
energy-storage molecules composed of carbon, hydrogen, and oxygen, with fewer oxygen atoms and more carbon and hydrogen atoms than carbohydrates

fatty acids
organic compounds made up of long hydrocarbon chains with carboxylic acid groups at one end

filtrate
liquid collected after filtration

filtration
separation of solid particles from a liquid by passing the mixture through a material that retains the solid particles

flocculation
formation of an insoluble material suspended in or precipitated from a solution

fluorescence
emission of visible light when exposed to radiant energy (usually ultraviolet radiation)

fluoridation
addition of small quantities of fluoride ion (F^-) to treated water supplies

food additive
substance added to food during processing and intended to increase its nutritive value or to enhance its storage life, visual appeal, or ease of production

formula unit
group of atoms or ions represented by a compound's chemical formula; simplest unit of an ionic compound

fraction (petroleum)
mixture of petroleum-based substances with similar boiling points and other properties; collected during distillation

fractional distillation
separating a mixture into its components by boiling and condensing the components sequentially

free radicals
atoms or molecules with unstable arrangements and numbers of electrons

frequency
number of waves that pass a given point each second; rate of oscillation

functional group
atom or group of atoms that imparts characteristic properties to an organic compound

fusion
See nuclear fusion

G

gamma ray
high-energy electromagnetic radiation emitted during decay of some radioactive elements

gaseous state
state of matter having no fixed volume or shape

gasohol
a fuel produced by blending small quantities of ethanol with gasoline

Geiger-Mueller counter
device used to detect and quantify radiation by collecting electrical signals produced by ionizing radiation

glycerol
simple three-carbon alcohol that is a component of triglycerides

gray (Gy)
unit that measures ionizing radiation delivered to tissue; equal to one joule per kilogram of body tissue

green chemistry
approach to chemistry that aims to improve industrial chemical products and processes by evaluating every aspect of an industrial approach to make production less hazardous to human health and to the environment

greenhouse effect
trapping and returning of infrared radiation to Earth's surface by atmospheric substances such as water and carbon dioxide

greenhouse gases
atmospheric substances that effectively absorb infrared radiation

groundwater
water that collects underground

group (Periodic Table)
vertical column of elements in the Periodic Table; also called a *family*; members of a group share similar properties

H

Haber-Bosch process
method of ammonia production in which nitrogen and hydrogen are combined directly under conditions of high temperature and pressure

half-cell
metal (or other electrode material) in contact with a solution of ions to form one half of a voltaic cell

half-life
time within which any particular radioactive atom has a 50:50 chance of undergoing radioactive decay

half-reaction
a reaction involving either the loss of electrons from an electrode (anode) or the gain of electrons by an electrode (cathode); an oxidation-reduction reaction can be expressed as the sum of two half-reactions

hard water
water containing relatively high concentrations of calcium (Ca^{2+}), magnesium (Mg^{2+}), or iron(III) (Fe^{3+}) ions

heat of combustion
quantity of thermal energy released when a specific amount of a material burns

heterogeneous mixture
mixture that is not uniform throughout

high-level nuclear waste
radioactive waste products associated with nuclear fission or products formed when neutrons are absorbed by nuclear fuel

histogram
graph indicating the frequency or number of instances of particular values (or value ranges) within a set of related data

homogeneous mixture
mixture that is uniform throughout

hydrocarbon
molecular compound composed only of carbon and hydrogen

hydrochlorofluorocarbons (HCFCs)
compounds used as coolants in place of chlorofluorocarbons (CFCs) that contain hydrogen and/or fluorine in place of some chlorine atoms

hydrofluorocarbons (HFCs)
compounds containing only hydrogen, fluorine, and carbon that may be used instead of hydrochlorofluorocarbons (HCFCs) and chlorofluorocarbons (CFCs); HFCs lack chlorine and decompose before reaching the stratosphere

hydrogenation
chemical reaction that adds hydrogen atoms to a molecule

hydrologic cycle
See water cycle

hydrometallurgy
metal-processing methods involving treatment of rocks or ores with reactants in water solution

hydronium ion
ion resulting from bonding of a hydrogen ion (H^+) to a water molecule; H_3O^+

I

ideal gas
gas that behaves under all conditions as predicted by kinetic molecular theory

intermolecular forces
forces of attraction between molecules

ions
electrically charged atoms or groups of atoms

ion exchange
process of purifying water, which may involve the exchange of hard-water ions for other ions such as sodium (Na^+)

ionic compound
substance composed of positive and negative ions

ionizing radiation
nuclear radiation and high-energy electromagnetic radiation with sufficient energy to produce ions by ejecting electrons from atoms and molecules

isomers
compounds with the same molecular formula but different structural formulas

isotopes
atoms of the same element with differing numbers of neutrons

K

kinetic energy
energy associated with the motion of an object

kinetic molecular theory (KMT)
model for gas behavior based on rapidly moving particles (molecules) that are relatively far apart and change direction only through collisions with each other or the container walls

kinetics
study of reaction rates

L

law of conservation of energy
energy can change form, but is not created or destroyed in a chemical reaction

law of conservation of matter
matter is not created or destroyed in a chemical reaction

LeChatelier's Principle
a system at equilibrium tends to shift to a new equilibrium position due to a change in temperature, concentration, or pressure; the direction of the shift partially offsets the imposed change

limiting reactant
starting substance that is used up first as a chemical reaction occurs

liquid state
state of matter with fixed volume but no fixed shape

low-level nuclear waste
radioactive waste products with relatively low levels of radioactivity, such as discarded protective clothing from nuclear laboratories

M

macromineral
any mineral essential to human life that occurs in relatively large quantities (at least 5 g) in the body

Magnetic Resonance Imaging (MRI)
noninvasive nuclear medical method of imaging soft tissues using nonionizing radiation (low-energy radio waves) to identify atoms in tissue without affecting the patient

malleability
property related to a material's ability to be flattened without shattering

mass number
sum of the number of protons and neutrons in the nucleus of an atom of a particular isotope

mean value
number obtained by dividing the sum of a set of values by the total number of values in the set; also referred to as the *average value*

mechanical filtering
pollution-control method in which combustion waste products pass into filters that trap particles too large to pass through

median value
within an ascending or descending set of values, the number that represents the middle value, with an equal number of values above and below it

metal
a material possessing properties such as luster, ductility, conductivity, and malleability

metalloid
a material with properties intermediate between those of metals and nonmetals

metastable
nuclear state resulting from alpha or beta decay in which resulting nuclei are energetically excited

mineral
a naturally occurring solid substance; commonly removed from ores to obtain a particular element of interest or value

mixture
combination of materials in which each material retains its separate identity

molar concentration
See molarity

molar heat of combustion
quantity of thermal energy released from burning one mole of a substance

molar mass
mass (usually in grams) of one mole of a substance or other chemical species

molar volume
volume occupied by one mole of a substance; at 0 °C and 1 atm, the molar volume of any gas is 22.4 L

molarity
solution concentration determined by dividing the total moles of solute by the volume of the solution (expressed in liters); molar concentration

mole (mol)
amount of substance or chemical species that is equal to 6.02×10^{23} units, where the unit may be atoms, molecules, formula units, electrons, or other specified entities; chemist's "counting" unit

molecule
smallest particle of a substance retaining the properties of the substance

molecular substance
substance composed of molecules, such as H_2O and CH_4

monomer
compound whose molecules can react to form the repeating units of a polymer

monosaccharide
simple-sugar molecule, usually containing five or six carbon atoms

monounsaturated fat
fat molecule containing only one carbon-carbon double bond in its fatty acid components

mutagen
material that causes mutations in DNA

mutation
changes in the structure of DNA that may result in the production of altered proteins

N

natural water
untreated water gathered from natural sources such as rivers, ponds, or wells

negative oxidation state
negative number assigned to an atom in a compound when that atom has greater control of bonding electrons than the control exerted by one or more atoms to which it is bonded

neutralization
the combination of an acid and a base in amounts that result in elimination of any excess acid or base

neutron
particle possessing no electrical charge; found within the nucleus of most atoms

nitrogen cycle
movement of atmospheric nitrogen atoms through Earth's ecosystems via collection by bacteria, conversion into ammonia or ammonium ions, conversion into nitrate ions, uptake by plants, passage through the food chain, release as ammonia or ammonium ions, and conversion back into atmospheric nitrogen

noble gas
very unreactive element belonging to the last (rightmost) group on the Periodic Table

nonconductor
material that does not allow electricity to flow through it

nonionizing radiation
electromagnetic radiation in the visible and lower-energy regions of the electromagnetic spectrum with insufficient energy to form ions when transferring energy to matter

nonmetal
a material possessing properties such as brittleness, lack of luster, and nonconductivity (it acts as an insulator)

nonrenewable resource
a resource that will not be replenished by natural processes over the time frame of human experience

nuclear fission
splitting an atom into two smaller atoms

nuclear fusion
combining of two nuclei to form a new, heavier nucleus

nuclear radiation
form of ionizing radiation resulting from changes in the nuclei of atoms

nucleon
nuclear particle, specifically a neutron or proton

nucleus, atomic
dense, positively charged central region in an atom; composed of protons and neutrons

O

octane number
rating indicating the combustion quality of gasoline compared to the combustion quality of isooctane

OIL RIG
mnemonic device for remembering the definitions of oxidation and reduction: **O**xidation **I**s **L**oss, **R**eduction **I**s **G**ain (of electrons)

ore
rock or other solid material from which it is profitable to recover a mineral containing a metal or other useful substances

organic chemistry
branch of chemistry dealing with hydrocarbons and their derivatives

oxidation
any process in which one or more electrons are lost by a chemical species

oxidation-reduction (redox) reaction
reaction in which oxidation and reduction occur

oxidizing agent
species that causes another atom, molecule, or ion to become oxidized

oxygenated fuel
fuel with oxygen-containing additives, such as methanol, that increase octane rating and reduce harmful emissions

P

paper chromatography
method for separating substances that relies on solution components having different attractions to solvent (mobile phase) and paper (stationary phase)

particulates
solid particles that enter the air from human activities or natural processes

parts per million (ppm)
an expression of concentration, indicating the number of units of solute found within a million units of solution

peptide bond
bond that links amino acids together in peptides and proteins

percent composition
percent by mass of each component found in a material

period (Periodic Table)
horizontal row of elements in the Periodic Table

Periodic Table
table in which elements are placed in order of increasing atomic number, such that elements with similar properties are located in the same vertical column (group)

petrochemical
any organic compound produced from petroleum or natural gas

photochemical smog
potentially hazardous mixture of secondary pollutants formed by the irradiation by sunlight of certain primary pollutants in the presence of oxygen

photon
energy bundle of electromagnetic radiation that travels at the speed of light

pH scale
a measure of the acidic or basic character of a solution; at 25 °C solutions with pH values lower than 7 are acidic, solutions above 7 are basic, and neutral solutions are pH 7; based on the solution's hydrogen ion (H^+ or H_3O^+) concentration

physical change
change in matter in which the identity of the material involved does not change

physical property
a property that can be observed or measured without changing the identity of the sample of matter

polar
having regions of positive and negative charge resulting from uneven distribution of electrical charge

polyatomic ion
ion containing two or more atoms

polymer
molecule composed of very large numbers of identical repeating units

polysaccharide
polymer composed of many monosaccharide units

polyunsaturated fat
triglyceride molecule containing multiple carbon–carbon double bonds in its fatty acid portions

positive oxidation state
positive number assigned to an atom in a compound when that atom has less control of its electrons than as a free element

Positron Emission Topography (PET)
nuclear medical procedure relying on radioactive isotopes that emit positrons that generate detectable gamma rays upon interaction with electrons

positron
positively charged subatomic particle with the same mass as an electron; the antimatter counterpart to the electron

post-chlorination
addition of a low level of chlorine to treated water to prevent later bacterial infestation

potential energy
energy associated with position

pre-chlorination
addition of chlorine early in the water-treatment process to kill organisms

precipitate
insoluble solid substance that has separated from a solution

primary air pollutant
contaminant that directly enters the atmosphere rather than being formed as a result of reactions of airborne substances

product nucleus
isotope produced in a nuclear-bombardment reaction

products
substances produced in a chemical reaction

projectile particle
particle fired at the target nucleus in a nuclear-bombardment reaction

protein
major structural component of living tissue made from many linked amino acids

proton
particle possessing positive electrical charge; found in the nucleus of all atoms

pure substance
See substance

pyrometallurgy
use of thermal energy (heat) to process metals or their ores

R

rad
unit that expresses the quantity of ionizing radiation absorbed by tissue; equal to one-hundredth of a gray

radioactive decay
change in an atom's nucleus due to emission of alpha, beta, or gamma radiation

radioactivity
spontaneous emission of nuclear radiation

range
difference between the highest and lowest values in a data set

reactants
starting substances in a chemical reaction

reaction rate
expression of how fast a particular chemical change occurs

redox reaction
See oxidation-reduction reaction

reducing agent
species that causes another atom, molecule, or ion to become reduced

reduction
any process in which electrons are gained by a chemical species

reference solution
solution of known composition used as a comparison in chemical tests

reflectivity
proportion of radiation that a material reflects rather than absorbs

regeneration
process for renewing a material; renewing used ion-exchange resin by replacing hard-water ions in the resin with ions such as sodium (Na^+)

rem
unit that expresses the ability of radiation to cause ionization in human tissue; equal to one-hundredth of a sievert

renewable resource
resource that can be replenished by natural processes over the time frame of human experience

Responsible Care
international program involving chemical manufacturing companies that voluntarily agree to public scrutiny and evaluation according to criteria focused on safety and environmental responsibility

reversible reaction
chemical reaction in which products form reactants at the same time that reactants form products

R_f value
ratio of the distance a substance moves to the distance the solvent moves in paper chromatography

S

sand filtration
separation of solid particles from a liquid by passing the mixture through sand

saturated fat
fat molecule containing only single carbon-carbon bonds within its fatty acid components

saturated hydrocarbon
hydrocarbon consisting of molecules in which each carbon atom is bonded to four other atoms

saturated solution
solution in which the solvent has dissolved as much solute as it can retain stably at a given temperature

scintillation counter
detector of ionizing radiation that measures light emitted by atoms that have been excited by ionizing radiation

scrubbing
pollution control method that removes particles and sulfur oxides from industrial combustion processes

secondary air pollutant
contaminant formed in the atmosphere through chemical reactions between primary air pollutants and/or natural components of air

semiconductor
solid substance, such as silicon, with electrical conductivity between that of conductors and nonconductors at normal temperatures

sievert (Sv)
unit that measures the ability of radiation to cause ionization in human tissue; one sievert represents the ionizing capability of one gamma ray

single covalent bond
bond in which two electrons are shared by the two bonded atoms

smog
potentially hazardous combination of smoke and fog

solid state
state of matter having a fixed volume and fixed shape

solid-state detector
device used in research laboratories to monitor changes in the movement of electrons through semiconductors as they are exposed to ionizing radiation

solubility
quantity of a substance that will dissolve in a given quantity of solvent to form a saturated solution

solute
dissolved substance in a solution, usually the component present in the smaller quantity

solution
homogeneous mixture of two or more substances

solution concentration
quantity of solute dissolved in a specific quantity of solvent or solution

solvent
dissolving agent in a solution, usually the component present in the larger quantity

species
in chemistry, atoms, molecules, ions, free radicals, or other well-defined entities

specific heat capacity
quantity of thermal energy needed to raise the temperature of 1 g of a material by 1 °C

states
forms—gas, liquid, and solid—in which matter is found

straight-chain alkane
alkane consisting of molecules in which each carbon atom is linked to no more than two other carbon atoms

strong force
force that holds protons and neutrons together in an atom's nucleus

structural formula
chemical formula showing the arrangement of atoms and covalent bonds in a molecule, in which each electron pair in a covalent bond is represented by a line between the symbols of two atoms

subatomic particles
particles that make up atoms, including neutrons, protons, and electrons

subscript
character printed below the line of type and used to indicate the number of atoms of a given element in a chemical formula; in H_2O, for example, the subscript 2 indicates the number of H atoms

substance
an element or a compound; a material with uniform, definite composition and distinct properties

substrate
reactant molecule that interacts with an enzyme to undergo a reaction

supersaturated solution
solution containing a higher concentration of solute than a saturated solution at the given temperature

surface water
water on the surface of Earth

suspension
mixture containing large, dispersed solid particles that can settle out or be separated by filtration

synergistic interaction
interaction where the combined effect of several factors is greater than the sum of their separate effects

synthetic substance
substance produced solely as the result of human activity; entirely human-made

T

tap water
water drawn from plumbing lines

target particle
stable isotope bombarded in nuclear-bombardment reactions

temperature inversion
atmospheric condition in which a cool air mass is trapped beneath a less dense warm air mass, often in a valley or over a city

tetrahedron
a regular triangular pyramid; the four bonds of each carbon atom in an alkane point to the corners of a tetrahedron

titration
common procedure used to determine the concentrations of substances in solutions

trace mineral
any mineral essential to human life that occurs in relatively small quantities (fewer than 5 g) in the body

transmutation
conversion of one element to another either naturally or artificially

transuranium element
any element with an atomic number greater than that of uranium, atomic number 92

triglyceride
common type of fat molecule made from a simple three-carbon alcohol (glycerol) and three fatty acid molecules

turbidity
cloudiness

Tyndall effect
visible pattern caused by reflection of light from suspended particles in a colloid

U

unsaturated compound
organic compound containing one or more double or triple bonds per molecule

unsaturated fat
fat molecule containing one or more carbon-carbon double bonds in its fatty acid components

unsaturated solution
solution containing a lower concentration of solute than a saturated solution at the given temperature

V

vitamin
biomolecule necessary for growth, reproduction, health, and life

vitrification
part of a common storage method for radioactive wastes involving conversion of wastes into a type of ceramic by heating

voltaic cell
electrochemical cell in which a spontaneous chemical reaction is used to produce electricity

W

water cycle
repeating processes of rainfall (or other precipitation), run-off, evaporation, and condensation that continuously circulate water between Earth's crust and atmosphere

water softener
apparatus containing an ion-exchange resin; used to treat water by removing ions that cause water hardness

water treatment
purification processes applied to water before it is distributed for consumption

wavelength
distance between the corresponding parts of successive waves

X

X-ray
high-energy electromagnetic radiation, unable to penetrate dense materials such as bone or lead, but able penetrate less dense materials

Z

zero oxidation state
the oxidation state of atoms of an element not chemically combined with any other element

ILLUSTRATION CREDITS

COVER:

top (*left*) Tom Stewart/The Stock Market;
top (*right*) Ken Karp;
bottom (*right*) Digital Stock/Corbis;
bottom (*left*) Pat O'Hara/Stone;

UNIT 1

2-3 (*background*) Digital Stock/Corbis;
3 (*left*) Matt Meadows/Peter Arnold;
3 (*right*) Peter Arnold;
6 Photo Researchers;
12 (*left*) Matt Meadows/Peter Arnold;
12 (*right*) The Stock Market;
19 Young-Wolff/PhotoEdit;
23 (*left*) Still Pictures/Peter Arnold;
23 (*right*) Peter Arnold;
25 (*left*) Michael Dalton/Fundamental Photographs;
25 (*center*) Michael Dalton/Fundamental Photographs;
25 (*right*) Michael Dalton/Fundamental Photographs;
33 Manfred Kage/Peter Arnold;
40 (*left*) Randy Brandon/Alaska Stock;
40 (*right*) James Holmes/Thomson Laboratories/Science Photo Library/Photo Researchers;
41 Rachel Epstein/PhotoEdit;
64 Frank Rossotto/The Stock Market;
78 Richard Megna/Fundamental Photographs;
80 Stan W. Elems/Visuals Unlimited;
83 (*top left*) Island Water Association, Florida, photo by Rusty Isler;
83 (*top right*) Island Water Association, Florida;
83 (*bottom*) Joe McDonald/Visuals Unlimited;
86 Daniel J. Schaefer/PhotoEdit;

UNIT 2

90-91 (*background*) Richard Megna/Fundamental Photographs;
90 (*top left*) Richard Megna/Fundamental Photographs;
90 (*top right*) John Madere/The Stock Market;
90 (*bottom*) Milton Heiberg;
91 (*top*) Charles D. Winters/Photo Researchers;
91 (*bottom*) Charles O'Rear/Corbis;
95 (*left*) Ken Coleman/PhotoEdit;
95 (*right*) John Madere/The Stock Market;
96 Richard Megna/Fundamental Photographs;
118 Milton Heiberg;
120 (*top*) Archivo Iconografico, S.A./Corbis;
120 (*center*) Jonathan Blair/Corbis;
120 (*bottom*) Archivo Iconografico, S.A./Corbis;
136 Richard Megna/Fundamental Photographs;
138 Fran Gale/NCPTT-National Center for Preservation Technology and Training;
139 Mark E. Gibson/Visuals Unlimited;
155 (*left*) Kristen Brochmann/Fundamental Photographs;
155 (*center*) Tom Pantages;
155 (*right*) Visuals Unlimited;
157 (*counter clockwise from top left*) Visuals Unlimited;
157 Bill Stanton/Rainbow;
157 Bill Gallery/Stock Boston;
157 Jeff Greenberg/Visuals Unlimited;
157 Kent Knudson/Stock Boston;
157 Mark Segal/Stock Boston/PictureQuest;
157 Chris Knapton/Science Photo Library/Photo Researchers;
163 Davis Barber/PhotoEdit;
164 Courtesy of John Langhans;
165 (*left*) Charles D. Winters/Photo Researchers;
165 (*right*) Charles O'Rear/Corbis;
166 (*bottom*) Davis Barber/PhotoEdit;
166 (*top*) Michael Newman/PhotoEdit;
167 Courtesy of Mercedes-Benz USA, Inc.;

UNIT 3

172-173 (*background*) Peter Menzel/Stock Boston;
173 (*top left*) David Woods/The Stock Market;
173 (*top right*) Joseph Nettis/Stock Boston;
173 (*top center*) Julie Houck/Stock Boston;
173 (*bottom*) Young-Wolff/PhotoEdit;
198 (*left*) Courtesy of Susan Landon;
198 (*right*) Courtesy of Susan Landon;
210 Alan Schein/The Stock Market;
218 (*clockwise from top left*) David Woods/The Stock Market;
218 Michael Townsend/Allstock/PictureQuest;
218 Tony Freeman/PhotoEdit;
218 Julie Houck/Stock Boston;
218 Paul Silverman/Fundamental Photographs;
218 Mark Burnett/Stock Boston;
218 Joseph Nettis/Stock Boston;
224 Richard Megna/Fundamental Photographs;
225 (*counter clockwise from top right*) J. Gerard Smith/Photo Researchers;
225 Corbis;
225 Peter Fisher/The Stock Market;

225 Michael Powers/Stock Boston;
225 Jeffrey Dunn/Stock Boston;
225 Leonard Harris/Stock Boston;
225 Paul Silverman/Fundamental Photographs;
225 Ray Coleman/Photo Researchers;
226 (*counter clockwise from top left*) Corbis;
226 Corbis;
226 Richard Megna/Fundamental Photographs;
226 Jeffrey W. Myers/Stock Boston;
230 Courtesy of DuPont;
231 Courtesy of DuPont;
237 Young-Wolff/PhotoEdit;
241 Peter Menzel/Stock Boston;

UNIT 4

244-245 (*background*) JSC/NASA;
245 John Maier, Jr./The Image Works;
255 The Photo Works;
257 (*left*) Russ Lappa/Photo Researchers;
257 (*right*) Richard Megna/Fundamental Photographs;
258 Arthur R. Hill/Visuals Unlimited;
269 Mark Richards/PhotoEdit;
279 (*left*) John Beatty/Tony Stone;
279 (*right*) Dennis Flaherty/Photo Researchers;
282 (*top*) Frank Zullo/Photo Researchers;
282 (*bottom*) Richark Hutchings/Photo Researchers;
298 John Maier, Jr./The Image Works;
299 Richard Megna/Fundamental Photographs;
303 Robert J. Llano
306 Richard Megna/Fundamental Photographs;
315 Barbara Alper/Stock Boston;
316 Biophoto Associates/Photo Researchers;
326 Courtesy of Hughes Environmental Systems Inc.
327 Chart Applied Technologies
331 TOMS/GSFC/NASA;

UNIT 5

338-339 (*background*) David Joel/Tony Stone
338 Grant Heilman/Grant Heilman Photography
344 (*top left*) Tom Tracy/The Stock Market
344 (*top right*) Gabriel Covian/Image Bank
344 (*center right*) John Gillmoure/The Stock Market
344 (*center left*) Randy Duchaine/The Stock Market
344 (*bottom left*) Chris Sorensen/The Stock Market
344 (*bottom right*) Ken Whitmore/Tony Stone
346 (*left*) Richard Megna/Fundamental Photographs
346 (*right*) Richard Megna/Fundamental Photographs
350 A & J Verkaik/The Stock Market
351 Grant Heilman/Grant Heilman Photography
368 Bilderberg/The Stock Market
376 Courtesy of Pfizer, Inc.
387 Bill Aron/PhotoEdit
399 Courtesy of Technic, Inc.

UNIT 6

408-409 (*background*) J. Pasachoff/LBL/Visuals Unlimited
410 (*top*) Will & Deni McIntyre/Photo Researchers
410 (*bottom*) Carlos Humberto T.D.C./Contact Press Images/PictureQuest
413 (*left*) Science Photo Library/Photo Researchers
413 (*right*) Science Photo Library/Photo Researchers
414 CNRI/Phototake
434 Hank Morgan/Photo Researchers
444 CERN/Science Photo Library/Photo Researchers
453 Will & Deni McIntyre/Photo Researchers
459 Courtesy of Darleane Hoffman
461 Courtesy of Beverly Buck
468 A.J. Copley/Visuals Unlimited
470 Carlos Humberto T.D.C./Contact Press Images/PictureQuest
473 LLNL/Science Source/Photo Researchers
476 Courtesy of Michael Arnold

UNIT 7

484-485 (*background*) Jom Foster/The Stock Market
484 Charles D. Winters/Photo Researchers
485 Matt Haas/American Chemical Society
486 (*top left*) Sonda Dawes/The Image Works
486 (*top right*) David Young-Wolfe/PhotoEdit
486 (*center left*) L. D. Gordon/Image Bank
486 (*center right*) Andy Levin/Photo Researchers
486 (*bottom left*) Charles D. Winters/Photo Researchers
486 (*bottom right*) Richard Falco/Black Star Publishing/PictureQuest
490 Bob Daemmrich/Stock Boston
494 Courtesy of Ken Kashian, Illinois Farm Bureau
496 (*left*) Charles D. Winters/Photo Researchers
496 (*right*) Felicia Martinez/PhotoEdit
525 Bill Aron/PhotoEdit
538 Matt Haas/American Chemical Society
552 Leonard Lessin/Peter Arnold

CREDITS

ChemCom is the product of teamwork involving individuals from all over the United States over more than ten years. The American Chemical Society is pleased to recognize all who contributed to *ChemCom*.

The team responsible for the fourth edition of *ChemCom* is listed on the copyright page. Individuals who contributed to the initial development of *ChemCom*, to the first edition in 1988, the second edition in 1993, and the third edition in 1998 are listed below.

Principal Investigator:
W. T. Lippincott

Project Manager:
Sylvia Ware

Chief Editor:
Henry Heikkinen & Conrad L. Stanitski

Contributing Editor:
Mary Castellion

Assistant to Contributing Editor:
Arnold Diamond

Editor of Teacher's Guide:
Thomas O'Brien & Patricia J. Smith

Revision Team:
Diane Bunce, Gregory Crosby, David Holzman, Thomas O'Brien, Joan Senyk, Thomas Wysocki

Editorial Advisory Board:
Joseph Breen, Glenn Crosby, James DeRose, I. Dwaine Eubanks, Lucy Pryde Eubanks, Regis Goode, Henry Heikkinen (*chair*), Mary Kochansky, Ivan Legg, W. T. Lippincott (ex officio), Steven Long, Nina McClelland, Lucy McCorkle, Carlo Parravano, Robert Patrizi, Max Rodel, K. Michael Shea, Patricia Smith, Susan Snyder, Conrad Stanitski, Jeanne Vaughn, Sylvia Ware (ex officio)

Writing Team:
Rosa Balaco, James Banks, Joan Beardsley, William Bleam, Kenneth Brody, Ronald Brown, Diane Bunce, Becky Chambers, Alan DeGennaro, Patricia Eckfeldt, Dwaine Eubanks (dir.), Henry Heikkinen (dir.), Bruce Jarvis (dir.), Dan Kallus, Jerry Kent, Grace McGuffie, David Newton (dir.), Thomas O'Brien, Andrew Pogan, David Robson, Amado Sandoval, Joseph Schmuckler (dir.), Richard Shelly, Patricia Smith, Tamar Susskind, Joseph Tarello, Thomas Warren, Robert Wistort, Thomas Wysocki

Steering Committee:
Alan Cairncross, William Cook, Derek Davenport, James DeRose, Anna Harrison (ch.), W. T. Lippincott (ex officio), Lucy McCorkle, Donald McCurdy, William Mooney, Moses Passer, Martha Sager, Glenn Seaborg, John Truxall, Jeanne Vaughn

Consultants:
Alan Cairncross, Michael Doyle, Donald Fenton, Conrad Fernelius, Victor Fratalli, Peter Girardot, Glen Gordon, Dudley Herron, John Hill, Chester Holmlund, John Holman, Kenneth Kolb, E. N. Kresge, David Lavallee, Charles Lewis, Wayne Marchant, Joseph Moore, Richard Millis, Kenneth Mossman, Herschel Porter, Glenn Seaborg, Victor Viola, William West, John Whitaker

Synthesis Committee:
Diane Bunce, Dwaine Eubanks, Anna Harrison, Henry Heikkinen, John Hill, Stanley Kirschner, W. T. Lippincott (ex officio), Lucy McCorkle, Thomas O'Brien, Ronald Perkins, Sylvia Ware (ex officio), Thomas Wysocki

Evaluation Team:
Ronald Anderson, Matthew Bruce, Frank Sutman (dir.)

Field Test Coordinator:
Sylvia Ware

Field Test Workshops:
Dwaine Eubanks

Field Test Directors:
Keith Berry, Fitzgerald Bramwell, Mamie Moy, William Nevill, Michael Pavelich, Lucy Pryde, Conrad Stanitski

Pilot Test Teachers:
Howard Baldwin, Donald Belanger, Navarro Bharat, Ellen Byrne, Eugene Cashour, Karen Cotter, Joseph Deangelis, Virginia Denney, Diane Doepken, Donald Fritz, Andrew Gettes, Mary Gromko, Robert Haigler, Anna Helms, Allen Hummel, Charlotte Hutton, Elaine Kilbourne, Joseph Linker, Larry Lipton, Grace McGuffie, Nancy Miller, Gloria Mumford, Beverly Nelson, Kathy Nirei, Elliott Nires, Polly Parent, Mary Parker, Dicie Petree, Ellen Pitts, Ruth Rand, Kathy Ravano, Steven Rischling, Charles Ross, Jr., David Roudebush, Joseph Rozaik, Susan Rutherland, George Smeller, Cheryl Snyder, Jade Snyder, Samuel Taylor, Ronald Tempest, Thomas Van Egeren, Gabrielle Vereecke, Howard White, Thomas Wysocki, Joseph Zisk

Field Test Teachers:
Vincent Bono, Allison Booth, Naomi Brodsky, Mary D. Brower, Lydia Brown, George Bulovsky, Kay Burrough, Gene Cashour, Frank Cox, Bobbie Craven, Pat Criswell, Jim Davis, Nancy Dickman, Dave W. Gammon, Frank Gibson, Grace Giglio, Theodis Gorre, Margaret Guess, Yvette Hayes, Lu Hensen, Kenn Heydrick, Gary Hurst, Don Holderread, Michael Ironsmith, Lucy Jache, Larry Jerdal, Ed Johnson, Grant Johnson, Robert Kennedy, Anne Kenney, Joyce Knox, Leanne Kogler, Dave Kolquist, Sherman Kopelson, Jon Malmin, Douglas Mandt, Jay Maness, Patricia Martin, Mary Monroe, Mike Morris, Phyllis Murray, Silas Nelson, Larry Nelson, Bill Rademaker, Willie Reed, Jay Rubin, Bill Rudd, David Ruscus, Richard Scheele, Paul Shank, Dawn Smith, John Southworth, Mitzi Swift, Steve Ufer, Bob Van Zant, Daniel Vandercar, Bob Volzer, Terri Wahlberg, Tammy Weatherly, Lee Weaver, Joyce Willis, Belinda Wolfe

Field Test Schools:
California: Chula Vista High, Chula Vista; Gompers Secondary School, San Diego; Montgomery High, San Diego; Point Loma High, San Diego; Serra Junior-Senior High, San Diego;

Southwest High, San Diego. Colorado: Bear Creek Senior High, Lakewood; Evergreen Senior High, Evergreen; Green Mountain Senior High, Lakewood; Golden Senior High, Golden; Lakewood Senior High, Lakewood; Wheat Ridge Senior High, Wheat Ridge. Hawaii: University of Hawaii Laboratory School, Honolulu. Illinois: Project Individual Education High, Oak Lawn. Iowa: Linn-Mar High, Marion. Louisiana: Booker T. Washington High, Shreveport; Byrd High, Shreveport; Caddo Magnet High, Shreveport; Captain Shreve High, Shreveport; Fair Park High, Shreveport; Green Oaks High, Shreveport; Huntington High, Shreveport; North Caddo High, Vivian; Northwood High, Shreveport. Maryland: Charles Smith Jewish Day School, Rockville; Owings Mills Junior-Senior High, Owings Mills; Parkville High, Baltimore; Sparrows Point Middle-Senior High, Baltimore; Woodlawn High, Baltimore. New Jersey: School No. 10, Patterson. New York: New Dorp High, Staten Island. Texas: Clements High, Sugar Land; Cy-Fair High, Houston. Virginia: Armstrong High, Richmond; Freeman High, Richmond; Henrico High, Richmond; Highland Springs High, Highland Springs; Marymount School, Richmond; Midlothian High, Midlothian; St. Gertrude's High, Richmond; Thomas Dale High, Chester; Thomas Jefferson High, Richmond; Tucker High, Richmond; Varina High, Richmond. Wisconsin: James Madison High, Madison; Thomas More High, Milwaukee. Washington: Bethel High, Spanaway; Chief Sealth High, Seattle; Clover Park High, Tacoma; Foss Senior High, Tacoma; Hazen High, Renton; Lakes High, Tacoma; Peninsula High, Gig Harbor; Rogers High, Puyallup; Sumner Senior High, Sumner; Washington High, Tacoma; Wilson High, Tacoma

Safety Consultant:
Stanley Pine & William H. Breazeale, Jr.

Social Science Consultants:
Ross Eshelman, Judith Gillespie

Art:
Rabina Fisher, Pat Hoetmer, Alan Kahan (dir.), Kelley Richard, Sharon Wolfgang

Copy Editor:
Martha Polkey

Production Consultant:
Marcia Vogel

Administrative Assistant:
Carolyn Avery

ACS Staff:
Rebecca Mason Simmons, Martha K. Turckes

Student Aides:
Paul Drago, Stephanie French, Patricia Teleska

ACS also offers thanks to the National Science Foundation for its support of the initial development of *ChemCom*, and to NSF project officers Mary Ann Ryan and John Thorpe for their comments, suggestions, and unfailing support.

Second Edition Revision Team

Project Manager:
Keith Michael Shea & Ted Dresie

Chief Editor:
Henry Heikkinen

Assistant to Chief Editor:
Wilbur Bergquist

Editor of Teacher's Guide:
Jon Malmin

Second Edition Editorial Advisory Board:
Diane Bunce, Henry Heikkinen (ex officio), S. Allen Heininger, Donald Jones (chair), Jon Malmin, Paul Mazzocchi, Bradley Moore, Carolyn Morse, Keith Michael Shea (ex officio), Sylvia Ware (ex officio)

Teacher Reviewers of First Edition:
Vincent Bono, New Dorp High School, New York; Charles Butterfield, Brattle Union High School, Vermont; Regis Goode, Spring Valley High School, South Carolina; George Gross, Union High School, New Jersey; C. Leonard Himes, Edgewater High School, Florida; Gary Hurst, Standley Lake High School, Colorado; Jon Malmin, Peninsula High School, Washington; Maureen Murphy, Essex Junction Educational Center, Vermont; Keith Michael Shea, Hinsdale Central High School, Illinois; Betsy Ross Uhing, Grand Island Senior High School, Nebraska; Jane Voth-Palisi, Concord High School, New Hampshire; Terri Wahlberg, Golden High School, Colorado.

Teacher Reviewers of Second Edition:
Michael Clemente, Carlson High School, Gibraltar, MI; Steven Long, Rogers High School, Rogers, AR; William Penker, Neillsville High School, Neillsville, WI; Audrey Mandel, Connetquot High School, Bohemia, NY; Barbara Sitzman, Chatsworth High School, Chatsworth, CA; Kathleen Voorhees, Shore Regional High School, West Long Branch, NJ; Debra Compton, Cy-Fair High School, Houston, TX; Christ Forte, York Community High School, Elmhurst, IL; Gwyneth D. Sharp, Cape Henlopen High School, Lewes, DE; Louis Dittami, Dover-Sherborn High School, Dover, MA; Sandra Mueller, John Burroughs School, St. Louis, MO; Kirk Soule, Sunset High School, Beaverton, OR; Sigrid Wiolkinson, Athens Area High School, Athens, PA; Millie McDowell, Clayton High School, Clayton, MO; Leslie A. Roughley, Steward School, Richmond, VA; Robert Houle, Bacon Academy, Colchester, CT; Robert Storch, Bishop Ireton High School, Alexandria, VA; Michael Smolarek, Neenah High School, Neenah, WI; Fred Nozawa, Timpview High School, Provo, UT; Michael Sixtus, Mar Vista High School, Imperial Beach, CA

Safety Consultant:
Stanley Pine

Editorial:
The Stone Cottage

Design:
Bonnie Baumann & P.C. & F., Inc.

Art:
Additional art for this edition by Seda Sookias Maurer

INDEX

Page numbers in **boldface** indicate pages with definitions.

The Chemistry i

created exclusively for this edition!

St

Wraparound Teache

0-7167-3918-6

The teacher's edition offers
support for teaching stude
different skill levels and lea

Activity Workbook

0-7167-3920-8

Developed by experienced
teachers, these activity she
students organize the mate
opporunities for additiona

Skills Building Hand

0-7167-3917-8

The skills book provides te
students a review of the m
for high school chemistry.
amplifies and clarifies topi

Printed Test Bank, 0

The test bank offers hundr
short answer questions at

Test Bank CD-ROM

one disc), 0-7167-3993-X;
The CD version of the test
resequence questions to su
access point for...

Diploma Online Tes

from the Brownstone Re

With Diploma, instructors
secure exams over a netwo
questions that incorporate
exercises.

Color Transparenci

Blackline Copy Mas

These masters of charts an
used as class handouts or

Meeting the Standards, Chemistry that Counts

Chemistry in the Community (ChemCom), Fourth Edition

W. H. Freeman and Company

American Chemical Society

UNFOLD FLAP FOR YOUR *CHEMCOM* EVALUATOR'S GUIDE

We're providing this evaluator's guide to highlight important changes in the Fourth Edition of *Chemistry in the Community.*
Unfold the flap to find page references for specific coverage as well as to examples of the following:

•***ChemQuandary*** – these brief (10-15 minute) activities challenge students to identify questions and concepts that guide inquiry

•***Laboratory Activity*** – with these activities (timed for approximately one lab per week), students conduct scientific inquiries, make decisions based on data, and communicate results

•***Making Decisions*** – in which students gather data for a problem, recognize and analyze alternative explanations and models, and propose a solution

•***Building Skills*** – here, students use technology and mathematics to master explanations of unit concepts

•***Modeling Matter*** – students critique as well as create visual representations of abstract chemical concepts, formulating and revising scientific explanations and models using logic and evidence

•***Chemistry at Work*** – these reveal how working chemists tackle real, from-the-headlines problems

•***Section Summaries*** – with *Reviewing the Concepts, Extending the Concepts, and Connecting the Concepts*, students can sharpen their higher-order thinking skills and monitor their understanding

•***Putting It All Together*** – each Unit summary offers a variety of scenarios for students to work through, applying what they have learned

•***Technology Icons*** – these margin references point to relevant resources on the *ChemCom* CD-ROM or on the Web.

The Elements (Values in parentheses are the mass numbers of the longest-lived isotopes.)

Element	Symbol	Atomic Number	Atomic Weight	Element	Symbol	Atomic Number	Atomic Weight
Actinium	Ac	89	(227)	Neodymium	Nd	60	144.24
Aluminum	Al	13	26.98	Neon	Ne	10	20.18
Americium	Am	95	(243)	Neptunium	Np	93	(237)
Antimony	Sb	51	121.76	Nickel	Ni	28	58.69
Argon	Ar	18	39.95	Niobium	Nb	41	92.91
Arsenic	As	33	74.92	Nitrogen	N	7	14.01
Astatine	At	85	(210)	Nobelium	No	102	(259)
Barium	Ba	56	137.33	Osmium	Os	76	190.23
Berkelium	Bk	97	(247)	Oxygen	O	8	16.00
Beryllium	Be	4	9.01	Palladium	Pd	46	106.42
Bismuth	Bi	83	208.98	Phosphorus	P	15	30.97
Bohrium	Bh	107	(264)	Platinum	Pt	78	195.08
Boron	B	5	10.81	Plutonium	Pu	94	(244)
Bromine	Br	35	79.90	Polonium	Po	84	(209)
Cadmium	Cd	48	112.41	Potassium	K	19	39.10
Calcium	Ca	20	40.08	Praseodymium	Pr	59	140.91
Californium	Cf	98	(251)	Promethium	Pm	61	(145)
Carbon	C	6	12.01	Protactinium	Pa	91	231.04
Cerium	Ce	58	140.12	Radium	Ra	88	(226)
Cesium	Cs	55	132.91	Radon	Rn	86	(222)
Chlorine	Cl	17	35.45	Rhenium	Re	75	186.21
Chromium	Cr	24	52.00	Rhodium	Rh	45	102.91
Cobalt	Co	27	58.93	Rubidium	Rb	37	85.47
Copper	Cu	29	63.55	Ruthenium	Ru	44	101.07
Curium	Cm	96	(247)	Rutherfordium	Rf	104	(261)
Dubnium	Db	105	(262)	Samarium	Sm	62	150.36
Dysprosium	Dy	66	162.50	Scandium	Sc	21	44.96
Einsteinium	Es	99	(252)	Seaborgium	Sg	106	(263)
Erbium	Er	68	167.26	Selenium	Se	34	78.96
Europium	Eu	63	151.96	Silicon	Si	14	28.09
Fermium	Fm	100	(257)	Silver	Ag	47	107.87
Fluorine	F	9	19.00	Sodium	Na	11	22.99
Francium	Fr	87	(223)	Strontium	Sr	38	87.62
Gadolinium	Gd	64	157.25	Sulfur	S	16	32.07
Gallium	Ga	31	69.72	Tantalum	Ta	73	180.95
Germanium	Ge	32	72.61	Technetium	Tc	43	(98)
Gold	Au	79	196.97	Tellurium	Te	52	127.60
Hafnium	Hf	72	178.49	Terbium	Tb	65	158.93
Hassium	Hs	108	(265)	Thallium	Tl	81	204.38
Helium	He	2	4.003	Thorium	Th	90	232.04
Holmium	Ho	67	164.93	Thulium	Tm	69	168.93
Hydrogen	H	1	1.008	Tin	Sn	50	118.71
Indium	In	49	114.82	Titanium	Ti	22	47.87
Iodine	I	53	126.90	Tungsten	W	74	183.84
Iridium	Ir	77	192.22	Ununnilium	Uun	110	(269)
Iron	Fe	26	55.85	Unununium	Uuu	111	(272)
Krypton	Kr	36	83.80	Ununbium	Uub	112	(277)
Lanthanum	La	57	138.91	Ununquadium	Uuq	114	(285)
Lawrencium	Lr	103	(262)	Ununhexium	Uuh	116	(289)
Lead	Pb	82	207.2	Ununoctium	Uuo	118	(293)
Lithium	Li	3	6.94	Uranium	U	92	238.03
Lutetium	Lu	71	174.97	Vanadium	V	23	50.94
Magnesium	Mg	12	24.31	Xenon	Xe	54	131.29
Manganese	Mn	25	54.94	Ytterbium	Yb	70	173.04
Meitnerium	Mt	109	(268)	Yttrium	Y	39	88.91
Mendelevium	Md	101	(258)	Zinc	Zn	30	65.39
Mercury	Hg	80	200.59	Zirconium	Zr	40	91.22
Molybdenum	Mo	42	95.94				